NON-CIRCULATING MATERIAL

DISCARDED

AIChE Symposium Series No. 323
Volume 96, 2000

Fifth International Conference on
FOUNDATIONS of COMPUTER-AIDED PROCESS DESIGN

Proceedings of the Fifth International Conference on
Chemical Process Design
Breckenridge, Colorado, July 19-24, 1999

Editors

Michael F. Malone
University of Massachusetts

James A. Trainham
E. I. DuPont Company, Inc.

Production Editor, CACHE Publications

Brice Carnahan
University of Michigan

CACHE
American Institute of Chemical Engineers

2000

© 2000
American Institute of Chemical Engineers (AIChE)
and
Computer Aids for Chemical Engineering Education (CACHE)

Neither AIChE nor CACHE shall be responsible for statements or opinions advanced in their papers or printed in their publications.

Library of Congress Cataloging-in-Publication Data

International Conference on Foundations of Computer-Aided Process Design (Fifth : 1999 : Breckenridge, Colorado)
    Fifth International Conference on Foundations of Computer-Aided Process Design
    Michael F. Malone, James A. Trainham, Brice Carnahan, volume editors.
    p. cm. – (AIChE Symposium Series; 323 = v.96)
    Includes bibliographical references and index.
    ISBN 0-8169-0826-5
        1. Chemical processes – Data processing – Congresses. I. Malone, Michael F., 1952.
    II. Trainham, James A., 1950. III. Carnahan, Brice, 1933. IV. Title.
    V. AIChE Symposium Series; no. 323.

       TP155.7. I58 1999
       660'.2815–dc21   00-042050

    All rights reserved whether the whole or part of the material is concerned, specifically those of translation, reprinting, re-use of illustrations, broadcasting, electronic networks, reproduction by photocopying machine or similar means, and storage of data in banks.

    Authorization to photocopy items for internal use, or the internal or personal use of specific clients, is granted by AIChE for libraries and other users registered with the Copyright Clearance Center Inc., 222 Rosewood Drive, Danvers, MA 01923. This consent does not extend to copying for general distribution, for advertising, for promotional purposes, for inclusion in a publication, or for resale.

    Articles published before 1978 are subject to the same copyright conditions. AIChE Symposium Series fee code: 0065-8812/1998. $15.00 per article.

PREFACE

FOCAPD 99

In the two decades since the first FOCAPD meeting, we have seen astounding progress in the size, difficulty, speed and fidelity of models and tools for Computer Aided Process Design. This conference brought together 150 researchers and practitioners to assess the state of the art and progress in the last five years, since FOCAPD 94. Of these 150 participants, 56% are affiliated with academic institutions, 39% from industry and 5% from vendors or government.

The conference included 19 invited papers, 2 oral presentations selected from the contributed papers, and an invited conference summary, all of which appear in this volume. These invited papers were in the following sessions:

Keynote Address
- Chairs: M. F. Malone and J. A. Trainham
- Author: A. E. Fowler

Challenge Problems
- Chair: M. F. Doherty
- Authors:
  - J. P. O'Connell and M. Neurock
  - S. Kim
  - R. W. Sylvester, W. D. Smith, and J. B. Carberry

Process Development in Agricultural Chemicals and Pharmaceuticals
- Chair: P. K. Basu
- Authors:
  - N. Shah, N. J. Samsatli, M. Sharif, J. N. Borland, and L. G. Papageorgiou
  - G. Stephanopoulos, A. Ali, A. Linninger, and E. Salomone

Optimization Methods, Tools, and Applications for Process Design
- Chairs: G. V. Reklaitis and I. E. Grossman
- Authors:
  - S. J. Wright
  - I. E. Grossmann and J. Hooker
  - C. A. Floudas and P. M. Pardalos

Design for Process Operability, Dynamics and Control
- Chair: J. D. Perkins
- Authors:
  - J. van Schijndel and E. N. Pistikopoulos
  - B. D. Tyreus and M. L. Luyben

Modeling and Simulation
- Chair: B. A. Finlayson
- Authors:
  - L. T. Biegler, D. Alkaya, and K. J. Anselmo
  - J. L. Sinclair
  - D. J. Quiram, K. F. Jensen, M. A. Schmidt, P. L. Mills, J. F. Ryley, and M. D. Wetzel

Design Education
- Chair: J. M. Douglas
- Authors:
  - M. F. Doherty, M. F. Malone, R. S. Huss, M. A. Alger, and B. A. Watson
  - B. A. Finlayson and B. M. Rosendall

Tools and Environments for Process Design
- Chair: L. T. Biegler
- Authors:
  - W. Marquardt, L. von Wedel, and B. Bayer
  - A. Shah, B. Nagy, and K. D. Ganesan
  - B. L. Braunschweig, C. C. Pantelides, H. Britt, and S. Sama

Two papers were selected from the contributed papers on *Recent Developments*
- Chair: M. F. Malone
- Authors:
  - J. Stefanovic and C. C. Pantelides
  - S. K. Bermingham, A. M. Neumann, H. J. M. Kramer, P. J. T. Verheijen, G. M. van Rosmalen, and J. Grievink

There were also 53 contributed papers, selected by review of abstracts. Of these, all but one appears in this volume. These were loosely categorized according to their main emphasis into six categories and presented in three poster sessions.

Session A (Chair: M. L. Luyben) included topics on:

Batch- and Bio-processing (7 papers),
Conceptual Design and Synthesis (6 papers),
Dynamics, Operability and Control (4 papers).

Session B (Chair: J. W. Ponton) included:

Environmental Design (6 papers),
Separation and Reaction Systems (11 papers).

Session C (Chair: M. M. El-Halwagi) covered:

Computing, Tools, and Modeling (18 papers).

A Conference Summary, authored by J. M. Douglas, J. A. Trainham, and A. W. Westerberg, appears in this volume, so we make no attempt to summarize the individual papers or sessions here.

The Conference Organizing Committee members were:

R. Agrawal, Air Products and Chemicals Inc.
L. T. Biegler, Carnegie Mellon University
H. I. Britt, Aspen Technology
A. R. Ciric, University of Cincinnati
M. F. Doherty, University of Massachusetts
J. M. Douglas, University of Massachusetts
C. A. Floudas, Princeton University
H. M. Gehrhardt, Amoco Chemical Company
I. E. Grossman, Carnegie Mellon University
I. Hashimoto, Kyoto University
M. L. Mavrovouniotis, Northwestern University
D .E. McKee, E. I. DuPont Company
Y. Natori, Mitsubishi Chemical Corporation
J. L. Robertson, Independent Consultant
G. V. Reklaitis, Purdue University
J. D. Seader, University of Utah
W. D. Seader, University of Pennsylvania
J. J. Siirola, Eastman Chemical Company
W. D. Smith, E. I. DuPont Company
G. Stephanopoulos, MIT
D. R. Vredeveld, Union Carbide Corporation
W. Wagner, Bayer AG
A. W. Westerburg, Carnegie Mellon University

We are grateful for their advice and participation.

The *CACHE Corporation* and the Computing and Systems Technology Division of the *American Institute of Chemical Engineers* sponsored this conference. Their continued sponsorship of conferences such as this has been decisive in advancing all aspects of computer-aided process engineering.

Financial support for the conference was provided by:

Bayer AG
E. I. DuPont Co., Inc.
Parke-Davis Pharmaceutical Research
National Science Foundation (Grant No. CTS-9909204)

Brice Carnahan and his students, Scott Couzens and Decker Ringo, at the University of Michigan continue a long and distinguished record of excellent work on the editing, assembly, printing, and distribution of CACHE conference publications. Special thanks are due Janet Sandy in the CACHE office and Robin Craven of Alliance, LLC, who handled all of the logistics and on-site arrangements.

Michael F. Malone
Amherst, Massachusetts

James A. Trainham
Wilmington, Delaware

TABLE OF CONTENTS

INVITED PAPERS

KEYNOTE SESSION

Process Development in the New Millenium – Hands on or Modeled, In-sourced or Out-sourced, Solo or Shared? 1
    Allan E. Fowler

CHALLENGE PROBLEMS

Trends in Property Estimation for Process and Product Design 5
    John P. O'Connell and Matthew Neurock

The Web and Flow of Information in Pharmaceutical R&D 23
    Sangtae Kim

Information and Modeling for Greener Process Design 26
    Robert W. Sylvester, W. David Smith, and John Carberry

PROCESS DEVELOPMENT IN AGRICULTURAL CHEMICALS AND PHARMACEUTICALS

Modeling and Optimization for Pharmaceutical and Fine Chemical Process Development 31
    Nilay Shah, Nouri J. Samsatli, Mona Sharif, John N. Borland, and Lazaros G. Papageorgiou

Batch Process Development: Challenging Traditional Approaches 46
    George Stephanopoulos, Shahin Ali, Andreas Linninger, and Enrique Salomone

OPTIMIZATION METHODS, TOOLS, AND APPLICATIONS FOR PROCESS DESIGN

Algorithms and Software for Linear and Nonlinear Programming 58
    Stephen J. Wright

Logic Based Approaches for Mixed Integer Programming Models and their Application in Process Synthesis 70
    Ignacio E. Grossmann and John Hooker

TABLE OF CONTENTS

Recent Developments in Deterministic Global Optimization and their Relevance to Process Design 84
    Christodoulos A. Floudas and Panos M. Pardalos

DESIGN FOR PROCESS OPERABILITY, DYNAMICS, AND CONTROL

Towards the Integration of Process Design, Process Control, and Process Operability: Current Status and Future Trends 99
    Jan van Schijndel and Efstratios N. Pistikopoulos

Industrial Plantwide Design for Dynamic Operability 113
    Bjorn D. Tyreus and Michael L. Luyben

MODELING AND SIMULATION

Multi-Solver Modeling for Process Simulation and Optimization 125
    L. T. Biegler, Dilek Alkaya, and Kenneth J. Anselmo

CFD Multiphase Flow: Research Codes and Commercial Software 138
    Jennifer Lynn Sinclair

Integrated Microchemical Systems: Opportunities for Process Design 147
    David J. Quiram, Klavs F. Jensen, Martin A. Schmidt, Patrick L. Mills, James F. Ryley, and Mark D. Wetzel

DESIGN EDUCATION

Decision-Making by Design: Experience with Computer-Aided Active Learning 163
    Michael F. Doherty, Michael F. Malone, Robert S. Huss, Montgomery A. Alger, and Brian A. Watson

Reactor/Transport Models for Design: How to Teach Students and Practitioners to Use the Computer Wisely 176
    Bruce A. Finlayson and Brigette M. Rosendall

TOOLS AND ENVIRONMENTS FOR PROCESS DESIGN

Perspectives on Lifecycle Process Modeling 192
    Wolfgang Marquardt, Lars von Wedel, and Birgit Bayer

Streamlining Projects through Integrated Front-End Automation Solutions 215
    Bert Nagy, K. D. Ganesan, and Ashish Shah

Open Software Architectures for Process Modeling: Current Status and Future Perspectives 220
    Bertrand L. Braunschweig, Constantinos C. Pantelides, Herbert I. Britt, and Sergei Sama

TABLE OF CONTENTS

RECENT DEVELOPMENTS

Towards Tighter Integration of Molecular Dynamics within Process and
Product Design Computations 236
    Jelena Stefanovic and Constantinos C. Pantelides

A Design Procedure and Predictive Models for Solution Crystallisation Processes 250
    Sean K. Bermingham, Andreas M. Neumann, Herman J. M. Kramer,
    Peter J. T. Verheijen, Gerda M. van Rosmalen, and Johan Grievink

CONFERENCE SUMMARY

Conference Summary and Challenges for the Next FOCAPD and Beyond 265
    J. M. Douglas, J. A. Trainham, and A.W. Westerberg

CONTRIBUTED PAPERS

BATCH- AND BIO-PROCESSING

Simultaneous Design and Layout of Batch Processing Facilities 279
    Ana Paula Barbosa-Póvoa, Ricardo Mateus, and Augusto Q. Novais

Integrated Intelligent Tools for Automated Batch Operating Record Synthesis 284
    Linas Mockus, Jonathan M. Vinson, and Prabir K. Basu

Design of Multiproduct Batch Plants with Semicontinuous Stages and Storage Tanks 289
    Jorge M. Montagna and Aldo R. Vecchietti

Computational Experience Solving the Design of Multiproduct Batch Plants
Modeled as Nonlinear Continuous/Discrete Problems 294
    Jorge M. Montagna and Aldo R. Vecchietti

Production Simulation for Pipeless Batch Process Design and Operation 298
    Hirokazu Nishitani and Tadao Niwa

Semi-Continuous Operation of a Middle-Vessel Distillation Column 302
    James R. Phimister and Warren D. Seider

Process Performance Models in the Optimal Design of Protein Production Plants 306
    José M Pinto, Juan A. Asenjo, Jorge M. Montagna, Aldo R. Vecchietti and Oscar A. Iribarren

TABLE OF CONTENTS

CONCEPTUAL DESIGN AND SYNTHESIS

There is More to Process Synthesis than Synthesising a Process 311
  David Glasser, Diane Hildebrandt, and Craig McGregor

Overall Process Design and Optimization - An Industrial Perspective ….....…............ 315
  J. S. Kussi, H. J. Leimkühler, and R. Perne

A New Concept in Process Synthesis: Maximising the Average Rate ....…...... 320
  Craig Mcgregor and Diane Hildebrandt

Conceptual Design of Processes for Structured Products 324
  F. Michiel Meeuse, Johan Grievink, Peter J. T. Verheijen, and Michel L. M. vander Stappen

Industrial Examples of Process Synthesis Analysis Driving Research and Development .. 329
  Daniel L. Terrill

DYNAMICS, OPERABILITY, AND CONTROL

Driving Force Cost and Its Use in Optimal Process Design 333
  Jianguo Xu

Incorporation of Prior Plant Knowledge into Nonlinear Dynamic Black Box
Models Based on Information Analysis 338
  Jun Hsien Lin, Shi-Shang Jang, M. Subramaniam, and Shyan-Shu Shieh

Conceptual Dynamic Models for the Design of Batch Distillations 342
  Enrique Salomone and José Espinosa

Plant Design Based on Economic and Static Controllability Criteria 346
  Panagiotis Seferlis and Johan Grievink

Information Models for Control Engineering Data 351
  Vernon A. Smith and Neil L. Book

ENVIRONMENTAL DESIGN

Design and Analysis of Optimal Waste-Treatment Policies 355
  Andreas A. Linninger and Aninda Chakraborty

Integrated Design and Operability Analysis for Economical Waste Minimization 360
  Usha Gollapalli, Mauricio Dantus, and Karen High

Industrially Applied Process Synthesis Method Creates Synergy between
Economy and Sustainability 364
  Jan Harmsen, Peter Hinderink, Jo Sijben, Axel Gottschalk, and Gerhard Schembecker

TABLE OF CONTENTS

Pollution Prevention via Mass Integration 367
    H. Dennis Spriggs and Mahmoud M. El-Halwagi

Modeling and Design of an Environmentally Benign Reaction Process 371
    Benito A. Stradi, Gang Xu, Joan F. Brennecke, and Mark A. Stadtherr

Simplified System Disturbance Propagation Models and Model-Based
Process Synthesis 376
    Q. Z. Yan, Y. H. Yang, and Y. L. Huang

Improving Efficiency of Distillation with New Thermally-Coupled
Configurations of Columns 381
    Rakesh Agrawal and Zbigniew T. Fidkowski

SEPARATION AND REACTION SYSTEMS

Experimental and Computational Screening Methods for Reactive Distillation 385
    Bernd Bessling

Optimum Design of Mass Exchange Networks Using Pinch Technology 389
    Duncan M. Fraser and Nick Hallale

Automated Generation of Reaction Products 393
    Haifa Wahyu, Ramachandran Lakshmanan, and Jack W. Ponton

Properties of Sectional Profiles in Reactive Separation Cascades 397
    Steinar Hauan, Kristian M. Lien, Jae W. Lee, and Arthur W. Westerberg

Feed Addition Policies for Ternary Distillation Columns 402
    Brendon Hausberger, Diane Hildebrandt, David Glasser, and Craig McGregor

Uncertainty Considerations in the Reduction of Chemical Reaction Mechanisms 406
    Marianthi G. Iepapetritou and Ioannis P. Androulakis

Computationally Efficient Dynamic Models for the Optimal Design and
Operation of Simulated Moving Bed Chromatographic Separation Processes 411
    Karsten-U. Klatt and Guido Dünnebier

Rigorous Optimal Design of a Pervaporation Plant 415
    J. I. Marriott, E. Sørensen, and I. D. L. Bogle

Integrating Chemistry Process Development with Downstream Considerations 419
    David C. Miller and James F. Davis

Synthesis of Optimal Distillation Sequences for the Separation of Zeotropic Mixtures
Using Tray-by-Tray Models 423
    Hector Yeomans and Ignacio E. Grossmann

TABLE OF CONTENTS

COMPUTING, TOOLS AND MODELING

Economic Risk Management for Design and Planning of Chemical Manufacturing Supply Chains 427
    G. E. Applequist, J. F. Pekny, and G. V. Reklaitis

Process Plant Ontologies Based on a Multi-dimensional Framework 433
    Rafael Batres and Yuji Naka

MODEL.LA: A Phenomena-Based Modeling Environment for Computer-Aided Process Design 438
    Jerry Bieszczad, Alexandros Koulouris and George Stephanopoulos

Information Models for the Electronic Storage and Exchange of Process Engineering Data 442
    Neil L. Book

An Object-Oriented Framework for Process Synthesis and Optimization 446
    E. S. Fraga, M. A. Steffens,, I. D. L. Bogle, and A. K. Hind

A Task and Version-Oriented Framework for Modeling and Managing the Process Design Process 450
    S. Gonnet, G. Mannarino, H. Leone, and G. Henning

The Value of Design Research: Stochastic Optimization as a Policy Tool 454
    Timothy Lawrence Johnson and Urmila M. Diwekar

Decomposition Algorithms for Nonconvex Mixed-Integer Nonlinear Programs 458
    Padmanaban Kesavan and Paul I. Barton

Metamodeling – A Novel Approach for Phenomena-Oriented Model Generation 462
    Andreas A. Linninger

Integrated Use of CAPE Tools - An Industrial Example 466
    H. H. Mayer and H. Schoenmakers

A General Framework for Considering Data Precision in Optimal Regression Models 470
    Mordechai Shacham and Neima Brauner

Redefining the Process Simulation Flowhseet: A Component Based Approach 474
    Carl Spears and Vince Tassone

Fuel-Additives Design Using Hybrid Neural Networks and Evolutionary Algorithms 478
    Anantha Sundaram, Prasenjeet Ghosh, Venkat Venkatasubramanian, James. M. Caruthers, and Daniel T. Daly

The Emerging Discipline of Chemical Engineering Info Transfer 482
    Thomas L. Teague and John T. Baldwin

A Principal Variable Approach for Batch Process Design Based on a Black Box Model .. 486
    Po-Feng Tsai, Shi-Shang Jang, M. Subramaniam, and Shyan-Shu Shieh

TABLE OF CONTENTS

A Sampling Technique for Correlated Parameters in Nonlinear Models for
Uncertainty and Sensitivity Analysis 490
    Victor R. Vasquez and Wallace B. Whiting

CHEOPS: A Case Study in Component-Based Process Simulation 494
    Lars von Wedel and Wolfgang Marquardt

A Layers Architecture Based Process Simulator 498
    Naava Zaarur and Mordechai Shacham

AUTHOR INDEX 503

SUBJECT INDEX 506

PROCESS DEVELOPMENT IN THE NEW MILLENNIUM - HANDS ON OR MODELED, IN-SOURCED OR OUT-SOURCED, SOLO OR SHARED?

Allan E. Fowler
The Dow Chemical Company
Midland, MI 48667

Abstract

This document describes a corporate snapshot of the dynamic changes facing the chemical industry. A relationship between mergers and acquisitions and their impact on process development and engineering is presented. Further, a comparison of the various methods to secure process development, i.e., modeling vs. hands-on, whether to out-source, or to share in an external collaboration is discussed.

Keywords

Process development, Modeling, Outsourcing, Mergers, Acquisitions.

Introduction

It has been said years ago "the times they are a changing"…and so it is today. Who could or would have predicted the rate of change would be so dramatic. If you are a glass is half empty type of person….these changes, and more importantly, the rate of change, would find you frustrated, scared, fearful, anxious, cautious and unprepared. If, however, you live a "glass is half full" kind of life, these are some of the most exciting times, challenge vs. fear, opportunity vs. frustration, risk-taking vs. caution, and so on.

I intend to show you how these changing times and the rate of change are inexorably linked to the titled "hands on vs. modeled, in-sourced vs. out-sourced, and solo vs. shared". You can use your judgment as to which are the dependent vs. independent variables.

One more introductory comment. I use the word millennium in the keynote title. I thought long and hard about it, (and certainly like the term better than "Bridge to the next century") it being so overused, misused, and the like. But I left it in, not to highlight this year as really any different than 1992, 1997 or 2009, rather more like a bookmark. The chemical industry has virtually all of its roots in the 20th century. In the grand scheme of things, we are pretty young indeed.

Solo or Shared?

Now, let's get on to the subject of Process Development. The first comparison I would like to address is that of solo vs. shared. To do that, we need to look at industry in general, and more importantly, at the chemical industry.

First, of the Fortune 500 companies in 1979, only 215 of them exist today. The remaining 285 have merged, been acquired or have gone out of business, a rate of 57% in just 20 years. It has been reported that 1998 was the "best" year for Mergers and Acquisitions involving a U.S. company with nearly 12,000 transactions totaling 1.7 trillion dollars. Note the term "best". Remember the half empty/half full glass analogy. The stakeholders in these transactions include the shareholders, the bankers, the employees, the communities they operate in, and of course, the customers. 1998 as a "best" year may not be the term used by all.

Let's look at some specifics in Figure 1. This is just a

snapshot of some of the activities in the chemical and related industries. Add to this, the M&A activity in areas like retail and banking, for example, and you can see what defines intense global competitive environment.

Mergers and Acquisitions

Mobil & Exxon
Rohm & Haas & Morton
ICI (Chem) → Huntsman
Amoco & BP…now plus Arco
Millennium & Lyondell → Equistar
Lyondell + Rexene + Citgo +
    Oxychem + Arco → Equistar
Henkel → Cognis
Sandoz Chemicals Div. + Hoechst Specialty
    Chemicals Bus → Clariant
Sandoz + Ciba → Novartis
Chrysler + Daimler-Benz → DiamlerChrysler

Divestiture

Hoechst - Celanese
DuPont - Conoco
Shell - Shell Chemical

And ??

Union Carbide
Monsanto

Figure 1. Activities in the chemical and related industries.

And on and on it goes. Even though my company, Dow Chemical, still has no name preceding it or succeeding it, significant restructuring of our portfolio has taken place. In the past five years, Dow has divested $10B worth of assets, and has acquired $10B of yet different assets. For a nominal $20 billion company, this is 50% change in our offering. For every M&A, usually some assets must be sold, primarily for either debt reduction, portfolio mix, or anti-trust issues. Furthermore, companies seek to underwrite the high cost of R&D by combining forces.

So how does this impact Process and Development and Design? It is an obvious redistribution of resources and likely a better critical mass. There is more sharing. Whether this is a conscious decision on the part of corporate America or the consequence of heavy M&A activity can be debated. At the end of the day, the formula for better/faster/cheaper must be satisfied.

Regardless of the many corporate machinations just described, the work product of Process Development must still be done. Whether your company name, logo, location or personnel have changed doesn't excuse you or your corporation from this critical need. Think about the value of common tools and standards. The industry needs them and are demanding them. They can satisfy at least two of the better/faster/cheaper requirements. Faster, because with the myriad of corporate changes, tools must be standard and fungible. Economy of scale implies cheaper. We're left with better. This is where many of you come in. Technological advancements in Process Development tools from a fundamental engineering or from an interchangeability standpoint will be ultimately rewarded. So, whether we are talking about Cape Open, Global Cape Open, AspenTech, Pavillion, gProms, etc., this is your challenge.

So to answer the question solo or shared? It will be increasingly more shared, but shared in different ways.

In-Sourced or Out-Sourced?

Let's move on to discuss the issue of in-sourced vs. out-sourced. There are parts of this discussion that may sound similar to the shared vs. solo discussion. The distinction I would focus on is this: in an out-sourcing decision, a company is making a conscious choice as to where and how they want something to be done, and are willing to pay someone else to do it. It is not the total enterprise, rather a portion of the work process deemed to be better served by an outside partner. The basic premise held by industry is in-source all your core competencies and out-source those items that others can do better than you do.

Classic examples are benefits plan, computer support, accounting, customer service centers, etc. In most cases, these programs have served industry well. The companies involved both would agree they "stuck to their knitting" and continued to focus their energy and resources on advancing their core competencies. But does it end here?

If the M&A activity wasn't confusing enough as to who you work for or with, the outsourcing during the past decade has truly muddied the water. How many people do you know that have changed companies and still have the same office? Are still doing the same job? It is difficult to tell when one company ends and the next company begins. How many ID badges can you find in the building where you work?

Back to the subject at hand. How does all this impact Process Development? As most of us work for a technology-based company, isn't that a core competency and therefore should be in-sourced? Well, we said the times they are a changing.

Recent reports state that 67% of organizations will out-source some of their R&D in 1999. What is driving this, even when core competencies are at stake? The past decade of flat budgets and downsizing in the chemical industry has certainly had an impact. I would argue the driver continues to point toward the better/faster/cheaper formula.

We are witnessing an industry where more engineering is being done offshore, particularly India and China and, more specifically, by the larger global companies. The rate of out-sourcing is increasing in all aspects of Process Development. In life sciences/biotech, many examples exist where the traditional or combinatorial chemistry is out-sourced. Speaking of combinatorial chemistry, I predict an increase in the use of this tool in the more industrial chemical sector, like catalysis, polymers, etc. If you look at the more traditional industrial or specialty chemicals, we see an increasing number of engineering companies doing process conceptualization/development and beyond.

How about chemical production? The industry may call it contract manufacturing, but this term contains a significant helping of Process Development. And every time you do this, you are outsourcing a portion of your technology development.

We've talked about the early stage development and the contract manufacturing scenarios. It should be noted that, while in-sourcing the chemical process technology, you may choose to out-source the design and construction of your pilot plants. Witness the growth of companies like Zeton and Xytel for this very purpose.

Recall in the first section I spoke of engineering tools, models and processes that can be common and shared across company, software, or business lines. Every time technology development is out-sourced, the need for a fungible tool increases.

To this point in the discussion, I have focused on industrial collaboration, M&As, and outsourcing. What about the industrial/academia interface? Certainly this is an environment where sharing and/or out-sourcing must be fertile. Witness the growth in industrial support of R&D shown in Figure 2 (source IRI). The increase between 1993 and 1998 was 33% and it has more than doubled since 1988. While this estimated 1998 value of $1.83 billion represents only 7.1% of the total $25.7 billion spent by universities and colleges, the share funded by industry has been increasing.

While goals may be in conflict (Research vs. Development, profit vs. science, intellectual property rights) the need to satisfy the demands of the market place encourages more collaboration. For Process Development, this can take many forms, from reaction engineering to Computational Fluid Dynamics, to dynamic modeling simulations to fully integrated process models.

So is the answer in-source or out-source Process Development? You should not be surprised that it's both...with the pendulum swinging more toward out-sourcing.

Since I mentioned core competencies, it should be noted that the chemical companies are not likely to out-source the crown jewels, nor are they likely to compromise on intellectual asset ownership. These are the first two variables that will be addressed, followed by our old formula for better/faster/cheaper.

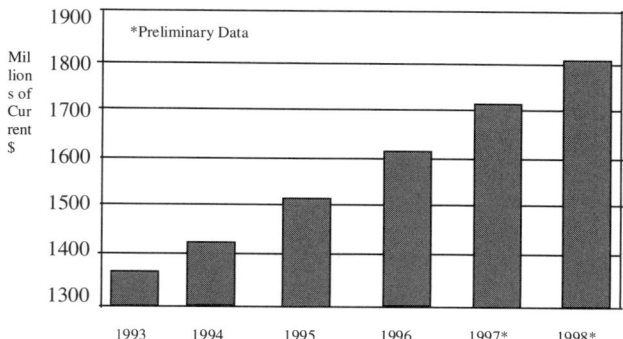

Figure 2. Industrial support of R&D at universities and colleges.

represents only 7.1% of the total $25.7 billion spent by universities and colleges, the share funded by industry has been increasing.

If you pay another company to do it, call it out-sourcing. If you pay a university to do it, call it collaboration or shared. In either event, the number of people involved continues to grow while the time to market gets shorter and shorter.

Hands On or Modeled?

The final comparison is hands on vs. modeled. I've saved this for last for two reasons. First, it has the most to do with science and technology, and second, when it comes to Process Development, it represents a sea change in what the future holds. If the petrochemical industry is celebrating its first 100 years, the progress on real vs. modeled is a great place to start the next 100 years.

Many of us talk about "Edisonian" research. That gave way to empirical research to statistical experimental design. Many were the process projects that demanded that we not only pilot plant every unit operation of the process, but we also had to have a fully integrated pilot plant to do it.

Then, when we solved that headache, we wrestled with the scale-up factors. These factors are found in Perry's, in company technology centers, academia and in the minds of some of the most senior and influential process experts. To violate either the need for an integrated pilot plant or a scale-up factor could be a career-limiting move.

Well, the modelers are prevailing, as they should be. It advances the technology, is faster, and is more cost effective.

The pharmaceutical and agricultural chemical sectors are very well established in the use of molecular models.

These models, based on fundamental chemical, physical, and scientific principles, have drastically reduced the time from ideation to synthesis, and subsequently to market.

Applications of these same modeling principles are finding their way into the larger, high volume, lower unit cost chemical sector. Using modeling to predict structure-property relationships in polymers, fundamental kinetic modeling of complex multi-path reactions, and Computational Fluid Dynamics with reaction are all on the rise, first in study and, ultimately, as an integrated part of the entire commercial process. The sheer number of experiments needed, the time to market, and the advancement of the computational tools continue to converge on modeling as the best enabler to success. But the models must be right and they must be founded in good scientific experiments. We need to understand that, when new knowledge is needed, we develop this knowledge with the right dose of experimentation and with a higher level of trust of the vendor.

The chemical industry is one of the few in the U.S. that enjoys a favorable balance of trade, with investment in the U.S. from abroad continuing to exceed U.S. investment abroad.

But where are the boundaries? The advent of advanced materials puts companies like Intel, Microsoft, GM and Ford squarely in the hunt for chemical process expertise. The rapid growth of life sciences from pharma to Ag to medicine to biomaterials blurs the interface on technology advancement within an "industry". Tools will be interchangeable, scale-up R&D may mean more microelectromechanical and microelectrochemical devices, and the Process Development work process will be further integrated between ideation to commercialization.

Much of the discussion has focused on Process Development in the context of R&D. There is much more under the Process Development umbrella, such as the use of modeling tools for optimization of existing assets, products, and processes. For sure, this may not be as exciting as new product development and commercialization of new technology; but rest assured, it is every bit as critical to our industry.

To those of you applying your skills to advancing pinch technology, basic unit operation optimization, process intensification, model based control, and integrated models, please keep it up. We may not be able to violate the fundamental laws of thermodynamics, physics, and chemical reactions, but there is waste to be minimized, energy to be saved, capital to be reduced, and yields to be increased.

So the question of hands on or modeled is more easily answered. Modeling, provided it is robust, technically correct, and applied with common sense, will always win. In the better/faster/cheaper equation, it should satisfy all three.

Conclusions

Let me close with these comments. At this conference in 1994, Larry Biegler of Carnegie Mellon and Michael Doherty of University of Massachusetts spoke of the challenge facing industry. The message was increased competition, greater regulatory pressure and uncertain prices for energy, raw materials and products.

Well, five years later the story is unchanged…and I see nothing on the horizon that is going to change this story line.

This state of affairs is coupled with the lament of virtually all chemical companies today, that of eroding prices, Asian turmoil, a strong U.S. dollar and structural cost reduction behind schedule or harder to achieve.

Ladies and gentlemen, this is true raw material for change. And, I again remind you that technology will be the catalyst, as well as the product. Anyone who satisfies the better/faster/cheaper formula will be allowed in the game.

TRENDS IN PROPERTY ESTIMATION FOR PROCESS AND PRODUCT DESIGN

John P. O'Connell and Matthew Neurock
University of Virginia
Charlottesville, VA 22903

Abstract

Physical properties and phase equilibria play essential roles in process design, simulation and optimization. They provide values of thermodynamic and transport properties, give target values for specification of process conditions, and delimit the solution space and suggest strategies for solving process design problems. Recent advances in theory, experiment, and microsimulation have significantly expanded options for property determination, enhancing their use at all stages of process design including chemical kinetics and mechanisms. A brief review is given of how properties are used and then some opportunities and limitations of molecular simulation and computational chemistry are considered. Future computer-aided process and product engineering will be able to utilize a greater richness of property model options, though measurement and judgment will always be required for reliable results.

Keywords

Physical properties, Property estimation methods, Molecular simulation, Computational chemistry, Thermodynamics.

Introduction

The initial series of the Foundations of Computer-Aided Process Design Conferences (Mah and Seider, 1981; Westerberg and Chien, 1984) showed that workers in Computer-Aided Process Engineering have long had an interest and appreciation for the importance of physical properties along with phase and reaction equilibria. Property values are used for the whole range of design activities from invention through conceptual design to equipment, optimization and control systems. Properties are most evident in process simulators where codes for mathematical models are used to provide values for conceptual quantities such as internal energy and fugacity in the fundamental equations for material, energy, entropy balances and equilibria. But properties also help inventors, innovators and designers recognize opportunities to exploit or avoid certain separation and reaction schemes, anticipate challenging or convenient phase and reaction behavior, and suggest optimal routes to determine process configuration and operation including for specified products (Gani and Constantinou, 1996; Gani and O'Connell, 1999; Harper, *et al.*, 1998; Joback and Stephanopoulos, 1990).

As with all technologies, major advances in property estimation have been made in the last few decades because of our ability to comprehensively assess the accuracy and reliability of various model formulations, to gather much data over ever wider ranges of type, time and space, and to increasingly make computations from molecular-level principles. This has been described well by Ishikawa (1999) and has even reached the level of creating a "super-chemist" for the chemical industry (Dohgane and Yuki, 1998). Useful comprehensive books covering this subject are by Horvath (1992) and Doucet and Weber (1996). The purpose of our review is to highlight the current situation in property methods for process and product design and to indicate directions for the future.

Diverse options are currently being pursued especially with high-speed computers. But traditional methods will be fully adequate for many process design needs and few designs will be completed without both experimental verification and engineering judgment. We do expect that molecular level computations will provide increased guidance about the behavior of chemical systems, such as at extreme conditions and for reaction details, leading to minimized experiment and maximized exploration of creative alternatives. In fact, practice has already reached the stage at some companies where "it is an almost unwritten rule that proposed experimental research and development be justified by calculational evidence." (Golab, 1997).

Structure and Roles of Properties in Process Design

Physical and chemical properties are the result of codifying certain fundamental behaviors of Nature into laws. As suggested in Fig. 1 (O'Connell, 1999), these are formulated into useful models by three different means: Theory, Experiment, and atomic/molecular (Micro)simulation. All of these endeavors, especially when used complementarily, yield quantitative expressions or models which are then coded into process simulation programs or made available for predicting properties of chemical products.

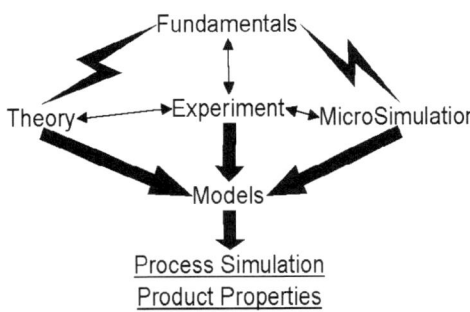

Figure 1. A structure for properties methodology.

This structure exists because thermodynamics, transport phenomena, and statistical and quantum mechanics are unusual disciplines; their fundamentals can be stated unequivocally in quantitative form and can be relied upon to hold as long as the system is adequately characterized. For macroscopic systems, the properties must be homogeneous over a suitable range of space. Then the conservation of mass and the 1st and 2nd Laws of Thermodynamics will lead to balance equations for material, energy and entropy. While mass and energy are conserved, entropy is generated in real processes. Yet the general balance equation form is still valid and becomes useful when entropy generation is made quantitative such as with the isentropic efficiency of turbines or obtained in detail from transport phenomena for fluxes in momentum, energy, and species. For statistical mechanics, after enumerating the feasible energy states, the fundamentals assert that each state contributes equally to a system's average value. There are different modes of averaging, which have been defined: partition functions, spatial and temporal distribution functions, and mean fluctuations. Though rigorous results must be the same, approximate methods will differ not only in ease of solution, but also in quantitative answer. In quantum mechanics, Schrödinger's equation is the fundamental quantum relationship for the distribution of energy in time and space. Its solution can yield all of the information of statistical and macroscopic properties, but so far, such descriptions have been found only approximately for common chemical processes.

The mathematical consequences of the macroscopic laws include phase and reaction equilibrium and the flux laws to achieve equilibrium by heat, mass and momentum transfer and by reaction; nothing new needs to be said about Nature to reach these conclusions. The transfers generally follow linear flux laws with transport coefficients being the proportionality constants for the proper driving forces (*e.g.*, chemical potential, not concentration, for diffusive mass transfer). Finally, in the cases where analytic solutions to the microscopic fundamentals can be found, they are rigorous.

Unfortunately, at the macroscopic level, the laws, consequences, and generalities are not very convenient for either properties or mechanisms. Invariably properties involve nonmeasurables or what can be called conceptuals. Thermo*statics* connects fundamental conceptuals such as entropy and fugacity to combinations of measurables such as temperature and composition and to more accessible conceptuals such as vapor pressures and activity coefficients. Then models for these conceptuals are formulated with parameters that can adequately represent property variations for the particular system or substance of interest. These same models can also be used to describe the tendencies for change via heat, momentum and mass transfer and/or reaction from a specified state whether mechanisms for the changes exist or not. Then thermo*dynamics* can be used to obtain the process independent changes of properties and species that could result if mechanisms exist. These set directions and limits on the possibilities. Finally, though we can infer some aspects of process dependent behavior by considering transition states in *pseudo*thermo*static* terms, full, detailed process evolution requires transport phenomena and chemical kinetics.

The conceptual/measurable property connections can be established via experiment on many systems, provided there is an empirically observed pattern to the behavior, which allows a statistically meaningful analysis of the raw data. But experiment is expensive, time-consuming and not always reliable or sensitive enough. So other methods must be employed.

Theories are approximate treatments of idealized images of the system involving parameterized expressions that are inspired by rigorous, but inaccessible formalisms (O'Connell, 1999). Thus, lattice models of fluids, despite obviously having too low entropy values, have been used because the regularity of lattice positions and energies leads to convenient approximation and parameterization of the partition function. Microsimulation of models of the atoms and molecules of the substances of interest can be more efficient and potentially more generalizable as we describe below. But these are subject to approximations whose effect is not always known, so their results should always be tested against data.

The goals of macroscopic modeling are well known (Reid, *et al.*, 1987; Sandler, 1993; Millat, *et al.*, 1996; O'Connell, 1984; Carlson, 1996; Dohrn, 1999). Data with descriptions of experimental techniques and careful evaluation are often published in specialized journals; some focus on giving the raw numbers or correlating equations for desired properties; others expect the data to support a theoretical analysis. Many data tabulations exist though not always of uniform quality. Sometimes these include modeled or estimated "data" or merely give model parameter values. Theory is normally developed within the context of statistical mechanics, both equilibrium and time dependent. The fundamental ideas can be found in statistical mechanics and statistical thermodynamics books (Reed and Gubbins, 1973; McQuarrie, 1975; Hill, 1962; Ben-Naim, 1992).

In order to appreciate how models can be utilized in chemical process and product design, Gani and O'Connell (1999) have articulated the ways in which properties are employed. They note that for the same process, models of different formulation, complexity and accuracy may be used to solve different process engineering problems. Thus the choices of process models depend on both the problem type and the process details. All process models use properties as intermediate variables; their values are obtained from property *models* chosen to mimic the behaviour of the process substances and the operating conditions. The information they provide to a process model is how the process description "knows'" differences among the processes to make liquid air, sulphuric acid, jet fuel and pharmaceuticals. The common use of physical property models is to implement the effects of a physical phenomenon (such as the constraints of phase equilibrium) or to provide values for an intermediate variable (such as the enthalpy of a stream). Typically the process model passes to the property model the process conditions of state: composition $\{x\}$, temperature (T), and pressure (P), and it returns the desired variable (property) values. As with process models, different property models may be used, depending on the problem, system and process.

Gani and O'Connell (1999) show three main roles that properties can play in process synthesis, design, optimization and control. First there is a *service* role where property models are used to provide the needed property values for process models when asked. Different property models, either in approximations for the same components and mixtures or different parameter values for different substances will usually give different results with the same process model. The impact of these differences depends on the sensitivity of the process model to properties; great care must be exercised in certain situations and unsuspected effects can arise (O'Connell, 1984; Whiting, 1999, Dohrn, 1999). It is common for property model complexities to be the cause for non-linear behaviour of the process model equations, causing difficulties in achieving convergence.

In the service role of process simulation, accuracy and generality of properties models are of greatest importance. However, Gani and O'Connell (1999) note that simulators are also beginning to be used for on-line optimization and control where greater speed and reliability are demanded. At this point, petroleum and other small molecule organic systems are adequately treated, but the expanding breadth of chemical systems such as aqueous electrolytes, polymers and biochemicals often cannot be reliably correlated or predicted with acceptable accuracy by current generalized simulator models. The future will require property models that apply to such complex substances and systems since simulations using unreliable property models are unlikely to have acceptable process uncertainties.

In addition to the *service* role, there is also the role of *advice* for properties. Often process design and synthesis set out *explicit* or *implicit target values* for properties where the problem solution achieves these values by manipulating unspecified intensive variables. A common case is solvent design where the properties have *explicit* target values such as selectivity for particular solutes at specified solution conditions (Joback and Stephanopoulos, 1990; Harper, *et al.*, 1998; Ahmad and Barton, 1999). The structure of candidate solvents is manipulated until the solution properties match the target values. Solving both of these process design/synthesis problems requires two steps (irrespective of whether they are solved sequentially or simultaneously): 1) an *advice* step where alternatives are generated (*e.g.*, T or solvent), and 2), a *service* step where properties are determined (number of phases or selectivity) and alternatives are verified. Strategies for the advice role attempt to eliminate infeasible solutions or at least reduce the size of the search space and/or mathematical problem by providing the valuable *insight*.

Finally, Gani and O'Connell (1999) describe the most comprehensive role for properties in the solution of CAPE problems: *solve*. In process/tools integration and graphical (visual) design techniques, properties actually can *define the solution strategy*. Historical examples are graphical distillation column design and pinch analysis for heat integration. More recently they include design and analysis of azeotropic mixtures via distillation boundaries and residue curves together with the phase equilibrium curves and even extensions to reactive systems. In all of these, the relations among thermodynamic properties and intensive variables generate the algorithm or solution strategy in addition to providing their values. "Intelligent" manipulation of the process and property model equations and variables using physicochemical and mathematical insights can identify a small number of intensive variables as the unknown process model variables to represent the essential information about the problem. This tools integration, which is most valuable in applying integrated algorithms for synthesis, design and/or control, can be enhanced by considering the functions of properties in process models.

There are further extensions to be anticipated. Prausnitz (1999) points out that simulation/design problems of future will require the simultaneous modelling of reaction kinetics and physical properties. Kleintjens (1999) asserts that the design and study of future organic materials will not be done from process thermodynamics but from a materials/product approach.

Microsimulation Overview

It is within this structure of methodology, roles and challenges that we describe how microsimulation is beginning to be utilized for physical and chemical properties. Currently, "chemistry modeling is considered an enabling technology that provides additional, alternative, nonexperimental, technical information upon which better business decisions can be based. . . . [It] is used as a tool to predict molecular properties as accurately as required by our business partners. . . . it is of interest to link [it] to computer aided process engineering modeling software." (Golab, 1997). To provide motivation, Table 1 shows a collection of industrial companies and cases compiled by Westmoreland (1997) for which microsimulation has been used. There was topical conference on the area of 157 papers at the AIChE Meeting in Miami Beach, FL, November, 1998.

We have coined the term microsimulation for the two principal types of computer modeling of atomic- and molecular-level systems. The first is molecular simulation which assumes the molecular structure and interaction forces and then computes average and transient molecular (microscopic) and local collective (mezoscopic) structure which can be used to obtain thermodynamic and transport (macroscopic) properties. Its programs use the fundamental equations of statistical mechanics to compute properties by sampling a system's molecular phase space of positions and energy (Monte Carlo) or positions and velocity (molecular dynamics) and then weighting the results to obtain average physical properties and structure. Since the molecular structure is assumed to be known, molecular simulation can only yield physical properties.

The other technique of microsimulation, commonly called computational chemistry, is the application of quantum mechanics to describe the formation and structure of molecular bonds and interactions. The technique solves the fundamental equations of quantum mechanics through a series of approximations to obtain the average or time dependent distribution of energy within systems at the subatomic level, especially for electrons associated with nuclei and in chemical bonds.

The results yield optimized molecular structures, electronic charge distributions and their energies. These can then be used to calculate a great number of additional physical and chemical properties of a system. Together with molecular simulation techniques for entropy and free energy (see below), a system's microscopic structure and physical properties can also be determined.

Table 1. Industrial Applications of Microsimulation (from Westmoreland, 1997).

Company	Application Area
Air Liquide	O_2/N_2 Separation Zeolites
Air Products	Adhesives, Adsorption
Albemarle	Flame Retardents, Lubes
Amoco	Catalysis, Thermochemistry
ARCO	Polymer Weathering
BASF	Surfactant Systems
BP	CH_4 Adsorption
Chevron	Gas Hydrates, Lubricants
Dow	Polymers, Thermochemistry
Dow Corning	Silicone Chemistry
DuPont	Catalysis, Thermochemistry
Exxon R&E	NO_x Kinetics, Zeolites
Hercules	Polysaccharide Rheology
IBM - Zurich	Polymers, Coatings
Lubrizol	Reactions, Corrosion
General Motors	Monolithic Catalysts
Phillips	Catalysts, Gas Hydrates
Procter & Gamble	Detergent Enzyme Design
Royal Dutch Shell	Porous Media Diffusion
Schlumberger	Setting of Drilling Cements
Shell USA	Solvent Seps, Catalysis
Union Carbide	Catalysis, Materials
Xerox	Photocopying Materials

While microsimulation was initially the domain of scientists, chemical and other engineers began to use molecular simulation methods in the 1970's as their familiarity with statistical mechanics suggested that simulation could overcome the limitations of approximate analytic treatments.

The applications of computational chemistry are somewhat less known in process design because it has many complexities and subtleties. Many of the initial attempts led to disappointing differences with experimental data or showed extreme sensitivity to certain approximations. Further, in some organizations, computational chemistry was considered "the method of last resort, with little planning, just before the termination of a project." (Golab, 1997). Ironically, the success of other efforts has not been widely advertised because of proprietary concerns.

The computer resources for computational chemistry can be exceedingly great, though the combination of newer techniques, more powerful computers, and efficient program packages are likely to make computational chemistry ever more accessible to application. So far, there have been few well-publicized "killer applications/solutions" to prove the technology as indispensable and provide direction for future profitable applications. Such expectations are probably unrealistic since microsimulation is, and probably should be, "rarely employed to single-handedly solve a problem." (Golab, 1997).

In any case, for some or all of these reasons, computational chemistry has been less apparent in chemical process engineering than has molecular simulation. Suggestive of the situation is that of the AIChE topical conference papers were about two to one of molecular simulation to computational chemistry themes.

Molecular Simulation

The "essence" of molecular simulation "is simply stated: numerically solve the N-body problem of classical mechanics." (Haile, 1991) A simulation "consists of three principal steps: (1) construction of a model, (2) calculation of molecular trajectories, and (3) analysis of those trajectories to obtain property values." Molecular dynamics (MD) involves numerically integrating the differential equations of motion resulting in a deterministic temporal development of the system either with fixed (equilibrium) or moving (nonequilibrium) boundaries. The Monte Carlo (MC) method uses Markov chains which are sequences of configurations generated stochastically. There are also hybrid methods in which both deterministic and stochastic elements are involved. As Kofke and Cummings (1997) point out, the utilization of energy simulations to obtain free energies and species chemical potentials is an especially important objective. Recent text and reference books on molecular simulation include those by Haile (1991), Allen and Tildesley (1993), Binder (1995), Frenkel and Smit (1996) and Sadus (1999). There are now occurring electronic conferences on the subject (Quirke, 1999).

The literature of this area began with the initial applications of computers to physical processes, principally in the 1950's. It has now reached the stage where it is considered equally important with theory and experiment for investigating Natural phenomena (Fig. 1). One can find both intensive research in methodology and diverse applications to systems where experiment is difficult, expensive or ambiguous for interpretation, and to test approximations to solve complex equations to validate theories (Cummings, 1996, 1999; Panagiotopoulos, 1996). For physical properties in chemical processing, molecular simulation offers particular opportunities in both service and advice roles. Note that it cannot be directly applied to reacting systems.

When the simulations are properly executed, the basic uncertainties relative to "real" systems arise from, 1) the approximate interaction energy models between and within molecules, and 2) the time scale over which processes can be simulated compared to those of Nature.

For the former, it is common to use convenient pairwise interactions among limited numbers of particles to model the potential energies, which can cause deviations from Natural systems. Simplification of potentials can ignore significant effects; for example, nonadditivity of polarizability turns out to be very important, especially in aqueous systems, but this was not realized until recently and still has not yet been treated satisfactorily in all aspects (Chialvo and Cummings, 1998; Kiyohara, et al., 1998). The time scale affects reliability for large molecules and networked systems which evolve over periods much longer than nanoseconds and microseconds. Increasing computer speed and capacity is diminishing these limitations somewhat, but biological and certain materials systems represent significant challenges even when the assumed potential energies might be reliable (Kollman, 1996).

Classical statistical thermodynamics formulates certain characteristic properties in terms of multidimensional integrals over the positions and orientations of the particles in the system. For example, in the canonical system the variables are the number of particles, N, the temperature, T, and the total volume, V, and the configurational portion of the helmholtz energy, A_c, arising from the interaction potential energy, U_N, is directly related to the configurational integral, Z

$$A_c(T,V,N) = -kT \ln Z \tag{1}$$

where

$$Z = \int ... \int \exp[-U_N(\mathbf{r}^N)/kt] d\mathbf{r}_1 ... d\mathbf{r}_N \tag{2}$$

The average value of any dependent thermodynamic property such as the configuration internal energy, E_c, is found either by thermodynamic manipulation of A_c such as

$$E_c = \left.\frac{\partial(A_c/T)}{\partial(1/T)}\right|_{V,N} = -k \left.\frac{\partial \ln Z}{\partial(1/T)}\right|_{V,N} \tag{3}$$

where a theoretical or model expression for Z is developed, or by averaging of the contributions of the configurations

$$E_c = \frac{1}{Z} \int ... \int U_N(\mathbf{r}^N) \exp[-U_N(\mathbf{r}^N)/kT] d\mathbf{r}_1 ... d\mathbf{r}_N \tag{4}$$

As described below, microsimulation uses MC and MD to obtain average property values. These can be used directly for volumetric (see, e.g., Neubauer, et al. 1999) and energetic properties. One method, applicable to both MC and MD, can be used to obtain changes in Gibbs energy when the system is evolved from one set of energy relations to another. It is known by various names: free energy perturbation, thermodynamic integration, coupling (or charging) parameter process,. An example is to find the solubility of methane in water from a series of simulations where one water molecule is changed gradually from interacting as a water to interacting as a methane (Kofke and Cummings, 1998). The basic equation connecting two systems 1 and 2 is

$$G_2 - G_1 = \sum_i \int_{\lambda_{i1}}^{\lambda_{i2}} \left\langle \frac{\partial U_N}{\partial \lambda_i} \right|_{N,T,P,\lambda_{j \neq i}} d\lambda_i \right\rangle \quad (5)$$

where U_N is written in terms of separation, orientation, etc., but the two systems are obtained from the same model with different values of a set of coupling or charging parameters, $\{\lambda\}$. That is,

$$U = U(\mathbf{r}^N, \ldots, \{\lambda\}) \quad (6a)$$

$$U_2 = U(\mathbf{r}^N, \ldots, \{\lambda_2\}) \quad (6b)$$

$$U_1 = U(\mathbf{r}^N, \ldots, \{\lambda_1\}) \quad (6c)$$

The integrands of Eq. (5) are the ensemble (MC) or time (MD) averages of derivatives of Eq. (6a). A convenient formulation is where the interaction such as for methane with water (state 2) is a linear difference from the interaction of water with water (state 1), i.e.,

$$U = U_1 + \lambda(U_2 - U_1) \quad (7)$$

Then there is only one term in the sum of Eq. (7) and the integrand is the same quantity over the integration (though its *value* depends on $\{\lambda\}$ through the averaging).

This method was originally conceived by Kirkwood (1935) and described by Hill (1960), but was not able to be applied until computers became powerful enough. Chialvo and Haile (1987), Chialvo (1993), Kollman (1996), and Shukla (1997ab) describe fundamentals and applications of this approach with references. Most recently in chemical process systems, it has been used to obtain infinite dilution activity coefficients by Slusher (1998; 1999) though the results cannot yet be relied on to be as accurate as experiment. Kollman (1996) describes its use in systems of biological interest. One of the most interesting opportunities to use this approach is for solvent effects on transition state species in reacting systems.

Monte Carlo Simulations

The MC technique, which is often used to evaluate multidimensional integrals in applied physics, samples the configurations in such a way that suitable average values can be obtained. Then the average property value is found from n configurations by

$$<A> = \frac{1}{n} \sum_{k=1}^{n} A_k \quad (8)$$

The integrand in Eqn. (2) indicates that high-energy configurations (repulsions) contribute little and so need not be included. Methods of biasing the sampling toward those configurations that do contribute are used to increase computational efficiency especially for dense systems (Panagiotopoulos, 1996).

The calculation consists of assigning the N particles to positions and computing the potential energy, U_N. To minimize calculation, N is made as small as possible, often from 100 to 1000, though current simulations use as many as 10^6-10^9 particles. To minimize small system (edge) effects, replicas of the system are placed at system boundaries (periodic boundary conditions) to provide interactions. Special treatments may be done to include long-range intermolecular energies such as in electrolytes. Typically, a pairwise model is assumed for the N-body potential:

$$U_N(r^N) = \sum_{i=1}^{N-1} \sum_{J>i}^{N} u_{ij}(r_{ij}) \quad (9)$$

A common pair potential is the Lennard-Jones 12-6 model

$$u_{ij}(r_{ij}) = 4\varepsilon_{ij} \left[\left(\frac{\sigma_{ij}}{r_{ij}} \right)^{12} - \left(\frac{\sigma_{ij}}{r_{ij}} \right)^{6} \right] \quad (10)$$

The procedure is to randomly select a particle, move it to a random new position and recompute the energy. There are rules for accepting the new configuration into the sample or not. Whether or not it is accepted, the random step process is repeated. Commonly, several million steps are used in order to obtain an average with reasonable uncertainties. The number of steps depends on the property of interest and the desired sampling uncertainty. There are many subtle effects to be considered in either doing simulations oneself or in analyzing the work of others (Binder, 1979; Haile, 1991; Doucet and Weber, 1996).

It is possible to change the independent thermodynamic variables of the system to more desirable quantities (Panagiotopoulos, 1996). Fixed pressure can be maintained by moving the boundary positions. This allows variations of the average density of a fluid. The chemical potential can be found from changes in Z when a particle is added to the system. This has been used to obtain Henry's constants of solutes in solvents (Cummings, 1999) and for ethylene/ethane adsorption separation modeling (Blas, *et al.*, 1998); it is also the basis for Gibbs ensemble Monte Carlo (GEMC) phase equilibrium simulations including near-critical regions (Potoff and Panagiotopoulos, 1998).

The outputs of MC simulations are average macroscopic (thermodynamic) and microscopic (molecular structure) properties. Except for truly stochastic processes which include certain model polymer motions (Binder, 1995), no dynamic behavior nor pathways can be found from MC because systems normally have deterministic rather than stochastic kinetic evolution (Binder, 1979). However, stochastic (Monte Carlo) simulation of reactions, which we do not cover, does reveal reaction pathways (see, for example, Even and Bertault, 1999, and its references).

Molecular Dynamics Simulations

The equilibrium form of MD simulations involves numerical solution to Newton's equations of motion for a set of particles in a system with specified thermodynamic constraints. In the isolated system case, the vector of the total force on a particle i, $\mathbf{F}_i$, is related to the second time derivative of the position, $\mathbf{r}_i$ and the first position derivative of the potential energy, U_N:

$$\mathbf{F}_i(t) = m\ddot{\mathbf{r}}_i(t) = -\frac{\partial U_N(\mathbf{r}^N)}{d\mathbf{r}_i} \qquad (11)$$

where m is the mass of particle i. When a model for $U_N(\mathbf{r}^N)$ such as the pairwise (ij) potential is used and the particles are put in some initial set of positions, $\mathbf{r}_0^N$, with initial momenta, $\mathbf{p}_0^N$, integration once yields momenta and integration twice yields positions for a time step, Δt. These integrations are usually repeated several tens of thousands of times for steps of the order of femtoseconds (10^{-15}) so that time periods much more than picoseconds (10^{-12}) can be covered. A set of steps is first taken to eliminate the effects of the initial conditions and then for n steps, an average property, $<A>$, can be estimated from the values computed at regular intervals, $k\Delta t$

$$<A> = \lim_{n \to \infty} \frac{1}{n} \sum_{k}^{n} A_k(k\Delta t) \qquad (12)$$

The property can be either an equilibrium property such as energy or pressure, or it can be a dynamic transport property such as the diffusion coefficient. The value's accuracy depends upon the number and size of the steps.

The ergodic hypothesis asserts that in the limit of infinite samples, the time average of Eqn. (8) would be the same as the ensemble average of Eqn. (5). This can provide a check on whether the MC and MD simulations have been done correctly or, at least, consistently.

It is also possible to do nonequilibrium molecular dynamics (NEMD) by applying an external force to the system so that steady conditions of shear, concentration difference and material and energy flow are established.

As a result, bulk and shear viscosity, thermal conductivity and diffusion coefficients have been obtained. The programming and running of such methods is more complex and demanding than is equilibrium MD but some new insights and relevant applications are being treated this way (Cummings, 1999; Gupta, *et al.*, 1998; Sarman, *et al.*, 1998).

Comparisons of MC and MD Methods

With regard to which molecular simulation method is most effective and accurate, MC has an advantage in systems with complex potential energy functions such as rigid bodies or multifunctional and orientation-dependent species, it is more easily adapted to a greater variety of constraints such as for phase behavior and adsorption, and it is easier to code. However, MD is more efficient in such properties as heat capacities, compressibilities and interfacial properties, and uniquely gives transport properties. It is also easier to determine how long to run and to detect coding errors. Finally it provides effective visualization of molecular processes for understanding behavior. Thus, as usual with powerful methodologies, the desired objective and resources at hand determine which is to be used. Workers should be prepared to employ or analyze results from all types of simulation.

Molecular Simulation of Chemical Process Systems

Gubbins and Panagiotopoulos (1989) provided an early description of the range of engineering contributions that molecular simulation could and would make to properties and structure. They note that simulation can be very effective in overcoming thermodynamic errors associated with systems of small numbers of molecules such as droplets, colloids, surfactants, liquid crystals, porous media, and nucleation as well as systems of great multifunctionality such as polymers and proteins. Examination of the papers at the AIChE topical conference showed that these and many more applications are occurring a decade later. For example, Cummings (1999) has compiled a set of properties used in process design for which molecular simulation has been used. While many of these can and have been determined from experiment, simulation has allowed exploration of situations not easily measured, aided in solving controversies, and suggested measurements to find unusual behavior. We explore some cases to show microsimulations in service and advice roles.

The first case is associated with the chain length dependence of the critical properties of pure normal alkanes, a service role. As described by Tsonopoulos and Tan (1993), there was an uncertainty from experiment about critical mass density when the number of carbons exceeded about 10 ($\rho_c \sim 0.23$ g cm^{-3}). Some earlier measurements with significant uncertainty suggested that the density continued to increase while later results showed a decrease. Though there was probably a basis for choosing the data with the extremum (Ambrose and Tsonopoulos, 1995), this choice was also confirmed by Siepmann, *et al.* (1993) who performed a series of GEMC calculations. Their results not only agreed with the experiments which were limited to $N_C < 20$, they were able to provide values up to $N_C = 48$. Figure 2 shows the different sources of data for critical density. In addition to this property that showed an extremum, the calculations confirmed the trends for T_c and P_c and gave full phase envelopes of vapor pressure and coexisiting phase densities.

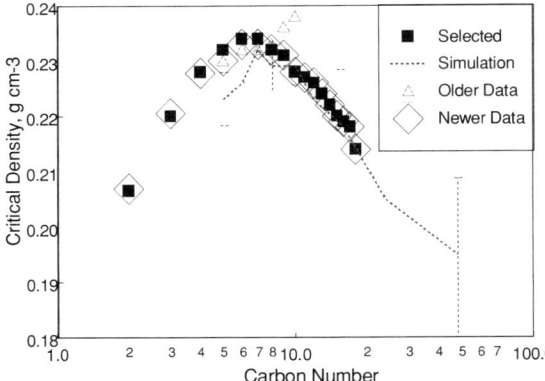

Figure 2. Critical mass densities of n-Alkanes; experiment and simulation aid selection of final values.

In the same vein, there have been simulations of the phase coexistence and critical behavior of pure perfluorocarbons. The results are quite consistent with experiment but provide information about a much wider range of these substances. The key to the fluorocarbons was to find inter- and intramolecular potential energy functions that would accurately reflect the systems. The major uncertainty was the torsional potential for rotation of about the C-C bond because CF_2 chains can form helices while CH_2 chains cannot. Cui, et al. (1996) used results from quantum mechanical calculations though further investigation (Mei, 1999) has shown that the potential model was very similar to that obtained from molecular measurements in solution over several decades. The sensitivity of many properties to the torsional potential is not great though molecular structure and dynamics can be significantly affected.

Another example of molecular simulation providing properties information is for fluids in porous media. Gubbins and coworkers (Blas, *et al.*, 1998; Gelb and Gubbins, 1999; McCallum, *et al.*, 1999; Sliwinska-Bartkowiak, *et al.*, 1997), among others, have examined a number of interesting systems to elucidate the effect of fluid confinement on properties as well as to simulate adsorption systems and relate common characterizations to molecular phenomena. For example, the critical temperature of simulated liquid-liquid systems is significantly decreased under confinement. Below the critical, two-phase systems show significant reduction of the solubility of the less strongly adsorbed component in the phase rich in the more strongly adsorbed component while the other liquid has a significantly higher concentration of the more strongly adsorbed component. Thus, the whole envelope for confined phases is shifted down and toward the more strongly adsorbed component.

These results suggest that adsorption purification and natural porous media processes may have large differences from nonconfined systems. In this case, the simulated properties are providing an advice role since a process design would require knowing the conditions necessary for obtaining the two phases for separation; the phase envelope and critical line define them.

Many more cases can be described. Extensive molecular simulation investigations of proteins, polymers, aqueous solutions at extreme conditions, etc. are ongoing (Cummings, 1999); many of these use commercial software which include computational chemistry as well as molecular simulation. An extensive listing of programs is maintained by ANTAS (1999). As this technology becomes more widespread, users will be challenged know enough of the state-of-the-art and the literature to decide how and when to utilize such methods.

Computational Chemistry

In Volume I of Reviews in Computational Chemistry the editors (Lipkowitz and Boyd, 1990) defined the subject to be "those aspects of chemical research which are expedited or rendered practical by computers." This is also implied in some of the professional press (Wilson, 1999). We have not taken this broad a view but rather focus on the aspects that pertain to chemical structure and reactivity. This does encompass *molecular mechanics* which optimizes the conformational energy in a molecular or macromolecular system by manipulating the internal structure, i.e. bond lengths, bond angles, torsion angles, and *quantum mechanics* which describes the electronic structure for a system of N-electrons.

Molecular mechanics (MM) is an empirically-based approach which was born out of efforts to predict structure and minimum energy conformations from known atomic connectivity of a molecule or macromolecular system. The expected structure or conformation of an N-atom system is postulated as that with the minimum potential energy calculated from a series of terms whose parameters are fitted to empirical information. The potential energy surface is also known as the force-field (Burket and Allinger, 1982).

$$V = V_{Stretch} + V_{Bend} + V_{Torsion} + V_{vdW} + V_{H-Bond} + V_{Cross} \quad (13)$$

Eq. (13) presents a typical force field that includes terms for harmonic bond stretching, angle bending, and torsional angle twisting as well as those for van der Waals repulsion (overlap) and dispersion (attraction) between atoms as in Eq. (10), for multipolar and non-bonding interactions such as hydrogen-bonding, and for interactions involving more than one source). For the basic organic systems that have been parametrized, molecular mechanics has proven to be quite effective in determining the relative energy differences between different conformations. The computational effort for systems with less than 50 atoms is minimal. One of the greatest advantages of MM has been its ability to explore very large systems such as proteins, polymers and oxides. The major limitation of molecular mechanics is its dependence on empirical force fields, its inability to model electron transfer, i.e. reactivity, and its difficulty in treating transition metal systems. Subsequent, and less accurate, calculations are needed to get entropies and free energies from MM.

Chemical reactivity, that is, bond breaking and formation, requires explicit treatment of electron transfer which is quantum mechanical in nature. Computational quantum chemistry finds solutions to the time-independent Schrödinger equation for an N-interacting *electron* system (see, for example, Doucet and Weber, 1996, chapter 9).

$$\hat{H}\Psi = E\Psi \quad (14)$$

where $\hat{H}$ is the Hamiltonian operator and Ψ is the wave function which defines the state of the N-electron system and the probability of finding the N-electron system with particular spatial and spin coordinates. This probability is found by integration and provides both the electronic and energetic structure of the system and therefore all properties of the system.

A series of basic approximations enables the N-particle problem to be solved as a series of n single-particle (electron) equations which are solved by the expansion of atomic orbital basis functions into a solution vector of molecular orbitals. The complete Hamiltonian operator describes the kinetic and potential energy terms for the single particle wave functions ϕ_i which contains terms for the kinetic energy of the electrons and of the nuclei, the repulsive nuclear-nuclear and attractive nuclear-electron interactions, and the repulsive electron-electron interactions.

$$H = -\frac{h^2}{8\pi^2 m_e}\sum \nabla_i^2 - \frac{h^2}{8\pi^2 m_e}\sum \frac{1}{M_\alpha}\nabla_\alpha^2 + \\ \sum\sum \frac{Z_\alpha Z_\beta}{|R_\alpha - R_\beta|} - \sum\sum \frac{Z_\alpha e}{|r_i - R_\alpha|} + \sum\sum \frac{e^2}{|r_i - r_j|} \quad (15)$$

Normally only the nuclei terms (second and third in Eq. 15) are considered separately. In addition to predicting the electronic state of the N-particle system, the molecular structure can be optimized to predict many other properties including energetic, spectroscopic, and chemical properties of the system.

The three basic approaches to solving the N-electron system are semiempirical quantum chemistry, *ab initio* molecular orbital theory, and density functional theory. All three approaches have their advantages and disadvantages and each has its own series of different variants (Head-Gordon, 1996; Irikura and Frurip, 1998). However, in essentially all cases, the greater the accuracy, the greater the expense. As a result, the best method depends strongly on the problem of interest.

Semiempirical Methods

Among the terms of Eq. (15), the most difficult and computationally expensive step involves the solution of the (last) electron-electron potential energy terms; it requires integrating the interactions among four electrons (two on two different atoms) over phase space or doing the *configuration interaction* (CI) calculations. Semiempirical methods replace the electron-electron repulsion integrals with empirically regressed parameters. This allows solutions for much larger macromolecular systems such as polymers, proteins and enzymes. The accuracy for simple hydrocarbon systems is usually quite good and the effectiveness of using semiempirical approaches for drug design efforts is high. Systems on the order of hundreds of atoms can routinely be optimized to determine structural, energetic, and electronic properties. The results from these efforts are typically used toward the development of simple structure-property relationships that can provide rapid screening, reduce experiments and design better products.

However, the empiricism that is introduced in these methods is often a severe limitation. The number of atoms that have been accurately parametrized is still rather small and obtaining reliable results for transition metals, for example, has proven to be extremely difficult.

Ab Initio Molecular Orbital Methods

There is a full suite of programs to expand electronic wave functions into a series of molecular orbitals and solve the basic Schrödinger's equations without the introduction of empirical information. They are typically called *ab initio* molecular orbital methods. They range from the very approximate Hartree-Fock (HF) methods to highly accurate couple cluster (CC) theory with size-consistent basis sets. Two of the major approximations required in solving the N-electron system are known to introduce various degrees of uncertainty in the accuracy of the solution. The first involves the neglect of electron correlation. The basic HF solution avoids solving all multi-centered electron integrals by approximating individual electron-electron interactions with a potential electronic field. Through a series of iterations, the field used is the on that interacts with the individual electrons. The solution to the many-body problem is then to find the ground state determinant. Electron-correlation effects can be added back to the solution in a systematic way, but this requires greater computational cost.

The second major approximation involves the application of molecular basis sets to describe the many-body wave function. The most rigorous solution would involve an infinite number of basis functions to solve. Most methods employ atomic orbital bases for the description of the molecular wave functions; by increasing the basis set completeness, accuracy can be increased. Also involved is the level of CI used. The exact solution to N-electron problem would require performing the intractable full-CI calculation with an infinite basis set. Instead, there are wave function recipes for increasing the accuracy of the electronic structure calculation for any given system. As described by Head Gordon (1996), the approach starts at the most basic level, *i.e.*, HF, and systematically increases both the level of CI and basis set completeness.

Dunning and co-workers (Peterson and Dunning, 1997ab; Peterson, *et al.*, 1997; Petersson, *et al.*, 1998) at Pacific Northwest Laboratories have demonstrated the state-of-the-art by performing very high-level CI calculations with high-level basis functions and extrapolating to the complete basis set limit. These are CC methods which include double and triple electron excitations [CCSD(T)] and size-consistent CI basis functions. They can predict heats of formation of small gas phase molecules to within 0.5-2 kJ mol^{-1} accuracy, which rivals even the best experiments. However, the computational expense is quite high, so the molecules are limited to the order of 10 atoms.

More realistic levels of computational expenditures involve performing many-body perturbation calculations which add back about 80% of the configuration energy. These allow much larger (25-50 atoms) systems than the very high level CI calculations. CC methods can also be performed; they treat the electronic ground state as well as single and double excitation (CCSD) determinants having greater fractions of CI but they demand more computation.

Two *ab initio* based approaches, the G2 by Pople and coworkers (Curtiss, *et al.*, 1991, 1993), and the complete basis set (CBS) methods by Petersson and coworkers (Petersson *et al.*, 1998) offer design strategies for increasing rigor in these methods. Both use a number of less taxing calculations along with a few high level calculations to determine what happens for many members of a series of species. This is known as applying *model chemistries*. Many recent industrial efforts to establish thermochemical properties databases of gas phase intermediates have employed such model-based chemistries. They cost about $3000-$5000 for a single heat of formation for molecules with less than 15 nonhydrogen atoms, a substantial savings compared to experiment (DOE, 1999).

Density Functional Theory (DFT)

Since the development of DFT by Nobel Prize winner Walter Kohn, an explosive growth has been seen in its use. The approach is fundamentally different from wave function methods in that it describes the energy of the ground state of an N-electron system as a functional of the energy density, $\rho(r)$ rather than of the wave function, Ψ. The rigorous energy equation is (Laird, *et al.*, 1996)

$$E_{el} = -\frac{1}{2}\sum_A \int \phi_i(r_1)\nabla^2 \phi_i(r_1)dr_1 + \sum_A \frac{Z_A}{|R_A - r_1|}\rho(r_1)dr_1$$
$$+ \frac{1}{2}\int \frac{\rho(r_1)\rho(r_2)}{|r_1 - r_2|}dr_1 dr_2 + E_{XC} \quad (16)$$

The problem to be solved is to find the function, $\rho(r)$, from

$$\left[-\frac{1}{2}\nabla_i^2 + \frac{Z_A}{|R_A - r_i|} + \int \frac{\rho(r_i)\rho(r_j)}{|r_i - r_j|}dr_j + \frac{\partial E_{XC}}{\partial \rho(r_i)}\right]\varphi_i(r_i) = \varepsilon_i \varphi_i(r_i) \quad (17)$$

The advantage of Eq. (17) is that the correlation effects are explicitly included in the exchange-correlation term, E_{XC}. The technique uses various sets of functional relations that are known to approximately describe electron correlation and exchange. DFT methods are much less costly since they scale as N^3 where N is the number of basis functions (or roughly the number of electrons) compared to N^5 or higher powers for advanced CI methods. DFT methods have proven to be invaluable for analyzing larger and more complex models of polymers, metals, metal oxides, peptides, proteins and other macromolecular systems (Laird, *et al.*, 1996; Truhlar and Morokuma, 1999).

Although DFT has proven to be applicable across a broad range of chemistries, there are areas where DFT has difficulties. The main limitation of DFT is that there is no systematic way improve the accuracy of the results from a particular method by modifying the functional expressions. And, like other methods, the prediction of van der Waals interactions and the properties of excited states are problematic for DFT. It should be noted that newer DFT methods attempt to solve the time-dependent problem which should increase our ability to address these latter problems (Runge and Gross, 1984).

The Reaction Environment - Mixing Molecular Simulation and Computational Chemistry

The greatest limitation of computational chemistry methods is the small size of systems that can be reliably modeled. For metals, the range is from a few to 30 atoms (Neurock, 1997). Using a single atom to describe the metal surface would be quite inadequate for systems such as benzene adsorption on a Ni(111) surface. So, efforts are being made to define *the model* as the reaction environment that surrounds the active site. This can include the nearest and next-nearest neighbor atoms on a surface, the presence of a solvent, or the local region of protein that surrounds an active site. The interaction of a molecule with a substrate, either liquid or solid, is clearly dependent upon its local environment. Many of the methods for extending *ab initio* treatments to more complex macromolecular systems involve coupling high level quantum chemical (QC) methods with lower level semiempirical QC algorithms or with molecular simulation methods. These include *ab initio*/MD, *ab initio*/MM, and *ab initio*/MC techniques.

Car and Parinello (1985) revolutionized the connection between electronic structure and molecular simulations when they demonstrated that molecular dynamics techniques could be coupled with DFT methods into "*ab initio*/MD" to simulate the changes in the electronic state as well as the dynamic changes in an N-particle system. In this approach, a series of fictitious masses are used to model the electrons; this enables the simulation of dynamic forces on both electrons and nuclei. These forces can subsequently be used to simulate the dynamic changes in atomic structure as a function of time and process conditions (see, *e.g.,* Blochel and Togni, 1996; Car and Parrinello, 1985; Curioni, *et al.*, 1997).

The development and application of *ab initio*/MM methods (*e.g.*, Matsubara, *et al.*, 1996) have grown exponentially over the past decade as scientists try to unravel behavior of macromolecules in solution. Most of these efforts have focused on understanding protein folding and solvent effects in biomolecular systems (*e.g.*, Truhlar and Morokuma, 1999). In *ab initio*/MM, the N-particle system is apportioned into two distinct regions. The first involves a small core area that incorporates the "active" site(s). This region, where electron transfer is carried out, requires rigorous QC calculations (DFT or advanced CI calculations) to treat bond breaking and formation processes. The second region is the environment that surrounds the active core. This region is much more extended. The sensitivity of the electron transfer to conditions is typically much less than that for the active core region. As a result, the *reaction environment* can be treated with a much lower level of theory. Empirical models such as MM offer an attractive solution in that they are computationally much easier to extend to longer length scales.

The main obstacle to reliable mixed system approaches involves accurately treating the QM/MM interface such as in solvent-based and materials systems. Morokuma (*e.g.*, Matsubara, *et al.*, 1996) has constructed a strategy for this issue that can be used to examine many different systems ranging from solvent effects to solid surfaces. For heterogeneous catalysis, Sauer (Eichler, *et al.*, 1997) has developed a similar method for zeolites. Whitten and Yang (1996) have used rigorous CI calculations to treat the site of chemisorption on a metal surface. In all these cases, the active core is then embedded into larger set of atoms that represent the environment which is modeled with single particle wave functions and a lower level theory.

The integration of *ab initio*/MC (static or dynamic) allows considerable expansion of length and time scales, though there have been relatively few efforts in this area. (Neurock and Hansen, 1998; Hansen and Neurock, 1999; Eichler, *et al.*, 1997; Hill and Sauer, 1995; Jorgenson and Ravimohan, 1985; Jorgenson and Gao, 1986; Jorgenson, *et al.*, 1994). Jorgenson's work has involved simulating the effect of aqueous solvation on reaction kinetics by including up to 500 water molecules. Quantum mechanics is used to map out the potential energy surface and structural MC simulations establish the conformations of solvents and their effects on the potential energy surface. This combination of quantum chemistry and MC simulation can model large systems with relatively modest CPU efforts.

Applications of Computational Chemistry To Chemical Processing

Computational chemistry has already begun to play an important role in process engineering. In particular, quantum chemistry is proving to be a valuable partner to experiment in predicting a broad range of properties important for process engineering models. Consistent with the service, advice and solve roles of properties in process design, computational chemistry can predict thermodynamic, kinetic, and material information.

Thermochemical and Thermophysical Properties

Computational chemistry prediction of thermodynamic properties has focused primarily on thermochemical and thermophysical properties of single phase systems. The accuracy is such that the prediction of thermochemical properties such as heats of formation, heats of reaction, and entropic terms can be computed with a fairly high level of accuracy (Dixon and Feller, 1999). Heats of formation within 0.5-1.5 kJ mol^{-1} of the experiment for molecular systems are possible for molecules with less than 12 atoms (Peterson and Dunning, 1997ab; Peterson, *et al.*, 1997) Not only are these values as accurate and less expensive than experiment, since toxic and explosive substances can be treated, there is no need to expose workers to potentially unsafe situations (Frurip, *et al.*, 1998).

Two examples for which QC played major service roles include DuPont's push to rapidly bring HFC-134A to market as a replacement for chlorofluorocarbons (CFCs) and Dow's development of more rigorous reaction/reactor models for thermal chlorination processes.

In the first case, the Montreal protocol which banned the production of CFCs placed serious demands on industry to quickly find suitable replacements for them. DuPont was able to capitalize on HF fluorination research efforts and conclude that (CF_3CH_2F) was one suitable replacement. However, very complex reaction pathways for producing HFC-134a were found as Figure 3 shows (Manzer, 1990).

The development of process models was limited by the few property data for the many possible intermediates. High level CI calculations were used to compute the geometries, vibrational and electronic energies which then led to the prediction of thermodynamic properties such as ideal gas heats of formation, equilibrium constants and heat capacities. Pair interactions between HFC-134a molecules and intermediates were computed and then fitted to the Leonard-Jones 12-6 potential (Eq. 10) to estimate fluid density and ideal gas transport properties of the proposed species (Dixon, *et al.*, 1995). The predicted heat capacity values matched the known experimental data quite well. This information was a critical complement to experiment in establishing the database used in the process design.

Kinetic Properties

In work at the Dow Chemical Company, Tirtowidjojo *et al.* (1995) used high-level G2 calculations in conjunction with transition state theory to calculate the reaction pathways, reaction energies, transition states and activation barriers for the chlorination of different C_2 and C_3 hydrocarbon intermediates. The kinetic results, along with some empirical information, were used to construct reactor models. The authors carried out a comprehensive set of calculations to develop a database for the general free radical reaction of R• + Cl_2•• RCl + Cl•. Many of their predicted rate constants were found to be in good agreement with data. The authors point out the difficulties in predicting accurate kinetics and show examples where G2 theory could not predict the barriers for some systems. Despite these limitations, they were able to construct reactor models that followed the computed kinetics.

Thus some rate constants can be obtained *a priori*, though there are still many for which calculations are significantly different from experiment. We expect that in the next 5 - 10 years, many of these issues will be solved with higher level calculations and better ways to incorporate the appropriate statistical mechanics. In particular, we should see more reliable QC predictions of kinetics for molecules with less than 20 atoms. Thus *ab initio* prediction of kinetics for small molecules will be able to function in the service and advice roles as process engineers seek pathways and conditions for maximum yield and selectivity. For systems of more than 20 atoms, the accuracy is likely to remain beyond that useful for quantitative precision and accuracy. Though some of this limitation will be the accuracy of the available methods, the more critical factor will be better modeling of the reaction environment. It will be necessary for experimental measurements to provide the proper atomic structure of the ground, product and transition states as the basis for the calculations.

Material Properties

The more critical questions for macromolecular systems such as catalysts, proteins, enzymes, polymers, and electronic materials are focused at the level of designing materials with specific properties, the advice role. Questions of material properties are directly tied to atomic structure, the properties and how they change with time or vary with processing conditions.

Pharmaceuticals

Computational chemistry has probably been most successful in the pharmaceutical industry where the largest fraction of computational chemists are employed. These companies view computational chemistry and combinatorial chemistry as "must have" technologies. Combinatorial chemistry involves the rapid synthesis and screening of target materials in order to map out the space for the possible combination of elements that lead to most active materials. Computational efforts can play an important role in these rapid scanning efforts by their ability to quickly identify candidate compounds with appropriate activities. This even affects business decisions such as Pharmecopia, a leading combinatorial synthesis company, purchasing Molecular Simulations, Inc. (Rogers, 1999).

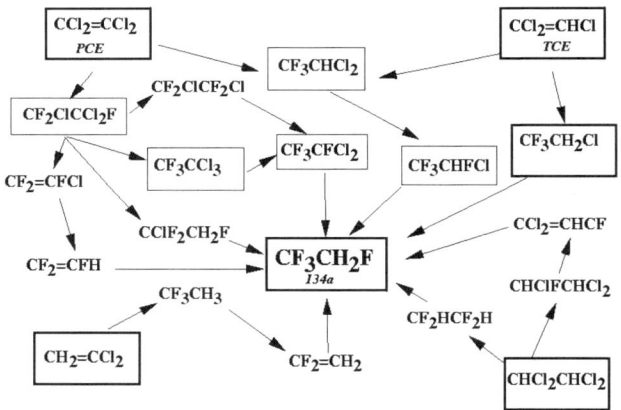

Figure 3. Fluorination routes of different chlorocarbons and chlorofluorocarbons to HFC-134a. Adapted from Manzer (1990).

The search for new drugs often starts with molecular mechanics and semiempirical methods which provide reasonable first candidates without detailed analysis of activity. It commonly involves more simple quantitative structure activity relationships (QSARS). Then as the range of phase space is narrowed, more advanced methods such as DFT, MP2 and even CI calculations are used for more accurate predictions which are then confirmed by experiment. Examples of the success of this methodology include the search for ibuprofen and the screening and identification of candidates for new HIV protease inhibitors.

Homogeneous Catalysts

Computational chemistry has been most successful when it has dealt with systems that are: 1) structurally well defined, 2) small in size, and 3) composed of first and second row atoms with simple electronic wave functions. Larger macromolecular structures such as homogeneous catalysts, are much more complicated than this. Thus, the number of atoms to describe a number of commercially relevant homogeneous catalysts are often in the hundreds. Although the structures may be well defined, modeling the complexity of the ligand environment and the influence of solvent can be difficult.

The exciting developments in the synthesis of novel metallocene-based olefin polymerization catalysts have recently opened up a fertile area for computational efforts. Much of the computational chemistry research that is currently being carried out in the chemical industry is directed toward developing novel homogeneous catalysts. Changes in the active metal center, the electronic and structural effects of the ligands and the solvation environment are all important design features. Industry has been working toward developing rapid screening methods which use a lower level of theory on idealize active complexes which mimic actual systems in order to identify promising leads. These efforts have provided a wealth of valuable advice to the experimentalist who are looking for new leads. Exxon, Union Carbide, Dow, DuPont and Phillips all have strong efforts for "intelligent screening" and modeling of homogeneous catalysts. Much of this work, unfortunately, is highly proprietary and is therefore not openly discussed in the literature.

A number of academic groups have also shown that computational chemistry, particularly DFT, is proving invaluable for understanding present generation homogeneous catalysts and generating new leads. Woo, *et al.* (1999), for example, have used DFT/MM to examine Brookhart Ni(II)diimine olefin polymerization catalysts which has the following general form of $(ArN=C(R)=C(R)=NAr)Ni-R'^+$ where the R is for organic ligands and Ar is for aromatic hydrocarbon ligands. The steric interactions between bulky ligands significantly influence the barriers for different elementary reaction steps. The DFT/MM embedding algorithm proved to be critical in describing energetic results in such a large system. Such breakthroughs have enabled these authors to examine design strategies for both ligand and metal centers.

Heterogeneous Catalysts

The application of computational methods to heterogeneous catalysis is significantly more challenging because of our inability to resolve the active atomic surface structure. So far, most of the QC efforts in this area have focused on *analysis* rather than *design* in order to develop techniques and demonstrate how computational chemistry can be used to predict reliable energies for elementary surface adsorption and reaction steps. For example, there have been a number of studies of chemisorption and surface reaction processes on idealized and well-defined model clusters or surfaces (see, *e.g.*, van Santen and Neurock, 1995; Whitten and Yang, 1996). DFT calculations treat transition metal, oxide and sulfide surfaces with similar accuracy to that of homogeneous complexes. The size and complexity of the model required to simulate an active site or region is best done with DFT methods. The results for well characterized surfaces typically report energies that are within about 20 kJ mol^{-1} of experiment.

Now, comprehensive efforts are beginning to move into designing surfaces with optimal properties. Besenbacher, *et al.* (1998) combined theory, ultrahigh vacuum experiments and novel synthesis methods to try to make a new steam reforming catalyst. The process involves dissociation of methane followed by the formation of carbon surface intermediates that can rapidly coke the catalyst surface. First-principle periodic slab calculations combined with ultrahigh vacuum scanning tunneling microscopy experiments probed different surface alloys and their effect on surface reactivity. Calculations and the STM studies suggested that there were novel Ni-Au surface compositions that might be resilient to coke formation. The next step was the synthesis and stabilization of these structures which performed as anticipated. This advice role of the calculations indicated which bimetallic surface structure would lower the metal-carbon interaction yet still promote the desired selective chemistry.

Neurock, *et al.* (1997) have studied the synthesis of vinyl acetate over Pd/Au catalysts. They showed that alloying palladium with gold at particular configurations shuts down the decomposition of acetic acid but helps to promote the addition of vinyl and acetate to form the vinyl acetate monomer. Their results provide a fundamental understanding as to why Pd/Au is much more selective than Pd for this chemistry. Figure 4 shows results for the C-H bond activation barrier for acetate. Figure 4A shows the energy of the barrier when gold is isolated in particular surface positions; the barrier energy is higher when gold is added. By way of illustration, Figure 4B shows the DFT transition state for the lowest energy barrier when no gold is present. Thus the calculations show how the addition of gold dramatically changes the barrier to nonselective decomposition routes. Current DFT methods now enable scientists to probe different bimetallics, different compositions, and different spatial arrangements on the surface. These might be used as "molecular handles" to begin to think about dialing in specific reactivity. The methods are therefore able to begin answering a number of "what if" type questions, thus a valuable role for advice.

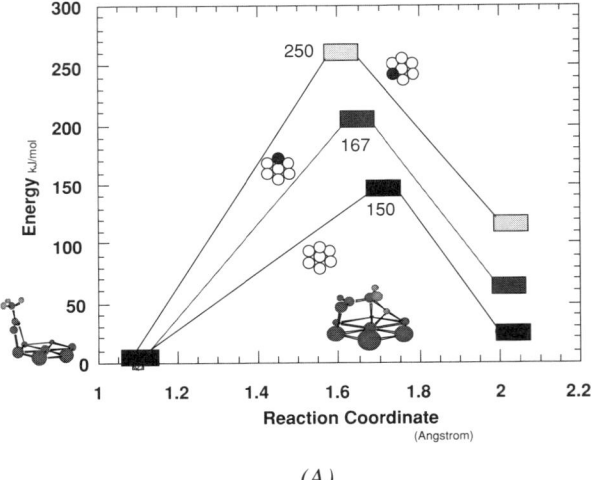

(A)

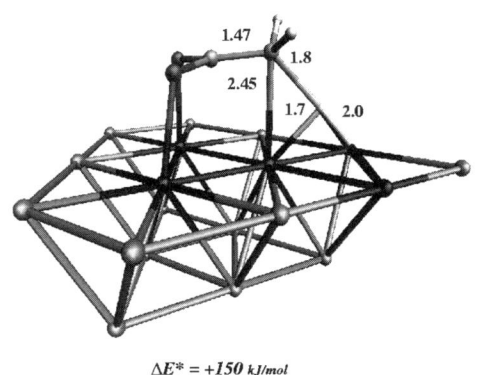

$\Delta E^* = +150 \; kJ/mol$

(B)

Figure 4. The effect of gold on activating the C-H bond of acetate for its decomposition on Pd(111). A) Representation of the DFT computed transition state; the energy to break the C-H bond is 150 kJ mol^{-1}. B) Adding gold at different locations changes the computed barrier.

Conclusions

The microsimulation fields of molecular simulation and computational chemistry have developed rapidly and are expected to grow in size and scale of activity and importance. However, given the roles of properties in computer-aided process engineering, this will probably be by evolution rather than a revolution. The evidence for this is twofold. First, the properties programs and methodology of process simulators use techniques which have long and strong traditions. While there are publications and codes which expand the simulators into new kinds of substances and conditions, there is yet to be seen, for example, any computational chemistry method directly accessible to process simulator users. Second, many of the equation of state and activity coefficient models included were first published decades ago. Carefully selection insures that they are quite adequate for many purposes. Thus, it will take some time before the service role of properties will directly involve microsimulation methods.

Rather, the first steps toward microsimulation meeting the service needs of process design will probably be to compute necessary data and validate them, followed by making them available to the process simulation routines in the same way as experimental information is currently included in their data bases.

However, it is more likely that the advice role will be implemented quickly since this normally happens in a more *ad hoc* fashion. The process will explore feasibility and ultimate limits by using thermodynamic and kinetic information. Further, microsimulation will yield more reliable mechanisms to guide understanding of how a process works. This can be used to make a better process.

The procedure needs to be done carefully. The problem must be phrased in an answerable fashion. The questions posed must be specific and realistic, the resources applied must be adequate, and the analysis must extract the appropriate results (Golab, 1997). It "would be foolish to model a binary vapor-pressure composition diagram *a priori*", but determining "the free energy of solvation for various organic solutes in water" is certainly appropriate. One might not expect to obtain the "absolute activity of a specific metallocene catalyst", but asking if a "ferrocene is better than a zirconocene" is reasonable even now. Not to be ignored are considerations of how to minimize the costs as well as how to articulate the results ways that lead to useful decisions and action. Thus the process should be directed toward information that is both valuable and timely. There should be influences on the issues of yield, selectivity, waste reduction. "Time to market" should be reduced. And, even if these are not found, when an innovation or an "aha!" occurs, the process is probably still of positive impact. The future should see microsimulation become routine for process innovation (Golab, 1997). The best way to utilize molecular simulation and computation chemistry experts is as essential and prominent members of a team. Their role is well-defined: it is to prepare the modeling strategy and validation plan for proposed concepts. They are to organize, interpret and present the results for the team as well as suggest further steps for possible for improved accuracy and for optimal use by managers in making decisions. Like the other process design team members, the microsimulation expert will be committed to the Technology Vision 2020 goals (AIChE, 1999) of 1) improving manufacturing processes for maximizing yields and minimizing environmental impact, and 2) to rationally designing new materials based on the retail properties of the final product and using longer-lived and more active catalysts. This mode of operation will be viable because there will be available so much more quantifiable, accurate and pertinent technical information. Further, younger workers will simply demand more modeling to support their ideas.

Microsimulation, especially as a part of comprehensive programs for investigation and innovation, will facilitate projects which move in many directions toward process design, optimization and control. We urge process design workers in both industry and academia to "get to know" their microsimulator colleagues so that research, education and practice can take advantage of these exciting opportunities for integration, efficiency and advancement.

Acknowledgments

J.P.O'C. is grateful to P.T. Cummings, R. Gani, and J. Golab, for sharing their presentation notes and to A.A. Chialvo for useful information. M.N. thanks the Donors of the Petroleum Research Foundation for Grant ACS PRF #31342-G5 and the National Science Foundation for Career Award Grant CTS-9702762.

References

Ahmad, B.S., and P.I. Barton (1999). Solvent recovery targeting. *AIChE J.*, **45**, 335-349.

AIChE (1999) Technology Vision 2020: The US chemical industry a call to action, innovation and change. http://www.aiche.org/vision2020.

Allen, M.P., and D.J. Tildesley (1993). *Computer Simulations in Chemical Physics.* NATO ASI Series 397, Kluwer, Dordrecht.

Ambrose, D. and C. Tsonopoulos (1995). Vapor-liquid critical properties of elements and compounds. 2. Normal alkanes. *J. Chem. Eng. Data*, **40**, 531-46.

ANTAS (1999) Computational chemistry software list and links. http://antas.agraria.uniss.it/software.html

Ben-Naim, A. (1992). *Statistical Thermodynamics for Chemists and Biochemists*, Plenum, NY.

Besenbacher, F., I. Chorkendorff, B.S. Clausen, B. Hammer, A.M. Molenbroek, J.K. Norskov, and I. Stensgaard (1998). Design of a surface alloy catalyst for steam reforming. *Science*, **279**, 1913-1915

Binder, K., (1979). Introduction: Theory and "Technical" Aspects of Monte Carlo Simulation. In K. Binder (Ed.), *Monte Carlo Methods in Statistical Physics.* Springer Verlag, Berlin.

Binder, K. (Ed.) (1995). *Monte Carlo and Molecular Dynamics Simulations in Polymer Science.* Oxford Press, New York.

Blas, F.J., L.F. Vega and K.E. Gubbins (1998). Modeling new adsorbents for ethylene/ethane separations by adsorption via π-complexation. *Fluid Phase Equil.*, **150-1**, 117-124.

Blochel, P., and A. Togni (1996). First-Principles Investigation of enantioselective catalysis: Asymmetric allylic amination with Pd complexes bearing P,N-ligands. *Organometallics*, **15**, 4125-4132.

Burket, J. L., and N.L. Allinger (1982). *Molecular Mechanics.* Am. Chem. Soc., Washington D.C.

Car, R., and M. Parrinello (1985). Unified approach for molecular dynamics and density functional theory. *Phys. Rev. Lett.*, **55**, 2471-2486.

Carlson, E.C., (1996). Don't gamble with physical properties for simulation, *Chem. Eng. Prog.*, **92**(10), 35-46.

Chemical and Engineering News (1997). Computation chemistry. May 12, 30-40; October 6, 26-32.

Chialvo, A.A., and J.M. Haile (1987). Determination of excess gibbs free energy from computer simulation: Multiple-parameter charging approach. *Fluid Phase Equil.*, **37**, 293-303.

Chialvo, A.A. (1993). Accurate calculation of excess thermal, infinite dilution, and related properties of liquid mixtures via molecular-based simulation. *Fluid Phase Equil.*, **83**, 23-32.

Chialvo, A.A., and P.T. Cummings (1996) Microstructure of ambient and supercritical water. Direct comparison between simulation and neutron scattering experiments. *J. Phys. Chem.*, **100**, 1309-16.

Chialvo, A.A., P.T. Cummings and H.D. Cochran (1998) Interplay between molecular simulation and neutron scattering in developing new insights into the structure of water. *IEC Res.*, **37**, 3021.

Cui, S.T., J.I. Siepmann, H.D. Cochran, and P.T. Cummings (1996). Intermolecular potentials and vapor-liquid phase equilibria of perfluorinated alkanes. *Fluid Phase Equil.*, **146**, 51-61.

Cummings, P.T., (1996). Molecular dynamics simulation of realistic systems. *Fluid Phase Equil.*, **116**, 237-47.

Cummings, P.T. (1999). Chemical engineering applications of molecular simulation and computational chemistry. Amer. Chem. Soc. meeting, Anaheim.

Curioni, A., M. Sprik, W. Andreoni, H. Schiffer, J. Hutter, and M. Parrinello (1997). Density Functional Theory-Based Molecular Dynamics Simulation of Acid-Catalyzed Chemical Reactions in Liquid Trioxane. , *J. Am. Chem. Soc.*, **119**, 7218-7229.

Curtiss, L. A., K. Raghavachari, G.W. Trucks, and J.A. Pople (1991). Gaussian-2 theory for molecular energies of first- and second-row compounds. *J. Chem. Phys.*, **94**, 7221-35.

Curtiss, L. A., K. Raghavachari, and J.A. Pople (1993). Gaussian-2 theory using reduced Moller-Plesset orders. *J. Chem. Phys.*, **98**, 1293.

Dixon, D. A., and D. Feller (1999). Computational Chemistry and Process Design, *Chem. Eng. Sci.*, In press.

Dixon, D., K. D. Dobbs, M. Neurock, J.J. Lerou, and T. Nakao (1995). The Prediction of Molecular Properties for Product Life Cycles. *Proc. of the 9th Int. Symp. on Large Chem. Plants*, 1-38.

DOE Workshop Report (1999). *Chemical Industry of the Future: Technology Roadmap for Computational Chemistry.* 1-73.

Dohgane, I. and K. Yuki (1998). The creation of a super-chemist by using computers in the chemical industry. *Internet J. Chem.* 1, article 35.

Dohrn, R., (1999). Thermophysical properties for chemical process design at Bayer. 17th European Seminar on Applied Thermodynamics, Villamoura, Portugal.

Doucet, J-P., and J. Weber (1996). *Computer-aided Molecular Design: Theory and Applications.* Academic Press, New York.

Eichler, U., M. Brandle, and J. Sauer (1997). Predicting Absolute and Site Specific Acidities for Zeolite Catalysts by a Combined Quantum Mechanics/Interatomic Potential Function Approach. *J. Phys. Chem. B*, **101**, 10035-10050.

Even, J., and M. Bertault (1999). Monte-Carlo simulations of chemical reactions in molecular crystals. *J. Chem. Phys.*, **110**, 1087-96.

Frenkel, D. and B. Smit (1996) *Understanding Molecular Simulation.* Academic Press, San Diego.

Frurip, D. J., N.G. Rondan, and J.W. Storer (1998). Implementation and application of computational thermochemistry to industrial process design at the Dow Chemical Company. In K.K. Irikura and D. J. Frurip, (Eds.); *Computational Thermochemistry: Prediction and Estimation of Molecular Thermodynamics*. Am. Chem. Soc.: Washington D.C., 319-340.

Gani, R., and L. Constantinou (1996). Molecular structure based estimation of properties for process design. *Fluid Phase Equil.*, **116**, 75-86.

Gani, R., and J.P. O'Connell (1999). Properties and CAPE: From present uses to future challenges. ESCAPE-9 Meeting, Budapest, May.

Gelb, L.D., and K.E. Gubbins (1998). Characterization of porous glasses: Simulation models, adsorption isotherms and the Brunauer-Emmett-Teller analysis method.

Golab, J. (1997). Better technical decisions with chemistry modeling. NSF Workshop, *Future directions in molecular modeling and simulation: Fundamentals and applications.*, Nov. 3-4.

Gubbins, K.E. and A.Z. Panagiotopoulos (1989). Molecular simulation. *Chem. Eng. Prog.*, **86**(10), 23-27.

Gupta, S.A., H.D. Cochran, and P.T. Cummings (1998). Nanorheology of liquid alkanes. *Fluid Phase Equil.*, **150-1**, 125-31.

Haile, J.M. (1991). *Molecular Dynamics Simulation.* Wiley Interscience, New York.

Hansen, E. W., and M. Neurock (1999). Modeling surface kinetics with first-principles-based molecular simulation. *Chem. Eng. Sci.*, **1999**, In press.

Harper, P.M., R. Gani, P. Kolar and T. Ishikawa (1998). Computer aided molecular design with combined molecular modeling and group contribution. Paper 222g, AIChE Annual Meeting, Miami Beach.

Head-Gordon, M., (1996). Quantum chemistry and molecular processes *J. Phys. Chem.*, **100**, 13213-13225.

Hill, T.L. (1960). *Introduction to Statistical Thermodynamics*, Addison-Wesley, Reading.

Hill, J., and J. Sauer (1985). Molecular mechanics potential for silica and zeolite catalysts based on *ab Initio* calculations. 2. Aluminosilicates. *J. Phys. Chem.*, **99**, 9536-9550.

Horvath, A.L. (1992), *Molecular Design: Chemical Structure Generation from the Properties of Pure Organic Compounds*, Elsevier, New York.

Irikura, K. K., and D.J. Frurip, (Eds.) (1998). *Computational Thermochemistry: Prediction and Estimation of Molecular Thermodynamics*. Am. Chem. Soc.: Washington D. C.

Ishikawa, T. (1998) *Traditional and New Chemical Products at Mitsubishi. 8th International Conference on Properties and Phase Equilibria for Product and Process Design*, Nordwijkerhout, Netherlands.

Joback, K.G., and G. Stephanopoulos (1990). Designing molecules possessing desired physical property values. In J.J. Siirola, I.E. Grossman and G. Stephanopoulos (Eds.), *Foundations of Computer-Aided Chemical Process Design*. CACHE, Elsevier, Amsterdam, 363-387.

Jorgensen, W. L., and C. Ravimohan (1985). Monte Carlo simulation of differences in free energies of hydration. *J. Chem. Phys.*, **83**, 3050-3054.

Jorgensen, W. L., and J. Gao (1986). Monte Carlo simulations of the hydration of ammonium and carboxylate Ions. *J. Phys. Chem.*, **90**, 2174-2182.

Jorgenson, W. L., J.F. Blake, D. Lim, and D.L. Severance (1994). Investigation of solvent effects on pericyclic reactions by computer simulation. *J. Chem. Soc. Far. Trans.*, **90**, 1727-1732.

Kirkwood, J.G. (1935). *J. Chem. Phys.*, **3**, 300. .

Kiyohara, K., K.E. Gubbins, and A.Z. Pangiotopoulos (1998). Phase Coexistence properties of polarizable water models. *Mol. Phys.*, **94**, 803-8.

Kleintjens, L.A.L. (1998). Thermodynamics of organic materials: A challenge for the next decades. *Fluid Phase Equil.*,**158-60**, 113 - 121.

Kofke, D.A., and P.T. Cummings (1997). Quantitative comparison and optimization of methods for evaluating the chemical potential by molecular simulation. *Mol. Phys.* **92**, 973-990.

Kofke, D.A., and P.T. Cummings (1998). Precision and accuracy of staged free-energy perturbation methods for computing the chemical potential by molecular simulation. *Fluid Phase Equil.*, **150-151**, 41-49.

Kollman, P.A., (1996). Advances and continuing challenges in achieving realistic and predictive simulations of the properties of organic and biological molecules. *Acc. Chem. Res.*, **29**, 461-9.

Laird, B., R. B. Ross, and T. Ziegler (Eds.) (1996). *Chemical Applications of Density Functional Theory*. Vol. 629. Am. Chem. Soc., Washington D.C.

Lipkowitz, K.B., and D.B. Boyd (1990). Preface on the meaning and scope of computational chemistry. *Rev. Comp. Chem.*, **I**, vii-xii.

Mah, R.S.H. and W.D. Seider (Eds.) (1981). *Foundations of Computer-Aided Chemical Process Design II.* CACHE, AIChE, New York, pp. 1-60.

Manzer, L. E. (1990). The CFC-ozone issue: Progress on the development of alternatives to CFCs. *Science*, **1990**, 31-35.

Matsubara, T., S. Sieber, and K. Morokuma (1996). A test of the new "Integrated MO+MM" (IMOMM) method for the conformational energy of ethane and n-butane, *Int. Jour. of Quant. Chem.*, **60**, 1101.

McCallum, C.L., T.J. Bandosz, and K.E. Gubbins (1999). A molecular model for adsorption of water on activated carbon: Comparison of simulation and experiment.

McQuarrie, D.A., (1975). *Statistical Mechanics*, Harper and Row, NY.

Mei, D. (1999). Personal Communication, Univ. Virginia.

Millat, J., J.H. Dymond and C.A. Nieto de Castro (1996). *Transport Properties of Fluids*, Cambridge U. Press, Cambridge, UK.

Neubauer, B., B. Tavitan, A. Boutin, and P. Ungerer (1999). Molecular simulations on volumetric properties of nautral gas. *Fluid Phase Equil.*, **161**, 45-62.

Neurock, M. (1997). Catalytic reaction pathways and energetics from first principles. In G. F. Froment and K.C. Waugh (Eds.), *Dynamics of Surfaces and Reaction Kinetics in Heterogeneous Catalysis*, Vol. 109. Elsevier, Amsterdam, 1-35.

Neurock, M., and E.W. Hansen (1998). First-principles-based molecular simulation of heterogeneous catalytic surface chemistry. *Comp. Chem. Eng.*, **22**, S1045-S1059.

O'Connell, J.P., (1984), Structure of thermodynamics in process calculations. In A.W. Westerberg and H.H. Chien (Eds.), *Proc. Sec. Int. Conf. on Foundations of Computer-Aided Chemical Process Design*. CACHE, Amer. Inst. Chem. Eng., New York. 219-250.

O'Connell, J.P. (1999). When theory might be useful: Connecting fundamentals to practice in physical properties models, AIChE meeting, Houston, TX.

Panagiotopoulos, A.Z., (1996). Current advances in Monte Carlo methods. *Fluid Phase Equil.*, **116**, 257-66.

Peterson, K. A., and T.H. Dunning (1997a). The CO molecule - The role of basis set and correlation treatment in the calculation of molecular properties *Theochem-Jour. Mol. Struct.*, **400**, 93-117.

Peterson, K. A., and T.H. Dunning, (1997b). Benchmark calculations with correlated molecular wave functions. 8. Bond energies and equilibrium geometries of the CHN and C2HN (N=1-4) series *J. Chem. Phys.*, **106**, 4119-4140.

Peterson, K. A., A.K. Wilson, D.E. Woon, and T.H. Dunning (1997). Benchmark calculations with correlated molecular wave functions. 12. Core correlation effects on the homonuclear diatomic molecules B-2-F-2 *Theo. Chem. Acc.*, **97**, 251-259.

Petersson, G. A., (Ed.) (1998). *Complete Basis Set Thermochemistry and Kinetics*; Am. Chem. Soc., Washington, D.C., 1998.

Petersson, G. A., D.K. Malick, W.G. Wilson, J.W. Ochterski, J.A. Montgomery Jr., and M. J. Frisch (1998). Calibration and comparison of the Gaussian-2, complete basis set and density functional methods for thermochemistry *J. Chem. Phys.*, **109**, 10570-10579.

Potoff, J.J., and A.Z. Panagiotopoulos (1998). Critical point and phase behavior of the pure fluid and a Lennard-Jones mixture. *J. Chem. Phys.*, **109**, 10914-10922.

Prausnitz, J.M. (1999). Thermodynamics and the other chemical engineering sciences; Old models for new chemical products and processes. *Fluid Phase Equil.*, **158-60**, 95-111.

Quirke, N. (1999) Applications and methodology of molecular simulation - An international conference in electronic format. http://molsim.vei.co.uk, April.

Reed, T.M., and K.E. Gubbins, *Applied Statistical Mechanics*, McGraw-Hill, NY 1973.

Reid, R.C., J.M. Prausnitz, and B.E. Poling (1987). *Properties of Gases and Liquids* 4th Ed. McGraw-Hill, New York. 5th Ed. in Preparation.

Rogers, R. (1999). MSI combines computers with chemistry. *Chem. and Eng. News*, **77**(17), 17-18. See also http://www.msi.com.

Runge, F., and E.K.U. Gross (1984). Density functional theory for time-dependent systems. *Phys. Rev. Lett.*, **52**, 997-1000.

Sadus, R.J., (1999). *Molecular Simulation of Fluids*. Elsevier, Amsterdam.

Sandler, S.I. (Ed.) (1993). *Models for Thermodynamic and Phase Equilibria Calculations*. M. Dekker, New York.

Sarman, S., D.J. Evans, and P.T. Cummings (1998). Recent developments in Non-Newtonian molecular dynamics. *Phys. Rep.*, **305**, 1-92.

Siepmann, J.I., S. Karaborni and B. Smit (1993). Simulating the critical behaviour of complex fluids. *Nature*, **365**, 330.

Shukla, K. (1997a) Thermodynamic properties of binary mixtures of atoms and diatomic molecules from computer simulation and perturbation theory. *Fluid Phase Equil.*, **127**, 1.

Shukla, K. (1997b) Thermodynamic properties of binary mixtures of atoms and quadrupolar diatomic molecules from computer simulation. *Fluid Phase Equil.*, **128**, 29 - 45, 47.

Shukla, K. (1997c) Total and excess properties of nonideal ternary fluid mixtures via isothermal-isobaric molecular dynamics simulation: size and energy parameter effects. *Fluid Phase Equil.*, **128**, 47 -

Slusher, J.T. (1998). Estimation of infinite dilution activity coefficients in aqueous mixtures via molecular simulation. *Fluid Phase Equil.*, **153**, 45-61.

Slusher, J.T. (1999). Infinite dilution activity coefficients in hydrogen-bonded mixtures via molecular dynamics: The water/methanol system. *Fluid Phase Equil.*, **154**, 181-192.

Sliwinska-Bartkowiak, M., R. Sikorski, and K.E. Gubbins (1997). Phase separations for mixtures in well-characterized porous materials: Liquid-liquid transitions. *Fluid Phase Equil.*, **136**, 93-100.

Tirtowidjojo, M. M., B.T. Colegrove, and J.L. Durant (1995). Transition state and G2Q analysis of hydrocarbon radical reactions with Cl_2. *IEC Res.*, **34**, 4202-4211.

Truhlar, D. G., and K. Morokuma, (Eds.) (1999). *Transition State Modeling for Catalysis*. Vol. 721. Am. Chem. Soc. Washington D.C.

Tsonopoulos, C. and Z. Tan (1993). The critical constants of normal alkanes from methane to polyethylene II. Application of the Flory theory. *Fluid Phase Equil.*, **83**, 127-138.

van Santen, R. A., and M. Neurock, (1995). Concepts in theoretical heterogeneous catalytic reactivity. *Catal. Rev. Sci. Eng.*, **37**, 557-698.

Westerberg, A.W., and H.H. Chien (Eds.) (1984). *Proc. Sec. Int. Conf. on Foundations of Computer-Aided Chemical Process Design.* CACHE-AIChE, New York. 219-302.

Westmoreland, P.R. (1997). NSF Workshop, *Future directions in molecular modeling and simulation: Fundamentals and applications.*, Nov. 3-4.

Whiting, W. B., and Y. Xin. (1999). Sensitivity and uncertainty of process simulation to thermodynamic data and models: Case studies. AIChE meeting, Houston.

Whitten, J. L., and H. Yang (1996). Theory of chemisorption and reactions on metal surfaces. *Surf. Sci. Rep.*, **24**, 55-124.

Wilson, E.K. (1999). Bridging Chemistry and Engineering. *Chem. and Eng News*, **77**(17), 24-32.

Woo, T. K., P. M. Margl, L. Deng, L. Cavallo, and T. Ziegler (1999). Towards more realistic computational modeling of homogeneous catalysis by density functional theory: Combined QM/MM and *ab initio* molecular dynamics. *Catal. Today,* **50**, 479.

THE WEB AND FLOW OF INFORMATION IN PHARMACEUTICAL R&D

Sangtae Kim
Parke-Davis Pharmaceutical Research
Warner-Lambert Company
Ann Arbor, MI 48105

Abstract

This paper considers how the world-wide web, in its internal (intranet) and external (internet) forms, has transformed the flow of information in pharmaceutical research and development. The opportunities and challenges are especially striking when the advances in information technology are overlayed against the backdrop of the changes in the profile of the discovery and development pipeline of the pharmaceutical industry.

Keywords

Information technology, Intranet, Internet, Information silo.

Introduction

The synergistic emergence of the web browser and the internet is one of the major (and some would argue that it is *the* major) development of the final decade of the second millennium. Entire industries have undergone great upheavals as e-commerce and "dot-coms" rearrange societal perceptions of market valuations. It is thus not surprising that these same forces are creating significant opportunities in pharmaceutical research. The main point of this presentation, however, is the interactions between the revolution in information technology (IT) and the dramatic changes in the profile of the discovery and development pipeline of the pharmaceutical industry.

A Changing Profile

The typical time line for pharmaceutical research and development, from the discovery of a new compound and assays against biological targets, to the phases of the clinical trials and the approval by the national regulatory agencies is shown in the last figure. The "funnel" in the first figure on the following page depicts the concept of attrition, i.e., starting from many and ending with a few, as candidates are winnowed for various reasons including toxicity, efficacy, stability and bioavailability. The high attrition rate is an important factor in the large sums invested in research by the pharmaceutical industry.

The current *raison d'etre* of the information technology organization in a pharmaceutical research center is cycle compression, as captured in the second figure on the following page by the compressed funnel overlayed on top of the prior (ostensibly IT-deficient) funnel. In addition to the positive financial implications, the more rapid introduction of an innovative new therapeutic product can be linked statistically with an enhancement in the quality of life or even number of lives saved. However, this conventional picture will change dramatically over the next decade with the introduction of new technologies in the discovery and preclinical development segments of the pipeline.

The advances in genomics will provide a new framework for a more fundamental understanding of the underlying mechanisms for diseases. The number of biological targets (i.e., the number of biological assays for the compound library) will increase dramatically. The number of entities in the compound library will also grow dramatically thanks to combinatorial chemistry and its associated technologies. New, automated, high throughput screening (HTS) technologies will accommodate the rapidgrowth in both the "rows" and "columns" of the matrix. From an IT perspective, the result is an explosion in the volumes of data generated in the discovery segments of the pipeline.

It is equally clear, however, that a matching growth will not occur in the number of clinical trials, as these are constrained by societal factors such as the number of patient volunteers. The implication of this reality is the final point to be drawn from the new profile shown in the second figure below. Advances in datamining are required to help filter the greater volumes of data and translate the new wealth of choices in the preclinical segments into a set of more robust candidates for the clinical trials, i.e., less attrition in the high-cost portion of the pipeline. So the message of the second figure below is a *changing profile* as well as the conventional goal of cycle-compression.

Last but not least, the final segment of the pipeline will also undergo a transformation as the regulatory review process moves to a "paperless", electronic review.

Concluding Remarks

The technological challenges in building the infrastructure to support the new profile and to facilitcycle compression are of themselves almost insurmountable. But an even greater challenge lies in the fact that the information storage in each segment of the pipeline may be "owned" by different departments (perhaps spread over multiple continents over the globe) within the R&D organization. The term "information silo" is often used to describe this phenomenon. The image that springs to mind is quite the opposite of a pipeline with a rapid flow of information from department to department. The situation is not an IT problem per se, but the IT organization nevertheless may be held accountable for "solving" the silo problem.

This brings me back to the title of my paper. Rather than expending huge monetary capital (not to mention political capital) in the cause of "taking over" or dismantling silos, the astute information technology organization of the new millennium will create incentives to foster the widespread adoption of "web portals" that link the silos together with access control lists (ACL) that reassure all "silo owners" that only those that "need to know" are sharing the information. Indeed, as we change our vantage point to a higher elevation to encompass data-sharing with external alliances, the internal and external portals (ACLs) appear as a self-similar hierarchy (fractal structure) of webs within webs.

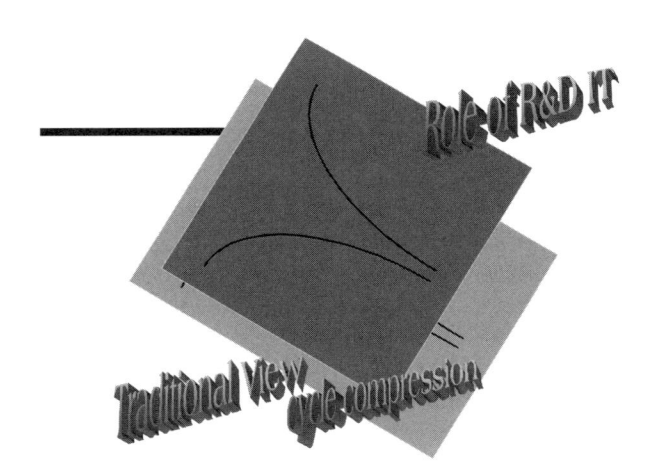

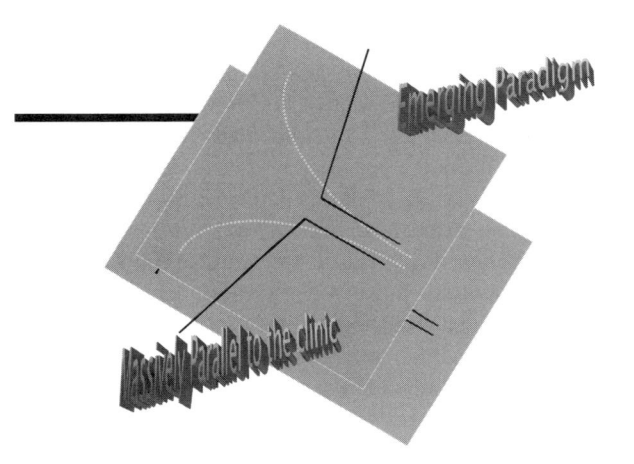

New Drug Development Timeline for USA

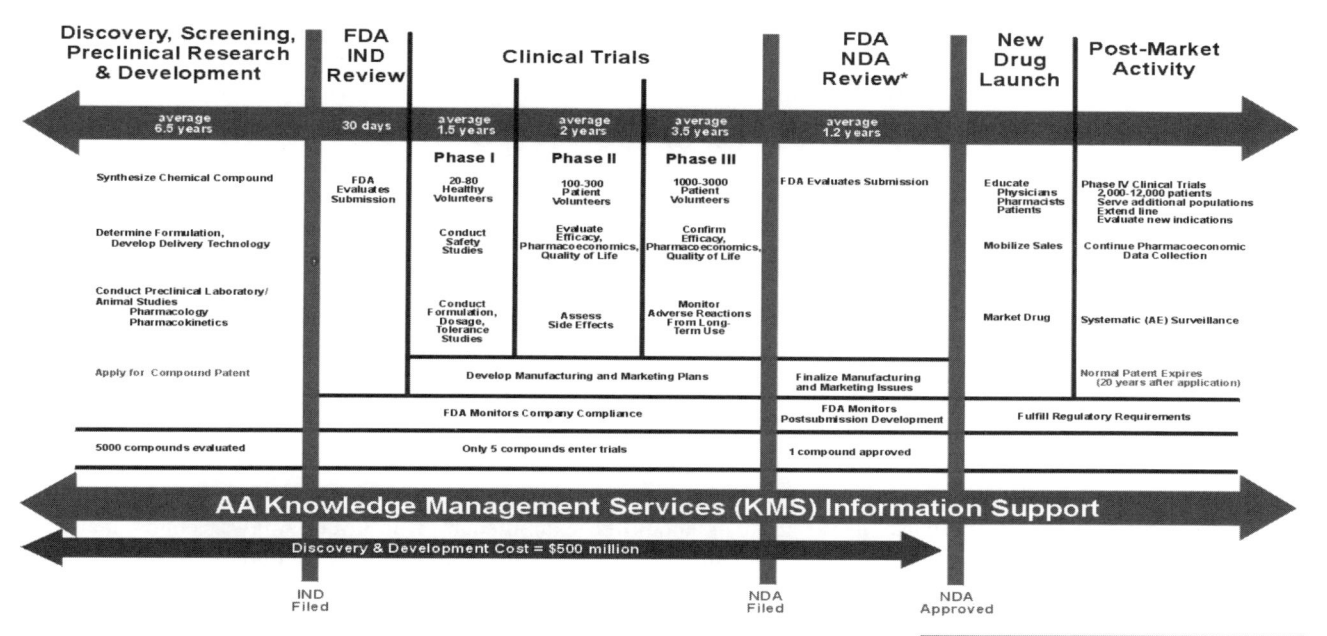

Discovery, Screening, Preclinical Research & Development	FDA IND Review	Clinical Trials			FDA NDA Review*	New Drug Launch	Post-Market Activity
average 6.5 years	30 days	average 1.5 years	average 2 years	average 3.5 years	average 1.2 years		
		Phase I	Phase II	Phase III			
Synthesize Chemical Compound	FDA Evaluates Submission	20-80 Healthy Volunteers	100-300 Patient Volunteers	1000-3000 Patient Volunteers	FDA Evaluates Submission	Educate Physicians Pharmacists Patients	Phase IV Clinical Trials 2,000-12,000 patients Serve additional populations Extend line Evaluate new indications
Determine Formulation, Develop Delivery Technology		Conduct Safety Studies	Evaluate Efficacy, Pharmacoeconomics, Quality of Life	Confirm Efficacy, Pharmacoeconomics, Quality of Life		Mobilize Sales	Continue Pharmacoeconomic Data Collection
Conduct Preclinical Laboratory/ Animal Studies Pharmacology Pharmacokinetics		Conduct Formulation, Dosage, Tolerance Studies	Assess Side Effects	Monitor Adverse Reactions From Long-Term Use		Market Drug	Systematic (AE) Surveillance
Apply for Compound Patent		Develop Manufacturing and Marketing Plans			Finalize Manufacturing and Marketing Issues		Normal Patent Expires (20 years after application)
		FDA Monitors Company Compliance			FDA Monitors Postsubmission Development	Fulfill Regulatory Requirements	
5000 compounds evaluated		Only 5 compounds enter trials			1 compound approved		

AA Knowledge Management Services (KMS) Information Support

Discovery & Development Cost = $500 million

IND Filed — NDA Filed — NDA Approved

* As mandated by the FDA Modernization Act of 1997, the FDA Performance Goals include reviewing Standard NDAs within 10 months by fiscal year 2002.

This chart was compiled by members of the Parke-Davis KMS Staff through published sources. It may be reproduced only for educational or research purposes. August 1998

Drug Development Timeline for USA
Bibliography

All data and timeline information on the Drug Development Timeline came from the published literature listed below.

US Drug and Biologic Approvals in 1996
Beary JF, Robillard LE, Woollett GR, Siegfried JD, White TX, Shriver DA
Drug Dev Res. 1997; 40: 275-291

Sustaining Innovation in US Pharmaceuticals: Intellectual Property Protection and the Role of Patents
Boston Consulting Group
Boston: Boston Consulting Group, 1996
pages 37-38

CDER 1997 Report to the Nation: Improving Public Health Through Human Drugs
Center for Drug Evaluation and Research
www.fda.gov/cder/reports/rptntn97.pdf
pages 3-5

Patent Fundamentals for Scientists and Engineers
Gordon, Thomas T. and Cookfair, Arthur S.
Boca Raton: CRC Lewis, 1995
page 5
Call #T211. G67 1995

New Drug Development: A Regulatory Overview
Mathieu, Mark P. and Evans, Anne G.
Waltham, MA: PAREXEL International Corporation
pages 6-16, 19, 129-130, 146-147
Call # HD9000.9.U5 N41 1997

Clinical Research in Pharmaceutical Development
Bleidt, Barry and Montague, Michael
New York: Dekker, 1996
pages 62-67
Call # RM301.27 .C576 1996

Drug Research: From the Idea to Product
Kuhlmann J
Int J Clin Pharmacol Ther. 1997; 35: 541-552

Multinational Pharmaceutical Companies: Principles and Practices
Spilker, Bert
New York: Raven Press, 1994
pages 511-517, 555-561
Call # HD9665 .S65 1994

Inside the Drug Industry
Spilker, Bert and Cuatrecasas, Pedro
Barcelona: Prous Science Publishers, 1990
page 45
Call # HD9665 S6

This bibliography was compiled by members of the Parke-Davis KMS Staff. It may be reproduced only for educational or research purposes. August 1998

INFORMATION AND MODELING FOR GREENER PROCESS DESIGN

Robert W. Sylvester, W. David Smith and John Carberry
E. I. DuPont de Nemours and Company, Inc.
Wilmington, DE 19880-0101

Abstract

The concept of "sustainable growth" is described from the viewpoint of the DuPont Company. Heuristics for estimating waste treatment costs and for solvent selection are reviewed, and open questions are identified. Sustainable products should be examined by Life Cycle Analysis, and the results of such an analysis for Nylon 66 in an automobile application are presented. Future information and modeling needs for developing greener processes are identified and discussed.

Keywords:

Sustainable, Wastes, Inherent safety, Pollution prevention, Life-cycle assessment, Process design.

Introduction

The chemical processing industry is challenged by customers, shareholders, employees, neighbors and regulators to minimize environmental impact in the broadest sense. This broad sense includes not only minimization of waste releases, but also remediation of past wastes, assurance that finite natural resources are appropriately stewarded, proof of the safety of products through downstream processing, use and disposal of wastes, and protection of employees and neighbors from explosions and emergency releases. The challenge of all members of the industry is to move from a defensive posture of regulatory compliance, through optimization of processes which minimize wastes and treatment costs, to a final state in which customers may see broader environmental performance as a competitive advantage. Within DuPont, this movement to greener processes is termed "sustainable growth."

Traditional industry practice has focussed on economic performance as the objective for process optimization. Techniques to optimize economic potential are applied in DuPont beginning with evaluation of research opportunities and including research management, conceptual design, process design, process integration, utility specification, operations, and technology retrofit. These techniques include MINLP (mixed integer, nonlinear programming), operations research, process design heuristics, heat and mass integration, process modeling and dynamic simulation. However, environmental performance and safety performance, not just economics, are objectives for sustainable processes. This paper examines both new developments and challenges to include these objectives in design.

Waste Treatment

Consider waste treatment broadly to include remediation of soil and contaminated groundwater, treatment of wastewater and solid wastes, and abatement of air emissions. Many processes, especially those involving molecular transformation, produce byproducts, which can not be economically recycled. An example is acetaldehyde, an undesired byproduct of high temperature polyester polymerization, created by the dehydration of ethylene glycol and generated from a modern polymerization process at the rate of perhaps 0.1% of the polymer production rate. The industry has no technology for economically producing polyester without generating acetaldehyde in small amounts. These amounts are far too small to permit profitable recovery. Treatment of the acetaldehyde may be in the form of absorption into water with biological oxidation in dilute wastewater or combustion – often with recovery of the fuel value. The challenge in selecting the treatment process is to identify the appropriate degree of control – in this case essentially complete destruction – and to rapidly identify the control technology with the lowest net present cost. Models of treatment costs conclude that the minimum cost for treatment of wastewater is an annual cost of $2,000 per pound per hour of biological oxygen demand and $300 per gpm. The extent of acetaldehyde destruction by biological oxidation may be limited by air stripping during treatment. If new investment is required to expand the water treatment facility, minimum incremental investment is $6,000 per pound per hour of BOD and $2,000 per gpm. Alternatively, minimum investment for incremental capacity of air pollution control equipment is $40 per cubic foot per minute of gas (air) capacity, and annual operating costs for dilute streams are $6 per cfm. (Dyer and Taylor) Optimization of treatment costs requires development of process models, flowsheets and approximate process design and consideration of full vs. incremental treatment costs. Providing the design engineer with up-to-date economic information, available spare capacities of treatment facilities, and a methodology for using the information is one of the challenges for the design process. (Mulholland and Dyer)

Many chemical process industry sites have concluded that catalytic and thermal oxidation are appropriate abatement technologies for volatile wastes like acetaldehyde. Also, they conclude that one central oxidizer connected to many tanks, loading spots, and process vents is far less expensive than multiple oxidizers. A frequent practice is to connect many pieces of equipment into a common vent collection system. Such a practice may lead to a tremendous increase in the risk of fire moving from one area throughout the entire plant – an example of possible conflict between environmental and safety objectives. Stringent design and operating practices must be implemented to minimize the risk of fire propagation through the vent collection system. (Clark and Sylvester).

While the literature provides many sources of information on the performance capabilities of environmental control systems, economics and safety, certain important questions seem to remain unanswered. These include:

- What is the range of typical investments in environmental treatment systems? What is the added process investment due to waste and requirements for internal recycle? The experience of the authors is that environmental treatment systems are 5 to 15% of total site investment for many chemical plants and that the additional processing costs for waste generation are at least a small percentage of the full process investment plus any dedicated equipment for recycle. That added process investment, including larger process equipment, additional separation steps, and utilities may equal the treatment system costs – but these are only "guesstimates". A broad study of these questions might be very enlightening.
- Treatment systems typically experience a maximum load during transients at startup or shutdown. Even processes designed for minimum waste generation often create waste from vessel cleaning prior to maintenance. How can these wastes be avoided? An equally important challenge for the designer is to determine how these transients can be quantified and then how these short time constant changes can be coupled in process modeling with the long time constant biological waste disposal facilities.

Source Reduction

No process designer sets out to produce wastes, which only add cost and investment. What are barriers to source reduction? An obvious barrier is the limitations of knowledge, such as how to produce polyester at economical rates without generating acetaldehyde. In some cases, the lack of knowledge is not so fundamental. An example is the selection of solvents for the batch synthesis of pharmaceuticals and agricultural chemicals. These chemicals are complicated, multifunctional molecules prepared from commercially available raw materials in six to ten reaction steps. Intermediate and final products are frequently solids if isolated and must be dissolved in solvents that do not interfere with the reaction steps. A development chemist who typically investigates choices within a class of solvents known to be non-reactive often does the actual solvent selection. The chemist seeks to obtain the highest yield of product in an inexpensive solvent. Whether partially recovered or used on a once through basis, (increasingly rare since the 1970's) solvent losses and exposures can be a major environmental burden and threat to sustainability. Economic impacts of the solvents include the costs of safe guards which protect employees and neighbors from toxic solvents; the management of fire and explosion hazards; solvent recovery; and control of emissions to meet corporate objectives and regulations. Impacts may vary widely among possible solvents, and even within a class of solvents.

DuPont recently developed a guide to solvent selection for internal use. One premise of the guide is that development chemists understand the necessary class of solvent and would value guidance on the relative safety and environmental impact of choices within that class. A second premise is that facilities will be designed to protect employees and the environment, and that these facilities will be less expensive if less hazardous solvents can be used with no yield penalty. Thus, the role of the solvent selection guide is to identify optimal solvents within a class. Key evaluation characteristics were selected to be easily available, readily understood, and related to the economics of risk management. A simple set of heuristics were identified:

- Relative costs of treatment to achieve regulatory compliance for air and water emissions. This is greatest for RCRA (Resource Conservation and Recovery Act) listed compounds, intermediate for hazardous air pollutants (HAP) and priority pollutants, and least for volatile organic compounds (VOCs). While classification is based on US regulations, similar criteria are widely used in other countries and reflect good environmental control practices.
- Relative costs for protection from toxic chemical exposure and loss in labor productivity associated with this protection. This is modeled as a function of threshold exposure as measured by TLV, Threshold exposure Limit Value. Highly bioaccumulative compounds are penalized.
- Risk from fire is modeled using the flammability classifications of the National Fire Protection Associate system, based on flash point and normal boiling point.
- Estimated costs of recovery from water.

Example applications of this system, focussing on differences among solvents within a class, are given in Tables 1-3. A surprising conclusion was that for many solvent classes a clearly preferred solvent might be identified. For commonly used aliphatic solvents, heptane is clearly preferable to n-hexane because of higher TLV. Higher boiling aliphatic compounds with a flash point above ambient temperature would be even more preferable unless the TLV is higher. Among aromatic solvents, xylene is preferred by all listed criteria. Only in the case of the normal alcohols is there a true tradeoff. Ethanol has a larger TLV, but requires special control unless contaminated with denaturants and azeotropes with water. Butanol might be preferable to methanol if dermal exposure is not a concern.

SmithKline Beecham independently derived environmental and safety criteria to create a broad guide to solvent selection – with very similar conclusions. (Curzons et al.) The challenge is to build on these early systems and to package them in a way that they can be incorporated into computer aided design systems.

Sustainable Products

Superior process design is not sufficient to define an environmentally preferable product. The experience within DuPont is that the full life cycle environmental impact of a product must be examined from extraction of minerals and hydrocarbons to the final disposal and ultimate fate of products. This examination is done by a life cycle assessment. Environmental impact included global impacts, such as ozone depletion and global warming potential, regional impacts, such as acid rain, and local impacts, such as water consumption and contamination. Life cycle assessments are particularly meaningful when used to compare process options, materials options, or reductions in environmental impact through time.

Table 1. Solvent Selection Guide, Alkanes.

	n-Hexane	Heptane
Environmental Class	HAP	VOC
TLV	50	400
CC flash point, C	-21	-1
Sep from water	No significant differences	

Clear incentive to move to higher molecular weight alkanes
HAP = Hazardous Air Pollutant, VOC = Volatile Organic Compound

Table 2. Solvent Selection Guide, Aromatics.

	Benzene	Toluene	Xylene
Environmental Class	RCRA	RCRA	HAP
TLV	0.5	50	100
CC flash point, C	-11	-4	29
Sep from water	No significant differences		

Table 3. Solvent Selection Guide, n-Alcohols.

	C1	C2*	C3	n-C4
Environmental Class	HAP	VOC	VOC	VOC
TLV, (proposed)	200	1,000	200	25
CC flash point, C	12	12	13	37
Sep from water	+	--	--	+

* Ethanol is subject to special controls

DuPont recently participated in the life cycle analysis of materials for the air intake manifold in an automobile. Four options were considered aluminum; recycled aluminum, Nylon 66 and post consumer recycled Nylon 66. (Keoleian and Kar, Spitzley and Keoleian) The sponsor, an automobile company, invited DuPont participation in data collection and limited participation in data critique and analysis.

DuPont learned several things from this study:

- If an emission is not regulated, it is rarely reported in commercial life cycle databases. Because hazardous air emissions are a US

designation, US oil wells report them, but North Sea oil wells do not. Similarly, initial data showed no global warming emissions from aluminum smelting, even though smelters are known sources of the potent global warming gas carbon tetrafluoride. This omission was subsequently corrected.
- Final fate is an important consideration for the global warming impact of organic compounds. Methane is a much more potent global warmer than carbon dioxide. If, at the end of useful life, organic compounds are disposed of in a landfill and are assumed to slowly decompose to methane, then the global warming contribution should include a contribution of about 22 of carbon dioxide equivalents from each molecule of carbon – a large contribution to the global warming burden of many products. Organic compounds burned directly for fuel and combusted to carbon dioxide, or disposed in a land fill with methane collection and combustion will have a smaller life time contribution to global warming.
- When all environmental impact data was compared, DuPont concluded that the effect of intake manifold weight on automobile fuel consumption dominated comparison among the four cases. Fortunately for DuPont, Nylon 66 was the lightest and had the lowest impact.
- While the difference between Nylon 66 and post consumer recycled Nylon 66 was quite small, our customer very much wanted to emphasize the recycle content of the final car. Thus, DuPont has proceeded with plans to develop mechanically recovered and depolymerized products to meet customer-marketing objectives. The overall advantage of Nylon 66 as determined in the life cycle assessment has led to a large growth for this application for both new and recovered Nylon.

Challenge: It will be increasingly important to get these types of decision systems into the hands of the designer. How do we develop user friendly decision tools that are also based on acceptably complete and reasonably up-to-date data?

Future Information and Modeling Needs

How could the academic research community help companies like DuPont develop even greener processes and products? Several needs seem to frequently recur:
- How to quickly identify, while doing conceptual design, appropriate extractants and adsorbents? This need is particularly acute for the removal of high boiling polar organic compounds from water containing proteins, biomass or salts. A related question is how to identify the optimum extractant or sorbant, short of exhaustive testing.
- How to create realistic targets for solid/ liquid separations? The rate and extent of solid/liquid separation is a function of both liquid and solid properties. For pharmaceutical processes, the centrifuge and filter system may require five times as much investment as the reactor. Separations may be improved in many ways, for example by seeding, by slowing the growth rate during nucleation, or by changing the agitation level. How can industry rapidly identify the economic stake for further process development to improve separation properties?
- How to deal with interesting problems presented by biotechnology routes to commodity chemicals? How best to couple a batch biosynthesis process to a continuous separations process? How to economically recovery low value products from dilute solutions? How to purify incoming and outgoing nutrient, air and waste streams?
- How to improve the quality of data available for life cycle analysis? How to avoid regional biases based on regulation differences? How to quickly estimate the benefits and impacts of post consumer recycle versus extended life?
- How to continually evaluate inherent safety and environmental impact during conceptual design and development? Are the inherent safety benefits of a nonflammable halocarbon solvent ever greater than the environmental impact of such a solvent?
- How to rapidly assess relative risks among competing commercial processes if forced to manufacture and use a highly toxic intermediate, such as phosgene?
- How can engineers better communicate process integration and optimization opportunities in ways, which are readily, understood by management and operations staff? How can engineers confidently speak to changes in startup characteristics, waste generation while transitioning products, and similar issues using language that creates confidence and consensus?

Conclusions

Our understanding of the characteristics of greener, more sustainable, and inherently safer processes and products continues to evolve with time. Public expectations for reduced environmental impact and inherent safety expand with time and now include product characteristics and consideration of ultimate fate. Wastes have not yet been eliminated from the chemical processing industry. Consequently, economic and effective waste treatment technologies are still required. Incorporation of these waste reduction and waste treatment technologies into overall processes and optimization of the whole remains an important, but presently unfulfilled need.

Pollution prevention is an important goal for process designers. Progress continues in process modeling and optimization and capabilities that were unknown or exploratory in the 1970s are widely utilized in the 1990s. New opportunities for process improvements continue to emerge, and expectations for further pollution prevention should remain high. These tend to be based on engineering experience. The experiences are, thus far, only modestly incorporated into design systems, and few are based on fundamentals. An example is that tools for greener and more inherently safe solvent selection are emerging. Another example is that engineers are still very much challenged to identify extractants and sorbents from theory.

The public further expects both industry processes and products to be sustainable. Life cycle assessment has emerged as a key tool in examining the sustainability of products. It can be used to compare products, to examine high impact steps in a product value chain, to better understand improvements with time, and to highlight opportunities for improvement. Environmental data bases and life cycle assessment tools are not fully developed to rapidly examine the relative benefits of materials recycle versus extended service.

Looking forward, the leading challenge seems to be how engineers can best combine the concepts of reduced environmental impact and inherent safety with greener process chemistry and codify these in a way that can be in-corporated into design systems. Additionally, how can engineers improve their ability to communicate the opportunities of process integration and further optimization to manufacturing personnel and financial leaders?

While industry can be pleased with its progress, substantial opportunities remain to improve the understanding and modeling of greener processes.

References

Clark, D. and R. W. Sylvester. Ensure process vent collection safety, *Chem. Eng. Prog.*, **92**(1):65-77.

Curzons, A., D. C. Constable, V. L. Cunningham (1999). Solvent selection guide: A guide to the integration of environmental, health and safety criteria into the selection of solvents. *Clean Products and Processes*, **2**.

Dyer, J. A., and W. C. Taylor, June 1994. Waste management: A balanced approach. *Proceedings of the Air & Waste Management Association*. 87th Annual Meeting and Exhibition, 94-RP122B.05, Cincinnati, OH.

Keoleian, G.A. and K. Kar. *Life Cycle Design of Air Intake Manifolds: Phase I: 2.0 L Ford Contour Air Intake Manifold*, National Risk Management Research Laboratory, Office of Research and Development, U.S. Environmental Protection Agency, (in press).

Mulholland, K. L., and J. A. Dyer (1999). *Pollution Prevention – Methodology, Technologies, and Practices.* American Institute of Chemical Engineers, New York.

Spitzley, D. and Keoleian, G.A. *Life Cycle Design of Air Intake Manifolds: Phase II: Lower Plenum of the 5.4 L F-250 Air Intake Manifold, Including Recycling Scenarios*, National Risk Management Research Laboratory, Office of Research and Development, U.S. Environmental Protection Agency, (in review).

MODELLING AND OPTIMISATION FOR PHARMACEUTICAL AND FINE CHEMICAL PROCESS DEVELOPMENT

Nilay Shah, Nouri J. Samsatli and Mona Sharif
Centre for Process Systems Engineering
Imperial College
London SW7 2BY

John N. Borland
John_Borland@compuserve.com

Lazaros G. Papageorgiou
Department of Chemical Engineering
University College London,
London WC1E 7JE

Abstract

The specialty chemical manufacturing industry is under increasing pressure to reduce costs and increase its speed of response to customer requirements. The response, currently, has been to stay with old manufacturing technology and to move to lower cost economies. The Britest Project is researching how to change design and operation radically in this sector to achieve significantly improved business performance.

Keywords

Batch process development, Plant design, Time-to-market, Modeling, Optimisation, Uncertainty.

Introduction

Manufacturing industry has always been under pressure to reduce costs and to minimise risks for new investments. The history of manufacturing has been one of increased productivity, by which we mean lower manpower and lower costs, and the continued search for new ways of achieving that improvement.

In the UK, manufacturing is an activity, which does not have a good public image and has been in steady decline for some considerable time. The Technology Foresight Report 9 (OST, 1995) points out that, as a proportion of GDP, manufacturing has been falling since 1800 and is now only 20% of GDP. Nationally there is regret at the loss of jobs but there is little commitment to changing attitudes to improve the status of manufacturing as a career opportunity or as an activity of national importance which makes significant contribution to wealth creation. Whenever the decline is discussed there is always an economist who points out that the economy can be successful without a high level of manufacturing and points to the comparable position in the US, where it is a lower percentage of GDP.

The chemical industry, which in overall terms has seen enormous growth over the last 50 years, is in a period of serious rationalisation. In Europe there is now a period of merger and change in the chemical industry. This is particularly so in the area of relatively low tonnage products, which are usually made by batch manufacture, and include pharmaceuticals and fine chemicals; we shall refer to the whole sector as specialty chemicals.

It is interesting to note that many companies are targeting the higher added-value sector of the market. This is probably most graphic in the massive change ICI is currently undergoing in its attempt to move from commodities to specialties. There is a danger here in that, with increased focus on this area, competitive pressures will force prices down and the level of added value will fall. It is worth noting that Dussage et al. (1992) define value added as: "the *perception* by customers of the product or service resulting from the utilisation of the technology. This perception and the value associated with it governs both the price level that can be applied and the size of the potential market." How much of the industry's success is based on that perception and how quickly can perceptions change?

Cost Reduction

Companies have responded to the challenge; in particular, cost cutting has been a continual feature for over 30 years to our certain knowledge. This has taken a number of forms and there has been considerable creativity in thinking of new ways to continue cost and 'numbers' reduction. Strategic reviews, cost cutting exercises, repositioning, core competencies, Business Process Re-Engineering, Supply Chain Management, customer focus, TQM, JIT … the list goes on. It is to the credit of the consultancy business that it continues to think of new names and different approaches to improving efficiency and reducing costs. There is no doubt that without these measures many more companies would have failed, but are these measures enough to secure the future?

Technical measures taken to reduce costs in the specialty chemicals sector have been to use mechanical handling, increase scale where possible (though in some sectors the relatively low demand levels mean that scale is still small) and improve control. Initially this was through pneumatic controls and is now through computer control, which has played a major part in reducing manpower and improving operational consistency. Counter to these, there has been the increase in containment standards for safety and environmental reasons. Though there have been considerable benefits from these changes, overall they have increased cost.

Where, then, is the next cost reduction to come from? Manufacturing in other sectors, most notably car manufacturing, has seen enormous changes, which have been well documented, where the move from mass production to lean production has reduced the assembly hours from 41 to 18 per unit, lowered inventory levels and brought many other improvements in quality.

The chemical industry generally has a bad press; its main publicity is when there is a fire, explosion or environmental incident. There are sections of the community which seek to stop some manufactures altogether, e.g. pesticides. So what is the way forward now when there are compelling reasons for moving the manufacture to low cost economies, such as China and India? Their costs are low and are easy to exploit to produce material at lower cost; avoiding the adverse publicity associated with manufacturing is an additional benefit.

Additional Pressures

In addition to the pressure on cost there are pressures to respond more quickly, through faster introduction of new products and faster response to changes in demand. The pharmaceutical industry is changing and considerable effort is being put into reducing the time-scale for the introduction of new products; this has a knock on effect on manufacturing. In other sectors, increased speed of response is being demanded, while at the same time business managers want to reduce inventory and working capital. In a capital-intensive industry, where production facilities are designed for specific products, there is pressure to reduce capital, particularly for new products where the risks are higher.

These improvements in business performance are demanded from a manufacturing process, which is inherently uncertain and incorporates significant variability.

Research

Research is to underpin the future prosperity of a company. Businesses with high research spending are the ones that are more likely to have a secure future. There is scope for debate about how much should be spent on research and in what areas.

Hutton (1996) and the Technology Foresight Report 9 (OST, 1995) have recorded research expenditure in the UK. Against a world average expenditure in R&D of 4.8% of sales the UK spends 2.4%. This is significantly lower than some developed countries, at 1.25% of GDP for the UK compared with over 2% in Japan. In 1993 this was £9.1 billion and over 90% of that expenditure was in 100 companies. The chemical industry accounts for a quarter of all industrial R&D expenditure at around 7% of sales, and can be seen as more forward looking than other industries, though there is a case for even higher spending. Expressing this when compared to dividends, the world's top companies' R&D spending is three times dividends whereas in the UK it is two thirds of dividends.

Much of this expenditure is in the pharmaceutical sector and is focussed on research into new products. There is no denying the importance of new product research but how much is spent on process research? If the UK Government's Department of Trade and Industry, Technology Foresight report No. 1 is to be believed, the future success of the industry depends on "research and innovation to develop more efficient processes and to shift to higher added-value products…" — process innovation and the customers perception.

How strong a base is that for the future of the industry?

Current Approach

The purpose of process development in the specialty batch manufacturing industry is primarily to achieve a cost-effective reproducible manufacturing route, where that effectiveness is determined by the value of the end product, and secondly to develop a process that can be operated in existing plant. This is particularly important in the pharmaceutical industry for new compounds and in the contract manufacturing industry, although the continued re-use of plants as products change is a feature of many batch processes. It is worth noting that the response to the faster introduction of new products is to *ensure* that processes are developed to fit existing plant. This has the merit of eliminating plant construction time.

The ability to fit processes into existing plants is an important feature of the training of new development chemists. They must be aware of plant limitations and therefore do not waste time in developing a process that cannot be readily accommodated in either existing plant or standard plant. Standard plant is that which is most similar to the laboratory, *i.e.* the glass lined mild steel (GLMS) vessel—equivalent to the laboratory glass flask. In many organisations, chemical engineers work with development chemists and this provides further safeguards to ensure that the process will be most readily accommodated in standard plant.

In pharmaceuticals, it is essential that all processes fit in standard plant because that gives most flexibility on available capacity, hence quicker response to demand changes and most importantly to meet regulatory requirements. It is necessary to manufacture in plant that corresponds to development plant and once fixed the process cannot be changed.

Chemists have been very innovative, bringing to large-scale manufacture many processes that have only been text book or laboratory possibilities. It is a tribute to their creativity that they have achieved high yields with all these constraints and have achieved complex syntheses with great ingenuity. It is only when standard plant will not operate under the only feasible conditions that different plant is necessitated, *e.g.* high pressure or high/low temperature. The importance of fitting processes to plant is well recognised. It is important for development chemists to know what changes in performance there will be in the process on scaling up from the laboratory to the plant. Indeed that is a significant part of their experience and a measure of their success. Knowing what will and will not work, allowing for the reduction in yield on scale up, ensuring that the process is robust enough to be operated on the large scale with all its variances are all important aspects of a development chemist's know-how.

Occasionally, for difficult reactions, kinetic models are developed. These usually take significant effort and can only be undertaken if the product potential is large enough and the problems warrant that effort. It is expected that the fastest way to full scale manufacture is to transfer from the laboratory as soon as possible so that scale-up problems can be quickly identified and resolved.

Experienced chemists and chemical engineers will resolve many problems at the development stage, many of which arise because of the limitations of the full-scale equipment and come under the heading of 'scale-up'. However, pressure to get into production and to contain expensive development costs plus limited facilities for carrying out experiments to investigate all aspects of scale up results in processes being operated at large scale with many problems still to be resolved and with significant potential for improvement. Technical effort to support, trouble-shoot and improve processes through 'learning by doing' is seen as an essential element.

Pisano (1997) has characterised process development as currently practised in the pharmaceutical industry, drawing the distinction between 'learning before doing' and 'learning by doing'. Through analysis of manufacturing problems, he concludes that 'learning by doing' is an essential part of any new manufacture. What is clear is that business managers, unwilling to spend more than a minimum sum on process development and new capital, accept both the need for significant support and low performance. When manufacture starts, they will pour resource and money in to resolve production problems.

That is an indictment of the batch processing industry and a statement that it does not really care about having efficient processes.

The question is whether continuing this approach is good enough to secure the future.

Process Development

The current development procedure is for a chemist to run the desired reaction in the laboratory, usually starting in a glass flask because that is what is most easily used. There are a series of improvement steps, built up from experience, of changing solvents, concentrations, temperature, addition patterns, *etc.* until a satisfactory performance has been achieved. This means avoiding certain conditions and critical operations and making allowances for loss of yield. These factors are well known and the good development chemists are the ones who take these difficulties into account with little loss of reaction performance and with a minimum potential for problems when the process is operated on the plant.

Scale up problems are experienced when the plant performance does not mirror laboratory performance, either in the operating conditions achieved or the consistency with which the operating conditions are achieved. Considerable expertise and ingenuity go into resolving these difficulties to achieve good yield and product quality with a robust process.

Significant cost reduction is achieved through process development. Initially this is to achieve a reasonable manufacturing cost when compared with the early process. Further development results in cost savings, though often this is improving plant performance towards the best laboratory results and does little more than counter the effects of inflation.

Batch manufacturing requires the use of large quantities of solvents and reagents such that the amount produced is many times less than the total material handled. Clearly that ratio depends on number of stages and which particular reactions are being followed, but it is generally considered good if the product is 10% of the materials handled in its manufacture. For a multi-stage manufacturing process, the overall yield may be only 20% and then the product may be less than 1% of materials handled. That material costs money: to buy, to handle, to process, to store, to separate, to dispose of or to recover. Reduction in these materials would result in savings throughout, capital productivity, working capital and operating costs.

Modern manufacturing looks at activities that add value to the product. Much effort goes into removing wasteful operations carried out by people: for example unnecessary paperwork. In batch chemical manufacturing, how much time does material spend on the works without undergoing an added-value activity? Activities such as storing, moving around *e.g.* from vessel to vessel, when material is being worked on slowly when it could be worked on quickly. How many operations are introduced because the one that would be best cannot be achieved? How often are reagents or solvents introduced to modify reactions because the desired operating conditions cannot be achieved? How often is process performance reduced on the plant because of scale-up issues? How often are processes modified from the best performance so as to improve robustness? A conservative estimate is that the manufacturing velocity, a measure of the time during which material is being worked on to effect, could be increased by an order of magnitude.

Current methods of investigating processes are very expensive in time and resource, and are only applied when the product is to be made at very large scale and for a long period of time so that new processes can be the basis of new productive capacity. Modelling reactions can take considerable effort and, unless there are over-riding reasons, that effort is usually best applied to carrying out experimentation.

Much has been written about 'learning by doing' contrasting with 'learning before doing' in which the conclusion is that for complex problems there is no alternative to learning by doing. This supports the main experience of development chemists and their approach to process development. It underpins the importance of gaining good plant operating experience and strong links between chemical engineers and chemists to ensure that processes developed in the laboratory fully take into account the limitations of the plant. Processes therefore fit into large-scale glass flasks (*i.e.* GLMS vessels) and batch cycle times are based around chemist's days with further time extensions due to longer full-scale operations. A GLMS vessel can be used as a mixer, reactor, heat exchanger, liquid-liquid extractor, crystalliser, still... what could be more versatile?

Once a process has been developed and product made for trials, there is often little reason to change. In fact, in the pharmaceutical industry the regulatory authorities seek to ensure that nothing changes.

The Challenge

So, given the history of process development and the increased pressure to improve, what are the options? Is it inevitable that manufacture will increasingly migrate to lower cost economies? There are downsides to this move in the loss of a manufacturing base and the loss of competence which goes with that, so that the capability for the future is seriously undermined.

Learning by doing is well established in complex situations. von Hippel and Tyre (1995), working in the semi-conductor industry, whilst concluding that 'learning by doing' is essential, show that around 80% of the information required to resolve problems experienced was known at the design stage. They indicate that the issue may be more how to structure the problem-solving activity.

Another factor, which is well recognised, is that the biggest impact on the cost of a project is at the early definition stage. For many batch plants this is seen as when the process flowsheet is being drawn and the number of stages, with intermediate storage, is being determined. The question is whether this is the best way forward, to continue to strive to do better what has been done for many years, seeking continual improvement slowly whittling away at costs. The collaborators in the *Britest Project* do not think so; they are committed to finding a better way.

Britest Project

The Britest Project is carrying out research into process and plant design with the aim of initiating radical change. Britest is an acronym for 'Batch Route Innovative Technology, Evaluation and Selection Techniques'—a title which attempts to capture something of the project's essence. It is a three-year project supported by the UK Government's Innovative Manufacturing Initiative with research work funded by the Engineering and Physical Sciences Research Council. There are thirteen companies and three universities collaborating in four areas of research, which are designed to address the problems outlined above.

One group is looking fundamentally at process design, trying to understand how to determine key factors which affect process performance for given chemical trans-formations. They are then seeking to map the required performance to equipment capable of delivering it. Another group is looking at plant design to assess why plants have to be uniquely designed and yet are converted to manufactures not envisaged at the design stage. They are also looking at whether it is necessary to build plants so that they are difficult to modify, as we do now, or whether they could be built for rapid reconfiguration to a new process.

Computer modelling can be a great help in gaining maximum benefit from minimum information. Another group is working to use modelling and process systems techniques to gain greater insights and understanding from limited experimentation as well as to investigate the implications of reconfigurable plant design.

There is no doubt that for this work to be successful, it does have to bring significant business benefit. It is general experience that the best option for the finance director is one in which the capital is the lowest. Current assessment methods look at rates of return on a discounted cash-flow basis but rarely investigate the implications of whole life effects and the implications of significantly different technical options. Risks are something the business manager wants the technologist to tell him will be 'alright'. An important part of Britest is to investigate this area and to develop a business model that will help understanding of how the technical advances of the other groups can bring real benefit to the different businesses.

One project will not solve all of the problems nor will it revolutionise the specialty chemical manufacturing industry but if it changes views, ends complacency and develops some new methods and tools to allow the forward thinkers to make progress, it will make a worthwhile contribution.

Britest has a public web site, www.britest.co.uk, and, although the main benefit will be to the collaborators, information will be published from time to time. We also hope to widen the debate, including specific technical issues, and would welcome contributions from anyone who shares the concerns and aspirations expressed in this paper.

Britest Project Collaborators

Academic: Centre for Process Systems Engineering, Imperial College; Department of Chemical Engineering, UMIST; Keyworth Manufacturing Institute and Leeds Business School, University of Leeds.

Industrial: Zeneca, Glaxo-Wellcome, ICI, Jacobs Engineering, Rhodia Chemicals, with a group supported by SOCSA a branch of the Chemical Industries Association, Bush Boake Allen, FMC Process Additives, Hampshire Chemicals, Holliday Dyes and Chemicals, Inspec Fine Chemicals, Macfarlan Smith, Pentagon Chemicals, Robinson Brothers.

Paper Outline

In the remainder of this paper we will describe the role that process systems engineering has played and will play in the Britest project. In the next two sections, we shall give an overview of the areas in which process systems engineering is seen to be of most relevance, and the areas in which it has been applied in the past literature. In the subsequent sections, we will describe in overview the specific applications that have been developed so far.

Process Systems Engineering in Batch Process Development

As discussed above, the aim of the Britest project is to bring radical change to the way that processes and plants are designed in the specialty manufacturing industry. At this stage, we feel that computer-aided process engineering techniques should play a pivotal role in design and development from the point at which a number of chemical route options have been generated to the final design of the plant (and beyond into operation). Future work may also include the automatic generation of chemical routes.

The main concern with current industrial practice is that processes are generally developed within the restrictions of standard plant items, and that standard plant items are designed to match closely the behaviour of laboratory equipment. In order to develop a significantly more efficient process, we need to examine how it behaves on a physicochemical basis alone. To this end, we can define a process only in terms of fundamental properties and the "conditions" that are required to allow the process to perform at its best (an "abstract" process). Depending on the level of abstraction, the conditions might be properties such as pressure, rates of heat transfer, rates of mass transfer, rates of mixing, *etc.* from which the behaviour (and hence performance) of the process can be calculated. We can then relate the required conditions for the best process to the capabilities of the equipment in which the process may be carried out. This will highlight any inadequacies in standard equipment and hopefully stimulate new equipment designs that have significantly greater ability to provide the right conditions for the process. At worst, if no novel equipment design is available, we will have a precise understanding of the effects on the process of using a standard plant and the ability to make the best use of existing plant without the problems associated with scale-up.

Similarly, processes are defined in terms of "unit operations" too early in development. This can stifle innovative thinking and lead to inefficient processes. If we use the above philosophy to define a process that is independent of equipment and any pre-conceived unit operations, we have what we term an "abstract" process, where individual component flows can be manipulated to generate the ideal conditions for the process. (This will be discussed in more detail later on.) From this abstract process, an idea of what types of operations are best for the process can be gained; as well as indicating which standard unit operations are suited to the process, more importantly it should indicate where novel techniques can be of benefit. The key aspect here is that the modelling activity should drive innovation towards new efficient processes and process operations.

The approach that we are developing in this project is one of a hierarchical series of procedures. It begins with a completely abstract description of the process (one with no notion of unit operations or equipment limitations), passes through various intermediate representations, and finally arrives at a concrete process and a plant design. This provides a mechanism that takes direct account of the effects of any decisions on the performance of the process.

The current hierarchical procedure is summarised in the following steps:

- "abstract" process design;
- "conceptual" process design;
- "concrete" process design and mapping to equipment (including consideration of novel operations and equipment);
- plant design.

Associated with the above hierarchy are a number of key issues, the first of which is the handling of data and its link to experimentation—this gives rise to an associated data hierarchy.

A key feature of our approach is the way in which experiments are driven by the process design activity in order to generate the necessary data. Currently, experiments are carried out with the goal of obtaining the "best" reproducible process in the laboratory. Any data that are available for modelling are almost purely as a consequence of this, and quite often the data are insufficient to build a reliable model of the process (because the right properties were not measured or because the data are valid only in a small region of operation, or more than likely both). In our approach, we want to use the model to predict the best process (because this can be done without equipment limitations) and discriminate quickly between alternative routes and use experimentation mainly to support and validate the modelling activity.

A second issue, one that is perhaps in conflict with the above, is that of rapid evaluation of process alternatives. Clearly, it is important to keep the development cycle as short as possible for business reasons. Another advantage with speeding up the evaluation of process alternatives is that it can stimulate more creative thinking. One of the aspects of the hierarchical design philosophy described above is that engineers will need to be creative in interpreting the results of the abstract designs in order to move to more concrete designs. With a faster evaluation procedure, it will be possible to assess more of the ideas generated from the design activity. This will have a knock-on effect, where the evaluated ideas can be fed back very quickly in order to drive further creativity, ultimately at the route generation level.

An issue that will be inevitable in any design procedure for fine chemical processes is that of uncertainty and robustness. There are obviously a number of uncertainties associated with the design procedure and the data used therein. Further uncertainties arise from the inability to run the process exactly as specified in the design, which will be due to raw material variability, imperfect control, *etc*.

The final issue discussed here is that of performance measures. There are many business and engineering metrics that can be used to give a quantitative measure of the performance of a process. One issue is, of course, which measures are appropriate for the process. A further issue is that of performance measures that can be used to distinguish between *abstract* processes. It is not clear which metric will be suited to abstract process design, particularly as many metrics depend on equipment characteristics (*i.e.* cost).

The approach is based on:

- representation of a candidate process at different levels of abstraction;
- developing a mathematical description of the key phenomena in each step;
- using dynamic modelling and optimisation techniques at each level to indicate the most promising alternatives means of operating the process in terms of performance requirements that need to be met at that level (*e.g.* reaction rates, mass transfer rates, degree of mixing/heat transfer, segregation of species *etc.*) rather than constraining operation based on notions of what standard equipment can deliver;
- using sensitivity analyses to highlight the main sources of variability;
- using the information on sensitivity to parameters or assumptions to drive experimentation; and
- using robust optimisation techniques to improve reproducibility in the face of such sources of variability.

One issue explored is the extent to which existing research can immediately be brought to bear on the problem and when new research needs to be undertaken. We will revisit this later.

Overview of Previous Related Work

Hierarchical Method (e.g. Douglas, 1988)

This approach, which perhaps best describes a great deal of existing practice is based on the use of models of increasing complexity and rules which evolves a series of candidate designs with evaluations at each stage to prune the decision space. It has been widely applied to continuous processes but is less immediately applicable to batch processes.

Superstructure Optimisation (e.g. Grossmann and Sargent, 1978)

This technique has again been widely applied to continuous processes, where relatively detailed steady-state models are used for a superset of interconnected equipment items and the required parts of the process and their operational details are determined through the solution of an optimisation problem (usually mixed integer non-linear).

Global/local Variable Decomposition (Tricoire and Malone, 1992)

This technique has been developed specifically for entire batch processes, whereby the decision space is divided into "global" variables important to the overall process (*e.g.* conversions, recoveries) and "local" variables which are only important for individual steps. The local variables define the operating policy of a step and therefore determine the values of the global variables.

Detailed/simple Model Iteration (e.g., Salomone et al., 1994)

This approach is based on an iterative procedure, which develops overall processes using simple models (based on parametric fits through experimentation or through detailed modelling) and then uses detailed models for simulation and scheduling to assess processes so derived more accurately

Hierarchical material-balance oriented approach (Linninger, 1996)

This group has developed a toolkit with a series of tools that explore and support different levels of the development hierarchy. The emphasis is on the use of knowledge bases and material balancing to choose and assess options. The process was particularly applied to the problem of developing low environmental impact processes, including route synthesis.

Functional Operator Method (Smith & Pantelides, 1995; Sharif, 1999)

This approach is applicable to batch, continuous and batch/continuous processes. The idea is that a number of functional operators (*e.g.* reaction, separation) are defined through detailed models. A series of mixing and splitting operations allows all operators to be interconnected in any possible way. Given a potential starting set of raw materials and some final product specifications, the technique uses optimisation to select the required operators, their operating characteristics and their interconnections. Smith and Pantelides (1995) also demonstrated that, provided the models fulfil a certain condition, the number of discrete alternatives that need to be considered is relatively small; *i.e.* combinatorial explosion is *not* a problem.

Single-level Detailed Model Approach (Charalambides et al., 1995)

This approach is used to develop the best operating policy and resourcing for a process with a fixed structure. It may therefore be used to assess different process structures (generated for example through the phase strategy) in the best possible light. It used detailed dynamic models to trade-off intensities of processing at each step as well as to determine operating policies of each step.

Plant Design and Scheduling

A very large body of work has considered the problem of how best to implement a fixed process, either in an existing plant, a new plant or a plant to be retrofitted. In this case, all details of the process (times, yields, recoveries, materials, *etc.*) are fixed, and the problem is to determine the type, number and capacities of the key equipment items (reactors, crystallisers, columns, storage tanks *etc.*).

Typical approaches include:

- heuristics,
- simulation,
- mathematical programming, and
- evolutionary approaches.

The approaches all take some account of production scheduling—efficient scheduling will result in a lower resource requirement.

Tools Overview

There is no single tool that supports the process development activity envisaged in Britest, hence the motivation for the project. There are a number of different tools that support elements of it, and a couple that aim eventually to support most of it. Relevant tools include:

i. Batch process recipe description, reaction path synthesis and material balancing tools;
ii. Reaction characterisation and calorimetry;
iii. Dynamic modelling, simulation and optimisation;
iv. Batch process scheduling and equipment selection.

Aspects of the work programme envisaged here are supported by existing tools, but we expect to have to develop methodologies to:

- generate alternative processing options;
- assess and improve options in the face of uncertainty;
- characterise existing and novel equipment;
- match equipment to process conditions and assess resulting matches; and
- define, assess and optimise versatility and robustness of ensuing designs.

Methodology and Components

The methodology that we apply to batch process development is one of linked hierarchical structures as shown in Fig. 2. The main hierarchy is that of the process development procedure, which consists of various levels of abstraction before finally arriving at a given process and plant design. This is illustrated in the right-hand element of Fig. 2. Linked to this hierarchy is an associated data hierarchy, driven by the needs of the process development activities. They are linked by a process of selecting a set of "conditions" that give rise to the best performance, which is then tested through experimentation that in turn leads to new and more detailed data sets.

Within any of the levels of process development, there may be scope for existing CAPE techniques or a need to develop new methods. Since the use of such techniques has not been adopted in the batch processing sector, it seems likely that there will be substantial scope for applying existing techniques in this area.

In the next subsection, we look at some of the existing techniques that have been applied to some case studies under the Britest project.

Use of Current CAPE Techniques in Process Development

The use of existing techniques has already played an important role in a number of Britest case studies. Perhaps the most useful of these is the State-Task-Network (STN) representation (Charalambides *et al.*, 1995), in which a process can be represented by a network of tasks and their operating policies that transform one or more states into one or more different states.

The STN representation has been used to analyse an industrial process. Here, using simple process models (experimental yields, mass balances), the whole process is described using the STN representation so that the interactions of the various process operations can be seen on an overall process performance metric. This approach where the process is analysed as a whole is not common practice in this sector.

As a result of the full-process modelling, even with simple models, it was possible to perform rapidly a series of sensitivity analyses that highlight which operations are critical to the performance of the thus creating a number of important opportunities.

First, this approach clearly shows where further experimental or modelling work should be focussed and is important in reducing the data requirements: based on an initial small set of data, additional data is obtained only where it is expected to be of greatest benefit. The results of these sensitivity analyses were somewhat surprising in that they indicated that most separation operations had a much lower influence than the reactions.

A particular reaction was highlighted as being the most significant. Again, the use of standard, if advanced, dynamic parameter estimation techniques were combined with structural analyses and a simple set of experimental data to identify an approximate kinetic model. A dynamic optimisation was then performed to identify the best way of operating the reaction in contrast to that devised through experimentation. A novel feeding strategy was identified which has the scope to improve the yield by about 30%. Experiments are underway to verify this. The overall effect of this improved yield can be quantified fairly accurately by resorting to the approximate whole process model again.

The results of the sensitivity analysis and optimisation can also be used in more traditional methods of process development, such as "brainstorming" sessions. In this case study, the results of such a session were two new chemical routes for a key process step. After a rapid analysis using a simplified model of the process (in STN form), the new routes were shown to be substantially better than the base case. In another industrial case study, a mathematical model was derived by the research group at UMIST from experimental data. Using standard dynamic optimisation, we were able to obtain an operating policy that resulted in a significant improvement of the process. A comparison of the existing and proposed operating policies can be seen in Fig. 1 for the reaction scheme $A + B \rightarrow C$; $A + C \rightarrow D$ where C is the desired product and D is a by-product. It can be seen that the C:D selectivity is greatly enhanced in the optimised case.

New Techniques for Process Development

As mentioned earlier, we consider a hierarchical approach which allows the evaluation of a number of alternative routes at a different levels of detail. This facilitates two things in particular: (i) the screening of alternatives without having to generate complete designs (a feature in common with the approach of Douglas (1988)); and (ii) the ability to define processes in a fashion that transcends traditional limitations and highlights fundamental limitations associated with the choice of route.

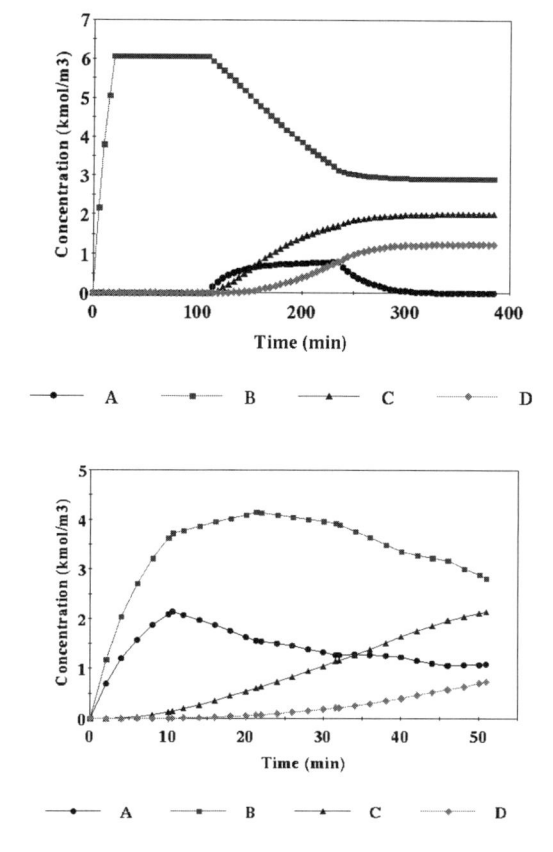

Figure 1. Comparison of base case with best conventional process.

The top-level activity attempts to design processes at a very abstract level, which is concerned with the spatial and temporal distribution of phases, species and conditions. At the next level, the abstract process is refined by considering the basic process operations that might realise it (*e.g.* heat and mass transfer). This level still operates at a degree of abstraction and is not concerned with specific equipment, and therefore bears some similarity with the mass/heat exchange module approach of Ismail *et al.* (1997) and co-workers. The conceptual level is currently being evolved by the research groups and will not be described further here. The concrete process design activity seeks to implement a given process in a plant superstructure (see, *e.g.* Sharif, 1999). Any compromises on the process performance will be identified here.

Finally, any potential manufacturing processes must be studied from the point of view of robustness and reproducibility. We introduce two concepts in later sections associated with robustness metrics and operational envelopes.

As mentioned earlier, a data hierarchy is also evolved as illustrated in Fig. 2. At each level, data are both required and generated. Initial experimentation will usually aim at route proof-of-concept (the "first set") and the determination of some fundamental parameters, typically that generated by development chemists in the course of identifying feasible routes. Then, as alternative good processes emerge, they should be used to drive experimentation, either to validate them, or if they indicate some degree of infeasibility or mismatch to improve the representation and data set.

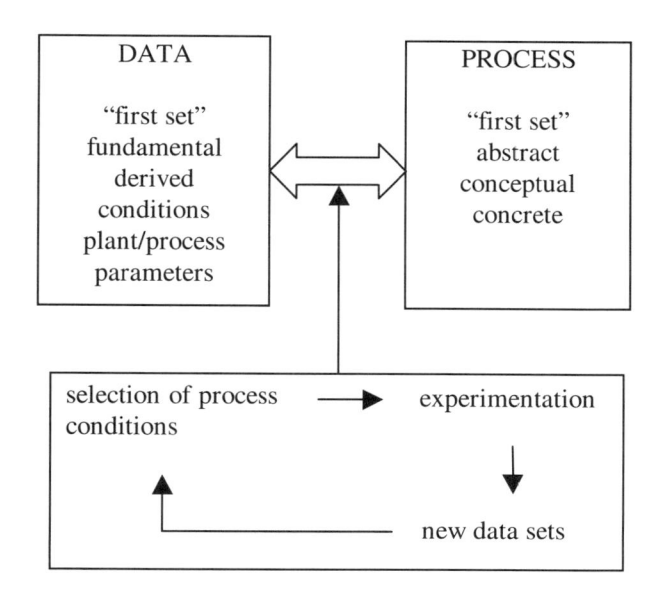

Figure 2. Schematic of dual-hierarchy methodology.

This contrasts with the traditional role of experimentation, where experiments are carried out only to optimise the process at the laboratory scale: the experiments are aimed to converge on a feasible process, whereas our approach is committed to using experimentation and modelling together in a "divergent" manner to explore the potential of the process systematically.

The next sections of the paper will describe the abstract level and the research on robustness and operational envelopes.

Abstract Process Design

As described briefly before, abstract process design is the technique we use to determine the ideal or perfect set of conditions for a process.

Our first attempt at this is based on the concept of a series of interconnected process cells as illustrated in Fig. 3 and Fig. 4.

Each cell is assumed to be well mixed and to operate in a dynamic fashion, with time-varying flows, component hold-ups and conditions. The only physico-chemical process that is modelled explicitly in a cell is chemical reaction. There are two types of flows (which are bidirectional). The solid flows are between pure component reservoirs and the cell; there is a time-varying flow for each component from or to a cell. These represent either feeding policies or hypothetical separation processes that are able to remove single species with perfect resolution. The dotted flows are between cells and are assumed to be at the prevailing compositions of the source cells.

It can be shown that in the absence of explicit treatment of multiple phases, a single cell is all that is required to describe multistage reaction-separation processes at a very abstract level. There may, however, be computational benefits to a multi-cell structure. The cell concept is utilised as follows:

i. a dynamic model is developed for each cell and reservoir, taking account of reaction stoichiometry and kinetics;
ii. a dynamic optimisation is performed whereby the cell conditions (*e.g. T, P*) and flows are varied dynamically to optimise a function based on the rate of value-added associated with the contents of each reservoir;
iii. the results of the abstract designs may be used to screen poor solutions (bearing in mind that at this level there is large uncertainty in the design);
iv. the results are used to attempt to identify possible operations to achieve the implied conditions.

An example is used below to illustrate this.

Single Reaction Example

Here, we consider an abstract involving the reaction $2A \rightarrow B \rightarrow C$. The desired product is B. We contrast it with the best possible conventional semi-batch reaction. In both cases, the objective is to maximise the rate of value-added. In the conventional case, this value is 2500 and the profiles are illustrated in the first graph of Fig. 5. In the abstract case, an alternative solution is found. The optimisation removes B from the cell into the B reservoir and generates a rate of value-added of 5400. The profiles are illustrated in the second graph of Fig. 5 and the hourly flow-rates out of the cell are shown in the final graph. The motivation for considering the types of operations to achieve such a process is clear and thus a new challenge is posed. There is no shortage of techniques to rise to such challenges (*e.g.* membrane reactors, extractive reaction, chemical conversion/inactivation *etc.*); the critical step is identifying the opportunity.

The bar chart in Fig. 6 shows the differences in rates of value added for conventional optimised semi-batch processes and the optimised abstract process for examples. The fourth example is an industrial case study. The work on the conceptual design level is somewhat preliminary and will not be reported further here. In contrast, the concrete level has been widely reported (see, *e.g.*, Sharif, 1999). In the next section we therefore focus on the work on robustness

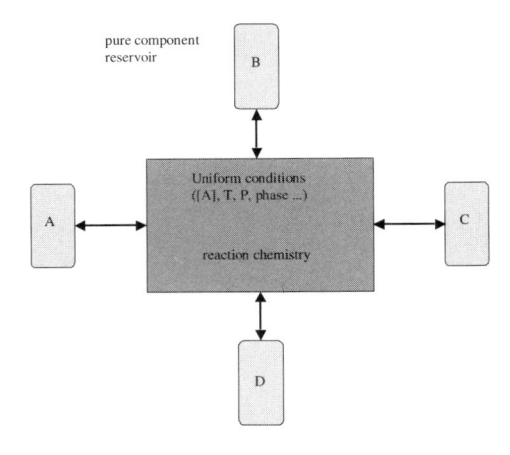

Figure 3. Single-cell abstract design.

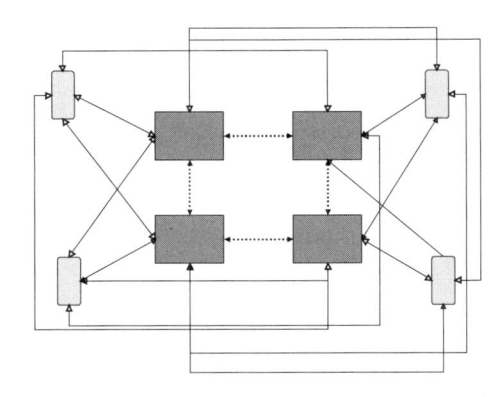

Figure 4. Multi-cell abstract design.

Robustness metrics

In this section we describe a method for measuring the ability of a process to cope with various sources of uncertainty. Although we focus on uncertainty in the model parameters, the approach may be useful in other areas such as market uncertainty, *etc*. This problem is particularly pertinent to the Britest project, where many processes are subject to significant uncertainty (as opposed to the petrochemical industries for example). The method is illustrated with two examples.

Introduction

The treatment of uncertainty in process design is not a new problem. Comprehensive reviews of the various approaches adopted have been presented by Grossmann *et al.* (1983), Pistikopoulos (1995) and Wets (1996).

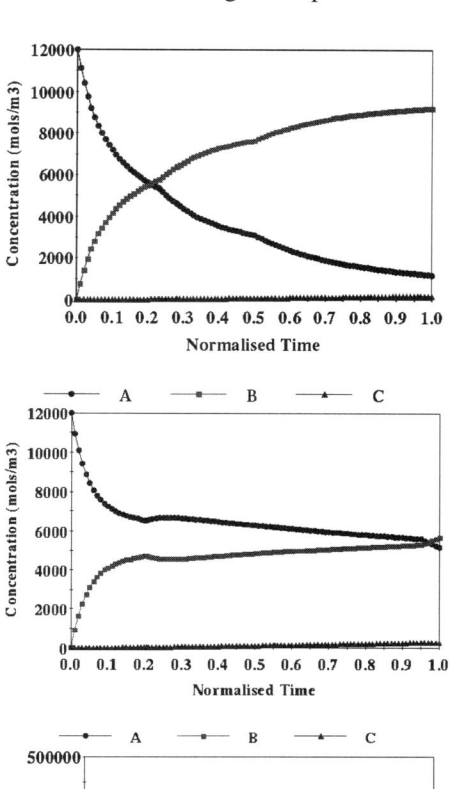

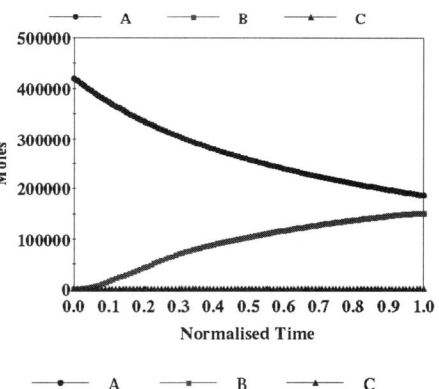

Figure 5. Comparison of conventional and abstract processes.

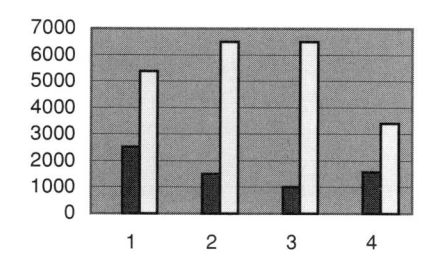

Figure 6. Comparisons of conventional and abstract processes

Most approaches involve modeling the uncertainty using parameters with probability distributions. Then, rather than optimising a single value of a performance metric (*e.g.* profit), the objective is to optimise the distribution of the performance metric that results from the uncertainty (*i.e.* probability distributions) in the parameters. Probability distributions are very often characterised using the mean (or expectation) and variance to indicate roughly the position of the distribution and its breadth respectively. Most approaches to design under uncertainty aim to optimise the mean of a given performance metric; others involve the variance or both. There are two potential pitfalls to this approach:

1. Other than perhaps profit, most performance metrics (*e.g.* those associated with product quality) are required to meet known target values.

2. Many performance requirements are not symmetric and may even be one sided (*e.g.* only values of the metric below a target value are problematic).

This shows that the mean and variance may not always be appropriate measures of robustness. A more detailed argument is presented in Samsatli *et al.* (1999).

There is also a subtle difference between robustness and reproducibility: robustness must be directly related to deviations from a desired value and need not depend on the breadth of a distribution, whereas reproducibility is directly related to the breadth of a distribution and is suitably characterised by the variance.

Clearly, a different approach may be necessary in a number of circumstances. For many performance metrics, we desire a robustness metric that is related only to deviations from a target value. In other cases (*e.g.* if the purity had to meet an exact target—or, in practice, be as close as possible), mean and variance constraints may be most appropriate. Whatever the requirements, it would be useful to have an alternative approach where positive deviations from the target value could be treated differently from negative ones, but also to be able to use conventional metrics where appropriate. In the next subsection, we describe such an approach.

New Robustness Metrics

The general robustness metric is defined in terms of deviations of a performance metric, z, from a desired value, z^*. Two functions of the deviation $(z - z^*)$ are defined for deviations above and below the desired value, z^*, and are combined to form a "general" deviation function q:

$$q = \begin{cases} f_1(z - z^*) & \text{if } z \geq z^* \\ f_2(z - z^*) & \text{if } z < z^* \end{cases} \quad (1)$$

The robustness metric is then defined as the expectation of the general deviation function, q, over the uncertain parameter space.

By selecting appropriate values of f_1 and f_2, various types of robustness metric can be defined for both equality and inequality constraints. For example, a robustness metric for an inequality constraint of the form $P \geq P^{LB}$ (*e.g.* the purity must be at least P^{LB}) can be defined by selecting $f_1 = 0$ and $f_2 = 1$, which calculates the probability that the constraint will be met. Various general deviation functions are shown in Fig. 7, where robustness metrics for both equality and inequality constraints are shown. The variance function is simply a special case.

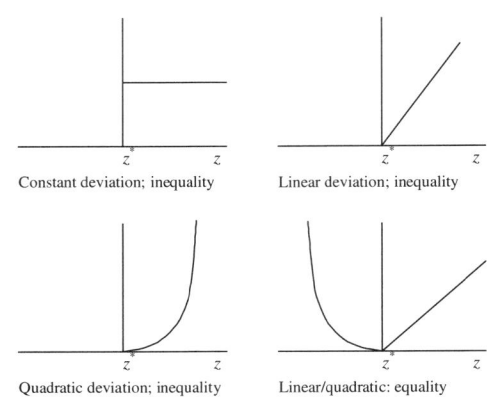

Figure 7. General deviation functions for equality and inequality constraints on z.

Solution Procedure

Since a large number of uncertain parameters may result in an intractable problem, an overall solution procedure is proposed that accounts for the most important uncertain parameters and neglects those of less importance. The procedure is first to simulate the process at the "nominal" values of the uncertain parameters and to perform sensitivity analyses on them to determine the critical ones. The multi-scenario model is then formulated using only the critical parameters. Once a solution is obtained, accurate values of the robustness metrics are generated from a stochastic simulation of the operating policy/design. The details of the solution procedure can be found in Samsatli *et al.* (1998).

Example Problem

We consider a two-stage reaction process and assume that the reaction rate parameters in the first reaction step (two reactions) are uncertain. Since the Arrhenius activation energies appear in an exponential term, they are deemed to be the most critical. The objective is to maximise the production rate subject to a lower bound constraint on the product concentration; the robustness metric is based on this. A comparison of the product concentration distributions for the nominal and robust operating policies is shown in Fig. 8. The key result of this example is that not only has the robust optimisation significantly reduced the probability that the concentration constraint will be violated but also that the variance of the distribution has been allowed to increase. Applying a variance-based constraint would have had the opposite effect and possibly resulted in a worse process.

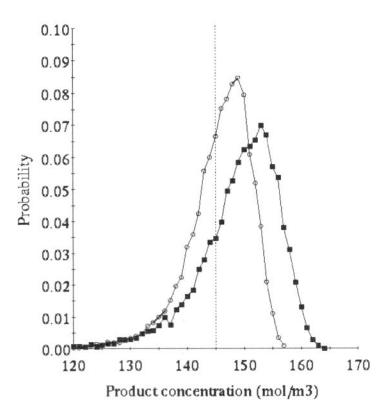

Figure 8. Comparison of product concentration distributions for nominal (circles) and robust (squares) operating policies.

Batch Process Design and Operation using Operational Envelopes

Currently, batch process design is usually undertaken with the objective of establishing a "fixed" operation, where the details (*e.g.* processing times/temperatures) are specified exactly. However, in operating practice, there are usually deviations from these, but while some deviations are usually acceptable, their limits are not normally known. It would make sense, therefore, to design processes with explicit "envelopes", *i.e.* allowable variations in the operating parameters, which will not affect the final outcome (*i.e.* a profitable and feasible process). This is particularly useful for processes whose operation has a high manual component. For example, a statement along the lines of "as long as the temperature remains between 95°C and 118°C and the processing time between 4.5 hours and 6 hours the process will work" would represent a more practical alternative to "maintain the temperature as close as possible to 100°C and run for 5 hours exactly".

The envelope concept becomes even more interesting when applied to multi-step processes. Here, it may be possible to derive multiple decoupled envelopes which ensure that the overall process will work as long as the constituent steps are operated within their envelopes. This introduces a form of "robustness" into the process operation, since no adaptation of downstream operating policies will be required depending on the performance of upstream stages. It is interesting to note that there is in fact a coupling between the dimensions of the different envelopes which is affected by the intensive and extensive properties of the materials joining adjacent stages.

Overall, we aim to optimise some properties of the operational envelopes of each step, for example a total scaled volume. We aim for simple geometric forms of the envelope where a hyper-rectangle is generated by determining a range for each parameter.

This concept bears some resemblance to that of operational flexibility indices as first introduced by

Swaney and Grossmann (1985). They partition the variables into design and operational variables, and attempt to find a set of design variables which give rise to the maximal coverage of a parameter space θ, where the operational variables may be adjusted for different values of θ. The parameters θ normally correspond to data uncertainties. Our problem formulation and procedure differs from similar approaches in the following regards:

- from the philosophical perspective, we are not trying to design a process that will be resilient to variations in unknown parameters, but rather selecting some fixed parameters that result in attractive operational envelopes in the other operational parameters;
- we have no adjustable control parameters that are set optimally upon the realisation of the uncertainties;
- we are interested in determining coverage over *ranges* of parameters (i.e. the ranges 120–150°C and 130–160°C would be equally desirable) rather than coverage over absolute values;
- we are interested in batch processes;
- our solution procedure does not presuppose a nominal point or expected deviations;
- our envelope region is of a simple hyper-rectangle form, which allows the formulation of straightforward operating procedures.

Mathematical Formulation

A typical mathematical formulation for dynamic process design consists of continuous variables representing differential state variables, algebraic state variables, operational variables and design variables (the latter two being degrees of freedom). The model of the process is represented by equations that relate the state variables to the degrees of freedom (and, for dynamic processes, to define the initial state of the system), and equality and inequality constraints to describe safety and operational limits. In most cases, such a model will represent a single point or trajectory of operation of the process, *i.e.* it assumes that the operational variables can be controlled precisely at the specified values or profiles.

In our formulation we want to account explicitly for variations in these operational variables. This means treating the operational variables as ranges rather than points or profiles and results in a hyper-rectangular envelope. We also want to ensure that the process is feasible and meets certain performance requirements anywhere within the ranges of the operational variables. The problem is formulated to maximise a measure of the envelope size, subject to satisfying the model equations, all constraints and the performance requirements at all points within the envelope.

The size of the envelope is defined as a weighted sum of the ranges of the "bounded" parameters. The weights are derived from the fact that: i) there will be a lower bound on the ranges, representing the limit to which the variables can be controlled; and ii) there will be an upper bound representing the level of control that is trivial to achieve; so that the maximum size of the envelope will be 1, corresponding to loose control, and the minimum size will be 0, corresponding to the tightest control possible.

The performance requirement is written in the form $\Phi \geq \Phi^*$ throughout the envelope. Φ would typically be the objective function in a standard optimisation problem (e.g. maximising profit) and Φ^* would be a suitable target value.

The details of the formulation are given in Samsatli *et al.* (1999).

Solution Procedure: Restricted Vertex Search

In this work, we apply a relatively simple solution procedure. We assume that, at least for the example problems presented here, ensuring feasibility of the constraints only at the vertices of the envelope will be sufficient to guarantee feasibility throughout. However, since the objective function is evaluated from the variables that define *every* vertex, the problem must be solved in a single level optimisation. This essentially means solving a multi-scenario problem where each scenario corresponds to a vertex of the operating envelope.

One issue arises immediately from the need to solve the problem in a single-level optimisation: the problem size increases exponentially with the number of "bounded" variables and thus industrial-scale problems would likely be intractable. In such cases, we can exploit the relationship between the operational envelope and the process envelope (feasible region) to reduce the number of scenarios that need be included in the optimisation. The solution procedure is based on the observation that a small proportion of the scenarios will be limiting the size of the operational envelope (*i.e.* the feasibility constraints will be active for only a few scenarios). This can be seen in Fig. 10, where only two of the four scenarios restrict the size of the operational envelope. If all of the limiting scenarios were know *a priori*, we would only need to solve a much smaller optimisation problem. Since this is obviously not possible, we use a heuristic procedure to select (or reject) scenarios for inclusion in a reduced-scenario optimisation problem. It is an iterative procedure using a full-scenario simulation to check for feasibility in all scenarios and to select the scenarios for the next optimisation.

The first step of the procedure is to select two scenarios, corresponding to all bounded variables at the lower bound and all bounded variables at the upper bound (any pair of opposite vertices will suffice), for inclusion in the first optimisation problem. Then, a full-scenario simulation is performed on the resulting operating policy to test for feasibility and to select an alternative pair of scenarios or additional scenarios for the next optimisation (*i.e.* at the current iteration, select the scenarios with the greatest infeasibility). The procedure is repeated, adding or replacing scenarios, until all scenarios are shown to be feasible in the full-scenario simulation step.

Example Problem 1

To illustrate our approach, we first consider a simple single-stage example process with only two envelope variables. This will then be extended to demonstrate the use of parametric optimisation to determine the trade-off between envelope size and process performance.

The process, for the first example, is a batch reaction in which the reactions $2A \to B \to C$ take place, where B is the desired product. The processing time and temperature profile may be adjusted to maintain the purity and production rate above certain levels (representing constraints of the form $\Phi \geq \Phi^*$). The formulation above was used to obtain operating ranges for the processing time and the average temperature, while all other operating parameters were kept constant. The results of the optimisation are shown in Fig. 9, where the operational envelope is compared with the feasible region of the process.

This example can be extended to generate a trade-off curve for envelope size versus process performance, by performing a parametric optimisation. As can be seen in Fig. 10, with quite a loose envelope the process can be "guaranteed" to generate a profit of up to 80% of the maximum possible profit. Achieving anything more than 80% of the maximum profit requires far tighter envelopes, and the limit would seem to be slightly less than 90%. This is naturally because an envelope size of zero relates to the smallest envelope achievable; not an operating point.

Example Problem 2

In this final example, we consider system in which there are seven bounded variables. This is a case where the full-scenario optimisation (consisting of 128 scenarios) would be at best difficult to solve.

The process consists of two reactors in which the following reaction scheme is carried out:

$$A + B \to 2C; \quad C \to D; \quad C + E \to F,$$

where F is the desired product and D is a by-product.

The seven "bounded" degrees of freedom represent the operating policies of the two reactors (e.g. the initial charges, cooling water flow-rate, feeding profiles, etc.). There are also two "fixed" degrees of freedom representing the volumes of both reactors.

The full-scenario simulation model comprises 11310 equations; some 1792 additional constraints would be required to perform a full-scenario optimisation. Clearly, it would be desirable not to have to solve the full-scenario optimisation. Here, we apply the solution procedure outlined before, using optimisation models that contain only two scenarios.

The first iteration results in an operational envelope that leads to only 35 infeasible scenarios when simulated using the full-scenario model. After selecting an alternative pair of scenarios for the second iteration, the resulting operational envelope is feasible in all 128 scenarios. Other optimisation problems for this system, with differing constraints, also behave similarly. These results show that quite complex problems can be solved with considerably less effort than might be expected.

The key results are illustrated in Fig. 11, where the first bars indicate the lower bounds on the ranges, the middle bars the actual values of the ranges and the last bars the upper bounds on the ranges.

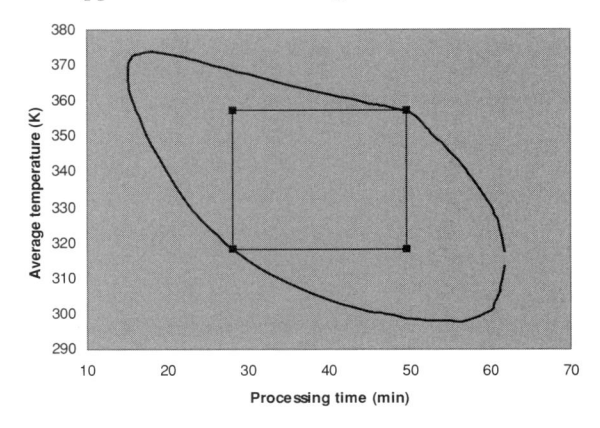

Figure 9. Comparison of the operating envelope with the feasible region.

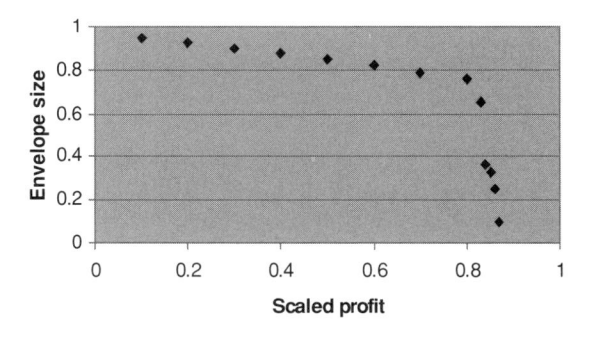

Figure 10. Result of parametric optimisation, example 1.

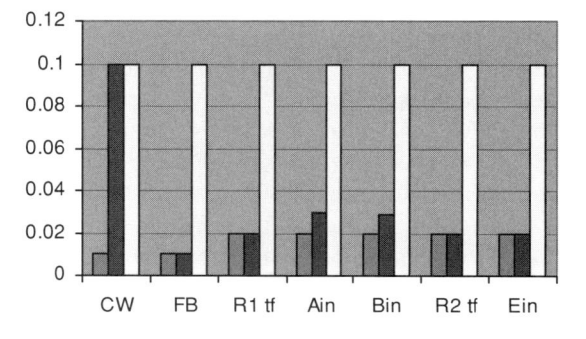

Figure 11. Range sizes for the second example.

Conclusions

We have highlighted a number of serious concerns in the way that pharmaceutical and fine chemical processes are developed in industry. This concern is also shared by many in the industry. Our approach to the problem is to take a radically different view than current practice. By relying more heavily on modelling and optimisation, we hope to exploit properties of the systems that would probably not be found using current practice.

The overall concept is to define how a process behaves based only on the conditions that it experiences, without specifying what unit operations are to be performed or what equipment will be used. This, we hope, will lead to more innovative process development and new equipment that are better suited to the process.

In this paper, we described a number of technologies that should facilitate the rapid development of better processes. Of key importance is that of "abstract" process design, the results of which can be used to determine the best set of operations (not necessarily standard) to carry out the process.

Once a more concrete view of the process has been defined, other important aspects such as robustness and operational envelopes can be considered. The robustness metrics described here have been shown to be particularly useful for product quality type constraints, but more work will be needed to find suitable solution procedures that are capable of solving industrial-scale problems. The technique for specifying the operating procedure of a process in terms of envelopes has been successfully demonstrated using a number of simple problems.

The next stage will be to look at more realistic problems, but given the apparent efficiency of the proposed solution procedure, we do not expect this to pose an insurmountable problem.

References

Charalambides M. S., N. Shah and C. C. Pantelides (1995). A software tool for optimal design of integrated batch processes. AIChE Annual meeting, Miami.

Douglas, J. M. (1988). *Conceptual Design of Chemical Processes*. McGraw-Hill Chemical Engineering Series. McGraw-Hill, New York.

Dussage P., S. Hart and B. Ramanantsoa (1992*). Strategic Technology Management.* John Wiley & Sons, 53.

Grossmann, I. E. and R. W. H. Sargent (1978). Optimum design of heat exchanger networks. *Comp. Chem. Eng.*, **2**, 1–7.

Grossmann, I. E., K. P. Halemane and R. E. Swaney (1983). Optimization strategies for flexible chemical processes. *Comp. Chem. Eng.*, **7**, 439.

Hutton W. (1996). *The State We're In*. Vantage, Jonathan Cape. London, p. 8.

Ismail, S.R., E.N. Pistikopoulos and K.P. Papalexandri (1997). Separation of nonideal mixtures based on mass/heat exchange principles. The entrainer selection and sequencing problem. *Comp. Chem. Eng.*, **21**, S211–S216.

Linninger, A. A., A. A. Shahnin and G. E. Stephanopoulos (1996). Knowledge-based validations and waste management of batch pharmaceutical process designs. *Comp. Chem. Eng.*, **20**, S1431–S1436.

Pisano G. P. (1997). *The Development Factory*. HBS Press. 34.

Pistikopoulos, E. N. (1995). Uncertainty in process design and operations. *Comp. Chem. Eng.*, **19**, S553.

Samsatli, N. J., L. G. Papageorgiou and N. Shah (1998). Robustness metrics for dynamic optimisation models under parameter uncertainty. *AIChE J.*, **44**, 1993–2006.

Samsatli, N. J., L. G. Papageorgiou and N. Shah (1999). Batch process design and operation using operational envelopes, *Comp. Chem. Eng.*, **23**, S877–S880.

Sharif, M. (1999). *Design of Integrated Batch Processes*, PhD thesis, University of London.

Salomone, H. E., J. M. Montagna and O. A. Iribarren (1994). Dynamic simulations in the design of batch processes. *Comp. Chem. Eng.*, **16**, 173–184.

Smith, E. M. B. and C. C. Pantelides (1995). Design of reaction separation networks using detailed models. *Comput. Chem. Engng*, **19**, S83–S88.

Technology Foresight, *Progress through Partnership Report No 1 - Chemicals*. Office of Science and Technology, London, 15–16.

Technology Foresight (1995). *Progress through Partnership Report No 9 - Manufacturing, Production and Business Processes (1995)*. Office of Science and Technology, London, 7.

Tricoire, B. and M. F. Malone (1992). Design of multiproduct batch processes for polymer production. Paper No. 134e in *Proceedings of the AICHE Annual Meeting*, Miami.

von Hippel, E. and M. J. Tyre (1995). *How learning by doing is done: Problem identification in novel process equipment. Research Policy,* **24**(1), 1–12.

Wets, R. J. B. (1996). Challenges in stochastic programming. *Math. Prog.*, **75**, 115.

BATCH PROCESS DEVELOPMENT: CHALLENGING TRADITIONAL APPROACHES

George Stephanopoulos and Shahin Ali
Massachusetts Institute of Technology
Cambridge, MA 02139

Andreas Linninger
University of Illinois - Chicago
Chicago, IL

Enrique Salomone
INGAR
Santa Fe, Argentina

Abstract

Batch process development is a fairly complex series of engineering tasks, overwhelmed by the lack of pivotal scientific information and the simultaneous pressures for short development cycle and production of small quantities of product for testing purposes. The broad range of scientific knowledge and engineering expertise that must be brought together, the effective management of a multitude of process development projects, the interaction between product design and the ensuing processing concepts, the multitude of metrics and constraints (e.g. economics, safety-health-environmental regulations, operability, flexibility) that need to be taken into account, are some of the features that call for different process development paradigms. In this paper we examine the nature of batch process development, its phases and the needs for each phase, the requisite knowledge, and its current weaknesses. We also discuss certain methodological approaches that have evolved over the last few years with emphasis on the work carried out by the authors, and outline needed approaches for further progress in this important field. In particular, we discuss the profound effect that the selection of the proper materials and reaction schemes have on the structure and behavior of the resulting process, the management concepts that need to be implemented for effective process development, as well as the computer-aided engineering tools and information systems that need to be in place, in order to support a wide spectrum of activities (e.g. from the process chemist's wet lab to the manufacturing plant) constituting the process of batch process development.

Keywords

Batch processes, Process synthesis, Computer-aided design, Environmental considerations, Batch process development, Solvent selection, Waste treatment.

Introduction

With a global consumer driven market place, the demand of high value-added pharmaceuticals, agricultural and specialty chemicals has risen sharply. These types of products are naturally suited for batch manufacturing due to factors such as, strict quality constraints, small production volumes, short product life, a large variety of products, and short time-to-market. However, while batch production is by far the more widely used and the much older of the two, the literature has been dominated by contributions pertaining to the design, control and operation of continuous processes.

Economic opportunities for "being first in the market" are extremely attractive. The continuous evolution in product specifications has led to increasingly shorter life-cycles for a number of chemicals, chemical compositions, and various types of materials, the large majority of which is manufactured through batch processes. Such perpetual product/process evolution, necessitates the deployment of rational and systematic research and development approaches, as well as the rapid execution of the chain of tasks, *"product development – process development – process design and engineering"* (Stephanopoulos et al., 1999).

The surge of interest in batch processing has led to work in all aspects of batch processing. There are many studies discussing one or more of these aspects such as, the *scheduling* of batch processes, the use of *intermediate storage*, the *sizing* of processing equipment and the *construction of task networks* given a complete process recipe and set of operations. All types of plants have been considered, from the *single process dedicated* plant, to the *multiproduct* plant, to the *multipurpose* plant.

However, while published work in batch process scheduling and the design of batch plants has grown exponentially, there has been very little work in the area of *conceptual design* or *process synthesis* for batch processes. Sharratt (1997) has underlined that "*...Academia has recognized the opportunity to contribute, but seems unclear as to what is required. Industrial research and development is also fragmented.*". In a series of articles, Basu and his collaborators (Basu, 1998; Basu, et al., 1997, 1998) have articulated very concisely the nature of process development and have made a very compelling case for the need of further improvements in pharmaceuticals process development.

The synthesis of a batch processing scheme with minimum production cost, satisfaction of regulatory constraints (safety, health, environmental), and inherent flexibility to fit existing multi-purpose batch plants, is a far more complicated task than its continuous-process counter-part. This complexity stems from a fundamental difference between the batch process and the continuous process: continuous processes are synthesized through *equipment-centered* approaches with a one-to-one mapping between operations and equipment, while in batch processes, the same piece of equipment is employed to perform many operations, leading to an *operations-centered* view for synthesizing, evaluating and deploying a batch process.

In addition, batch processes employ a wide variety of processing materials, e.g. solvents, which can have a profound effect on the structure and behavior of the ensuing process. Identifying the best set of materials is the pivotal question to be answered during batch process development. Furthermore, the significant molecular size and complexity of the materials encountered in batch processes puts the model-based estimation of reaction rates and physical properties beyond the current capabilities of the available engineering tools. This weakness offers one of the most exciting opportunities in defining new and effective *molecular modeling-based approaches* in process development.

It should be emphasized that progress in batch product and process development is profoundly affected by new scientific developments. Innovation is at the core of batch process development. Chasing the innovation wave is essential for continued growth of industries based on batch manufacturing (Thayer, 1999). For example, molecular biology is revolutionizing the character of biochemical operations, while the synthesis of new nano-sized materials is introducing new types of reactors and separators, never encountered before. In other words, *batch process development cannot afford to limit its perspectives to existing operations or unit operations*.

Finally, it has to be recognized at the outset of this paper that, *managing process development activities is at least as important as any scientific or technical issue involved*. Pisano (1996) in his book *"The Development Factory"*, has provided a very convincing case for the significance of effective management in fostering innovation and deploying new batch products and processes for the pharmaceutical industry. Ayers (1999) has also presented very compelling arguments for the role of management tools in process development.

The remainder of this paper is structured as follows. The next section presents a case study for the development of a pharmaceutical process. It illustrates the phases of process development and identifies the issues that need to be resolved. The subsequent section summarizes these observations, thus laying the general framework for the various activities encountered in batch process development. Managing the activities of batch process development, the interaction among the various tasks, metrics of effectiveness, constitute the subject matter of the following section, while the last section of the paper deals with the computer-aided tools that are needed to support batch process development.

Developing a Pharmaceutical Process: A Case Study

Consider the 9-stage manufacturing chain of Fig. 1 for the production of the intermediate pharmaceutical component, Mark VIII. Several questions arise regarding the efficiency of the above scheme:

a) Is there a better set of reactions, using different raw materials, to produce the same final product?
b) The main reaction at each stage is carried out in a specific solvent. Is the selected solvent the best from a reaction rate and selectivity (against competing reactions) points of view?

c) What are the competing by-products and how are they viewed from an environmental point of view?

d) What are the consequences of the employed reaction solvents on the recovery and purification of the main intermediate product?

e) The reactions in Stage-1 (Fig. 2) require the use of a complex quench mixture for the formation and stabilization of the main product. Is the employed quench mixture the best for the complete recovery of the product, and what are the environmental consequences of its use?

f) The Grignard agent, CH3MgBr, in Stage-1 is stored in Et2O. Is this the best storage medium, and what are the consequences on the downstream processing efficiencies and wastes generation?

2. Select the best solvents, homogeneous catalysts, and quench mixtures.

3. Construct the best processing scheme, using selected chemicals and chemical reactions.

It is clear that the simultaneous solution of the above design issues leads to an enormous combinatorial problem, whose solution defies the capabilities even of the best combinatorial optimization algorithms. For the time being it is important to examine the characteristics of the design issues raised above.

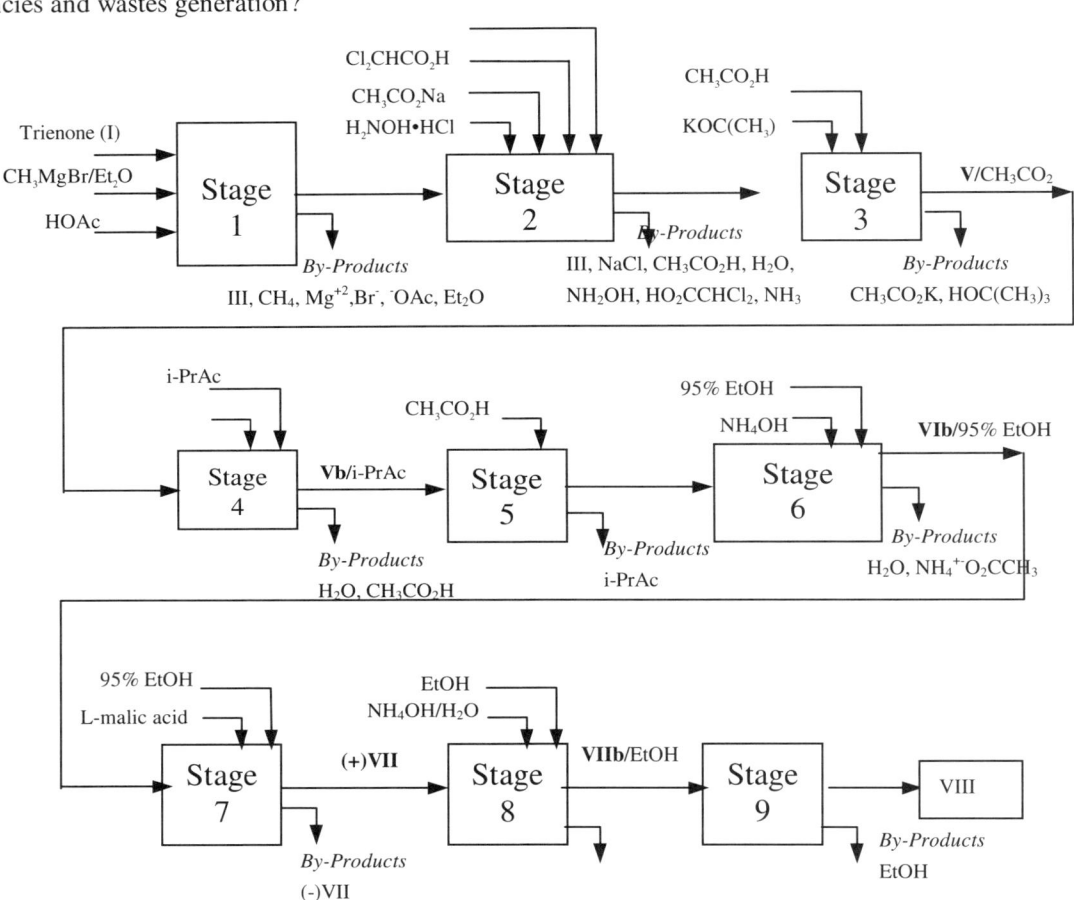

Figure 1. The 9-stage plant depicts the multistage production of 5-Methyl-10,11-dihydro-5H-dibenzo[a,d]cyclohepten-5,10-imine.maleate (Mark VIII).

If we focus our attention on any specific stage, we can raise a number of different questions, each of which is related to the role, the processing efficiency, and the environmental consequences from the use of any of the materials employed in the stage.

All of the above questions require the simultaneous treatment of the following design issues:

1. Select the best set of raw materials and associated network of reactions for the production of the desired chemical product.

Identification of the Pharmaceutical "Targets"

The intermediate molecule, Mark VIII, is not the final pharmaceutical product, which for confidentiality reasons will remain unnamed. The following question though arises; *how was the desired pharmaceutical molecule identified?*

Modern practices in the pharmaceutical industry for the identification of pharmaceutical targets and the design of appropriate molecules (drugs) have advanced significantly as a result of the revolution brought by genomics, proteomics, and the use of molecular modeling techniques. For example, the known sequence of a genome leads to the identification of the relevant gene and of the associated protein. Search of extensive libraries of known chemicals, systematic combinatorial synthesis of alternatives to natural analogs, and evaluation of their binding strength to the given protein through molecular mechanics and estimation of binding energies, have revolutionized the science and technology in target identification. However, for our purposes, the point is that several molecules may exhibit similar estimates of anticipated medical efficacy, in which case how does one select the most promising of the various alternative targets? It is clear that once the target molecule has been selected, the scope of the ensuing allowable reactions and processing schemes has been greatly reduced. No systematic approach is presently available for answering the above question, and pharmaceutical companies rely on the subjective evaluation of very skilled synthetic chemists. Typically, a small number of molecules may be selected for the continuation of process development along a small number of parallel paths.

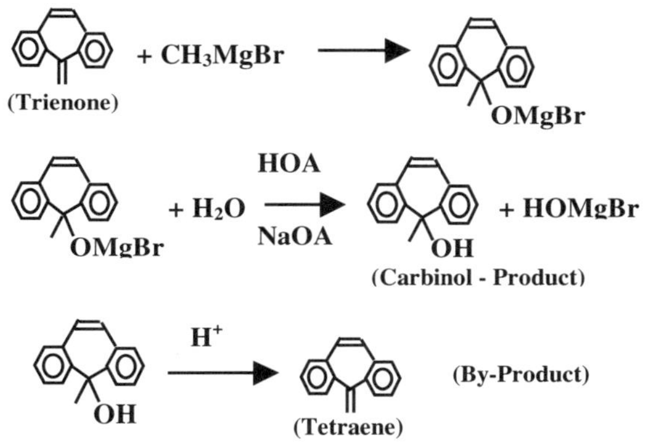

Figure 2. The set of reactions for stage-1.

Raw Materials and the Reaction Network

Starting with the desired final product one can employ the well-established *retrosynthetic* approaches in order to identify the potential *precursor-intermediate* products, as shown in Fig. 3. The synthetic algorithms of Mavrovouniotis and Stephanopoulos (Mavrovouniotis and Stephanopoulos, 1992; Mavrovouniotis and Bonvin, 1995) can also be used to ascertain the completeness and consistency of the generated reaction pathways.

The feasibility and attractiveness of each alternative sequence of chemical reactions depends on the following considerations:

a) Overall equilibrium conversion yields from the raw materials to the desired product.
b) Relative reaction rates to the desired product (intermediate or final) and the various by-products, i.e. the relative reaction sensitivities.
c) The prices of the various raw materials.
d) The ecological characteristics of all chemicals; raw materials, intermediates, and by-products.

Although the selection of the best does involve a large combinatorial problem, the pivotal impediment is the estimation of the relative reaction rates in various solvents, and the estimation of the ecological properties.

Synthesis of the Batch Processing Scheme

Fig. 4 shows a diagrammatic depiction of the series of operations, which implement the set of reactions in Fig. 2 for the production of the intermediate carbinol in Stage-1. Note that only the operations involved in the production of the principal intermediate are shown. The ancillary operations for the recovery of raw materials, solvents, and other processing materials, as well as the generation of wastes need treatment are discussed in a later paragraph. It can be easily seen that the building blocks of the processing scheme are elementary operations such as, *heat, charge, concentrate, decant, etc.* No specification of the processing vessels has been made, and several of the operations shown in Fig.4 will be eventually carried out in the same vessels of a manufacturing plant, after the plant has been selected and the operations have been scheduled in the most appropriate equipment.

The synthesis of a batch processing scheme, such as the one shown in Fig.4, corresponds to the *synthesis of an operating procedure*. It involves the selection of the appropriate elementary processing steps (the *operations* of the procedure) and their sequencing (the *ordering* of the operations). In a subsequent paragraph we will discuss a specific planning approach for the synthesis of a batch processing scheme, using *means-ends analysis* and concepts of *monotonic planning*.

Plant-Wide Selection of Solvents

Solvents are used in a batch-processing scheme to: (a) Store raw materials. (b) Enhance reaction rates and selectivities. (c) Quench a reaction. (d) Stabilize the formation of the desired product. (e) Extract the desired product or by-products from the reacting mixture. (f) Provide the medium for the crystallization of a product. (g) Prepare filters for the filtering operation. (h) Wash the crystals removed from a filter. (i) Clean process vessels from impurities. (j) Break azeotropes. (k) Act as transfer media for potentially hazardous compounds.
Traditionally, the selection of solvents has been addressed within the scope of specific process operations. As a result, a solvent, which is excellent for the enhancement of reaction rates and selectivities, could create insurmountable problems for the downstream processing operations. Consequently, only a plant-wide scope for the selection of the best solvents could lead to consistent and optimal results. All solvents should have properties, which satisfy three groups of desired specifications; (i) processing, (ii) environmental, (iii) safety, and health related. The processing requirements are dictated by the

role that a particular solvent is called to serve. Table 1 provides a partial list of such requirements for the solvents needed in the Stage-1 of the processing scheme shown in Figure 2 (see Modi, Aumond and Stephanopoulos, 1996).

The selection of the solvents, which satisfy all the desired physical and ecological constraints, is a fairly complex combinatorial problem, and is hampered by missing data, or methodologies for the estimation of certain physical properties. To overcome the combinatorial complexity of the solvent selection problem one may pursue a hierarchical approach, which proceeds as follows:

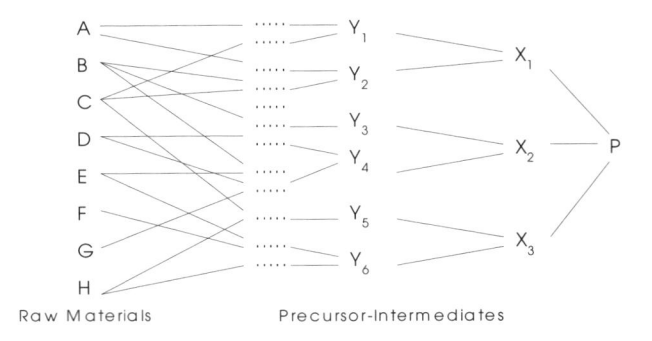

Figure 3. Schematic of alternative reaction chains leading to the same final product.

Level-1: Given a bench-scale description of a batch-processing scheme, identify all the solvent-based processing tasks.

Level-2: Screen-out solvents, which violate task-level constraints; processing or ecological.

Level-3: Screen-out solvents, which violate plant-wide constraints.

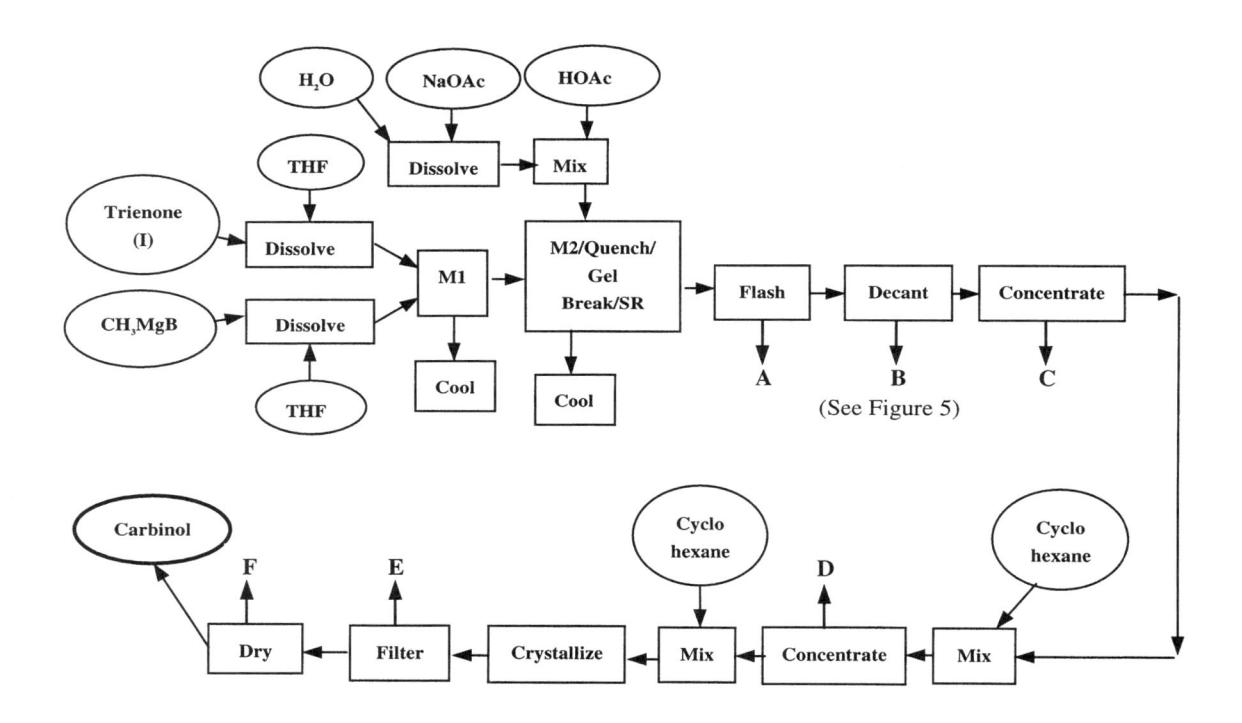

Figure 4. The main line of processing operations leading to the production of intermediate Mark VIII.

Synthesis of a Waste Treatment System

The manufacturing of specialty chemicals by batch processing schemes, produces *very large amounts of wastes per unit product*. As ecological regulations become more stringent, one needs to depart from existing practices. End-of-pipe treatment of wastes, or/and relocation of manufacturing facilities to places with more permissive environmental regulations, are temporary and in some respect counter-productive solutions. Selection of treatment options should be viewed as an integral part of the synthesis of batch processing schemes. Linninger et al., (1995a, b; 1997) have developed such an approach, which is based on the concept of *Augmented Planning Networks* and as such offers a natural setting for integration with the MEA-NMP approach, employed for the synthesis of batch processes. Fig. 5 shows the lines of processing operations associated with the treatment of the "wastes" generated by the various operations for the production of Mark VIII in Stage-1. For the production of 25 Kg of intermediate Carbinol, 985 Kg of wastes are generated, i.e. almost 25 Kg of wastes per Kg of intermediate product. If we consider the whole sequence of 9 stages, shown in Fig. 1, then the ratio of wastes to final product approaches 500, clearly an unacceptable situation.

Facilities Allocation

The problem of allocating a batch processing scheme, or equivalently, a *processing recipe*, into a multipurpose production facility is a major activity in batch process engineering. Like many other engineering activities it is characterized by the reconciliation of requirements with availability of resources. During the life cycle of a batch process development, the problem is approached with different objectives.

A. *The Facilities Allocation Problem During the Conceptual Synthesis of Batch Processing Schemes.* During the synthesis of conceptual processing schemes, preliminary studies on the allocation of a production sequence into existing manufacturing facilities are conducted. At this stage, focus is on feasibility, evaluation of processing alternatives and the definition of requirements. The problem is posed at a higher level of abstraction. Decisions taken at this stage will usually have long term and deep impact on the process under development. Typical questions at this level are:
a) Do we have the necessary equipment to realize this process?
b) Among several production facilities, which are the best for allocating this process?
c) Can we identify the main pros and cons of assigning this process to a given production facility?
d) Given the inventory of equipment in a production facility, which is the attainable scale of production?

Being able to address the above questions, while still it is possible to introduce modifications to the production recipe is in at the core of the modern concepts of process development. To answer them requires to bring together data and knowledge traditionally segregated. Therefore, consistent integration of information is the key. The nature of this integration is largely evolutionary, so the interaction among designers and information is also a critical issue.

B. *The Facilities Allocation Problem at the Manufacturing Stage.* Later on, when the process reaches the manufacturing stage, the problem of allocating facilities is considered again, but at this stage, focus is on scheduling of tasks, distribution of limited resources and performance improvement.

In this second case, operational level decisions are taken in usually short periods of time. Efficient automation of these decisions is the key. This is the stage where mathematical formulations of the problem and optimization algorithms are in order for quantitatively sound results. Scheduling of tasks is by and large the problem dealing with the allocation of production facilities that has received most attention, but it should not be concluded that the problem of allocating production facilities is the problem of scheduling.

Having recognized the hierarchical nature of the problem of facilities allocation it follows that a framework for its representation is needed. This representation has to be consistent with any level of abstraction and to serve as a common repository of information available all along the evolution of the process development.

Such a computer-integrated framework for the representation of the problem of allocation of production facilities has been developed by Salomone and Stephanopoulos (1997) and is part of the *BatchDesign-Kit* (*BDK*), described in a subsequent section. Based on this framework, a systematic approach has been implemented for solving the problem of preliminary allocation of production facilities during the conceptual design stage.

Summary of Batch Process Development Tasks

The development of the batch process proceeds through a series of phases before its implementation, i.e. it goes from chemist to process developer, to process engineer. At each of these phases, the process must be reviewed in order to insure a feasible design and, generally, once a design leaves a particular phase, that portion of the design is considered complete. Minor iterations and backtracking do not substantially change this mode of conception, development and engineering of a batch process. This current approach is not conducive in allowing changes to the previous levels of design. Ideally, the system of process development should allow for the rapid iteration between the process developers, engineers and chemists, facilitating rapid communication between the groups and rapid revisions of design alternatives. The rapid iterative process proposed is equivalent to placing feedback loops into the development process.

Table 1. The Explicit Requirements for the Solvents of Stage-1 in the Process of Figure 1.

Processing Task	Current Solvent	Explicit Processing Constraints
Reaction	Tetra hydrofuran	1. High capacity for CH$_3$MgBr, trienone and carbinol 2. High boiling point 3. Low vapor pressure 4. Low reactivity with trienone, carbinol, bromide 5. Large effect on reaction rate 6. Low viscosity
Quench	Aqueous solution of glacial acetic acid and sodiumacetate	1. Low capacity for carbinol 2. High reactivity with bromide 3. High boiling point 4. Low vapor pressure 5. Low density
Crystallization Of Carbinol	Cyclohexane	1. High capacity at $T=T_{ambient}$ 2. Low capacity at $T<T_{ambient}$ 3. Low vapor pressur 4. Low reactivity; carbinol 5. Low tendency form solvates 6. Low viscosity
Solids Wash	Cyclohexane	1. High capacity for impurities 2. High boiling point 3. Low freezing point 4. Low vapor pressure 5. Low reactivity with carbinol 6. Low viscosity 7. Low density
Extraction	Brine	1. Low capacity for carbinol 2. High boiling point 3. Low freezing point 4. Low vapor pressure 5. Low viscosity 6. Low density

In principle, batch process development goes through the following stages:

1. *Molecule Discovery*, during which a molecule with desired properties is identified, synthesized in lab-scale facilities, and tested for function. Whether this is the best molecule for the intended market function or not, is not examined until much later. At the beginning simple feasibility is sufficient.

2. *Selection of the Reaction Pathway*, that will produce the desired molecule from economically available raw materials. During this phase, process chemists do explore a limited number of alternative pathways, but the pressures of process development outweight a more extensive search. It is possible that an ill-conceived set of reactions may lead to an unacceptable process, an outcome that will not be visible until later on.

3. *Synthesis of Conceptual Processing Schemes*, an activity which proceeds in parallel with the lab-scale production of the desired molecule. Reaction conditions, product separation and purification operations, reaction solvents and catalysts, mass separating agents, are all integral elements of the process definition. Rudimentary economics and materials' assessment from an environmental, health, and safety point of view are normally considered to avoid major pitfalls of the evolving process. For pharmaceuticals, agricultural chemicals and other specialties, which are subject to efficacy testing, toxicity and other health hazards assessment, this stage of process development doubles as a production phase for the molecules to be tested.

4. *Selection of Treatment Options*, to process or/and dispose the wastes generated by the evolving manufacturing scheme. Negligence to consider these issues early on may lead to premature abortion of the specific batch process development activities, or rejection of the resulting process due to high cost in ensuring acceptable environmental impact and health risks.

5. *Allocation of Manufacturing Facilities*, where the processing concept is implemented on existing multi-purpose or new dedicated plant. This decision is very closely linked with major decisions regarding the timetable of product manufacturing and thus to the planning and scheduling of production facilities. Consequently, allocation of a batch process to specific facilities creates an inherent integration of the process under development with the rest of the processes employed by a given company.

Synthesis of Conceptual Batch Processing Schemes

Conceptual synthesis of a batch processing scheme is defined as the selection of process operations and the interconnections between theses operations to transform a set of raw materials into the final product. Planning, as an activity, is defined as the construction of action sequences to achieve a goal. By comparing these two definitions, we see that *the synthesis of batch processing schemes is a planning activity*. Thus, the process developer is creating an *operating plan* for the process, i.e. selecting and ordering the process operations and determining what the operating conditions should be. While linear, or equivalently, monotonic planning was the first paradigm advanced for the compilation of plans, non-linear or non-monotonic planning has evolved to be a far more effective approach.

Nonmonotonic Planning in Batch Process Synthesis

Nonmonotonic planning is the construction of plans through the refinement of *partial plans*. A partial plan is an incomplete plan, which leaves some information unspecified. The unspecified information could be either the temporal order of the operators in the plan. For example, it has been determined that a process requires an azeotropic separation, but the exact location of this operation has yet to be determined. "A single partial plan actually describes a large number of completions." (Ali, 1998). For different values of the unspecified information,

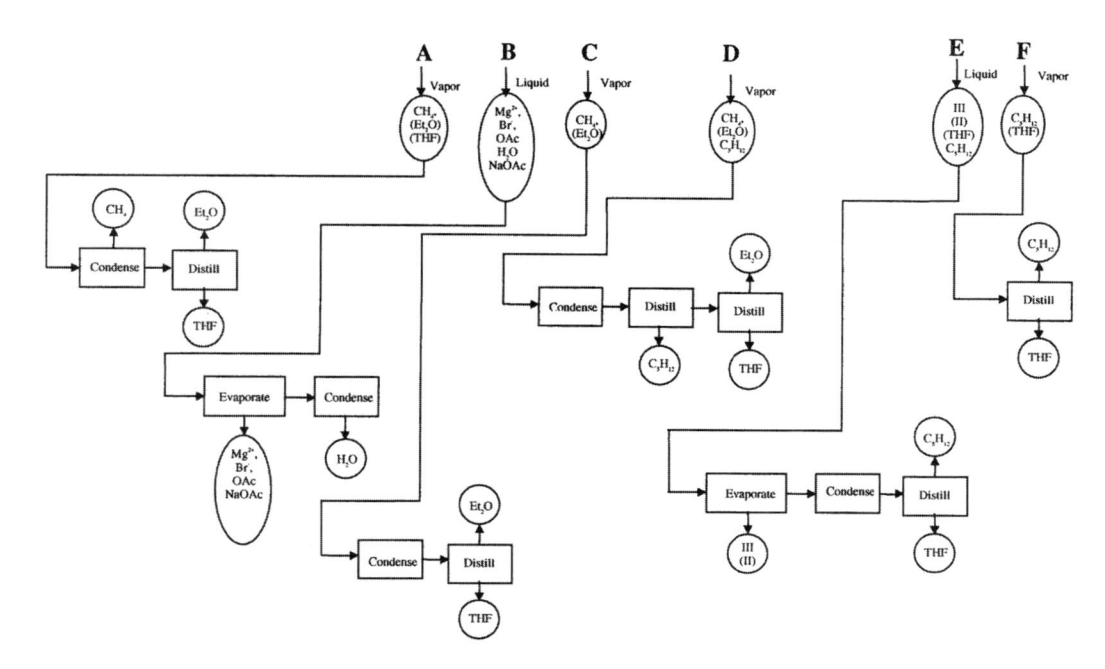

Figure 5. The sequence of ancillary operations for the treatment of wastes generated in stage-1.

alternative complete plans can be developed. For example, depending upon how the previous operations have modified the state of the current mixture, an azeotropic split using a mass separating agent may be chosen over a pressure distillation giving us two possible completions.

A major advantage of nonmonotonic over monotonic planning is that the former can reduce, in principle, the need for backtracking. Since information can be left unspecified, the method of *constraint posting without backtracking* can be employed. Thus, instead of backtracking, when a constraint violation, called a *clobberer*, is detected, nonmonotonic planning attempts to identify an operator, called a *white knight*, that would negate the preconditions leading to the constraint violation. The white knight is applied before the operation that causes the constraint violation, leading to the so-called *promotion* of the white knight. The resulting posting of a new constraint, determines the temporal ordering of the operations in the plan, i.e. the white knight must be applied before the clobberer.

Another form of promotion of operators, which also results in constraint posting without backtracking, is the promotion of an infeasible operation. If we are unable to identify a white knight operation, before we remove an operation from consideration, we attempt to promote the infeasible operation in order to determine if reordering of the operations will remove the precondition violation. The infeasible operation is placed before the preceding operation in the plan and its preconditions are evaluated.

In order to use this type of promotion, the preconditions of the promoted operation must not be violated or violations must be removed through white knight identification and the resulting post-conditions from the application of the promoted operation must not cause the preconditions of the demoted operation to be violated.

The Means-Ends-Analysis and Non-Monotonic-Planning (MEA-NMP) Approach to Batch Synthesis

Ali and Stephanopoulos (Ali, 1998) have combined non-monotonic planning ideas with the classical means-ends-analysis, in order to compose a systematic approach for the synthesis of batch processing schemes. This approach allows for the systematic and simultaneous selection and ordering of processing operations. It consists of the following three main elements:

1. Means-ends analysis is used to identify differences between the current and the

desired final states of the batch process, and to determine the class of operators, which could alleviate the differences.
2. Non-monotonic planning is used to identify partial sequences of process operations in bridging the gaps between current and desired states of the process.
3. Constraint-posting without backtracking allows for the identification of missing tasks or/and determines the relative ordering of these tasks, in order to remove constraint violations.

The following example illustrates the approach:

Illustration of the MEA-NMP Ideas

Let's assume that we have the following reaction, $A + B \rightarrow C$, where A and B are the raw materials and are available as solids. Our goal is to produce compound C. We first apply MEA to identify differences. In our example the difference is quite obvious, the molecular structure of A and B differ from the molecular structure of C. The next step is to identify an operation that can resolve this difference. In this case, the task is chemical reaction:

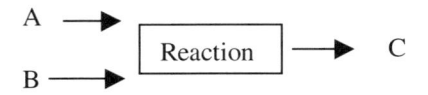

```
A ─┐
   ├─→[ Reaction ]─→ C
B ─┘
```

From the figure it would appear that we are finished, but before we can apply the operation we must first check the preconditions associated with the operation. Here, in order to have a reaction, the reactants must come into close contact and since our reactants are both solids this is not possible. At this point a direct application of MEA would suggest marking the reaction as infeasible. MEA would find no solution. Through the application of the NMP, we try to identify operations (*white knights*) to eliminate the precondition violations, before we attempt to apply the *react* operation.

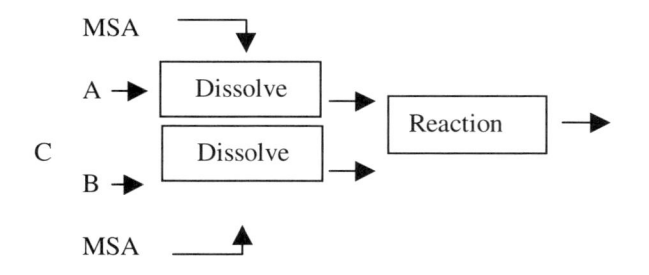

```
       MSA ─┐
            ↓
A ─→[ Dissolve ]─┐
                 ├─→[ Reaction ]─→
C ─→[ Dissolve ]─┘
B ─→
       MSA ─┘
```

The procedure for identifying the white knights is given in detail in Ali (1998). For this example, the white knight, which removes the close contact violation, is the *dissolve* operation. Since we are eliminating a precondition violation, the temporal ordering generated, places the white knight before the reaction operation, i.e. the *dissolve* operation is placed before the reaction. Assuming that the preconditions of the dissolve operations have been met, one can proceed and apply the white knights, generate the new material states and then apply the *reaction* operation.

The MEA-NMP approach is implemented in eight iteratively and recursively applied steps, as follows:

1. Identify the differences between the current state and the desired goal state.
2. Screen the current state and the differences identified in Step 1 for *critical features*.
3. Apply MEA for task category and primary operation identification.
4. Evaluate the preconditions of the proposed operation.
5. Identify white knights to remove precondition violations.
6. Promote operator if white knight cannot be found.
7. Apply the operators and generate the new current state.
8. Check the plan for abstract operators.

Managing Batch Process Development

As the process moves from Stage-1 to Stage-5 (see Section on "Summary on Batch Process Development Tasks", above), the "added value" or "lost opportunity" decreases by several orders of magnitude, e.g. by a factor of 10,000 to 1,000,000. Indeed, an "optimally" selected set of reactions offers far greater value than an "optimal" process based on a complex and cumbersome reaction pathway, which employs a large number of solvents and catalysts. Similarly, judicious selection or complete absence of solvents would lead to far more economical process than an optimal facilities allocation of a poorly conceived processing scheme.

Although this is fairly well known and appreciated by process developers, there have been only limited attempts to overcome the consequences through innovative ways for process management. There are many reasons for this inherent weakness:

1. *The Economics of Being First in the Market* overwhelm any gains that may come from more lengthy process development. This has been a pivotal and all-determining driver in process development.
2. *Parallel Process Development and Product Manufacturing.* In the pharmaceutical industry, a process chemist not only attempts to synthesize a lab-scale processing scheme for the production of a pharmaceutical, but at the same time he/she must produce small quantities of the molecule for toxicological tests. A process development engineer not only attempts to scale-up the process and resolve manufacturing issues at the pilot

plant, but she/he must produce sufficient amounts of product for clinical testing. Consequently, the pressures of interim production are intrinsically linked to issues of process conceptualization, and lead to premature "freezing" of process options.

3. *Dispersion of Knowledge and Expertise.* The range of scientific knowledge and engineering expertise required by the activities in the five stages of the development process defy the abilities of any known form of collaborative teams. The requisite knowledge and expertise may not even be part of a company's intellectual assets. Even when it is available within a given company, the scientific knowledge and expertise rest with a large number of individuals or in dispersed databases (not necessarily in electronic form), while by their nature are quite fragmented and poorly codified. The capacity of a company to accelerate batch process development and improve the quality of the resulting process is directly linked to its ability to master all available scientific knowledge and engineering expertise, while accumulating its newly derived intellectual assets into easily codifiable electronic form.

4. *Complexity in Decision-Making.* Batch process development is essentially a very complex decision-making process, which in its entirety involves all decisions from the selection of the best molecule to the optimal coordination of a product's manufacturing operations with the manufacturing of a range of other products. A frontal attack through an integrated optimization, as if all decisions were of equal importance, is not realistic neither desirable. An inherent dominance of certain decisions, e.g. set of reactions, over others, e.g. separation operations, implies a lexicographic ordering of these decisions. However, no systematic methodology has been as yet proposed to handle the overall process.

5. *Insufficient Codification of Scientific Knowledge.* It is surprising how difficult is still the scale-up of a batch process, despite the enormous progress in the scientific understanding of the essential physico-chemical phenomena occurring in batch processes. The basic reason is the lack of effective codification of the existing scientific knowledge, or to use a more common term, the lack of effective models to describe a number of typical batch operations, e.g. crystallization. This is a major obstacle and it should be resolved quickly. Furthermore, it is also surprising how ineffective are the existing methods for predicting (a) the complete array of reaction by-products in a quantitatively satisfactory manner, or (b) the physical properties of mixtures of complex and large molecules.

Pisano (1997), in a very thorough study of the process development activities for the pharmaceutical industry, outlined the series of interacting tasks, which shape the effectiveness of the development process. Specifically, the activities on *Product Development* proceed in the following steps:

1. Discovery of Lead Compound
2. Pre-clinical development
3. Clinical trials

The tasks of Process Development

1. Process research,
2. Pilot development, and
3. Transfer to manufacturing plant and start-up.

These two, seemingly separate and independent lines of development, are in essence tightly coupled and interacting. However, the perception that the added value during Product Development outstrips the added value during Process Development, has led to a serial execution of the tasks, where the Product Development activities dominate the course of action and define the specifications within which the process is developed.

Thus, the efficient progression of activities in the development of a pharmaceutical process requires the synergistic interaction among a variety of interrelated tasks, and is characterized by the following features:

1. Several reaction paths must be explored in parallel for the generation of a lead compound.
2. The generation of sufficient quantities of the lead compound for safety and medical efficiency testing, dictates that the process chemist (at the bench-scale lab) and the process development engineer (at the pilot-scale plant), are concerned not only with the definition of the chemical production routes (generation and isolation of lead compound), but also the production of sufficient quantities of the lead compound. Thus, process development is coupled with "production" requirements.
3. The selection of process materials (e.g. catalysts, solvents) has a very profound effect on the structure and behavior of the resulting process, thus rendering a very tight coupling between, (i) product and process development, and (ii) the various phases of process development.
4. The prevailing uncertainty in scientific knowledge about reactions and materials'

properties, introduces an overwhelming need for setting the proper trade-offs using qualitative knowledge (from experts) and making use of the accumulated corporate knowledge from past development projects.

Pisano (1997) has evaluated the significance of all of the above and has suggested a management structure for the efficient evolution of development activities. Most of these suggestions have focused on the criticality of new science and the effective communication among various experts in resolving the series of inevitable tradeoffs. A summary of his recommendations is beyond the scope of this paper, but the interested reader is encouraged to consult his book for further details.

Information Processing and the *BatchDesign-Kit* in Batch Process Development

In the previous paragraphs we have indicated the pivotal role that the appropriate information plays is securing the proper and optimal decision-making during the various phases of product and process development. A significant portion of effective management of all these activities depends on the availability of the requisite information to the decision-makers at the critical junctures of the whole process. A typical example of a computer-aided system which has attempted to answer some of the information processing needs during batch process development is the *BatchDesign-Kit (BDK)*, developed by the co-authors and their collaborators at MIT.

The *"BatchDesign-Kit"(BDK)* (Linninger et al., 1995a,b) is an integrated set of software components, which provide the computer-aided implementation of the methodologies sketched in the previous paragraphs. The original version of the system was developed at the Laboratory for Intelligent Systems in Process Engineering (LISPE) at MIT, and its more advanced version was developed for commercialization by Hyprotech Ltd.

BDK has pioneered some new metaphors for the description of batch processes, e.g. the *Process_Sequence_Diagram*, a two-dimensional depiction of the material flows in time and among processing facilities, the *BatchSheet* a textual description of the recipe with the various operations in a batch process, the *Materials_Model* which constitutes the basis for the state-task network of materials transformations occurring in a batch process, etc.

Fig. 6 depicts the essential components of BDK, which have been designed to offer the following capabilities in support of batch process development, as outlined in earlier sections:

Production Route Planner, which assists in the formulation and early economic evaluation of alternative production routes for a particular chemical.

Process Synthesizer, is the facility that implements the MEA-NMP approach, described in earlier section, for the synthesis of conceptual processing schemes (production recipes).

Simulator-Evaluator, is the set of components, which implement the operations-oriented language of BDK for the definition and simulation of production recipes, and their evaluation in terms of overall process economics, waste generation, batch cycles, and facilities allocation.

Materials Selection, encompasses the methodologies for the selection of solvents for chemical reactions and separations, as described in earlier section.

Materials Assessment, is an integrated set of expert systems, which encompass the US Federal regulations for the environmental, health and safety considerations by a chemical process. It provides an early warning of high-risk or unacceptable situations, during the evolution of a processing scheme.

Facilities Allocation, implements the ideas described earlier for the early identification of possible allocation scenarios, and estimates the corresponding batch cycles and costs.

Databases, in relational forms, capture the information used by the various facilities of BDK, e.g. chemicals, their properties, reactions, plants, equipment, solvents, regulations, etc.

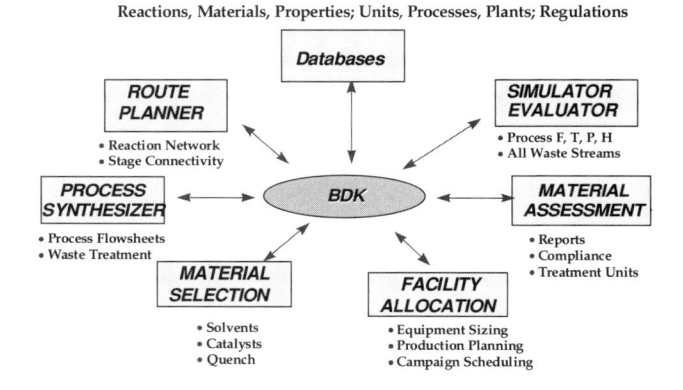

Figure 6. The structure and components of the BatchDesign-Kit

Conclusions

Batch Process Development for the manufacturing of pharmaceuticals, agricultural chemicals, and other specialties is an extremely important activity, which is shaping today the industrial structure for the manufacturing of a very broad range of chemicals. Its significance cannot be over-emphasized. The scientific, technological, and management issues that govern it, are complex, inter-related, and at times quite cumbersome. Nevertheless, significant progress has been accomplished towards its rationalized evolution.

The most pressing issues for further progress are the following:

(a) Integrated approaches for a parallel product and process development are of paramount importance.
(b) New computational approaches for the generation and evaluation of alternative chemical production

routes are essential for the selection of the best route.
(c) Early integration of process engineering considerations, e.g. process synthesis, selection of processing materials, assessment of materials, treatment of wastes, is critical for setting the proper specifications, within which the subsequent steps of process development will be carried out.
(d) Effective modeling approaches for the estimation of physical properties for large and complex molecules, as well as the description of basic processing steps, e.g. crystallization, are sorely needed for the efficient scale-up of elementary operations.
(e) New management structures are necessary for the effective and well-orchestrated integration of a variety of tasks.
(f) New breeds of computer-aided tools for efficient information processing and decision-making are essential and should be developed and deployed.

These challenges can be met if research activities in academia and industry, related to the core batch process development tasks, are properly focused and directed towards the generic problems, discussed in earlier sections of this paper. So far, this has taken place with limited success.

References

Ali, S.A. (1998). Ph.D. Thesis, MIT.
Ali, S. A., E. Stephanopoulos, A. A. Linninger, and G. Stephanopoulos (1994). *LISPE Report*, MIT.
Ayers, J.B. (1999). *Chem. Eng. Prog.,* **95**(2), 31.
Basu, P.K (1998). *Chem. Eng. Prog.,* Sept., 75.
Basu, P.K., J. Quaadgras, J.E. Holleman, R.A. Mack, and A.R. Noren (1997). *Chem. Eng. Prog.,* June, 66.
Basu, P.K., R.L. Buchman, and A.L. Campbell (1998). *Chem. Eng. Prog.,* Dec, .52.
Linninger, A., S. Ali, E. Stephanopoulos, C. Han, and G. Stephanopoulos (1995a). *AIChE Symp. Series*, No.303, **90**, 46-58.
Linninger, A., E. Stephanopoulos, S. Ali, C. Han, and G. Stephanopoulos (1995b). *Comp. Chem. Eng.,* **19**, S7-S14.
Linninger, A., S. Ali, and G. Stephanopoulos (1996). *Comp. Chem. Eng.,* **20**, S1431-S1436.
Linninger, A., E. Salomone, S. A. Ali, E. Stephanopoulos, and G. Stephanopoulos (1997). *Waste Management*, **17**, 165-173.
Mavrovouniotis, M.L., and G. Stephanopoulos (1992). *Ind. Eng. Chem. Res.* **31**, 1625-1637.
Mavrovouniotis, M.L. and Bonvin, D. (1995). *FOCAPD-'94, AIChE Symp. Ser.*
Modi, A., J.P. Aumond, M. Mavrovouniotis, and G. Stephanopoulos (1996). *Comp. Chem. Eng.,* **20**, S375-380.
Pisano, G.P. (1997). *The Development Factory*, Harvard Business School Press, Boston.
Salomone, E. and G. Stephanopoulos (1997). *LISPE Report*, MIT.
Sharratt, P.N. (1997). *Handbook of Batch Process Design*, Blakie, London.
Siirola, J.J.(1996). *Comp. Chem. Eng.,* **19**.
Stephanopoulos, G., S. Ali, A. Linninger, and E. Salomone (1999). *Comp. Chem. Eng.,* **23**, .S975.
Thayer, A.M. (1999). *Chem. Eng. News*, Feb. 8, 17.

ALGORITHMS AND SOFTWARE FOR LINEAR AND NONLINEAR PROGRAMMING

Stephen J. Wright
Mathematics and Computer Science Division
Argonne National Laboratory
Argonne IL 60439

Abstract

The past ten years have been a time of remarkable developments in software tools for solving optimization problems. There have been algorithmic advances in such areas as linear programming and integer programming which have now borne fruit in the form of more powerful codes. The advent of modeling languages has made the process of formulating the problem and invoking the software much easier, and the explosion in computational power of hardware has made it possible to solve large, difficult problems in a short amount of time on desktop machines. A user community that is growing rapidly in size and sophistication is driving these developments. In this article, we discuss the state of the art in algorithms for linear and nonlinear programming and its relevance to production codes. We describe some representative software packages and modeling languages and give pointers to web sites that contain more complete information. We also mention computational servers for online solution of optimization problems.

Keywords

Optimization, Linear programming, Nonlinear programming, Integer programming, Software.

Introduction

Optimization problems arise naturally in many engineering applications. Control problems can be formulated as optimization problems in which the variables are inputs and states, and the constraints include the model equations for the plant. At successively higher levels, optimization can be used to determine setpoints for optimal operations, to design processes and plants, and to plan for future capacity.

Optimization problems contain the following key ingredients:

1. Variables that can take on a range of values. Variables that are real numbers, integers, or binary (that is, allowable values 0 and 1) are the most common types, but matrix variables are also possible.

2. Constraints that define allowable values or scopes for the variables, or that specify relationships between the variables;

3. An objective function that measures the desirability of a given set of variables.

The optimization problem is to choose from among all variables that satisfy the constraints the set of values that minimizes the objective function.

The term "mathematical programming", which was coined around 1945, is synonymous with optimization. Correspondingly, linear optimization (in which the constraints and objective are linear functions of the variables) is usually known as "linear programming," while optimization problems that involve constraints and have nonlinearity present in the objective or in at least some constraints, are known as "nonlinear programming" problems. In convex programming, the objective is a convex function and the feasible set (the set of points that satisfy the constraints) is a convex set. In quadratic programming, the objective is a quadratic function while the constraints are linear. Integer programming problems are those in which some or all of the variables are required to take on integer values.

Optimization technology is traditionally made available to users by means of codes or packages for specific classes of problems. Data is communicated to the software via simple data structures and subroutine argument lists, user-written subroutines (for evaluating nonlinear objective or constraint functions), text files in the standard MPS format, or text files that describe the problem in certain vendor-specific formats. More recently, modeling languages have become an appealing way to interface to packages, as they allow the user to define the model and data in a way that makes intuitive sense in terms of the application problem. Optimization tools also form part of integrated modeling systems such as GAMS and LINDO, and even underlie spreadsheets such as Microsoft's Excel. Other "under the hood" optimization tools are present in certain logistics packages, for example, packages for supply chain management or facility location.

The majority of this paper is devoted to a discussion of software packages and libraries for linear and nonlinear programming, both freely available and proprietary. We emphasize in particular packages that have become available during the past 10 years, that address new problem areas or that make use of new algorithms. We also discuss developments in related areas such as modeling languages and automatic differentiation. Background information on algorithms and theory for linear and nonlinear programming can be found in a number of texts, including those of Luenberger (1984), Chvatal (1983), Bertsekas (1995), Nash and Sofer (1996), and the forthcoming book of Nocedal and Wright (1999).

Online Resources and Computational Servers

As with so many other topics, a great deal of information about optimization software is available on the world-wide web. Here we point to a few noncommercial sites that give information about optimization algorithms and software, modeling issues, and operations research. Many other interesting sites can be found by following links from the sites mentioned below.

The NEOS Guide at www.mcs.anl.gov/otc/Guide contains

- A guide to optimization software containing around 130 entries. The guide is organized by the name of the code, and classified according to the type of problem solved by the code.
- An "optimization tree" containing a taxonomy of optimization problem types and outlines of the basic algorithms.
- Case studies that demonstrate the use of algorithms in solving real-world optimization problems. These include optimization of an investment portfolio, choice of a lowest-cost diet that meets a set of nutritional requirements, and optimization of a strategy for stockpiling and retailing natural gas, under conditions of uncertainty about future demand and price.

The NEOS Guide also houses the FAQs for Linear and Nonlinear Programming, which can be found at www.mcs.anl.gov/otc/Guide/faq/. These pages, updated monthly, contain basic information on modeling and algorithmic issues, information for most of the available codes in the two areas, and pointers to texts for readers who need background information.

Michael Trick maintains a comprehensive web site on operations research topics at mat.gsia.cmu.edu. It contains pointers to most online resources in operations research, together with an extensive directory of researchers and research groups and of companies that are involved in optimization and logistics software and consulting.

Hans Mittelmann and Peter Spellucci maintain a decision tree to help in the selection of appropriate optimization software tools at plato.la.asu.edu/guide.html. Benchmarks for a variety of linear and nonlinear programming codes can be found at the site plato.la.asu.edu/bench.html, which is also maintained by Mittelmann. Arnold Neumaier maintains the page solon.cma.univie.ac.at/~neum/glopt.html, which emphasizes global optimization algorithms and software.

The NEOS Server at www.mcs.anl.gov/neos is a computational server for the remote solution of optimization problems over the Internet. By using an email interface, a Web page, or an xwindows "submission tool" that connects directly to the Server via Unix sockets, users select a code and submit the model information and data that define their problem. The job of solving the problem is allocated to one of the available workstations in the Server's pool on which that particular package is installed, then the problem is solved and the results returned to the user.

The Server now has a wide variety of solvers in its roster, including a number of proprietary codes. For linear programming, the BPMPD, HOPDM, PCx, and XPRESS-MP/BARRIER interior-point codes as well as the XPRESS-MP/SIMPLEX code are available. For nonlinear programming, the roster includes LANCELOT, LOQO, MINOS, NITRO, SNOPT, and DONLP2. Input in the AMPL modeling language is accepted for many of the codes.

The obvious target audience for the NEOS Server includes users who want to try out a new code, to benchmark or compare different codes on data of relevance to their own applications, or to solve small problems on an occasional basis. At a higher level, however, the Server is an experiment in using the Internet as a computational, problem-solving tool rather than simply an informational device. Instead of purchasing and installing a piece of software for installation on their local hardware, users gain access to the latest algorithmic technology (centrally maintained and updated), the hardware resources needed to execute it and, where necessary, the consulting services of the authors and maintainers of each software package. Such a means of delivering problem-solving technology to its customers is an appealing option in areas that demand access to huge amounts of computing cycles (including, perhaps, integer programming), areas in which extensive hands-on consulting services are needed, areas in which access to large, centralized, constantly changing data bases, and areas in which the solver technology is evolving rapidly.

Linear Programming

In linear programming problems, we minimize a linear function of real variables over a region defined by linear constraints. The problem can be expressed in standard form as

$$\min c^T x \quad \text{subject to} \quad Ax = b, \; x \geq 0,$$

In linear programming problems, we minimize a linear function of real variables over a region defined by linear constr

where x is a vector of n real numbers, while $Ax = b$ is a set of linear equality constraints and $x \geq 0$ indicates that all components of x are required to be nonnegative. The dual of this problem is

$$\max b^T \lambda \quad \text{subject to} \quad A^T + s = c, \; s \geq 0,$$

where λ is a vector of Lagrange multipliers and s is a vector of dual slack variables. These two problems are intimately related, and algorithms typically solve both of them simultaneously. When the vectors $x^*, \lambda^*,$ and s^* satisfy the following optimality conditions:

$$A^T \lambda^* + s^* = c,$$
$$Ax^* = b,$$
$$x^* \geq 0, s^* \geq 0,$$
$$(x^*)^T s^* = 0,$$

then x^* solves the primal problem and (λ^*, s^*) solves the dual problem.

Simple transformations can be applied to any problem with a linear objective and linear constraints (equality and inequality) to obtain this standard form. Production quality linear programming solvers carry out the necessary transformations automatically, so the user is free to specify upper bounds on some of the variables, use linear inequality constraints, and in general make use of whatever formulation is most natural for their particular application.

The popularity of linear programming as an optimization paradigm stems from its direct applicability to many interesting problems, the availability of good, general-purpose algorithms, and the fact that in many real-world situations, the inexactness in the model or data means that the use of a more sophisticated nonlinear model is not warranted. In addition, linear programs do not have multiple local minima, as may be the case with nonconvex optimization problems. That is, any local solution of a linear program—one whose function value is no larger than any feasible point in its immediate vicinity—also achieves the global minimum of the objective over the whole feasible region. It remains true that more (human and computational) effort is invested in solving linear programs than in any other class of optimization problems. A fine reference on linear programming, with an emphasis on the simplex method, is the book of Chvatal (1983).

Prior to 1987, all of the commercial codes for solving general linear programs made use of the simplex algorithm. This algorithm, invented in the late 1940s, had fascinated optimization researchers for many years because its performance on practical problems is usually far better than the theoretical worst case. A new class of algorithms known as interior-point methods was the subject of intense theoretical and practical investigation during the period 1984—1995, with practical codes first appearing around 1989. These methods appeared to be faster than simplex on large problems, but the advent of a serious rival spurred significant improvements in simplex codes.

Today, the relative merits of the two approaches on any given problem depend strongly on the particular geometric and algebraic properties of the problem, and the picture continues to change significantly as improvements are made to both simplex and interior-point codes. In general, however, good interior-point codes continue to perform as well or better than good simplex codes on larger problems when no prior information about the solution is available. When such "warm-start" information is available, however, as is often the case in solving continuous relaxations of integer linear programs in branch-and-bound algorithms, simplex methods are able to make much better use of it than interior-point methods. Interior-point methods also have the advantage on multiprocessor architectures. Their basic operation of factoring a sparse matrix can be implemented in parallel with reasonable efficiency, at least on a modest number of processors, while parallelization of the simplex method is not possible except on problems with special structure. Finally, we note a logistical advantage for the interior-point approach: A number of excellent interior-point codes are freely available, at least for purposes of research and evaluation, while the few freely available simplex codes are not competitive with the best commercial codes.

A comparison of some interior-point codes and the simplex code CPLEX 3.0 was performed by Andersen et al. (1996), but since both CPLEX and the interior-point codes have changed considerably since that time, the conclusions are somewhat dated. Up-to-date comparisons of many of the most popular interior-point and simplex codes can be found at Mittelmann's benchmark page plato.la.asu.edu/bench.html.

The simplex algorithm generates a sequence of feasible iterates x^k for the primal problem, where each iterate typically has the same number of nonzero (strictly positive) components as there are rows in A. We use this iterate to generate dual variables λ and s such that two other optimality conditions are satisfied, namely,

$$A^T \lambda^k + s^k = c, \left(x^k\right)^T s^k = 0.$$

If the remaining condition $s^k \geq 0$ is also satisfied, then the solution has been found and the algorithm terminates. Otherwise, we choose one of the negative components of s^k and allow the corresponding component of x to increase from zero. To maintain feasibility of the equality constraint $Ax = b$, the components that were strictly positive in x^k will change. One of them will become zero when we increase the new component to a sufficiently large value. When this happens, we stop and denote the new iterate by x^{k+1}.

Each iteration of the simplex method is relatively inexpensive. It maintains a factorization of the submatrix of A that corresponds to the strictly positive components of x–a square matrix B known as the *basis*–and updates this factorization at each step to account for the fact that one column of B has changed. Typically, simplex methods converge in a number of iterates that is about two to three times the number of columns in A. A good discussion of the elements of the simplex algorithm is given by Murtagh (1981). An online Java applet that demonstrates the operation of the simplex method on small user-defined problems can be found at www.mcs.anl.gov/otc/Guide/CaseStudies/simplex/.

Interior-point methods proceed quite differently, applying a Newton-like algorithm to the three equalities in the optimality conditions and taking steps that maintain strict positivity of all the x and s components. It is the latter feature that gives rise to the term "interior-point" –the iterates are strictly interior with respect to the inequality constraints. Each interior-point iteration is typically much more expensive than a simplex iteration, since it requires refactorization of a large matrix of the form $AS^{-1}XA^T$, where S and X are diagonal matrices whose diagonal elements are the components of the current iterates s and x, respectively. The solutions to the primal and dual problems are generated simultaneously. Typically, interior-point iterates converge in between 10 and 100 iterations. Wright (1997) summarizes the theoretical properties of primal-dual interior-point methods–a class that includes the methods implemented in practical codes–and discusses some of the issues that arise in turning the algorithm into a reliable piece of software.

Codes can differ in a number of important respects, apart from the different underlying algorithm. All practical codes include presolvers, which attempt to reduce the dimension of the problem by determining the values of some of the primal and dual variables without applying the algorithm. As a simplex example, suppose that the linear program contains the constraints

$$10x_3 - 4x_{10} + x_{12} = -4,$$
$$x_3 \geq 0, 0 \leq x_{10} \leq 1, x_{12} \geq 0,$$

then the only possible values for the three variables are

$$x_3 = 0, x_{10} = 1, x_{12} = 0.$$

These variables can be fixed and deleted from the problem, along with the three corresponding columns of A and the three components of c. Presolve techniques have become quite sophisticated over the years, though little has been written about them because of their commercial value. An exception is the paper of Andersen and Andersen (1995).

For information on specific codes, refer to the online resources mentioned earlier; in particular, the NEOS Software Guide, the Linear Programming FAQ, and the benchmarks maintained by Hans Mittelmann.

Modern, widely used commercial simplex codes include CPLEX and the XPRESS-MP. Both these codes accept input in the industry-standard MPS format, and also in their own proprietary formats. Both have interfaces to various modeling languages, and also a "callable library" interface that allows users to set up, modify, and solve problems by means of function calls from C or FORTRAN code. Both packages are undergoing continual development. Freely available simplex codes are usually of lower quality, with the exception of SOPLEX. This is a C++ code, written as a thesis project by Roland Wunderling, that can be found at www.zib.de/Optimization/Software/Soplex.

The code MINOS is available to nonprofit and academic researchers for a nominal fee. Commercial interior-point solvers are available as options in the CPLEX and XPRESS-MP packages. However, a number of highly competitive codes are available free for research and noncommercial use, and can for the most part be obtained through the Web. Among these are BPMPD, PCx, COPLLP, LOQO, HOPDM, and LIPSOL. See Mittelmann's benchmark page for comparisons of these codes and links to their web sites. These codes mostly charge a license fee for commercial use, but it is typically lower than for fully commercial packages. All can read MPS files, and most are interfaced to modeling languages. LIPSOL is programmed in Matlab (with the exception of the linear equations solver), while the other codes are written in C and/or FORTRAN.

Modeling Languages

From the user's point of view, the efficiency of the algorithm or the quality of the programming may not be the critical factors in determining the usefulness of the code. Rather, the ease with which it can be interfaced to his particular applications may be more important; weeks of person-hours may be more costly to the enterprise than a few hours of time on a computer. The most suitable interface depends strongly on the particular application and on the context in which it is solved. For users that are well acquainted with a spreadsheet interface, for instance, or with MATLAB, a code that can accept input from these sources may be invaluable. For users with large legacy modeling codes that set up and solve optimization problems by means of subroutine calls, substitution of a more efficient package that uses more or less the same subroutine interface may be the best option. In some disciplines, application-specific modeling languages allow problems to be posed in a thoroughly intuitive way. In other cases, graphical user interfaces may be more appropriate.

For general optimization problems, a number of high-level modeling languages have become available that allow problems to be specified in intuitive terms, using data structures, naming schemes, and algebraic relational expressions that are dictated by the application and model rather than by the input requirements of the optimization code. Typically, a user starting from scratch will find the process of model building more straightforward and bug free with such a modeling language than, say, a process of writing FORTRAN code to pack the data into one-dimensional arrays, turning the algebraic relations between the variables into FORTRAN expressions involving elements of these arrays, and writing more code to interpret the output from the optimization routine in terms of the original application.

The following simple example in AMPL demonstrates the usefulness of a modeling language (see Fourer, Gay, and Kernighan (1993), page 11). The application is to a steel production model, in which the aim is to maximize profit obtained from manufacturing a number of steel products by choosing the amount of each product to manufacture, subject to restrictions on the maximum demands for each product and the time available in each work week to manufacture them. The following file is an AMPL "model file" that specifies the variables, the parameters that quantify aspects of the model, and the constraints and objective.

```
set PROD;
param rate {PROD} >0;
param avail >= 0;
param profit {PROD};
param market{PROD};

var Make {p in PROD} >= 0, <= market[p];
maximize total_profit: sum {p in PROD} profit[p] *Make[p];
subject to Time: sum {p in PROD} (1/rate[p]) * Make[p] <= avail;
```

PROD is the collection of possible products that can be manufactured, while rate, profit and market are the rate at which each product can be manufactured, the profit on each product, and the maximum demand for each product, respectively. avail represents the total time available for manufacturing. Make is the variable in the problem, representing the amount of each product to be manufactured. In its definition, each element of Make is constrained to lie between zero and the maximum demand for the product in question. The last two lines of the model file specify the objective and constraint in a self-evident fashion.

The actual values of the parameters can be assigned by means of additional statements in this file, or in a separate "data file." For instance, the following data file specifies parameters for two products, bands and coils:

```
set PROD := bands coils;

param: rate   profit market :=
  bands 200    25     6000
  coils 140    30     4000;
param avail := 40;
```

These statements specify that the market[bands] is 6000, profit[bands] is 25, and so on. An interactive AMPL session would proceed by invoking commands to read these two files and then invoking an option solver command to choose the linear programming solver to be used (for example, CPLEX or PCx) together with settings for parameters such as stopping tolerances, etc, that the user may wish to change from their defaults. A solve command would then solve the problem (and report messages passed through from the underlying optimization code). Results can be inspected by invoking the display command. For the above example, the command display Make invoked after the problem has been solved would produce the following output:

```
Make [*] :=
bands  6000
coils      0
;
```

Note from this example the intuitive nature of the algebraic relations, and the fact that we could index the parameter arrays by the indices bands and coils, rather than the numerical indices 1 and 2 that would be required if we were programming in FORTRAN. Note too that additional products can be added to the mix without changing the model file at all.

Of course, the features of AMPL are much more extensive than the simple example above allows us to demonstrate. The web site www.ampl.com contains a great deal of information about the language and the optimization software to which it is linked, and allows users to solve their own simple models online.

Numerous other modeling languages and systems can be found on the online resources described above, particularly the NEOS Software Guide and the linear and nonlinear programming FAQ's. We mention in particular AIMMS (Bisschop and Entriken (1993)) which has a built in graphical interface; GAMS (www.gams.com), a well established system available with support for linear, nonlinear, and mixed-integer programming and newly added procedural features; and MPL, a Windows-based system whose web site www.maximal-usa.com contains a comprehensive tutorial and a free student version of the language.

Other Input Formats

The established input format for linear programming problems has from the earliest days been MPS, a column oriented format (well suited to 1950s card readers) in which names are assigned to each primal and dual variable, and the data elements that define the problem are assigned in turn. Test problems for linear programming are still distributed in this format. It has significant disadvantages, however. The format is non-intuitive and the files are difficult to modify. Moreover, it restricts the precision to which numerical values can be specified. The format survives only because no universally accepted standard has yet been developed to take its place.

As mentioned previously, vendors such as CPLEX and XPRESS have their own input formats, which avoid the pitfalls of MPS. These formats lack the portability of the modeling languages described above, but they come bundled with the code, and may be attractive for users willing to make a commitment to a single vendor.

For nonlinear programming, SIF (the standard input format) was proposed by the authors of the LANCELOT code in the early 1990s. SIF is somewhat hamstrung by the fact that it is compatible with MPS. SIF files have a similar look to MPS files, except that there are a variety of new keywords for defining variables, groups of variables, and the algebraic relationships between them. For developers of nonlinear programming software, SIF has the advantage that a large collection of test problems–the CUTE test set–is available in this format. For users, however, formulating a model in SIF is typically much more difficult than using one of the modeling languages of the previous section.

For complete information about SIF, see www.numerical.rl.ac.uk/lancelot/sif/sifhtml.html

Nonlinear Programming

Nonlinear programming problems are constrained optimization problems with nonlinear objective and/or constraint functions. However, we still assume that all functions in question are smooth (typically, at least twice differentiable), and that the variables are all real numbers. If any of the variables are required to take on integer values, the problem is a (mixed-) integer nonlinear programming problem, a class that we will not consider in this paper. For purposes of description, we use the following formulation of the problem:

$$\min \quad f(x) \quad \text{subject to} \quad c(x) \leq 0, h(x) = 0,$$

where x is a vector of n real variables, f is a smooth real-valued function, and c and h are smooth functions with dimension m_E and m_I, respectively.

Algorithms for nonlinear programming problems are more varied than those for linear programming. The major approaches represented in production software packages are sequential quadratic programming, reduced gradient, sequential linearly constrained, and augmented Lagrangian methods. (The latter is also known as the method of multipliers.) Extension of the successful interior-point approaches for linear programming to the nonlinear problem is the subject of intense ongoing investigation among optimization researchers, but little production software for these approaches is yet available.

The use of nonlinear models may be essential in some applications, since a linear or quadratic model may be too simplistic and therefore produce useless results. However, there is a price to pay for using the more general nonlinear paradigm. For one thing, most algorithms cannot guarantee convergence to the global minimum, i.e., the value x^* that minimizes f over the entire feasible region. At best, they will find a point that yields the smallest value of f over all points in some feasible neighborhood of itself. (An exception occurs in convex programming, in which the functions f and $c_i, i = 1, 2, \ldots, m_I$, are convex, while $h_i, i = 1, 2, \ldots, m_E$, are linear. In this case, any local minimizer is also a global minimizer.

Note that linear programming is a special case of convex programming.) The problem of finding the global minimizer, though an extremely important one in some applications such as molecular structure determination, is very difficult to solve. While several general algorithmic approaches for global optimization are available, efficient implementations are obtained only when the special properties of the underlying application are exploited heavily. Even so, computational requirements are often large. We refer to Floudas and Pardalos (1992) and the journal *Global Optimization* for information on recent advances in this area. The paper by Floudas and Pardalos (1999) in these Proceedings discusses applications of global optimization techniques to process design problems.

A second disadvantage of nonlinear programming over linear programming is that general-purpose software is somewhat less effective because the nonlinear paradigm encompasses such a wide range of problems with a great number of potential pathologies and eccentricities. Even when we are close to a minimizer x^*, algorithms may encounter difficulties because the solution may be degenerate, in the sense that certain of the active constraints become dependent, or are only weakly active. Curvature in the objective or constraint functions (a second-order effect not present in linear programming) can cause difficulties for the algorithms, especially when second derivative information is not supplied by the user or not exploited by the algorithm. Finally, some of the codes treat the derivative matrices as dense, which means that the size of the problems they can handle is somewhat limited. However, most of the leading codes, including LANCELOT, MINOS, SNOPT, and SPRNLP are able to exploit sparsity, and are therefore equipped to handle large-scale problems.

Algorithms for special cases of the nonlinear programming problem, such as problems in which all constraints are linear or the only constraints are bounds on the components of x, tend to be more effective than algorithms for the general problem because they are more able to exploit the special properties. (We discuss a few such special cases below.) Even for problems in which the constraints are nonlinear, the problem may contain special structures that can be exploited by the algorithm or by the routines that perform linear algebra operations at each iteration. An example is the optimal control problem (arising, for example, in model predictive control), in which the equality constraint represents a nonlinear model of the plant, and the inequalities represent bounds and other restrictions on the states and inputs. The Jacobian (matrix of first partial derivatives of the constraints) typically has a banded structure, while the Hessian of the objective is symmetric and banded. Linear algebra routines that exploit this bandedness, or dig even deeper and exploit the control origins of the problem, are much more effective than general routines on such problems.

Local solutions of the nonlinear program can be characterized by a set of optimality conditions analogous to those described above for the linear programming problem. We introduce Lagrange multipliers λ and π for the constraints $c(x) \leq 0$ and $h(x) = 0$, respectively, and write the Lagrangian function for this problem as

$$L(x,\lambda,\pi) = f(x) + \lambda^T c(x) + \pi^T h(x).$$

The first-order optimality conditions (commonly known as the KKT conditions) are satisfied at a point x^* if there exist multiplier vectors λ^* and π^* such that

$$\nabla_x L(x^*,\lambda^*,\pi^*) = 0,$$
$$h(x^*) = 0,$$
$$c(x^*) \leq 0, \lambda^* \geq 0,$$
$$(\lambda^*)^T c(x^*) = 0.$$

The active constraints are those for which equality holds at x^*. All the components of h are active by definition, while the active components of c are those for which $c_i(x^*) = 0$.

When the constraint gradients satisfy certain regularity conditions at x^*, the KKT conditions are necessary for x^* to be a local minimizer of the nonlinear program, but not sufficient. A second-order sufficient condition is that the Hessian of the Lagrangian, the matrix $\nabla_{xx} L(x^*,\lambda^*,\pi^*)$, has positive curvature along all directions that lie in the null space of the active constraint gradients, for some choice of multipliers λ^* and π^* that satisfy the KKT conditions. That is, we require

$$w^T \nabla_{xx} L(x^*,\lambda^*,\pi^*) w > 0,$$

for all vectors w such that $\nabla h(x^*)^T w = 0$ and $\nabla c_i(x^*)^T w = 0$ for all active indices i.

The *sequential quadratic programming (SQP)* approach has been investigated extensively from a theoretical point of view and is the basis of several important practical codes, including NPSOL and the more recent SNOPT. It works by approximating the nonlinear programming problem by a quadratic program around the current iterate x^k, that is,

$$\min_d \quad \nabla f(x^k)^T d + \frac{1}{2} d^T H_k d, \quad subject\ to$$
$$c(x^k) + \nabla c(x^k)^T d \leq 0,$$
$$h(x^k) + \nabla h(x^k)^T d = 0,$$

where H_k is a symmetric matrix (usually positive definite) that contains exact or approximate second-order information about the objective and constraint functions. There are many modifications of this basic scheme. For instance, a trust-region bound limiting the length of the step d may be added to the model, or the linear constraints may be adjusted so that the current step is not required to remedy all the infeasibility in the current iterate x^k.

The approximate Hessian H_k can be chosen in a number of ways. Local quadratic convergence can be proved under certain assumptions if this matrix is set to the Hessian of the Lagrangian, that is, $\nabla_{xx} L(x^k,\lambda^k,\pi^k)$, evaluated at the primal iterate x^k and the current estimates λ^k, π^k of the Lagrange multiplier vectors. The code SPRNLP allows users to select this value for H_k, provided that they are willing to supply the second derivative information. Alternatively, H_k can be a quasi-Newton approximation to the Lagrangian Hessian. Update strategies that yield local superlinear convergence are well known, and are implemented in dense codes such as NPSOL, DONLP2, NLPQL, and are available as an option in a version of SPRNLP that does not exploit sparsity. SNOPT also uses quasi-Newton Hessian approximations, but unlike the codes just mentioned it is able to exploit sparsity and is therefore better suited to large-scale problems. Another quasi-Newton variant is to maintain an approximation to the reduced Hessian, the two-sided projection of this matrix onto the null space of the active constraints. The latter approach is particularly efficient when the dimension of this null space is small in relation to the number of components of x. Such is the case in many process control problems, where the most of the active constraints are equality constraints that represent the dynamics of the process. Biegler, Nocedal, and Schmid (1995) describe this approach, and Biegler and Ternet (1998) describe software that implements this method.

To ensure that the algorithm converges to a point satisfying the KKT conditions from any starting point, the basic SQP algorithm must be enhanced by the addition of a "global convergence" strategy. Usually, this strategy involves a *merit function*, whose purposes is to evaluate the desirability of a given iterate x by accounting for its objective value and the amount by which it violates the constraints. The commonly used ℓ_1 penalty function simply forms a weighted average of the objective and the constraint violations, as follows:

$$P(x,\rho) = f(x) + \rho \|h(x)\|_1 + \rho \|c(x)^+\|_1,$$

where $c(x)^+$ is the vector of length m_I whose elements are $\max(0, c_i(x))$ and ρ is a positive parameter. The simplest algorithm based on this function fixes ρ and insists that all steps produce a "sufficient decrease" in the value of P. Line search or trust region strategies are applied to ensure that steps with the required property can be found whenever the current point x^k does not satisfy the KKT conditions. More sophisticated strategies contain mechanisms for adjusting the parameter ρ and for ensuring that the fast local convergence properties are not compromised by the global convergence strategy.

We note that the terminology can be confusing–"global convergence" in this context refers to convergence to a KKT point from any starting point, and *not* to convergence to a global minimizer.

For more information on SQP, we refer to the review paper of Boggs and Tolle (1996), and Chapter 18 of Nocedal and Wright (1999).

A second algorithmic approach is known variously as the *augmented Lagrangian method* or the *method of multipliers*. Noting that the first KKT condition, namely, $\nabla_x L(x^*, \lambda^*, \pi^*) = 0$, requires x^* to be a stationary point of the Lagrangian function L, we modify this function to obtain an augmented function for which x^* is not just a stationary point but also a minimizer. When only equality constraints are present (that is, c is vacuous), the augmented Lagrangian function has the form

$$A(x, \pi; \rho) = f(x) + \pi^T h(x) + \frac{\rho}{2} \|h(x)\|_2^2 ,$$

where ρ is a positive parameter. It is not difficult to show that if π is set to its optimal value π^* (the value that satisfies the KKT conditions) and ρ is sufficiently large, that x^* is a minimizer of A. Intuitively, the purpose of the squared-norm term is to add positive curvature to the function L in just those directions in which it is needed–the directions in the range space of the active constraint gradients. (We know already from the second-order sufficient conditions that the curvature of L in the *null* space of the active constraint gradients is positive.)

In the augmented Lagrangian method, we exploit this property by alternating between steps of two types:

Fixing π and ρ, and finding the value of x that approximately minimizes $A(x, \pi; \rho)$;

Updating π to make it a better approximation to π^*.

The update formula for π has the form

$$\pi \leftarrow \pi + \rho h(x),$$

where x is the approximate minimizing value just calculated. Simple constraints such as bounds or linear equalities can be treated explicitly in the subproblem, rather than included in the second and third terms of A. (In LANCELOT, bounds on components of x are treated in this manner.) Practical augmented Lagrangian algorithms also contain mechanisms for adjusting the parameter ρ and for replacing the squared norm term $\|h(x)\|^2$ by a weighted norm that more properly reflects the scaling of the constraints and their violations at the current point.

When inequality constraints are present in the problem, the augmented Lagrangian takes on a slightly more complicated form that is nonetheless not difficult to motivate. We define the function $\varphi(t, \sigma; \rho)$ as follows:

$$\varphi(t, \sigma; \rho) = \begin{cases} \sigma t + (\rho/2)t^2 & \text{if } t + \sigma/\rho \geq 0, \\ -\sigma^2/(2\rho) & \text{otherwise.} \end{cases}$$

The definition of A is then modified to incorporate the inequality constraints as follows:

$$A(x,\lambda,\pi;\rho) = f(x) + \pi^T h(x) + \frac{\rho}{2}\|h(x)\|_2^2$$
$$+ \sum_{i=1}^{m_I} \varphi(c_i(x), \lambda_i, \rho)$$

The update formula for the approximate multipliers λ is

$$\lambda_i \leftarrow \max(\lambda_i + \rho c_i(x), 0), \quad \text{all } i = 1, 2, \ldots, m_I.$$

See the references below for details on derivation of this form of the augmented Lagrangian.

The definitive implementation of the augmented Lagrangian approach for general-purpose nonlinear programming problems is LANCELOT. It incorporates sparse linear algebra techniques, including preconditioned iterative linear solvers, making it suitable for large-scale problems. The subproblem of minimizing the augmented Lagrangian with respect to x is a bound-constrained minimization problem, which is solved by an enhanced gradient projection technique. Problems can be passed to Lancelot via subroutine calls, SIF input files, and AMPL.

For theoretical background on the augmented Lagrangian approach, consult the books of Bertsekas (1982, 1995), and Conn, Gould, and Toint (1992), the authors of LANCELOT. The latter book is notable mainly for its pointers to the papers of the same three authors in which the theory of Lancelot is developed. A brief derivation of the theory appears in Chapter 17 of Nocedal and Wright (1999). (Note that the inequality constraints in this reference are assumed to have the form $c(x) \geq 0$ rather than $c(x) \leq 0$, necessitating a number of sign changes in the analysis.)

Interior-point solvers for nonlinear programming are the subjects of intense current investigation. An algorithm of this class, known as the *sequential unconstrained minimization technique (SUMT)* was actually proposed in the 1960s, in the book of Fiacco and McCormick (1968). The idea at that time was to define a barrier-penalty function for the NLP as follows:

$$B(x;\mu) = f(x) - \mu \sum_{i=1}^{m_I} \log(-c_i(x))$$
$$+ \frac{1}{2\mu} \sum_{i=1}^{m_E} h_i^2(x),$$

where μ is a small positive parameter. Given some value of μ, the algorithm finds an approximation to the minimizer $x(\mu)$ of $B(x,\mu)$. It then decreases μ and repeats the minimization process. Under certain assumptions, one can show that $x(\mu) \to x^*$ as $\mu \downarrow 0$, so the sequence of iterates generated by SUMT should approach the solution of the nonlinear program provided that μ is decreased to zero. The difficulties with this approach are that all iterates must remain strictly feasible with respect to the inequality constraints (otherwise the log functions are not defined), and the subproblem of minimizing $B(x,\mu)$ becomes increasingly difficult to solve as μ becomes small, as the Hessian of this function becomes highly ill conditioned and the radius of convergence becomes tiny. Many implementations of this approach were attempted, including some with enhancements such as extrapolation to obtain good starting points for each value of μ. However, the approach does not survive in the present generation of software, except through its profound influence on the interior-point research of the past 15 years.

Some algorithms for nonlinear programming that have been proposed in recent years contain echoes of the barrier function B, however. For instance, the NITRO algorithm (Byrd, Gilbert, and Nocedal (1996)) reformulates the subproblem for a given positive value of μ as follows:

$$\min_{x,s} f(x) - \mu \sum_{i=1}^{m_I} \log s_i \quad \text{subject to}$$
$$c(x) + s = 0, \quad s > 0, \quad h(x) = 0.$$

NITRO then applies a trust-region SQP algorithm for equality constrained optimization to this problem, choosing the trust region to have the form

$$\|(x, Ds)\| \leq \tau,$$

where the diagonal matrix D and the trust-region radius τ are chosen so that the step $(\Delta x, \Delta s)$ does not violate strict positivity of the s components, that is, $s + \Delta s > 0$. NITRO is available through the NEOS Server at www.mcs.anl.gov/neos. The user is required to specify the problem by means of FORTRAN subroutines to evaluate the objective and constraints. Derivatives are obtained automatically by means of ADIFOR.

An alternative interior-point approach is closer in spirit to the successful primal-dual class of linear programming algorithms. These methods generate iterates by applying Newton-like methods to the equalities in the KKT conditions. After introducing the slack variables s for the inequality constraints, we can restate the KKT conditions as follows:

$$\nabla f(x^*) + \sum_{i=1}^{m_E} \pi_i \nabla h_i(x^*) + \sum_{i=1}^{m_I} \lambda_i \nabla c_i(x^*) = 0,$$

$$h(x^*) = 0,$$

$$c(x^*) + s^* = 0,$$

$$\Lambda^* S^* e = 0,$$

$$s^* \geq 0, \lambda^* \geq 0,$$

where Λ^* and S^* are diagonal matrices formed from the vectors λ^* and s^*, respectively, while e is the vector $(1,1,\ldots,1)^T$. We generate a sequence of iterates $(x^k, s^k, \pi^k, \lambda^k)$ satisfying the strict inequality $(s^k, \lambda^k) > 0$ by applying a Newton-like method to the system of nonlinear equations formed by the first four conditions above. Modification of this basic approach to ensure global convergence is the major challenge associated with this class of solvers; the local convergence theory is relatively well understood. Merit functions can be used, along with line searches and modifications to the matrix in the equations that are solved for each step, to ensure that each step at least produces a decrease in the merit function. However, no fully satisfying complete theory has yet been proposed.

The code LOQO implements a primal-dual approach for nonlinear programming problems. It requires the problem to be specified in AMPL, whose built-in automatic differentiation features are used to obtain the derivatives of the objective and constraints. LOQO is also available through the NEOS Server at www.mcs.anl.gov/neos/Server/ , and or can be obtained for a variety of platforms.

The reduced gradient approach has been implemented in several codes that have been available for some years, notably, CONOPT and LSGRG2. This approach uses the formulation in which only bounds and equality constraints are present. (Any nonlinear program can be transformed to this form by introducing slacks for the inequality constraints and constraining the slacks to be nonnegative.) Reduced gradient algorithms partition the components of x into three classes: basic, fixed, and superbasic. The equality constraint $h(x) = 0$ is used to eliminate the basic components from the problem by expressing them implicitly in terms of the fixed and superbasic components. The fixed components are those that are fixed at one of their bounds for the current iteration. The superbasics are the components that are allowed to move in a direction that reduces the value of the objective f. Strategies for choosing this direction are derived from unconstrained optimization; they include steepest descent, nonlinear conjugate gradient, and quasi-Newton strategies. Both CONOPT and LSGRG2 use sparse linear algebra techniques during the elimination of the basic components, making them suitable for large-scale problems. While these codes have found use in many engineering applications, their performance is often slower than competing codes based on SQP and augmented Lagrangian algorithms.

Finally, we mention MINOS, a code that has been available for many years in a succession of releases, and that has proved its worth in a great many engineering applications. When the constraints are linear, MINOS uses a reduced gradient algorithm, maintaining feasibility at all iterations and choosing the superbasic search direction with a quasi-Newton technique. When nonlinear constraints are present, MINOS forms linear approximations to them and replaces the objective with a projected augmented Lagrangian function in which the deviation from linearity is penalized. Convergence theory for this approach is not well established–the author admits that a reliable merit function is not known–but it appears to converge on most problems.

The NEOS Guide page for SNOPT contains some guidance for users who are unsure whether to use MINOS or SNOPT. It describes problem features that are particularly suited to each of the two codes.

Obtaining Derivatives

One onerous requirement of some nonlinear programming codes has been their requirement that the user supply code for calculating derivatives of the objective and constraint functions. An important development of the past 10 years is that this requirement has largely disappeared. Modeling languages such as AMPL contain their own built-in systems for calculating first derivatives at specified values of the variable vector x, and supplying them to the underlying optimization code on request. Automatic differentiation software tools such as ADIFOR (Bischof et al. (1996)), which works with FORTRAN code, have been used to obtain derivatives from extremely complex "dusty deck" function evaluation routines. In the NEOS Server, all of the nonlinear optimization routines (including LANCELOT, SNOPT, and NITRO) are linked to ADIFOR, so that the user needs only to supply FORTRAN code to evaluate the objective and constraint functions, not their derivatives. Other high quality software tools for automatic differentiation include ADOL-C (Griewank, Juedes, and Utke (1996)), ODYSSEE (Rostaing, Dalmas, and Galligo (1993)), and ADIC (Bischof, Roh, and Mauer (1997)).

Object-Oriented Software Design

Most existing optimization software is programmed in Fortran, or occasionally C, and does not make use of modern software principles such as object-oriented design. Increasingly, however, optimization researchers are coming to understand the potential benefits of utilizing these techniques, and it is a safe bet that the next generation of optimization software will be made up largely of object-oriented codes consisting of reusable modules for linear algebra and basic arithmetic operations.

Apart from its aesthetic appeal, object-oriented design has practical advantages that will lead to more flexible, maintainable, and efficient software. For example, it allows the major operations in an algorithm to manipulate data objects in a way that is independent of the particular structure in which they are stored, so that the higher-level code can be identical for a wide range of applications with vastly different structures. The need for recoding much of the algorithm for each new application is eliminated, so maintainability of the code is enhanced. Efficiency is ensured by tailoring the lower-level to take advantage of the structure of each application.

A second advantage is reusability. Some operations are common to more than one algorithm, and can therefore be encoded in a single module that can be reused in a number of different contexts. An example is the line search operation in a nonlinear optimization code, which finds a step length to move along a given direction that satisfies certain criteria of sufficient reduction and approximate minimization. Such an operation is needed in both constrained and unconstrained optimization codes, regardless of whether the step is generated by a Newton, quasi-Newton, sequential quadratic programming or some other scheme.

In some cases, such as simplex codes for general linear programming, most linear algebra operations are highly specialized, so it is not particularly important for the software modules that implement them to be reusable in other applications.

The Role of Heuristics and Simulated Annealing

We close with some remarks about the role of approaches such as simulated annealing, genetic algorithms, and other heuristics. Some applications experts have tended to rely too much on techniques such as these in situations where algorithmic approaches such as those described above are more appropriate, and much more efficient. In solving a constrained optimization problem, for instance, nonlinear programming algorithms such as SQP or augmented Lagrangian are appropriate when the functions are reasonably smooth. There is no need to rely on simulated annealing or other heuristics unless it is critical to find the global minimum, or unless the problem has some special properties for which a suitable heuristic has been designed. Given the availability of automatic differentiation tools, the inconvenience or impracticality of computing the derivative information required by most algorithms is no longer an excuse for not using these approaches.

Heuristics have proved to be effective in many combinatorial optimization problems, and are often used in conjunction with algorithms such as branch-and-bound to obtain good feasible solutions or to make branching decisions. Indeed, for some problems, there are heuristics that often find the optimum in very little time, leaving the much more expensive branch-and-bound process to the subordinate role of verifying optimality. Since combinatorial and discrete problems are beyond the scope of this paper, we have omitted discussion of heuristics of this type.

Problems in which the objective function is nonsmooth or discontinuous, or in which random measurement or calculation errors produce a "noisy" objective function, give other cases in which alternatives to the algorithms described above are needed. Part II of the book by Kelley (1999) and the articles cited therein describe direct search and implicit filtering approaches that are useful in this context.

References

Andersen, E. D. and K. D. Andersen (1995). Presolving in linear programming. *Mathematical Programming Series A,* **71**, 221-245.

Andersen, E. D., Gondzio, J., Meszaros, C., and X. Xu (1996). Implementation of interior-point methods for large-scale linear programming. In *Interior-Point Methods for Mathematical Programming* (T. Terlaky, Ed.), 189-252. Kluwer.

Bertsekas, D. P. (1982). *Constrained Optimization and Lagrange Multiplier Methods.* Academic Press, New York.

Bertsekas, D. P. (1995). *Nonlinear Programming.* Athena Scientific.

Bielger, L. T., Nocedal, J., and C. Schmid (1995). A reduced Hessian method for large-scale constrained optimization. *SIAM Journal on Optimization,* **5**, 314-347.

Biegler, L. T. and Ternet, D. J. (1998). Recent improvements to a multiplier-free reduced Hessian successive quadratic programming algorithm. *Comp. Chem. Eng.,* **22**, 963.

Bischof, C., Carle, A., Khademi, P., and A. Mauer (1996). ADIFOR 2.0: Automatic differentiation of FORTRAN programs. *IEEE Comp. Sci. Eng.,* **3**, 18-32.

Bischof, C., Roh, L., and A. Mauer (1997). ADIC: An extensible automatic differentiation tool for ANSI-C. *Software-Practice and Experience,* **27**, 1427-1456.

Bisschop, J. and R. Entriken (1993). *AIMMS: The Modeling System.* Available from AIMMS web site at http://www.paragon.nl

Boggs, P. T. and J. W. Tolle (1996). Sequential quadratic programming, *Acta Numerica,* **4**, 1-51.

Byrd, R. H., Gilbert, J.-C., and Nocedal, J. (1996). *A Trust-region Algorithm Based on Interior-point Techniques for Nonlinear Programming.* OTC Technical Report 98/06, Optimization Technology Center. (Revised, 1998.)

Chvatal, V. (1983). *Linear Programming.* Freeman, New York.

Conn, A. R., Gould, N. I. M., and P. L. Toint (1992). *LANCELOT: A FORTRAN Package for Large-Scale Nonlinear Optimization (Release A).* Volume 17, Springer Series in Computational Mathematics, Springer-Verlag, New York.

Czyzyk, J., Mesnier. M. P., and J. J. More' (1998). The NEOS server. *IEEE J. Comp. Sci. Eng.,* **5**, 68-75.

Fiacco, A. V. and G. P. McCormick (1968). *Nonlinear Programming: Sequential Unconstrained Minimization Techniques.* John Wiley and Sons, New York. (Reprinted by SIAM Publications, 1990.)

Floudas, C. and P. Pardalos, Eds. (1992). *Recent Advances in Global Optimization.* Princeton University Press, Princeton.

Floudas, C. and P. Pardalos (1999). Recent developments in global optimization methods and their relevance to process design. These Proceedings.

Fourer, R., Gay, D. M., and B. W. Kernighan (1993). *AMPL: A Modeling Language for Mathematical Programming.* The Scientific Press, San Francisco.

Griewank, A., Juedes, D., and J. Utke (1996). ADOL-C, A package for the automatic differentiation of algorithms written in C/C++. *ACM Trans. Math. Software*, **22**, 131-167.

Kelley, C. T. (1999). *Iterative Methods in Optimization.* SIAM Publications.

Luenberger, D. (1984). *Introduction to Linear and Nonlinear Programming.* Addison Wesley.

Murtagh, B. (1981). *Advanced Linear Programming: Computation and Practice.* McGraw-Hill.

Nash, S. and Sofer, A. (1996). *Linear and Nonlinear Programming.* McGraw-Hill.

Nocedal, J. and S. J. Wright (1999). *Numerical Optimization.* Springer.

Rostaing, N., Dalmas, S., and A. Galligo (1993). Automatic differentiation in Odyssee. *Tellus*, **45a**, 558-568.

Wright, S. J. (1997). *Primal-Dual Interior-Point Methods.* SIAM Publications.

LOGIC BASED APPROACHES FOR MIXED INTEGER PROGRAMMING MODELS AND THEIR APPLICATION IN PROCESS SYNTHESIS

Ignacio E. Grossmann[1] and John Hooker[2]
[1]Department of Chemical Engineering
[2]Graduate School of Industrial Administration
Carnegie Mellon University
Pittsburgh, PA 15213

Abstract

This paper presents an overview of logic based optimization methods, a new exciting direction that has emerged recently in the superstructure optimization of process systems. A brief review is first presented of the approach that is based on mixed-integer nonlinear programming techniques. The paper discusses general issues related to the solution of logic-based models, emphasizing the importance of relaxations and inference techniques, as well as establishing connections with constraint programming. Recent methods based on generalized disjunctive programming, and that can explicitly handle nonlinear process models, are reviewed. Finally to compare their performance with existing MINLP methods, three synthesis problems are presented: a small process network, a vinyl chloride flowsheet, and a distillation sequencing problem.

Keywords

Mixed integer programming, Nonlinear programming, Process synthesis, Optimization methods.

Introduction

The mathematical programming approach to process design and synthesis problems has emerged as a major methodology for addressing these problems. As will be shown in this paper, significant advances have taken place with this methodology, as there has been considerable progress in mathematical programming techniques, particularly in the solution of mixed-integer nonlinear programming problems. These models, however, are still largely limited by the size of problems that they can handle, as well as by the difficulties in effectively handling nonlinearities.

It is the objective of this paper to present an overview of logic based optimization techniques that have the potential of overcoming or at least reducing difficulties that are normally experienced with algebraic mixed-integer linear and nonlinear programming techniques. The paper is organized as follows. We first present an overview of superstructures and models for process design and synthesis, and their general formulation as mixed-integer linear and nonlinear problems (MILP, MINLP), and generalized disjunctive programming problems (GDP). We next discuss specific logic based optimization techniques for linear and nonlinear models.

Superstructures and Models

The application of mathematical programming techniques to design and synthesis problems involves three major steps. The first is the development of a representation of alternatives from which the optimum solution is selected. The second is the formulation of a mathematical program that generally involves discrete and continuous variables for the selection of the configuration and operating levels, respectively. The third is the solution of the optimization model from which the optimal solution is determined. Whether one uses a high level aggregated model, or a fairly detailed model, it is always necessary to postulate a superstructure of alternatives. Most of the previous work has relied on representing the superstructure for each particular problem at hand, but without following some general principles. There are two major issues that arise in postulating a superstructure. The first is, given a set of alternatives that are to be analyzed, what are the major types of representations that can be used, and what are the implications for the modeling. The second is, for a given representation that is selected, what are all the feasible alternatives that must be included to guarantee that the global optimum is not overlooked.

As for types of superstructures, Yeomans and Grossmann (1999a) have characterized two major types of representations. The first is the State-Task Network (STN) which is motivated by the work in scheduling by Kondili, Pantelides and Sargent (1994). The basic idea here is that the representation makes use of two types of nodes: states and tasks (see Fig. 1.a). The assignment of equipment is dealt implicitly through the model. Both the cases of one-task one-equipment [OTOE] or variable task equipment assignment [VTE] can be considered. The second representation is State Equipment Network (SEN) which is motivated by recent work of Smith (1995), and where the basic idea is to work with two types of nodes: states and equipment (see Fig. 1.b). The tasks in this case are treated implicitly through the model. This representation considers the case of variable task equipment assignment [VTE]. Yeomans and Grossmann (1999a) have developed GDP models for each of the two different types of representations. These can then be used for solution with a GDP algorithm, or they can be used for reformulation as MILP or MINLP problems.

As for the issue on how to systematically generate the superstructure that includes all the alternatives of interest, Friedler et al. (1993) have proposed a novel graph theoretic approach that has polynomial complexity to find all the interconnections in process networks, given that nodes for processes and chemicals are specified. This procedure has been successfully applied for synthesizing process networks for waste minimization (Friedler at al., 1995). These authors have also used these ideas to perform more efficiently the search in the optimization (Friedler et al., 1996).

Closely related to the selection of the superstructure, is the selection of level of detail of the optimization model. A common misconception about the mathematical programming approach is that models are always detailed and require a great deal of information. This, however, is not necessarily true. In general mathematical programming models can be classified into three main classes:

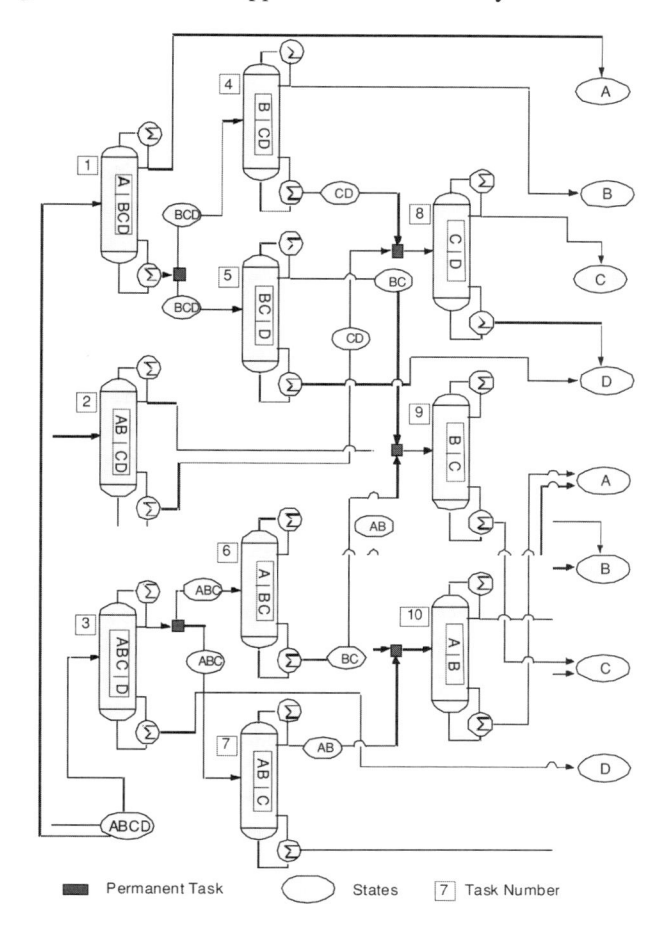

(a) State-task network

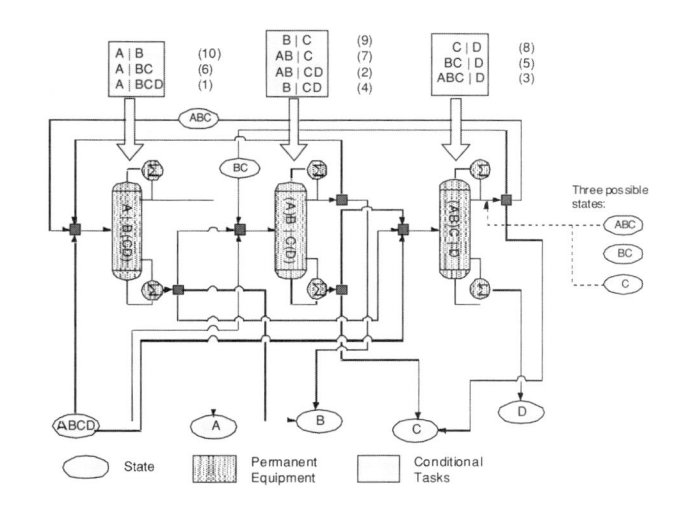

(b) State-equipment network

Figure 1. Alternative superstructure representations for ternary distillation.

As for the issue on how to systematically generate the superstructure that includes all the alternatives of interest, Friedler et al. (1993) have proposed a novel graph theoretic approach that has polynomial complexity to find all the interconnections in process networks, given that nodes for processes and chemicals are specified. This procedure has been successfully applied for synthesizing process networks for waste minimization (Friedler at al., 1995). These authors have also used these ideas to perform more efficiently the search in the optimization (Friedler et al., 1996).

Closely related to the selection of the superstructure, is the selection of level of detail of the optimization model. A common misconception about the mathematical programming approach is that models are always detailed and require a great deal of information. This, however, is not necessarily true. In general mathematical programming models can be classified into three main classes:

a) *Aggregated models*. These refer to high level representations in which the design or synthesis problem is greatly simplified by an aspect or objective that tends to dominate the problem at hand. Examples of aggregated models include the transshipment model for minimum utility and minimum number of units in heat exchanger networks (Papoulias and Grossmann, 1983) and mass exchanger networks (El-Halwagi and Maniousiouthakis, 1989a), the set of heat integration constraints based on the pinch location method (Duran and Grossmann, 1986a; Grossmann et al, 1998), distillation models for minimizing cost of utilities (Caballero and Grossmann, 1998), reactor network models for maximizing yield (Balakrishna and Biegler, 1992). All these models are specific to the corresponding problem at hand. An exception is for instance the work by Papalexandri and Pistikopoulos (1996) which considers heat and mass transfer building blocks, that allow integration of functions (e.g. reactive distillation). Daichendt and Grossmann (1998) have outlined a theoretical framework for deriving aggregated models, which however, must be adapted to each particular application.

b) *Short cut models*. These refer to fairly detailed superstructures that involve cost optimization (investment and operating costs), but in which the performance of the units is predicted with relatively simple nonlinear models in order to reduce the computational cost, and/or for exploiting the algebraic structure of the equations, especially for global optimization. Examples of such models include synthesis models for heat exchanger networks (Yee at al., 1990; Ciric and Floudas, 1991), distillation sequences (Aggrawal, and Floudas, 1990; Yeomans and Grossmann, 1999b), and process flowsheets (Kocis and Grossmann, 1989; Türkay and Grossmann, 1996).

c) *Rigorous models*. These also rely on detailed superstructures, but involve rigorous and complex models for predicting the performance of the units. The area of synthesis of distillation sequences (ideal and non-ideal) is perhaps the one that has received the most attention for developing rigorous models. Examples are the work by Bauer and Stichlmair (1996,1998) and Smith and Pantelides (1995).

It should be noted that aggregated models give rise to simpler types of optimization models. They are often LP, NLP or MILP models of modest size, that are simpler to solve than larger MINLP models. In contrast, both short cut and detailed models give rise almost exclusively to MINLP problems, which as mentioned above, can also be formulated as GDP problems. The important point to realize here is that mathematical programming can accommodate models of various degree of complexity.

Mathematical Programming Models

Design and synthesis problems give rise to discrete/continuous optimization problems, which when represented in algebraic form, correspond to mixed-integer optimization problems that have the following form:

$$\min Z = f(x, y)$$
$$st. \quad h(x, y) = 0$$
$$g(x, y) \leq 0 \quad \text{(MIP)}$$
$$x \in X, \ y \in \{0,1\}$$

where $f(x, y)$ is the objective function (e.g. cost), $h(x, y) = 0$ are the equations that describe the performance of the system (mass and heat balances, design equations), and $g(x, y) \leq 0$ are inequalities that define the specifications or constraints for feasible choices. The variables x are continuous and generally correspond to the state or design variables, while y are the discrete variables, which generally are restricted to take 0-1 values to define the selection of an item or an action. Problem (MIP) corresponds to a mixed-integer nonlinear program (MINLP) when any of the functions involved are nonlinear. If all functions are linear it corresponds to a mixed-integer linear program (MILP). If there are no 0-1 variables, the problem (MIP) reduces to a nonlinear program (NLP) or linear program (LP) depending on whether or not the functions are linear.

The formulation and solution of major types of mathematical programming problems can be effectively performed with modeling systems such as GAMS (Brooke at al., 1992), and AMPL (Fourer et al., 1992). While these require that the model be expressed explicitly in algebraic form, they have the advantage that they automatically interface with codes for solving the various types of problems. They also perform automatic differentiation and allow the use of indexed equations, with which large scale models can be readily generated. It should also be noted that these modeling systems now run mostly on desktop and PC computers, making their use and application widely available.

A state-of-the-art review of methods for solving LP and NLP problems, can be found in the paper by Wright in this conference. It should be noted that standard NLP methods are only guaranteed to find the global optimum if the problem is convex (i.e. convex objective function and constraints). When the NLP is nonconvex a global optimum cannot be guaranteed. One option is to try to convexify the problem, usually through exponential transformations, although the number of cases where this is possible is rather small. Alternatively, one could use rigorous global optimization methods, which over the last few years have made significant advances (see Floudas and Pardalos this conference).

MILP methods rely largely on simplex LP-based branch and bound methods (Nemhauser and Wolsey, 1988) that consists of a tree enumeration in which LP subproblems are solved at each node, and eliminated based on bounding properties. These methods are being improved through cutting plane techniques (Balas et al., 1993), which produce tighter lower bounds for the optimum. LP and MILP codes are widely available. The best known include CPLEX (1999), OSL (1999) and XPRESS-MP (1999), all which have achieved impressive improvements in their capabilities for solving problems. It is worth noting that since MILP problems are NP-complete it is always possible to run into time limitations when solving problems with large number of 0-1 variables, especially if the integrality gap is large.

Major methods for MINLP problems include Branch and Bound (BB) (Gupta and Ravindran, 1985; Nabar and Schrage, 1991; Borchers and Mitchell, 1992; Stubbs and Mehrotra, 1996; Leyffer, 1998), which is a direct extension of the linear case, except that NLP subproblems are solved at each node. Generalized Benders Decomposition (GBD) (Benders, 1962; Geoffrion, 1972), and Outer-Approximation (OA) (Duran and Grossmann, 1986a; Yuan, Zhang, Piboleau and Domenech, 1988; Fletcher and Leyffer, 1994; Ding-Mai and Sargent, 1992), are iterative methods that solve a sequence of alternate NLP subproblems with all the 0-1 variables fixed, and MILP master problems that predict lower bounds and new values for the 0-1 variables. The difference between the GBD and OA methods lies in the definition of the MILP master problem; the OA method uses accumulated linearizations of the functions, while GBD uses accumulated Lagrangian functions parametric in the 0-1 variables. The LP/NLP based branch and bound by Quesada and Grossmann (1992) essentially integrates both subproblems within one tree search, while the Extended Cutting Plane Method (ECP) (Westerlund and Pettersson, 1995) does not require the solution of NLP subproblems as it relies exclusively on successive linearizations. All these methods assume convexity to guarantee convergence to the global optimum. Nonrigorous methods for handling nonconvexities include the equality relaxation algorithm by Kocis and Grossmann (1987) and the augmented penalty version of it (Viswanathan and Grossmann, 1990). A review of these methods and how they relate to each other can be found in Grossmann and Kravanja (1997). The only commercial code for MINLP is DICOPT (GAMS, 1999), although there are a number of academic versions, such as MINOPT (Schweiger and Floudas, 1999), α-ECP by Westerlund and co-workers).

In recent years a new trend that has emerged in process synthesis is the formulation and solution of discrete/continuous optimization problems through a model that is known as Generalized Disjunctive Programming (GDP) (Raman and Grossmann, 1994). The basic idea in GDP models is to use Boolean and continuous variables, and formulate the problem with an objective function, and subject to three types of constraints: (a) global inequalities that are independent of discrete decisions; (b) disjunctions that are conditional constraints involving an OR operator; (c) pure logic constraints that involve only the Boolean variables. More specifically, the problem is as follows:

$$\min Z = \sum_{k \in K} c_k + f(x)$$
$$s.t. \quad g(x) \leq 0$$
$$\bigvee_{j \in I_k} \begin{bmatrix} Y_{jk} \\ h_{jk}(x) \leq 0 \\ c_k = \gamma_{jk} \end{bmatrix}, \quad k \in K \quad \text{(GDP)}$$
$$\Omega(Y) = true$$
$$x \in X, \, c_k \geq 0, \, Y_{jk} \in \{true, false\}$$

where x are continuous variables and y are the Boolean variables. The objective function involves the term $f(x)$ for the continuous variables (e.g. operating cost) and the charges c_k that depend on the discrete choices. The equalities/inequalities $g(x) \leq 0$ must hold regardless of the discrete conditions, and $h_{jk}(x) \leq 0$ are conditional constraints that must be satisfied when the corresponding Boolean variable Y_{jk} is True for the j'th term of the k'th disjunction. The set I_k represents the number of choices for each disjunction defined in the set K. Also, the fixed charge c_k is assigned the value γ_{jk} for that same variable. Finally, the constraints $\Omega(Y)$ involve logic propositions in terms of Boolean variables.

In the context of synthesis problems the disjunctions in (GDP) typically arise for each unit i in the following form:

$$\begin{bmatrix} Y_i \\ h_i(x) \leq 0 \\ c_i = \gamma_i \end{bmatrix} \vee \begin{bmatrix} \neg Y_i \\ B^i x = 0 \\ c_i = 0 \end{bmatrix} \quad i \in I \quad (2)$$

in which the inequalities h_i apply and a fixed cost γ_i is incurred if the unit is selected (Y_i); otherwise ($\neg Y_i$) there is no fixed cost and a subset of the x variables is set to zero with the matrix B^i.

It is important to note that any problem posed as (GDP) can always be reformulated as an MINLP of the form of problem (MIP), and any problem in the form of (MIP) can be posed in the form of (GDP). For modeling purposes, however, it is advantageous to start with model (GDP) as it captures more directly both the qualitative (logical) and quantitative (equations) part of a problem. As for the transformation from (GDP) to (MIP), the most straightforward way is to replace the Boolean variables Y_{ik} by binary variables y_{ik}, and the disjunctions by "big-M" constraints of the form,

$$h_{ik}(x) \leq M(1 - y_{ik}), \ i \in D_k, \ k \in SD$$
$$\sum_{i \in D_k} y_{ik} = 1, \ k \in SD \quad (3)$$

where M is a large valid upper bound. Finally, the logic propositions $\Omega(Y)=True$, are converted into linear inequalities as described in Williams (1985) and Raman and Grossmann (1991).

In the next sections we discuss recent modeling techniques and methods for addressing logic-based optimization problems of the form of (GDP). We first consider the general issues, with an emphasis on linear models.

Solution Methods

Problem (GDP) can be solved either by using specific logic-based optimization methods, or else by using MILP or MINLP methods which in turn require appropriate reformulations of the GDP problem. Specific logic-based optimization methods include branch and bound, and decomposition techniques. In the next section we describe the basic elements for branch and bound methods, in which the importance of relaxations and inference techniques is highlighted. In the section that follows we discuss decomposition techniques, and emphasize reformulations and the handling of nonlinearities.

Logic-based Branch and Bound Methods

The disjunctive formulation (GDP) not only provides a more natural modeling framework than a mixed integer formulation (MIP) for many problems, but it can lead to more efficient solution as well.

A disjunctive problem can be solved in a manner that is somewhat parallel to the traditional branch-and-bound solution of mixed integer problems (Hooker 1994; Hooker and Osorio 1997). A search tree is created by branching on the logical variables Y_{jk}. At the beginning of the search (the root node of the tree), one branches on a selected Y_{jk} by setting $Y_{jk} = True$ and $Y_{jk} = False$. (There are other branching schemes as well, which may be more efficient.) This creates two successor nodes of the root node, at which the process repeats.

At each node of the search tree, one solves a continuous relaxation of the problem, whose optimal value Z is a lower bound on the optimal value of the original problem (GDP). If Z is greater than or equal to the value of the incumbent solution (the best feasible solution found so far), there is no need to branch further at the current node, and the search backtracks to a higher node in the tree. When all nodes have been explored, the incumbent solution is optimal.

The continuous relaxation at a given node is determined as follows. Certain logical variables Y_{jk} have been fixed to True or False. By applying logical inference to the logical constraints $\Omega(Y) = True$, one may be able to determine the values of still more logical variables. The continuous relaxation consists of the constraints $g(x) \leq 0$, $x \in X$, and the constraints $h_{jk}(x) \leq 0$, $c_k = \gamma_k$ for each Y_{jk} that is fixed to True. The relaxation is solved by a linear or nonlinear programming algorithm. If (a) the relaxation has a solution (x,c), (b) every Y_{jk} has a truth value, and (c) at least one disjunct of every disjunction is satisfied (i.e., for every k, $Y_{jk} = True$ for at least one j), then (x,c,Y) is a feasible solution of the original problem.

It is often possible to identify a feasible solution even if not all Y_{jk}'s are assigned values. This can be done by using a heuristic or other search method to assign values to the unfixed Y_{jk}'s so that $\Omega(Y) = True$, and $Y_{jk} = False$ whenever the solution (x,c) of the continuous relaxation violates $h_{jk}(x) \leq 0$ or $c_k = \gamma_{jk}$.

This approach has at least four potential advantages over the traditional branch-and-bound solution of mixed integer models.

(a) The continuous relaxations are smaller and therefore easier to solve. Traditional relaxations of mixed integer problems (MIP) contain all of the 0-1 integer variables y_j, with the 0-1 restriction replaced by $0 \leq y_j \leq 1$. The logic-based relaxations usually contain none of the discrete variables.

(b) Inference methods can fix discrete variables by processing the logical constraints $\Omega(Y) = True$, thus reducing the amount of branching. Preprocessing achieves some of this effect in mixed integer programming, but it lacks the full power of complete logical inference methods (explained below).

(c) In traditional branch-and-bound, some of the 0-1 variables can have fractional values in the relaxation when the solution is in fact feasible, which results in unnecessary branching. This does not occur in a logic-based approach.

(d) In many design problems, singularities can result when variables are fixed to zero to represent the absence of a component. In a logic-based setting, the absence of a component is treated as a separate case in a disjunction which the expressions causing the singularity are simply omitted.

As indicated above, an inference method can infer the truth values of some logical values Y_{jk} from the logical constraints $\Omega(Y) = \text{True}$ and from the truth values that have already been fixed. If the method is *complete*, it fixes all values that can possibly be inferred. Although complete methods may be impractical in applications with a large number of discrete variables, they are often appropriate for process design problems. This is because in these problems, it is generally expensive to solve the (often nonlinear) continuous relaxations at the nodes of the search tree. It is worth a substantial investment in logical inference to minimize the number of nodes.

The simplest complete inference method for logical variables is the *resolution* method. It requires that the logical constraints be written in *conjunctive normal form*, which is a conjunction of disjunctions of *literals* (logical variables Y_{jk} or their negations $\neg Y_{jk}$). This is readily accomplished by converting implications such as $Y_1 \Rightarrow (Y_3 \vee Y_4 \vee Y_5)$ to disjunctions $\neg Y_1 \vee Y_3 \vee Y_4 \vee Y_5$ and by using other simple rules (see Hooker and Osorio 1997). Resolution proceeds by finding two disjunctions between which exactly one variable Y_j changes sign. The resolvent is a disjunction of all of the literals in the two clauses except Y_j and $\neg Y_j$. For example, the third disjunction below is the resolvent of the first two.

$$Y_1 \vee Y_2 \vee \neg Y_3$$
$$\neg Y_1 \vee Y_2 \vee Y_4$$
$$Y_2 \vee \neg Y_3 \vee Y_4$$

Any disjunctions in the constraint set that are implied by the resolvent are removed. The process continues until no resolvents can be generated that are not implied by other constraints. When resolution generates a single-term disjunction Y_j or $\neg Y_j$, Y_j is fixed to *True* or *False*, respectively. If resolution generates an empty disjunction, the constraints are unsatisfiable, and the search backtracks.

Although the models discussed here use two-valued logical variables, in many applications multivalued discrete variables are convenient and lead to more efficient logic processing. Suppose for example that each of m inputs is to be assigned to a processing unit, and each unit can process at most one input. The assignments can be indicated in either of two ways.

(a) Bivalent logical variables Y_{jk} are set to 1 when input k is assigned to unit j. In this case one must impose assignment constraints $\vee_j Y_{jk}$ for all k and $\neg Y_{jk} \vee \neg Y_{jk'}$ for all j, k, k' ($k \neq k'$).

(b) Multivalued discrete variables Y_k have a value indicating the unit to which input k is assigned. Here it suffices to use a single constraint, all-different$(Y_1,...,Y_m)$, which requires each unit to be assigned at most one input.

Multivalued variables have the advantage that *constraint propagation* algorithms (a form of logical inference) may be able to reduce their domains efficiently. The *domain* of a variable is the set of values it may assume. Suppose that in the assignment problem just mentioned, $Y_1 \in \{1\}$, $Y_2 \in \{2,3,5\}$, $Y_3 \in \{1,2,3,5\}$, $Y_4 \in \{1,5\}$, and $Y_5 \in \{1,2,3,4,5\}$. That is, input 1 must be assigned to unit 1, input 2 may be assigned to unit 2, 3 or 5, and so forth. It can be deduced from this information that input 2 cannot be assigned to unit 5. In fact, the domains can be reduced to $Y_2 \in \{2,3\}$, $Y_3 \in \{2,3\}$, $Y_4 \in \{5\}$, and $Y_5 \in \{4\}$. This reduces the amount of branching on each variable and even fixes two of the variables. This deduction can be accomplished rapidly using an algorithm related to maximum cardinality bipartite matching (Régin 1993).

The constraint programming community has developed applications and propagation algorithms for a number of useful constraints in addition to all-different (Marriott and Stuckey 1998; Tsang 1993; Van Hentenryck 1988). They are embodied in such constraint programming packages as CHIP (Dincbas et al. 1988) and the ILOG Solver (Puget 1994). These techniques have seen most of their commercial success in scheduling and logistics but have been applied to some engineering design problems as well. These include analog circuitry synthesis (Simonis and Dincbas 1987), transistor amplifier design (Heintze et al. 1992), analysis and partial synthesis of truss structures (Lakmazaheri and Rasdorf 1989), designing gear boxes (Sthanusubramonian 1991), and other mechanical systems (Subramanian and Wang 1993).

A potential weakness of the logic-based approach is that its continuous relaxations may be weaker than the relaxations obtained in classical mixed integer programming. The logic-based relaxations are smaller and easier to solve, but because they tend to have fewer constraints, their optimal values may be smaller. This can be offset when necessary by adding relaxations of disjunctive constraints to the continuous relaxation at each node of the search tree.

The simplest relaxation of a disjunction is obtained by replacing the 0-1 constraints $Y_{ik} \in \{0,1\}$ with $0 \le Y_{ik} \le 1$ in the big-M model (3). One can also use the convex hull formulation (RDP), to be discussed in the next section. Both of these approaches, however, introduce variables Y_{ik} into the relaxation, and the latter introduces additional continuous variables.

In some cases a disjunction can be given a useful relaxation by adding one or a few constraints, with no additional variables. Consider, for example, a disjunction of the following form,

$$\bigvee_{i \in D} \begin{bmatrix} Y_i \\ h_i(x) \le 0 \end{bmatrix}$$

where $h_i(x) \le 0$ is a single inequality and $0 \le x \le m$. This disjunction has a relaxation that involves no additional variables, consists of a single inequality, and is equivalent to the big-M relaxation (Beaumont 1990):

$$\sum_{i \in D} \frac{h_i(x)}{M_i} \le |D| - 1$$

where

$$M_i = \max\{h_i(x) \mid 0 \le x \le m\}$$

If $h_i(x) = b^i x - \beta_i$ is linear, then the relaxation is linear:

$$\left(\sum_{i \in D} \frac{b^i}{M_i} \right) x \le \sum_{i \in D} \frac{\beta_i}{M_i} + |D| - 1$$

where

$$M_i = \sum_j \max\{0, b_j^i\} m_j - \beta_i$$

The disjunction (2) can be similarly relaxed. Let $h_i(x) \le 0$ be a system of m inequalities $h_{ip}(x) \le 0$ and let each $B^i = I$. Then the relaxation is

$$\frac{h_{ip}(x)}{\gamma_i} + \frac{c_i}{\gamma_i} \le 1, \, p = 1\ldots,m$$

$$\frac{\sum x_j}{\sum m_j} \le \frac{c_i}{\gamma_i}$$

where

$$M_{ip} = \max\{h_{ip}(x) \mid 0 \le x \le m\}$$

When relaxations obtained as above are too large or otherwise unsatisfactory, one can use the traditional big-M or convex hull relaxation. Another option is to use *optimal separating inequalities* as described in Hooker and Osorio (1997), which are logic-based analogs of the lift-and-project cuts of Balas et al. (1996).

When the continuous relaxation is nonlinear, any of several linearization techniques may be used. Two approaches suitable for certain types of design problems that involve bilinearities are discussed in Bollapragada et al. (1998). More general techniques are presented in the next section.

General-purpose software that implements the methods described above does not exist at this writing but is likely to emerge in the near future. Hooker et al. (1998,1999) are preparing a system of this kind. Fourer and Gay have written an experimental version of AMPL that accepts logic-based formulations (Fourer 1998). It generates input for the ILOG Solver, however, and so does not take full advantage of mathematical programming software. Several investigators have proposed schemes for combining mathematical programming with constraint programming and in some cases have obtained computational results. They include Bockmayr and Kasper (1998), Darby-Dowman and Little (1998), and Rodosek et al. (1997). The modeling language OPL can invoke either a linear programming (CPLEX) or a constraint programming (ILOG) solver but does not integrate them as described above (Van Hentenryck and Lustig 1999).

Logic-based Decomposition and Reformulation Methods.

We focus in this section on the solution of logic based optimization models that can be posed in the form of problem (GDP), and involve nonlinear process models. We consider first the corresponding OA and GBD algorithms for process network problems expressed in the form of (GDP). As described in Turkay and Grossmann (1996) for fixed values of the Boolean variables, $Y_{\hat{i}k} = true$ and $Y_{ik} = false$ for $\hat{i} \ne i$, the corresponding NLP subproblem is as follows:

$$\min \; Z = \sum_{k \in SD} c_k + f(x)$$
$$s.t. \quad g(x) \le 0$$
$$\left. \begin{array}{l} h_{ik}(x) \le 0 \\ c_k = \gamma_{ik} \end{array} \right\} \text{ for } Y_{ik} = true \quad \hat{i} \in D_k, k \in SD$$
$$\left. \begin{array}{l} B^i x = 0 \\ c_k = 0 \end{array} \right\} \text{ for } Y_{ik} = false \quad i \in D_k, i \ne \hat{i}, k \in SD$$
$$x \in R^n, \; c_i \in R^m \qquad \text{(NLPD)}$$

Note that for every disjunction $k \in SD$ only constraints corresponding to the Boolean variable Y_{ik} that is true are imposed. Also, fixed charges γ_{ik} are only applied to these terms. Assuming that K subproblems (NLPD) are solved in which sets of linearizations $l = 1,...K$ are generated for subsets of disjunction terms $L_{ik} = \{ l \mid Y^l_{ik} = true \}$, one can define the following disjunctive OA master problem:

$$\text{Min} \quad Z = \sum_k c_k + f(x) \quad \text{(MGDP)}$$

s.t.
$$\left. \begin{array}{l} \alpha \geq f(x^l) + \nabla f(x^l)^T (x - x^l) \\ g(x^l) + \nabla g(x^l)^T (x - x^l) \leq 0 \end{array} \right\} l = 1,...,L$$

$$\bigvee_{i \in D_k} \left[\begin{array}{l} Y_{ik} \\ h_{ik}(x^l) + \nabla h_{ik}(x^l)^T (x - x^l) \leq 0, l \in L_{ik} \\ c_k = \gamma_k \end{array} \right], k \in SD$$

$$\Omega(Y) = True$$
$$\alpha \in R, \; x \in R^n, \; c \in R^m, \; Y \in \{true, false\}^m$$

It should be noted that before applying the above master problem it is necessary to solve various subproblems (NLPD) so as to produce at least one linear approximation of each of the terms in the disjunctions. As shown by Turkay and Grossmann (1996), selecting the smallest number of subproblems amounts to the solution of a set covering problem, which is of small size and easy to solve. In the context of flowsheet synthesis problems, another way of generating the linearizations in (MGDP) is by starting with an initial flowsheet and suboptimizing the remaining subsystems as in the modeling/decomposition strategy (Kocis and Grossmann, 1989; Kravanja and Grossmann, 1990).

The above problem (MGDP) can be solved by the methods described by Beaumont (1991), Raman and Grossmann (1994), and Hooker and Osorio (1997). For the case of process networks, Turkay and Grossmann (1996) have shown that if the convex hull representation of the disjunctions in (2) is used in (MGDP), then by converting the logic relations $\Omega(Y)$ into the inequalities $Ay \leq a$, leads to the MILP problem,

$$\text{Min } Z = \sum_k c_k + f(x) \quad \text{(MIPDF)}$$

s.t.
$$\left. \begin{array}{l} \alpha \geq f(x^l) + \nabla f(x^l)^T (x - x^l) \\ g(x^l) + \nabla g(x^l)^T (x - x^l) \leq 0 \end{array} \right\} l = 1,...,L$$

$$\nabla_{x_{Zi}} h_i(x^l)^T x_{Zi} + \nabla_{x_{Ni}} h_i(x^l)^T x^1_{Ni}$$
$$\leq [-h_i(x^l) + \nabla_x h_i(x^l)^T x^l] y_i$$
$$l \in K^i_L, i \in I$$
$$x_{N_i} = x^1_{N_i} + x^2_{N_i}$$
$$0 \leq x^1_{N_i} \leq x^U_{N_i} y_i$$
$$0 \leq x^2_{N_i} \leq x^U_{N_i} (1 - y_i)$$
$$Ay \leq a$$

$$x \in R^n, \; x^1_{N_i} \geq 0, x^2_{N_i} \geq 0, \; y\{0,1\}^m$$

where the vector x is partitioned into the variables (x_{Z_i}, x_{N_i}) for each disjunction i according to the definition of the matrix B^i. The linearization set is given by $K_L^i = \{ l \mid Y_{il} = True, l = 1,...,L\}$ that denotes the fact that only a subset of inequalities were enforced for a given subproblem l. It is interesting to note that the logic-based Outer-Approximation algorithm represents a generalization of the modeling/decomposition strategy Kocis and Grossmann (1989) for the synthesis of process flowsheets.

Turkay and Grossmann (1996) have also shown that although a logic-based Generalized Benders method (Geoffrion, 1972) cannot be derived as in the case of the OA algorithm, one can exploit the following property. For MINLP problems, performing one Bender's iteration (Turkay and Grossmann, 1996) on the MILP master problem of the OA algorithm, is equivalent to generating a Generalized Benders cut. Therefore, a possible logic-based version of the Generalized Benders method consists of performing one Benders iteration on the MILP master problem (MIPDF) based on the property that the cut obtained from performing one Benders iteration on the MILP master of outer-approximation is equivalent to the cut obtained from the GBD algorithm. Finally, it should also be noted that slacks can be introduced to (MGDP) and to (MIPDF) to reduce the effect of nonconvexities as in the augmented-penalty MILP master problem (Viswanathan and Grossmann, 1990).

In an alternative approach for solving problem (GDP), Lee and Grossmann (1999) have shown, based on the work by Stubbs and Mehrotra (1994), that the convex hull of the disjunction $k \in SD$ in the Generalized Disjunctive Program (GDP), is given by the following set of equations and inequalities,

$$x = \sum_{i \in D_k} v_{ik}, \quad \sum_{i \in D_k} \lambda_{ik} = 1$$
$$c_k = \sum_{i \in D_k} \lambda_{ik} \quad \text{(CH}_k\text{)}$$
$$\lambda_{ik} h_{ik}(v_{ik} / \lambda_{ik}) \leq 0 \quad i \in D_k$$

where $h_{ik}(x)\leq 0$ are assumed to be convex inequalities. The derivation of the above equations can be performed by an exact linearization in which each term in a disjunction k is multiplied by the non-negative variables λ_{ik}. Mehrotra and Stubbs (1994) proved that $\lambda h(v/\lambda)$ is a convex function if $h(x)$ is a convex function. In (CH$_k$) the variables v_{ik} can be interpreted as disaggregated variables that are assigned to each disjunctive term, while λ_{ik}, can be interpreted as weight factors that determine the validity of the inequalities in the corresponding disjunctive term. Note also that (CH$_k$) reduces to the result by Balas (1985) for the case of linear constraints, $h_{ik}(x)\leq 0$.

By using the equations in (CH$_k$), the following nonlinear programming can be defined for problem (GDP),

$$\min Z^L = \sum\sum \gamma_{ik}\lambda_{ik} + f(x)$$

$$\text{st} \quad g(x) \leq 0 \quad\quad\quad\quad (\text{RDP})$$

$$x = \sum_{i\in D_k} v_{ik}, \quad \sum_{i\in D_k} \lambda_{ik} = 1 \quad k \in SD$$

$$\lambda_{ik} h_{ik}(v_{ik}/\lambda_{ik}) \leq 0 \quad i \in D_k, k \in SD$$

$$A\lambda \leq a$$

$$x \in R^n, \; v_{ik} \geq 0, \; 0 < \lambda_{ik} \leq 1, \; i\in D_k, k\in SD$$

The relaxation problem (RDP), which yields a valid lower bound to the solution of problem (GDP), can be used as a basis to construct a special purpose branch and bound method as has been proposed by Lee and Grossmann (1999). Alternatively, problem (RDP) can be used to reformulate problem (GDP) as a tight MINLP problem of the form,

$$\min Z^L = \sum\sum \gamma_{ik}\lambda_{ik} + f(x)$$

$$\text{st} \quad g(x) \leq 0$$

$$x = \sum_{i\in D_k} v_{ik}, \quad \sum_{i\in D_k} \lambda_{ik} = 1 \quad k \in SD$$

$$(\lambda_{ik}+\varepsilon)h_{ik}(v_{ik}/(\lambda_{ik}+\varepsilon)) \leq 0 \quad i \in D_k, k \in SD$$

$$-U_{ik} \leq v_{ik} \leq U_{ik}$$

$$A\lambda \leq a$$

$$x \in R^n, \; v_{ik} \geq 0, \; \lambda_{ik}=\{0,1\}, \; i\in D_k, k\in SD \quad (\text{MIP-DP})$$

In the above model λ_{ik} are binary variables that represent the Boolean variables Y_{ik}, and ε is a small tolerance to avoid numerical difficulties in the nonlinear inequalities. All the algorithms that were discussed in the section on MINLP methods can be applied to solve this problem. For the case when the disjunctions have the form of (2), Lee and Grossmann (1999) noted that there is the following relationship of problem (MIP-DP) with the logic-based outer-approximation by Turkay and Grossmann (1996). If one considers fixed values of λ_{ik} this leads to an NLP subproblem of the form (NLPD). If one then peforms a linearization on problem (MIP-DP), this leads to the MILP problem (MIPDF). Hence, for the case of disjunctions with two terms (one for true to impose equations, and one for false to set zero values), the reformulation (RDP) when solved by the outer-approximation method, reduces to the method by Turkay and Grossmann (1996). It should also be noted that the methods described in this section are being implemented in the LOGMIP code by Vecchietti and Grossmann (1997).

Applications

Synthesis of a Process Network

We first present numerical results of an example problem dealing with the synthesis of a process network that was originally formulated by Duran and Grossmann (1986a) as an MINLP problem, and later by Turkay and Grossmann (1996) as a GDP problem. Fig. 2 shows the superstructure that involves the possible selection of 8 processes. The model, which involves convex functions, is given in Turkay and Grossmann (1996). The Boolean variables Y_j denote the existence or non-existence of processes 1-8. The global optimal solution is $Z^* = 68.01$, consists of the selection of processes 2,4,6, and 8.

As seen in Table 1, the branch and bound (BB) algorithm by Lee and Grossmann (1999) finds the optimal solution in only 5 nodes compared with 17 nodes of standard branch and bound method when applied to the MINLP formulation with big-M constraints. A major difference in these two methods is the lower bound predicted by the relaxed NLP. Clearly the bound at the root node in the proposed BB method, which is given by problem (RDP), is much stronger (62.48 vs. 15.08). Table 2 shows the comparison with other algorithms when the problem is reformulated as the tight MINLP problem (MIP-DP). Note that the proposed BB algorithm and the standard BB yield the same lower bound (62.48) since they start by solving the same relaxation problem. The difference in the number of nodes, 5 vs. 11, lies in the branching rules, which are better exploited in the special branch and bound method by Lee and Grossmann (1999). The OA, GBD and ECP methods start with initial guess $Y^0 = [1,0,1,1,0,0,1,1]$. Note that in GBD and OA methods, one major iteration consists of one NLP subproblem and one MILP master problem. As predicted by the theory, the logic-based OA method yields the lower bound 8.541, which is stronger than the one of the GBD method.

Therefore, OA requires 3 major iterations versus 8 from GBD. The ECP method requires 7 iterations, each involving the solution of an MILP. Thus, these results show the improvements that can be obtained through the logic based formulation, such as with the generalized disjunctive program (GDP). It also shows that the OA algorithm requires fewer major iterations than the GBD and ECP methods.Superstructure of Vinyl Chloride Process

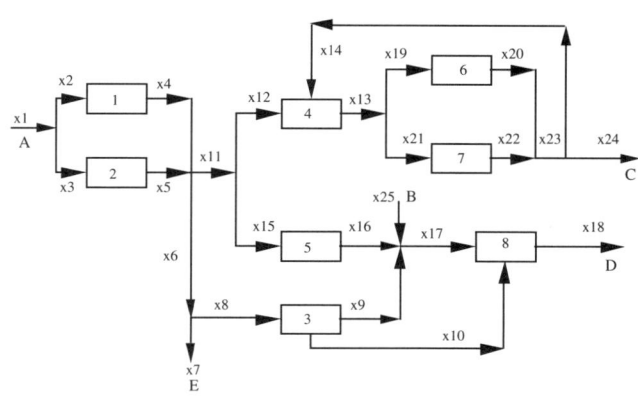

Figure 2. Superstructure for process network example.

Table 1. Comparison of Branch and Bound Methods.

Optimum Solution	Model Method	(BM) Standard BB	MIP-DP Proposed BB Algorithm
-	No. of nodes	17	5
68.01	Relaxed Optimum	15.08	62.48

Table 2. Comparison of Several Algorithms with Reformulation MIP-DP.

Method*	No. of Nodes / Iterations	Relaxed Optimum
Standard BB	11 (Nodes)	62.48
Proposed BB	5 (Nodes)	62.48
OA (Major)	3 (Iter.)	8.541
GBD (Major)	8 (Iter.)	-551.4
ECP (Major)	7 (Iter.)	-5.077

*All methods solve the reformulated MINLP problem (MIP-DP).

The superstructure for vinyl chloride production proposed by Turkay and Grossmann (1998) and shown in Fig. 3, consists of three sections: direct chlorination reaction, oxychlorination reaction, and pyrolysis and separation sections. The superstructure contains a total of 32 process units that give rise to 96 feasible flowsheet configurations. The objective function, profit, includes raw material costs, costs associated with utility consumptions (electricity, cooling water, and steam), annualized investment costs for equipment, and revenues from the sales of vinyl chloride product and hydrogen chloride by-product. Discontinuous functions were used to model the cost of units. Different correlations over several ranges of sizes, with pressure and temperature factors, were modeled using disjunctions and converted into mixed-integer constraints using the equations in (CH_k). The models used for the flowsheet were simplified or short-cut models: isentropic compression of an ideal gas for compression, the Fenske-Underwood-Gilliland correlation for distillation, Raoult's Law for phase equlibrium. For the direct chlorination reactor a CSTR reactor was used, a PFR for the oxychlorination, and a PFR for the pyrolysis reactor. The objective was to select the flowsheet configuration, as well as operating conditions (pressures, temperatures, recycle rates) to maximize the annual profit.

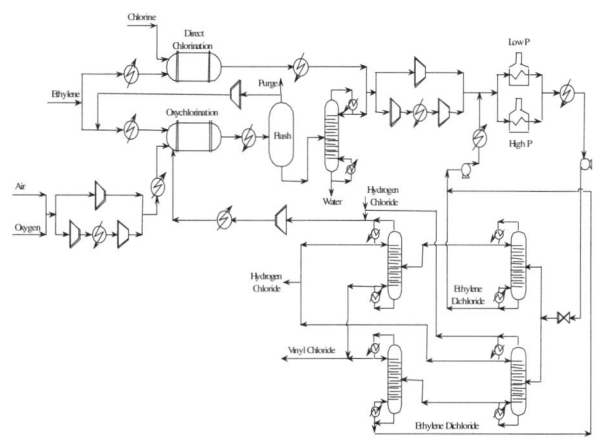

Figure 3. Superstructure of vinyl chloride process.

The algorithm by Turkay and Grossmann (1996) required the solution of two initial flowsheets in order to derive outer-approximations for each unit which were used in the first master problem. The MINLP problem for the first flowsheet involved 162 binary variables, 879 continuous variables, and 699 constraints, and was solved with DICOPT++ (Viswanathan and Grossmann, 1990), yielding a profit of $27,678,000/yr. The MINLP for the second flowsheet was somewhat larger, involving 168 binary variables, 883 continuous variables, and 703 constraints, and yielded a profit of $75,283,000/yr. Next, the first master problem was derived, involving 259 binary variables, 1715 continuous variables and 1750 constraints. The MILP master problem is solved with GAMS/OSL (Brooke et al., 1992) requiring the enumeration of only 6 nodes. The master problem predicts the flowsheet in Fig. 5, with a potential profit of $82,763,000/yr. The corresponding MINLP for the flowsheet calculates an optimal profit for this flowsheet of $75,283,000/yr, which corresponds to a lower bound. The second master problem is modeled with the linearizations from the first, second, and third flowsheets. This master problem predicts a potential profit of $65,262,000/yr corresponding to the upper bound on the profit of the remaining alternatives. Since the upper bound on the profit is less than the current lower bound, the algorithm terminates confirming that the optimal flowsheet corresponds to the second flowsheet in Fig. 4. The total CPU time required to optimize this flowsheet was less that 4 minutes on a IBM RS-6000. It is worth noting that a direct solution of this problem as an MINLP failed with DICOPT.

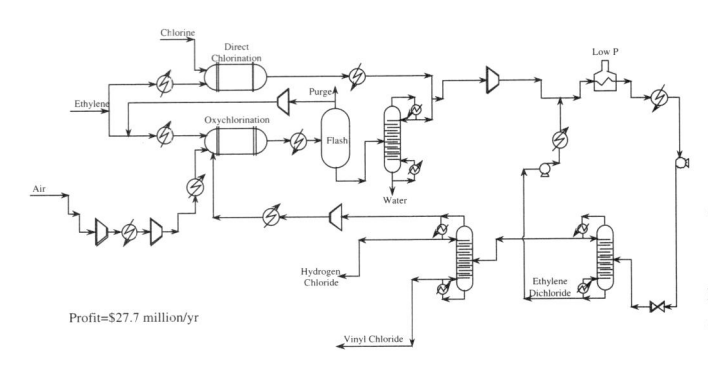

Figure 4. Optimal flowsheet configuration.

Synthesis of Distillation Sequences

Finally, we present an example by Yeomans and Grossmann (1999c), dealing with the synthesis of a distillation sequence for separating a mixture of butane (C4), pentane (C5) and hexane (C6) into pure components. An SEN superstructure of two columns with a maximum number of trays of 30 is used, with a tray-by-tray model and ideal equilibrium. The objective function selected was the present cost of the equipment and utility costs. The problem was formulated as a GDP problem, involving 56 Boolean variables, 1962 continuous variables, and 2037 constraints. The problem was solved with an extension of the algorithm by Turkay and Grossmann (1996), and required 42 min CPU-time. Figure 5 shows the optimal configuration, which has a cost of $2,694,000.

It is worth noting that the optimal solution removes the most abundant component last, violating a well-known heuristic. If the heuristic configuration is used, this leads to a cost of $3,462,000, which is 22% more expensive. Also, an important feature in the solution of this problem is that the NLP subproblems only included columns with the actual number of trays, with which the problem of singularities was entirely avoided. A direct solution as an MINLP of this problem did not even yield a feasible solution.

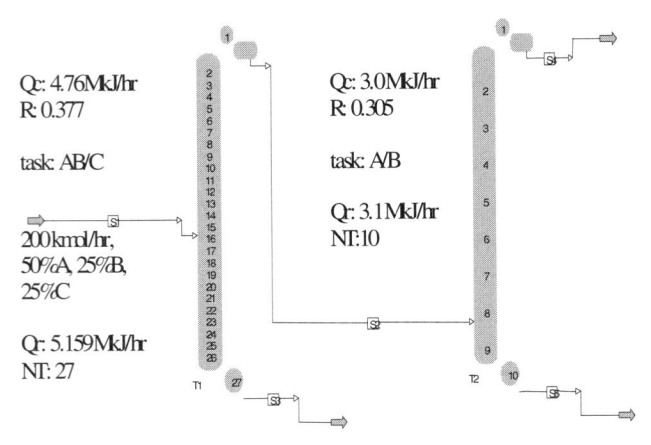

Figure 5. Optimal configuration of ideal hydrocarbon mixture with tray-by-tray models.

Conclusions

This paper has presented a general overview on recent developments in the area of logic based optimization for mixed-integer programming. As has been shown, the first important feature of these techniques is that they consist of declarative models that involve Boolean and continuous variables, in equations, disjunctions and logic propositions, as is the case in generalized disjunctive programming. A number of general issues were discussed regarding their solution, including the questions of constructing relaxations, and of performing inferences with the logic. The connection with the emerging area of constraint logic programming was also pointed out. The paper then reviewed specific algorithms for nonlinear problems, including the logic based version of the outer-approximation method, and methods that use the relaxation based on the convex hull on each disjunction. Three examples were presented, a small process network problem, a flowsheet synthesis problem, and the synthesis and design of a separation sequence. In all three cases the logic based techniques offered significant advantages in terms of robustness and computer time compared to the algebraic MINLP models and their solution methods.

Acknowledgement

Financial support from the National Science Foundation under grant CTS-9710303 is gratefully acknowledged.

References

AlphaECP, Mixed-Integer Process Optimization Group, Abo Akademi, Finland; website at http://at8.abo.fi/PSE/Software.shtml

Balakrishna, S. and L.T. Biegler (1992a). A constructive targeting approach for the synthesis of isothermal reactor networks. *Ind. Eng. Chem. Res.* **31**(9), 300.

Balas, E. (1985). Disjunctive programming and a hierarchy of relaxations for discrete optimization problems. *SIAM J. Alg. Disc. Meth.*, **6**, 466-486.

Balas, E., Ceria, S. and G.A. Cornuejols(1993). Lift-and-project cutting plane algorithm for mixed 0-1 programs, *Math. Prog.*, **58**, 295-324.

Balas, E., S. Ceria and G. Cornuéjols. Mixed 0-1 programming by lift-and-project in a branch and cut framework, *Management Sci.*, **42**, 1229-1246.

Bauer, M.H. and J. Stichlmair (1998). Design and economic optimization of azeotropic distillation processes using mixed-integer nonlinear programming. *Comp. Chem. Eng.*, **22**(9) 1271.

Bauer, M.H. and J. Stichlmair (1996). Superstructures for the mixed integer optimization of nonideal and azeotropic distillation processes. *Comp. Chem. Eng.*, S25.

Bazaraa, M.S., H.D. Sherali and C.M. Shetty (1994). *Nonlinear Programming*, John Wiley.

Beaumont, N. (1991). An algorithm for disjunctive programs, *European J. Ops. Res.*, **48**, 362-371.

Benders J.F. (1962). Partitioning procedures for solving mixed-variables programming problems. *Numer. Math.*, **4**, 238-252.

Biegler, L.T., I.E. Grossmann and A.W. Westerberg (1997). *Systematic Methods for Chemical Process Design*, Prentice-Hall.

Bockmayr, A., and T. Kasper (1998). Branch-and-infer: A unifying framework for integer and finite domain constraint programming, *INFORMS J. Comp.*, **10** 287-300.

Bollapragada, S., O. Ghattas and J. N. Hooker (1998). Optimal design of truss structures by logic-based branch and cut, to appear in *Ops. Res.* Available at http://ba.gsia.cmu.edu/jnh.

Borchers, B. and J.E. Mitchell (1994). An improved branch and bound algorithm for mixed integer nonlinear programming. *Comp. Ops. Res.*, **21**, 359-367.

Brooke, A., Kendrick, D. and A. Meeraus (1992). *GAMS - A User's Guide*. Scientific Press, Palo Alto.

Caballero, J.A. and I.E. Grossmann (1999). Aggregated models for integrated distillation systems. to appear *in Ind. Eng. Chem. Res.*

Ciric, A.R. and C.A. Floudas (1991). Heat exchanger network synthesis without decomposition," *Comp. Chem. Eng.*, **15**, 385-396.

CPLEX (1999). A Division of ILOG, web site at http://www.cplex.com.

Daichendt, M.M. and I.E. Grossmann (1994b). Preliminary screening for the MINLP synthesis of process systems I: Aggregation and decomposition techniques. *Comp. Chem. Eng.* **18**, 663.

Dakin, R.J. (1965). A tree search algorithm for mixed-integer programming problems. *Comp. J*, **8**, 250-255.

Darby-Dowman, K., and J. Little (1998). Properties of some combinatorial optimization problems and their effect on the performance of integer programming and constraint logic programming, *INFORMS J. Comp.*, **10** 276-286.

Dincbas, M., P. Van Hentenryck, H. Simonis, A. Aggoun, T. Graf, F. Bertier (1998). The constraint programming language CHIP, *Proceedings on the International Conference on Fifth Generation Computer Systems FGCS-88*, Tokyo.

Ding-Mei and R.W.H. Sargent (1992). *A Combined SQP and Branch and Bound Algorithm for MINLP Optimization*, Internal Report, Centre for Process Systems Engineering, Imperial College, London.

Duran, M.A. and I.E. Grossmann (1986a). An outer-approximation algorithm for a class of mixed-integer nonlinear programs. *Math. Prog.* **36**, 307.

Duran, M.A. and I.E. Grossmann (1986b). Simultaneous optimization and heat integration of chemical processes. *AIChE J.*, **32**, 592.

Fletcher, R. and S. Leyffer (1994). Solving mixed integer nonlinear programs by outer approximation. *Math Prog.*, **66**, 327.

Flippo, O.E. and A.H.G. Rinnoy Kan (1993). Decomposition in general mathematical programming. *Math. Prog.*, **60**, 361-382.

Fourer, R. (1998). Extending a general-purpose algebraic modeling language to combintorial optimization: A logic programming approach, in D. L. Woodruff, ed., *Advances in Computational and Stochastic Optimization, Logic Programming, and Heuristic Search,* Kluwer, 31-37.

Friedler, F., K., J.B. Varga, E. Feher and L.T. Fan (1996). Combinatorially accelerated branch and bound method for solving the MILP model of process network synthesis. in *Nonconvex Optimization and its Applications*, Eds. C.A. Floudas and P.M. Pardalos, 609-626, Kluwer Academic Publishers, Nowell, MA.

Friedler, F., Tarjan, K., Huang, Y.W., L.T. Fan (1993). Graph-theoretic approach to process synthesis: Polynomial algorithm for maximal structure generation. *Comp. Chem. Eng.*, **17**, 929.

Friedler, F., J.B. Verga, L. T. Fan (1995). Algorithmic approach to the integration of total flowsheet synthesis and waste minimization.. *Chem. Eng.. Sci.*, **50**, 1218.

GAMS (1999). *DICOPT MINLP Solver*, web site at http://www.gams.com/solvers/diopt/main.htm.

Geoffrion, A. M. (1972). Generalized Bender's decomposition. *J. Optimization Theory and Applications*, **10**(4), 237-260.

Grossmann, I.E. (1990). Mixed-integer nonlinear programming techniques for the synthesis of engineering systems. *Res. Eng. Des.*, **1**, 205.

Grossmann, I.E. (1996a). Mixed-integer optimization techniques for algorithmic process synthesis. *Adv. Chem. Eng.,*, **23**, *Process Synthesis*, 171-246.

Grossmann, I.E. (ed.) (1996b). *Global Optimization in Engineering Design*. Kluwer, Dordrecht.

Grossmann, I.E., J.A. Caballero and H. Yeomans (1999). Advances in mathematical programming for automated design, integration and operation of chemical processes. *Proceedings of the International Conference on Process Integration*, Copenhagen.

Grossmann, I.E. and M.M. Daichendt (1996). New trends in optimization-based approaches for process synthesis. *Comp. Chem. Eng.*, **20**, 665-683.

Grossmann, I.E. and Z. Kravanja (1997). Mixed-integer nonlinear programming:: A survey of algorithms and applications. The IMA Volumes in Mathematics and its Applications, **93**, *Large-Scale Optimization with Applications. Part II: Optimal Design and Control* (eds, Biegler, Coleman, Conn, Santosa) 73-100, Springer Verlag.

Grossmann, I.E., H. Yeomans and Z. Kravanja (1998). A rigorous disjunctive optimization model for simultaneous flowsheet optimization and heat integration. *Comp. Chem. Eng.*, **22**, S157-S164.

Gupta, O. K. and V. Ravindran, V. (1985). Branch and bound experiments in convex nonlinear integer programming. *Management Sci.*, **31**(12), 1533-1546.

Heintze, N., S. Michaylov and P. Stuckey (1992). CLP(R) and some electrical engineering problems. *J. Automated Reasoning*, **9**, 231-260.

Hooker, J.N. and M.A. Osorio. Mixed logical linear programming. to appear in *Discrete Applied Math.*, available at http://ba.gsia.cmu.edu/jnh

Hooker, J. N., Hak-Jin Kim and G. Ottosson (1998). A declarative modeling framework that integrates solution methods, to appear in *Annals of Ops. Res.* special issue on Modeling Languages and Approaches.

Hooker, J. N. (1994). Logic-based methods for optimization. in Borning et al., eds., *Principles and Practice of Constraint Programming, Lecture Notes in Computer Science*, **874**, 336-349.

Hooker, J. N., G. Ottosson, E. S. Thorsteinsson and H.J. Kim (1999). On integrating constraint propagation and linear programming for combinatorial optimization, *AAAI99 Proceedings*, to appear.

Hooker, J. N., and M. A. Osorio (1997). Mixed logical/linear programming, to appear in *Discrete Applied Math.*

Horst, R. and P.M. Pardalos (eds.) (1995). *Handbook of Global Optimization*, Kluwer.

Kelley Jr., J.E. (1960). The cutting-plane method for solving convex programs. *J. SIAM*, **8**, 703-712.

Kocis, G.R. and I.E. Grossmann (1987). Relaxation strategy for the structural optimization of process flowsheets. *Ind. Eng. Chem. Res.*, **26**, 1869.

Kondili E,. Pantelides C.C. and Sargent R.W.H (1993). A general algorithm for short-term scheduling of batch operations: I. MILP formulation. *Comp Chem.Eng.*, **17**(2), 211-227.

Lakmazaheri, S., and W. Rasdorf (1989). Constraint logic programming for the analysis and partial synthesis of truss structures. *Art. Intel. Eng. Des. Anal. Manuf.* **3** 157-173.

Lee, S. and I.E. Grossmann (1999). Nonlinear convex hull reformulation and algorithm for generalized disjunctive programming. submitted for publication.

Leyffer, S. (1993) *Deterministic Methods for Mixed-Integer Nonlinear Programming*. Ph.D. thesis, Department of Mathematics and Computer Science, University of Dundee, Dundee.

Leyffer, S. (1998). Integrating SQP and branch and bound for mixed Integer nonlinear programming. submitted for publication.

Magnanti, T. L. and R. T. Wong (1981). Acclerated benders decomposition: algorithm enhancement and model selection criteria. *Operations Research*, **29**, 464-484.

Marriott, K., and P. J. Stuckey (1998). *Programming with Constraints: An Introduction*, MIT Press, Cambridge.

Mawengkwang, H. and B.A. Murtagh (1986). Solving monlinear integer programs with large-scale optimization software. *Annals of Operations Research*, **5**, 425-437.

Nabar, S. and L. Schrage, L. (1991). Modeling and solving nonlinear integer programming problems. AIChE meeting, Chicago.

Nemhauser, G. L. and L. A. Wolsey (1988). *Integer and Combinatorial Optimization*. Wiley-Interscience, New York.

OSL (1999). The IBM Optimization Subroutine Library, web site at http://www.research.ibm.com/osl, 1999.

Papoulias, S.A. and I.E. Grossmann (1983). A structural optimization approach in process synthesis.: Part II: heat recovery networks. *Comp. Chem. Eng.*, **7**, 707.

Paules, G. E. and C. A. Floudas (1989). APROS: Algorithmic development methodology for discrete-continuous optimization problems. *Ops. Res.*, **37**, 902-915.

Puget, J.-F. (1994). A C++ implementation of CLP, in *Proceedings of SPICIS'94*, Singapore.

Quesada, I. and I.E. Grossmann (1992). An LP/NLP based branch and bound algorithm for convex MINLP optimization problems. *Comp. Chem. Eng.*, **16**, 937-947.

Raman, R. and I.E. Grossmann (1991). Relation between MILP modelling and logical inference for chemical process synthesis. *Comp. Chem. Eng.*, **15**, 73.

Raman, R. and I.E. Grossmann (1993). Symbolic integration of logic in mixed integer linear programming techniques for process synthesis. *Comp. Chem. Eng.*, **17**, 909.

Raman, R. and I.E. Grossmann (1994). Modelling and computational techniques for logic based integer programming. *Comp. Chem. Eng.*, **18**, 563.

Régin, J.-C. (1993). *A Filtering Algorithm for Constraints of Difference in CSPs*, Research Report No. LIRMM 93-068, Laboratoire d'Informatique, de Robotique et de Microélectronique de Montpellier, Université Montpellier, France.

Rodosek, R., M. Wallace and M. Hajian (1999). A new approach to integrating mixed integer programming and constraint logic programming, *Ann. Ops. Res.*, **86**, 63-87.

Schweiger, C. A., and C. A. Floudas (1999). MINOPT, A modeling language and algorithmic framework for linear, mixed-integer, nonlinear, dynamic, and mixed-integer nonlinear optimization. web site at http://titan.princeton.edu/MINOPT.

Simonis, H., and M. Dincbas. (1987). Using an extended prolog for digital circuit design, in *IEEE International Workshop on AI Applications to CAD Systems for Electronics*, Munich, Germany, 165-188.

Smith, E.M. and C.C. Pantelides (1995). Design of reaction/separation networks using detailed models. *Comp. Chem. Eng.*, **19**, S83.

Sthanusubramonian, T. (1991). *A Transformational Approach to Configuration Design*. Master's thesis, Engineering Design Research Center, Carnegie Mellon University.

Stubbs, R.A. and S.Mehrotra (1996). A branch and cut method for 0-1 mixed convex programming. INFORMS meeting, Washington.

Subramanian, D., and C. S. Wang (1993). Kinematic synthesis with configuration spaces, in D. Weld, ed., *Proc. Qualitative Reasoning,* 228-239.

Tsang, E. (1993). *Foundations of Constraint Satisfaction,* Academic Press.

Turkay, M. and I.E. Grossmann (1996) A logic based outer-approximation algorithm for MINLP optimization of process flowsheets. *Comp. Chem. Eng.,* **20**, 959-978.

Turkay, M. and I.E. Grossmann (1998). Structural flowsheet optimization with complex investment cost functions. *Comp. Chem. Eng.,* **22**, 673-686.

Van Hentenryck, P. (1989). *Constraint Satisfaction in Logic Programming*. MIT Press (Cambridge.

Van Hentenryck, P., and I. Lustig. *The OPL Optimization Programming Language.* MIT Press, Cambridge.

Vecchietti, A. and I.E. Grossmann (1997). LOGMIP: A discrete continuous nonlinear optimizer. *Comp. Chem. Eng.,* **21**, S427-S432.

Viswanathan, J. and I.E. Grossmann (1990). A combined penalty function and outer-approximation method for MINLP ptimization. *Comp. Chem. Eng.,* **14**, 769.

Westerlund, T. and F. Pettersson (1995). A cutting plane method for solving convex MINLP problems. *Comp. Chem. Eng.,* **19**, S131-S136.

Williams, H.P. (1999) *Mathematical Building in Mathematical Programming.* John Wiley, Chichester.

XPRESS-MP (1999). Dash Optimisation Company, web site at http://www.dash.co.uk.

Yee, T.F. and I.E. Grossmann (1990). Simultaneous optimization models for heat integration, II: Heat exchanger network synthesis. *Comp. Chem. Eng.,* **14**, 1165.

Yeomans, H. and I.E. Grossmann (1999a). A systematic modeling framework of superstructure optimization in process synthesis. *Comp. Chem. Eng.,* **23**, 709-731.

Yeomans, H. and I.E. Grossmann (199b). Disjunctive programming models for the synthesis of heat integrated distillation sequences. To appear in *Comp. Chem. Eng.*

Yeomans, H. and I.E. Grossmann (1999c). Nonlinear disjunctive programming models for the optimal design of distillation columns and separation sequences. submitted for publication.

Yuan, X., S. Zhang, L. Piboleau and S. Domenech (1988). Une methode d'optimisation nonlineare en variables mixtes pour la conception de procedes. *RAIRO,* **22**, 331.

RECENT DEVELOPMENTS IN DETERMINISTIC GLOBAL OPTIMIZATION AND THEIR RELEVANCE TO PROCESS DESIGN

Christodoulos A. Floudas
Princeton University
Princeton, NJ 08544-5263

Panos M. Pardalos
University of Florida
Gainesville, FL 32611

Abstract

This paper presents an overview of the recent advances in global optimization and their relevance to process design applications. The review focuses on (i) twice differentiable nonlinear optimization problems, (ii) mixed-integer nonlinear optimization problems, (iii) discrete optimization problems, and (iv) the enclosure of all solutions of constrained nonlinear equations. Current and future research challenges are outlined.

Keywords

Deterministic global optimization, Integer programming, All solutions, Noncovexities, Process design, Phase equilibrium, Parameter estimation, Data reconciliation; Azeotropes.

Introduction

Global optimization addresses the computation and characterization of global optima (i.e., minima and maxima) of nonconvex functions constrained in a specified domain. Given an objective function f that is to be minimized and a set of equality and inequality constraints S, *Deterministic Global Optimization* focuses on the following important issues:

1. Determine a global minimum of the objective function f (i.e., f has the lowest possible value in S subject to the set of constraints S.
2. Determine *lower* and *upper bounds* on the global minimum of the objective function f on S that are valid for the whole feasible region S.
3. Enclose all solutions of the set of equality and inequality constraints S.

The nonconvex mathematical models that we review in this paper from the deterministic global optimization point of view belong to three classes. The first class takes the form:

$$\min_{\mathbf{x}} f(\mathbf{x})$$
$$\text{s.t. } \mathbf{h}(\mathbf{x}) = \mathbf{0}$$
$$\mathbf{g}(\mathbf{x}) \leq \mathbf{0} \quad (1)$$
$$\mathbf{x} \in \mathbf{X} \subseteq \Re^n$$

where f, $\mathbf{h}$ and $\mathbf{g}$ are the objective function, and the vectors of the equality and inequality constraints, respectively, and $\mathbf{x}$ is a vector of continuous variables of size $\mathbf{n}$.

Within this first class of global optimization models (1) we have (i) the biconvex and/or bilinear mathematical structure; (ii) the signomial mathematical structure, and (iii) the general mathematical structure of f, $\mathbf{h}$ and $\mathbf{g}$ belonging to C^2, the set of twice continuously differentiable functions.

The second class corresponds to mixed-integer nonlinear MINLP optimization problems that take the form :

$$\min_{x,y} f(x,y)$$
$$\text{s.t.} \quad h(x,y) = 0$$
$$g(x,y) \leq 0 \quad (2)$$
$$x \in X \subseteq \Re^n$$
$$y \in Y \text{ integer,}$$

where x represents a vector of n continuous variables and y is a vector of integer variables. Within this class we also have discrete optimization problems.

The third class of mathematical models focuses on enclosing all solutions of a nonlinear system of equations subject to inequality constraints and variable bounds :

$$h_j(x) = 0, j \in N_E$$
$$g_k(x) \leq 0, k \in N_I \quad (3)$$
$$x^L \leq x \leq x^U,$$

where N_E is the set of equalities, N_I the set of inequality constraints, and x the vector of n continuous variables.

Chemical Engineering Contributions: 1994-99

In the last decade we have experienced a dramatic growth of interest in Chemical Engineering for new methods of global optimization and their application to important process design and synthesis problems. In this section, we discuss the chemical engineering contributions during the last five years. The books of Floudas and Pardalos (1996) and Grossmann (1996) and the review papers Floudas (1997), (1999a), (1999b), and Floudas et al (1999b) discuss a number of recent chemical engineering and computational chemistry contributions. A detailed account of the recent approaches and contributions can be found in the forthcoming book by Floudas (2000).

Maranas and Floudas (1994a), (1994b) proposed a novel branch and bound method combined with a difference of convex functions transformation for the global optimization of molecular conformation problems that arise in computational chemistry. Vaidyanathan and El-Halwagi (1994) proposed an interval analysis based global optimization method. Quesada and Grossmann (1995) combined convex underestimators in a branch and bound framework for linear fractional and bilinear programs. McDonald and Floudas (1995a), (1995b), (1995c) addressed the fundamental problems of (i) minimization of the Gibbs free energy and (ii) the tangent plane stability criterion that arise in phase and chemical reaction equilibrium as global optimization problems for the first time. They proposed decomposition based approaches for biconvex problems that result from the use of the NRTL equation, and branch and bound approaches for the UNIQUAC, UNIFAC, ASOG, and TK-Wilson equations. Psarris and Floudas (1995) proposed a global optimization approach based on the GOP for the robust stability of control systems under real parametric uncertainty. Ryoo and Sahinidis (1995) suggested the application of reduction tests within the framework of branch and bound methods and they proposed a branch and reduce global optimization approach (Ryoo and Sahinidis (1996). Dorneigh and Sahinidis (1995) proposed a global optimization approach for the chip layout and compaction problem Androulakis et al. (1995) proposed the global optimization method αBB that addresses general continuous optimization problems with nonconvexities in the objective function and/or constraints.

This approach classifies the nonconvexities as special structure (e.g., bilinear, signomial, univariate) or generic structure and is based on convex relaxations and a branch and bound framework. Maranas and Floudas (1995) proposed a new approach for enclosing all ε-feasible solutions of nonlinearly constrained systems of equations. This approach transforms the problem into a min-max form and corresponds to enclosing all multiple global optima via the αBB global optimization approach.

Adjiman and Floudas (1996) proposed novel approaches for the rigorous determination of the α parameters that are employed in the αBB global optimization approach. These methods are based on a modified Kharitonov theorem for interval polynomials, interval analysis of the Hessian matrices and on calculations of rigorous bounds on the minimum eigenvalue for general twice differentiable problems. Staus et al. (1996) formulated the combined adaptive controller design and estimation problem as a nonconvex problem with convex objective and bilinear constraints, and proposed a branch and bound global optimization method which is based on the McCormick underestimators. Visweswaran et al. (1996) addressed bilevel linear and quadratic programming problems to global optimality by employing the basic principles of the GOP and developing additional theoretical properties that exploit the underlying mathematical structure of such problems.

Androulakis et al (1996) developed a distributed version of the GOP, and discussed the key theoretical and implementation issues along with extensive computational results on large scale indefinite quadratic and pooling/blending problems. Shectman and Sahinidis (1996) proposed a finite algorithm for separable concave programs, discussed the design of such branch and bound approaches, and presented computational results employing domain reduction tests. McKinnon et al. (1996) addressed the global optimization in phase and chemical reaction equilibrium using interval analysis coupled with the tangent plane stability criterion of Gibbs. Hua et al. (1996) applied an interval analysis method for the phase stability computations of binary and ternary mixtures using activity coefficient models. Mockus and

Reklaitis (1996) proposed a continuous formulation for the short term batch scheduling problem, and developed a global optimization approach based on the Bayesian heuristic.

Lucia and Xu (1996) studied nonconvexity issues in sparse successive quadratic programming, and Banga and Seider (1996) introduced a stochastic global optimization approach which they applied to the optimal design of a fermentation process, phase equilibrium problems and optimal control problems. Epperly and Swaney (1996a), (1996b) proposed a branch and bound method with a new linear programming underestimating problem, and provided extensive computational studies. This approach is applicable to NLPs in factorable form which includes quadratic objective and constraints, and twice differentiable transcendental functions. Visweswaran and Floudas (1996a) proposed new formulations for the GOP algorithm which are based on a branch and bound framework. These allow for the implicit solution of the relaxed dual problems, which are formulated in a single MILP model, and feature a linear branching scheme.

Visweswaran and Floudas (1996b) discussed the implementation issues of the GOP and provide extensive computational experience on a variety of chemical engineering problems. Byrne and Bogle (1996a), (1996b) and Vaidyanathan and Elhalwagi (1996) proposed global optimization methods that are based on interval analysis for constrained NLPs and MINLPs respectively. Liu et al (96) proposed an approach for planning of chemical process networks which is based on global concave minimization. The approach is based on their earlier work on finite global optimization approaches and their computational studies revealed the efficiency of the proposed approach. Ierapetritou and Pistikopoulos (1996) studied the global optimization of stochastic planning, scheduling and design problems, applied the decomposition-based approach, GOP, and demonstrated that significant reductions in the number of relaxed dual problems can be achieved by exploiting the mathematical structure further. Iyer and Grossmann (1996) extended the global optimization approach of Quesada and Grossmann (1993) to the multiperiod heat exchanger networks that feature fixed configuration, linear cost functions, arithmetic mean driving forces, and isothermal mixing. Quesada and Grossmann (1996) studied further the use of alternative bounding approximations and applied them to a variety of engineering design problems that include structural design, batch processes, layout design, and portfolio problems. Smith and Pantelides (1996) proposed a symbolic manipulation algorithm for the automatic reformulation of algebraic constraints and introduced a spatial type branch and bound approach within the gPROMS framework.

Maranas and Floudas (1997) proposed a global optimization approach for generalized geometric programming problems that have many applications in robust control and engineering design problems. McDonald and Floudas (1997) proposed the combination of the Gibbs free energy minimization problem with the stability criterion problem, developed a special purpose program GLOPEQ, and performed an extensive computational study on difficult phase equilibrium problems. Harding et al. (1997) proposed novel formulations and global optimization methods for the enclosure of all homogeneous azeotropes. This approach offers a theoretical guarantee of enclosing all homogeneous azeotopes and it was applied to a variety of activity coefficient models. Androulakis et al (1997a) proposed a novel global optimization approach based on atomistic level modeling and the αBB principles for the structure prediction of oligopeptides. Androulakis et al (1997b) proposed a decomposition based global optimization approach for the peptide docking problem of amino acids into the binding sites of the HLA-DR1 protein.

Harding and Floudas (1997) proposed novel global optimization approaches for the batch design under uncertainty of multiproduct and multipurpose plants. These approaches are based on the exact calculation of the α parameters within the αBB global optimization framework and they result in the tightest possible underestimators. Epperly et al (1997) proposed a global optimization approach for stochastic batch design problems that is based on a reduction of the number of variables needed for branching as suggested in their reduced space branch and bound approach Epperly and Pistikopoulos (1997). Sriniwas and Arkun (1997) introduced a global optimization approach based on the GOP principles for the nonlinear model predictive control problems with polynomial ARX models. Vanantwerp et al (1997) introduced a global optimization approach based on branch and bound for the design of a robust controller to time-varying nonlinear plant perturbations. Smith and Pantelides (1997) studied further their reformulation/spatial branch and bound approach for the global optimization of MINLP problems.

Klepeis et al. (1998) and Klepeis and Floudas (1999a) proposed new global optimization approaches for the structure prediction of solvated peptides using area and volume accessible to the solvent models. A review of the global optimization activities in the areas of protein folding and peptide docking can be found in Floudas et al. (199b). Esposito and Floudas (1998) proposed the first global optimization method for the parameter estimation and data reconciliation of nonlinear algebraic models using the principles of the αBB approach. Mckinnonn and Mongeau (1998) proposed a generic global optimization approach for the phase and chemical reaction equilibrium problem that is based on interval analysis and combines the stability criterion with the minimization of the Gibbs free energy. Hua et al. (1998)applied an interval analysis method for the phase stability computations of binary and ternary mixtures using equation of state models. Maier and et al. (1998) applied an interval analysis based approach for the enclosure of homogeneous azeotropes. They employed the formulations proposed by Harding and et al (1997) and studied systems with activity coefficient and equation of state models. Meyer and Swartz (1998) proposed a new approach for testing convexity for phase equilibrium problems that can be coupled with the global optimization methods of McDonald and Floudas (1994), (1995a), (1995b), (1995c), (1997). Shectman and Sahinidis (1998) proposed a finite global optimization method for separable concave problems.

Yamada and Hara (1998) proposed a global optimization approach based on the triangle covering for H-infinity control with constant diagonal scaling. Androulakis and Floudas (1998) studied the parallel computation issues that arise using the αBB global optimization approach. Adjiman et al (1998a) proposed several new rigorous methods for the calculation of the α parameters for (i) uniform diagonal shift of the Hessian matrix and (ii) non-uniform diagonal shift of the Hessian matrix, and they established their potential trade-offs. Adjiman *et al.* (1998b) presented the detailed implementation of the αBB approach and computational studies in process design problems such as heat exchanger networks, reactor-separator networks, and batch design under uncertainty. Westerlund et al. (1998) proposed an extended cutting plane approach for the global optimization of pseudoconvex MINLP problems. Zamora and Grossmann (1998a) proposed a hybrid branch-and-bound and outer approximation method for the global optimization of heat exchanger networks with no stream splits.

Zamora and Grossmann (1998b) introduced a deterministic branch and bound approach for structured process systems that have univariate concave, bilinear and linear fractional terms. They proposed several properties of the contraction operation, embedded them in the global optimization algorithm and studied the contraction effects on several applications. Adjiman *et al.* (1998d) proposed two novel global optimization approaches for nonconvex mixed-integer nonlinear programming problems. The first approach is for separable continuous and integer domains while the second approach addresses general mixed-integer nonlinear problems. Klepeis and Floudas (1999b) proposed a novel deterministic global optimization approach for free energy calculations of peptides. Westerberg and Floudas (1999) introduced a global optimization framework for the enclosure of all transition states of potential energy hypersurfaces, and studied the reaction pathways and dynamics of protein folding. Floudas *et al.* (1999) introduced a handbook of global optimization test problems that cover a wide spectrum of engineering and applied science applications.

In this paper, we will focus on the deterministic global optimization method denoted as αBB, and present the key contributions in addressing (a) twice-differentiable NLPs, (b) mixed-integer nonlinear problems, (c) discrete optimization problems, and (d) the enclosure of all solutions of nonlinear systems of equations.

The αBB Aproach for General NLPs

The αBB algorithm is based on novel underestimation schemes and a branch-and-bound framework and addresses nonconvex minimization problems of the formulation (1). The theoretical properties of the algorithm guarantee that such a problem can be solved to global optimality with finite ε-convergence.

Each iteration of the algorithm consists of a *branching* step and a *bounding* step. In the *step*, a lower bound is obtained by constructing valid convex underestimators for the functions in the problem and solving the resulting convex NLP to global optimality. An upper bound is calculated either by solving the original problem locally over each subdomain of the solution space or by performing a problem evaluation at the solution of the lower bounding problem. The identification of the global optimal solution depends on the validity of the lower bounding problems as well as the construction of increasingly tight lower bounding problems for successive partitions of the solution space. Such properties lead to the generation of a *nondecreasing sequence* of lower bounds which progresses towards the optimal solution.

An important step in the convexification strategy is the decomposition of each nonlinear function into a sum of terms belonging to one of several categories: linear, bilinear, trilinear, fractional, fractional trilinear, convex, univariate concave or general nonconvex. It is also possible to construct customized underestimators for other mathematical structures such as signomial expressions, as shown by Maranas and Floudas (1997). In constructing a convex underestimator for the overall function, it is first noted that the linear and convex terms do not require any transformation. The convex envelope of the bilinear, fractional, and univariate concave terms can be constructed by following simple rules. For a more detailed exposition the reader is directed to Maranas and Floudas (1995,1997) and Adjiman et al. (1998a,b).

Bilinear Terms: In the case of a bilinear term xy, Al-Khayyal and Falk (1983) showed that the tightest convex lower bound over the domain $[x^L, x^U] \times [y^L, y^U]$ is obtained by introducing a new variable w_B which replaces every occurrence of xy in the problem and satisfies:

$$w_B \geq x^L y + y^L x - x^L y^L, \\ w_B \geq x^U y + y^U x - x^U y^U. \quad (4)$$

An upper bound can be imposed on w to construct a better approximation of the original problem (McCormick, 1976). This is achieved through the addition of two linear constraints:

$$w_B \leq x^U y + y^L x - x^U y^L,$$
$$w_B \leq x^L y + y^U x - x^L y^U. \quad (5)$$

Fractional Terms: Fractional terms of the form x/y are underestimated by introducing a new variable w_F and two new constraints (Maranas and Floudas, 1995) which depend on the sign of the bounds on x.

$$w_F \geq \begin{cases} x^L/y + x/y^U - x^L/y^U & \text{if } x^L \geq 0 \\ x/y^U - x^L y/y^L y^U + x^L/y^L & \text{if } x^L < 0 \end{cases}$$

$$w_F \geq \begin{cases} x^L/y + x/y^U - x^L/y^U & \text{if } x^U \geq 0 \\ x/y^L - x^U y/y^L y^U + x^U/y^U & \text{if } x^U < 0 \end{cases} \quad (6)$$

Univariate Concave Terms: Univariate concave functions are underestimated by their linearization at the lower bound of the variable range. Thus the convex envelope of the concave function $ut(x)$ over $[x^L, x^U]$ is the linear function of x:

$$ut(x^L) + \frac{ut(x^U) - ut(x^L)}{x^U - x^L}(x - x^L). \quad (7)$$

General Nonconvex Terms: For the most general nonconvexities, a slightly modified version of the underestimator proposed by Maranas and Floudas (1994b) is used. A function $f(x) \in C^2(\Re^n)$ is underestimated over the entire domain $[x^L, x^U]$ by the function $L(x)$ defined as

$$L(\mathbf{x}) = f(\mathbf{x}) + \sum_{i=1}^{n} \alpha_i (x_i^L - \mathbf{x})(x_i^U - \mathbf{x}) \quad (8)$$

where the α_i's are positive scalars. Since the summation term in Equation (8) is negative over the entire region $[x^L, x^U]$, $L(\mathbf{x})$ is a guaranteed underestimator of $f(x)$. Furthermore, since the quadratic term is convex, all nonconvexities in the original function $f(x)$ can be overpowered given sufficiently large values of the α_i parameters: $L(x)$ is therefore a *valid convex underestimator*. The following theorem can then be used to ensure that $L(x)$ is indeed a convex underestimator:

Theorem 1: $L(x)$, as defined in Equation (8), is convex if and only if $H_f(x) + 2 = H_f(x) + 2 \text{ diag}(\alpha_i)$ is positive semi-definite for all $\mathbf{x} \in [x^L, x^U]$.

For general twice-differentiable functions, the elements of the Hessian matrix $H_f(x)$ are likely to be nonlinear functions of the variables. The difficulty arising from the presence of the variables in the convexity condition can be overcome through the transformation of the exact x-dependent Hessian matrix to an interval matrix $[H_f]$ such that $H_f(x) \subseteq [H_f], \forall x \in [x^L, x^U]$. The elements of the original Hessian matrix are treated as independent when calculating their natural interval extensions (Neumaier, 1990). The interval Hessian matrix family is then used to formulate a theorem in which the α calculation problem is relaxed.

Theorem 2: Consider a general function f(**x**) with continuous second-order derivatives and its Hessian matrix $H_f(\mathbf{x})$. Let $L(\mathbf{x})$ be defined by Equation (8). Let $[H_f]$ be a real symmetric interval matrix such that $H_f(x) \subseteq [H_f], \forall x \in [x^L, x^U]$.
If the matrix $[H_L]$ defined by $[H_L] = [H_f] + 2\Delta = [H_f] + 2 \text{ diag}(\alpha_i)$ is positive semi-definite, then $L(\mathbf{x})$ is convex over the domain $[\mathbf{x}^L, \mathbf{x}^U]$.

A rigorous method for the calculation of the α parameters which is very effective is the Scaled Gerschgorin Theorem presented below. This method is based on the non-uniform diagonal shift matrix and the details of its proof as well as several other methods are in Adjiman et al. (1998a).

Theorem 3: For any vector **d** > 0 and a symmetric interval matrix [A], define the vector α as

$$\alpha_i = \max\left\{0, -\frac{1}{2}\left(\underline{a}_{ii} - \sum_{j \neq i}|a|_{ij}\frac{d_j}{d_i}\right)\right\}$$

where $|a|_{ij} = \max\left\{|\underline{a}_{ij}|, |\overline{a}_{ij}|\right\}$.

Then, for all $A \in [A]$ the matrix $A_L = A + 2\Delta$ with $\Delta = \text{diag}(\alpha_i)$ is positive semi-definite.

Overall Valid Convex Underestimator

A convex underestimator for any given twice-differentiable function can now be obtained through a decomposition approach. A function $f(x)$ with continuous second-order derivatives can be written as:

$$f(x) = LT(x) + CT(x) + \sum_{i=1}^{bt} b_i x_{B_{i,1}} x_{B_{i,2}}$$
$$+ \sum_{i=1}^{ft} f_i \frac{x_{F_{i,1}}}{x_{F_{i,2}}} + \sum_{i=1}^{ut} UT_i(x) + \sum_{i=1}^{nt} NT_i(x)$$

where $LT(x)$ is a linear term; $CT(x)$ is a convex term; bt is the number of bilinear terms, $x_{B_{i,1}}$ and $x_{B_{i,2}}$ denote the two variables that participate in the ith bilinear term and b_i is its coefficient; ft is the number of fractional terms, $x_{F,1}$ and x_{F2} denote the two variables that participate in the ith fractional term and f is its coefficient; ut is the number of univariate concave terms, $UT_i(x^i)$ is the ith univariate concave term, x^i denotes the variable that participates in UT_i; nt is the number of general nonconvex terms, $NT_i(x)$ is the ith general nonconvex term.

The corresponding lower bounding function is

$$L(x,w) = LT(x) + CT(x)$$
$$+ \sum_{i=1}^{bt} b_i w_{Bi} + \sum_{i1}^{ft} f_i w_{Fi}$$
$$+ \sum_{i=1}^{ut} \left(UT_i(x^{i,L}) + \frac{UT_i(x^{i,U}) - UT_i(x^{i,L})}{x^{i,U} - x^{i,L}} (x - x^{i,L}) \right) \quad (9)$$
$$+ \sum_{i=1}^{nt} \left(NT_i(x) + \sum_{j=1}^{n} \alpha_{ij} (x_j^L - x_j)(x_j^U - x_j) \right)$$

where α_{ij} corresponds to term i and variable j. The w_{Bi} variables are defined by Equations (4) and (5). The w_{Fi} variables must satisfy constraints of the forms given by (6).

Mixed-Integer Nonlinear Problems

A wide range of chemical engineering problems can effectively be framed as Mixed—Integer Nonlinear Problems (MINLP) as this approach allows the simultaneous optimization of the continuous variables pertaining to a certain structure, and of the structure itself which is modeled via binary variables (Floudas, 1995; Grossmann 1990,1996a). Such a mathematical framework has been proposed for a variety of process synthesis problems (e.g., heat recovery networks, separation systems, reactor networks), process operations problems (e.g., scheduling and design of batch processes), molecular design problems and synthesis of metabolic pathways. A number of these applications are described in Floudas (1995), Grossmann (1996b) and Floudas (2000). The degree of nonconvexity of the participating functions is generally arbitrary and nonlinearities can be identified in the continuous, the integer, or joint domains. The difficulties in solving these MINLPs therefore stem not only from the combinatorial characteristics of the problem which are a direct result of the presence of the integer variables, but also from the presence of nonconvexities Floudas and Grossmann (1995).

A number of global MINLP algorithms that have been developed to address different types of nonconvex MINLPs that belong to (2). These include: the Branch-and-Reduce Approach, (Ryoo and Sahinidis, 1995); the Interval Analysis based Approach, Vaidyanathan and El Holwagi (1996); the Extended Cutting Plane Approach, (Westerlund et al., 1994; Westerlund et al., 1998); the Reformulation/Spatial Branch-and-Bound Approach, (Smith and Pantelides 1997); the Hybrid Branch-and-Bound and Outer Approximation Approach, (Zamora and Grossmann (1998a); the SMIN-αBB Approach, (Adjiman et al 1997c; Adjiman et al., 1998d); and the GMIN-αBB Approach, (Adjiman et al., 1997c; Adjiman et al 1998d).

In the sequel, we will briefly discuss the GMIN-αBB approach.

The GMIN-αBB Algorithm. The GMIN-αBB algorithm is a global optimization algorithm for general MINLPs of the form (2).

The key idea of the GMIN-αBB algorithm is to embed the αBB algorithm within a branch and bound framework which handles the binary variables. At each node of the branch and bound tree a continuous relaxation of the original problem is being solved, with some of the binary variables fixed to 0 or 1, according to the branching rules that are discussed in a subsequent section. The most important consequence of this approach is that the continuous relaxation at each node is a non-convex NLP whose global optimum solution can provide a guaranteed lower bound to the MINLP problem. It is therefore crucial to be able to solve each branch and bound node efficiently to global optimality. Note that any lower bound on the global solution of these non-convex NLPs is a valid lower bound for the global solution of the original MINLP problem. The αBB algorithm is employed so as to provide valid lower bounds for the outer branch and bound algorithm. A detailed description of the three major components of the GMIN-αBB approach, namely the lower bound generation, the selection of branching variables, and the selection of branching nodes can be found in Adjiman et al. (1997), (1998d).

Discrete Optimization Problems

Discrete (or combinatorial) optimization problems, that is, problems with a discrete feasible domain and/or a discrete domain objective function, model a large spectrum of applications in computer science, operations research and engineering.

Solution methods for discrete optimization problems can be classified into combinatorial and continuous approaches. A typical combinatorial approach generates a sequence of states, which represent a partial solution, drawn from a discrete finite set. Continuous approaches for solving discrete optimization problems are based on different equivalent characterizations in a continuous space. These characterizations include equivalent continuous formulations, or continuous relaxations (including semidefinite programming), that is, embeddings of the discrete domain in a larger continuous space.

There are many ways to formulate discrete problems as equivalent continuous problems or to embed the discrete feasible domain in a larger continuous space (relaxation). The surprising variety of continuous approaches reveal interesting theoretical properties which can be explored to develop new algorithms for computing (sub)optimal solutions to discrete optimization problems.

Mixed Integer Programming and LCP

The simplest nonconvex constraints are the 0 − 1 integer constraints. Integer constraints are equivalent to continuous nonconvex constraints. Another example would be as followed: $z \in \{0,1\} \Leftrightarrow z + w = 1, z \geq 0, w \geq 0, zw = 0$ or in another approach $z \in \{0,1\} \Leftrightarrow z - z^2 = z(1-z) = 0$. Therefore, it seems that there is no significant difference between discrete and continuous optimization. However, there is a considerable difference (in terms of problem complexity) between convex and nonconvex optimization problems. Next we show that the mixed integer feasibility problem is equivalent to the complementarity problem. The complementarity conditions which are present in optimality conditions reveal deep connections with discrete optimization Horst *et al* (1995), Horst and Pardalos (1995).

We consider the general linear complementarity problem (LCP) of finding a vector $x \in R^n$ such that $Mx + q \geq 0, x \geq 0, x^T M x + q^T x = 0$ (or proving that such an x does not exist) where M is an $n \times n$ rational matrix and $q \in R^n$ is a rational vector. For given data M and q, the problem is generally denoted by LCP(M,q). The LCP unifies a number of important problems in operations research. In particular, it generalizes the primal-dual linear programming problem, convex quadratic programming, and bimatrix games.

For the general matrix M, where $S = \{x : M_x + q \geq 0, x \geq 0\}$ can be bounded or unbounded, the LCP can always be solved by solving a specific zero-one, linear, mixed-integer problem with n zero-one variables. Consider the following mixed zero-one integer problem (MIP):

$$\max_{\alpha, y, z} \alpha$$
$$\text{s.t. } 0 \leq My + \alpha q \leq e - z \quad (10)$$
$$\alpha \geq 0, 0 \leq y \leq z$$
$$z \in \{0,1\}^n$$

Theorem: Let (α^*, y^*, z^*) be any optimal solution. If $\alpha^* > 0$, then $x^* = y^*/\alpha^*$ solves the LCP. If in the optimal solution $\alpha^* = 0$, then the LCP has no solution.

In fact, every feasible point (α, y, z) of MIP, with $\alpha > 0$, corresponds to a solution of LCP. On the other hand, the mixed integer feasibility problem can be formulated as an LCP.

Given matrices $A_{n \times n}, B_{n \times 1}$ and a vector $b \in R^n$ with rational entries, the mixed integer feasibility problem is to find (x, z), such that $x \in R^n, x \geq 0, z \in \{0,1\}^l$ that satisfy $Ax + Bz = b$.

The condition $z_i \in \{0,1\}$ is equivalent to:

$$z_i + w_i = 1, z_i \geq 0, w_i \geq 0, z_i w_i = 0.$$

With this transformation z_i is a continuous variable and for each z_i a new continuous variable w_i is introduced. In addition, let $s, t \in R^n$ be such that

$$s = Ax + Bz - b \geq 0, \quad t = -Ax - Bz + b \geq 0.$$

The only way for these two inequalities can be satisfied is to have $s = t = 0$, which implies that $Ax + Bz = b$. Then, the mixed integer feasibility problem can be reduced to the problem of finding a solution of the LCP: Find v, y such that

$$v \geq 0, y \geq 0, v^T y = 0, v = My + q.$$

where

$$y = \begin{pmatrix} z \\ x \\ \theta \end{pmatrix}, v = \begin{pmatrix} w \\ s \\ t \end{pmatrix}, M = \begin{pmatrix} -I & 0 & 0 \\ B & A & 0 \\ -B & -A & 0 \end{pmatrix}, q = \begin{pmatrix} e \\ b \\ -b \end{pmatrix},$$

where $\theta \in R^n$ and $e \in R^l$ is the vector of all 1's.

Satisfiability Problems

The satisfiability problem (SAT) is central in mathematical logic, computing theory, and many industrial application problems. Problems in computer vision, VLSI design, databases, automated reasoning, computer-aided design and manufacturing, involve the solution of instances of the satisfiability problem. Furthermore, SAT is the basic problem in computational complexity. Developing efficient exact algorithms and heuristics for satisfiability problems can lead to general approaches for solving combinatorial optimization problems (Pardalos (1994), Pardalos (1996)).

Traditional methods treat the SAT problem as a constrained decision problem. In recent years, researchers have applied global optimization methods to solve the SAT problem. In this approach, universal satisfiability models, are formulated that transform a discrete SAT problem on Boolean space $\{0,1\}^m$ into a continuous SAT problem on real space R^m. Thus, this decision problem is transformed into a global optimization problem which can be solved by existing global optimization methods.

Let $C, C_2, \ldots, C_m$ be m clauses, involving n Boolean variables $x_1, x_2, \ldots, x_n$, which can take on only the values `true` or `false` (1 or 0). Define clause i to be

$$C_i = \bigvee_{j=1}^{n_i} l_{ij},$$

where the literals $l_{ij} \in \{x_i, \bar{x}_i \mid i = 1,...,n\}$.

In SAT (*CNF* formula) one is to determine the assignment of truth values to the *n* variables that satisfy all clauses $C_1, C_2,..., C_m$.

Continuous Unconstrained Formulations. There are many possible formulations in this category. A simple formulation, *UniSAT*, (Universal SAT Problem Models), suggests:

$$\min_{y \in E^n} f(\mathbf{y}) = \sum_{i=1}^{m} c_i(\mathbf{y})$$

where

$$c_i(\mathbf{y}) = \prod_{j=1}^{n} q_{i,j}(y_j)$$

$$q_{i,j}(y_i) = \begin{cases} (y_j - 1)^2 & \text{if } x_j \in C_i \\ (y_j + 1)^2 & \text{if } \bar{x}_j \in C_i \\ 1 & \text{otherwise} \end{cases}$$

where *T* and *F* are positive constants.

Two special formulations to $q_{i,j}(y_j)$ exist. In the *UniSAT5* model

$$q_{i,j}(y_i) = \begin{cases} (y_j - 1)^2 & \text{if } x_j \in C_i \\ (y_j + 1)^2 & \text{if } \bar{x}_j \in C_i \\ 1 & \text{otherwise} \end{cases}$$

UniSAT5 can be solved with efficient, discrete, greedy local search algorithms. *UniSAT7* requires computationally expensive continuous optimization algorithms, rendering them applicable to only small formulas.

Continuous Constrained Formulation

This generally involves a heuristic objective function that measures the quality of the solution obtained (such as the number of clauses satisfied). One formulation similar to the formulation proposed for the unconstrained case is as follows.

$$\begin{cases} \min_{y \in E^n} f(y) = \sum_{i=1}^{m} c_i(y) \\ \text{s.t. } c_i(y) = 0 \ \forall i \in \{1,2,...,m\} \end{cases}$$

As an example, consider the *CN*F formula

$$F(\mathbf{x}) = (x_1 \lor \bar{x}_2) \land (\bar{x}_1 \lor x_2 \lor x_3)$$

is translated into

$$f(\mathbf{y}) = |y_1 - 1||y_2 + 1| + |y_1 + 1||y_2 - 1||y_3 - 1| \text{ and}$$
$$f(\mathbf{y}) = (y_1 - 1)^2(y_2 + 1)^2 + (y_1 + 1)^2(y_2 - 1)^2(y_3 - 1)^2$$

respectively. The solution to the SAT problem corresponds to a set of global minimum points of the objective function. To find a true value of F(*x*) is equivalent to find the minimum value, i.e., 0, of f(y).

The translation of SAT problem into a nonlinear program is different from the integer program approach. Although nonlinear problems are intrinsically more difficult to solve, an unconstrained optimization problem is conceptually simple and easy to handle. Many powerful solution techniques have been developed to solve unconstrained optimization problems, which are based primarily upon calculus, rather than upon algebra and pivoting, as in the simplex method.

Depending on the global optimization strategy used, the objective function can be minimized in one dimension or in multiple dimensions. Presently many families of global optimization algorithms have been developed for the UniSAT problem models.

Enclosure of All Solutions

The problem of enclosing all solutions of a nonlinear system of equations subject to inequality constraints and variable bounds is formulated as in (3). Note that in formulation (3), the total number of variables is allowed to be different than the total number of equalities so as neither the existence nor the uniqueness of a solution of (3) is postulated. Therefore, overspecified and underspecified systems are included in the present investigations. Note that a number of important problems naturally arise as special instances of formulation (3). Eliminating all inequality constraints, (3) corresponds to a system of nonlinear equations. Eliminating all equality constraints (3) checks the existence of feasible points for the given inequality constraint set.

Introducing a single slack variable *s*, formulation (3) can be written as the following optimization problem (11):

$$\min_{\mathbf{x}, s \geq 0} s$$

$$\begin{aligned} \text{s.t. } & h_j(\mathbf{x}) - s \leq 0, & j \in N_? \\ & -h_j(\mathbf{x}) + s \leq 0, & j \in N_? \\ & g_k(\mathbf{x}) \leq 0, & k \in N_I \\ & \mathbf{x}^L \leq \mathbf{x} \leq \mathbf{x}^U \end{aligned} \quad (11)$$

Notice that there is a one to one correspondence between multiple global minima (x^*, s^*) of (11) for which $s^* = 0$ and solutions of (3). This means that if the global minimum of (11) involves a nonzero slack variable s^* then the original problem (3) has no solutions. Note that, unless the functions $h_j(\mathbf{x})$ are *linear* and $g_k(\mathbf{x})$ are *convex*, formulation (11) corresponds to a *nonconvex* optimization problem. This implies that if a local optimization approach is used to solve (11) one might miss some of the multiple global minima of (11) or even erroneously deduce that there are no solutions for (3).

A lower bound on the solution of (11) is found by first replacing each nonconvex constraint in (11) with a convex underestimation of it and the subsequent solution of the convex relaxation of (11). This approach naturally partitions the constraints of formulation (11) into convex constraints for which no relaxation is required and nonconvex ones. This partitioning yields the following alternative formulation (12)

$$\begin{aligned}
\min_{x, s \geq 0} \quad & s \\
\text{s.t.} \quad & h_j^{nonc}(x) - s \leq 0, \quad j \in N_{noncE} \\
& -h_j^{nonc}(x) + s \leq 0, \quad j \in N_{noncE} \\
& g_k^{nonc}(x) \leq 0, \quad k \in N_{noncI} \quad (12) \\
& h_j^{lin}(x) = 0, \quad j \in N_{linE} \\
& g_k^{conv}(x) \leq 0, \quad k \in N_{convI} \\
& x^L \leq x \leq x^U
\end{aligned}$$

where N_{noncE}, N_{linE} are the sets of nonconvex and linear equality constraints respectively, and N_{noncI}, N_{convE} are the sets of nonconvex and convex inequality constraints,

$$N_E = N_{noncE} \cup N_{linE}, \quad N_I = N_{noncI} \cup N_{convI}$$

A convex relaxation (13) of (12) of the form:

$$\begin{aligned}
\min_{\mathbf{x}, s \geq 0} \quad & s \\
\text{s.t.} \quad & \hat{h}_{+,j}^{nonc}(\mathbf{x}) - s \leq 0, \quad j \in N_{noncE} \\
& \hat{h}_{-,j}^{nonc}(\mathbf{x}) + s \leq 0, \quad j \in N_{noncE} \\
& \hat{g}_k^{nonc}(\mathbf{x}) \leq 0, \quad k \in N_{noncI} \quad (13) \\
& h_j^{lin}(\mathbf{x}) = 0, \quad j \in N_{linE} \\
& g_k^{conv}(\mathbf{x}) \leq 0, \quad k \in N_{convI} \\
& \mathbf{x}^L \leq \mathbf{x} \leq \mathbf{x}^U
\end{aligned}$$

can be obtained by replacing the original nonconvex functions with some convex tight lower bounding functions

$$\hat{h}_{+,j}^{nonc}(\mathbf{x}), \quad \hat{h}_{-,j}^{nonc}(\mathbf{x}), \quad \hat{g}_k^{nonc}(\mathbf{x})$$

An efficient convex lower bounding of the nonconvex functions appearing in formulation (12) is based on the principles outlined in the section of the αBB approach.

Procedure for Enclosing All Solutions

The multiple ε - global minima of (12) (if any) can then be localized based on a branch and bound procedure involving the successive refinement of convex relaxations (13) of the initial problem (12). Formulation (13) will always underestimate the global minimum of (12) within the same box constraints. Therefore, if the solution of (13) inside some rectangular region is strictly positive, then the solution of (12) inside the same rectangular domain will also be strictly positive. A strictly positive solution for (12) implies that the slack variable s cannot be driven to zero, and thus (3) is guaranteed not to have any solutions inside the rectangular region at hand. This provides a mechanism for fathoming (eliminating) parts of the target region which are guaranteed not to contain any solutions. If on the other hand, the global minimum of (13) is negative then (12) may or may not involve a solution with a zero slack variable and therefore no deduction can be drawn regarding the existence or not of solutions for (3) inside the current rectangular domain. In this case, further partitioning of the current rectangular region is required until the global minimum of (13) becomes positive or a feasible point for (12) is found.

Remark. Even if the number of multiple global minima of (12) is finite (Hansen *et al* 1992b) have shown that no algorithm can locate all of them with a finite number of function evaluations. Furthermore, a corollary of this result (Hansen *et al* 1992b) is that no algorithm can always localize, with a finite number of function evaluations, all globally optimal points by compact subintervals in one-to-one correspondence with them. Therefore, a more reasonable aim than finding all global minima of (12) is to find arbitrarily small disjoint subintervals containing all globally optimal points of (12) if any.

Relevance to Process Design

Global optimization problems are ubiquitous in chemical process synthesis and design. Important design and synthesis applications include: pooling and blending operations, phase equilibrium calculations such as minimization of the Gibbs free energy and stability, synthesis of heat exchanger networks; design and synthesis of nonideal distillation columns and sequences of columns, synthesis of complex reactor networks, parameter estimation, optimal control, synthesis of pump networks, design, scheduling and planning of batch, semi continuous and continuous processes, trim loss minimization, layout and nesting, and design under uncertainty (see Floudas, (2000)).

In this section, four classes of problems are briefly discussed to illustrate the relevance of the deterministic global optimization approaches for the process design applications.

Minimization of Gibbs Free Energy: The minimization of the Gibbs free energy problem is a fundamental problem in thermodynamics and more specifically in phase and chemical reaction equilibrium. It can be stated as:

Given a set of components $C = \{i\}$, the elements that constitute these components $E = \{e\}$, the set of phases $P = \{k\}$ consisting of vapor and liquid phases, denoted as P_V and P_L respectively, the liquid phase mol fractions $\mathbf{x} \equiv \{x_i\}$, and the vapor phase mol fractions $\mathbf{y} \equiv \{y_i\}$, we would like to determine the mol vector n that minimizes the value of the Gibbs free energy while also satisfying the appropriate material balance constraints.

For simultaneous phase and chemical equilibrium where reaction does occur, conservation of the constituent atoms must be satisfied:

$$\sum_{i \in C}\sum_{k \in P} a_{ei} n_i^k = b_e \quad \forall \quad e \in E \quad (14)$$

where a_{ei} represents the number of gram-atoms of element e in component i, and b_e the total number of gram-atoms of element e in the system.

For phase equilibrium without chemical reaction, conservation over the components need only hold:

$$\sum_{k \in P} n_i^k = n_i^T \quad \forall \quad i \in C \quad (15)$$

where n_i^T is the total number of moles of component i in the beginning mixture.

A physically realizable solution requires that

$$0 \leq n_i^k \leq n_i^T \quad \forall \quad i \in C \quad k \in P \quad (16)$$

where n^T is the total number of mols in the system.

The Gibbs free energy expression for the case of an ideal vapor phase and liquid phases modeled using the NRTL activity coefficient expression and after application of the theoretical properties developed by McDonald and Floudas (1994) can be written as a combination of a convex part and a nonconvex part:

$$\min \hat{G}(n) = \sum_{k \in P} C_N^k + \sum_{i \in C}\sum_{k \in P_L} n_i^k \left\{ \sum_{j \in C} \frac{G_{ij}\tau_{ij}n_j^k}{\sum_{l \in C} G_{lj} n_l^k} \right\}$$

where C_N^k is defined as follows:

$$C_N^k = \sum_{i \in C} n_i^k \left\{ \frac{\Delta G^{k,f}}{RT} + \ln \frac{n_i^k}{\sum_{i \in C} n_i^k} \right\} \forall k \in P$$

and it is a convex function. This mathematical formulation features nonconvexities in the objective function.

The tangent plane distance function is defined as the distance between the Gibbs surface and the tangent plane constructed to this surface at **z**. If we define:

$$g(\mathbf{x}) = \sum_{i \in C} x_i \mu_i(\mathbf{x}), \quad T(\mathbf{x}) = \sum_{i \in C} x_i \mu_i^0(\mathbf{z}),$$
$$F(\mathbf{x}) = g(\mathbf{x}) - T(\mathbf{x})$$

then the postulated solution corresponds to a global minimum of the Gibbs free energy as long as the tangent plane lies completely below the Gibbs surface. The formulation for the tangent plane stability problem is:

$$\min F(\mathbf{x}) = g(\mathbf{x}) - \sum_{i \in C} x_i \mu_i^0(\mathbf{z})$$

$$\text{s.t.} \quad \sum_{i \in C} x_i = 1$$

$$0 \leq x_i \leq 1$$

Note that the tangent plane distance function is explicitly minimized. If a nonnegative global optimum solution is obtained, then the stability of the postulated solution is guaranteed. This model has nonconvexities in the objective function. McDonald and Floudas (1994), (1995a), (1995b), (1997) provided analytical properties that exploit the mathematical structure and conducted extensive computational studies on the aforementioned fundamental phase equilibrium problems for activity coefficient models. Recently, Harding and Floudas (1999) developed deterministic global optimization approaches for equations of state models and demonstrated their excellent performance for systems up to eight components.

Parameter Estimation: Mathematical models which accurately predict physical phenomena are essential in many fields of engineering and applied sciences. These models frequently contain adjustable parameters which need to be determined from available experimental data. In chemical engineering, mathematical models form the basis for the design, synthesis, control, operation, and optimization of process systems.

Such models take the form of an implicit algebraic system of equations

$$\mathbf{f}(\theta, \mathbf{z}) = \mathbf{0} \quad (17)$$

where θ is a vector of p unknown parameters, **z** is a vector of *n* experimentally measured variables, and **f** represents the system of *l* algebraic functions.

All experimentally measured variables are affected to some extent by error. The measurements are related to the true values through:

$$z_\mu = \zeta_\mu + e_\mu \quad \mu = 1,...,m \quad (18)$$

where ζ_μ is a vector of unknown true values of the experimentally measured variables, z_μ, at the μ data point, and e_μ is the vector of additive error.

We define the likelihood function *L* as the probability of the observed errors in all data points occurring given a set of statistical parameters (φ) for the distribution used to represent those errors. Assuming that (i) the measurement errors from different data points (e.g., experiments) are considered as uncorrelated or independent, (ii) the joint probability density can be expressed as a product of the individual probability densities, (iii) the errors follow a normal distribution with zero mean and covariance matrix **V**, the ln *L* likelihood function becomes:

$$\ln L = -\frac{nm}{2}\ln 2\pi - \frac{1}{2}\sum_{\mu=1}^{m}\ln|V_\mu| - \frac{1}{2}\sum_{\mu=1}^{m}e_\mu^T v_\mu^{-1} e_\mu.$$

Assuming further that the errors in each experiment are known, equal, and independent, that is, the covariance matrix is diagonal with elements v_i, then the objective function becomes:

$$\min \sum_{\mu=1}^{m}\sum_{i=1}^{n} e_{\mu,i}^2 v_i^{-1} \quad (19)$$

where $e_{\mu,i}$ is the *i*th component on the vector e_μ that is, the error associated with the *i*th variable in the μth experiment.

Defining v_i as σ_i^2 where σ_i is the standard deviation assigned to the *i*th variable in all experiments, substituting the definition of $e_{\mu\mu,i}$ from (19), and subjecting the minimization to the model equations, we have the following optimization problem, known as the *error-in-variables* formulation:

$$\min_{\hat{z}_\mu, \theta} \sum_{\mu=1}^{r}\sum_{i=1}^{m} \frac{(\hat{z}_{\mu,i} - z_{\mu,i})^2}{\sigma_i^2} \quad (20)$$

s.t.

$$\mathbf{f}(\hat{z}_\mu, \theta) = 0 \quad \mu = 1...m.$$

The optimization variables, $\hat{z}_\mu$, are referred to as the fitted data variables. These are approximations, obtained through the optimization, of the true values of the experimental data, ζ_μ.

This formulation has a convex the objective function but since the model equations enter into the formulation as equality constraints, nonlinear models result in nonconvex optimization problems. Esposito and Floudas (1998) proposed the first global optimization approach for the parameter estimation and data reconciliation of algebraic models. Recently, Esposito and Floudas (1999) extended the approach to differential algebraic systems and optimal control problems.

Enclosure of All Azeotropes: The ability to predict whether a given mixture will form one or more homogeneous, heterogeneous and reactive azeotropes and to calculate the conditions and compositions of each azeotrope is essential in modeling separation processes.

Considering here the case of homogeneous azeotropes (i.e., one liquid phase), there are three thermodynamic conditions which a system must meet in order for a homogeneous azeotrope to exist. These conditions are: (i) equilibrium, (ii) the composition of vapor phase must be identical to the overall composition of the liquid phase, and (iii) the mole fractions of the components in each phase must sum to unity and must be non-negative.

The equilibrium condition requires that the chemical potential of each component must be the same in all phases. Since an homogeneous azeotropic system contains a vapor phase (V), and only one liquid phase (L), this condition can be written:

$$\mu_i^V = \mu_i^L \quad \forall i \in N \quad (21)$$

where μ_i^V and μ_i^L are the chemical potential of component *i* in the vapor and liquid phases. Assuming that the vapor phase can be modeled as an ideal gas, we have:

$$\frac{y_i}{x_i} = \frac{\gamma_i^L P_i^{sat}}{P} \quad (22)$$

The saturated vapor pressure is calculated using the Antoine equation:

$$\ln P_i^{sat} = a_i - \frac{b_i}{T + c_i} \quad (23)$$

where a_i, b_i, and c_i, are constant parameters with a_i and b_i, being positive, while c_i may be positive or negative but $|c_i| < T$.

Condition (ii) takes the form:

$$y_i = x_i \quad \forall i \in N. \quad (24)$$

Condition (iii) requires that the mole fractions in each phase sum to unity and have values between 0 and 1.

$$\sum_{i \in N} y_i = \sum_{i \in N} x_i = 1$$
$$0 \leq y_i, x_i \leq 1 \quad \forall i \in N, \quad (25)$$

This mathematical model corresponds to a nonlinear constrained system of equations that exhibits multiplicity of solutions and each solution is an homogeneous azeotrope. Harding *et al.* (1997) proposed the first deterministic global optimization approach for the enclosure of all homogeneous azeotropes and demonstrated that this can be achieved efficiently for activity coefficient models. Hua *et al.* (1998) proposed an interval analysis approach for both activity coefficient and equation of state models. Recently, Harding and Floudas (1999a) proposed novel formulations and global optimization approaches for the enclosure of heterogeneous and reactive azeotropes.

Current and Future Challenges

This review paper presented the recent advances in global optimization and their relevance to process design. The focal point was on novel deterministic global optimization methods for (i) general twice differentiable NLPs, (ii) mixed integer nonlinear problems, (iii) continuous approaches for discrete problems, and (iv) the enclosure of all solutions of nonlinear constrained systems of equations. From this review and the one in the 1994 FOCAPD meeting, it is evident that during the last decade (a) the area of Global Optimization has received significant attention in Chemical Engineering, (b) new theoretical and algorithmic developments have taken place, (c) new application areas of global optimization have been identified and research efforts have been initiated, and (d) computational tools have been developed. At this point, a variety global optimization problems of medium size (e.g., up to 100 to 200 variables) have been solved to global optimality. These correspond to the application areas that are reviewed in the Chemical Engineering contributions section. Large scale global optimization problems, have been addressed (see Harding and Floudas, 1997), but these correspond to special structure models. These exciting developments are complemented by a number of new challenges that include:

1. Global optimization approaches for bilevel nonlinear and multilevel nonlinear models,
2. Continuous approaches for mixed integer nonlinear and discrete models that have many applications in process synthesis problems,
3. New global optimization methods for black and gray box models with applications in design and parameter estimation problems,
4. Global optimization methods for differential and algebraic models,
5. Global optimizations approaches for partial differential and algebraic models,
6. Global optimizations methods for optimal control problems,
7. Global optimization approaches for design under uncertainty, the feasibility test and the flexibility index problems,
8. Global optimization methods for nonconvex parametric MINLPs,
9. Global optimization approaches for deterministic and stochastic uncertainty,
10. New rigorous methods for the design and synthesis of nonideal and hybrid separations,
11. Global optimization approaches for the design and scheduling of batch and continuous processes
12. New methods for the facility location-allocation models,
13. Global optimization approaches for the plant layout and nesting problems,
14. Novel global optimization frameworks for the structure prediction in protein folding, the peptide docking and the reaction pathways in the dynamics of peptide folding,
15. Distributed computing methods and efficient implementations of the global optimization approaches that will address medium to large problems.

Acknowledgements

Christodoulos A. Floudas gratefully acknowledges support from the National Science Foundation, Air Force Office of Scientific Research, the National Institutes of Health, Mobil Technology Co., General Motors Corp. and Delphi Corporation. Panos M. Pardalos is grateful for the support from the National Science Foundation.

References

Adjiman, C. S., Androulakis, I. P., and C. A. Floudas (1997c). Global optimization of MINLP problems in process synthesis and design. *Comp. Chem. Eng.*, **21**: S445-S450.

Adjiman, C. S., Androulakis, I. P., and Floudas, C. A. (1998b). A global optimization method, αBB, for general twice-differentiable NLPs - II. Implementation and computational results. *Comp. Chem. Eng.*, **22(9)**: 1159-1179.

Adjiman, C. S., Androulakis, I. P., and Floudas, C. A. (1998d). Global optimization of mixed-integer nonlinear problems. Submitted for publication.

Adjiman, C. S., Dallwig, S., Floudas, C. A., and A. A. Neumaier (1998). Global optimization method, αBB, for general twice-differentiable NLPs - I. Theoretical advances. *Comp. Chem. Eng.*, **22**(9), 1137-1158.

Adjiman, C. S., and Floudas, C. A. (1996). Rigorous convex underestimators for general twice-differentiable problems. *J. Global Opt.*, **9**, 23-40.

Androulakis, I. P., and Floudas, C. A. (1998). Distributed branch and bound algorithms in global optimization. In *Parallel Processing of Discrete Problems*, (Ed. P. M. Pardalos, **106**, *IMA Volumes in Mathematics and Its Applications*, Springer-Verlag,, 1-36.

Androulakis, I. P., Maranas, C. D., and Floudas, C. A. (1997a). Global minimum potential energy conformations of oligopeptides. *J. Global Opt.*, **11**: 1-34.

Androulakis, I. P., Nayak, N. N., Ierapetritou, M. G., Monos, D. S., and C. A. Floudas (1997b). A predictive method for the evaluation of peptide binding in pocket 1 of HLA-DRB1 via global minimization of energy interactions. *Proteins*, **29**, 87-102.

Androulakis, I. P., Visweswaran, V. and C. A. Floudas (1996). Distributed decomposition-based approaches in global optimization. In *Proceedings of the State of the Art in Global Optimization: Computational Methods and Applications* (Eds. C.A. Floudas and P.M. Pardalos), Book Series on Nonconvex Optimization and Its Applications, 285-302. Kluwer Academic Publishers.

Androulakis, I. P., Maranas, C. D. and C. A. Floudas (1995). αBB: a global optimization method for general constrained nonconvex problems. *J. Global Opt.*, **7**, 337-363.

Banga, J. R., and W. D. Seider (1996). Global optimization of chemical processes using stochastic algorithms. In In *Proceedings of the State of the Art in Global Optimization: Computational Methods and Applications* (Eds. C.A. Floudas and P.M. Pardalos), Book Series on Nonconvex Optimization and Its Applications, 285-302. Kluwer Academic Publishers.

Byrne, R. P. and Bogle, I. D. L. (1996a). An accelerated interval method for global optimization. *Comp. Chem. Eng.*, **20**, S49-S54.

Byrne, R. P. and I. D. L. Bogle (1996b). Solving nonconvex process optimization problems using interval subdivision algorithms. In I. E. Grossmann, Book Series on Nonconvex Optimization and Its Applications, 155-174. Kluwer Academic Publishers., Chapter 5.

Dorneigh, M. C. and N. V. Sahinidis (1995). Global optimization algorithms for chip layout and compaction. *Eng. Opt.* **25**(2), 131-154.

Epperly, T. G. W., Ierapetritou, M. G., and Pistikopoulos, E. N. (1997). On the Global and Efficient Solution of Stochastic Batch Plant Design Problems. *Comput. chem. engng.*, **21**: 1411-1431.

Epperly, T. G. W. and E. N. Pistikopoulos (1997). A reduced space branch-and-bound algorithm for global optimization. *J. Global Opt.*, **11**, 287-311.

Epperly, T. G. W. and R. Swaney (1996a). Branch and bound for global NLP: new bounding LP. In I. E. Grossmann, editor, *Global Optimization in Engineering Design*, Book Series on Nonconvex Optimization and Its Applications, 1-36. Kluwer Academic Publishers. Chapter 1.

Epperly, T. G. W. and R. Swaney (1996b). Branch and bound for global NLP: iterative LP algorithm and results. In I. E. Grossmann, editor, *Global Optimization in Engineering Design,* Kluwer Book Series on Nonconvex Optimization and Its Applications, 37-74. Kluwer Academic Publishers. Chapter 2.

Esposito, W. R. and C. A. Floudas (1999). Global optimization in parameter estimation of differential algebraic systems. *I&EC Res.*, submitted for publication.

Floudas, C. A. (1995). *Nonlinear and Mixed Integer Optimization: Fundamentals and Applications*. Oxford University Press, New York, 1995.

Floudas, C. A. (1997). Deterministic global optimization in design, control, and computational chemistry. In L. T. Biegler, T. F. Coleman, A. R. Conn, and F. N. Santosa, editors, *IMA Volumes in Mathematics and its Applications: Large Scale Optimization with Applications, Part II,* **93**, 129-184, Springer-Verlag, New York.

Floudas, C. A. (1999a). Global optimization in design and control of chemical process systems. *J. Process Control,* in press.

Floudas, C. A. (1999b). Recent advances in global optimization for process synthesis, design and control: enclosure of all solutions. *Comp. Chem. Eng.*, 963-973.

Floudas, C. A. (2000). *Deterministic Global Optimization: Theory, Algorithms and Applications. Nonconvex Optimization and its Applications*. Kluwer Academic Publishers.

Floudas, C. A. and I. E. Grossmann (1995). Algorithmic approaches to process synthesis: logic and global optimization. In *FOCAPD'94,* **91** *AICHE Symposium Series,* 198-221.

Floudas, C. A., Klepeis, J. L., and P. M. Pardalos (1999b). Global optimization approaches in protein folding and peptide docking. In M. Farach-Colton, F. S. Roberts, M. Vingron, and M. Waterman (Eds.), editors, *DIMACS Series In Discrete Mathematics and Theoretical Computer Science,* **47**, 141-171.

Floudas, C. A., and P. M. Pardalos (1996). State of the art in global optimization*: Computational Methods and Applications*. Kluwer Academic Publishers.

Floudas, C. A., Pardalos, P. M., Adjiman, C. S., Esposito, W. R., Gümüs, Z. H., Harding, S. T., Klepeis, J. L., Meyer, C., and C. A. Schweiger (1999). *Handbook of Test Problems in Local and Global Optimization*. Kluwer Academic Publishers.

Grossmann, I. E. (1996). Global optimization in engineering design. *Nonconvex Optimization and Its Applications*. Kluwer Academic Publishers, Dordrecht.

Grossmann, I. E. (1996b). Mixed integer optimization techniques for algorithmic process synthesis. In J.L. Anderson, editor, *Advances in Chemical Engineering, Process Synthesis,* **23**, 171-246. Academic Press.

Hansen, P. Jaumard, B., and S. Lu (1992b). Global optimization of univariate Lipschitz functions: II. New algorithms and computational comparison. *Math. Prog.*, **55**, 273.

Harding, S. T. and C. A. Floudas (1997). Global optimization in multiproduct and multipurpose batch design under uncertainty. *Ind. Eng. Chem. Res.,* **36**(5), 1644-1664.

Harding, S. T. and C. A. Floudas (1999). Phase stability for equations of state: a global optimization approach. submitted for publication.

Harding, S. T. and C. A. Floudas (1999a). Locating all heterogeneous and reactive azeotropes in multicomponent mixtures. *Ind. Eng. Chem. Res.,* in press.

Harding, S. T., Maranas, C. D., McDonald, C. M., and C. A. Floudas (1997). Locating all heterogeneous and reactive azeotropes in multicomponent mixtures. *Ind. Eng. Chem. Res.,* **36**(1), 160-178.

Horst, R. and P. M. Pardalos (1995) *Handbook of Global Optimization.* Kluwer Academic Publishers, Dordrecht.

Horst, R., Pardalos, P. M., Thoai, N. V. (1995). *Introduction to Global Optimization.* Nonconvex Optimization and its Applications. Kluwer Academic Publishers.

Hua, Z., Brennecke, J. F., and M. A. Stadtherr (1996). Reliable prediction of phase stability using an interval-Newton method. *Fluid Phase Equilibria,* **116**, 52-59.

Hua, Z., Brennecke, J. F., and M. A. Stadtherr (1998). Enhanced interval analysis for phase stability: Cubic equation of state models. *Ind. Eng. Chem. Res.,* **37**, 1519-1527.

Ierapetritou, M. G. and E. N. Pistikopoulos (1996). Batch plant design and operations under uncertainty. *Ind. Eng. Chem. Res.,* **35**(2),: 772-787.

Iyer, R. R. and I. E. Grossman (1996). Global optimization for heat exchanger networks with fixed configuration for multiperiod design. In I. E. Grossman, editor, *Global Optimization in Engineering Design,* Kluwer Book Series in Nonconvex Optimization and its Applications, 289-308, Chapter 9.

Klepeis, J. L., Androulakis, I. P., Ierapetritou, M. G., and C. A. Floudas (1998). Predicting solvated peptide conformations via global minimization of energetic atom to atom interactions. *Comp. Chem. Eng.,,* **22**(6),: 765-788.

Klepeis, J. L. and C. A. Floudas (1999a). A comparative study of global minimum energy conformations of hydrated peptides. *J. Comp. Chem.,* **20**(13), 1354-1370.

Klepeis, J. L. and Floudas, C. A. (1999b). Free energy calculations for peptides via deterministic global optimization. *J. Chem. Phys.,* **110**(15), 7491-7512.

Liu, M. L., Sahinidis, J. P., and J. P. Shectman (1996). Planning of chemical process networks via global concave minimization. In I. E. Grossmann, editor, *Global Optimization in Engineering Design,* Kluwer Book Series in Nonconvex Optimization and its Applications, 195-230. Chapter 7.

Lucia, A. and J. Xu (1996). Nonconvexity and descent in nonlinear programming. In *Proceedings of the State of the Art in Global Optimization: Computational Methods and Applications.* (Eds. C.A. Floudas and P.M Pardalos), Kluwer Book Series on Nonconvex Optimization and Its Applications. Kluwer Academic Publishers.

Maier, R. W., Brennecke, J. F., and M. A. Stadtherr (1998). Reliable computation of homogeneous azeotropes. *AIChE J.,* **44**, 1745-1755.

Maranas, C. D. and C. A. Floudas (1994a). A deterministic global optimization approach for molecular structure determination. *J. Chem. Phys.,* **100**(2), 1247-1261.

Maranas, C. D. and C. A. Floudas (1994b). Global minimum potential energy conformations of small molecules. *J. Global Opt.,* **4**, 135-170.

Maranas, C. D. and C. A. Floudas (1995). Finding all solutions of nonlinearly constrained systems of equations. *J. Global Opt.,* **7**(2), 143-182.

Maranas, C. D. and C. A. Floudas (1997). Global optimization in generalized geometric programming. *Comp. Chem. Eng.,* **21**(4), 351-370.

McDonald, C. M. and C. A. Floudas (1994). Decomposition based and branch and bound global optimization approaches for the phase equilibrium problem. *J. Global Opt.,* **5**, 205-251.

McDonald, C. M. and C. A. Floudas (1997). Global optimization for the phase and chemical equilibrium problem: application to the NRTL equation. *Comp. Chem. Eng.,* **19**(11), 1111-1141.

McDonald, C. M. and Floudas, C. A. (1997). Global optimization for the phase and stability problem. *AIChE J.,* **41**(7), 1798-1814.

McDonald, C. M. and C. A. Floudas (1997). GLOPEQ: A new computational tool for the phase and chemical equilibrium problem. *Comp. Chem. Eng.,* **21**, 1-23.

McDonald, C. M. and C. A. Floudas (1997). Global optimization and analysis for the Gibbs free energy function using the UNIFAC, Wilson and ASOG equations. *Ind. Eng. Chem. Res.,* **34**, 1674.

McKinnon, K. and M. Mongeau (1998). A generic global optimization algorithm for the chemical phase equilibrium problem. *J. Global Opt.,* **12**, 325-351.

McKinnon, K. I. M., Millar, C. and M. Mongeau (1996). Global optimization for the chemical and phase equilibrium problem using interval analysis. In *Proceedings of the State of the Art in Global Optimization: Computational Methods and Applications,* (Eds. C.A. Floudas and P.M. Pardalos). Kluwer Book Series on Nonconvex Optimization and Its Applications. Kluwer Academic Publishers.

Meyer, C. A. and C. L. E. Swartz (1998). A regional convexity test for global optimization: Application to the phase equilibrium problem. *Comp. Chem. Eng.,* **22**, 1407-1418.

Mockus, L. and G. V. Reklaitis (1996). A new global optimization algorithm for batch process scheduling. In C.A. Floudas and P. M. Pardalos, eds., *State of the Art in Global Optimization: Computational Methods and Applications,* (Eds. C.A. Floudas and P.M. Pardalos). Kluwer Book Series on Nonconvex Optimization and Its Applications. Kluwer Academic Publishers.

Pardalos, P. M. (1994). On the passage from local to global optimization. In J.R. Birge and K. G. Murty, eds., *Mathematical Programming: State of the Art 1994,* 220-247. The University of Michigan.

Pardalos, P. M. (1996). Continuous approaches to discrete optimization. In G. Di Pillo and F. Giannessi, eds., *Nonlinear Optimization and Applications,* 318-328. Plenum Publishing.

Psarris, P. and C. A. Floudas (1995). Robust stability analysis of systems with real parameter uncertainty: A global optimization approach. *Int. J. Robust Nonlinear Control,* **6**, 699-717.

Quesada, I. and I. E. Grossmann (1993). Global optimization algorithm for heat exchanger networks *Ind. Eng. Chem. Res.,* **32**, 487.

Quesada, I. and I. E. Grossmann (1993). Global optimization algorithm linear fractional and bilinear programs. *J. Global Opt.,* **6**, 39-76.

Quesada, I. and I. E. Grossmann (1996). Alternative bounding approximations for the global optimization of various engineering design problems. In I.E. Grossmann, editor, *Global Optimization in Engineering Design,* Kluwer Book Series in Nonconvex Optimization and its Applications, 309-332, Chapter 10.

Ryoo, H. S. and N. V. Sahinidis (1995). Global optimization of nonconvex NLPs and MINLPs with applications in process design. *Comp. Chem. Eng.,* **19**(5), 551-566.

Ryoo, H. S. and N. V. Sahinidis (1996). A branch and reduce approach to global optimization. *J. Global Opt.,* **8**(2), 107-139.

Shectman, J. P. and N. V. Sahinidis (1996). A finite algorithm for global optimization of separable concave programs. In *Proceedings of the State of the Art in Global Optimization: Computational Methods and Applications* (Eds. C.A. Floudas and P.M. Pardalos), Kluwer Book Series on Nonconvex Optimization and Its Applications. Kluwer Academic Publishers.

Shectman, J. P. and N. V. Sahinidis (1998). A finite algorithm for global optimization of separable concave functions. *J. Global Opt.,* **12**, 1-36.

Smith, E. M. B. and C. C. Pantelides (1996). Global optimization of general process models. In I. E. Grossman, ed., *Global Opt. Eng. Design,* 355-386. Kluwer Academic Publishers.

Smith, E. M. B. and C. C. Pantelides (1997). Global optimization of nonconvex MINLPs. *Comp. Chem. Eng.,* **21**, S791-S796.

Sriniwas, G. R. and Y. Arkun (1997). A global solution to the nonlinear model predictive control algorithms using polynomial ARX models. *Comp. Chem. Eng.,* **21**, 431-439.

Staus, G. H., Biegler, L. T., and B. E. Ydstie (1996). Adaptive control via nonconvex optimization. . In *Proceedings of the State of the Art in Global Optimization: Computational Methods and Applications* (Eds. C.A. Floudas and P.M. Pardalos), Kluwer Book Series on Nonconvex Optimization and Its Applications.

Vaidyanathan, R. and M. El-Halwagi (1994). Global Optimization of nonconvex nonlinear programs via interval analysis. *Comp. Chem. Eng.,* **18**, 889-897

Vaidyanathan, R. and M. El-Halwagi (1996). Global optimization of nonconvex MINLP's by inerval analysis. In I.E. Grossman, ed., *Global Optimization in Engineering Design,* 175-193. Kluwer Academic Publishers.

VanAtwerp, J. G., Braatz, R. D., and N. V. Sahinidis (1997).. Globally optimal robust control for systems with time-varying nonlinear perturbations. *Comp. Chem. Eng.,* **21**, S125-130.

Visweswaran, V. and C. A. Floudas (1996a). New formulations and branching strategies for the GOP algorithm. In I. E. Grossmann, ed., *Global Optimization in Engineering Design,* Kluwer Book Series in Nonconvex Optimization and its Applications, 75-110, Chapter 3.

Visweswaran, V. and C. A. Floudas (1996b). Computational results for an efficient implementation of the GOP algorithm and its variants. In I. E. Grossmann, ed., *Global Optimization in Engineering Design,* Kluwer Book Series in Nonconvex Optimization and its Applications, 110-154, Chapter 4.

Visweswaran, V. and Floudas, C. A., Ierapetritou, M. G., and E. N. Pistikopoulos (1996). A decompostion-based global optimization approach for solving bilevel linear and quadratic programs. In C. A. Floudas and P. M. Pardalos, eds., *State of the Art in Global Optimization: Computational Methods and Applications,* Kluwer Book Series on Nonconvex Optimization and Its Applications. Kluwer Academic Publishers.

Westerberg, K. M. and C. A. Floudas (1999) Locating all transition states and studying the reaction pathways of potential energy surfaces. *J. Che. Phys.,* **110**(18), 9259-9296.

Westerlund, T., Pettersson, F., and I. E. Grossman (1994). Optimization of pump configuration problems as a MINLP problem. *Comp. Chem. Eng.,* **18**(9), 845-858.

Westerlund, T., Skrifvars, Harjunkoski, I. and Pörn (1998). An extended cutting plane method for a class of non-convex MINLP problems. *Comp. Chem. Eng.,* **22**(3),: 357-365.

Yamada, Y. and S. Hara (1998). Global optimization for H-infinity control with constant diagonal scaling. *IEEE Trans. Auto. Control,* **43**, 191-203.

Zamora, J. M. and I. E. Grossmann (1998a). A global MINLP optimization algorithm for the synthesis of heat exchanger networks with no stream splits. *Comp. Chem. Eng.,* **22**(3), 367-384.

Zamora, J.M. and I. E. Grossmann (1998b). Continuous global optimization of structured process systems models. *Comp. Chem. Eng.,* **22**(12), 1749-1770.

TOWARDS THE INTEGRATION OF PROCESS DESIGN, PROCESS CONTROL, AND PROCESS OPERABILITY:
CURRENT STATUS AND FUTURE TRENDS

Jan van Schijndel
Shell International Chemicals
1031 CM Amsterdam, The Netherlands

Efstratios N. Pistikopoulos
Centre for Process Systems Engineering
Imperial College
London SW7 2BY, U.K.

Abstract

In this paper, we discuss critical issues and requirements for the integration of process design, process control and process operability. A paradigm for such integration, based on dynamic modeling and optimization principles, is presented along with its application to two typical industrial processes. Future research directions and needs towards establishing a holistic approach to Process Design for Operability are identified.

Keywords

Process design, Process control, Process operability, Process optimization.

Introduction

In one of the invited sessions of the last FOCAPD Meeting, on the topic of 'Design for Operations and Control', a number of important developments and contributions in the area were reviewed, both academic and industrial perspectives were given and future needs were identified. Some of the valuable observations and recommendations, which were put forward, included statements such as: "what is ultimately of interest is the closed loop behavior of the system once it is operating and subjected to disturbances" (Morari and Perkins, 1995); "an ideal method for designing a process with sufficient controllability is yet to be developed" (Hashimoto and Zafiriou, 1995); "maintaining acceptable variability is an operability issue of increasing prominence", "linkage of technology that can be used to manage process variability with the process design function is believed to lead to significant savings" (Downs and Ogunnaike, 1995).

Five years later, in this paper, by assessing the current status in the area of 'Process Design for Control and Operability', some of the above issues and concerns will be discussed and addressed in an attempt to establish the foundations of a holistic approach towards the integration of process design, process control and process operability. It will be shown that, while many issues still remain open, a lot of progress has been made in at least substantially bridging the gap between academia and industry, which was considered "an urgent need" (Hashimoto and Zafiriou, 1995). A paradigm for such an integration of operability objectives in process design and optimization, and its application to two typical industrial problems, will then be briefly presented. Future trends and research needs will finally be identified.

Process Design for Operability – an Overview

Operability in Process Design, including flexibility, controllability, robustness, availability, as well as process design and optimization under uncertainty, is a broad research subject which has been receiving considerable and increasing attention, in particular over the last two decades. While not exhaustive, the overview that follows intends to provide, from the authors' perspective, a classification of the main research trends and a roadmap of the key developments.

In the area of 'Operability Criteria in Process Design', with over 70 publications since 1985, the works can be broadly classified as shown in Table 1 (see Appendix). There are three main strands: (a) analysis methods for comparing alternative designs, (b) methods for designing operable systems and (c) applications of these methods to specific processes. While the works differ in the specific criterion used (such as flexibility, robustness, controllability indicators), Table 1 clearly depicts that almost all of the works in this area use steady-state or linear dynamic models, open-loop control considerations, with no apparent evidence of any (simultaneous or sequential) process and control optimization strategy.

In the area of 'Interactions of Design and Control', with over 50 publications since 1982, two international workshops (1992, 1994) and many dedicated invited sessions at international conferences (including FOCAPD' 94, the Annual AIChE meeting in 1997, DYCOPS-5), a classification attempt, as the one shown in Table 2 (see Appendix), reveals some interesting general trends: (i) the bulk of the works has been based on either steady-state or simple (often linear) dynamic models, with only a recent shift towards more rigorous dynamic models, (ii) most works use analysis and/or simulation tools to compare candidate process designs, (iii) either no or simple control schemes have been used in most cases and (iv) in only a handful of works are both process and control design optimized, mostly in a sequential fashion.

In the related areas of 'Plant-wide Control' and 'Control Structure Selection' (including 'Multivariable Controller Synthesis'), with over 110 publications in the last ten years, the classification of Table 3 (see Appendix) shows: (i) no direct consideration of economics is the norm in the multivariable controller synthesis area, (ii) steady-state or simple dynamic models and heavy reliance on analysis and dynamic simulation dominate the plant-wide control literature and (iii) analysis is also the norm for control structure selection, with only few works considering economics within an optimization framework.

As another example, in the related field of 'Availability and Maintenance' for chemical process systems, with almost 100 publications, despite the wealth of traditional risk assessment techniques, it is interesting to note, as shown in Table 4 (see Appendix), that very few works have even attempted to formalize the interactions of process design, process operation, uncertainty and maintenance for general classes of processes and models – furthermore, optimization attempts are at best sporadic.

From an optimization theory and technology perspective, the classifications presented in Tables 5 and 6 (see Appendix), merely support the argument why it is *not* surprising that progress so far has been limited in establishing a convincing strategy for the application of optimization tools for process design, process control and process operability in the process industry. With over 200 relevant publications cited, it becomes apparent that: (a) methods and computational tools to address optimization issues in process systems described by dynamic models in the presence of stochastic/parametric uncertainty are either non-existent or at their infancy (see Table 5) and (b) the area of dynamic optimization with discrete decisions has only recently started to attract some attention, with a few theoretical developments gradually emerging, but without any reported industrial applications (see Table 6).

While caution should be exercised in drawing any general conclusions from such classifications, it is the authors' personal view that bridging the gap between academic theory and industrial practice in the field of Process Design for Operability, cannot only be an adventure in cultural, managerial and organizational perspective shift, nor can it only be the result of technology push developments (especially if these rely on either oversimplified models/problems or lack any realism in the problem definition/representation). It requires a coordinated effort between industry and academia in properly defining the needs and demonstrating the benefits to key business/industrial applications, thereby setting the targets of what such technology should be offering – a market pull approach. Such an approach will enable re-focussing on the key, most relevant issues to make this technology work, thereby paving the way to significant developments in all the areas briefly covered above. In the next section, a paradigm towards this endeavor is presented.

Process Design for Operability – a Paradigm

Figures 1 and 2 schematically show two typical industrial process systems, a double effect distillation and an azeotropic ternary distillation system with a side column. Both systems have very tight product specifications and are subjected to disturbances and uncertainty/variability, such as for example in the feed composition and flow-rate. The objective is to design the columns and control scheme at minimum annualized (capital and operating) cost, able to maintain feasible operation over a finite time horizon of interest, i.e. meeting the tight product quality specifications and other operating constraints, subject to given parametric and/or stochastic uncertainty and disturbances. While the details of these studies are described in detail elsewhere (Bansal et al, 1999; Ross et al, 1999), here some key issues and basic features will only be highlighted.

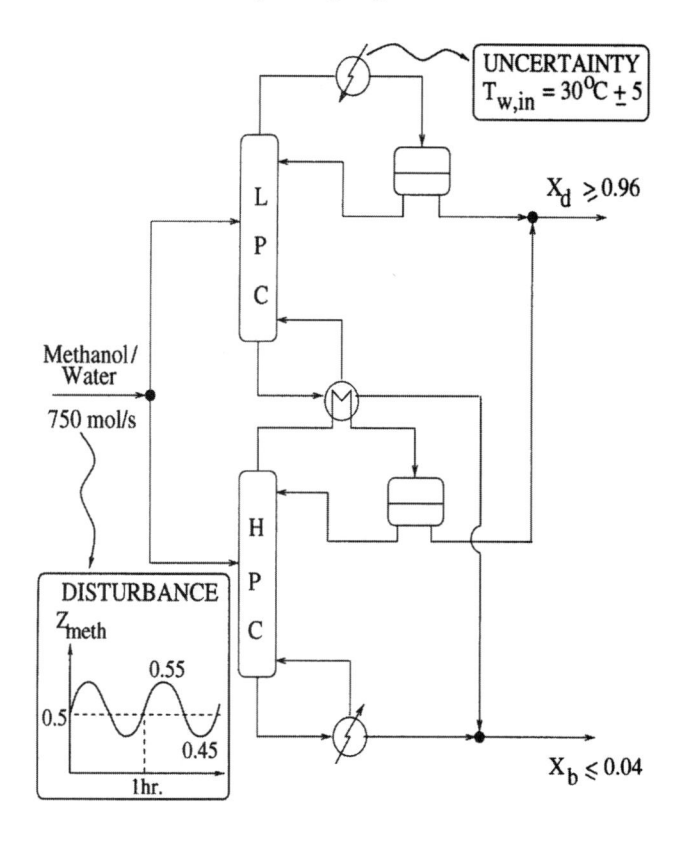

Figure 1. Double-effect distillation system.

In both systems, a large number of possible process and control configurations exist related to the process structure itself, the control structure and controller type, as well as the detailed design (of the process – equipment sizing, material flow-rates – and of the controller – tuning parameters). Furthermore, the control scheme clearly depends on the process structure and the design itself, intensifying the need for studying the crucial interactions of design and control in an integrated way, in the additional presence of uncertainty and variability.

The necessity for accurately capturing the process dynamics and identifying potential process/operability bottlenecks is another critical issue. For example, in the double effect distillation system, pressure dynamics are important as are modeling provisions for the identification of flooding and related minimum column diameter requirements. For the azeotropic distillation system, the dramatic effect of feed disturbances on the temperature and composition profiles within the first column, has a further devastating effect on the feed composition to column II – this can result in either overloading column II with water or flooding the lower part of column I. Such complex phenomena call for a high-fidelity dynamic modeling representation of the inherent non-linear process behavior of any industrially relevant process system.

To address these needs, the problem of the integration of process design, process control and process operability is conceptually posed as follows:

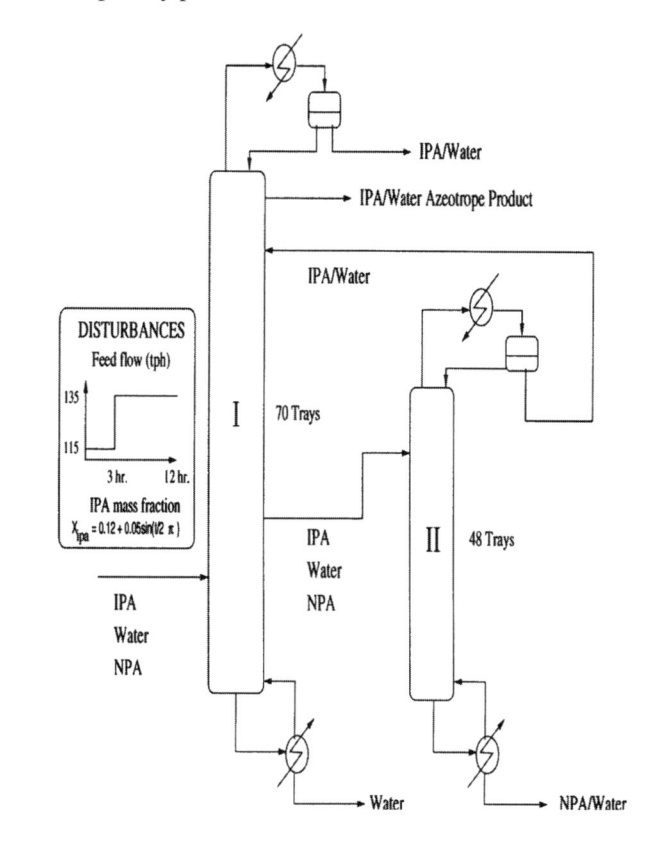

Figure 2. Industrial azeotropic distillation system.

Minimize   Expected Total Annualized Cost   **(P)**

Subject to   Differential-Algebraic Process Model
Inequality Path Constraints
Control Scheme Equations
Process Design Equations
Feasibility of Operation (over time)
Process Variability Constraints

To determine Process and Control Design

Note that (P) attempts to encapsulate most of the essential ingredients of the problem at hand: high-fidelity dynamic models; explicit considerations for feasibility in process performance over time; full account of the interactions between process design and control design as well as provisions for the effective management of the process variability within the process design formalism. However, it is not a mere philosophical statement of the authors' wishful thinking – it does provide a sound mathematical and process engineering basis for the formal representation of a novel paradigm to Process Design for

Operability as well as for the development of appropriate strategies for its solution (Mohideen et al, 1996, 1997a). As will be discussed in the final section, it also helps in identifying needs and requirements for future developments.

From a theoretical perspective, (P) corresponds to a large-scale stochastic/parametric mixed integer dynamic optimization problem – a class of problems for which, as noted in Table 5, there are not yet established rigorous theoretical and computational tools. Nevertheless, for special cases of (P), computational strategies have been developed, such as the one schematically outlined in Figure 3. It corresponds to the solution of a variant of (P), (see Mohideen et al, 1996, for details), in which a PI control scheme is included together with a PI controller superstructure model, no explicit variability

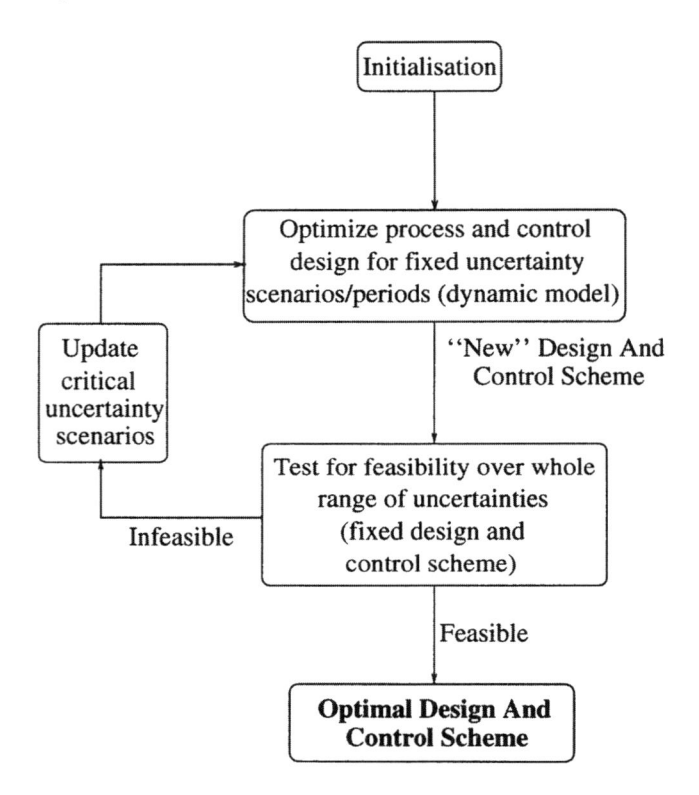

Figure 3. Simultaneous design and control strategy.

constraints are present and uncertainty is modeled via parametric upper and lower bounds. The solution strategy involves an iterative decomposition algorithm between (a) a multi-period simultaneous design and control sub-problem for a given uncertainty and disturbance profile and (b) a feasibility over time analysis step (which may also involve a stability test, see Mohideen et al, 1997b), which identifies further operability bottlenecks to be included in (a). The algorithm determines a process design and PI control scheme, which is feasible to cope with the specified ranges of uncertainty and disturbances at minimum average investment and operating costs.

From a process engineering viewpoint, the significance of applying such a simultaneous optimization strategy to the solution of the two systems shown in Figures 1 and 2 is amply demonstrated by the results summarized in Table 7, in which a comparison is given of the integrated design with that obtained using a conventional, sequential approach (process design first followed by control design). A striking feature of these results is that the application of the advanced simultaneous optimization strategy enables improved economic and control performance to be achieved. For the double effect distillation system, annual cost savings of $100,000 are reported (Bansal et al, 1999), with significant simultaneous improvements in control performance (15% less ISE). For the azeotropic distillation system, a 6% ($300,000 per annum) cheaper design and control system is reported (Ross et al 1999), with a new distillation system design proposed featuring a reduced capacity in column I (7% smaller diameter, 9% lower reboiler duty) and an increased capacity in column II (1% larger diameter, 24% higher reboiler duty)!

From a computational viewpoint, these problems, even for the case considered here in which the structure (process and control) is fixed (i.e., no integer variables present in (P)), correspond to very large-scale dynamic optimization models. For the double effect distillation column, the model involves over 4000 variables, of which 180 are differentiable state variables and 20 are optimization variables, and 20 inequality path constraints. For the azeotropic system, approximately 10,000 variables (without including 17,000 physical property variables), of which about 600 are differentiable states, 6 process design variables (two column diameters, both reboiler duties and the two flow-rates between columns I and II, see Figure 2), 2 control design variables (the gain and reset time of the PI controller in the temperature-product flow-rate loop) and 7 inequality path constraints describing feasible operation of the process (minimum column diameter/flooding constraint and a fractional entrainment limit for each column, composition constraints in the azeotrope product near the top of column I and in the bottom product of column II) are involved in the mathematical model. The *gOPT* dynamic optimization interface for *gPROMS* (Process Systems Enterprise Ltd., 1998) was used in these studies.

To our knowledge, as also supported by the review in Table A.6, these are amongst the largest dynamic process optimization problems ever solved – it is interesting to note that the solution of such large models only five years ago was not even foreseeable! Despite this success and the very encouraging results obtained, the computational efficiency and mathematical rigor of such optimization models still poses some formidable theoretical and computational challenges (as will be discussed later in this paper).

While the paradigm briefly described above may not fully incorporate all the components and dimensions of the Process Design for Operability problem, it does go some way in overcoming a number of previously perceived limitations and offers enough hope for further exciting developments in the near future. Some of its advantages are listed below.

1. The use of, often over-simplified, linear-based process models for control optimization, along with the use of, often misleading, controllability indicators, in the context of simultaneous process and control design optimization, is effectively avoided.
2. The need for, often extensive and computationally expensive, dynamic simulations to verify the results obtained, which is the standard practice of any conventional process design and control strategy to-date, is essentially alleviated.
3. Control optimization is performed *simultaneously* with design optimization in an integrated way, with the control/dynamic performance requirements included as a set of appropriate constraints in the design problem.
4. Other operability criteria, such as robustness and variability, can also be included in the strategy outlined in Figure 3, by appropriately defining the corresponding statistical terms of interest (for example, six-sigma, signal-to-noise ratio, variance).
5. From a computational point, the tools for Process Design for Operability are being linked to state-of-the-art commercial software for dynamic simulation and optimization, in a seamless, to the user, way – this will significantly boost its potential for industrial applicability and acceptance.
6. From a philosophical perspective, the paradigm outlined above is a fundamental departure from the dogma "Design via Analysis and Simulation", which clearly constitutes the current industrial and academic thinking and practice 'status quo'. Instead, it flags the beginning of a new era of "Design via Synthesis and Optimization" in the eve of the new millenium.

Process Design for Operability – Future Trends

The significant progress achieved over the last few years in the field of Process Design for Operability should be a source for optimism: an essential first step towards establishing the foundations of a truly holistic approach for the integration of process design, process control and process operability has been made. At the same time, a critical assessment of what has been achieved versus the actual industrial needs provides enough ammunition for further research and development requirements, an opportunity that the FOCAPD community will, undoubtedly, not miss as part of its agenda for the new millenium. Some of these challenges for 'Process Design for Operability in 2010' are briefly discussed below.

Towards ... linking conceptual process synthesis and plant-wide control.

The paradigm proposed in the previous section can, in principle, be applied at the process synthesis and plant-wide control level. However, there is a clear need to reduce as much as possible the potentially very large number of process and control configurations, for example through a rigorous screening prior to the optimization phase. In this context, advances in thermodynamic and phenomena based representations for process synthesis (see, for example, Hauan and Lien, 1998, Papalexandri and Pistikopoulos, 1996) and process control (Ydstie and Alonso, 1997, Hangos et al, 1999, Tyreus, 1999), proper plant-wide analysis of thermophysical properties (Gani and O'Connell, 1999) and hybrid hierarchical-thermodynamic-intelligent approaches (Douglas and Stephanopoulos, 1995) can be of particular relevance.

Towards ... more emphasis on more realistic applications, involving complex, multi-component, highly non-ideal and nonlinear systems.

This will further 'bridge the gap' between academia and industry and substantially expand the boundary of applications of Process Design for Operability. Here, advances in azeotropic and reactive separation, possibly with multiple steady-states (see for example, Malone and Doherty, 1995, Bekiaris and Morari, 1996, Westerberg and Wahnschafft, 1996), in systems with unsteady state operation (see for example, Seider et al, 1991) and batch processes (see for example, Furlonge et al, 1999), perhaps incorporated in an optimization strategy similar to the one in Figure 3, can provide the basis for such development.

Towards ... the grand unification of operability and environomics in process design and optimization.

While the paradigm in (P) allows for flexibility requirements, robustness and dynamic performance to be simultaneously considered with economics, issues related to reliability and availability (see for example Thomaidis and Pistikopoulos, 1995), safety (see for example, Dimitriadis et al., 1997) and environmental impact minimization (see for example, El-Halwagi, 1997, Cano-Ruiz and McRae, 1998,) are of equal importance in the design stage. A simultaneous optimization strategy to link all these components together, while a formidable task, will enable for the first time the 'optimal' design of 'operable' and 'environmentally conscious' process systems.

Towards ... even more emphasis on high-fidelity modeling and optimization based applications.

Here, the challenge is to use the same detailed nonlinear dynamic optimization model from the process and control design step (featuring general multi-variable control), all the way to real time optimization (for example, by linking the framework in Figure 3 to the

MPC/RTO framework of Loeblein and Perkins, 1996) - with the model reduction, if and where needed, being carried out at the level of the mathematical model itself, *not* being the result of a process simplification or approximation (such as, for example, a linear model representation).

Towards ... more advanced optimization solution strategies and software tools.

One of the main challenges in the successful application of optimization-based paradigms, such as (P), is the *rigorous* and *efficient* solution of the underlying large-scale stochastic/parametric mixed integer dynamic optimization problem. Here, recent theoretical and computational advances in mixed integer dynamic optimization (see for example, Allgor and Barton, 1999, Ross et al., 1998), stochastic and parametric mixed integer optimization (for example, Acevedo and Pistikopoulos, 1997), global optimization (for example, Floudas and Grossmann, 1995) provide an excellent starting point for future developments.

Towards ... new processing concepts.

Another exciting application route and source for future growth for the Design for Operability field concerns with the design and operation of new processing concepts, where truly integrated designs will be the norm. Here, responsive plants (Rowe et al, 1997), plant on a chip (Santini et al, 1999), invisible control (Benson, 1999), down-the-pipe-plant for oil and gas reservoir management (Maitland, 1999) as well as fuel cell technology in automobile applications, are some of the novel processing concepts, for which Design for Operability will not be a luxurious add-on feature but a strict technological requirement – a survival kit.

Acknowledgements

Financial support from EPSRC (IRC grant), the Centre for Process Systems Engineering (IRC Industrial Consortium), European Union (JOE3-CT97-0085) and Shell International Chemical B.V. is gratefully acknowledged. The authors would also like to acknowledge the valuable contributions of Professor John Perkins, Dr. Roderick Ross and Mr. Vikrant Bansal at Imperial College and Mr. Gerard Koot at Shell. ENP is also grateful to Vik Bansal, Rod Ross, Vivek Dua and Costas Vassiliadis for their help in preparing the tables.

References

Acevedo, J. and E.N. Pistikopoulos (1997). A multiparametric programming approach for linear process engineering problems under uncertainty. *Ind. Eng. Chem. Res.*, **36**, 717-728.

Allgor, R.J. and P.I. Barton (1999). Mixed-integer dynamic optimization I: problem formulation. *Comp. Chem. Eng.*, **23**, 567-584.

Bansal, V., R. Ross, J.D. Perkins and E.N. Pistikopoulos (1999). The interactions of design and control: double-effect distillation. *J. Process Control*, in print.

Bekiaris, N. and M. Morari (1996). Multiple steady states in distillation –infinity/infinity predictions, extensions and implications for design, synthesis and simulation. *Ind. Eng. Chem. Res.*, **35**, 4264-4280.

Benson, R. (1999). Trends in process control. Presented at Industrial Research Consortium, Spring Meeting, Centre for Process Systems Engineering, Imperial College.

Cano-Ruiz, J.A. and G.J. Mc Rae (1998). Environmentally conscious process design. *Annu. Rev. Energy Environ.*, **23**, 499-536.

Dimitriadis, V.D., N. Shah and C.C. Pantelides (1997). Modeling and safety verification of discrete/continuous processing systems. *AIChE J.*, **43**, 1041-1059.

Douglas, J.M. and G. Stephanopoulos (1995). Hierarchical approaches in conceptual process design: framework and computer aided implementation. In L.T. Biegler and M.F. Doherty (Eds), *Foundations of Computer-Aided Process Design*, AIChE Symposium Series, No. 304, Vol. 91, CACHE-AIChE, pp. 183-197.

Downs, J.J. and B. Ogunnaike (1995). Design for control and operability: an industrial perspective. In L.T. Biegler and M.F. Doherty (Eds), *Foundations of Computer-Aided Process Design*, AIChE Symposium Series, No. 304, Vol. 91, CACHE-AIChE, 115-123.

El-Halwagi, MM. (1997). *Pollution prevention through process integration: systematic design tools*. Academic Press, San Diego, CA.

Floudas, C.A. and I.E. Grossmann (1995). Algorithmic approaches to process synthesis: logic and global optimization. In L.T. Biegler and M.F. Doherty (Eds), *Foundations of Computer-Aided Process Design*, AIChE Symposium Series, No. 304, Vol. 91, CACHE-AIChE, 198-221.

Furlonge, H.I., C.C. Pantelides and E. Sorensen (1999). Optimal operation of multivessel batch distillation columns. *AIChE J.*, **45**, 781-801.

Gani, R. and J.P. O'Connell (1999). Properties and CAPE: from present uses to future challenges. *Proc. ESCAPE-9*, Budapest, Hungary, in print.

Hangos, K.M., A.A. Alonso, J.D. Perkins and B.E. Ydstie (1999). Thermodynamic approach to the structural stability of process plants. *AIChE J.*, **45**, 802-816.

Hashimoto, I. and E. Zafiriou (1995). Design for operations and control. In L.T. Biegler and M.F. Doherty (Eds), *Foundations of Computer-Aided Process Design*, AIChE Symposium Series, No. 304, Vol. 91, CACHE-AIChE, 103-104.

Huan, S. and K.M. Lien (1998). A phenomena based design approach to reactive distillation. *Trans. IChemE*, **76**, 396-407.

Loeblein, C. and J.D. Perkins (1996). Economic analysis of different structures of on-line process optimization systems. *Comp. Chem. Eng.*, **20**, S551-S556.

Maitland, G.C. (1999). Real-time oil and gas reservoir management – from pore to pour. Presented at Industrial Research Consortium, Spring Meeting, Centre for Process Systems Engineering, Imperial College.

Malone, M.F. and M.F. Doherty (1995). Separation system synthesis for nonideal liquid mixtures. In L.T. Biegler and M.F. Doherty (Eds), *Foundations of Computer-Aided Process Design*, AIChE Symposium Series, No. 304, Vol. 91, CACHE-AIChE, 9-18.

Mohideen, M.J., J.D. Perkins and E.N. Pistikopoulos (1996). Optimal design of dynamic systems under uncertainty. *AIChE J.*, **42**, 2251-2272.

Mohideen, M.J., J.D. Perkins and E.N. Pistikopoulos (1997a). Towards an efficient numerical procedure for mixed integer optimal control. *Comp. Chem. Eng.*, **21(S)**, S457-S462.

Mohideen, M.J., J.D. Perkins and E.N. Pistikopoulos (1997b). Robust stability considerations in optimal design of dynamic systems under uncertainty. *J. Process Control*, **7**, 371-386.

Morari, M. and J.D. Perkins (1995). Design for operations. In L.T. Biegler and M.F. Doherty (Eds), *Foundations of Computer-Aided Process Design*, AIChE Symposium Series, No. 304, Vol. 91, CACHE-AIChE, 105-114.

Papalexandri, K.P. and E.N. Pistikopoulos (1996). Generalized modular representation framework for process synthesis. *AIChE J.*, **42**, 1010-1032.

Process Systems Enterprise Ltd (1998). *gPROMS Advanced User's Guide*.

Ross, R., V. Bansal, J.D. Perkins and E.N. Pistikopoulos (1998). A mixed-integer dynamic optimization approach to simultaneous design and control. Paper 230e, AIChE Annual Meeting, Miami Beach.

Ross, R., V. Bansal, J.D. Perkins, E.N. Pistikopoulos, G.L.M. Koot and J.M.G. van Schijndel (1999). Optimal design and control of an industrial distillation system. *Proc. ESCAPE-9*, Budapest, Hungary, in print.

Rowe, D.A., J.D. Perkins and S.P.K. Walsh (1997). Integrated design of responsive chemical manufacturing facilities. *Comp. Chem. Eng.*, **21**, S101-S106.

Santini, J.T., M.J. Cima and R. Langer (1999). A controlled-release microchip. *Nature*, **397**, 335-339.

Seider, W.D., D.D. Brengel and S. Widagdo (1991). Nonlinear analysis in process design. *AIChE J.* **37**, 1-38.

Thomaidis, T.V. and E.N. Pistikopoulos (1995). Flexibility, reliability and maintenance in process design. In L.T. Biegler and M.F. Doherty (Eds), *Foundations of Computer-Aided Process Design*, AIChE Symposium Series, No. 304, Vol. 91, CACHE-AIChE, 260-265.

Tyreus, B.D. (1999). Advances in plant-wide control. Presented at Industrial Research Consortium, Spring Meeting, Centre for Process Systems Engineering, Imperial College.

Westerberg, A.W. and O. Wahnschafft (1996). Synthesis of distillation-based separation processes. In J.L. Anderson (Ed.), *Advances in Chemical Engineering*, Vol 23, Academic Press, New York, 63-170.

Ydstie, B.E. and A.A. Alonso (1997). Process systems and passivity via the Clausius Plank inequality. *Syst. Control Lett.*, **30**, 352-263.

Appendix

The following seven pages contain seven tables with the following table titles:

Table 1. Operability in Process Design.

Table 2. Interactions of Design and Control.

Table 3. Related Control Papers

Table 4. Reliability-Maintenance in Process Design.

Table 5. Optimization under Parametric and Stochastic Uncertainty.

Table 6. Dynamic Optimization.

Table 7. Results of Process Design for Operability Studies.

References	General Description	Type of Process Model	Operability Criterion	Type of Control Scheme	Design Optimiz[n.] Process	Control
1–35	Analysis methods for comparing alternatives	All steady-state (SS) or linear dynamic (LD) except 10 and 30	1–10: Flexibility 11: Robustness 12–35: Controllability	None	No	N/A
36–51	Methods for designing operable systems	All SS except 50	36–47: Flexibility 48–49: Controllability 50–51: Robustness	None	Yes	N/A
52–75	Applic[n.] of analysis & design methods to specific processes	Mainly SS 59–60, 71: LD Others: non-LD	52–58, 62–64: Flexibility Others: Controllability	Mainly open-loop. Closed-loop simul[ns.] sometimes applied.	No	N/A

1. Swaney, R.E. and I.E. Grossmann (1985a,b). *AIChE. J.* **31**, 621–641.
2. Kabatek, U. and R.E. Swaney (1992). *Comp. Chem. Eng.* **16**, 1063–1071.
3. Grossmann, I.E. and C.A. Floudas (1987). *Comp. Chem. Eng.* **11**, 675–693.
4. Chacon-Mondragon, O.L. and D.M. Himmelblau (1988). *Comp. Chem. Eng.* **12**, 383–387.
5. Kubic, W.L, and F.P. Stein (1988). *AIChE. J.* **34**, 583–601.
6. Pistikopoulos, E.N. and T.A. Mazzuchi (1990). *Comp. Chem. Eng.* **14**, 991–1000.
7. Straub, D.A. and I.E. Grossmann (1990). *Comp. Chem. Eng.* **14**, 967–985.
8. Grossmann, I.E. and D.A. Straub (1991). In: *Computer-Oriented Process Engineering* (L. Puigjaner and A. Espuna, Eds.). pp. 49–59. Elsevier.
9. Ostrovsky, G.M., Y.M. Volin, E.I. Barit and M.M. Senyavin (1994). *Comp. Chem. Eng.* **18**, 757–767.
10. Dimitriadis, V.D. and E.N. Pistikopoulos (1995). *Ind. Eng. Chem. Res.* **34**, 4451–4462.
11. Georgiadis, M.C. and E.N. Pistikopoulos (1999). *Ind. Eng. Chem. Res.* **38**, 133–143.
12. Morari, M. (1983). *Chem. Eng. Sci.* **38**, 1881–1891.
13. Holt, B.R. and M. Morari (1985a,b). *Chem. Eng. Sci.* **40**, 59–74, 1229–1237.
14. Skogestad, S. and M. Morari (1987). *Chem. Eng. Sci.* **42**, 1765–1780.
15. Fisher, W.R. and J.M. Douglas (1985). *Comp. Chem. Eng.* **9**, 499–515.
16. Perkins, J.D. and M.P.F. Wong (1985). *Chem. Eng. Res. Des.* **63**, 358–362.
17. Russell, L.W. and J.D. Perkins (1987). *Chem. Eng. Res. Des.* **65**, 453–461.
18. Psarris, P. and C.A. Floudas (1990). *Chem. Eng. Sci.* **45**, 3505–3524.
19. Psarris, P. and C.A. Floudas (1991a,b). *Chem. Eng. Sci.* **46**, 2691–2728.
20. Skogestad, S., M. Hovd and P. Lundstrom (1991). *Modeling Ident. Control.* **12**, 159–177.
21. Hovd, M. and S. Skogestad (1992). In: *Interactions between Process Design and Process Control* (J.D. Perkins, Ed.), pp. 49–54. IFAC Workshop, London. Pergamon.
22. Wolff, E.A., S. Skogestad, M. Hovd and K.W. Mathisen (1992). In: *Interactions between Process Design and Process Control* (J.D. Perkins, Ed.), pp. 127–132. IFAC Workshop, London. Pergamon.
23. Skogestad, S. and E.A. Wolff (1996). *Modeling Ident. Control.* **17**, 167–181.
24. Fararooy, S., J.D. Perkins and T.I. Malik (1992). *Comp. Chem. Eng.* **16**, S517–S522.
25. Fararooy, S., J.D. Perkins, T.I. Malik, M.J. Oglesby and S. Williams (1993). *Comp. Chem. Eng.* **17**, 617–625.
26. Cao, Y., D. Biss and J.D. Perkins (1996). **20**, 337–346.
27. Cao, Y. and D. Biss (1996). *J. Proc. Control* **6**, 37–48.
28. Weitz, O. and D.R. Lewin (1996). *Comp. Chem. Eng.* **20**, 325–335.
29. Zafiriou, E. and H.W. Chiou (1996). *Comp. Chem. Eng.* **20**, 347–355.
30. Soroush, M. (1996). *Comp. Chem. Eng.* **20**, 357–363.
31. Swartz, C.L.E. (1996). *Comp. Chem. Eng.* **20**, 365–371.
32. Young, J.C.C., C.L.E. Swartz and R. Ross (1996). *Comp. Chem. Eng.* **20**, S677–S682.
33. Bahri, P.A., A. Bandoni and J.A. Romagnoli (1996). *Comp. Chem. Eng.* **20**, S787–S792.
34. Chenery, S.D. and S.P.K. Walsh (1998). *J. Proc. Control.* **8**, 165–174.
35. Zheng, A. and R.V. Mahajanam (1999). *Ind. Eng. Chem. Res.* **38**, 999–1006.
36. Halemane, K.P. and I.E. Grossmann (1983). *AIChE. J.* **29**, 425–433.
37. Grossmann, I.E. and M. Morari (1983). In: *Proc. Second Int. Conf. FOCAPD*. (A.W. Westerberg and H.H. Chien, Eds.). pp. 931–1030. CACHE.
38. Chacon-Mondragon, O.L. and D.M. Himmelblau (1996). *Comp. Chem. Eng.* **20**, 447–452.
39. Pistikopoulos, E.N. and Grossmann (1988a,b,c). *Comp. Chem. Eng.* **12**, 719–731, 841–843, 1215–1227.
40. Pistikopoulos, E.N. and Grossmann (1989a,b). *Comp. Chem. Eng.* **13**, 1003–1016, 1087–1096.
41. Straub, D.A. and I.E. Grossmann (1993). *Comp. Chem. Eng.* **17**, 339–354.
42. Pistikopoulos, E.N. and M.G. Ierapetritou (1995). *Comp. Chem. Eng.* **21**, 317–325.
43. Thomaidis, T.V. and E.N. Pistikopoulos (1994). *Comp. Chem. Eng.* **18**, S259–S263.
44. Thomaidis, T.V. and E.N. Pistikopoulos (1995). *Comp. Chem. Eng.* **19**, S687–S692.
45. Varvarezos, D.K., I.E. Grossmann and L.T. Biegler (1995). *Comp. Chem. Eng.* **19**, 1301–1316.
46. Ostrovsky, G.M., Y.M. Volin and M.M. Senyavin (1997). *Comp. Chem. Eng.* **21**, 317–325.
47. Bansal, V., J.D. Perkins and E.N. Pistikopoulos (1999). *Submitted for Publication. AIChE. J.*
48. Palazoglu, A., B. Manousiouthakis and Y. Arkun (1985). *Ind. Eng. Chem. Proc. Des. Dev.* **24**, 802–813.
49. Barton, G.W., M.A. Padley and J.D. Perkins (1992). In: *Interactions between Process Design and Process Control* (J.D. Perkins, Ed.), pp. 95–100. IFAC Workshop, London. Pergamon.
50. Samsatli, N.J., L.G. Papageorgiou and N. Shah (1998). *AIChE. J.* **44**, 1993–2006.
51. Bernardo, F.P. and P.M. Saraiva (1998). *AIChE. J.* **44**, 2007–2017.
52. Saboo, A.K., M. Morari and D.C. Woodcock (1985). *Chem. Eng. Sci.* **40**, 1553–1565.
53. Saboo, A.K., M. Morari and R.D. Colberg (1987a,b). *Comp. Chem. Eng.* **11**, 399–408, 457–468.
54. Colberg, R.D., M. Morari and D.W. Townsend (1989). *Comp. Chem. Eng.* **13**, 821–837.
55. Fisher, W.R., M.F. Doherty and J.M. Douglas (1985). *Ind. Eng. Chem. Res.* **24**, 593–598.
56. Terrill, D.L. and J.M. Douglas (1987). *Ind. Eng. Chem. Res.* **26**, 691–696.
57. Calandranis, J. and G. Stephanopoulos (1986). *Chem. Eng. Res. Des.* **64**, 347–364.
58. Floudas, C.A. and I.E. Grossmann (1987). *Comp. Chem. Eng.* **11**, 319–336.
59. Mathisen, K.W., S. Skogestad and E.A. Wolff (1992). *Comp. Chem. Eng.* **16**, S263–S272.
60. Skogestad, S. (1996). *Comp. Chem. Eng.* **20**, 373–386.
61. Kim, Y.H. and W.L. Luyben (1994). *Chem. Eng. Comm.* **128**, 65–94.
62. Djouad, S., P. Floquet, L. Piboleau and S. Domenech (1994). In: *Proc. 5th Int. Conf. PSE.* (E.S. Yoon, Ed.), pp. 303–306. Kyongju, S. Korea. KIChE.
63. Li, G.Q., B. Hua, B.L. Liu and G.R. Wu (1994). In: *Proc. 5th Int. Conf. PSE.* (E.S. Yoon, Ed.), pp. 407–413. Kyongju, S. Korea. KIChE.
64. Aguilera, N.E. and G. Nasini (1996). *Comp. Chem. Eng.* **20**, 1227–1239.
65. Uzturk, D., A.E.S. Konukman, C. Boyaci and U. Akman (1996). *Comp. Chem. Eng.* **20**, S943–S948.
66. Ohno, H. and E. Nakanishi (1985). *IChemE Symposium Series*. No. 92, 469–479.
67. Annakou, O., A. Meszaros, Z. Fonyo and P. Mizsey (1996). *Hung. J. Ind. Chem.* **24**, 155–160.
68. Koggersbol, A., B.R. Andersen, J.S. Nielsen and S.B. Jorgensen (1996). *Comp. Chem. Eng.* **20**, S853–S858.
69. Zhelev, T.K., P.S. Varbanov and I. Seikova (1998). *Hung. J. Ind. Chem.* **26**, 81–88.
70. Correa, D.J. (1998). *Heat Transfer Eng.* **19**, 22–28.
71. Lewin, D.R. and D. Bogle (1996). *Comp. Chem. Eng.* **20**, S871–S876.
72. Kuhlmann, C., D. Bogle and Z. Chalabi (1997). *Bioprocess Eng.* **17**, 367–374.
73. Russo, L.P. and B.W. Bequette (1997). *Comp. Chem. Eng.* **21**, S571–S576.
74. Russo, L.P. and B.W. Bequette (1997). *Chem. Eng. Sci.* **53**, 27–45.
75. Flores-Tlacuahuac, A. and A. Silva-Beard (1998). *Comp. Chem. Eng.* **22**, S703–S706.

References	General Description	Type of Process Model	Type of Control Scheme	Design Optimization Process	Design Optimization Control
1–4	Design of economically optimal, operable, HENs	Steady-state (SS)	None	Yes	N/A
5–11	Bi-objective optimiz^n economics vs. controllability	5–8: Simple dynamic 9–11: SS	5: P, 8: MPC 6–7, 9–11: None	Yes	5, 8: Yes 6–7, 9–11: N/A
12–24	Economic optimization using dynamic models	Dynamic	12–13: None 15–17: Feedforward, PI 14, 18–23: PI 24: PI, multivariable	Yes	12–13: N/A 15–17: Yes 18–23: Yes 14, 24: No
25–49	Use of analysis tools and/or dynamic simulation to compare alternatives	25–26, 33, 47: SS Others: (linear) dynamic	25–26, 29–31, 33, 46–48: None Others: mainly PI(D) 44: multivariable 49: non-linear	All no except 29–31	No
50–56	Workshops and review papers				

1. Marselle, D.M., M. Morari and D.F. Rudd (1982). *Chem. Eng. Sci.* **37**, 259–270.
2. Georgiou, A. and C.A. Floudas (1990). In: *Proc. Third Int. Conf. FOCAPD*. CACHE.
3. Papalexandri, K.P. and E.N. Pistikopoulos (1994a,b). *Ind. Eng. Chem. Res.* **33**, 1718–1755.
4. Boyaci, C., D. Uzturk, A.E.S. Konukman and U. Akman (1996). *Comp. Chem. Eng.* **20**, S775–S780.
5. Lenhoff, A.M. and M. Morari (1982). *Chem. Eng. Sci.* **37**, 245–258.
6. Palazoglu, A. and Y. Arkun (1986). *Comp. Chem. Eng.* **10**, 567–575.
7. Palazoglu, A. and Y. Arkun (1987). *Comp. Chem. Eng.* **11**, 205–216.
8. Brengel, D.D. and W.D. Seider (1992). *Comp. Chem. Eng.* **16**, 861–886.
9. Abbas, A., P.L. Yue and P.E. Sawyer (1992). Vol. 1, pp. A1–A15. IChemE.
10. Abbas, A., P.E. Sawyer and P.L. Yue (1993). *Chem. Eng. Res. Des.* **71**, 453–456.
11. Luyben, M.L. and C.A. Floudas (1994a,b). *Comp. Chem. Eng.* **18**, 933–993.
12. Nishida, N. and A. Ichikawa (1975). *Ind. Eng. Chem., Proc. Des. Dev.* **14**, 236–242.
13. Nishida, N. Y.A. Liu and A. Ichikawa (1975). *AIChE. J.* **22**, 539–549.
14. Soroush, M. and C. Kravaris (1993a,b). *Ind. Eng. Chem. Res.* **32**, 866–893.
15. Walsh, S.P.K. and J.D. Perkins (1994a). *Comp. Chem. Eng.* **18**, S183–S187.
16. Walsh, S.P.K. and J.D. Perkins (1994b). *Proc. Safety and Environ. Protection* **72**, 102–108.
17. Walsh, S.P.K. and J.D. Perkins (1996). In: *Advances in Chem. Eng.* Vol. 23. pp. 301–402. Academic Press.
18. Rowe, D.A. (1997). Integrated Design of Distributed Chemical Manufacturing Facilities. PhD Thesis. University of London.
19. Mohideen, M.J., J.D. Perkins and E.N. Pistikopoulos (1996). *AIChE. J.* **42**, 2251–2272.
20. Mohideen, M.J., J.D. Perkins and E.N. Pistikopoulos (1997). *J. Proc. Control.* **7**, 371–385.
21. Ross, R., V. Bansal, J.D. Perkins, E.N. Pistikopoulos, G.L.M. Koot and J.M.G. van Schijndel (1999). *Comp. Chem. Eng.* **23**.
22. Bansal, V., R. Ross, J.D. Perkins and E.N. Pistikopoulos (1999). *J. Proc. Control.* **9**.
23. Schweiger, C.A. and C.A. Floudas (1997). In: *Optimal Control: Theory, Algorithms and Applications* (W.W. Hager and P.M. Pardalos, Eds.). pp. 388–435. Kluwer.
24. Bahri, P.A., J.A. Bandoni and J.A. Romagnoli (1997). *AIChE. J.* **43**, 997–1015.
25. Fisher, W.R., M.F. Doherty and J.M. Douglas (1985). *AIChE. J.* **31**, 1538–1547.
26. Fisher, W.R., M.F. Doherty and J.M. Douglas (1988a,b,c). *Ind. Eng. Chem. Res.* **27**, 597–615.
27. Herman, D.J., G.R. Sullivan and S. Thomas (1985). *Chem. Eng. Res. Des.* **63**, 373–377.
28. Handogo, R. and W.L. Luyben (1987). *Ind. Eng. Chem. Res.* **26**, 531–538.
29. Barton, G.W., W.K. Chan and J.D. Perkins (1991). *J. Proc. Control.* **1**, 161–170.
30. Narraway, L.T., J.D. Perkins and G.W. Barton (1991). *J. Proc. Control.* **1**, 243–250.
31. Lear, J.B., G.W. Barton and J.D. Perkins (1995). *J. Proc. Control.* **5**, 49–62.
32. Bouwens, S.M.A.M. and P.H. Kosters (1992). In: *Interactions between Process Design and Process Control* (J.D. Perkins, Ed.), pp. 75–86. IFAC Workshop, London. Pergamon.
33. Laing, D.M. and J.W. Ponton (1992). In: *Interactions between Process Design and Process Control* (J.D. Perkins, Ed.), pp. 87–93. IFAC Workshop, London. Pergamon.
34. Bekkers, P. and J.E. Rijnsdorp (1992). *Comp. Chem. Eng.* **16**, S523–S530.
35. Grassi, V.G. (1993). *ISA Trans.* **32**, 323–332.
36. Helget, A., M. Groebel and E.D. Gilles (1994). In: *Proc. 5th Int. Conf. PSE* (E.S. Yoon, Ed.), pp. 1111–1116. Kyongju, S. Korea. KIChE.
37. Gong, J.P., G. Hytoft and R. Gani (1997). *Comp. Chem. Eng.* **19**, S489–S494.
38. Gong, J.P., G. Hytoft, C. Jaksland and R. Gani (1995). *Comp. Chem. Eng.* **21**, 1135–1146.
39. Elliott, T.R. and W.L. Luyben (1995). *Ind. Eng. Chem. Res.* **34**, 3907–3915.
40. Elliott, T.R. and W.L. Luyben (1996). *Ind. Eng. Chem. Res.* **35**, 3470–3479.
41. Elliott, T.R., W.L. Luyben and M.L. Luyben (1997). *Ind. Eng. Chem. Res.* **36**, 1727–1737.
42. Luyben, M.L., B.D. Tyreus and W.L. Luyben (1996). *Ind. Eng. Chem. Res.* **35**, 758–771.
43. Lyman, P.R., W.L. Luyben and B.D. Tyreus (1996). *Ind. Eng. Chem. Res.* **35**, 3484–3497.
44. van Dijk, J.F.M., S. DeWolf, R. Postma, L.C. Zullo, M. Guenin and P.H. Gustafson (1996). *Comp. Chem. Eng.* **20**, 387–393.
45. Gross, F., E. Baumann, A. Geser, D.W.T. Rippin and L. Lang (1998). *Comp. Chem. Eng.* **22**, 223–237.
46. Hopkins, L., P. Lant and B. Newell (1998). *J. Proc. Control* **8**, 57–68.
47. Gal, I.P., J.B. Varga and K.M. Hangos (1998). *J. Proc. Control* **8**, 251–263.
48. Lababidi, H.M.S., I.M. Alatiqi and R. Banares-Alcantara (1996). *Comp. Chem. Eng.* **20**, S207–S212.
49. Kumar, A. and P. Daoutidis (1999). *AIChE. J.* **45**, 51–68.
50. Interactions between Process Design and Process Control (J.D. Perkins, Ed.), IFAC Workshop, London, 7–8 Sept. 1992. Pergamon.
51. Morari, M. (1992). In: *Interactions between Process Design and Process Control* (J.D. Perkins, Ed.), pp. 3–16. IFAC Workshop, London. Pergamon.
52. IFAC Workshop on Integration of Process Design & Control, IPDC '94, Baltimore, Maryland, 27–28 June 1994. Selected Papers in *Comp. Chem. Eng.* **20**(4) (E. Zafiriou, Ed.).
53. Hashimoto, I. and E. Zafiriou (1995). In: *Proc. Fourth Int. Conf. FOCAPD* (L.T. Biegler and M.F. Doherty, Eds.), pp. 103–104. Snowmass, Colorado, July 10–14, 1994. CACHE.
54. Morari, M. and J.D. Perkins (1995). In: *Proc. Fourth Int. Conf. FOCAPD* (L.T. Biegler and M.F. Doherty, Eds.), pp. 105–114. Snowmass, Colorado, July 10–14, 1994. CACHE.
55. Downs, J.J. and B. Ogunnaike (1995). In: *Proc. Fourth Int. Conf. FOCAPD* (L.T. Biegler and M.F. Doherty, Eds.), pp. 115–123. Snowmass, Colorado, July 10–14, 1994. CACHE.
56. Downs, J.J. and J.J. Siirola (1997). *Paper 189a. Presented at AIChE Annual Meeting*, Los Angeles, Nov. 1997.

References	General Description	Remarks
1–18	Plantwide control	Steady-state or simple dynamic models. Heuristics and/or controllability analysis. Rely on dynamic simulations (no optimization).
19–73	Control structure selection	Almost all use analysis tools to "gain insight". Only 50–52 and 71 directly consider economics within an optimization framework.
74–111	Multivariable controller synthesis	No direct consideration of economics.

1. Price, R.M. and C. Georgakis (1993). *Ind. Eng. Chem. Res.* **32**, 2693–2705.
2. Price, R.M., P.R. Lyman and C. Georgakis (1994). *Ind. Eng. Chem. Res.* **33**, 1197–1207.
3. Fonyo, Z. (1994). *Comp. Chem. Eng.* **18**, S483–S492.
4. Chou, Y.S., C.Y. Lee and C.T. Liou (1995). *J. Chinese Inst. of Eng.* **18**, 343–352.
5. Yi, C.K. and W.L. Luyben (1995). *Ind. Eng. Chem. Res.* **34**, 2393–2405.
6. Luyben, M.L and W.L. Luyben (1995). *Ind. Eng. Chem. Res.* **34**, 3885–3898.
7. Luyben, M.L, B.D. Tyreus and W.L. Luyben (1997). *AIChE. J.* **43**, 3161–3174.
8. Belanger, P.W. and W.L. Luyben (1997). *Ind. Eng. Chem. Res.* **36**, 706–716.
9. Belanger, P.W. and W.L. Luyben (1998a,b,c). *Ind. Eng. Chem. Res.* **37**(2), 516–546.
10. Banerjee, A. and Y. Arkun (1996). *Comp. Chem. Eng.* **19**, 453–480.
11. Storz, M., M. Friedrich and E.D. Gilles (1996). *Comp. Chem. Eng.* **20**, 437–445.
12. Wu, K.L. and C.C. Yu (1996). *Comp. Chem. Eng.* **20**, 1291–1316.
13. Wu, K.L. and C.C. Yu (1997). *Ind. Eng. Chem. Res.* **36**, 2239–2251.
14. Dimian, A.C., A.J. Groenendijk and P.D. Iedema (1996). *Comp. Chem. Eng.* **20**, S805–S810.
15. Dimian, A.C., A.J. Groenendijk, S.R.A. Kersten and P.D. Iedema (1997). *Comp. Chem. Eng.* **21**, S291–S296.
16. Lausch, H.R., G. Wozny, M. Wutkewicz and H. Wendeler (1998). *Chem. Eng. Res. Des.* **76**, 185–192.
17. McAvoy, T. (1998). *Comp. Chem. Eng.* **22**, 1543-1552.
18. McAvoy, T. and R. Miller (1999). *Ind. Eng. Chem. Res.* **38**, 412–420.
19. Bristol, E.H. (1966). *IEEE Trans. Aut. Control.* **AC-11**, 133–134.
20. Niederlinski, A. (1971). *Automatica.* **7**, 691–701.
21. Witcher, M.F. and T.J. McAvoy (1977). *ISA Trans.* **16**, 35–41.
22. Morari, M., Y. Arkun and G. Stephanopoulos (1980). *AIChE. J.* **26**, 220–232.
23. Morari, M. and G. Stephanopoulos (1980a,b). *AIChE. J.* **26**, 232–260.
24. Arkun, Y. and G. Stephanopoulos (1980). *AIChE. J.* **26**, 975–991.
25. Arkun, Y. and G. Stephanopoulos (1981). *AIChE. J.* **27**, 779–793.
26. Stephanopoulos, G. (1983). *Comp. Chem. Eng.* **7**, 331–365.
27. Calandranis, J. and G. Stephanopoulos (1988). *Comp. Chem. Eng.* **12**, 651–669
28. Ng, C.S. and G. Stephanopoulos (1996). *Comp. Chem. Eng.* **20**, S999–S1004.
29. Arkun, Y. and S. Ramakrishnan (1984). *Chem. Eng. Sci.* **39**, 1167–1179.
30. Manousiouthakis, V., R. Savage and Y. Arkun (1986). *AIChE. J.* **32**, 991–1003.
31. Arkun, Y. (1988). *AIChE. J.* **34**, 672–675.
32. Reeves, D.E. and Y. Arkun (1989). *AIChE. J.* **35**, 603–613.
33. Chiu, M.S. and Y. Arkun (1990). *Ind. Eng. Chem. Res.* **29**, 369–373.
34. Chiu, M.S. and Y. Arkun (1991). *Automatica.* **27**, 419–421.
35. Lee, T.K., M.S. Chiu and Y. Arkun (1998). *Ind. Eng. Chem. Res.* **37**, 4734–4739.
36. Lau, H. J. Alvarez and K.F. Jensen (1985). *AIChE. J.* **31**, 427–439.
37. Grosdidier, P., M. Morari and B.R. Holt (1985). *Ind. Eng. Chem. Fundam.* **24**, 221–235.
38. Shimizu, K., B.R. Holt, M. Morari and R.S.K. Mah (1985). *Ind. Eng. Chem. Proc. Des. Dev.* **24**, 852–858.
39. Grosdidier, P. and M. Morari (1986). *Automatica.* **22**, 309–319.
40. Grosdidier, P. (1990). *Comp. Chem. Eng.* **14**, 687–689.
41. Bartee, J.F., K.F. Bloss and C. Georgakis (1989). Design of Nonlinear Reference System Control Structures. *Presented at AIChE Annual Meeting, San Francisco, USA.*
42. Skogestad, S. and M. Morari (1987). *AIChE. J.* **33**, 1620–1635.
43. Skogestad, S., P. Lundstrom and E.W. Jacobsen (1990). *AIChE. J.* **36**, 753–764.
44. Skogestad, S. and M. Morari (1992). *Modeling Ident. Control.* **13**, 113–125.
45. Skogestad, S. and M. Hovd (1990). *Proc. American Control Conference, San Diego, USA.* pp. 2133–2139.
46. Hovd, M. and S. Skogestad (1992). *Automatica.* **28**, 989–996.
47. Hovd, M. and S. Skogestad (1993). *AIChE. J.* **39**, 1938–1953.
48. Hovd, M. and S. Skogestad (1994). *Ind. Eng. Chem. Res.* **33**, 2134–2139.
49. Skogestad, S. and K. Havre (1996). *Comp. Chem. Eng.* **20**, S1005–S1010.
50. Narraway, L. and J.D. Perkins (1993). *Ind. Eng. Chem. Res.* **32**, 2681–2692.
51. Narraway, L. and J.D. Perkins (1994). *Comp. Chem. Eng.* **18**, S511–S515.
52. Heath. J.A. and J.D. Perkins (1994). *Proc. IChemE Research Event.* pp. 841–843
53. Zhu, Z.X. and A. Jutan (1993). *Chem. Eng. Sci.* **48**, 2337–2343.
54. Zhu, Z.X. (1996). *Ind. Eng. Chem. Res.* **35**, 4091–4099.
55. Zeghal, S. and A.N. Palazoglu (1993). *Comp. Chem. Eng.* **17**, S335–S341.
56. Ponton, J.W. and D.M. Laing (1993). *Chem. Eng. Res. Des.* **71**, 181–188.
57. Karlsmose, J., A. Koggersbol, N. Jensen and S.B. Jorgensen (1994). *Comp. Chem. Eng.* **18**, S465–S470.
58. Lin, X.G., M.O. Tade and R.B. Newell (1994). *Chem. Eng. Res. Des.* **72**, 26–37.
59. Lin, X., R.B. Newell, P.L. Douglas and S.K. Mallick (1994). *Int. J. of Sys. Sci.* **25**, 1437–1459.
60. Campo, P.J. and M. Morari (1994). *IEEE Trans. Aut. Control.* **39**, 932–943.
61. Lee, J.H., R.D. Braatz, M. Morari and A. Packard (1995). *Automatica.* **31**, 229–235.
62. Braatz, R.D., J.H. Lee and M. Morari (1996). *Comp. Chem. Eng.* **20**, 463–468.
63. Cao, Y. and D. Rossiter (1997). *Comp. Chem. Eng.* **21**, 563–569.
64. Cao, Y., D. Rossiter and D. Owens (1997). *Comp. Chem. Eng.* **21**, S403–S408.
65. Cao, Y. and D. Rossiter (1998). *J. Proc. Control.* **8**, 175–183.
66. Semino, D. and G. Giuliani (1997). *Comp. Chem. Eng.* **21**, S273–S278.
67. Stichlmair, J. (1995). *Chem. Eng. Proc.* **34**, 61–69.
68. Haggblom, K.E. (1997). *Comp. Chem. Eng.* **21**, 1441–1449.
69. Lee, J.H., P. Kesavan and M. Morari (1997). *IEEE Trans. Control Sys. Tech.* **5**, 402–416.
70. Hansen, J.E., S.B. Jorgensen, J. Heath and J.D. Perkins (1998). *J. Proc. Control.* **8**, 185–195.
71. Walsh, S.P.K., S. Chenery, P. Owen and T.I. Malik (1997). *Comp. Chem. Eng.* **21**, S391–S396.
72. Kookos, I.K. and A.I. Lygeros (1998). *Chem. Eng. Res. Des.* **76**, 458–464.
73. Wisnewski, P.A. and F.J. Doyle (1998). *J. Proc. Control.* **8**, 487–495.
74. Doyle, J.C. and G. Stein (1981). *IEEE Trans. Aut. Control.* **26**, 4–16.
75. Arkun, Y., V. Manousiouthakis and P. Putz (1984). *Int. J. Control.* **40**, 603–629.
76. Nwokah, O.D.I. (1984). *Int. J. Control.* **40**, 1189–1206.
77. Ohm, D.Y., J.W. Howze and S.P. Bhattacharyya (1985). *Automatica.* **21**, 35–55.
78. Garcia, C.E. and M. Morari (1985a,b). *Ind. Eng. Chem. Proc. Des. Dev.* **24**, 472–494.
79. Mandler, J.A., M. Morari and J.H. Seinfeld (1986). *Ind. Eng. Chem. Fundam.* **25**, 645–655.
80. Rivera, D.E. and M. Morari (1987). *Int. J. Control.* **46**, 505–527.
81. Zafiriou, E. and M. Morari (1987). *Int. J. Control.* **46**, 2087–2111.
82. Zafiriou, E. and M. Morari (1988). *Comp. Chem. Eng.* **12**, 757–765.
83. Zafiriou, E. and M. Morari (1991). *Int. J. Control.* **54**, 665–704.
84. Hsiao, F.H. and B.S. Chen (1988). *J. Dyn. Sys. Meas. Control - Trans. ASME.* **110**, 336–340.
85. Hsieh, J.G. and F.H. Hsiao (1989). *Int. J. Sys. Sci.* **20**, 2515–2537.
86. Tzouanas, V.K., W.L. Luyben, C. Georgakis and L.H. Ungar (1990). *Ind. Eng. Chem. Res.* **29**, 382–389.
87. Chang, J.W. and C.C. Yu (1992). *AIChE. J.* **38**, 521–534.
88. Meadowcroft, T.A., G. Stephanopoulos and C. Brosilow (1992). *AIChE. J.* **38**, 1254–1278.
89. Grimble, M.J. (1993). *IEE Proc. - Control Theory and Applications.* **140**, 353–363.
90. Hui, S. and S.H. Zak (1993). *Int. J. Robust and Nonlinear Control.* **3**, 115–132.
91. Chan, J.T.H. (1994). *Automatica.* **30**, 1479–1483.
92. Chan, J.T.H. (1995). *Automatica.* **31**, 781–786.
93. Huang, H.P., M.L. Roan and H.C. Sheu (1994). In: *Proc. 5th Int. Conf. PSE.* (E.S. Yoon, Ed.), pp. 1099–1104. Kyongju, S. Korea. KIChE.
94. Ramchandran, S. and R.R. Rhinehart (1995). *J. Proc. Control.* **5**, 115–128.
95. Kothare, M.V., V. Balakrishnan and M. Morari (1996). *Automatica.* **32**, 1361–1379.
96. Rosenvasser, Y.N. (1996). *Autom. Rem. Control.* **57**, 1091–1107.
97. Daoutidis, P. and C. Kravaris (1992). *Chem. Eng. Sci.* **47**, 1091–1107.
98. Daoutidis, P. and C. Kravaris (1994). *Chem. Eng. Sci.* **49**, 433–447.
99. Daoutidis, P. and A. Kumar (1994). *AIChE. J.* **40**, 647–669.
100. Christofides, P.D. and P. Daoutidis (1997). *J. Proc. Control.* **7**, 313–328.
101. Kravaris, C. and Soroush, M. (1990). *AIChE. J.* **36**, 249–264.
102. Soroush, M. and C. Kravaris (1993). *AIChE. J.* **39**, 1920–1937.
103. Soroush, M. and S. Valluri (1998). *Comp. Chem. Eng.* **22**, 1065–1088.
104. Hovd, M., R.D. Braatz and S. Skogestad (1997). *Automatica.* **33**, 433–439.
105. Haddad, W.M. and D.S. Bernstein (1991). *Systems & Control Letters.* **17**, 191–208.
106. How, J.P., E.G. Collins and W.M. Haddad (1996). *IEEE Trans. Control Sys. Tech.* **4**, 200–207.
107. Haddad, W.M. and V. Kapila (1997). *Int. J. Robust and Nonlinear Control.* **7**, 675–710.
108. Haddad, W.M. and V. Kapila (1998). *Int. J. Robust and Nonlinear Control.* **8**, 567–583.
109. Haddad, W.M. and V. Kapila (1998). *IEEE Trans. Automatic Control.* **43**, 987–992.
110. Aleksandrov, A.G. and Y.N. Chestnov (1998a,b). *Autom. Rem. Control.* **59**, 973–983, 1153–1164.
111. Pantas, A.V. (1998). Robustness and Economic Considerations in the Design of Multivariable Optimal Controllers. PhD Thesis. University of London.

	rigorous optimization of maintenance	Interactions of R/M with process design	Interactions of process operations with R/M	optimization of system effectiveness criteria	maintenance modelling flexibility	incorporation of process uncertainty
continuous processes	Ref.: 1-77	Ref.: 15-31	Ref.: 1-17	Ref.: 15-18 32-43	Ref.: 15-18 44-49	Ref.: 29,44	Ref.: 11-13 15-18 35, 40, 42
batch processes	Ref.: 78-86	Ref.: 86	Ref.: 85	Ref.: 78-83, 85	Ref.:	Ref.: 85	Ref.:

1. de la Mare, R., *Chemistry and Industry*, 3, 366-369 (1975).
2. Fahner, J. and W. Neumann, *Chemische Technik*, pages 319-322 (1997).
3. Henley, E. J. and H. Hoshino, *Ind. Eng. Chem. Fundam.*, 16, 439-443 (1977).
4. Konoki, K., *J. Chem. Eng. Japan*, 4(1), 77-80 (1971).
5. Ridell, H. S., *Terotechnica*, 2, 9-21 (1981).
6. Rudd, D. F., *Ind. & Eng. Chem. Fund.*, 1, 138-143 (1962).
7. Malathronas, J. P., J. D. Perkins, and R. L. Smith, *IIE Transactions*, 15, 195-201 (1983).
8. SchmidtTraub, H and M Koster and T Holkotter and N Nipper, *Comp. Chem. Eng.*, S22 S499-S504 (1998).
9. Georgiadis, M.C. and S. Macchietto, *Comp. Chem. Eng.*, S21 S337-S342 (1997).
10. Calabria, R., G. Pulcini, and M. Rapone, *Quality and Reliability Engineering International*, 11(3), 183-188 (1995).
11. Thomaidis, T. V. and E. N. Pistikopoulos, *Foundation of Computer Aided Process Design*, FOCAPD '94 (1994a).
12. Thomaidis, T. V. and E. N. Pistikopoulos, *Comp. Chem. Eng.*, 18, 259-263 (1994b).
13. Thomaidis, T. V. and E. N. Pistikopoulos, *IEEE Transactions on Reliability*, 44(2), 243-250 (1995).
14. Boukas, E. K., Q. Zhang, and G. Yin, *Lecture Notes in Control and Information Sciences*, 214, 197-222 (1996b).
15. Vassiliadis, C. G. and E. N. Pistikopoulos, *European Symposium on Safety and Reliability*, Trondheim, Norway, Vol.1, 255-272 (1998a).
16. Vassiliadis, C. G. and E. N. Pistikopoulos, *Comp. Chem. Eng.*, 22, 521-528 (1998b).
17. Vassiliadis, C. G. and E. N. Pistikopoulos, *Annual Reliability and Maintainability Symposium*, 18-21 January, Washington, Washington DC, pages 78-83 (1999).
18. Pistikopoulos, E. N. and C. G. Vassiliadis, *Conf. on Foundation of Computer Aided Operations*, FOCAP-O. Snowmass, Utah (1998).
19. Alkhamis, T. M. and J. Yellen, *Applied Mathematical Modelling*, 19, 543 (1995).
20. Arueti, S. and D. Okrent, *Reliability Engineering and System Safety*, 30, 93-114 (1990).
21. Aven, T. and R. Dekker, *Reliability Engineering and System Safety*, 58, 61-67 (1998).
22. Boukas, E. K. and H. Yang, *IEEE Transactions on Automatic Control*, 41(6), 881-885 (1996).
23. Boukas, E. K., J. Yang, G. Yin, and Q. Zhang, *Journal of Optimization Theory and Applications*, 91(2), 347-361 (1996a).
24. Boukas, E. K., Q. Zhang, and G. Yin, *IEEE Transactions on Automatic Control*, 40(4), 1098-1102 (1995).
25. Chien, T. W., C. H. Lin, and G. Sphicas, *Quality and Reliability Engineering International*, 13(4), 225-233 (1997).
26. D C Dietz, M. R., *IIE Transactions*, 29(5), 423-433 (1997).
27. Dekker, R., *Reliability Engineering and System Safety*, 51, 229-240 (1996).
28. Dekker, R. and P. A. Scarf, *Reliability Engineering and System Safety*, 60, 111-119 (1998).
29. Tan, J. S. and M. A. Kramer, *Comp. Chem. Eng.*, 21(12) (1997).
30. Tseng, S. T., *IIE Transactions*, 28, 687-694 (1996).
31. Vatn, J., P. Hokstad, and L. Bodsberg, *Reliability Engineering and System Safety*, 51, 241-257 (1996).
32. Ashayeri, S. W. and A Teelen and W Selen, *International Journal of Production Research*, 34(12), 3311-3326 (1996).
33. Pintelon, L.M.A. and P.L.B VanOuyvelde and L.F. Gelders, production process. *International Journal of Production Research*, 33(8), 2111-2123 (1995).
34. Procaccia, H., R. Cordier, and S. Muller, *Reliability Engineering and System Safety*, 55, 143-149 (1997).
35. E N Pistikopoulos, T A Mazzuchi, C. D. M. and T. V. Thomaidis, In *Proceedings of 11th Advances in Reliability University of Liverpool, UK*,18-20 April 1990, pages 284-297 (1990).
36. AlSultan, K. S. and M. A. AlFawsan, deterioration. *Production Planning and Control*, 9(1), 66-73 (1997).
37. Grievink, J. K., K. Smit, R. Dekker, and C. F. H. van Rijn, *Conf. on Foundation of Computer Aided Operations*, FOCAP-O, Crested Butte, Colorado (1993).
38. Gupta, S. M. and Y. A. Y. AlTurki, *Production Planning and Control*, 9(4), 349-359 (1998).
39. Kramer, F. J. and S. X. Bai, *Optimal Control Applications and Methods*, 17(4), 281-307 (1996).
40. Pistikopoulos, E. N., T. A. Mazzuchi, C. Maranas, and T. V. Thomaidis, In *Proceedings of 4th Int. Symposium PSE'91*, Quebec Canada (1991).
41. Song, D. P. and Y. X. Sun, *Automatica*, 34(9), 1047-1060 (1998).
42. Straub, D. A. and I. E. Grossmann, *Comp. Chem. Eng.*, 29(3), 967 (1990).
43. Nelson, S.C. and M.J. Haire and J.C. Schryver, in *Proceedings: Annual Reliability and Maintainability Symposium*, 150-156 (1992).

44. Brunelle, R. D. and K. C. Kapur, *Proc. Annual Reliability and Maintainability Symposium*, Anaheim CA, pages 286-292 (1998).
45. Lamb, R. G., *Hydrocarbon Processing*, 75(1), 51-55 (1996).
46. Narahari, Y. and N. Viswanadham, *IEEE Transactions on Robotics and Automation*, 10(2), 230-244 (1994).
47. Pujadas, W. and F. F. Chen, *Comp. Ind. Engng*, 31, 241-244 (1996).
48. van Rijn, C. F. H., *Conf. on Foundation of Computer Aided Operations*, FOCAP-O, Park City, Utah. (1987).
49. Woodhouse, J., *6th National Conference on Computers for Maintenance Management* (1986).
50. Falscheer, F. C., *Comp. Chem. Eng.*, S18, S731-S735 (1994).
51. Mauney, R. A. and M.E.G. Hunter, *Process Safety Progress*, 16(4), 262-268 (1997).
52. Bloch, H. P., *Hydrocarbon Processing*, pages 83-86 (1996).
53. Boukas, E. K., J. Yang, G. Yin, and Q. Zhang, *Stochastic Analysis and Applications*, 15(3), 269-293 (1997).
54. de la Mare, R. F. and K. J. Ball, *Reliability Engineering*, 2, 289-302 (1981).
55. Harnly, J. A., *Process Safety Progress*, 17(1), 32-38 (1998).
56. Hassett, T. F. and D. L. Dietrich, *Journal of Manufacturing Systems*, 14(6), 427-438 (1995).
57. Langseth, H., K. Haugen, and H. Sandtorv, *Reliability Engineering and System Safety*, 60, 103-110 (1998).
58. Lee, J., *Comp. Ind. Eng.*, 28(4), 793-811 (1995).
59. Michelsen, O., *Reliability Engineering and System Safety*, 60, 179-181 (1998).
60. Nahara, K., *Conf. on Foundation of Computer Aided Operations*, POCAP-O, pages 111-132 (1993).
61. Prickett, P., *Industrial Management and Data Systems*, 97(3-4), 143-151 (1997).
62. Richet, D., N. Cotaina, M. Gabriel, and K. Oreilly, *Control Engineering Practice*, 3(7), 1029-1034 (1995).
63. Sherwin, D. J. and P. P. Lees, *Proc Instn Mech Engrs*, 194, 301-307 (1980a).
64. Sherwin, D. J. and P. P. Lees, *Proc Instn Mech Engrs*, 194, 308-319 (1980b).
65. Tamaki, A., *Terotechnica*, 2, 3-4 (1981).
66. Tan, J. S. and M. A. Kramer, *AIChE Annual Meeting*, San Fransisco, November '94 (1994).
67. Tan, J. S. and M. A. Kramer, Hazard function modeling using cross validation: from data collection to model selection. *Reliability Engineering and System Safety* (1995).
68. van der Schaaf, T. W., *Microelectronics and Reliability*, 35, 1233-1243 (1995).
69. van Noortwijk, J. M., R. Dekker, R. M. Cooke, and T. A. Mazzuchi, *IEEE Transactions on Reliability*, 41(3), 427-432 (1992).
70. Wu, B., J. J. M. Seddon, and W. L. Currie, *International Journal of Production Research*, 30(11), 2683-2696 (1992).
71. Zakarian, A. and A. Kusiak, *IEEE Transactions on Robotics and Automation*, 13(2), 161-168 (1997).
72. Zeng, S. W., *Reliability Engineering and System Safety*, 55(2), 151-162 (1997).
73. Gaal, Z., *Chemische Technik*, 43(1) 33-35 (1991).
74. Gunn, D. J., *Terotechnica*, 2(4), 281-288 (1981).
75. Hedden, H., *Chemie ingenieur technik. Chemische Technik*, 38(9), 794-799 (1986).
76. Lauenstein, G., E. Nowotnick, and K. Renger, *Chemische Technik*, 36(8), 349-350 (1984).
77. Mann, L., *Maintenance Management International*, 40(10), 77-84 (1988).
78. Bassett, M. H., J. F. Pekny, and G. V. Reklaitis, *Com. Chem. Eng.*, 21, S1203-S1208 (1997).
79. Bassett, M. H., J. F. Pekny, and G. V. Reklaitis, *Industrial and Engineering Chemistry Research*, 36(5), 1717-1726 (1997).
80. Dedopoulos, I. T. and N. Shah, *Industrial & Engineering Chemistry Research*, 34, 192-201 (1995).
81. Sanmarti, E., A. Espuna, and L. Puigjaner, *Comp. Chem. Eng.*, page 1157 (1997).
82. Sanmarti, E., A. Espuna, and L. Puigjaner, *Comp. Chem. Eng.*, S19, S565-S570 (1995).
83. Kelly, A. and G. Seddon, *Maintenance Management International*, 10(4), 58-64 (1984).
84. Rotstein, G. E., R. Lavie, and D. R. Lewin, *Comp. Chem. Eng.*, 20, 201 (1996).
85. Malygin, E.N. and T.A. Prolova and M.N. Krasnyanskii, *Theoretical Foundations of Chemical Engineering*, 32(5), 518-525 (1998).
86. Dedopoulos, I. T. and N. Shah, *Trans IChemE*, 74, 307-320 (1996).

CLASSIFICATION	ROLE	SPECIFIC FEATURES							
				Linear	Convex Nonlinear	Mixed-Integer Linear	Mixed-Integer Nonlinear	Non-convex [*,#]	Parallel Algorithms [*]	Dynamic Systems
Parametric Programming/ Sensitivity Analysis [1-44]	Obtain profile of optimal solution as a function of uncertain parameters	Single Parametric	Theory	✓ [26,35]	✓ [9]	✓ [21,31]	✓ [1,29,31]	✗ [43]	✗	✗
			Computational Tools	✓ [16]	✓ [9]	✗	✗	✗	✗	✗
		Multi-Parametric	Theory	✓ [12,13]	✓ [7]	✓ [2,3,8]	✓ [7]	✗ [44]	✗	✗
			Computational Tools	✗	✗	✗	✗	✗	✗	✗
Stochastic Programming [45-129]	Optimal solution obtained as an expected economic or performance criteria		Theory	✓ [45-48,59-74,77,79-80,83-87,89-90, 92,95-96,98-99,102-115]	✓	✓	✓	✗ [125-127]	✗ [46]	✗
			Computational Tools	✓ [70]	✗	✗	✗	✗	✓ [46]	✗

✓ - Yes
✗ - No

[*] Relatively fewer references and/or for problems with specific structure
[#] [43-44] for parametric programming and [122-129] for stochastic programming

1. Acevedo, J. and E.N. Pistikopoulos (1996). *Ind. Eng. Chem. Res.* **35** 147-158.
2. Acevedo, J. and E.N. Pistikopoulos (1997). *Ind. Eng. Chem. Res.* **36** 717.
3. Acevedo, J. and E.N. Pistikopoulos (1998). *Op. Res. Letters.* Accepted for publication.
4. Bailey, M.G. and B.E. Gillet (1980). *J. Oper. Res.* **31** 257-262.
5. Carstensen, P.J. (1983). *Math. Prog.* **26** 64-75.
6. Cooper, M.W. (1981). *Naval Res. Log. Quarterly* **28** 301-307.
7. Dua, V. and E.N. Pistikopoulos (1998). *Comp. Chem. Eng. (Suppl.)* **22** S995.
8. Dua, V. and E.N. Pistikopoulos (1999). *Annals of Op. Res..* Accepted for publication.
9. Fiacco, A.V. (1983). *Introduction to Sensitivity Analysis in Nonlinear Prog.* (Academic Press, Inc., New York).
10. Fiacco, A.V. and J. Kyparisis (1986). *J. Optim. Theory Appl.* **48** 95-126.
11. Fiacco, A.V. and J. Kyparisis (1988). *Math. Prog.* **40** 213-221.
12. Gal, T. (1995). *Postoptimal Analyses, Parametric Prog., and Related Topics* (de Gruyter, New York).
13. Gal, T. and J. Nedoma (1972). *Management Sci.* **18** 406-422.
14. Ganesh, N. and L.T. Biegler (1987). *AIChE J.* **33** 282-296.
15. Geoffrion, A.M. and R. Nauss (1977). *Management Sci.* **23** 453.
16. Greenberg, H.J. (1996). *Annals of Op. Res.* **65** 91.
17. Holm, S. and D. Klein (1978). *Europ. J. Oper. Res.* **2** 50-53.
18. Holm, S. and D. Klein (1984). *Math. Prog. Study* **21** 97.
19. Jansen, B., J.J. de Jong, C. Roos and T. Terlaky (1997). *Europ. J. Oper. Res.* **101** 15-28.
20. Jenkins, L. (1982). *Management Sci.* **28** 1270-1284.
21. Jenkins, L. (1990). *Annals of Op. Res.* **27** 77.
22. Klein, D. and S. Holm (1979). *Management Sci.* **25** 64-72.
23. Marsten, R.E. and T.L. Morin (1977). *Annals of Discr. Math.* **1** 375-390.
24. McBride, R.D. and J.S. Yormark (1980). *Management Sci.* **26** 784-795.
25. Marty, K.G. (1980). *Math. Prog.* **19** 213-219.
26. Marty, K.G. (1983). *Linear Prog.* (John Wiley and Sons, New York).
27. Nishida, N. and Y. Ohtake (1986). *Comp. Chem. Eng.* **10** 119-122.
28. Ohtake, Y. and N. Nishida (1985). *Oper. Res. Lett.* **4** 41.
29. Papalexandri, K.P. and T.I. Dimkou (1998). *Comp. Chem. Res.* **37** 1866-1882.
30. Pertsinidis, A. (1992). Ph.D. Dissertation, Carnegie-Mellon University, Pittsburgh.
31. Pertsinidis, A. I.E. Grossmann and G.J. McRae (1998). *Comp. Chem. Eng.* **22(SS)** S205-S212.
32. Piper, C.J. and A.A. Zoltners (1976). *Management Sci.* **22** 759-765.
33. Roodman, G.M. (1972). *Naval Res. Log. Quarterly* **19** 435-447.
34. Roodman, G.M. (1974). *Naval Res. Log. Quarterly* **21** 595.
35. Roos, C., T. Terlaky and J.-Ph. Vial (1997). *Theory and Algorithms for Linear Optimisation: An Interior Point Approach* (John Wiley and Sons, Chichester).
36. Seferlis, P. and A.N. Hrymak (1996). *Comp. Chem. Eng.* **20** 1177.
37. Shapiro, J.F. (1977). *Annals of Discr. Math.* **1** 467-477.
38. Skorin-Kapov, J. (1989). *Math. Prog.* **44** 67-75.
39. Williams, A.C. (1989). *Bull. London Math. Soc.* **28** 311-316.
40. Williams, H.P. (1996). *J. Opt. Theory Appl.* **90** 257-278.
41. Williams, H.P. (1995). *Bull. London Math. Soc.* **28** 311-316.
42. Wolbert, D., X. Joulia, B. Koehret and L.T. Biegler (1994). *Comp. Chem. Eng.* **18** 1083.
43. Benson, H.P. (1982). *J. Opt. Theory Appl.* **38** 319-340.
44. Fiacco, A.V. (1990). *Annals of Op. Res.* **27** 381-396.
45. Acevedo, J. and E.N. Pistikopoulos (1997). *Ind. Eng. Chem. Res.* **36** 2262-2270.
46. Acevedo, J. and E.N. Pistikopoulos (1998). *Comp. Chem. Eng.* **22** 647-671.
47. Ahmed, S. and N.V. Sahinidis (1998). *Ind. Eng. Chem. Res.* **37** 1883-1892.
48. Ahmed, S., N.V. Sahinidis and E.N. Pistikopoulos (1998). *Ind. Eng. Chem. Res.* Submitted for publication.
49. Bienstock, D. and J.F. Shapiro (1988). *Management Sci.* **34** 215-229.
50. Birge, J.R. (1982). *Math. Prog.* **24** 314.
51. Birge, J.R. (1985). *Math. Prog.* **25** 31.
52. Birge, J.R. (1985). *Oper. Res.* **59** 1-18.
53. Birge, J.R. and M.A.H. Dempster (1995). *J. Global Opt.* **9** 417-451.
54. Birge, J.R. and R. Wets (1989). *Math. Prog.* **43** 131.
55. Bloom, J.A. (1983). *Oper. Res.* **31** 84-100.
56. Bodrov, V.I., S.I. Dvoretskii and D. Dvoretskii (1997). *Theor. Found. Chem. Eng.* **31** 494-499.
57. Bok, J.K., H. Lee and S. Park (1998). *Comp. Chem. Eng.* **22** 1037-1049.
58. Borision, A.N., P.A. Morris and S.S. Oren (1984). *Oper. Res.* **32** 1052-1068.
59. Chaudhuri, P.D. and U.M. Diwekar (1996). *AIChE J.* **42** 742-752.
60. Clay, R.L. and I.E. Grossmann (1994). *Chem. Engng. Res. Des.* **72** 415-419.
61. Clay, R.L. and I.E. Grossmann (1997). *Comput. Chem. Engng.* **21** 751-774.
62. Dantzig, G.B. and P.W. Glynn (1990). *Annals of Op. Res.* **22** 1-21.
63. Diwekar, U.M. and E.S. Rubin (1991). *Comp. Chem. Eng.* **15** 105-114.
64. Floudas, C.A. and I.E. Grossmann (1987). *Comp. Chem. Eng.* **11** 319.
65. Gassmann, H.I. (1998). *Annals Oper. Res.* **82** 107-137.
66. Grossmann, I.E. and K.P. Halemane (1982). *AIChE J.* **28** 686.
67. Grossmann, I.E. and C.A. Floudas (1986). *Comp. Chem. Eng.* **11** 675-693.
68. Grossmann, I.E., K.P. Halemane and R.E. Swaney (1983). *Comp. Chem. Eng.* **7** 439-462.
69. Halemane, K.P. and I.E. Grossmann (1983). *AIChE J.* **29** 425.
70. IBM Stochastic Solutions, http://www6.software.ibm.com/es/oslv2/features/stoch.htm.
71. Ierapetritou, M.G., J. Acevedo and E.N. Pistikopoulos (1996). *Comp. Chem. Eng.* **20** 703-709.
72. Ierapetritou, M.G. and E.N. Pistikopoulos (1994). *Ind. Eng. Chem. Res.* **33** 1930-1942.
73. Ierapetritou, M.G., E.N. Pistikopoulos and C.A. Floudas (1996). *Comp. Chem. Eng.* **20** 1499-1516.
74. Johns, W.R., G. Marketos and D.W.T. Rippin (1978). *Trans. IChemE.* **56** 249-257.
75. Jonsbraten, T.W. (1998). *J. Oper. Res. Soc.* **49** 811-818.
76. Jonsbraten, T.W., R.J.B. Wets and D.L. Woodruff (1998). *Annals Oper. Res.* **82** 83-106.
77. Kall, P. and S.W. Wallace, *Stochastic Prog.* (John Wiley and Sons, Chichester, 1994).
78. Kubic, W.L. and F.P. Stein (1988). *AIChE J.* **34** 583.
79. Liu, M.L. and N.V. Sahinidis (1996). *Ind. Eng. Chem. Res.* **35** 4154-4165.
80. Maranas, C.D. (1997). *AIChE J.* **43** 1250.
81. Modiano, E.M. (1987). *Oper. Res.* **35** 185-197.
82. Mulvey, J.M., R.J Vanderbei and S.A. Zenios (1995). *Oper. Res.* **43** 264-281.
83. Pai, C.C.D. and R.R. Hughes (1987). *Comp. Chem. Eng.* **11** 695-706.
84. Painton, L.A. and U.M. Diwekar (1994). *Comp. Chem. Eng.* **18** 369-381.
85. Pan, W., M.A. Tatang, G.J. McRae and R.G. Prinn (1997). *J. Geophysical Res.* **102** 21915.
86. Pan, W., M.A. Tatang, G.J. McRae and R.G. Prinn (1998). *J. Geophysical Res.* **103** 3815.
87. Paules IV, G.E. and C.A. Floudas (1992). *Comp. Chem. Eng.* **16** 189-210.
88. Pereira, M.V.F. and L.M.V.G. Pinto (1991). *Math. Prog.* **52** 359-375.
89. Petkov, S.B. and C.D. Maranas (1997). *Ind. Eng. Chem. Res.* **36** 4864.
90. Phenix, B.D., J.I. Dinaro, M.A. Tatang, J.W. Tester, J.B. Howard and G.J. McRae (1998). *Combustion and Flame* **112** 132.
91. Pistikopoulos, E.N. and I.E. Grossmann (1988a). *Comp. Chem. Eng.* **12** 719-731.
92. Pistikopoulos, E.N. and I.E. Grossmann (1988b). *Comp. Chem. Eng.* **12** 1215-1227.
93. Pistikopoulos, E.N. and I.E. Grossmann (1989a). *Comp. Chem. Eng.* **13** 1003-1016.
94. Pistikopoulos, E.N. and I.E. Grossmann (1989b). *Comp. Chem. Eng.* **13** 1087.
95. Pistikopoulos, E.N. and M.G. Ierapetritou (1995). *Comp. Chem. Eng.* **19** 1089-1110.
96. Pistikopoulos, E.N. and T.A. Mazzuchi (1990). *Comp. Chem. Eng.* **14** 991-1000.
97. Pistikopoulos, E.N. and T.V. Thomaidis (1992). Presented at AIChE Annual Meeting, Miami Beach, Florida.
98. Pistikopoulos, E.N. and T.V. Thomaidis (1996). *Comp. Chem. Eng.* **20** S1209-S1214.
99. Rippin, D.W.T. *Proceedings of the PSE'94*, 881 (1994).
100. Rockafellar, R.T. and R.J.B. Wets (1991). *Math. Oper. Res.* **16** 119-147.
101. Saboo, A.K., M.Morari and D.C. Woodcock (1983). *Comp. Chem. Eng.* **40** 1553-1565.
102. Shah, N. and C.C. Pantelides (1992). *Ind. Eng. Chem. Res.* **31** 1325.
103. *Stochastic Prog. - Part I & II, Annals of Op. Res.*, **30 & 31** (1991).
104. *Stochastic Prog., Annals of Op. Res.*, **36** (1995).
105. *Stochastic Prog. Algorithms and Models, Annals of Op. Res.*, **64** (1996).
106. *Stochastic Prog., State of the Art, 1998, Annals of Op. Res.*, **85** (1998).
107. Straub, D.A. and I.E. Grossmann (1990). *Comp. Chem. Eng.* **14** 967-985.
108. Straub, D.A. and I.E. Grossmann (1992). *Comp. Chem. Eng.* **16** 69.
109. Straub, D.A. and I.E. Grossmann (1993). *Comp. Chem. Eng.* **17** 339-354.
110. Swaney, R.E. and I.E. Grossmann (1985). *AIChE J.* **31** 621-630.
111. Subrahmanyam, S., J.F. Pekny and G.V. Reklaitis (1994). *Ind. Eng. Chem. Res.* **33** 2688-2701.
112. Subrahmanyam, S., M.H. Bassett, J.F. Pekny and G.V. Reklaitis (1994). *Ind. Eng. Chem. Res.* **33** 2688-2701.
113. Subrahmanyam, S., G.K. Kudva, M.H. Bassett and J.F. Pekny (1994). *Ind. Eng. Chem. Res.* **33** 2688-2701.
114. Tatang, M.A., W. Pan, R.G. Prinn and G.J. McRae (1997). *J. Geophysical Res.* **102** 21925.
115. Varvarezos, D.K., I.E. Grossmann and L.T. Biegler (1992). *Ind. Eng. Chem. Res.* **31** 1466.
116. Vidal, H. (1998). *European J. Oper. Res.* **108** 653-670.
117. Vladimirou, H. and S.A. Zenios (1997). *European J. Oper. Res.* **101** 177-192.
118. Walkup, D.W. and R. Wets (1970). *SIAM J. Appl. Math.* **14** 1143.
119. Wallace, S.W. (1987). *Math. Prog.* **38** 133.
120. Wellons, H.S. and G.V. Reklaitis (1989). *Comp. Chem. Eng.* **13** 115-126.
121. Wets, R.J.B. (1996). *Math. Prog.* **75** 115-135.
122. Epperly, T.G.W., M.G. Ierapetritou and E.N. Pistikopoulos (1997). *Comp. Chem. Eng.* **21** 1411-1431.
123. Ierapetritou, M.G. and E.N. Pistikopoulos (1995). *Comp. Chem. Eng.* **19** S627-S632.
124. Ierapetritou, M.G. and E.N. Pistikopoulos (1996). *Ind. Eng. Chem. Res.* **35** 772-787.
125. Ierapetritou, M.G. and E.N. Pistikopoulos (1996). *In Global Optimization in Engineering Design* (I.B. Grossmann, Ed.)
126. Petkov, S.B. and C.D. Maranas (1998). *AIChE J.* **44** 896-911.
127. Schmidt, C.W. and I.E. Grossmann (1996). *Comp. Chem. Eng.* **20** S1239-S1243.
128. Sen, S. (1992). *Oper. Res. Lett.* **11** 81-86.
129. Wagner, J.M., U. Shamir and H.R. Nemati (1992). *Water Resources Res.* **28** 1233-1246.

Ref.	Dynamic Model	Solution Strategy	Comments / Algorithmic Features		Problem Size (no. eq$^{ns.}$)
1–10	Continuous	Dynamic prog.	Recent extensions allow more general problem formulations		< 100
11–18		Indirect methods	Classical necessary optimality conditions → TPBVP		< 100
19–32		Direct: NLP-based	Simultaneous:	complete discretization (collocation)	> 100
33–40			Sequential:	control vector parametrization sensitivity-based gradients	up to 10000
41–50				adjoint-based gradients	up to 3000
51		Attainable region	Gain process insight but limited to very small problems		< 10
52–60		Non-gradient	Stochastic search, genetic algorithms and hybrid methods		> 100
61–67	Multi-stage	Direct	Extension to systems described by connected sets of DAEs		< 100
68–72	Hybrid	Decomposition	Integer variables reflect natural process discontinuities		< 100
73–81	Mixed-Integer	Decomposition	Integer variables reflect strucural alternatives. Iterate between primal, adjoint and master problems. Use direct methods for primal dynamic optimization.		100s

1. Bellman, R. (1957). *Dynamic Programming*, Princeton University Press.
2. Dadebo, S.A. and K.B. Mcauley (1995). *Comp. Chem. Eng.* **19**(5), 513.
3. Lin, J.S. and C. Hwang (1998). *Ind. Eng. Chem. Res.* **37**(6), 2469-2478.
4. Luus, R. (1993). *Ind. Eng. Chem. Res.* **32**(1), 859-865.
5. Luus, R. (1995). *Ind. Eng. Chem. Res.* **32**(5), 859-865.
6. Luus, R. (1996). *Chem. Eng. Res. Des.* **74**(A1), 55-62.
7. Luus, R. (1997). *Hungarian J. Ind. Chem.* **25**(4), 293-297.
8. Mekarapiruk, W. and R. Luus (1997a). *Ind. Eng. Chem. Res.* **36**(5), 1686.
9. Mekarapiruk, W. and R. Luus (1997b). *Can. J. Chem. Eng.* **75**, 806-811.
10. Wang, F.S. and T.L. Shieh (1995). *Ind. Eng. Chem. Res.* **36**(6), 2279-2286.
11. Agrawal, S.K. and N. Faiz (1998). *J. Opt. Theor. and Applics.* **97**, 11-28.
12. Agrawal, S.K. and X. Xu (1997). *J. of Vibration and Control.* **3**(4), 379-396.
13. Chen, Y. and J. Huang (1996). *Int J. Control* **65**(1), 177-194.
14. Isobe, T. (1987). *IEEE Trans. Autom. Cont.* **AC-32**(1), 63-67.
15. Lew, J.S. and D.J. Mook (1992). *Proc. American Control Conf.* **1**, 172-176.
16. Miele, A. and T. Wang (1993). *J. Opt. Theor. Applics.* **79**(1), 5-29.
17. Storen, S. and T. Hertzberg (1995). *Comp. Chem. Eng.* **19**, S495-S500.
18. Wright, S.J. (1992). *SIAM J. Sci. Stat. Comput.* **13**(3), 742.
19. Betts, J.T. and P.D. Frank (1994). *J. Opt. Theor. Applics.* **82**, 543.
20. Biegler, L.T. (1984). *Comp. Chem. Eng.* **8**(3/4), 243-248.
21. Cervantes, A. and L.T. Biegler (1998). *AIChE J.* **44**(5), 1038-1050.
22. Cuthrell, J.E. and L.T. Biegler (1987). *AIChE J.* **33**(8), 1257-1270.
23. Cuthrell, J.E. and L.T. Biegler (1998). *AIChE J.* **44**(5), 1038-1050.
24. Logsdon, J.S. and L.T. Biegler (1989). *Ind. Eng. Chem. Res.* **28**, 1628.
25. Logsdon, J.S. and L.T. Biegler (1992). *Chem. Eng. Sci.* **47**(4), 851-864.
26. Neuman, C.P. and A.A. Sen (1973). *Automatica* **9**, 601-603.
27. Reddien, G.W. (1979). *SIAM J. Control and Optim.* **17**(2), 298-306.
28. Tanartkit, O. and L.T. Biegler (1996). *Comp. Chem. Eng.* **20**(6,7), 735.
29. Tanartkit, O. and L.T. Biegler (1997). *Comp. Chem. Eng.* **21**(12), 1353.
30. Tsang, T.H., D.M. Himmelblau and T.F. Edgar (1975). *Int. J. Control* **21**, 763-768.
31. Vasantharajan, S. and L.T. Biegler (1990). *Comp. Chem. Eng.* **14**(10), 1083-1100.
32. Wang, F.S. and J.P. Chiou (1997). *Engineering Optimisation* **28**, 273-298.
33. Caracotsios, M. and W.E. Stewart (1985). *Comp. Chem. Eng.* **9**, 369.
34. Feehery, W.F. and P.I. Barton (1997). *Comp. Chem. Eng.* **22**(9), 1241.
35. Feehery, W.F., J.E. Tolsma and P.I. Barton (1997). *Applied Numerical Mathematics* **25**(1), 41-54.
36. Kraft, D. (1994). *ACM Trans. Math. Software* **20**, 262-281.
37. Maly, T. and L.R. Petzold (1996). *Appl. Num. Math.* **20**, 57-79.
38. Pantelides, C.C., R.W.H. Sargent and V.S. Vassiliadis (1994). *Int. Series of Numerical Mathematics* **115**, 177-191.
39. Ross, R., V. Bansal, J.D. Perkins, E.N. Pistikopoulos, G.L.M. Koot and J.M.G. van Schijndel (1999). *Comp. Chem. Eng.* **23**(SS).
40. Vassiliadis, V.S. (1993). *PhD Thesis*, Imperial College, London.
41. Bloss, K.F., L.T. Biegler and W.E. Schiesser (1999). *Ind. Eng. Chem. Res.* **38**(2), 421-432.
42. Colantonio, M.C. and R. Pytlak (1998). Submitted to: *Int. J. Control*.
43. Hasdorff, L. (1976). *Gradient Optimization and Nonlinear Control*, Wiley-Interscience, New York.
44. Goh, C.J. and K.L. Teo (1988a). *Adv. Engng. Soft.* **10**, 90.
45. Goh, C.J. and K.L. Teo (1988b). *Automatica* **24**, 3-18.
46. Jang, S.S., B. Joseph and H. Mukai (1987). *AIChE J.* **33**, 26.
47. Pytlak, R. (1998). *J. Opt. Theor. Applics.* **97**(3), 675-705.
48. Sargent, R.W.H. and G.R. Sullivan (1979). *Ind. Eng. Chem. Proc. Des. Dev.* **18**(1), 113.
49. Sullivan, G.R. (1977). *PhD Thesis*, Imperial College, London.
50. Teo, K.L. and C.J. Goh (1989). *J. Opt. Theor. Applics.* **63**, 1-22.
51. Godorr, S., D. Hildebrandt, D. Glasser and C. McGregor (1999a,b). *Ind. Chem. Eng. Res.* **38**(3), 638-651, 652-659.
52. Carrasco, E.F. and J.R. Banga (1997). *Ind. Eng. Chem. Res.* **36**(6), 2252-2261.
53. Banga, J.R., A.A. Alonso and R.P. Singh (1997). *Biotechnology Progress.* **13**(3), 326-335.
54. Lopez-Cruz, I. and J. Goddard (1996). 6th *Int. Conf. on Computers in Agriculture*, St. Joseph, MI, USA.
55. Mitra, K., K. Deb and S.K. Gupta (1998). *J. Applied Polymer Science* **69**(1), 69-87.
56. Pham, Q.T. (1998) *Comp. Chem. Eng.* **22**(7-8), 1089-1097.
57. Banga, J.R., IrizarryRivera, R. and W.D. Seider (1998). *Comp. Chem. Eng.* **22**(4-5), 603-612.
58. Park, Y.M., J.B. Park and J.R. Won (1998). *Int. J. Electrical Power & Energy Systems* **20**(4), 295-303.
59. Wang, F.S. and J.P. Chiou (1997). *Ind. Eng. Chem. Res.* **36**(12), S348-S357.
60. Weigand, W.A. and Q. Chen (1994). *AIChE J.* **40**(9), 1488-1487.
61. Babad, H.R. (1995). *J. Opt. Theor. Applics.* **86**(3), 529-552.
62. Clarke, F.H. and R.B. Vinter (1989a,b). *SIAM J. Cont. Opt.* **27**(5), 1048-1071, 1072-1091.
63. Gutenbaum, J. (1996). *Archives of Control Sciences* Vol. **5**(41)(3-4), 173-183.
64. Morison, K.R. and R.W.H. Sargent (1984). In: 4th *IIMAS Workshop on Numerical Analysis*, Guanajuato, Mexico.
65. Pantelides, C.C., R.W.H. Sargent and V.S. Vassiliadis (1994). *Int. Series of Numerical Mathematics* **115**, 177-191.
66. Vassiliadis, V.S., R.W.H. Sargent and C.C. Pantelides (1994a). *Ind. Eng. Chem. Res.* **33**, 2111-2122.
67. Vassiliadis, V.S., R.W.H. Sargent and C.C. Pantelides (1994b). *Ind. Eng. Chem. Res.* **33**, 2123-2133.
68. Avraam, M.P., N. Shah and C.C. Pantelides (1998a). *Comp. Chem. Eng.* **22**, S221-S228.
69. Avraam, M.P., N. Shah and C.C. Pantelides (1998b). Presented at AIChE Annual Meeting, Los Angeles, USA.
70. Galán, S. and P.I. Barton (1998). *Comp. Chem. Eng.* **22**, S183-S190.
71. Nishikawa, Y. and M. Okudaira (1978). *Memoirs of the Faculty of Engineering, Kyoto University*, Japan, **40**(2), 62-77.
72. Pantelides, C.C. In: Workshop on "Analysis and Design of Event-Driven Operations in Process Systems", London.
73. Allgor, R.J. (1997). *PhD Thesis*, Massachusetts Institute of Technology, Cambridge MA.
74. Allgor, R.J. and P.I. Barton (1999). *Comp. Chem. Eng.* **23**, 567-584.
75. Barton, P.I., R.J. Allgor, W.F. Feehery and S. Galán (1998). *Ind. Eng. Chem. Res.* **37**(3), 966-981.
76. Mohideen, M.J. (1997). *PhD Thesis*, Imperial College, London.
77. Mohideen, M.J., J.D. Perkins and E.N. Pistikopoulos (1996a). *AIChE J.* **42**, 2251-2272.
78. Mohideen, M.J., J.D. Perkins and E.N. Pistikopoulos (1997). *Comp. Chem. Eng.* **21**, S457-S462.
79. Ross, R., V. Bansal, J.D. Perkins and E.N. Pistikopoulos (1998). AIChE Annual Meeting: Paper 220e, Miami Beach, Florida.
80. Schweiger, C.A. and C.A. Floudas (1997). In: *Optimal Control: Theory and Applications*, (W.W. Hager and P.M. Pardalos, Eds.). pp. 388-435. Kluwer.
81. Sharif, M., N. Shah and C.C. Pantelides (1998). *Comp. Chem. Eng.* **22**(SS), S569-S576.

Table 7. Results of Process Design for Operability Studies.

Economics ($M/yr)	Double-Effect System		Industrial Azeotropic System	
	Sequential	Simultaneous	Sequential	Simultaneous
Capital Cost	0.74	0.72	0.63	0.58
Operating Cost	2.77	2.70	4.37	4.12
Total Cost	3.51	3.42	5.00	4.70
Features				
Weighted ISE	100	85	40	70
Q_{reboil} (Col. I) (MJ/s)			19.54	18.21
Diam. (Col. I) (m)			3.17	2.96
Draw-off to Col II (tph)			1.69	2.43
Q_{reboil} (Col. II) (MJ/s)			0.87	1.09
Diam. (Col. II) (m)			0.87	0.89

INDUSTRIAL PLANTWIDE DESIGN FOR DYNAMIC OPERABILITY

Bjorn D. Tyreus and Michael L. Luyben
E. I. du Pont de Nemours and Company, Inc.
Wilmington, DE 19880

Abstract

This paper explores the implications of process design decisions on the dynamic operability and controllability of a chemical plant. To achieve a plant's design objectives, the control system must provide stable operation, handle constraints and recycles, satisfy material and energy balances, and meet economic objectives. Process design must be used to ensure that the process is operational and that enough manipulators are available to achieve the control objectives. The hierarchical, thermodynamic, and optimization approaches to process synthesis are examined in terms of their impact on effective plantwide control. The strengths and weaknesses of each method for considering dynamic operability are discussed. The bridge between design and control engineers, who in industry are typically separated organizationally, can potentially be a dynamic process model to help choose among design alternatives.

Keywords

Process design, Plantwide control, Partial control, Dominant variables, Design and control, Conceptual design.

Introduction

The art and science of designing industrial process flowsheets have consumed an important part of research activities over the last few decades. Our objective is not to summarize here all the previous work, so we would refer to reviews of process synthesis by Nishida, Stephanopoulos, and Westerberg (1981) and Grossmann and Daichendt (1996) for perspective.

Three general approaches have been identified for tackling the synthesis of a complex, integrated process flowsheet: (1) methods that use heuristics, evolutionary techniques, and hierarchical decompositions, (2) methods that use superstructures, mathematical programming, and optimization, and (3) methods that involve the use of thermodynamic targets, process integration, and pinch analysis. Because the currency of business drives decisions within the chemical industry, all three approaches focus on synthesizing flowsheets that optimize economics.

At the same time, it has long been recognized that the design of a chemical process plays a significant role in its dynamic operation and control. And in fact we often encounter tradeoffs between a less capital-intensive plant design (good steady-state economics) and a more operable plant (good dynamic controllability). In the face of inevitable disturbances (in equipment performance, raw material quality, utility supply, etc.) a plant's design objectives can be met *only* if it can be operated safely and achieve its production, quality, and environmental goals. The interaction of design and control has been a widely accepted and acknowledged theme within the process design and control research communities. It has even been the subject of two workshops within the last decade (Perkins, 1992, and Zafiriou, 1994).

Within each of the three general approaches toward process synthesis, key decisions are made about the flowsheet design that have a bearing on the operability characteristics of the plant. For example, in a hierarchical procedure we will make decisions about whether the plant is batch or continuous, what types of reactors are used, how material is recycled, what methods and sequences of separation are employed, how much energy integration is involved, etc. In a thermodynamic pinch analysis, we typically start with some flowsheet information, but we must then decide what streams or units to include in the analysis, what level of utilities are involved, what thermodynamic targets are used, etc. In an optimization approach, we must decide the scope of the superstructure to use, what physical data to include, what constraints to apply, what disturbances or uncertainties to consider, what objective function to employ, etc.

These key decisions are characteristics of the particular flowsheet design method chosen. Yet these decisions, both implicitly and explicitly, affect the process behavior well beyond the steady-state economics. As process control engineers involved in process design, we encounter these effects regularly. The objective of this paper is to analyze for each of the three methodologies what implications the design decisions have on dynamic operability and controllability for continuous processes.

Heuristic and hierarchical methods certainly are the predominant tools in industrial practice today, and for many strong reasons are useful for generating initial flowsheet alternatives. And while these methods may not be perfect because they cannot consider the interaction among design variables at various levels, they are at least feasible. However, many of the design decisions can be negative for dynamic controllability unless the design engineer is aware of their implications. Looming on the horizon of industrial practice, though, are the mathematical programming approaches. From the start, these methods may be perfect because they consider all of the design variables simultaneously, but their solution may not yet be feasible. Hence we intend also in this paper to discuss both the strengths and the weaknesses of the methods for considering dynamic operability.

Thermodynamic Foundations

In this paper we make frequent references to the concepts of dominant variables and partial control. Tyreus (1999a) has recently shown that there is a strong connection between these concepts and thermodynamics. We thus start with a brief outline of some of the important aspects of thermodynamics that are relevant for describing dynamic systems and their operability.

It is well known from classical thermodynamics that the equilibrium state of a process system is completely defined by the mole numbers of each component, the system volume, and total internal energy. Since these variables all depend on the size of the system and since they are additive over sub-parts of the system, they are said to be extensive. Furthermore, the mole numbers and the internal energy behave in a manner that can be described by fundamental balance equations. We make use of these concepts when we construct first principles models of process systems.

It is also well known from classical thermodynamics that the state of equilibrium is attained when the entropy has reached its maximum for a fixed internal energy, system volume, and mole numbers. An equivalent description of equilibrium is also to specify the entropy of the system in place of the internal energy. The internal energy is then viewed as the derived property as expressed in the Gibbs fundamental relation.

It is much less convenient to use entropy as a state variable in place of internal energy since entropy is not a conserved property. This forces us to use a production term in the balance equation. However, from a conceptual viewpoint the entropy description is superior. The reason is that entropy behaves like mole numbers, electric charge, and momentum when it comes to describing physical systems. Specifically, all the aforementioned properties have a unique measure of their intensity or potential. For example, the intensity of a chemical component is called the chemical potential, whereas the potential of an electric charge is its voltage. The intensity of entropy is absolute temperature. The intensities of these quantities are not additive like extensive variables. Rather, they belong to the class of intensive variables as defined in classical thermodynamics.

An interesting and useful aspect of the fundamental variables (i.e. mole numbers, charge, momentum, and entropy) is that they can be viewed as *energy carriers*. The amount of energy carried with the flow is the product of its flow rate and intensity. For example, the product of an entropy flux and absolute temperature is equivalent to a

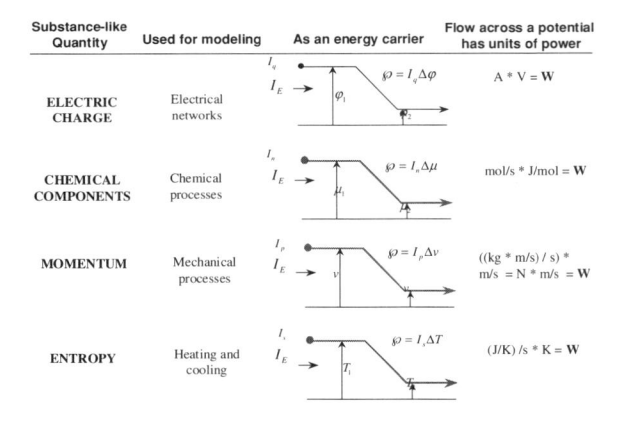

Substance-like Quantity	Used for modeling	As an energy carrier	Flow across a potential has units of power
ELECTRIC CHARGE	Electrical networks	I_q, I_E, φ_1, φ_2	$\wp = I_q \Delta\varphi$ — A * V = W
CHEMICAL COMPONENTS	Chemical processes	I_n, I_E, μ_1, μ_2	$\wp = I_n \Delta\mu$ — mol/s * J/mol = W
MOMENTUM	Mechanical processes	I_p, I_E, v_1, v_2	$\wp = I_p \Delta v$ — ((kg * m/s) / s) * m/s = N * m/s = W
ENTROPY	Heating and cooling	I_s, I_E, T_1, T_2	$\wp = I_s \Delta T$ — (J/K) /s * K = W

Figure 1. Fundamental variables for modeling physical systems.

heat flux, whereas the flow rate of chemical components and their chemical potentials account for the free energy in the stream. Furthermore, when a flow of these energy-carrier quantities changes its intensity or potential, energy is released or absorbed to account for the change in the stream's energy content (see Fig. 1). We will later show how the energy release relates to dominant variables and partial control.

Process Control Objectives

Luyben, Tyreus, and Luyben (1999) have shown that there are five major objectives of a plantwide control system. First, the control system must be able to stabilize the process. Second, the control system must cope with the constraints imposed on the operation. Third, the inventory of material and heat must be balanced. Fourth, the economic objectives of the plant must be met. Finally, the recycle structure must be controlled. Given that virtually every process has far fewer control degrees of freedom than objectives for the control system, we are inevitably forced to assign some priority to the various control objectives. In the past, when process control was an activity often performed fairly late in a design project, it became customary to assign priority to the objectives roughly in the order listed above. This often meant that after the stability, constraint, and inventory loops were assigned there were not enough manipulated variables left to meet the economic objectives adequately or to handle the recycle flows in the system. The result was that the control system fell short of helping the process meet its specifications. Worse yet, the control system sometimes contributed to the operational problems by allowing recycle flows to change in response to inventory variables when instead the recycle flow should have been controlled independently, as discussed in Chapter 2 of Luyben, Tyreus, and Luyben (1999).

In observing the shortcomings of past control system designs, it has become clear that more emphasis should be placed on helping the process meet its economic objectives. However, the need for stability, constraint, and inventory controls has not changed and serves as prerequisites for proper economic control. Thus the only way to break away from the traditions of the past is to use process design to ensure that the process is operational and that there are a sufficient number of manipulated variables available to meet all the control objectives. It is the influence of process design on controllability that forms the main theme of this paper. However, before we describe how various design decisions affect the controllability of the plant, we first examine in detail what is required to meet each of the five main control objectives.

Strategies for Plantwide Control

In this section we examine five different process design and control strategies aimed at helping the process meet the key control objectives. Since our main focus is to ensure proper control of the economic variables, we consider techniques to solve this task first. Next, we look at stability and constraint handling followed by recycle control. Last, we examine how construction of the inventory control system has to be consistent with the rest of the control strategies.

Control of the Economic Variables

In this area we are indebted to Shinnar (1981) for developing a most useful technique called *partial control*. Partial control is aimed at controlling the economic variables in the process, which are the quantities that bring significant financial value to the operation. For example, they may include production rate, yield, conversion, selectivity, purity, utility consumption, molecular weight, composition, etc. These variables are seldom directly measurable on-line. They also are often associated with the steady-state behavior of the process rather than the short-term, dynamic behavior that we think of for feedback control. Shinnar (1981) observed that the number of economic variables for a process usually exceeds the number of available manipulated variables or degrees of freedom, especially after loops have been assigned to meet the other four control objectives. Fortunately, Shinnar (1981) found that his idea of partial control provides a solution to this control dilemma. Partial control evolved from the recognition that it is seldom necessary to control all the economic variables to specific setpoints but it often is sufficient to control them within some acceptable ranges. For example, while we usually require that the production rate is controlled to setpoint, it is less important to hold the conversion and yield at exact values. We can thus understand the use of the word "partial" in the sense of approximate or range control.

Partial control can also be understood from the standpoint of controlling a small subset of all the variables that would have to be controlled if we tried to hold all the economic variables at their setpoints. Instead, Shinnar (1981) noticed that only a few key variables determine the dynamic behavior and steady-state productivity of many units. He appropriately coined the term *dominant variables* to denote these. He also found that it is sufficient to control the dominant variables with feedback controllers to achieve regulation or partial control of all the economic variables in the system. A typical example of partial control is the feedback regulation of temperature in an exothermic reactor. This single control loop simultaneously affects reactor stability, productivity, selectivity, and yield. Similarly, the control of one sensitive tray temperature in a conventional distillation column affects the purity of both products as well as the unit's operating cost.

Finding the dominant variables in a process is thus a main requirement for partial control. However, there must also be suitable manipulated variables to allow proper control. So, once the dominant variables have been identified they should be measured as frequently as possible, preferably continuously. Then, a matching number of independent, rapidly-acting manipulated variables must be identified. In an exothermic reactor the cooling rate is often used as a manipulator, whereas distillation columns frequently use steam rate to the reboiler as the manipulated variable. The dominant variables are paired with the manipulated variables in an appropriate fashion and the loops are closed with conventional PI or PID controllers.

An interesting feature of partial control is that the dominant variables in a system seldom have intrinsic economic value nor are they usually part of the list of economic objectives for the plant. For example, we don't assess the quality of a refined product in terms of a distillation column tray temperature and we don't measure productivity and yield with a reactor temperature. However, the dominant variables, by their nature, influence most other variables in the system including those that bring financial value to the operation. Additionally, establishing the control limits for dominant variables is essential to ensure on-aim quality control and the ability to run product-by-process. For example, reactor temperature often has a dominant effect on production rate, selectivity, and yield, all of which are important to costs and revenues. A partial control strategy takes advantage of this relationship by allowing the setpoints of the dominant feedback controllers to vary in order to meet the economic objectives. For example, when the overhead product from a distillation column contains excessive amounts of high boilers, we lower the setpoint of the tray temperature controller. This reduces the steam to the reboiler and eventually shifts the column's composition profile in a favorable direction. Similarly, when a high production rate in the plant is needed we raise the reactor temperature controller setpoint.

The usefulness of partial control hinges on three requirements. First, we must be able to identify the dominant variables. Second, we must establish how many dominant variables are sufficient to reach our partial control objectives. Finally, we have to be able to provide a matching number of manipulated variables to control the required set of dominant variables effectively.

Tyreus (1999a) has recently made advances in this area by using ideas from irreversible, non-equilibrium thermodynamics. Specifically, he observed that most of the economic objectives in a process are related to the internal process rates. For example, reaction rates determine productivity, yield, and selectivity and the steam rate to a distillation column determines its ability to separate the feed mixture. In searching for a common rate descriptor, Tyreus (1999a) selected the rate of energy transfer between the internal energy carriers in a process as these carriers undergo a change in their potentials (see Fig. 1).

With this in mind it is useful to consider a process or a unit operation as an exchanger or transceiver of energy between different internal process streams. For example, consider an isothermal exothermic reactor where a flow of free energy is carried into the reactor by the reactants. Some of the inflowing energy is released inside the reactor by lowering the chemical potentials of the reactants to those of the products at a rate equal to the reaction rate. The released energy per unit time (power) equals the heat flow leaving with the cooling water. This heat can also be seen as an entropy production rate multiplied by the absolute temperature in the reactor. The reactor thus serves as an exchanger of energy between a chemical component stream and a heat stream.

A second example of processes as energy tranceivers is a distillation column. Here heat (entropy at an elevated temperature) enters the column through the reboiler. The influx of entropy added to the entropy created inside the column leave the system through the condenser at a lowered temperature. The flow of heat across the temperature drop inside the column releases a certain amount of free energy. Part of this is used to elevate the chemical potentials of the components in the feed stream to those in the product streams. The remaining energy is used to create additional entropy to account for mixing processes and pressure drop over the trays.

When a unit operation is viewed as an energy transceiver we have a process stream serving as a power transmitter and another stream acting as a power receiver. Of particular interest from a control viewpoint are the variables affecting the rate of power transmission. Some of those variables are shown in Fig. 1. Specifically, these variables relate to the flow rate of an energy-carrier quantity and the associated intensities. Tyreus (1999a) noticed that these same variables were also used in control loops that had been found to be successful in controlling the unit operation. He thus proposed that the dominant variables for partial control could be found among the variables affecting the rate of internal power release in a process unit.

Control for Process Stability

Process stability is of vital concern before we can even attempt to consider other control objectives. It may thus seem surprising that we deal with this topic after control of the economic variables. However, as Shinnar (1981) and Arbel et al. (1995, 1996, 1997) have demonstrated for different systems, our ability to achieve process stability is a matter of being able to provide good control of the dominant variables in the system. This should not be surprising given that the dominant variables, by definition, have a dominating influence over the static and dynamic behavior of the process. So, while we have a strategy for dealing with instability through control, we may ask why process stability occasionally becomes an issue in the first place.

Instability is created from feedback, specifically positive feedback where the total gain in the feedback loop is greater than unity. Notice that the feedback has to be positive for instability to occur. Negative feedback provides stabilization. We may thus think that negative feedback arrangements are guaranteed to be trouble free. This would be true as long as the overall loop dynamics are less than third-order and the loop is free of non-minimum phase elements such as dead time and inverse response. However, with unfavorable loop dynamics negative feedback can exhibit instability while positive feedback loops are predestined to be unstable unless the loop gain is less than one.

Before we leave this discussion on stability, we should look at an example where a process design decision introduces feedback. The classic example is the choice of reactor type for a process. A perfect, adiabatic plug flow reactor has no feedback at all whereas a continuous stirred tank reactor has built-in feedback through the mixing process. However, this does not mean that a CSTR is always more difficult to control. For example, the mixing process provides negative feedback with negligible dynamics for the reactants participating in all non-autocatalytic reactions. The feedback in an isothermal CSTR thus provides a stabilizing effect on the operation. However, when we consider the influence of feedback of heat on the rate of reaction the picture is different. Since heat raises the reactor temperature and higher temperature raises the rate of reaction and hence the rate of heat evolution, we are now dealing with a positive feedback loop. For highly exothermic reactions with reasonably large activation energies, the CSTR will have unstable operating points. However, since temperature is a dominant variable for these systems, feedback control of the temperature will stabilize the system since we are now back to an isothermal reactor.

Constraint Control

Constraints are imposed on a process for a variety of reasons. Many times we encounter constraints on the intensities of fundamental process variables to ensure the integrity of the processing equipment or the safety of process operation. For example, pressure is part of the intensity for chemical components and must often be controlled below a certain limit dictated by the construction of a vessel. Similarly, temperature is the intensity for entropy and must also be controlled within ranges to prevent undesirable effects from taking place.

Unlike the economic variables of a process, the constraint variables are often directly measurable. The reason is that they belong to the intensive variables of the system. From the thermodynamic identification of dominant variables we also found that these are intensive. This has two important consequences opposite constraint control. First, it is possible that some of the constraints apply to the dominant variables. Since the dominant variables are already controlled in the partial control structure, no further considerations need be given to the control of those constraints. Second, it is possible that some constraints are limits or ranges. It is quite possible that the constraint variables satisfy those limits and ranges when the dominant variables are controlled to setpoint. In this case as well, the constraint control has no further implications beyond partial control. It is only when the constraint is not controlled by the actions of the partial control structure that we need to configure new loops with independent degrees of freedom (control valves).

Recycle Control

Luyben, Tyreus, and Luyben (1999) emphasize the importance of controlling the recycle streams in a process. They also offer several reasons why this should be done in order to achieve good plantwide control. We will not repeat the arguments here but instead offer another viewpoint from which recycle control can be justified.

Consider Fig. 2. Here we show the input-output structure of an arbitrary process. From a design standpoint this input-output structure says nothing about the recycle structure. In fact, the process could consist of a single reactor with no recycle at all. It is not until we use *design degrees of freedom* to create a recycle structure that we would actually have one as indicated in Fig. 2.

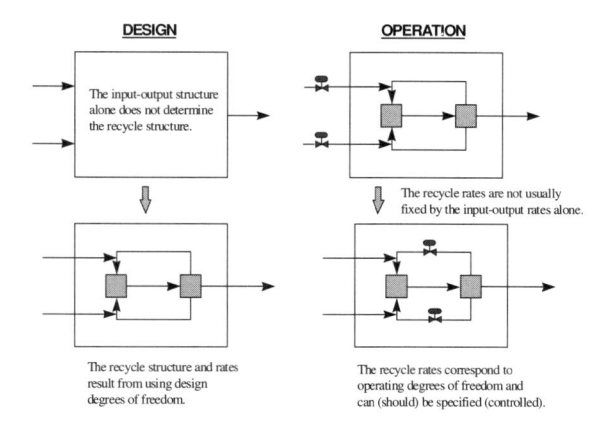

Figure 2. Design and control of material recycle structures.

When the process is operated we apply a similar line of argument around the recycle structure. Now, when we specify the input flow rates it does not follow that the recycle flows are uniquely determined. Instead, we must use independent *control degrees of freedom* to achieve the desired objectives set forth when the recycle structure was incorporated into the process during the design phase. This could amount to fixing the rate of the recycle flow or adjusting the flow to attain a desired recycle to feed ratio or to satisfy a process constraint.

Material and Heat Balance Control

As we pointed out earlier, physical systems can be described by applying the generalized balance equation to one or more energy-carrier quantities that characterize the system. In the case of chemical processes it is usually sufficient to apply the balance equation to each component in the system and to the system entropy. Since the entropy balance and the component balance equations contain production terms in addition to the inflow and outflow terms, it becomes necessary for the control system to match all three terms to achieve steady state operation.

Fortunately, this task is simplified when we consider material and heat balance controls as the last action item on the list of plantwide control strategies. The reason is that the significant production terms in the material and entropy equations are already controlled by the partial control structure. Recall that the thermodynamic approach for the identification of dominant variables targets the rate processes in the system. What remains to be done is then to match inflows and outflows of the various components to the rates determined by the partial control system. We can do this by applying the idea of self-consistent control as suggested by Price, Lyman, and Georgakis (1994). Figure 3 illustrates their idea. Here we have identified a step in the process where the dominant variables set the rate of production and consumption of various components. The control system matches the rates by bringing in reactants

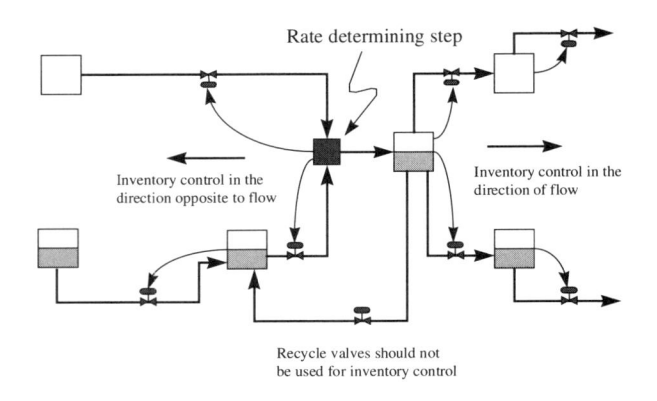

Figure 3. Arranging inventory control system to be consistent with the rate-determining step.

by inventory controllers arranged to work in the direction opposite to flow. The products are forced to exit the process by the use of inventory controllers operating in the direction of flow.

Occasionally there is a constraint on either the external feeds to the process or the product flows. This could occur for example when a process receives one reactant from another plant and there is negligible inventory between the two plants. This situation is illustrated in Fig. 4. For this process to work properly we must control the level in the recycle tank by adjusting the setpoints of the rate-controlling loop(s) in the partial control structure.

It is important not to fall for the temptation of using the valve associated with the recycle flow to control the inventory of the recycle tank. Such an erroneous assignment would lead to a mismatch between the flow rate of the constrained reactant and its consumption rate in the rate-controlling step. Such a control system would not help the process reach a steady state without frequent operator intervention.

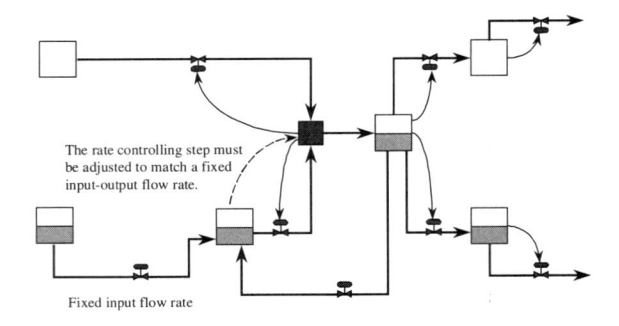

Figure 4. Arranging inventory controllers to comply with constraints on feed or product flows.

Hierarchical Approach

With the preceding discussion having set the stage for understanding dynamic operability, we are now in a position to describe how process design decisions affect the controllability of a plant and what specific control strategies have to be applied in order to ensure economic operation.

We start with a hierarchical approach to process design based upon the conceptual framework of decisions proposed by Douglas (1988).

1. Batch versus continuous
2. Input-output structure of the flowsheet
3. Recycle structure of the flowsheet
4. General structure of the separation system
    a. Vapor recovery system
    b. Liquid separation system
5. Heat-exchanger network

Batch versus Continuous

This decision has profound implications for the controllability and control system of the process. Batch process control systems are generally very different than control systems for continuous plants. In batch systems, controller setpoints often change based upon time trajectories, operations are performed in defined sequences, multiple operations are done in the same vessel, etc. This is not typical of continuous systems. Most of our experience stems from work with continuous processes, so we shall focus our attention on these. How design decisions affect the dynamic operability in batch processes we leave for future work.

Input-output Structure of the Flowsheet

The input-output design decisions revolve around measures that have to be taken when the feeds are not pure and when the process has reversible side reactions. Douglas (1988) provides a list of heuristics to help make the appropriate design decisions. Several of these heuristics, while beneficial for the steady state economics, can have negative effects on the dynamics and control of the plant. For example, if a feed contains a small amount of a low boiling inert, it is recommended to process the inert and remove it with a purge flow from the gas recycle stream where it will concentrate.

There are two problems with this approach, both associated with the desire to concentrate the impurity to the largest extent before it is allowed to leave the process and take valuable reactant with it. The first problem is that during the course of concentrating the purge stream in the inert component, we may build up a sizable inventory of this component within the process. A large inventory and a small external flow mean the time to purge the system becomes very long. This has implications for how the purge flow controller can be tuned and for the expected dynamic performance of the control loop.

The second problem with letting a low boiler build up in the gas recycle is the limit it imposes on our ability to change other components in the system. For example, if the recycle contains a component that influences a dominant variable in the reactor, a high concentration of the inert limits how much and how quickly we can influence the dominant variable. Tyreus (1999b) gives an example of this in his study of the Tennessee Eastman challenge problem. For both of these reasons, the better alternative for dynamics may be to have a lower concentration of impurity in the gas recycle loop.

Recycle Structure of the Flowsheet

The third step in the heuristic design procedure deals with the recycle structure of the plant. The specific questions Douglas (1988) has us ask are:

1. How many reactor systems are required?
2. How many recycle streams are required?
3. Do we want to use an excess of one reactant at the reactor inlet?
4. Is a gas compressor required?
5. Should the reactor be operated adiabatically, with direct heating or cooling, or is a diluent or heat carrier required?
6. Do we want to shift the equilibrium conversion?
7. How do the reactor costs affect the economic potential?

There is no question that the decisions made around the recycle structure of the plant have a great impact on the dynamic operability of the process. Not only is the time constant of the combined plant larger than the sum of the time constants for the individual units, but the chance of creating and retaining disturbances has also increased. The recycle flows may also affect the dominant variables in the process such that economic control becomes difficult. Finally, process stability is always an issue when we create feedback in the system.

Let us illustrate some of these points with a simple example. Consider a reaction where two components, A and B, are converted isothermally to component C as shown in Fig. 5. A straightforward economic analysis indicates that it is beneficial to install a medium size reactor and operate the reactor with a slight excess of both reactants. The excess must be recovered and recycled back to the reactor. Figure 6 shows that the economic optimum calls for roughly the same concentration of the reactants in the reactor and that this concentration is over 10 mole %.

Our ability to provide good control of this process hinges on identifying its dominant variables and making sure that we have a sufficient number of effective manipulated variables to close the loops in a partial control scheme. Tyreus (1999a) has shown how the dominant variables in the reactor can be identified. In this example the concentrations of both reactants and the reactor level are all potentially dominant. The degree of dominance (or the sensitivity) is inversely proportional to the steady-state value as shown in Fig. 5. It is clear from Fig. 6 that we have a conflict here. The only way that we can make the reactor concentration low in both reactants is by making the reactor very large. This would, however, make the system insensitive to level changes. Alternatively, we could try to make one reactant dominant at the expense of the other for a reasonably sized reactor. Such a strategy would require an investment cost larger than the economic optimum, but it provides added flexibility for control by making two variables dominant in the reactor (one composition and the level). The partial control scheme is shown in Fig. 7 along with the necessary recycle control and self-consistent inventory controllers to complete the plantwide control scheme. Notice that only the large B-recycle flow is controlled. The small recycle of A is allowed to vary to control the level in the second column's reflux drum. There is no risk of uncontrolled buildup of component A in this scheme since the reactor concentration is controlled by varying the fresh feed of component A.

General Structure of the Separation System

The structuring of the separation system deals with alternatives to sequencing distillation columns and selecting alternate methods for performing the required separation of reactants from products and of products from by-products. The control of a conventional distillation column (one feed and two products) is usually not particularly difficult and as long as we adhere to sound strategies for each unit there should not be any major controllability concern regarding the sequencing of the columns. However, complex columns are a different story. In these systems we attempt to combine several columns or operations into one and we produce several product streams for each feed stream. The difficulty we face is that a complex column may have as many dominant variables as there are significant separations but there is not a corresponding set of manipulated variables due to elimination of reboilers and condensers.

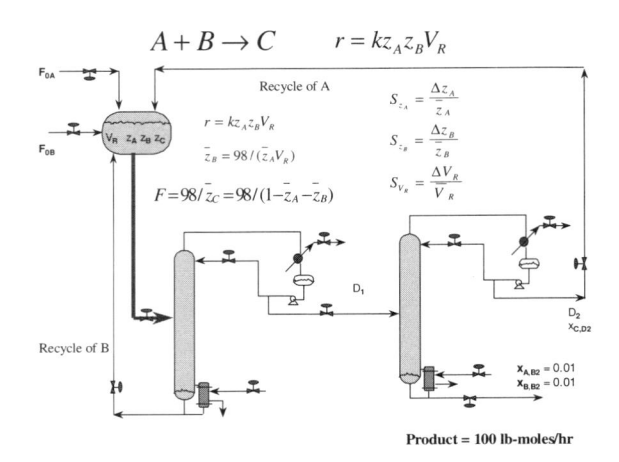

$$A + B \rightarrow C \qquad r = kz_A z_B V_R$$

Figure 5. Recycle flows affect the sensitivities (S_x) of the dominant variables in the reactor.

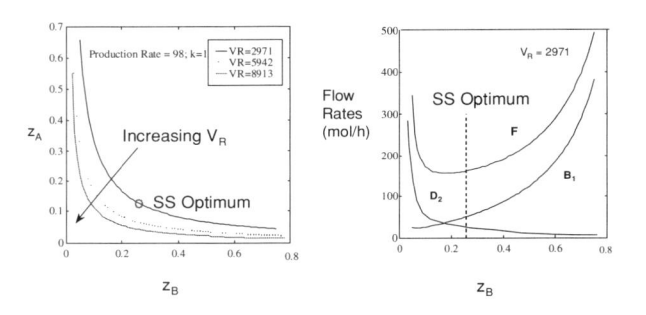

Figure 6. The steady-state economic optimum favors a medium sized reactor with incomplete conversion of both reactants.

This causes problems in constructing a sufficient partial control scheme with the result that control of the economic variables in the process may suffer. This deficiency is not difficult to remedy if it is analyzed at the conceptual design stage. As soon as we know what the dominant variables are from a thermodynamic analysis of the unit, we can consider what it would take to provide manipulated variables to move the dominant variables over wide enough ranges that we can adequately meet the economic objectives of the unit. The cost of these additions must factor into the investment cost of the process. On the benefit side we look at improved project economics due to better control of the economic objectives over the lifetime of the process.

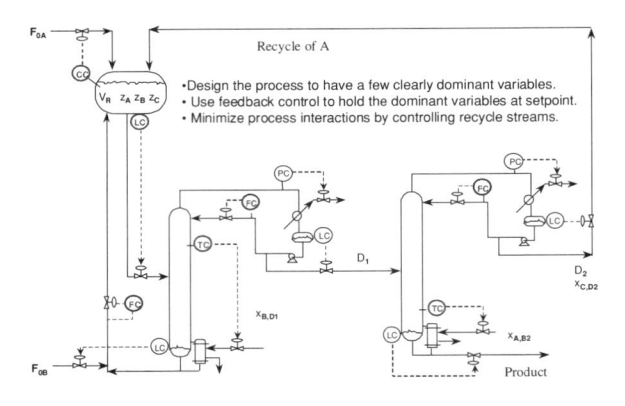

Figure 7. Plantwide strategy using partial, recycle, and self-consistent inventory controls.

Heat Exchanger Network

The design of heat exchanger networks is the last step in the hierarchical design process. The objective is to improve the thermodynamic efficiency of the plant and thereby reduce the steady-state operating utility costs. One of the most common applications of this type of design improvement is when reactor effluent heat is used to preheat the feeds as shown in Fig. 8. This process section is borrowed from the HDA plant described by Douglas (1988).

In addition to the feed-effluent heat exchanger (FEHE), there is a furnace and a quench stream. The quench stream reduces the reactor exit stream temperature down to the inlet temperature to prevent coke deposits in the FEHE (a constraint). The furnace is required because there is not enough temperature driving force left in the quenched stream to bring the reactor feed up to the required reactor inlet temperature.

A thermodynamic analysis of the heat exchanger and the furnace will readily show that there are three dominant variables in these two operations, namely the three temperatures shown in Fig. 8. A partial control scheme would use the quench flow to hold the inlet temperature to the exchanger (which then also takes care of the constraint control) and the fuel to the furnace to control the furnace temperature. This partial control scheme is sufficient to stabilize the process but would not be as flexible as a scheme where all three dominant variables were controlled. The question is how to provide a third degree of freedom? Figure 9 provides a possible solution to the problem. We have now made the feed preheater somewhat larger and provided a cold bypass stream around the exchanger. The bypass flow provides the extra degree of freedom. This design change will again bring about a higher investment cost but the operational flexibility may more than offset this investment over the lifetime of the process.

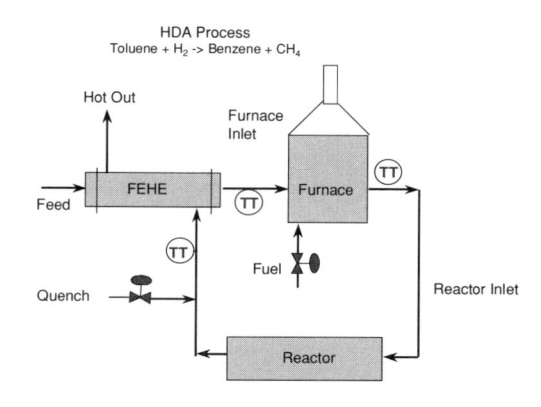

Figure 8. Exothermic, adiabatic reactor with a feed-effluent exchanger, a furnace, and a quench.

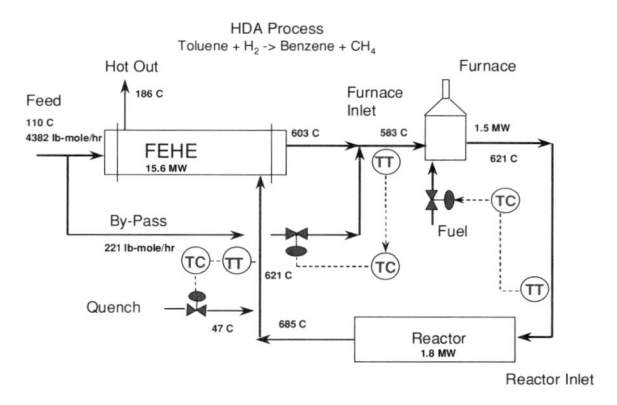

Figure 9. Addition of extra degrees of freedom for control provides operational flexibility.

Thermodynamic Approach

The thermodynamic approach to process design is motivated by the reduced operating cost achievable when the process is made thermodynamically efficient. According to Linhoff (1994), the classical use of pinch technology for the design of heat exchanger networks has evolved into a broad-based methodology for reducing capital costs, emissions, and energy consumption, spanning process design and total site planning. We will not deal with all the different facets of pinch technology in this paper but merely focus on the original use of the method for designing heat exchanger networks.

The basic idea of pinch analysis is to reuse internal heat in the process to avoid having to supply utility heat and cooling whenever a temperature change is required. Alternatively we can look at the use of heat exchangers for heat recovery as a way to allow an internal (or external) source of entropy to give up as much of its potential as possible inside the process. Recall that each time an entropy flow drops in potential (temperature), there is a certain amount of free energy liberated that can be used to perform useful work (e.g. separation work). Figure 10 shows a schematic of the HDA process with a number of additional heat exchangers inserted for efficient recovery of the heat of reaction.

The operational problems we encounter in a process with heat integration stem from the lack of manipulated variables for effective partial control and from the effect on driving forces for heat transfer. Each time we use a process-to-process heat exchanger for heat recovery, we eliminate a valve to a utility stream. This means that we loose ability to adjust the amount of entropy flow through a unit operation or its inlet temperature (both of which happen to be dominant variables for distillation systems). The use of process-to-process heat exchangers tends to make the process more "reversible" in the thermodynamic sense. As the driving forces for heat transfer become smaller, disturbances are more likely to propagate.

The only way we can restore flexibility and regain operability of a heat-integrated system is to provide through design the necessary degrees of freedom for effective partial control. Figure 11 gives an example of how this can be accomplished for the HDA process. Here we have provided controlled bypasses around all the process-to-process heat exchangers. In addition to a control valve in the bypass streams, we also have provided auxiliary coolers in the bypasses around the process-to-process reboilers on the columns. The reason for this is to regulate the amount of heat reaching the column without affecting the temperature level of the heating stream reaching the downstream equipment. Finally, we have provided auxiliary reboilers on the distillation columns to ensure that enough heat is available at all times to meet the economic objectives of the column.

Optimization Approach

In contrast to the hierarchical method outlined above, mathematical programming approaches toward process synthesis solve for all the design and flowsheet variables simultaneously. This means there is no decomposition of decisions concerning the input-output structure, the recycle structure, the separation system, and the heat exchanger network. This has interesting implications for dynamic operability considerations and for our plantwide control strategies.

The starting point for the optimization approach involves the construction of a superstructure incorporating all possible design alternatives of interest. This superstructure is then translated into a mathematical model, which typically contains both discrete variables (0-1), y, about the existence of units and continuous variables, x, about the values of the process parameters (flows, compositions, temperatures, pressures, vessel sizes, etc.). In principle, these continuous variables can involve both steady state and dynamic quantities.

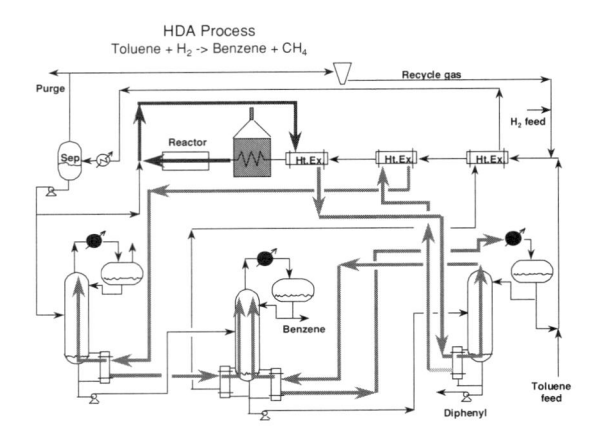

Figure 10. Entropy generated in the reactor and furnace is allowed to lower its potential in several stages before leaving the process.

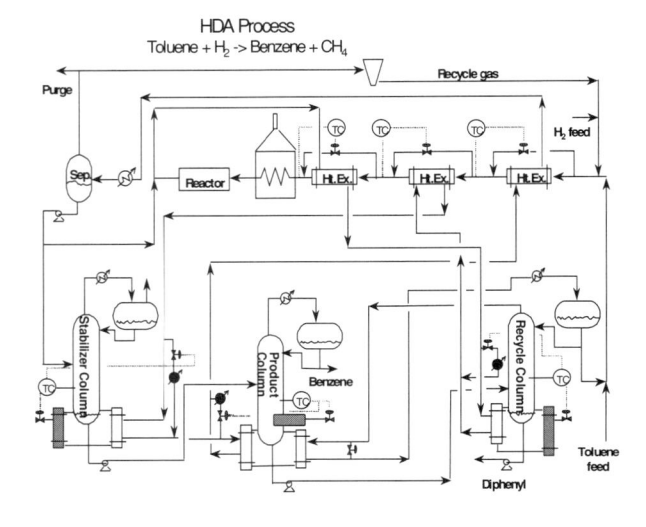

Figure 11. Heat integrated processes require design modifications to restore lost degrees of freedom for control.

Only recently, though, have researchers begun to incorporate dynamics, control, and uncertainty into the superstructure formulation. In this way, they can design both the process itself and the control structure that will be optimal in the face of disturbances and uncertainty. For example, Mohideen, Perkins, and Pistikopoulos (1996) presented an approach that includes simultaneously process and control system design as part of the optimization. It considers both parametric uncertainty and time-varying disturbances. Schweiger and Floudas (1997, 1999) incorporated dynamics and controllability within the superstructure to synthesize both the process flowsheet and the control structure. Finally, van Schijndel and Pistikopoulos (1999) lay out ideas for an optimization framework along with future trends.

In its most general form, then, the nonconvex, nonlinear mathematical model can be cast in the following form:

$$\begin{aligned} \min\ & f(x,y) \\ \text{subject to}\ & h(x,y) = 0 \\ & g(x,y) \ldots 0 \end{aligned} \quad (1)$$

This includes an objective function $f(x,y)$, equality constraints $h(x,y)$, and inequality constraints $g(x,y)$ that can contain algebraic and differential equations, sets of path and end-point constraints for feasible dynamic operation, and design and control constraints. Solving such mixed-integer nonlinear optimal control problems requires special decomposition algorithms that have been developed as part of this research work.

What then are the implications of these design methods on dynamic operability? In principle, this approach should be extraordinarily useful because it allows us to include within the optimization problem our knowledge about partial control, dominant variables, constraint and recycle control, and material and heat balance control. We can also ensure that we have sufficient degrees of freedom to be able to achieve effective partial control. However, the design decisions must be made in a way that allows an adequately rich problem formulation. This means that the superstructure, the choice of controlled and manipulated variables, the disturbances, the unit operations, etc. must cover the spectrum of plantwide control strategies. Without these, all is lost because the optimization problem will never be able to consider anything else.

Consider, for example, the vinyl acetate process shown in Fig. 12. Luyben and Tyreus (1998) presented the design details of this process as a realistic example for academic researchers to use in simulation, design, and control studies. In this process, ethylene, oxygen, and acetic acid react to form vinyl acetate, with water and carbon dioxide formed as byproducts.

If we used an optimization method to synthesize an optimal flowsheet for this process, we would formulate a mathematical model involving many possible unit operations, nonideal vapor-liquid equilibrium behavior, kinetic rate expressions involving partial pressures in both numerator and denominator terms, and other challenging tasks. Assuming that the formulation and solution to such a problem were feasible, then we would be in an excellent position to include control issues within the optimization. We would be able to specify the economic control objectives for the process. These include vinyl acetate production, reactor selectivity, catalyst life, safe reactor operating conditions, vinyl acetate product quality, acetic acid and ethylene yield losses, vinyl acetate composition in the acetic acid recycle stream, etc. We would then be able to ensure through the process design that we had effective manipulated variables to control the dominant variables. Some of these dominant variables, shown in

Fig. 12, are the reactor temperature, the reactor oxygen feed composition, and a tray temperature in the distillation column where the vinyl acetate composition profile breaks. We also could consider some of the expected disturbances to the process, including changes in production rate, changes in the ethylene feed composition, loss of catalyst activity, among others.

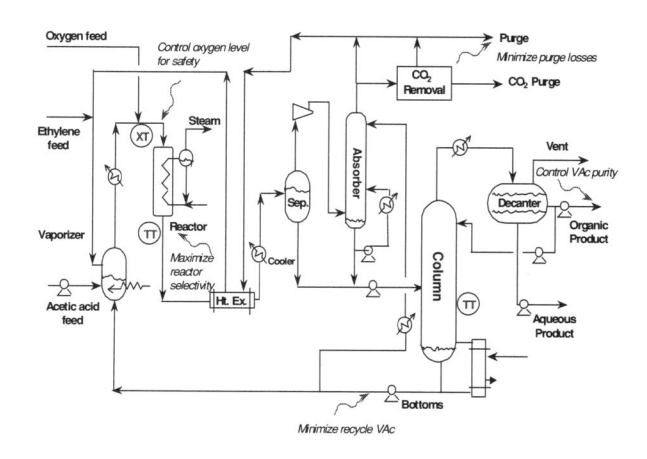

Figure 12. Economic control objectives and dominant variables in vinyl acetate process.

All of this, of course, is beyond the current capability of the optimization methods. Yet impressive progress continues to be made with these approaches. These considerations (the inclusion of control for economic objectives, stability, recycle, constraints, and material and energy balance) offer some potential focus for future research activities in this area. The analysis of dominant variables and effective manipulators is another consideration that can be brought into the formulation of the problem superstructure.

Conclusions

We have described in this paper the notion of using process design to ensure proper economic control for chemical plants. This means understanding what variables dominate the productivity of units in the process and having suitable manipulators available to implement an effective partial control strategy for the plant.

In our examination of the three general approaches to process synthesis, we highlighted how design decisions in each method can affect the dynamic operability of the resulting process design. Without question, heuristic and hierarchical methods predominate today in industrial practice. By decomposing the problem, they are extremely powerful at handling the many complexities of flowsheet alternatives, particularly in generating initial possibilities. The design engineer needs to be aware of the control implications of choosing particular types of reactors or separation systems, if they have unusual dynamic behavior or require certain manipulated variables to handle any specific safety, constraint, or economic objectives. Further, the recycle structure and energy integration must be carefully constructed to minimize the chances of disturbances propagating through the system and affecting the plant's economic objectives. The downside of the hierarchical methods is their inability to consider simultaneously the effects of all the design variables. This means that we can expect the methods to be feasible, but they may overlook choices that could improve steady-state economics or lead to better dynamic controllability.

The thermodynamic methods in process synthesis typically start with some flowsheet structure and then apply integration techniques. We again expect these methods to be feasible in terms of reaching a solution, but must caution against the downside of excessive amounts of integration. These include the loss of manipulated variables, the lowering of driving forces for heat transfer, and the potential for propagating disturbances throughout the entire plant.

Research efforts have made a great deal of progress on the practical application of optimization methods in process synthesis. At this point, these approaches seem more appropriate at a later stage of the conceptual design activity, when we have adequate information to formulate a meaningful process superstructure. We then may also be able to incorporate variables involving both the process design and the control system design. The clear advantage of an optimization formulation is the ability to consider all variables simultaneously. The fundamental question, though, about these methods is their feasibility. In principle, process synthesis can involve many nonlinear, nonconvex equations with both static and dynamic variables. Solving these problems is a challenging task indeed, particularly if we are also designing the best control system to reject specified disturbances. The incorporation of ideas from partial control and dominant variables may be a helpful contribution for these approaches.

It is an unfortunate but true fact that the dominant organizational structure in the chemical and petroleum industries has separate groups for design and control engineers. The current predominant use of heuristic and hierarchical methods in industrial plant design does not naturally facilitate better ways to do process design so that we account for the controllability issues we have highlighted in this paper. However, there is just one process. It happens to have both steady state and dynamic characteristics, and we must somehow reconcile the issues that arise in looking at the design and operation of a process. Certainly involving control engineers in the process design would contribute substantially to this. The use of dynamic process modeling could also be an effective mechanism for design and control engineers to communicate about picking the best design among several alternatives. Dynamic models show the effects of the design decisions on controllability, and so they act as a bridge for communicating about effective economic control. As industrial plant design begins to adopt more of the optimizaton methods, it will force the currently distinct design and control groups to work more closely together because this will be necessary for properly formulating the problem.

It is clear, though, that we foresee a substantial amount of future research work still ahead in understanding the best way to use process design to ensure effective economic control of chemical processes (independent of the synthesis method).

References

Arbel, A., I. H. Rinard, R. Shinnar, and A. V. Sapre (1995). Dynamics and control of fluidized catalytic crackers. 2. Multiple steady states and instabilities. *Ind. Eng. Chem. Res.*, **34**, 3014-3026.

Arbel, A., I. H. Rinard, and R. Shinnar (1996). Dynamics and control of fluidized catalytic crackers. 3. Designing the control system: choice of manipulated and measured variables for partial control. *Ind. Eng. Chem. Res.*, **35**, 2215-2233.

Arbel, A., I. H. Rinard, and R. Shinnar (1997). Dynamics and control of fluidized catalytic crackers. 4. The impact of design on partial control. *Ind. Eng. Chem. Res.*, **36**, 747-759.

Douglas, J. M. (1988). *Conceptual Design of Chemical Processes*. McGraw-Hill, New York.

Grossmann, I. E. and M. M. Daichendt (1996). New trends in optimization-based approaches to process synthesis. *Comp. Chem. Eng.*, **20**, 665-683.

Linnhoff, B. (1994). Use pinch analysis to knock down capital costs and emissions. *Chem. Eng. Prog.*, Aug. 32-57.

Luyben, M. L. and B. D. Tyreus (1998). An industrial design/control study for the vinyl acetate monomer process. *Comp. Chem. Eng.*, **22**, 867-877.

Luyben, W. L., B. D. Tyreus, and M. L. Luyben (1999). *Plantwide Process Control*. McGraw-Hill, New York.

Mohideen, M. J., J. D. Perkins, and E. N. Pistikopoulos (1996). Optimal design of dynamic systems under uncertainty. *AIChE J.*, **42**, 2251-2272.

Nishida, N., G. Stephanopoulos, and A. W. Westerberg (1981). A review of process synthesis. *AIChE J.*, **27**, 321-351.

Perkins, J. D. (Ed.), (1992). *Proceedings of IFAC Workshop on Interactions Between Process Design and Process Control*, Pergamon Press.

Price, R. M., P. R. Lyman, and C. Georgakis (1994). Throughput manipulation in plantwide control structures. *Ind. Eng. Chem. Res.*, **33**, 1197-1207.

Schweiger, C. A. and C. A. Floudas (1997). Interaction of design and control: optimization with dynamic models. In W. W. Hager and P. M. Pardalos (Eds.) *Optimal Control: Theory, Algorithms, and Applications*. Kluwer Academic Publishers. pp 388-435.

Schweiger, C. A. and C. A. Floudas (1999). Optimization framework for the synthesis of chemical reactor networks. *Ind. Eng. Chem. Res.*, **38**, 744-766.

Shinnar, R. (1981). Chemical reactor modeling for the purposes of controller design. *Chem. Eng. Comm.*, **9**, 73-99.

Tyreus, B. D. (1999a). Dominant variables for partial control. 1. A thermodynamic method for their identification. *Ind. Eng. Chem. Res.*, **38**, 1432-1443.

Tyreus, B. D. (1999b). Dominant variables for partial control. 2. Application to the Tennessee Eastman challenge process. *Ind. Eng. Chem. Res.*, **38**, 1444-1455.

van Schijndel, J. and E. N. Pistikopoulos (1999). Towards the integration of process design, process control and process operability – current status & future trends. These proceedings.

Zafiriou, E. (Ed.), (1994). *Proceedings of IFAC Workshop on Integration of Process Design and Control*, University of Maryland.

MULTI-SOLVER MODELING FOR PROCESS SIMULATION AND OPTIMIZATION

L. T. Biegler and Dilek Alkaya
Carnegie Mellon University
Pittsburgh, PA 15213

Kenneth J. Anselmo
Air Products and Chemicals, Inc.
Allentown, PA 18195

Abstract

With the increasing size and complexity of process simulation and optimization problems, exploitation of the process model becomes increasingly important. While most researchers recognize the need to exploit the problem structure, for instance at the linear algebra level, this study explores the case when multiple model solvers are required for simulation and optimization of the overall process system. While this approach is standard for process flowsheeting, we need to consider how we can take advantage of sophisticated simultaneous solution and optimization strategies for large-scale optimization. Here we discuss both open form and closed form models, and demonstrate that both are needed for different types of problems. We then consider an approach where closed form or 'black box' models can be 'opened up' to achieve simultaneous optimization without disturbing the inherent structure of the model's solver. In addition, several applications, including process flowsheets, dynamic optimization, PDE models and process integration are highlighted. Finally, we close with some challenges and areas for future work for both modeling environments and optimization algorithms.

Keywords

Process optimization, SQP algorithms, Open form (equation oriented) modeling, Closed form modeling, Tailored optimization, Process integration.

Introduction

With increasing complexity and size of chemical process models, there remains a constant challenge to develop more efficient and robust algorithms for simulation and optimization. Here model developers and engineers are faced with conflicting objectives. First, there is a need to develop robust general purpose modeling platforms and numerical algorithms that allow the rapid development of large process models. This paradigm allows a straightforward and seamless extension to very efficient optimization algorithms as well. And modeling, simulation and optimization tools that are built on these concepts allow us to conduct large-scale optimization studies with the same ease as solving the process model itself. Excellent illustrations in this conference include the papers by Wright (1999), Floudas and Pardalos (1999), and Grossmann and Hooker (1999).

On the other hand, increasing complexity of these models requires us to consider the exploitation of specific model structure and the development of specialized solvers for different kinds of systems. This approach has its roots in the early history of process simulation. For instance, equilibrium stage models exploit the block tridiagonal structure of the MESH equations and related systems. Moreover, further solver decompositions like the inside-out algorithm must be applied in order to isolate and control highly nonlinear characteristics of some columns. Moreover, the exploitation of structure and specialized solvers is especially acute in the solution of

PDE-based (Partial Differential Equation) models, where spatial and temporal discretizations create systems with millions of variables and equations. Here the 'one size fits all' solver paradigm breaks down. Excellent illustrations at this conference include the papers by Sinclair (1999) and Jensen (1999). However, for these systems, the extension of these models to optimization and integration with other process systems still remains a serious challenge.

This study attempts to bridge between highly developed process simulation and optimization tools and complex, diverse models that have yet to be considered in these environments. In the next section we begin by tracing the evolution of different process optimization strategies. In particular, we highlight the merits of simultaneous optimization strategies, both for efficiency and the ability to integrate different tasks and subsystems. Following this section, we highlight benefits and challenges required for optimization applied to process integration.

Section 4 then discusses the current state of open form and closed form systems, i.e., models expressed in a declarative, equation oriented way, vs. models that are closely coupled and exploited by a special purpose solver. The latter case also includes the presence of legacy codes and also covers a wide variety of detailed models that are not widely considered in process flowsheeting or real-time optimization. As will be shown in this section, both open and closed form paradigms are essential for their respective model types. Section 5 develops a tailored optimization strategy that extends the benefits of simultaneous optimization to a wide variety of closed form models. Derived as a variation of the Successive Quadratic Programming (SQP) algorithm, this approach has been demonstrated on a number of different process systems. Further challenges for nonlinear programming algorithms are discussed in section 6 and section 7 briefly summarizes and concludes the paper.

A Hierarchy of Optimization Approaches

In a recent lecture at the SIAM Optimization Conference, Betts (1999) provided a personal summary on the evolution of optimization applications in the aerospace industry. The main stages of this evolution can be classified through the use of *surrogate, black box* (with exact derivatives) and *simultaneous* optimization strategies applied to complex engineering models. In the *surrogate* approach, a simplified optimization problem is formed from easily obtained information from a detailed model. This problem is updated and solved repeatedly as the optimization proceeds and there is only a weak connection between the detailed model and the optimization process. In the *black box approach*, the detailed model is called repeatedly by the optimization algorithm in order to obtain accurate function (and gradient) information. Finally, the *simultaneous* optimization strategy leads to solution of the optimization problem and the complex model at the same time.

As one advances from the surrogate to the simultaneous strategies, it is clear that both the efficiency of the optimization approach and the scale of the application can be increased greatly. Also, the interactions between the detailed model and the optimization algorithm become much tighter, and considerably more effort is required to formulate the optimization problem. On the other hand, detailed models cannot all be adapted to optimization formulations in the same way. For some models, the application of simultaneous strategies can be difficult and possibly counterproductive with current approaches. For instance, at Boeing Corporation, orbit trajectory optimization is a well established and highly efficient simultaneous optimization procedure (Betts and Frank, 1994), while more complex and detailed structural and fluid flow optimizations for aircraft design are currently handled by more primitive surrogate methods (Booker et al., 1998).

Similarly, it is instructive to consider the evolution of optimization methods in process engineering. The *surrogate approach* applies to any strategy that forms a simplified model of the system, which can be optimized directly. In this context we assume that this simplification can be obtained quickly and cheaply from the complex model (e.g., with perhaps a few function evaluations). For these approaches, one can cite pattern searches developed in the 60s (Hooke-Jeeves, Nelder-Mead, EVOP, etc.) and related stochastic strategies such as simulated annealing and genetic algorithms. Interestingly, many process synthesis approaches such as the use of residue curve maps, attainable regions and pinch curves for heat or mass integration can also be viewed as surrogate optimization methods. These methods are easy to apply and require little interaction between the rigorous, detailed process model and the optimization method. As a result, these methods are very popular for 'one-shot' optimization studies. On the other hand, their limitations arise in dealing with highly constrained problems and in large-scale and routine applications where efficiency is crucial.

The *black box* approach using gradient-based methods is a popular procedure both for flowsheet optimization (e.g., ASPEN/Plus, ProSim, etc.) as well as the optimization of dynamic systems (gProms, ABACUSS). For efficient optimization strategies, a crucial component is the efficient and accurate calculation of derivative information. Nowadays, this task can be addressed through the application of efficient automatic differentiation (AD) and sensitivity strategies. For steady state models, AD approaches can be applied directly to the source code, so that exact derivatives with respect to any number of input variables are obtained with only a small multiple of the solution cost. AD (Griewank, 1989) has been applied to a wide variety of large-scale finite element and process models by replicating the calculation tree with corresponding derivative calculations. Nevertheless, this approach needs to be approached carefully to avoid differentiating through fixed point loops in internal calculations. Similarly, efficient sensitivity calculations have been adapted to parallel the solution of DAE systems (Feehery et al., 1998; Li et al., 1999). As a result, accurate gradients for Newton-type convergence and optimization can be obtained, often with only a small multiple of the solution cost.

Finally, *simultaneous methods* have seen widespread use in real-time optimization of petrochemical plants. Often used with 'open form' or equation based models, this strategy requires close collaboration of the modeler and the developers and users of the optimization strategy. Moreover, to obtain the performance benefits of this optimization approach, the development, analysis and implementation are much more difficult and time consuming than with the previous ones. Perhaps the greatest advantage of these approaches is their transparency to sophisticated general-purpose numerical algorithms. Derivatives are calculated in an efficient and accurate manner and, in principle, the solver has full access to all variables, equations and derivative information.

Table 1. Simulation Time Equivalents on Evolution of Methods.

Surrogate Models

•Friedman and Pinder (1972)	75-150
•Gaines and Gaddy (1976)	300

Black Box (with derivatives)

•Parker and Hughes (1981)	64
•Biegler and Hughes (1981)	13

Partially Simultaneous, Black Box

Biegler and Hughes (1982)	
Chen and Stadtherr (1985)	
Kaijaluoto (1984)	
ASPEN+, PRO/II, HYSYS	10-30
Wolbert et al (1994)	3-10

Simultaneous

•Locke and Westerberg (1983)	< 5
•Stephenson and Shewchuk (1986)	2
•RTOPT, NOVA, etc.	~1

Table 1 summarizes the performance characteristics for flowsheet optimization using a few selected studies as benchmarks. Using a process flowsheet with about ten decision variables, we see requirements of hundreds (or even thousands) of simulation time equivalents for surrogate strategies. On the other hand, with simultaneous strategies, optimization requires virtually the same effort as the solution of the process model. Also, we observe reductions of over an order of magnitude when passing from one strategy to the next. Note that in the case of partially simultaneous, black box methods, where derivatives are calculated very efficiently (Wolbert et al., 1994), there is some overlap in performance with simultaneous strategies.

It is clear from this discussion that simultaneous strategies provide clear performance benefits. Moreover, these benefits translate into the ability to model and solve much larger optimization problems. As discussed in the next section, the ability to integrate multiple process subsystems and multiple design tasks (e.g., economics, operability, controllability, safety, energy recovery) leads to more important benefits than increase in performance alone.

Optimization - A Tool for Integrated Process Engineering

Over the past decade the integration of tools and process design environments has become a major activity in the process industries. The impacts of this integration on standardization of work processes and incorporating design and operation issues into the supply chain (Ramage, 1998; van Schijndel and Pistikopoulos, 1999) are widely recognized as key corporate activities. Marquardt and Nagl (1998) surveyed the development of standards for tool integration and classified these as: *Presentation integration* – integrated tool set with common look and feel presented to user; *Data integration* – sharing and managing relations among data objects; *Control integration* – notification and activation of tools using e.g., message passing and *Platform integration* – execution of integrated suite of tools on heterogeneous, distributed computer network. Examples of these standards include STEP and PDXI for data integration as well as CAPE-OPEN for data and control integration. Moreover, commercial examples of tool integration include VTPLAN (Bayer), Plantelligence (Aspen Tech) and SimSight/Simulation Manager (SimSci/Bayer). Academic projects in this area include efforts to support conceptual, design and front-end engineering and are exemplified by the *n*-dim (Carnegie Mellon), KBDS/epee (Edinburgh), and the IMPROVE and CHEOPS (Aachen) systems.

More recently, Marquardt and coworkers (Backx et al., 1998; Helbig et al., 1998; Marquardt, 1999) demonstrated the importance of optimization tools within this integrated framework. Clearly the ability to model and optimize over entire systems and over multiple attributes leads to far superior solutions. Moreover, the integration of optimization formulations has been a fruitful activity in process systems engineering over the past decade. Studies include integration of batch process design and scheduling (Birewar and Grossmann, 1989; Voudouris and Grossmann, 1993), design under uncertainty, (Ierapetritou et al., 1996; Pistikopoulos, 1997), design and dynamic performance (Logsdon and

Biegler, 1993), design, scheduling and dynamic performance (Bhatia and Biegler, 1996), scheduling and dynamic performance (Mujtaba and Macchietto, 1993), interactions of energy, separation and reactor subsystems (Balakrishna and Biegler, 1996; Duran and Grossmann, 1986), interactions of control and design (Luyben and Floudas, 1994; Morari and Perkins, 1994; Walsh and Perkins, 1994), process design and planning (Pinto and Grossmann, 1994; Sahinidis and Grossmann, 1991) and safety, design and performance (Abel et al., 1998).

To show the quantitative benefits of integration, we briefly consider the design, operation and scheduling of a small batch process. By combining these problem aspects, we hope for synergies that lead, for instance, to shorter processing times, shorter planning horizons and higher quality batches. Two key aspects that aid in this case study are the use of simultaneous dynamic optimization and a simplified scheduling formulation. The former is due to the discretization of the differential equations (DAEs) and state and control profiles to form a large-scale optimization problem. For the latter we consider a continuous variable scheduling formulation adapted from Birewar and Grossmann (1986) that deals with the sequencing of tasks, products and equipment using idealized Unlimited Intermediate Storage (UIS) and Zero Wait (ZW) transfer policies. In both cases, a nonlinear program is formulated and the solution directly yields a minimum cycle time operating schedule for multiple products and stages. The detailed integrated formulation is given in Bhatia and Biegler (1996).

To demonstrate this approach we consider the process example given in Bhatia and Biegler (1996), which includes a batch reactor, batch distillation and several transfer units. We consider the sequencing of three products of varying purity with possible manipulations of the temperature profile in the reactor and the reflux profile in the column. Here optimal dynamic profile cases are R1 and C1 for the reactor and column, respectively, while best constant profiles are R0 and C0.

To compare formulations, we first consider the *sequential design* where the equipment design, operation and schedule are optimized one after another. The *intermediate case* deals with final time states fixed by a unit optimization followed by simultaneous optimization of design and scheduling. Finally, we consider a *fully simultaneous approach* with optimized final states. The results of this case study are summarized in Figure 1 for the ZW case.

Note that significant improvements result from the integrated formulation. Here we see that there is strong improvement due to a variable reflux ratio (from C0 to C1) but the greatest improvements are simply due to the integration of the design, operation and scheduling tasks in the optimization problem. The reason for this is apparent from the production schedule. The sequential solution still has long cycle times and long slack times in the ZW schedule. The intermediate solution reduces both of these considerably and the fully simultaneous approach eliminates the slack times altogether and creates a schedule with the shortest cycle time. This allows for much greater equipment utilization and significantly improves the profit.

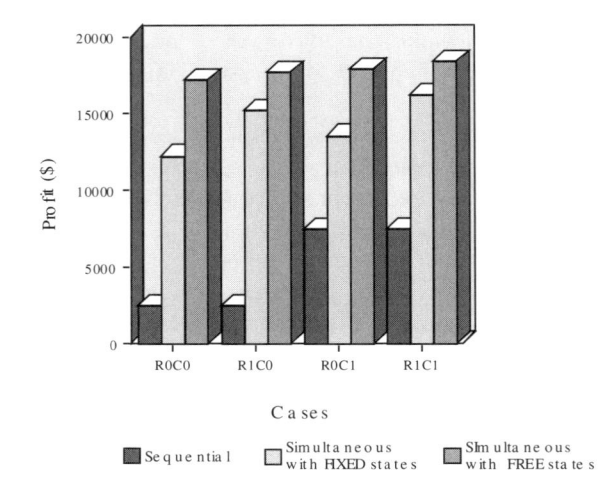

Figure 1. Integrated batch optimization.

The message of this example is that key aspects of the dynamic operation, equipment design and scheduling need to be modeled accurately and optimized simultaneously. Without these features, even the most detailed unit optimizations (along with separately optimized schedules) would realize less than half of the benefits of the overall optimization, for this example.

Most of the above studies on integrated optimization were performed using open form models. However, as we shall see next, these models are not always appropriate. In the next section we contrast the benefits and drawbacks of open and closed form models. This sets the stage for alternative simultaneous strategies in Section 5.

Open Form *vs.* Closed Form Models

In process engineering, open form models (also known as equation oriented or equation based models) are characterized by a strategy whereby all of the stream and unit equations are assembled, solved and optimized as one large system. Additional calls to procedures are kept to a minimum, although procedural calls to physical property routines are still considered essential. There are several key advantages to this approach for process simulation and optimization. In particular, performance in solving large flowsheets can be accelerated considerably, and highly integrated flowsheets can be converged and optimized very efficiently. These models have been extremely successful in the real time optimization of petrochemical processes (Perkins, 1998; Marlin and Hrymak, 1997).

On the other hand, closed form models (also known as procedural, modular or black box models) are tightly integrated to a special purpose solver and the user can only access a restricted set of information from the model. As a result, optimization algorithms cannot access function and gradient information about individual model equations and variables; frequently the equations are not even represented explicitly within the model. Instead, in an optimization environment, such models are currently solved repeatedly and this leads to considerable computational cost. Closed form models are prevalent in process simulation, design and analysis, especially for detailed reaction and fluid flow models. In this section, we contrast these two approaches from a number of different perspectives.

Declarative vs. Procedural Modeling

The open form approach decouples modeling of the process from the solution algorithm. This allows a lot of freedom in formulating the model and also allows the user to concentrate on the proper problem definition for the task at hand. Moreover, it is conceptually easier to extend and modify declarative models because the solution procedure does not need to be modified at the same time. This declarative capability is the main strength of platforms like ASCEND, SPEEDUP, gPROMS and GAMS. Note that this is fundamentally different from a specific procedural model or even an EXCEL spreadsheet where the user is intimately focused on the procedure of solving as well as modeling.

This declarative form of modeling offers a number of advantages when extending and modifying closed form models. Provided that powerful procedures are available for this effort (including automatic differentiation of the model equations), this modeling strategy greatly shortens development time and provides for easier maintenance.

Implicit in the declarative approach is that a generic solver (usually, Newton-Raphson) will be able to solve the derived model. Moreover, this solver will take limited advantage of the physics of the model, except in the exploitation of sparsity in the linearized equations. This means that a number of specialized closed form algorithms that involve nested loops, bootstrapping solution strategies and interpolation to avoid expensive calculations are not straightforward to implement with open form modeling. These may be needed to avoid overly large systems, improve convergence or minimize expensive physical property calls (e.g., the inside-out algorithm in RADFRAC).

Newton-like Convergence Behavior

The open form approach allows for a simultaneous convergence strategy since the most efficient generic solver is Newton's method. Therefore, convergence should be Q-quadratic under ideal assumptions (Kantorovich, 1948). This means that properly formulated and initialized open form models can be solved one or two orders of magnitude faster than modular models. For problem formulations of comparable size, it is unlikely that an alternative approach can beat this performance. Moreover, this approach provides a general, coordinated convergence strategy, regardless of problem structure. This is in contrast to flowsheeting programs where convergence behavior strongly depends on the tear stream selection and unit sequencing. On the other hand, Newton's method needs to be stabilized far from the solution, even with open form models. Here a generic stabilization strategy (line search or trust region) will not take advantage of specific problem features and the only recourse is for the user to reformulate the model. This becomes more serious when the problem becomes ill-conditioned or singular at isolated points. Under these circumstances, Newton's method has serious difficulties that should be tackled most effectively with problem-specific formulations. These can be remedied easily through additional safeguards in closed form approaches.

However, for open form models we need to resort to well-known line search and trust region strategies for solving nonlinear equations. In equation based environments these strategies are applied as a single damping factor for a trust region or line search. This can produce a 'conservative' stepsize that severely impacts overall convergence of the process system and offsets the expected quadratic convergence.

Toy Problem

To illustrate the problem of small stepsizes more closely, consider the following equations:

$$f_1 = 2x_1 + x_2 - 3 = 0$$
$$f_2 = \exp(-m(x_1-3)^2) - 1 = 0$$
$$x_1^0 = 1, \ x_2^0 = 1$$

First we consider the simultaneous solution of these equations with the above starting point using Newton iterations, for various values of m. As shown in Figure 2, a stepsize (α) equal to one is acceptable (using the Armijo criterion) for m = 0.5 at the first Newton step. However, for m = 5 only stepsizes below 1.3×10^{-7} will lead to a sufficient decrease in $\|f(x)\|$ along the first Newton direction. Figure 1 plots $\|f(x)\|$ for various values of m. Note that for m = 2 and above, $\|f(x)\|$ decreases only in very small regions around the ordinate. This will be true of *any open form model* that contains f_2. The table shows the number of MINPACK function evaluations to solve this problem. Here a very reliable solver (using QR factorizations, trust regions and analytic Jacobians) needs to be used.

In contrast, a procedural solution of this example is a trivial. Once x_1 is fixed, f_1 is solved immediately as a linear system. As shown above, f_2 can be solved using a simpler stabilized Newton method (S-N) where only a few iterations are needed. Of course, in both approaches, f_2 can easily be reformulated. But even with reformulation, the procedural approach offers the advantage of isolating the nonlinearities from the rest of the system.

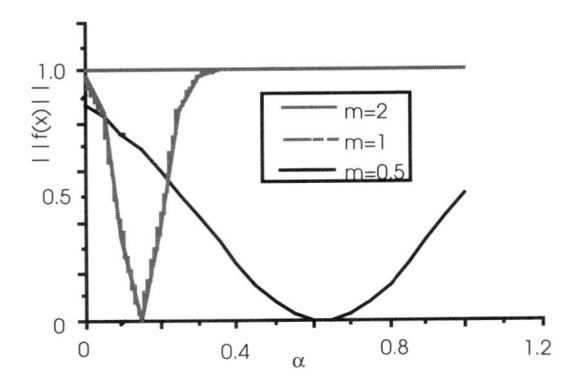

MINPACK		α_{max}	S-N	
0.5	77		1.0	10
1.0	83		0.32	10
2.0	83		0.007	9
5.0	65		1.3e-7	12

Figure 2. Stepsizes for toy problem.

The effect of the "conservative" stepsize has been observed on a number of process applications. For instance, simultaneous solution of countercurrent gas-gas heat exchanger models can be very unstable with very small damping factors required for convergence. Instead, a closed form reformulation of this system leads to a much more reliable method.

Open vs. Closed Form Structures

In addition to the advantages of declarative models, the open form of the model equations allows easy implementation and efficient calculation of exact derivative information. The ready availability of derivative information has led, along with Newton's method, to fast performance for open form approaches. Nevertheless, as discussed in the previous section, this can also be done in closed form using automatic differentiation strategies.

On the other hand, the open form necessarily requires the user to consider larger systems which are more difficult to set up, analyze, debug and get to solve. In essence, successful users *model declaratively but debug procedurally* by breaking down the problem and solving smaller problems. A closed form approach provides a natural, object-oriented strategy for this analysis. As a result, the process solver must give the user access to smaller subproblems to accomplish this. This can also be done, in principle, with an open form approach (e.g., in the ASCEND system), but currently a lot more experience is required by the user to do this. Here most diagnostics deal directly with open form variables and equations, rather than diagnostic messages that relate to a physical model.

A related problem is the initialization of the open form model. This task requires substantial user intervention and often more time and effort than in obtaining the solution itself. For future development, the open form approach will need to be improved, not just with better user interfaces, but also with better tools that allow more flexible problem decomposition and reformulation.

Moreover, the fully open form approach is resistant to incorporating legacy codes in closed form. These may involve specialized process models or solution algorithms, which are essential for a particular application (e.g., CFD systems with an iterative solver). This is especially difficult for ODE and PDE-based models. A closed form approach can incorporate these specialized models into a larger process model much more readily than open form packages can. Often, these codes must be reformulated entirely as declarative models in the open form environment. In future development, we need to allow the integration of closed models that may provide derivative information, which will be used by the solver. An interesting development in this direction is the use of *foreign objects* in the gProms environment (Kakhu et al., 1998).

Finally, the issue of accurate discretization of models using differential equations in space must be dealt with. For instance, closed form models solve ordinary differential equations with automatic step sizes to control error. That is, the integration step size is not known *a priori and* usually varies from step to step. In the open form approach, it is not at all clear how these step sizes are to be chosen prior to discretization. In academic studies, accurate discretization and error control has been demonstrated for small open form models (e.g., Tanartkit and Biegler, 1997). However, this may be difficult for commercial open form codes with many more stepsize variables. Currently, stepsize selection is done by trial and error by the modeler.

Conditional Models

With procedural approaches, it is straightforward to include IF-THEN relations triggered by inputs or calculated values. For instance, with simple flash calculations, decisions are based on dew and bubble point equations and the proper set of equations is then selected and solved. This approach is then embedded within much larger systems, which are successfully solved and optimized. The basic hope with closed form approaches, is that the solvers are able to 'jump over' these derivative and function discontinuities; this is often the case, even though it is easy to construct counterexamples.

There are more rigorous ways to deal with discontinuities and these are better suited to open form approaches. They are also more expensive than simple equation solving, as they involve combinatorial elements. Two examples are complementarity relations and integer variables. In both cases, the conditional relation must be reformulated by an experienced user to form a set of conditional equations or inequalities. The resulting model then must be solved with an advanced solver (nonlinear complementarity or mixed integer nonlinear program (MINLP), respectively).

Formulating Optimization Problems

Characteristics of open form modeling allow a natural extension of Newton-based solvers to nonlinear programming using algorithms like Successive Quadratic Programming (SQP), although it should be noted that *large-scale* versions of SQP are now required for simultaneous convergence. In contrast, much smaller optimization problem formulations can be considered for the same problem with the closed form approach.

Also, by extension, the same open form initialization effort required for model solution is needed for optimization. But, because more degrees of freedom are involved, the open form optimization may converge less reliably than the open form simulation problem. To promote convergence, safeguards that are easily built into the closed form model are not as straightforward in the open form approach.

Finally, the concept of formulating an optimization problem through the evaluation of procedures is often absent in the open form approach. In the modular approach, the user can visualize the optimization problem via a set of case studies applied to the (expected smooth) process model; these cases are then automated with an efficient, gradient-based (or even a direct search) optimization algorithm.

With proper safeguards, this automated case study approach leads to 'quick and dirty' optimization procedures, which are often successful, and lead to good, if not optimal, results. An open form approach to these problems requires complete rethinking of this concept and often results in a very different declarative model (e.g., MINLP formulation for feedtray optimization). In fact, the need to incorporate legacy codes and other closed form models has motivated several leading NLP researchers (Booker et al., 1998; Audet and Dennis, 1999; Kelley, 1999) and corporations (IBM, Boeing, Sandia Laboratories) to develop and analyze derivative free optimization methods.

Simultaneous Optimization with Existing, Closed-Form Models

From the previous section it is clear that both open and closed form models are required for different simulation and optimization applications. Still, a commonly held belief is that simultaneous gradient-based optimization is not possible with closed form models. As a result, there is considerable effort to extrapolate the open form successes from on-line optimization to many other areas of process engineering. As mentioned above, this effort will lead to a number of benefits in the creation, maintenance and efficient solution of many new process models. On the other hand, this approach does not incorporate legacy models well nor does it allow the use of (often essential) specialized solvers that exploit the model's structure.

Can we develop a simultaneous optimization approach using closed-form models? This section explores a hybrid strategy that allows the direct use of existing closed-form models within a simultaneous convergence and optimization strategy. Here we can also further exploit the simultaneous strategy for closed form models with Newton-based solvers. Note that such models may still be difficult to represent in open form. They include complex column models, two point boundary value models that involve reaction, separation or heat transfer and the solution of nonlinear partial differential equations (PDEs). For these models we assume that *neither the equations nor the Jacobian matrix are available directly*. Instead, we expect to access the Newton step and the 'sensitivity' of this step to input variables. With this information we can derive the following optimization algorithm based on SQP concepts in the reduced space (rSQP).

Derivation of a Tailored rSQP Algorithm

Consider the Nonlinear Programming Problem (NLP):

$$\text{Min } F(x) \\ \text{s.t. } h(x) = 0 \\ a \leq x \leq b \quad (1)$$

where $x \in R^n$ and $h(x): R^n \rightarrow R^m$. To obtain a local optimum we need to find a point that satisfies Karush Kuhn Tucker (KKT) conditions. The first order KKT conditions can be given by:

$$g(x^*) + \nabla \bar{h}(x^*) \lambda^* = 0 \\ \bar{h}(x^*) = 0 \quad (2)$$

where $g = \nabla F$ and $\bar{h}(x^*) = 0$ is the appropriate active constraint set at the solution x^*. To satisfy these equations (and choose the correct active set) we consider an extension of Newton's method by solving the QP subproblem at iteration k:

$$\text{Min } g(x_k)^T d + 1/2\, d^T Q d \\ \text{s.t. } h(x_k) + A(x_k)^T d = 0 \\ a \leq x_k + d \leq b \quad (3)$$

where $A = \nabla h$, $Q = \nabla_{xx} L(x, \lambda)$ (the second derivative matrix of the Lagrange function with respect to x) or its approximation and d is the search direction for x. The solution of the QP is given by the linear system at iteration k:

$$\begin{bmatrix} Q & \bar{A} \\ \bar{A}^T & 0 \end{bmatrix} \begin{bmatrix} d \\ \lambda \end{bmatrix} = - \begin{bmatrix} g \\ h \end{bmatrix} \quad (4)$$

(the k subscript is suppressed for brevity). We now consider the exploitation of the structure of the QP. In many process optimization problems n ~ m >> (n-m). As a result, the n x n matrix Q can either be considered in a large, sparse form or approximated in the reduced space. Because second derivatives are not available for closed form models (or in most commercial packages that use open form models), we choose the latter option.

Here we develop a simplified decomposition in the reduced space using only equality constraints and defer the explicit treatment of variable bounds until later. To solve the QP in the reduced space, let:

$$x^T = [\, y^T \mid z^T \,], \quad A^T = [\, C \mid N \,]$$

and select an n x n nonsingular matrix:

$$H = [\, Y \mid Z \,], \text{ where } A^T Z = 0.$$

Here Z and Y form null and range space bases for the linearized equality constraints. We similarly partition the search direction into range (p_Y) and null space (p_Z) components: $d = Y p_Y + Z p_Z$.

The former component deals with the dependent variables of the model, while the latter component determines the search direction for decision variables of the optimization problem. With this representation we can write:

$$H = \begin{bmatrix} I & -C^{-1}N \\ 0 & I \end{bmatrix} \quad H^{-1} = \begin{bmatrix} I & C^{-1}N \\ 0 & I \end{bmatrix} \quad (5)$$

Defining the linear system from the QP as M x = f gives:

$$\begin{bmatrix} Q & A \\ A^T & 0 \end{bmatrix} \begin{bmatrix} d \\ \lambda \end{bmatrix} = - \begin{bmatrix} g \\ h \end{bmatrix} \quad (6)$$

Now defining $X = \text{diag}\,[\,[\,Y \mid Z\,],\,I\,]$, we can consider the equivalent system $X^T M X z = X^T f$ as:

$$\begin{bmatrix} Y^T Q Y & Y^T Q Z & Y^T A \\ Z^T Q Y & Z^T Q Z & 0 \\ A^T Y & 0 & 0 \end{bmatrix} \begin{bmatrix} p_Y \\ p_Z \\ \lambda \end{bmatrix} = - \begin{bmatrix} Y^T g \\ Z^T g \\ h \end{bmatrix} \quad (7)$$

Because p_Y and p_Z are calculated from the last two rows, we can ignore the $Y^T Q Y$ and $Y^T Q Z$ terms and approximate $Z^T Q Y\, p_Y$ by w_k to get:

$$\begin{bmatrix} 0 & 0 & Y^T A \\ 0 & Z^T Q Z & 0 \\ A^T Y & 0 & 0 \end{bmatrix} \begin{bmatrix} p_Y \\ p_Z \\ \lambda \end{bmatrix} = - \begin{bmatrix} Y^T g \\ Z^T g + w_k \\ h \end{bmatrix} \quad (8)$$

Note that neglecting these terms does not affect the search direction, only the multiplier estimates λ. As the search direction converges to zero, λ also converges to its correct value. Also we see that these assumptions lead to a simple block diagonal decomposition for the rSQP strategy.

To relate the range and null space representation back to the original variables, we define the above system as: M x = f, and define:

$$X = \text{diag}\,[\,[\,Z \mid Y\,]^{-1},\,I\,].$$

This allows us to write $X^T M X z = X^T f$ as:

$$\begin{bmatrix} 0 & 0 & C^T \\ 0 & B_k & N^T \\ C & N & 0 \end{bmatrix} \begin{bmatrix} d_y \\ d_z \\ \lambda \end{bmatrix} = - \begin{bmatrix} g_y \\ g_z + w_k \\ h \end{bmatrix} \quad (9)$$

where $B_k \sim Z^T Q Z$. Now the C matrix is the combined Jacobian of the closed form multiple models and is assumed to be nonsingular. If not, we assume the closed form solver can be suitably modified to yield a nonsingular C. As a result we can modify the reduced system to yield:

$$\begin{bmatrix} 0 & 0 & I \\ 0 & B_k & N^T C^{-T} \\ I & C^{-1} N & 0 \end{bmatrix} \begin{bmatrix} d_y \\ d_z \\ C^T \lambda \end{bmatrix} = - \begin{bmatrix} g_y \\ g_z + w_k \\ C^{-1} h \end{bmatrix} \quad (10)$$

Note that if the model (h(x) = 0) is solved with a Newton-based procedure we may not be able to extract gradient information for the model equations. In many closed form models the Jacobian is not constructed explicitly and specialized decompositions are incorporated directly within the model (e.g., block tridiagonal decomposition in distillation models, condensation in collocation models). Nevertheless for many of these models, we can easily extract the Newton step, $p_y = -C^{-1} h$; the *model sensitivity matrix*, $D_z = -C^{-1} N$, which reflects the sensitivity of the Newton step to the model inputs can also be calculated. As a result, the KKT matrix for the QP is equivalent to:

$$\begin{bmatrix} 0 & 0 & I \\ 0 & B_k & -D_z^T \\ I & -D_z & 0 \end{bmatrix} \begin{bmatrix} d_y \\ d_z \\ \hat{\lambda} \end{bmatrix} = - \begin{bmatrix} g_y \\ g_z + w_k \\ -p_y \end{bmatrix} \quad (11)$$

and all of the explicit Jacobian information from the model disappears. Here the gradient of the objective function is usually specified directly by the user in terms of dependent and independent variables. The correction term w_k and the reduced Hessian B_k are approximated by reduced gradients. Moreover, D_z (the 'sensitivity' of p_y to decision variables, z) can be obtained by:

- calculating $\partial h / \partial z$ within model and solving multiple right hand sides for $-C^{-1} N$
- perturbing h(x) with respect to z and solving multiple right hand sides for $-C^{-1} N$, or

- executing an additional Newton step with a perturbed value of z.

Given the Newton steps and model sensitivity matrices for each unit, it is easy to create p_y and D_Z for the entire flowsheet, e.g., A" = [I | D_Z]. The tailored rSQP algorithm can then be represented as in Figure 3. where the necessary model matrices are collected and decision variables are updated for the entire process. QPKWIK (Schmid and Biegler, 1994b) is used to solve the reduced space QP problem.

Finally, we see that the actual multipliers λ are not calculated; only $C^T \lambda$ is available and this precludes the use of conventional line search strategies for SQP. Instead we apply a 'multiplier-free' line search where the penalty parameter is estimated by a lower bound on the multipliers. A global and local convergence analysis related to this rSQP approach is given in Biegler et al. (1997). Examples on the application of this tailored approach to closed form models can be found in Schmid and Biegler (1994a) and Tanartkit and Biegler (1996). The details of extending the tailored approach to flowsheet optimization and related decomposition strategies are given in Alkaya et al. (1999).

Flowsheet Optimization Study

To illustrate this tailored approach we briefly summarize the flowsheet optimization study, described in Tanartkit and Biegler (1996). The process contains several distillation columns as well as an exothermic reactor model solved with a Newton based collocation routine (with fixed stepsizes). Here we compare a "simultaneous" closed form approach with the tailored optimization approach. In this closed form approach, the tear equations are converged efficiently as equality constraints in the optimization problem. Also, we found that an open form approach solved with an rSQP method and the same initialization performs about the same as the tailored approach.

Five cases were considered where shortcut (i.e., split fraction) models are substituted by detailed Newton based models for the reactor (using COLDAE) and distillation column (using UNIDIST) in the process flowsheet:

1. Product Column (UNIDIST)
2. Reactor Model (COLDAE)
3. Recycle Column (UNIDIST)
4. Both Columns (UNIDIST)
5. Reactor and Product Column (UNIDIST and COLDAE)

The performance of the tailored vs. the modular approach is shown in Figure 3. In each of these cases the tailored approach leads to far better performance than in the conventional closed form (modular) approach. In the last case where both models need to be considered, the tailored approach leads to a four-fold increase in performance, but without requiring open form models to be used.

This approach has also been extended to dynamic systems as described in Alkaya et al. (1998). Here an optimal control problem for a binary batch column model is coupled to a batch reactor. Both units are self-contained

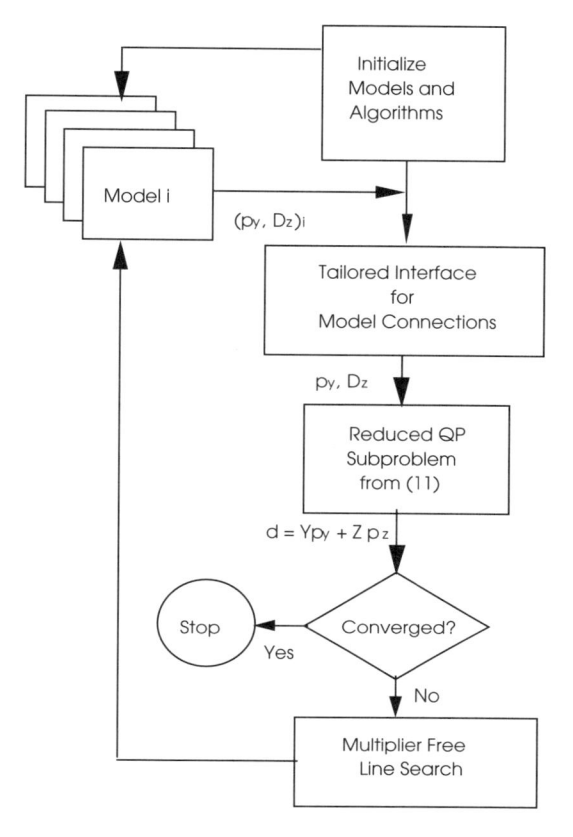

Figure 3. Tailored process optimization algorithm.

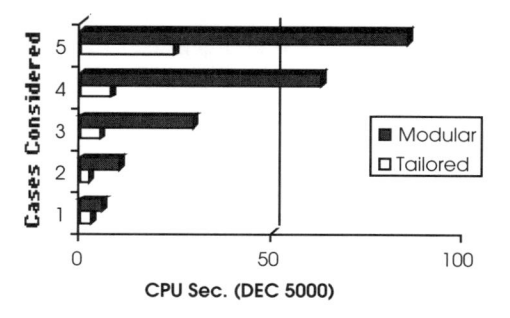

Figure 4. Flowsheet optimization with the tailored approach.

collocation models and coupling these models over time is performed using the above decomposition procedure. The optimization of this system with the tailored approach generally leads to the same performance as with open form models. Here finding the optimal reflux ratio to maximize product requires about 29 rSQP iterations.

Further Challenges for Nonlinear Programming Algorithms

The tailored approach represents only a preliminary effort to exploit models with closed form structure and multiple solvers. In this section we review a number of essential research issues for large-scale nonlinear programming.

Handling bound constraints

The development of the tailored approach does not explicitly deal with bound constraints. These have been handled in many process optimization applications (for both open form and tailored models) by the rSQP subproblem shown in Figure 3. For our studies we have relied on a reduced space QP algorithm (QPKWIK) to handle inequalities but this has clear limitations for larger problems. To deal with large-scale decomposition strategies, the KKT matrix should be exploitable and should not interfere with updating of bound constraints. This can be done by exploring both active set and interior point (IP) methods.

On problems where (n-m) is large (e.g., optimal control problems and multiperiod problems) interior point methods have a tremendous performance advantage in active set selection (Albuquerque et al., 1999). Figure 5 compares the number of active set iterations with the interior point method. Note that the number of IP iterations (six) remains independent of (n-m). Solving QP subproblems with an IP strategy has yielded significant benefits for process applications in process identification, multiperiod optimization and model predictive control.

To increase the usefulness of this approach we are currently exploring IP methods to solve the NLP problem directly. This further eliminates the overhead of solving the QP with an IP approach and can lead to an order of magnitude performance improvement. This approach has seen much recent work (Vanderbei and Shanno, 1999; Byrd, Gilbert and Nocedal, 1998; Conn, Gould and Toint, 1999) but many open questions still remain. Nevertheless, preliminary results show this approach is very successful for optimal control problems. For instance, with an IP NLP approach, the dynamic optimization of an air separation column with 462 DAEs and over 55,000 variables requires less than 5 CPU minutes on a DEC Alpha workstation (Cervantes et al., 1999).

On the other hand, for problems where (n-m) is small, the number of active bounds and the search complexity are also small. Therefore the overhead associated with IP methods does not compete with efficient active set methods (Ternet and Biegler, 1999). In particular, the application of Schur complement methods (Betts and Frank, 1994) has the advantage that the KKT matrix can be decomposed separately and updates to the active set are very cheap. This approach can be applied directly to (10) or (11) and appears to be about three times faster than QPKWIK (Bartlett, 1999).

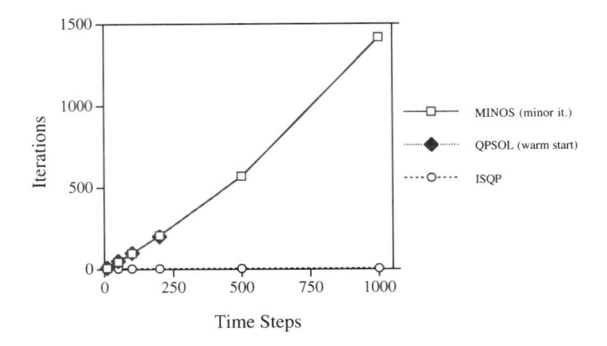

Figure 5. Comparison on interior point (ISQP) vs. active set (QPSOL, MINOS) iterations when (n-m) is large.

Iterative Linear Solvers

Similar concerns in exploiting the structure of the KKT matrix arise when dealing with very large NLPs arising from PDE applications. Large finite element models for fluid flow or structures are also solved with Newton-based procedures but are too large to attack with direct, sparse linear solvers. Instead iterative linear solvers (Preconditioned Krylov (PK) methods) are essential and the extension to simultaneous SQP methods poses a severe challenge. Recently, Biros and Ghattas (1999) used the rSQP decomposition and an approximated reduced Hessian to precondition the full space KKT matrix; the PK solver MINRES was used in a parallel environment to solve the KKT system. For a 3-D finite element model of the Navier-Stokes equations (with 90,000 state variables, 1200 decision variables) they obtained optimal solutions with less than 200 wallclock seconds on Cray T3E. This led to an average reduction of 34 times over rSQP method mentioned above and showed excellent scalability of the method on parallel processors. Moreover, on problems with few decision variables, Ghattas and coworkers have applied rSQP concepts to optimize finite element models with over 3 million state variables.

In addition to the above research directions, there are a number of specific issues related to handling increasingly large process optimization problems. Here we also need to consider:

Decomposition of large nonlinear systems for problems that exhaust the limitations of current tools. Included here are the handling of (multiple) RTO models for refinery wide optimization and multiple plant design models, which may run on different platforms, exploit different structure and could be developed under different modeling environments.

Coupling of multiple design models where mass and energy balance models need to be coupled with detailed equipment design and costing. Also included are important process aspects such as operability, controllability and safety, which can be hard to formulate for the optimization problem (often due to discontinuities in their calculation).

Multiperiod models reflect life cycle considerations for the process and allow the design to consider different operating scenarios, consideration of uncertainty and evolution of the process over time. These models lead to very large bordered block diagonal (or almost block diagonal) KKT matrices, where each diagonal block could represent the KKT matrix of any of the process optimization problems discussed thus far.

Most studies that respond to the challenges in integration assume that all process models are available in an open form environment. In this environment, discontinuities and nonsmoothness can be overcome, in principle, through additional constraints or MINLP formulations.

However, if process models need to remain in closed form, can the barriers to optimization be overcome? Here we can handle nonsmooth problems through reformulation or direct handling with bundle methods. More severe problems have prompted recent activity in derivative free optimization (DFO) methods in the math programming community, even to solve MINLP problems (Audet and Dennis, 1999). Also, convergence properties that link gradient based methods and DFO methods for unconstrained optimization, have recently been analyzed by Kelley (1999). In performance, these methods are clearly inferior to gradient based methods used with open form models. However, their robust application to closed form models could avoid the daunting expense of model conversion. Given the resurgence of these methods, an open question is whether they can be extended rigorously to constrained optimization problems with simultaneous convergence.

Finally, recent progress in the development of powerful methods for global optimization and their application to process problems (see Floudas and Pardalos, 1999) raises similar questions about their applicability to closed form models. It also focuses an awareness of the importance of local vs. global solutions in practice and better appreciation of the capabilities of optimization tools.

Conclusions

The widespread diversity of process models needs to be recognized and their individual structure needs to be exploited both for integration and optimization of these models. In the case of open form (equation based) models, extremely efficient optimization algorithms have been developed. Their influence is felt in the rapid solution of large-scale process optimization models, particularly for real-time optimization. Moreover, this capability has led to very creative NLP and MINLP problem formulations that allow engineers to model and exploit a number of difficult process attributes (e.g., integration of operability, controllability, safety constraints) in an efficient, quantitative manner.

However, more complex models need to be considered for process optimization and integration. And not all of these models can be represented in an open form, to be solved with a single generic solver. Difficult nonlinear features need to be isolated, particular equation structures need to be exploited more efficiently and specialized unit solvers are essential for many applications. To deal with these kinds of models, we outline a tailored optimization approach and illustrate its simultaneous capabilities on both flowsheeting and DAE models. This approach represents only a preliminary step toward exploiting these models. Moreover, a number of emerging areas in the optimization community are lending new algorithms and fundamental concepts to deal with these problems.

Finally, the ability to deal with specialized models is important not only for performance but also for the challenges posed for the integration of multiple process models, process attributes and for corporate-wide activities including design, control, operation, scheduling and planning. Application of optimization strategies is a key component of this integration and the most benefits are obtained *only if* accurate process models can be considered in the overall problem formulation.

Acknowledgements

Research funding from Mobil Research Corporation, Air Products and Chemicals, Inc. and the National Science Foundation is gratefully acknowledged. The authors would also like to thank Omar Ghattas, Wolfgang Marquardt, Peter Piela, Purt Tanartkit, Dimitrios Varvarezos and Sriram Vasantharajan for helpful discussions and comments.

References

Abel, O., A. Helbig, and W. Marquardt (1999). Productivity optimization of an industrial semi-batch polymerization Reactor under safety constraints. *J. Process Control*, to appear.

Albuquerque, J., V. Gopal, G. Staus, L. T. Biegler, and B. E. Ydstie (1999). Interior point SQP strategies for large-scale, structured process optimization problems. *Comp. Chem. Eng.*, **23**, 543.

Alkaya, D., and S. Vasantharajan, and L. T. Biegler (1988). Generalization and application of tailored approach for flowsheet optimization. AIChE Meeting, Miami Beach.

Alkaya, D., and L. T. Biegler (1999). Exploiting chemical process model structure for nonlinear programming. SIAM Annual Meeting, Atlanta.

Audet, C. and J.E. Dennis (1999). *On the Convergence of Mixed Integer Pattern Search Algorithms*. Technical Report, CAAM, Rice University.

Backx, T., O. Bosgra, and W. Marquardt (1998). Towards intentional dynamics in supply chain conscious process operations. *FOCAPO '98 Proceedings*, Snowbird, Utah.

Balakrishna, S. and L. T. Biegler (1999). Chemical reactor network targeting and integration - an optimization approach. Adv. Chem. Eng., 23, 248.

Bartlett, R. A. (1999). PhD Proposal, Carnegie Mellon University, Pittsburgh.

Betts, J. T. (1999). What's the answer? An industrial perspective on optimization. SIAM Conference on Optimization, Atlanta.

Betts, J. T., and P. Frank (1994). A sparse nonlinear optimization algorithm. J. Opt. Theo. Appl, 82, 3.

Bhatia, T., and L. T. Biegler (1996). Dynamic optimization in the design and scheduling of multiproduct batch plants. *Ind. Eng. Chem. Res.*, 35(7), 2234.

Bhatia, T. and L. T. Biegler (1999). Multiperiod design and planning with interior point methods. *Comp. Chem. Eng.*, accepted for publication.

Biegler, L.T. and R.R. Hughes (1981). Approximation programming of chemical processes with Q/LAP. *Chem. Eng. Prog.*, 76, April.

Biegler, L.T. and R.R. Hughes (1982). Infeasible path optimization for sequential modular simulators. *AIChE J.*, 28(6), 994.

Biegler, L. T., Claudia Schmid, and David Ternet (1997). A multiplier-free, reduced Hessian method for process optimization. *IMA Volumes in Mathematics and Applications,* Springer Verlag, 93, 101,

Birewar, D. and I. E. Grossmann (1989). Efficient algorithms for the scheduling of multiproduct batch plants. *Ind. Eng. Chem. Res.*, 28, 1333.

Biros, G. and O. Ghattas (1999). Parallel domain decomposition methods for optimal control of viscous incompressible flows. Proceedings of Parallel CFD '99, Williamsburg, VA.

Booker, A. J., Paul D. Frank, J.E. Dennis, Douglas W. Moore, and David B. Serafini (1998). *Managing Surrogate Objectives to Optimize a Helicopter Rotor Design - Further Experiments*. Tech. Report, CAAM, Rice University.

Byrd, R. H., J-C. Gilbert and J. Nocedal (1998). *A Trust Region Method based on Interior Point Techniques for Nonlinear Programming*. Technical Report, ECE Dept., Northwestern University.

Cervantes, A., A. Waechter and L. T. Biegler (1999). A decomposed barrier strategy for optimal control. SIAM Conference on Optimization, Atlanta.

Chen, H-S., M. A. Stadtherr (1985). A simultaneous modular approach to process flowsheeting and optimization - I, II, III. *AIChE J.*, 31(11), 1843.

Conn, A. R., N. Gould, D. Orban (1999). Global and local convergence of a primal-dual barrier algorithm for nonlinear optimization. SIAM Conference on Optimization, Atlanta.

Duran M. A., and I. E. Grossmann (1986). Simultaneous optimization and heat integration of chemical processes. *AIChE J.*, 32, 123.

Feehery, W. F., J. Tolsma, P. Barton (1997). Efficient sensitivity analysis of large-scale DAEs. *Appl. Num. Math.*, 25, 41.

Floudas, C. A. and C. Pardalos (1999). Recent developments in global optimization methods and their relevance to process design. These proceedings..

Friedman, P. and K. L. Pinder (1972). Optimization of a simulation model of chemical plants. *Ind. Eng. Chem. Proc. Des. Dev.*, 11(4), 512.

Gaines, L. D. and J. L. Gaddy (1976). Process optimization by flowsheet simulation. *Ind. Eng. Chem. Proc. Des. Dev.*, 15(1), 206.

Griewank, A. (1989). On automatic differentiation. in *Mathematical Programming: Recent Developments and Applications,* eds. M. Iri, K. Tanabe, Kluwer Academic Publishers, 83-108.

Grossmann, I. E., & J. Hooker (1999). Logic based approaches for mixed integer programming models. These proceedings.

Helbig, A., O. Abel, W. Marquardt (1998). Structural Concepts for Optimization Based Control of Transient Processes," *Proceedings of Nonlinear Model Predictive Control Workshop*, Ascona, Switzerland.

Ierapetritou, M.G.; Pistikopoulos, E.N.; and C. A. Floudas (1996). Operational planning under uncertainty. *Comp. Chem. Eng.*, 20(12),1499-1516

Jensen, K. F. (1984). Integrated microchemical systems: opportunities for process design. These proceedings.

Kaijaluoto, S. (1984). Process optimization by flowsheet simulation. Technical Research Center of Finland, Publication No. 20.

Kakhu, A., B. Keeping, Y. Lu and C. Pantelides (1998). An open software architecture for process modeling and model-based applications. *FOCAPO Proceedings, AIChE Symposium Series*, No. 320, 518.

Kantorovich, L. V. (1948). Functional analysis and applied mathematics. *Uspekhi Mat. Nauk*, 3, 89.

Kelley, C. T. (1999). *Iterative Methods for Optimization,* SIAM.

Li, S., L. Petzold and W. Zhu (1999). *Sensitivity Analysis of Differential-algebraic Equations: a Comparison of Methods on a Special Problem*. Tech. Report, Computer Science Department, University of California, Santa Barbara.

Locke, M.H.; and A. W. Westerberg (1983). The ASCEND-II system-a flowsheeting application of a successive quadratic programming methodology," *Comp. Chem. Eng.*, 7(5),615.

Logsdon, J. S. and L. T. Biegler (1993). Accurate determination of optimal reflux policies for the maximum distillate problem in batch distillation. *Ind. Eng. Chem. Res.*, 32(4), 692.

Luyben, M.L.; Floudas, C.A. (1994). Interaction between design and control in heat-integrated distillation synthesis. *Integration of Process Design and Control (IPDC '94) IFAC Workshop*, Oxford, Elsevier; 251; 79-84

Marlin and A. Hrymak (1997). Realtime operations optimization of continuous processes. *Chem. Proc. Control V*, J C Kantor, C E Garcia and B Carnahan (eds.), 217.

Marquardt, W. (1999). Perspectives on life cycle process modeling. These proceedings.

Marquardt, W., and M. Nagl (1998). Tool integration via interface standardization? computer application in process and plant engineering. Papers of the 36th DECHEMA-Monographie **135**, 95-126.

Morari, M. and J. Perkins (1994). Design for operations. Proceedings of FOCAPD '94, CACHE-AIChE, Biegler and Doherty (eds.), 105.

Mujtaba, I. and S. Macchietto (1993). Optimal operation of multicomponent batch distillation: multiperiod formulation and solution. *Comp. Chem. Eng.*, 17, 1191.

Parker, A. L. and R. R. Hughes (1981). Approxiamtion programming of chemical processes – I. *Comp. Chem. Enr.*, **5**(3), 123.

Perkins, J. (1998). Plantwide ptimization - opportunities and challenges. Proceedings of FOCAPO '98, Snowbird, Utah.

Pinto, J.M., and I. E. Grossmann (1994). Optimal cyclic scheduling of multistage continuous multiproduct plants. *Comp. Chem. Eng.* **18**(9), 797-816.

Pistikopoulos, E., (1997). Uncertainty in process design and operations. *Comp. Chem. Eng.*, **19**, S553.

Ramage, M. (1998). Computation and competitiveness: managing technology in the information age. *Proceedings of FOCAPO '98*, Snowbird, Utah.

Sahinidis, N. V. and I. E. Grossmann (1991). Multiperiod investment model for process networks with dedicated and flexible plants. *Ind. Eng. Chem. Res.*, **30**(6), 1165.

Schmid, C. and L. T. Biegler (1994a). Quadratic programming methods for tailored reduced Hessian SQP. *Comp. Chem. Eng.* **18**(9), 817.

Schmid, C. and L. T. Biegler (1994b). A Simultaneous approach for process optimization with existing modeling procedures. *Chem. Eng. Res. Des., Part A*, **72**, 382.

Shanno, D., and R. Vanderbei (1999). An interior-point algorithm for nonconvex nonlinear programming. SIAM Conference on Optimization, Atlanta.

Sinclair, J. L. (1999). CFD for multiphase flow: research codes and commercial software. These proceedings.

Stephenson, G. R., and C. F. Shewchuk (1986). Reconciliation of process data with process simulation. *AIChE J.*, **32**(2), 247.

van Schijndel, J. and S. Pistikopoulos (1999). Towards the integration of process design, process control and process operability - current status and future trends. these proceedings.

Tanartkit, P., and L. T. Biegler (1996). Reformulating ill-conditioned DAE optimization problems. *Ind. Eng. Chem. Res.*, **35**(6), 1853.

Tanartkit, P. and L. T. Biegler (1997). A nested, simultaneous approach for dynamic optimization problems-II: The outer problem. *Comp. Chem. Eng.*, **21**(12), 1365.

Ternet, D. and L. T. Biegler (1999). Interior point methods for reduced Hessian successive quadratic programming. *Comp. Chem. Eng*, accepted for publication.

Voudouris, V. T., and I. E. Grossmann (1993). Optimal synthesis of multiproduct batch plants with cyclic scheduling and inventory consideration. *Ind. Eng. Chem. Res.*, **32**(2), 1962.

Walsh, S. and J. Perkins (1994). Application of integrated process and control system design in waste water neutralization. *Comp. Chem. Eng.*, **8**, S183.

Wolbert, D., X. Joulia, B. Koehret and L. T. Biegler (1994). Flowsheet optimization and optimal sensitivity analysis using exact derivatives. *Comp. Chem. Eng.*, **18**, 1083.

Wright, S. (1999). Algorithms and software for linear and nonlinear programming. these proceedings.

CFD FOR MULTIPHASE FLOW: RESEARCH CODES AND COMMERCIAL SOFTWARE

Jennifer Lynn Sinclair
School of Chemical Engineering
Purdue University
West Lafayette, IN 47907-1283

Abstract

This paper gives an overview of the advances in development of computational fluid dynamics codes for multiphase, gas-solid flows. Both dilute-phase and dense-phase flows are discussed. Reliable prediction of dilute-phase flow requires an accurate description for the modulation of gas-phase turbulence by the presence of particles. In dense-phase flow, interactions between individual particles and clusters of particles, or particle-phase turbulence, are important phenomena that need to be accurately described. Experimental needs for validation and improvement of CFD codes are highlighted; other key challenges such as describing a distribution of particle sizes are discussed.

Keywords

Multiphase, Gas-solid, Fluidization, CFD turbulence.

Introduction

Throughout industry, fluid-solid unit operations are some of the most troublesome to operate. While other unit operations have a down-time less than 10%, fluid-solid unit operations have a down-time of approximately 40 to 50% (Ray Cocco and Bruce Hook, Dow Chemical, personal communication). The flow behavior of these two-phase systems is too complex in order to reliably scale-up, design or optimize them. Typically, experimental testing is performed on expensive, large-scale units for "effective" new designs or new units mimic existing units.

For the chemical industry, this situation, more often than not, directs engineers to implement more expensive, labor-intensive mechanical unit operations. For example, although fluidized bed reactors are well suited for reaction paths that are temperature sensitive, many times these same reactions are carried out in tubular reactors instead due to the uncertainties about designing or operating fluidized bed reactors.

The development of reliable computational fluid dynamic (CFD) models would limit the need for expensive large-scale test facilities. Evaluation of key variables and flow parameters such as particle concentration, particle size and size distribution, inlet flow conditions and geometry, etc. could be explored without elaborate testing equipment. In addition, turn-around time for trouble shooting an existing unit could be reduced and more efficient operation could be achieved. CFD modeling for multiphase flow based on a two-fluid framework is rapidly becoming a typical tool used for the prediction of flow behavior in fluidized beds, circulating fluidized beds, transport lines, etc. This paper presents a current status on CFD model development for multiphase flow.

Modeling Overview

In order to treat flows in systems of practical size where the solids loading ratio (ratio of solids mass flux to gas mass flux) exceeds one, particle flow models are constructed in the Eulerian framework. Research work in my group and several other groups [IIT (Gidaspow, Arastoopour), Princeton (Sundaresan), FETC (Syamlal, O'Brien), Los Alamos (Kashiwa)] has focused on the development of such models for dilute and dense-phase gas-solid flows. In many gas-solid systems, particle-particle interactions are significant and the resulting stresses can be described based on an analogy between gaseous molecules and particulate solids (Lun *et al.* 1984). Similar to molecules in a gas, the collisions between solid particles give rise to a random motion, which is superimposed, on the bulk motion of flow. Furthermore, the intensity of this random particle motion, referred to as the pseudothermal or granular temperature, affects the pressure and viscosity of the solid phase, just as the stresses in the gas phase depend on the true thermal temperature. Unlike the true thermal energy, however, the granular energy is dissipated by inelastic particle-particle and particle-wall collisions. The application of this kinetic theory analogy to fluidized systems (Sinclair and Jackson, 1989) allows for a fundamental description of solid-phase stress, as opposed to earlier descriptions which involved empirical input for the solids viscosity and solids stress modulus. A solid stress description based on concepts from kinetic theory only requires information on particle-particle and particle-wall collision properties, such as normal and tangential coefficients of restitution ($e = 1$ for perfectly elastic collisions, $e = 0$ for perfectly inelastic collisions).

Dilute-Phase Flow Modeling

In dilute-phase flow, an accurate description for particle modulation of gas-phase turbulence, as well as particle-particle interactions, is key to accurate flow prediction. Louge *et al.* (1991) and Bolio and Sinclair (1995) have included the effects of gas-phase turbulence and the modulation of gas-phase turbulence by the presence of particles. Louge *et al.* (1991) used a one-equation closure turbulence model and Bolio and Sinclair (1995) used the Myong and Kasagi (1990) k-ε model, both turbulence models modified for the presence of a dilute particle phase. Both works found that particle collisions are significant even in dilute-phase flow, contributing up to 25% of the overall pressure drop required to convey a given gas-solid mixture. Dilute-phase models are able to capture the following flow features: a non-uniform slip velocity over the tube cross-section, a change in sign in the slip velocity near the tube wall, a decrease in the slip velocity with decreasing particle size and increasing solids loading, a flattening of the mean and fluctuating gas velocity profile with the addition of solids, and particle velocity fluctuations which exceed gas velocity fluctuations. Validation of dilute-phase flow models is typically based on four key LDV data sets for vertical pipe flow (Sheen *et al.* (1993), Tsuji *et al* (1984), Lee and Durst (1982) and Maeda *et al* (1980)). Of these four data sets only Sheen *et al.* (1993) report both the axial and radial gas-phase fluctuating velocity components. This has led to some discrepancies in the literature between comparisons of model predictions and experimental data made at the level of the axial fluctuating component only, instead of the total turbulent kinetic energy (see Figures 1 and 2).

Dense-Phase Flow Modeling

In dense-phase gas-solid flows, the flow predictions from the treatment of Sinclair and Jackson (1989) exhibit an undue sensitivity to the coefficient of restitution e. To illustrate this behavior, Figure 3 shows the radial profile of the solid volume fraction obtained using the Sinclair and Jackson model in the case of fully developed flow in a vertical pipe (riser). For $e = 1.0$ and $e_{wall} = 0.9$, the model predicts a high concentration of particles near the wall since the only dissipation of granular energy is due to particle-wall collisions. At $e = 0.99$ and $e_{wall} = 0.9$, however, the segregation of particles now appears at the core of the pipe due to the inelastic particle-particle collisions. Such behavior where the particles are concentrated at the core is contrary to that observed experimentally. Evidently, not all of the mechanical interactions which are present in this dense-phase system were properly accounted for in the Sinclair and Jackson model.

Both experiments and computer simulations have indicated that densely-concentrated, gas-solid flows are characterized by spatial inhomogeneities and time-dependent solid-phase stresses, a behavior which is similar to the motion of eddies in a single-phase turbulent fluid. Based on an analogy with single-phase turbulence, two regimes of gas-solid flow can be identified. In the "laminar", dilute-phase flow regime, only the random motion of individual particles is significant; these random motions are described by kinetic theory concepts. In the dense-phase, "turbulent" flow regime, particle clustering and the random motion of these clusters, in addition to the random motion of individual particles, are significant. Both of these types of particle velocity fluctuations are important and contribute to solid mixing patterns in fluidized systems. Hence, a complete model for dense-phase flows should account for velocity fluctuations at both levels. Dasgupta *et al.* (1994) first introduced this concept of "particle-phase turbulence" in which the magnitude of the random velocity associated with collections of particles is defined by the turbulent kinetic energy of the particle phase (k_s), whereas the magnitude of the random velocity associated with individual particles is given by the granular temperature (T). Based on limiting forms of the model equations, it can be shown that solids volume fraction is inversely proportional to the granular temperature in the laminar regime, as is the case with the Sinclair and Jackson model since only the laminar contribution to the particle-phase is considered (Figure 3). Similarly, in the turbulent regime, the solids volume fraction is inversely proportional to k_s.

Figure 1a. Profiles of gas turbulent kinetic energy, clear gas.

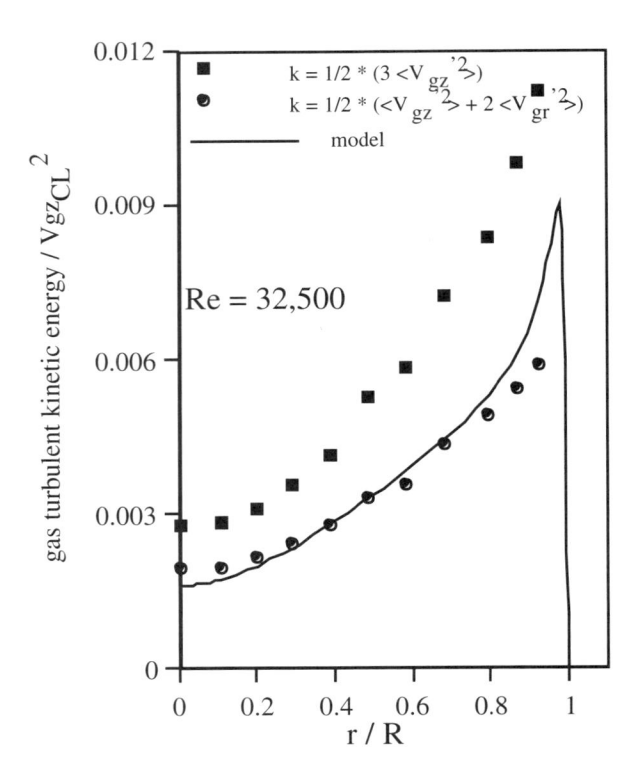

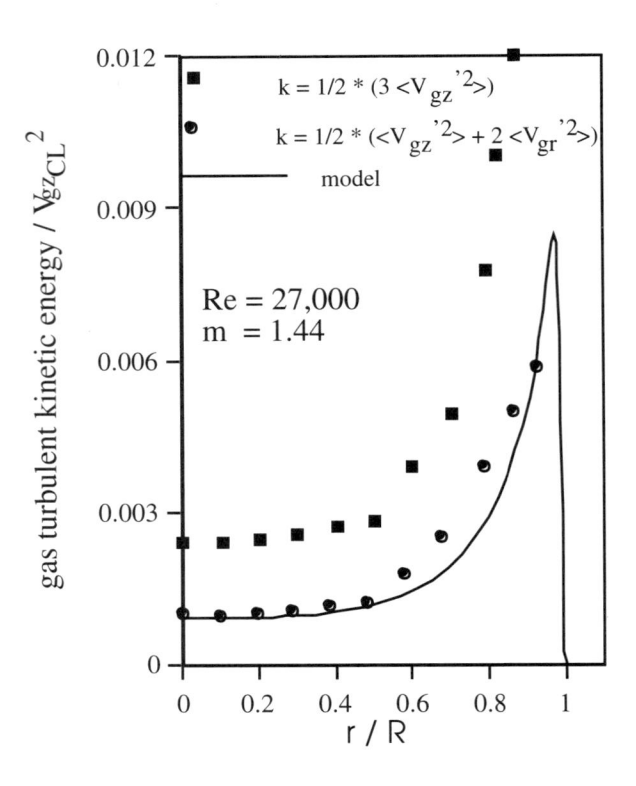

Figure 1b. Profiles of gas turbulent kinetic energy, with 450 μm polystyrene particles (data of Sheen et al. - 1993).

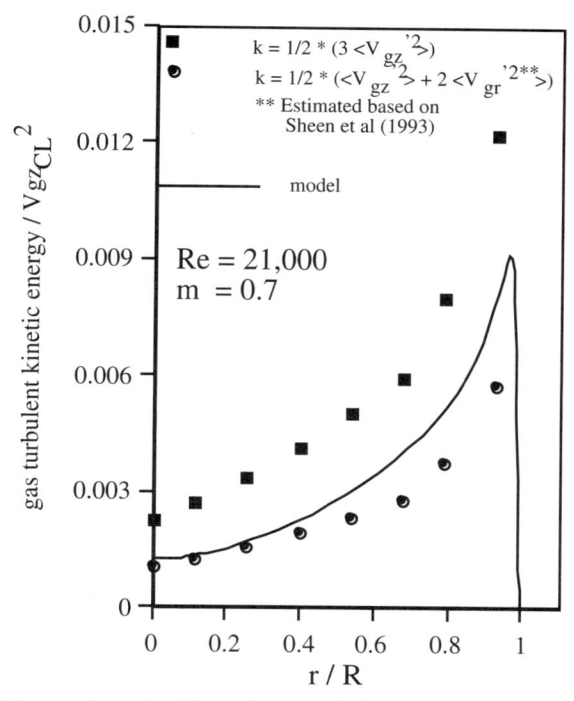

Figure 2a. Profiles of gas turbulent energy.

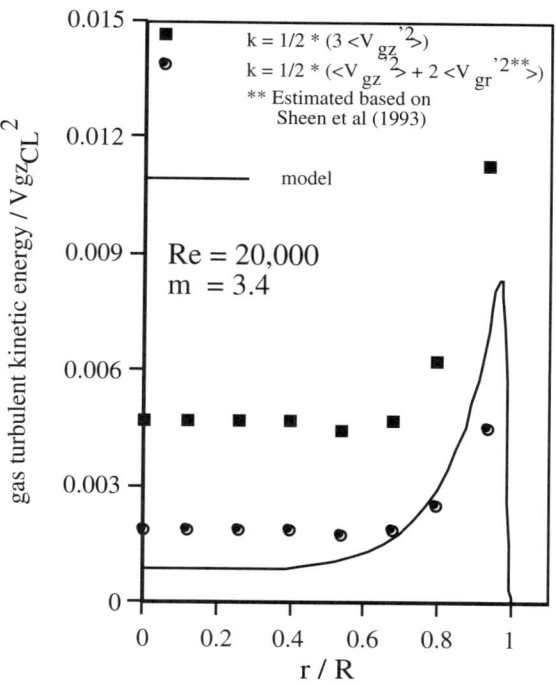

Figure 2b. Profiles of gas turbulent kinetic energy, with 500 μm polystyrene particles (data of Tsuji et al. - 1984).

Hrenya and Sinclair (1997) first investigated a model formulation in which only the governing equations (not the kinetic theory closures) were time-averaged. The results indicated that although the flow predictions from the particle-phase turbulent model were less sensitive to the coefficient of restitution than the laminar model, the sensitivity had not been completely eliminated (see Figure 4). The sensitivity to the coefficient of restitution remained because the *laminar* contribution to the solid-phase pressure was greater than the turbulent contribution. Hence, they found that a consideration of the effects of particle-phase turbulence on the governing equations alone was not sufficient to describe the flow; a linearization and time-averaging of the kinetic theory closures was needed. However, since this additional time-averaging gave rise to a number of complicated turbulence correlations with difficult physical interpretation, they focused attention on the dissipation rate of granular energy due to inelastic particle-particle collisions. It is this dissipation term which is responsible for the huge drop in the granular temperature associated with inelastic particle collisions.

When the dissipation rate was linearized and time-averaged, an additional term arose that contained a correlation between the fluctuating solids concentration and granular temperature. A physical analysis of this turbulence correlation revealed that it is always positive in sign; hence, the magnitude of the actual dissipation rate is less than in the case when particle phase turbulence is not taken into account. Hrenya and Sinclair (1997) took the turbulence correlation between the fluctuating solids concentration and granular temperature to be proportional to the intensity of solid-phase turbulence. Figure 5 displays the results of this improved model for the test case of fully-developed riser flow. Note that these results were obtained using a value for the particle-particle coefficient of restitution of 0.9 (e=0.9).

By including the effects of particle phase turbulence, very good flow predictions, both qualitatively and quantitatively, can be obtained in dense-phase systems for realistic values of the coefficient of restitution. An overall picture of the flow behavior for a FCC particle system in a vertical pipe with e = 0.9 is displayed in Figure 6. As shown, the Hrenya and Sinclair (1997) model is able to predict many of the salient qualitative features of gas-solid flows: cocurrent upflow, cocurrent downflow, and countercurrent flow regimes (1st, 2nd, and 3rd quadrants, respectively); existence of a flooding envelope (2nd quadrant); and the existence of multiple steady-state gas velocities for a given solids velocity at a fixed pressure gradient (1st and 3rd quadrants). The Zenz operating diagram of Figure 7 illustrates the ability of the model to predict additional features characteristic of gas-solid flows. In particular, the model exhibits the expected U-shape for systems of large solids concentration and a multiplicity of operating conditions. As depicted in Figure 8, the region of multiplicity disappears as the tube radius decreases, which implies that the scale-up of such systems is by no means straightforward. As shown in Figure 9, the progressive segregation of particles towards the wall of the riser with increasing solids flux at fixed gas velocity (Unpublished data, PSRI, 1995) is captured using the dense-phase model.

Heat Transfer in Dilute and Dense-Phase Gas-Solids Flow

Mallo and Sinclair (1998) and Mallo (1997) have coupled the above dilute and dense-phase gas-solid flow model with thermal energy balances for the gas and particle-phases to predict heat transfer rates in these multiphase systems. They have shown that the observed complex heat transfer behavior in gas-solid systems can be directly related to the hydrodynamic behavior.

In dilute-phase gas-solid flow with heat transfer, the Reynolds analogy is applied to both the gas and solid-phases (Wang and Campbell, 1990), and the interphase heat transfer coefficient is described as in the case of flow past a single sphere. The wall-to-bed heat transfer coefficient increases with increasing Reynolds number as in single-phase flow, and it increases with decreasing particle size due to an increase in the interphase heat transfer coefficient which is dependent on the surface area to volume ratio for the particle. Increases or decreases in the wall-to-bed heat transfer coefficient with increasing solids loading can be related to the competition between gas turbulence modulation which affects the magnitude of the thermal conductivity for the gas phase and the effect of particle-particle collisions which affects the magnitude of the thermal conductivity for the particle phase.

In dense-phase gas-solid flow with heat transfer, the effective thermal conductivity of the particle phase is augmented by particle velocity fluctuations not only at the level of individual particles (as in dilute-phase flow) but also by fluctuations at the level of clusters of particles. Mallo and Sinclair (1998) predict that the heat transfer coefficient increases with increasing solids flux and/or decreasing superficial gas velocity due to an increase in the effective thermal conductivity of the particle phase. The level of increase with increasing solids flux and/or decreasing superficial gas velocity is highly dependent on the solid segregation patterns and the solids volume fraction at the wall.

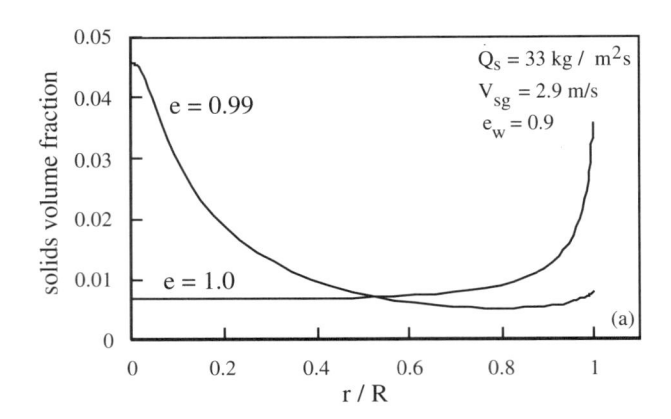

Figure 3. Sinclair and Jackson model (1989): sensitivity of solids volume fraction profile to e.

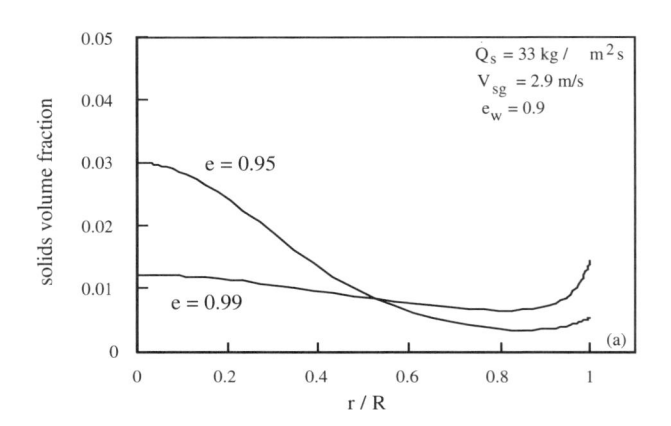

Figure 4. Sensitivity of solids volume fraction profile to e (effect of particle phase turbulence on governing equations only).

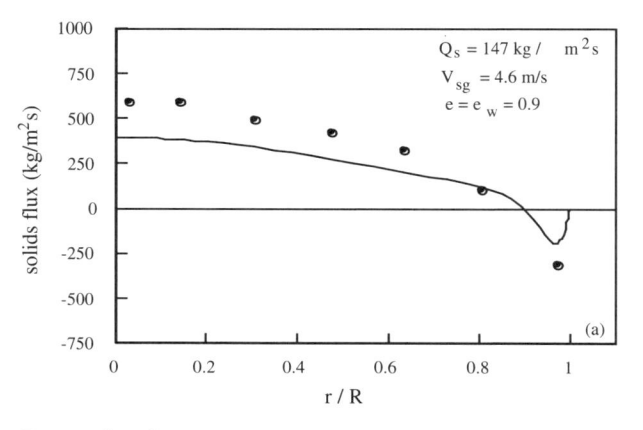

Figure 5. Comparison with the experimental data of Bader et al.

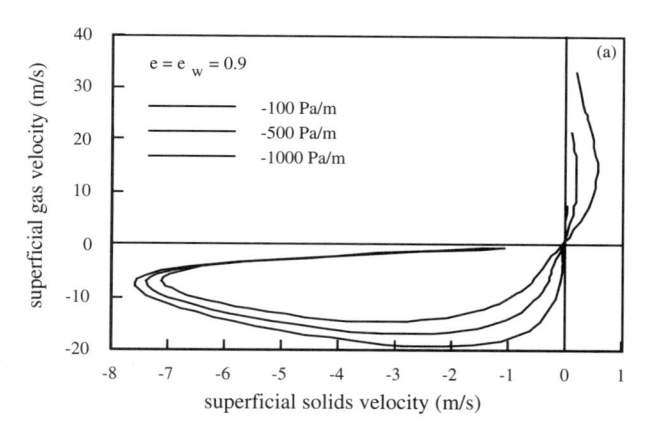

Figure 6. Contours of various pressure gradients in V_{ss}-V_{sg} plane.

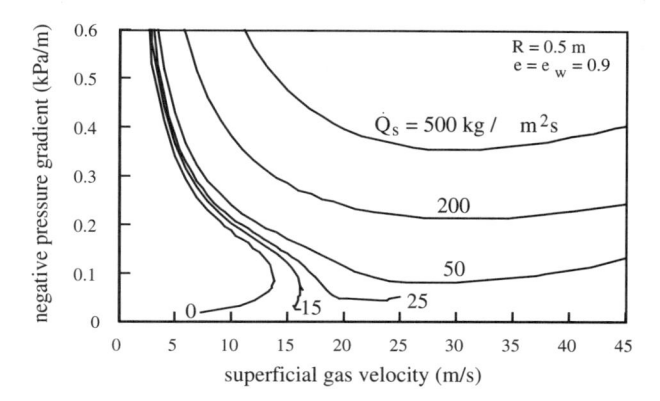

Figure 7. Zenz diagram for various overall solids flux.

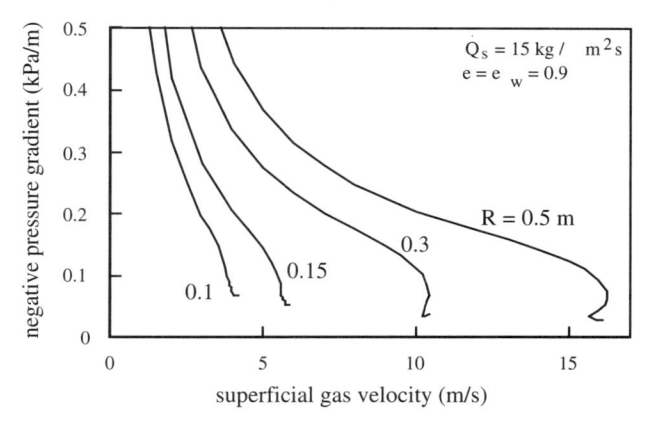

Figure 8. Zenz diagram for various pipe radii.

Commercial CFD Codes

Two-fluid models for multiphase flows are available in the industry-standard commercial codes CFX (AEA Technology) and Fluent (Fluent, Inc.). These two commercial codes dominate the multiphase CFD code market. The Bolio and Sinclair (1995) and Hrenya and Sinclair (1997) models have just recently been incorporated into the standard CFD Fluent package in their newest Version 4.5. Figure 10 presents, as an example, flow predictions in a vertical riser with the Sinclair model in the Fluent Version 4.5 code. In this flow simulation, the particles are FCC, the pipe diameter is 20 cm, the solids mass flux is 489 kg/m^{2}s, the superficial gas velocity is 5.2 m/s, the inlet uniform solids volume fraction is 0.02, and the inlet solids velocity is 14.4 m/s. The first simulation on the left is the full Sinclair model, including gas-phase turbulence, the second simulation is the Sinclair model neglecting gas-phase turbulence, and the third simulation is the Sinclair model neglecting particle-phase turbulence. By comparison of the first and second simulations, it is clear that the effect of gas-phase turbulence is negligible in dense-phase flow. The third simulation shows how particles move towards the center of the tube, contrary to experimental observations, when particle-phase turbulence is neglected. (The white region along the tube annulus indicates very dilute regions where the solids volume fractions is less than 0.02.)

The CFX multiphase flow code, unlike Fluent, allows for much flexibility by the user and the type of physical models previously described can be incorporated into user-defined subroutines. However, they are not part of the standard CFX package and must be input by the user. User-defined subroutines employing the kinetic theory concept have been developed by Van Wachem *et al.* (1999) and have been used to predict bed expansion and bubble size in fluidized beds.

Current Issues in CFD for Multiphase Flow

Lack of Detailed Non-Intrusive Flow Measuremen

Further model refinement and thorough model validation necessitates the acquisition of detailed, non-intrusive flow measurements. For example, there are no gas-solid flow measurements in the literature that report the local solids volume fraction, the mean velocities, the turbulence intensity for both phases simultaneously at one set of operating conditions. Some techniques are available for non-intrusive measurements in dense-phase flows. These include Gamma densitometry tomography for solids volume fraction, radioactive particle tracking techniques, capacitance imaging, cinematography, and NMR, to name a few. Each of these techniques, however, has limitations in terms of cost, speed of measurement, indirect measurement, and/or resolution. These needed flow measurements should to be made in simple flow geometries and should report inlet conditions. There is also a critical need to explore, in a controlled fashion, the factors which are known to influence flow behavior such as particle size distribution, particle shape and roughness. A DOE-OIT multiphase flow consortium involving several academic researchers, industrial companies and national laboratories has recently been formed to address this issue.

Variety of Two-Fluid Governing Equations Found in the Literature

Many two-fluid CFD models have been put forth by academic researchers, government laboratories, and commercial vendors. Not only do these models differ in terms of the closure relations but also there is often disagreement between the models even in terms of the basic two-fluid governing equations. This situation has resulted in much confusion in the literature. Many researchers put forth two-fluid governing equations in their work without citing a reference for the derivation of their origin. When the original derivation of the governing equations is cited, the two-fluid balances typically follow either the work of Anderson and Jackson (1967) or that of Ishii (1975). In Anderson and Jackson, a formal mathematical derivation of the local mean variables is used to translate the point Navier-Stokes equations for the fluid and the Newtonian equations of motion for the particles directly into continuum equations representing momentum balances for the fluid and solid phases. Local space averages are defined using a weighting function $g(r)$ which is a monotonic decreasing function of the separation between pairs of points in space, denoted by r. The integral of g over all space is normalized to unity, and the radius of g is chosen such that it is large with respect to the particle diameter but small with respect to the dimensions of the complete system. In Ishii, the derivation of the two-fluid model is based on a time-averaging procedure applied to the balance of a quantity (mass, momentum, energy) in a unit volume of a continuum. The original application for the derivation of Ishii was gas-liquid or liquid-liquid flows in which local instant fluctuations of variables are caused by rapidly moving and deforming interfaces. However, many researchers have adopted the approach of Ishii for gas-solid flows (Enwald *et al.*, 1996).

The key fundamental difference between the Anderson and Jackson (1967) formulation and the Ishii (1975) formulation has to do with the nature of the dispersed phase. As pointed out by Anderson and Jackson, when the dispersed phase is a rigid solid, the motion of each particle is completely determined by the translation and rotation about its center, and, correspondingly, the distribution of stress within each particle is not needed to determine the motion; only the resultant force and the resultant moment are relevant. Hence when the resultant forces due to fluid traction acting on the surfaces of the particles are evaluated, they are treated differently in the particle-phase momentum balance than in the fluid-phase momentum balance. In the particle-phase momentum balance, the resultant forces due to the fluid traction acting everywhere on the surface of the particles are first calculated, then these are averaged using values of the weighting function at the *particle centers*. In the fluid-phase momentum balance, the traction forces at all elements of fluid-solid interface are calculated, then these are averaged using values of the weighting function at the location of the *surface elements*. However, if the dispersed phase is a fluid, as in Ishii's formulation, such a distinction between the momentum balances for the two phases need not be made. In Ishii (1975), the distribution of stress within the dispersed phase is needed and "jump" conditions are used to determine the interfacial momentum transfer.

Effect of Particle Size Distribution

Previous experimental work has shown the significant role of particle size distribution in fluid-particle flows. A small concentration of fines, for example, can significantly influence the flow of coarser particles in a fluid. In Gilbert's (1914) experiments on the flow of various bimodal particle mixtures, he observed that when fines were added to a system of larger particles, the flow became smoother and "a quality of smoothness appealed to the eye". Matheson *et al.* (1949) showed that the viscosity of a suspension of larger FCC particles decreased by a factor of twelve upon the addition of FCC fines. Grace and Sun (1991) have also documented the increase in flow uniformity and smoothness upon the addition of fines to a gas-solid suspension. Horio, Morishita and Tachibana (1986) showed that the addition of fines to larger particles in riser flow increased the circulation rate and decreased the holdup of the coarser particles. The significant role of PSD has also been seen in the flow of dry granular materials involving mixtures of larger sized particles. Peciar, Buggisch and Renner (1994) showed that a widening of the particle size distribution produced a higher shearing capability in Couette flow. Particle diffusion and mixing increased as the particle size distribution broadened. Savage and Sayed (1984) showed that for mixtures of 600 micron and 1650 micron (mean diameter 1340 micron) polystyrene particles sheared in a Couette flow cell, the generated stresses were five times lower than those in a monodispersed system of 1340 micron particles.

Although particle size distribution plays a significant role in fluid-particle flows, accounting for a particle size distribution in the two-fluid framework is one of the next key challenges in CFD for multiphase flows. Separate particle-phase continuity and momentum balances can be easily formulated for individual particle-phase "bins", yet this approach will greatly increase the computational effort necessary for flow prediction.

In addition, a fundamental description for solid-phase stress in the case of a wide particle size distribution is difficult although the kinetic theory analogy has been extended by Jenkins and Mancini (1989) to consider the case of a bimodal particle size distribution. One possible approach is to employ the bimodal kinetic theory and invoke the assumption of pairwise additivity. Another approach is to combine population balancing with CFD.

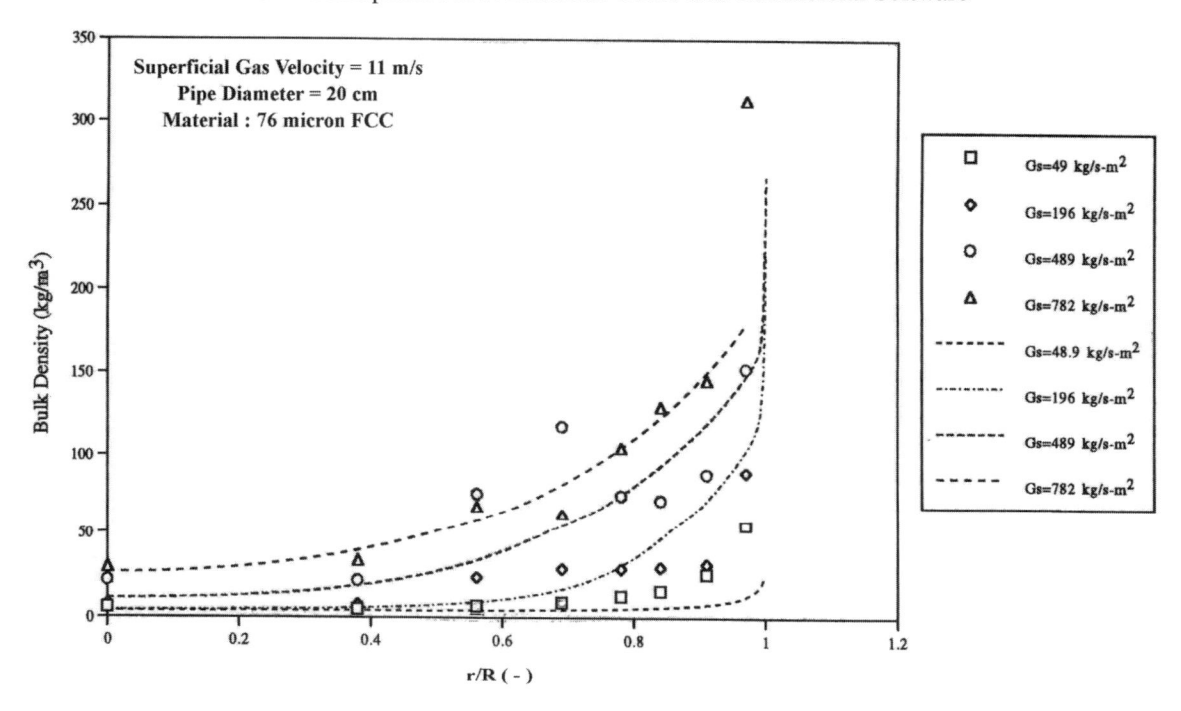

Figure 9. Particle segregation patterns in FCC risers.

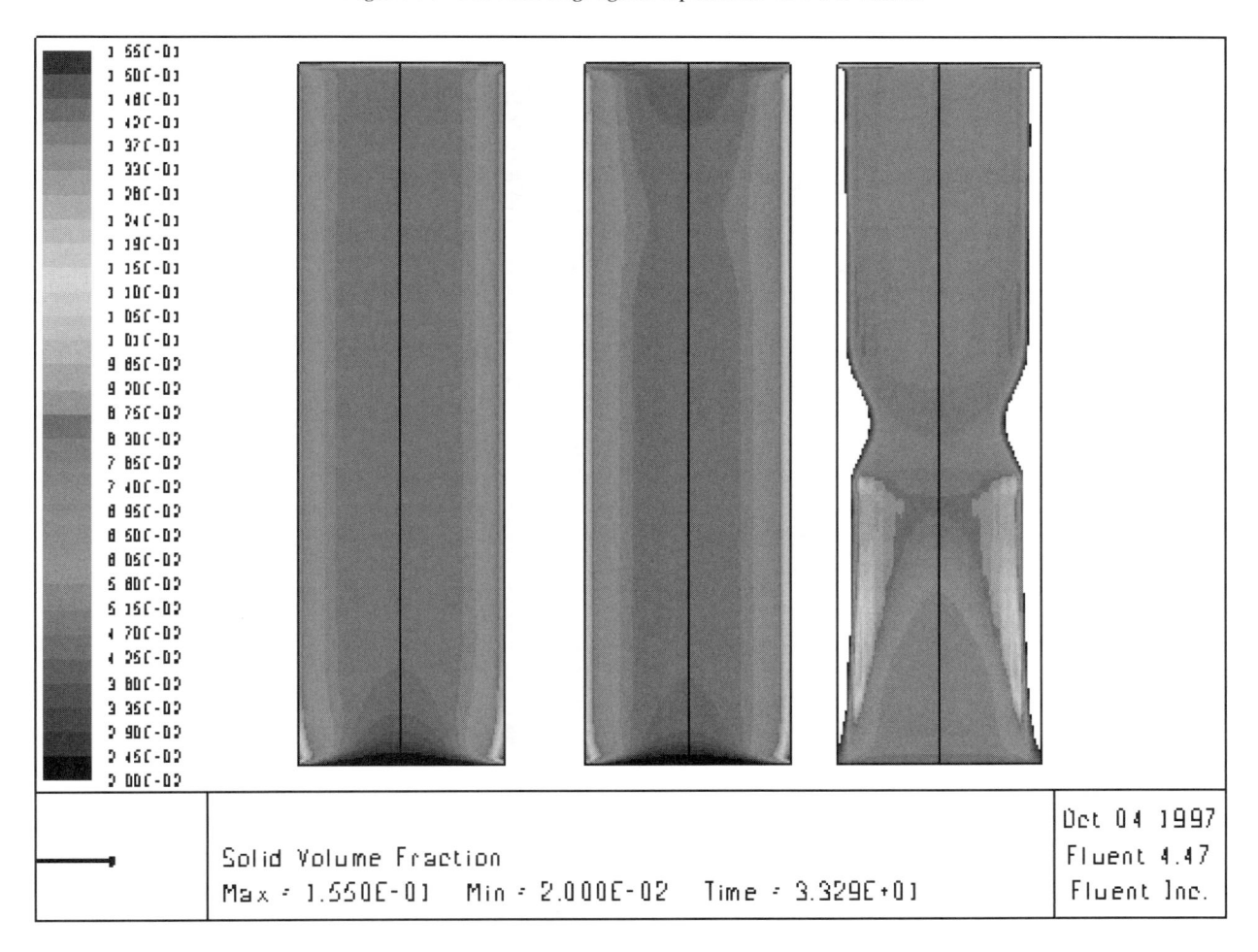

Figure 10. Riser flow predictions using Sinclair model in Fluent Version 4.5.

Concluding Remarks

CFD for multiphase flow has made significant strides in recent years and is rapidly becoming a common tool for system design, scale-up and optimization. Commercial codes, Fluent and CFX, are working to adopt the latest physical models although there is still much disparity in the literature even in terms of the basic governing equations. However, even with the restrictive assumptions of spherical particles and a monodispersed PSD, the proper physical models appear to capture many of the salient flow features. A comprehensive validation, though, relies on detailed, non-intrusive data which are lacking at present for dense-phase flows. An effective way in which to describe particle size distribution effects is a key future challenge.

References

Anderson, T.B. and Jackson, R. (1967). *Ind. Eng. Chem. Fund.*, **6**, 527.

Barlow, R. and C. Morrison (1990). *Experiments in Fluids*, **9**, 93.

Bolio, E and J. Sinclair 1996). *Int. J. Multiphase Flow*, **21**, 985.

Enwald, H., Peirano, E., and A.E. Almstedt (1996). *Int. J. Multiphase Flow*, **22**, 21.

Gidaspow, D. (1994). *Multiphase Flow and Fluidization*, Academic Press, San Diego.

Gilbert, G.K. (1914). U.S. Geological Survey Professional Paper No. 86.

Grace, J.R. and G. Sun (1991). *Can. J. Chem. Eng.*, **69**, 1126.

Horio, M., Morishita, K. and O. Tachibana, O. (1986). *Proceedings of the World Cong. II of Chem. Eng.*, 81-258.

Hrenya, C. and J. Sinclair (1997). *AIChE J.*, **43**, 853.

Ishii, M. (1975). *Thermo-Fluid Dynamic Theory of Two-Phase Flow*, Direction des Etudes et Recherches d'Electricit\'e de France, Eyrolles, Paris, France.

Jenkins, J. and F. Mancini (1989). *Phys. Fluids A.*, **12**, 2050.

Louge, M., Mastorakos, E. and J.T. Jenkins (1991). *J. Fluid Mech.*, **231**, 345.

Lun, C., Savage, S., Jeffery, D. and N. Chepurniy (1984). *J. Fluid Mech.*, **140**, 223.

Mallo, T. and J. Sinclair (1998). Heat transfer rate predictions in dilute and dense-phase gas-solid flows. in preparation, 1998.

Mallo, T. (1997). *Heat Transfer Rate Predictions in Dilute and Dense-phase Gas-solid Flows*. PhD Thesis, Carnegie Mellon University.

Matheson, G.L., Herbst, W.A. and P.H. Holt (1949). *Ind. Eng. Chem.*, **41**, 1099.

Myong, H.K. and N. Kasagi (1990). *JSME Int. J. Series II*, **33**, 63.

Peciar, M., Buggisch, H. and M. Renner (1994). *Chem. Eng. Prog.*, **33**, 39.

Savage, S. and M. Sayed (1993). *J. Fluid Mech.*, **142**, 391.

Sheen, H.J., Chang, Y. and Y. Chang (1993). *Proc. Natl. Sci. Counc. ROC(A)*, **17**, 200.

Sinclair, J.L. and R. Jackson, (1989). *AIChE J.*, **35**, 1473.

Van Wachem, B., Schouten, J., Krishna, R., Van den Bleek, C., and J. Sinclair (1999). CFD modeling for gas-solid systems: A comparative analysis of the various treatments. ASME Fluids Engineering Meeting, San Francisco, July.

Wang, D. and C. Campbell (1992). *J. Fluid Mechanics*, **244**, 52.

INTEGRATED MICROCHEMICAL SYSTEMS: OPPORTUNITIES FOR PROCESS DESIGN

David J. Quiram, Klavs F. Jensen[†], and Martin A. Schmidt
Massachusetts Institute of Technology
Cambridge, MA 02139

Patrick L. Mills, James F. Ryley, and Mark D. Wetzel
DuPont Company, Experimental Station
Wilmington, DE 19880-0304

Abstract

The current status of microfabricated chemical systems are reviewed with emphasis on applications relevant to the chemical process industry, specifically synthesis and catalyst research. The use of detailed simulations to understand the behavior of microreactors is summarized to demonstrate the value of reactor modeling. Researchers have made considerable progress in developing microreaction technology for industrial use, but few efforts have focused on integrating the technology into industrial production or laboratory test stations. Systems engineering tools are needed to evaluate potential microreactor applications and to design complex processes utilizing the technology. Experience at MIT and DuPont in building a multiple reactor test station has resulted in a design strategy for microchemical systems. Opportunities for system engineering in design and analysis of integrated microchemical systems are discussed.

Keywords

Microreactor, Microchemical, Microfluidic, Microelectromechanical systems, MEMS, Microfabrication, Minichemical plant.

Introduction

MicroElectroMechanical Systems (MEMS) emerged in the early 1970's with the application of microfabrication technology used by the IC industry to the development of mechanical sensors and actuators, such as pressure transducers (Petersen, 1982). Since then the field of MEMS has expanded from producing ink jet nozzles and pressure transducers to a broad range of applications. It now has the potential to be a leading growth sector of the world economy in the coming century (Gross and Port, 1998; Mandel, 1998; Amato, 1998). Although the first commercial products were mechanical in nature, recent trends suggest that MEMS will become pervasive not only in mechanical but in fluidic devices as well (Marshall, 1999). The current increase in companies devoted to producing lab-on-a-chip systems indicates the tremendous opportunities available by miniaturizing traditional laboratory analysis (Freemantle, 1999; Marshall, 1999; Marsili, 1999). Simultaneously, the development of combinatorial techniques for High Throughput Screening (HTS) is generating new interest in using MEMS as a discovery tool for pharmaceuticals, electronics materials, polymers, and catalysts.

The common theme of all MEMS devices is the potential advantage that simultaneous miniaturization and integration offer when compared to traditional devices. For example, microfabricated accelerometers for airbag

---

[†] Author to whom correspondence should be addressed.

release systems are not only smaller and cheaper than their macroscale counterparts, but have increased sensitivity and reliability (Jones et al., 1995). Similarly, micofabricated reactors offer several advantages over traditional reactors not only for catalyst testing, but chemical production as well (Ehrfeld et al., 1996; Lerou et al., 1996). Although the development of experimental microreactors has progressed considerably in the past few years, the development of systems utilizing this technology presents challenges that have not been encountered in traditional chemical engineering.

MIT, DuPont, and other groups from Europe and the United States have developed various microfabricated reactor designs (Brenchley and Wegeng, 1998; Baselt et al., 1998; Wegeng and Drost, 1998). As an example, Fig. 1 shows the MIT-DuPont microreactor. The main differences in these devices result from the target fluidic application (gas or liquid), level of sensor integration, and level of throughput. Their common feature is the use of reaction channel widths less than 1 mm. This size is large enough to include a variety of machining techniques for creation of the reaction channel (Ehrfeld et al., 1997). As a result, a wide range of fabrication materials and reaction systems can be explored.

Mathematical modeling has already been successfully applied to understanding microfabricated reactors (Jensen et al., 1997). Further work has also shown its value for design purposes (Quiram et al., 1998). These efforts now need to be extended to understand system level behavior. Future design efforts for complex microchemical systems will require these new simulation tools.

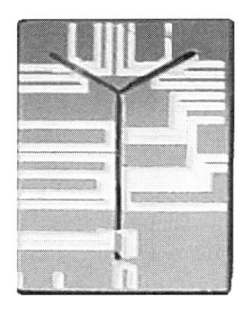

Figure 1. MIT-DuPont Y microreactor.

Motivation for Microchemical Systems

The two main advantages of a microfabricated reactor over a traditional reactor are its small volume and its high surface-to-volume ratio. The small volume is an important advantage for use in catalyst testing and biological assays, but it also limits reactor output. If microchemical production is going to be a viable technology, the process advantages of operating a microscale reactor must outweigh the low production capacity per reactor. For these systems, production is increased by building arrays of microreactors operating in parallel instead of the traditional scale-up approach of using a larger reactor (Ehrfeld et al., 1996; Lerou et al., 1996). Thus, the fabrication methods for these reactors must allow for inexpensive mass production to be competitive.

Benson and Pontn proposed the idea of the minichemical plant in 1993. Although they did not provide implementation details, they outlined the advantages these production units would have over the traditional large-scale chemical plants that is commonly associated with the chemical processing industry. At the first DECHEMA conference on microreaction technology, microfabricated reactors were proposed as a means to mass produce small chemical production units that would be operated on-site and on-demand (Ehrfeld et al., 1996; Lerou et al., 1996). At this conference, the following motivating concepts for minichemical plants were applied to microchemical production:

1) Inherent safety in small-scale production
2) On-site and on-demand production
3) Economy of mass production

Each of these issues is discussed below.

Safety

An important advantage of microchemical production is the inherent safety in having reactor volumes so small that the reactor system could be designed to contain even the worst case accident scenarios. Microreactors, with their narrow channel size, hold such a minor quantity of chemicals that a total failure would merely require temporary shut-down and reactor replacement. The small quantity of chemicals released by a failure would not pose an environmental threat. For example, a failure of the Y reactor for one minute before shutdown would release ~30 mg of chemicals. The design of a containment system to handle a hazardous release on this level would be straightforward.

Operating reactions in a microreactor is also safer because of the high surface-to-volume ratio that increases the net heat transfer rate away from the reaction zone. For example, microchannel heat exchangers/reactors have been built with heat transfer coefficients twenty times greater than conventional heat exchangers (Jäckel, 1996). Furthermore, the high surface-to-volume ratio also acts to quench free-radical chain mechanisms that induce a runaway reaction (Srinivasan et al., 1997). Researchers have already reported operating reactions in microreactors that would be considered unsafe or impossible to run on a larger scale because of the explosion potential (Hagendorf et al., 1998; Srinivasan et al., 1997).

On-site and On-demand Production

The point-of-use minichemical plant has advantages over a larger-scale manufacturing facility since it eliminates the need for hazardous chemical storage and transportation (Benson and Ponton, 1993). For example, methyl isocyanate production has already been demonstrated in a microfabricated reactor (Lerou et al., 1996). Using minichemical plants would avoid chemical storage by designing them to have flexible production schedules and by designing them for quick startup and shutdown. Parallel arrays of microreactors offer extraordinary flexibility for production levels since the reactors can be individually turned on or shutdown to increase or decrease production as necessary. This allows an individual microreactor to be designed for one particular set of conditions instead of compromising the design to operate over a variety of flow rates and feed compositions.

Economy of Mass Production

Traditionally, the chemical industry has used the economies of scale to reduce production costs. The minichemical plant takes advantage of the economies of mass production—manufacturing a large number of plants reduces cost (Benson and Ponton, 1993). This scaling law is applicable to microreactors since many of the techniques used for reactor fabrication come directly from the production of integrated circuits and electronic components. Using those industries as a model suggests that it is possible to mass produce individual reactors at a cost that makes assembling thousands of reactors in parallel economically attractive.

The minimization of plant size may be more economically competitive since it changes the nature of the chemical industry. A moderate sized chemical plant requires a capital investment of over $100 million. This limits the ability of countries with developing economies to enter the chemical industry because of their high cost of capital. By reducing plant size and plant cost, the competitive advantage of these regions with cheaper labor forces can be used to reduce production cost. Furthermore, smaller manufacturing facilities allow foreign companies to enter new countries while limiting their exposure to risk. Thus, the philosophy of large central manufacturing may be challenged as the globalization of the world's economies changes the chemical industry.

Development Path

Development of MEMS fluidic technology for chemical manufacturing will build upon advances in the areas of biological and chemical analysis, where MEMS offers immediate applications. Fig. 2 shows application areas for which microfluidic devices are being developed including combinatorial approaches. Research is in progress in all these fields with the most intense efforts in the fields promising immediate payoffs. Some examples of these areas are summarized in the following paragraphs.

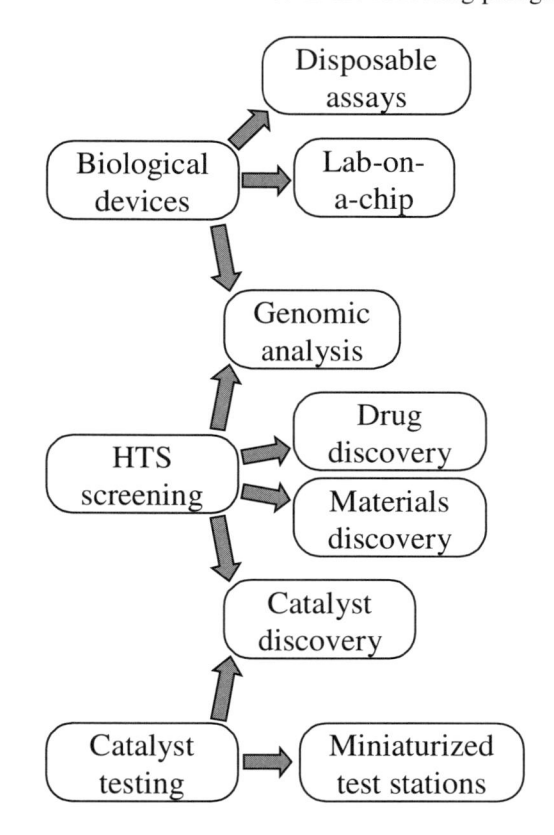

Figure 2. Development areas for microfluidic devices.

Microfabricated Biological Devices

In biological analysis, smaller is better for a number of reasons (Service, 1998; Marshall, 1999; Borman, 1999). The foremost factor is the possible reduction of lab analysis cost by developing systems that are not only less capital intensive but require little operator supervision and training. Similarly, analyses that use expensive reagents will be cheaper to perform because of the smaller quantities required. Furthermore, miniaturization reduces testing time in some analytical methods, such as electrophoresis, and this also reduces cost. In this case, cost may be a secondary factor when considering that life-saving information can be obtained quickly while a patient is being treated. One eventual goal for MEMS analytical devices is to enable physicians to perform most analytical testing in their offices and, in some cases, by bedside in hospitals. One European Commission study suggests that the capitalized annual growth rate of the disposable microfluidic assay device market will be 33%

with a $2.8 billion market by the year 2002 (Marshall, 1999).

Although the development of lab-on-a-chip and MEMS disposable assay devices are not directly applicable to microchemical production, the information gained from product development and manufacturing of these devices will be crucial. One key area of interest is the automation techniques used to control these devices. Specifically, the automation of a lab-on-a-chip device requires minimal operator training and intervention to run the device. Similarly, the operation of a minichemical plant should require only low level maintenance and intervention.

Applications in High Throughput Screening Techniques

The growing field of HTS techniques for genomics and drug discovery is generating significant interest in MEMS devices (Konrad *et al.*, 1997; Service, 1998; Marshall, 1999; Borman, 1999). Microfabrication and integration of analytical techniques with chemical systems–the "laboratory-on-a-chip" or "micro-total analysis system (μTAS)"–is an established and rapidly developing area (Berg and Harrison, 1998). The combination of MEMS technology with analysis and library creation is creating exciting new methods for combinatorial discovery. Already, devices are being designed that include gene-chips for miniature genetic analysis systems (Anderson *et al.*, 1998; Service, 1998). Other researchers are pursuing devices that are directly used in the screening process for new pharmaceutical compounds (Salimi-Moosavi *et al.*, 1998; Borman, 1999; Hicks, 1999). As an example of these efforts, Orchid Biocomputer has developed a multilayer microfluidic glass chip that interfaces with multiwell plates used in drug discovery (DeWitt, 1999)

The interest in combinatorial methods for biological applications has expanded to develop HTS techniques for materials and catalyst discovery (Dagani, 1998). In particular, Symyx Technologies has generated considerable industrial support for the development of combinatorial methods for materials and catalyst discovery (Jandeleit *et al.*, 1998; Weinberg *et al.*, 1998) The application of MEMS devices in these techniques is limited at this early stage. Catalyst screening with microreactors consisting of stacked microchannel plates sequentially sampled by mass spectrometry has recently been demonstrated (Zech *et al.*, 1999). The technique allows evaluation of different catalyst compositions by varying catalysts between plates, but it is limited by sequential sampling and the lack of individual plate control. MEMS reactors will have the largest impact on catalyst screening when devices can be individually controlled and integrated with microfabricated chemical sensors for product detection. Chemical detection is the rate-limiting step in most methods since detailed product information must be obtained using sequential screening. The ongoing merging of μTAS techniques with microreaction technology promises to yield a wide range of novel devices for high throughput screening.

Microreactors for Catalyst Testing

Besides the potential applications in combinatorial catalyst screening, individually controlled microreactors would be useful for catalyst performance testing. They could be used to obtain kinetic and catalytic activity information over a range of operating conditions in a relatively short time period. This use is not substantially different from the small tubular reactors, with volumes on the order of a few cm^3, currently used in catalyst testing. However, a microfabricated reactor would be favorable for safety reasons, the small amount of catalyst needed for testing, and improved thermal control (Ehrfeld *et al.*, 1996; Jäckel, 1996; Lerou *et al.*, 1996). Furthermore, the integration of a microfabricated reactor with other microfluidic components would substantially reduce the size of testing systems. Such systems would use MEMS technology not only for the reactors, but for the gas manifold, analytical method, and other key sub-systems as well. This may allow automated catalyst testing facilities to be mass produced, which would substantially reduce their cost. Typically, a commercially available fully automated catalyst test system with multiple reactor tubes costs around $200,000. A miniaturized system that fits on a desktop would be desirable, especially if it can be produced at a substantial cost reduction through large-scale manufacturing.

Current Microreactors

The most common design used for gas phase heterogeneous reactions is a microchannel device that can be integrated with a heat exchange layer for highly exothermic reactions. Figure 3 shows an example of such a device developed by the IMM and BASF (Jäckel, 1996). Several groups have independently developed this design to take advantage of the high heat transfer rate afforded by the small channel dimensions, which are typically around 0.1 mm wide and deep (Richter *et al.*, 1998; Wießmeier and Hönicke, 1996; Wegeng *et al.*, 1996). The difficulty of operating such devices is the limited capability for control and sensing of the individual channels in the reactor. However, their high throughput may make them more applicable for chemical production. Initial work using these devices has also shown enhanced performance compared to traditional reactor designs (Weissmeier *et al.*, 1997, Wiessmeier and Honicke, 1998, Jackel, 1996, Worz *et al.*, 1998).

The MIT-DuPont microreactor is an alternative design that has a single channel with integrated heating, temperature sensing, and flow sensing. Figure 4 shows the top and cross-sectional views of this reactor with its two reactant inlets and one product outlet that have the appearance of a "Y". The branches of the Y structure are the feed channels, which have integrated flow sensors. The gas feeds contact each other at the center of the Y, and the reaction takes place down the length of the main channel in the catalyst region. The reaction channel is 0.5 mm wide and 0.5 mm deep. It has two separate zones for heating and four temperature sensors along its length. Each heater is 2.8 mm in length and the temperature sensors are 0.8 m long.

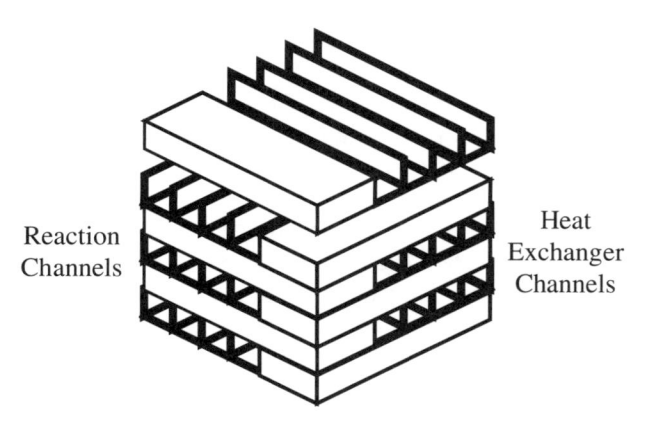

Figure 3. BASF microchannel/heat exchanger reactor(Jäckel, 1996).

The cross-section shown in Fig. 4 illustrates the use of a silicon nitride membrane to separate the heaters and temperature sensors from the reaction channel. The 1 μm thick membrane is impermeable to gas flow but provides an intimate thermal contact between the catalyst on one side of the membrane and the heaters on the opposite side. The reaction channel is sealed from the other side by an anodically bonded Pyrex® 7740 wafer. The Pyrex wafer has holes drilled at the feed inlets and the product outlet. The fabrication method for this device is similar to the "T" reactor described previously (Srinivasan et al., 1997). The main difference in the fabrication process is that a deep reactive ion etch is used to form the complex curved shape of the Y reactor. The main advantage of this device is that the integrated heaters and temperature sensors offer very localized temperature measurement and control. On the other hand, the single channel design results in lower throughput than the microchannel devices discussed above. Multiple channel versions of the integrated device can be designed, but the channel density of a microchannel device cannot be reached without sacrificing the advantages granted by sensor integration.

The use of a membrane also enables the integration of separation with chemical reaction, as in macroscopic membrane reactors. Such devices are capable of beyond equilibrium conversions, since the products and reactants are continuously separated. - Figure 5 shows an example of a novel palladium membrane microreactor (Franz *et al.*, 1999). This membrane microreactor can be used for a variety of hydrogenation and dehydrogenation reactions, as well as for hydrogen purification. The design includes integrated heaters and temperature sensors in the device. Microfabricated palladium membrane reactors could enable new applications in hydrogen purification because of their small size, fast thermal response times, and high efficiency.

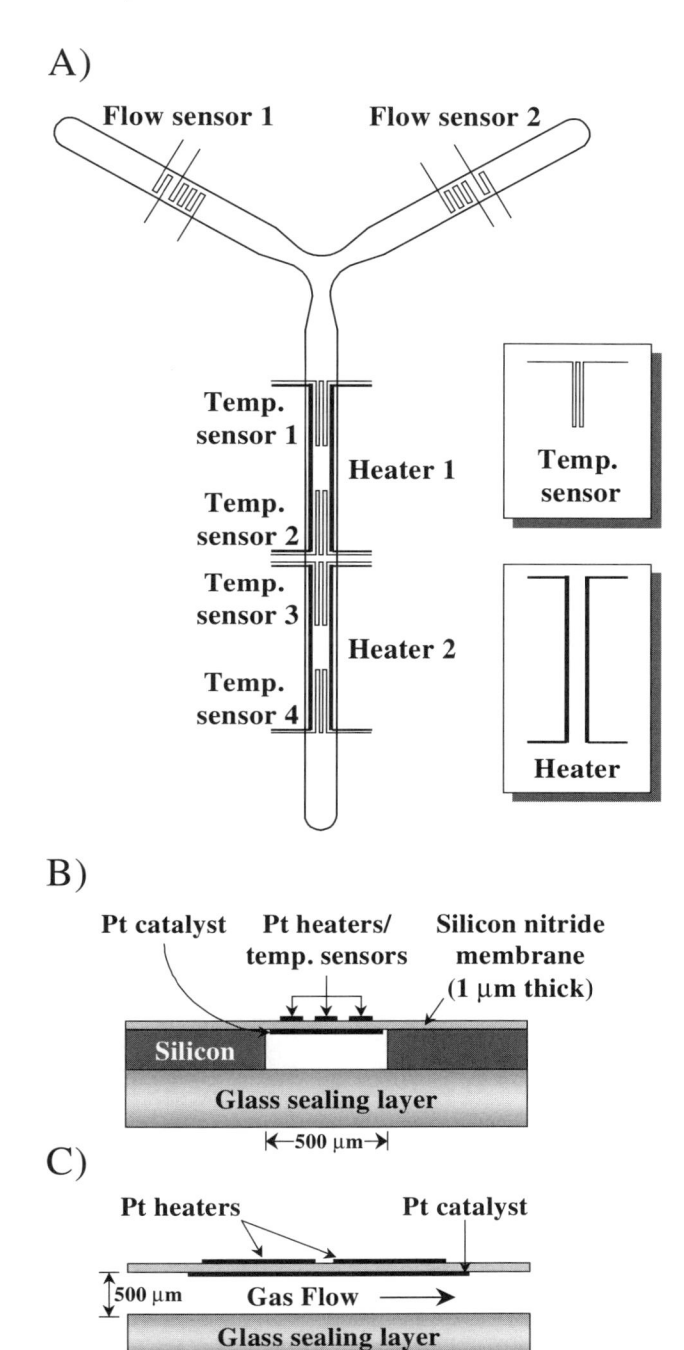

*Figure 4. Schematic of the Y microreactor.
A) Top View. B) Cross-section perpendicular to flow. C) Cross-section parallel to flow.*

The previous reactor designs are for gas phase chemistries, but there is also a significant effort to develop liquid phase, gas-liquid phase, and multiphase microreactors. In these cases, pressure drops and mixing effects become critical in reactor design. Mixing has to be achieved by diffusion because of the small dimensions and low flow rates in microfluidic devices.

Mixing in microfluidic devices is generally accomplished by the repeated lamination of the two streams to be mixed in order to increase the contact area and reduce diffusion lengths. In this "static mixer" scheme, the two fluids are brought into contact and the combined stream is separated into two flows along a line perpendicular to the mixing interface. These two streams are then recombined with a resultant doubling of the fluid interface and halving of the diffusion length.

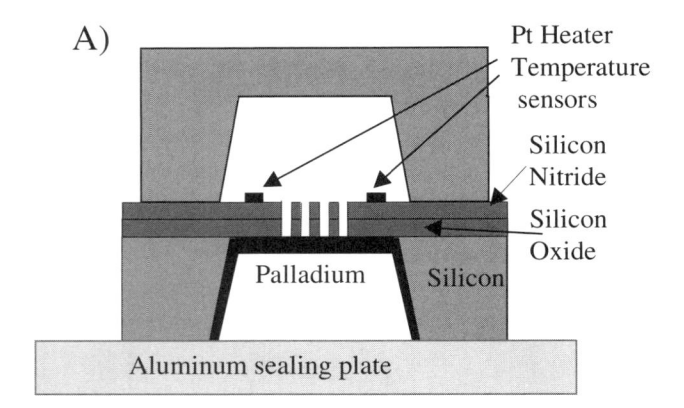

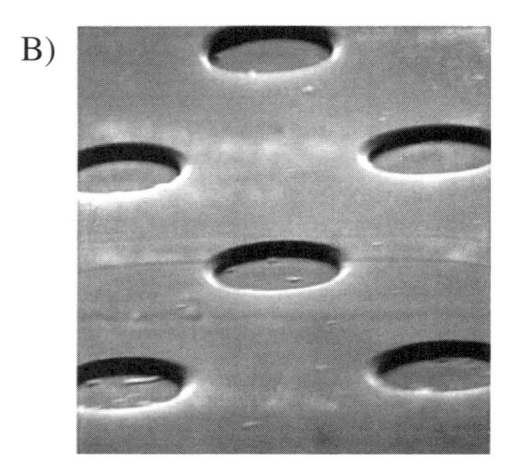

Figure 5. Palladium membrane microreactor. A) Schematic of cross-section perpendicular to flow. B) SEM of membrane - holes are 4 μm wide free standing palladium films (Franz et al., 1999).

By repeating the process several times, very short diffusion distances are obtained, thereby producing rapid mixing (Mobius et al., 1995; Ehrfeld et al., 1999). Ultimately, the design of a micromixing unit is a trade-off between mixing speed, pressure drop, volume flow, feasibility of microfabrication, and integration with chemical detection devices. The static mixing devices, in particular those developed by the IMM, have formed the basis for a number of successful liquid-liquid reaction studies (Wörz et al., 1998; Stoldt et al., 1999). This work indicated that the microfabricated design had superior performance compared to the production process and lab-bench reactors.

Figure 6 shows an example of a microfabricated liquid phase reactor that integrates laminar mixing, hydrodynamic focusing, heat transfer, and temperature sensing. This device is designed for the production of hazardous specialty chemicals, including organic peroxides (Floyd et al., 1999). The integrated reactor was fabricated in silicon using standard photolithography techniques and a deep reactive ion etch (Ayon et al., 1999) yielding channels 25-50 μm wide and ~500 μm deep. The wafer was capped by an anodically bonded Pyrex® 7740 wafer. The reactor achieves complete mixing in ~10 ms depending on the fluid properties and flow rates. Testing also shows that the microfabricated heat exchanger channels have a high heat transfer coefficient, 1500 W/(m^2·C°). The pressure drop is ~75 kPa at the 1.0 ml/min design flow rate.

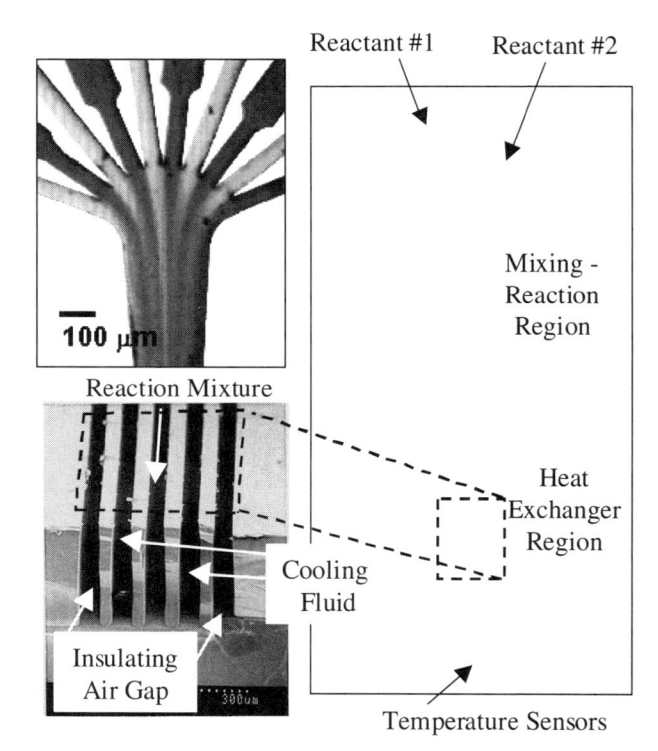

Figure 6. Liquid phase reactor with lamination of fluid streams and hydrodynamic focusing for mixing. Heat exchangers and temperature sensors in down stream reaction zone. Upper left hand insert shows mixing of acid base mixture in gray tone image of indicator color (Floyd et al., 1999).

Similar microfabricated devices, also designed for the lamination of fluid streams, have also been developed for extraction and reactive extraction processes. For example, two immiscible streams (e.g., organic and aqueous phase) can be contacted to allow phase transfer of a species or a phase transfer reaction. The short diffusion lengths result in fast mass transfer and the streams can easily be separated at the end of the reaction channel (Brody *et al.*, 1996; Burns *et al.*, 1997; Burns *et al.*, 1998).

Gas-liquid-solid reactions are ubiquitous throughout the chemical industry and provide unique opportunities for microfabrication. The high surface-to-volume ratios, controlled reactant distribution and contacting pattern, improved thermal management, and fast mass transfer rates attainable in microfabricated multiphase systems could result in performance advantages relative to conventional macroscopic systems. In addition, the microfabrication approach allows the integration of sensing and control, which provides an efficient mechanism for testing and controlling process performance. Figure 7 shows a multiphase packed bed reactor microfabricated using deep reactive-ion etch technology and standard photolithography tools (Losey *et al.*, 1999). A reaction chamber holds standard porous catalyst particles as reactants are fed continuously in a co-current fashion and distributed at the inlet of the fixed-bed.

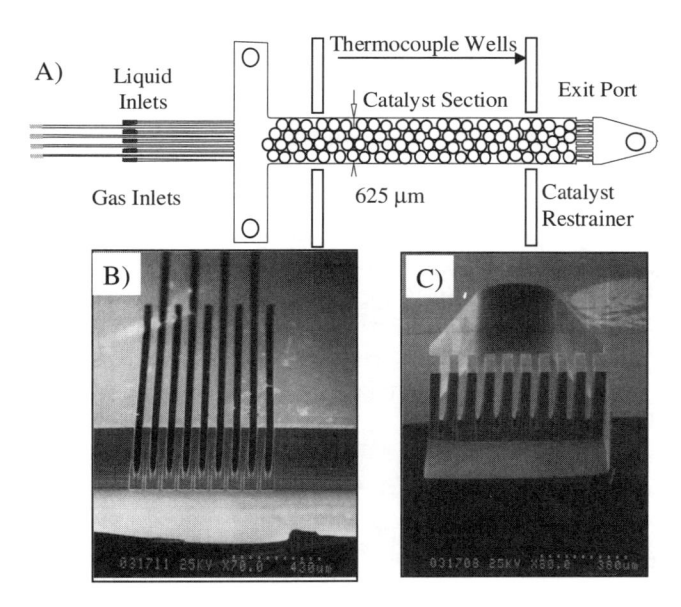

Figure 7 Example of multiphase packed bed reactor. A) Schematic, B) SEM of microfabricated inlet section, and C) SEM of microfabricated catalyst restrainer and fluid exit port (Losey et al., 1999).

The gas and liquid reactant streams are brought into contact by a series of interleaved, high-aspect ratio (25 μm wide 300 μm deep) inlet channels. A particle filter is formed within the reaction channel by etching a series of posts 40 μm wideseparated by 25 μm gaps. Reactions such as the hydrogenation of α-methylstyrene have been used to characterize the reactor.

Despite the development of microreactors for a variety of chemistries, relatively little work has focused on designing systems that take advantage of this technology. The reasons for this include the uncertainty that this technology will ever be a viable alternative to conventional reactors and the disruptive nature of microreaction technology. It is easier to justify small-scale laboratory experiments than the development of processes based on MEMS technology. However, the state-of-the-art suggests that designing chemical processes based solely on MEMS devices is feasible.

The critical unit operations of reaction, separation, and heat transfer have all been demonstrated in MEMS devices. Furthermore, fluidic control devices, including pumps and valves, have also been microfabricated. Separation devices that have been microfabricated include chromatographs and extractors (Martin *et al.*, 1998; Robins *et al.*, 1997; TeGrotenhuis *et al.*, 1998; Terry *et al.*, 1979; Frye-Mason *et al.*, 1998). Previous research has already shown that microfabricated heat exchangers are extremely effective because of their high surface-to-volume ratio in small channels (Jäckel, 1996; Wegeng et al., 1996). Microfabricated proportional valves are already commercially available from Redwood Microsystems[1] and TiNi Alloy Company[2]. Numerous designs for microfabricated pumps have been proposed, but pumps capable of generating high pressures have only been reported for liquids (Paul *et al.*, 1998). Currently, multiple MEMS components have only been integrated into devices for biological and chemical analysis. Considerable work is still needed to produce the complex systems required for chemical production.

Microreactor Modeling

The major difficulties in designing microfabricated systems are the lack of both experience and modeling tools for system development. A microfabricated device is typically constructed by past experience, engineering analysis, and through literature searches by the designer. In some cases, detailed simulation tools can be used, but

---

[1] Redwood Microsystems, Inc., 959 Hamilton Ave., Menlo Park, CA 94025.

[2] TiNi Alloy Company, 1621 Neptune Dr., San Leandro, CA 94577.

they are not universally applicable. MEMCAD®[3] is an example of one popular software package for the design of MEMS devices. It includes the capability to model moving structures, electric fields, heat transfer, and fluid flow in one package. It is moving towards the simulation of reactive flows, but this is difficult even for applications specifically designed for this purpose such as CFD-ACE+®[4] and Fluent®[5]. The experience with simulations in the MIT-DuPont microreactor project has been invaluable for understanding the reactors, but difficulties remain in modeling complex reactor behavior such as bifurcation phenomena and reactor dynamics.

During the fabrication and testing process of the original T microreactor, a simulation methodology was developed to understand reactor behavior and suggest design improvements (Quiram et al., 1998; Hsing et al., ; Jensen et al., 1997). Initially, the mathematical model was validated by comparing the thermal analysis results with experimental data. Subsequent simulations were run to demonstrate the ability to replicate the bifurcation behavior observed in the T reactor with highly exothermic partial oxidation reactions. The simulation tools were then used to analyze the effect of design changes before modifications were implemented in the laboratory. The use of modeling not only improved understanding of the microreactor but was critical in the evaluation of design changes.

The gas phase microreactor is well-suited for modeling because of its small characteristic dimension, which results in low Reynolds numbers. Typically, reactions are run with a flow rate of 10-20 cm^3/min at standard conditions (sccm), and this results in Reynolds numbers between 10 and 50, depending on the gas and the reactor temperature. Since the flow is typically in the laminar regime, models for approximating turbulence are not needed. The MIT-DuPont microreactor has the additional advantage of using a thin-film catalyst instead of a packed bed. This allows the classical conservation equations to be directly applied. However, there is a high degree of coupling between the species, momentum, and energy balances in the microreactor as in traditional reactors. Thus, these conservation equations cannot be solved separately, which drastically increases the problem complexity. Furthermore, the bifurcation behavior present makes modeling even more difficult. Further aspects of microreactor modeling are provided below.

Model Description

The first step in the model formulation was the transformation of the physical domain to a computational domain. A generalized procedure was developed that utilized ICEM-CFD®[6], which is a commercial, mesh generation software package. ICEM-CFD was used to create a CAD representation of the reactor that was then meshed using a Cartesian grid. The boundary conditions for the problem were added to the generated mesh using a preprocessor that finished the creation of the computational domain.

Discretization of the general conservation equations of mass, momentum, energy, and species over the computational domain was performed using the Galerkin Finite Element Method (GFEM). This scheme was chosen because of its broad generality and excellent convergence properties in handling elliptic problems. This general discretization process allows a wide variety of reactor types to be modeled with the only limitations placed by the computing power available.

The large set of non-linear algebraic equations generated by the discretization procedure was solved using Newton's method. Subsequently, the set of linear algebraic equations created by each Newton iteration was solved using a frontal algorithm. Although, Newton's method converges quadratically, the initial guess must be sufficiently close to the real solution. This becomes difficult for the highly nonlinear behavior exhibited by reaction systems that display multiple steady states. To address this problem, Keller's pseudo-arc length continuation scheme was used to trace the solution branches around the limit points of the ignition/extinction S curve. Other numerical problems arise when the Reynolds number becomes sufficiently large (Re > 15) so that the problem is convection dominated. To retain the convergence properties of the GFEM approach, the Brooks-Hughes Streamline Upwind Petrov-Galerkin scheme was implemented. More details concerning the model formulation and the numerical issues involved in microreactor simulation can be found in the work by Hsing et al. (1999).

Thermal Modeling

The first use of this model was to understand the temperature profiles in the T microreactor. They demonstrated the extremely localized heating produced by the microfabricated resistive heating devices, which was the original design intent. To validate the model, temperature measurements were taken in the heated region for a range of gas flow rates (Re < 150) and heater

---

[3] Microcosm Technologies, Inc., 5511 Capital Center Dr., Suite 104, Raleigh, NC 27606.

[4] CFD Research Corporation, 215 Wynn Dr., Huntsville, AL 35805.

[5] Fluent Inc., 10 Cavendish Ct., Lebanon, NH 03766.

---

[6] ICEM CFD/CAE, Version 3.3.3 D, ICEM CFD Engineering, 2855 Telegraph, Suite 501, Berkeley, CA 94705.

powers. These measurements were successfully reproduced in the simulation results.

Prediction of Bifurcation Behavior

Although predicting the reactor temperature profile is important, the simulations are more valuable if they successfully model reactor operation. Thus, an attempt was made to reproduce the experimentally observed ignition/extinction behavior, which occurs for fast, highly exothermic reactions. Bifurcation behavior is exhibited because of the limited rate of heat removal to the surrounding atmosphere and through the silicon nitride membrane. Although a simple 1-D heat transfer analysis can be used to qualitatively predict bifurcation behavior in the MIT-DuPont microreactor, the experimentally observed behavior was too complex for this analysis. During a partial oxidation reaction, ignition occurred near the end of the heater segment and the ignition front then moved to the beginning of the heater segment. This was visually observed since high temperature causes the membrane to buckle, which can be seen by the CCD camera used to observe testing.

This phenomenon was modeled by first starting with a low heater power in the simulation; continuing with this parameter until ignition was first observed; and then increasing the heater power to track the movement of the reaction front. The model showed that ignition first occurred at the end of the heater segment and moved to the front of the heater segment as more power was added (Jensen et al., 1997). This also suggested that the presence of multiple steady-states could be eliminated by increasing the rate of heat removal from the reaction zone.

Use as a Design Tool

The success of the mathematical model in reproducing experimental results led to its use as a tool for studying the effect of reactor design changes. This was first applied to studying the effect that changing the membrane material would have on the rate of heat removal from the reaction zone. The goal was to increase the heat transfer rate enough to prevent ignition/extinction behavior. However, if the rate was too high, the heaters would not provide a sufficient temperature rise because they would fail by electromigration. Simulations were used to predict the temperature versus heater power curve for the new microreactor design, which used a 2.6 µm silicon membrane instead of a 1 µm silicon nitride membrane. This predicted curve was extremely close to the experimental results that were obtained after the device was fabricated. The remarkable success in this trial led to the use of the modeling tools to evaluate more complex design changes.

Design of a New Flow Sensor

The success of the model in thermal analysis led to the design of a new flow sensor for the next generation microreactor. The original flow sensor used in the T microreactor was designed to operate by measuring the time-of-flight of a thermal pulse. This simplifies the design of the device, but it is difficult to operate because of the extremely fast time resolution required in the measurements. To simplify the measurement components for the flow sensor, a flow anemometer was designed using the simulation tools. This type of sensor indirectly measures flow rate by measuring the change in the temperature profile around a heating device caused by changes in gas flow rate.

The microfabricated flow anemometers were intended to be similar to the heaters in the reaction channel, but shorter in length. The simulations were used to design an optimal flow sensor by analyzing the effect of the heater segment length, the effect of heater power, and placement of the temperature sensors. In addition, the model was also used to evaluate the effect of membrane thermal conductivity and the thermal diffusivity of the gas on sensor performance.

Figure 8 shows a simulation of the temperature profile down the centerline of the channel for three gas flow rates. To better illustrate how the temperature profile changes with flow rate, Figure 9 shows the delta temperature profile. This is the difference between the temperature

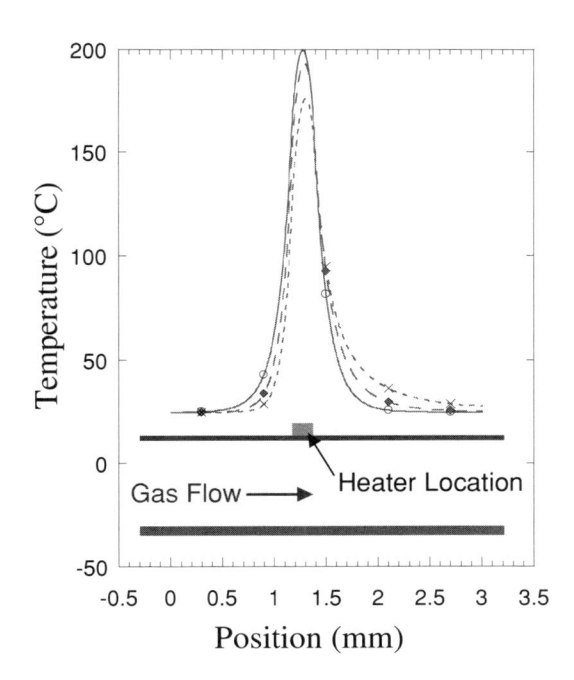

Figure 8. Temperature profile along the center of the membrane around the flow sensor heater.
—◦— *No O_2 flow;*
--◆-- *10 sccm O_2;* ⋯✕⋯ *30 sccm O_2.*

profile in the presence and absence of gas flow. The figure clearly indicates that the most sensitive configuration for this type of anemometer is to place the temperature sensor slightly upstream of the heater by 90 µm. This result is not obvious and shows the value of this type of detailed modeling.

After the ideal placement location for the temperature sensor placement was identified, the effect of the heater segment length and of the heater power were explored. The simulations gave the expected result that increasing the heater segment length yields temperature difference peaks that are taller and broader. Increasing the heater power had a similar effect, but it was more dramatic than increasing the heater length. The designer thus has the choice to increase sensitivity by either increasing the heater power or by simply making a longer heater. An optimal solution could be found in terms of minimum power requirement, but in this case the design goal was to achieve maximum sensitivity with a heater temperature of 200°C. This limit was set to avoid problems with electromigration in the heater element and to maintain its temperature well below the membrane fracture point, which is about 500°C for the current design.

The simulations were also extremely valuable in determining the effect of other parameters, which are less controllable, on the sensor performance. First, the effect of membrane thermal conductivity was explored because of the possibility of future changes in the fabrication process. Moreover, the thermal conductivity of the silicon nitride membrane is uncertain since its value depends strongly on the deposition process used. To address this issue, a series of simulations were performed to explore the range of reasonable thermal conductivity values of thin film silicon nitride. The simulations showed that changing this parameter has no effect on the optimum placement for the temperature sensor. This is extremely important, since it means that using a new membrane material does not require a redesign of the flow sensor. However, increasing the membrane thermal conductivity does decrease the overall performance of the sensor.

The effect of the gas on sensor performance was also investigated. This was done by choosing gases with a wide range of thermal diffusivities, from 1.4×10^{-5} m^2/s to 1.3×10^{-4} m^2/s. Gases with a higher thermal diffusivity had a relatively short and narrow delta temperature peak that was slightly farther away from the heater. A two-thirds loss of sensitivity was observed by switching from ammonia to hydrogen, the latter having the highest thermal diffusivity. If the flow sensor is used for hydrogen, the heater power must be increased to retain the sensor performance.

After the design simulations were completed, new microreactors were fabricated with the predicted optimal sensor design. The sensor performance was experimentally evaluated. The results are shown in Fig. 10, and they demonstrate that it is extremely sensitive to changes in flow rate between 0 and 10 sccm, which was the design goal. Sensitivity decreases as the flow rate increases further due to the development of a thermal boundary layer. The remarkable result from this figure is the excellent agreement between the experimental flow sensor performance and the performance predicted by the simulation. For these results, the experimental data were first used to determine the appropriate value of the thermal conductivity of the silicon nitride membrane since this is dependent on the deposition procedure. This value was found to be 13.1 W/(m·K). This is in the range of thermal conductivities for bulk silicon nitride, which are reported from 5 W/(m·K) to 30 W/(m·K). Considering this was the only fitted parameter in the model, the agreement between the experimental results and the simulations is excellent.

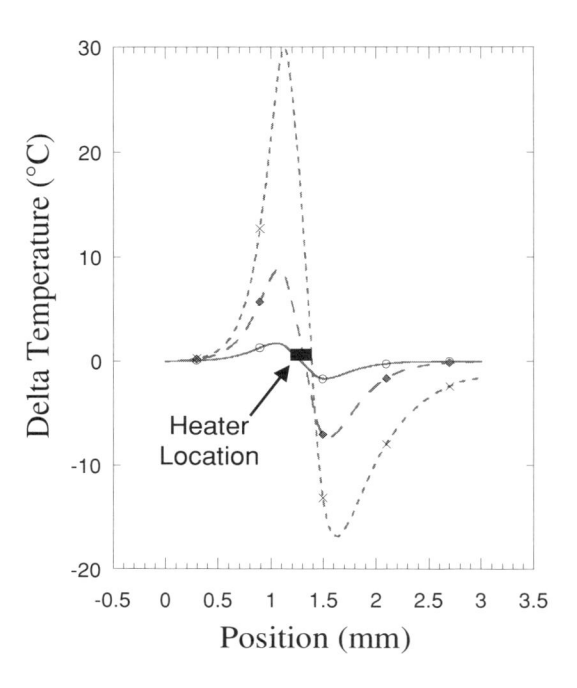

Figure 9. Delta temperature profile along the center of the membrane around the flow sensor heater. ──◯── *1 sccm O_2;* --◆-- *5 sccm O_2;* ·······✕······· *20 sccm O_2.*

Changing the Heater Design

The original T microreactor heater design consisted of a 50 µm platinum line that meandered across the 300 µm heater width down the length of the channel. This design is effective in providing uniform power for the heated zone, but it results in a non-uniform temperature across the heater. To reduce this effect, the heater was split into two symmetrically placed segments along the length of the channel as illustrated in Fig. 10. The simulation tools were then used to evaluate the effectiveness of the design and to determine whether electromigration would occur. Figure 12 shows the temperature profile obtained for a silicon nitride membrane reactor. The split heater design significantly improves the uniformity, but now there is a local temperature minimum in the center of the membrane.

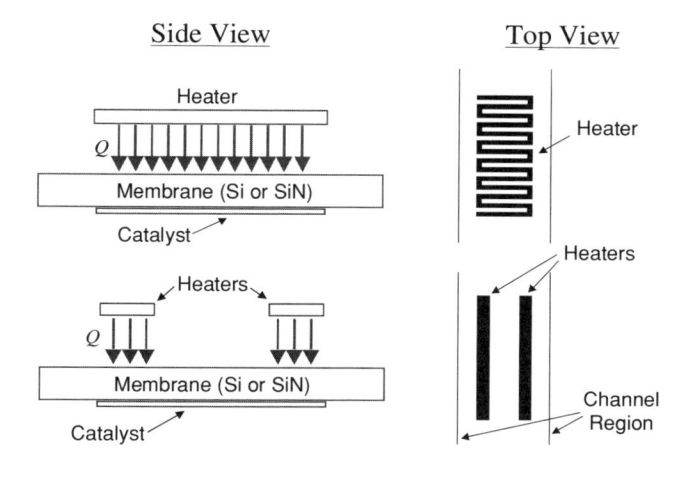

Figure 10. Top and side view schematics of the meandering and split heater designs.

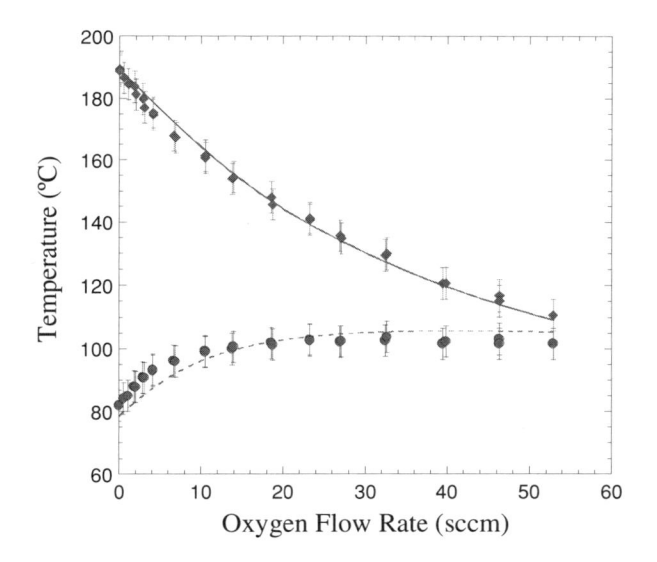

Figure 11. Upstream and downstream temperature sensor measurements for varying flow rates of oxygen. • Upstream (experimental) —(simulation); ♦ Downstream (experimental) ···(simulation)

The above results were obtained for non-reacting, stagnant gas conditions. The effect of adding a highly exothermic reaction on the thermal uniformity was studied using ammonia oxidation as the model system. Figure 13 shows the temperature profile across the channel for various NH_3 concentrations that vary from the ignition to the extinction region. For the cases shown, only the 5% NH_3 in air is operating at the lower steady state of the S curve. Because the reaction provides an additional heat source in the center of the channel, a reactor operating in the ignition regime has a much higher level of thermal uniformity.

Design of a Minichemical Plant

Because the parallel scale-up of microchemical reactors differs from the traditional scale-up paradigm, the design approach currently used for the latter is not completely applicable. Although obtaining greater production rates appears to be simple, parallelization presents new challenges that have not been addressed in previous system efforts. Particularly, the areas of reactor monitoring and control become increasingly complex as the parallel array size grows to potentially thousands of reactors. Questions also arise as to how such systems will perform as a whole, whether they can be designed in such a way to be financially competitive with conventional technology, and whether the technology can be safely practiced with a high reliability factor. The development of a systematic design methodology is needed, but the evolving nature of the technology makes the establishment of such a methodology difficult.

The construction of a complete packaged microfluidic system can be broken down into the following tasks: the individual device, the device packaging, the unit integration scheme, and the system-wide integration scheme. Figure 14 depicts the relationship between each of these tasks in the design strategy. In this case, the process design strategy changes from focusing on the entire system to focusing on one operational unit of components–each unit performs a designated process. For example, in the construction of a reactor unit, the key components could include the valves, pressure regulators, pressure sensors, flow sensors, and the reactor. Once the unit design is completed, then the design of a system to integrate the units can begin. To build even more complex systems, the number of structural levels can be increased so that the system is composed of units that are, in-turn, composed of sub-units. This leads to a self-similar design strategy to achieve greater complexity.

Traditional chemical plant design uses a top-down design strategy that adds increasing detail at each level. For example, the design of a plant begins with a production goal set by the market demand curve. The basic process flowsheet is then developed and an increasing level of detail is added as the design procedure progresses. This breadth-first search strategy may require the detailed study of several alternatives to produce the optimal design. Even then, the cost-effectiveness of the resulting design is limited by the ability to accurately estimate construction, startup, and production costs.

In the case of the minichemical plant, the design starts with individual units that will be later integrated to form the final system. Because each unit will be replicated tens to thousands of times, this design must be extremely thorough and include the construction of prototypes. The amount of detail placed in each unit depends on the operational requirements. For example, a reactor unit may consist of five reactors with one mass flow controller, one product analyzer, and one controller. The reactor behavior determines the necessary level of instrumentation. The trade-off is more instrumentation results in better performance, but at higher equipment costs. The designer would like to keep each unit as simple as possible and have units share expensive equipment such as product analyzers and programmable logic controllers.

As an example, consider the design of the data acquisition and control system for the MIT-DuPont microreactor. This poses a challenging problem because of the fast thermal and flow time constants in this microreactor. The thermal time constant for the heaters is ~15 ms and the residence time in the reaction zone is typically between 3 ms and 80 ms, depend on the flow rate used. Theoretically, very good control should be possible, but the control rate needed exceeds or is close to the limitations of even the most modern Programmable Logic Controllers (PLC). Thus, the requirement for extremely fast control may determine the PLC needed and how many reactors can be associated with one PLC before it is overloaded.

Design of a Multiple Reactor Test Station

To address these issues in microreactor scale-up, MIT and DuPont have begun construction of a multiple reactor test station for gas phase systems. The fabrication of this system serves as a test case for the development of more complex microreactor based systems. Unlike the integrated lab-on-a-chip devices discussed previously, the multiple reactor test station must be designed to operate in a continuous mode over a large range of gas flow rates. This requires a greater focus on interfacing the microfabricated components with conventional technology and on robustness for long-term continuous operation.

This test station will be capable of simultaneous operation of four microfabricated reactors and two conventional packed-bed microreactors. The conventional reactors will serve as a basis for comparing reactor performance since there is little quantitative data on the scale for chemical production. Additionally, reactor system life testing will be done to understand the long-term behavior of these devices. Although the first design will consist mainly of conventional components, it will serve as the first step to constructing a packaged multiple microreactor test station made entirely from microfabricated devices.

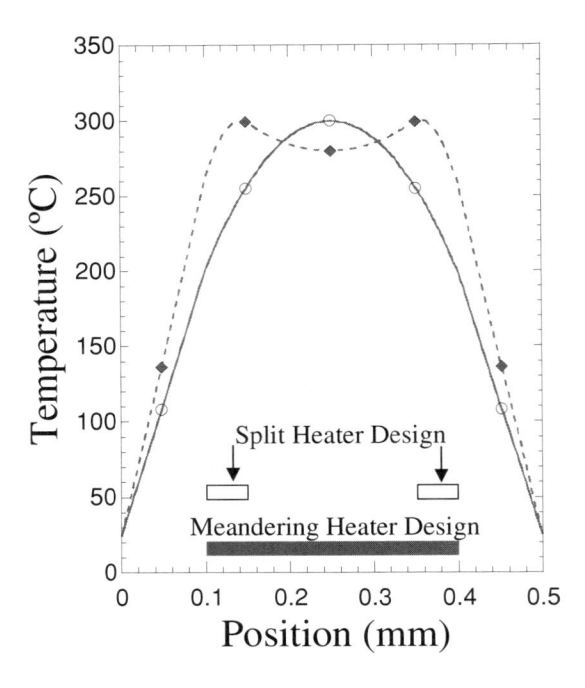

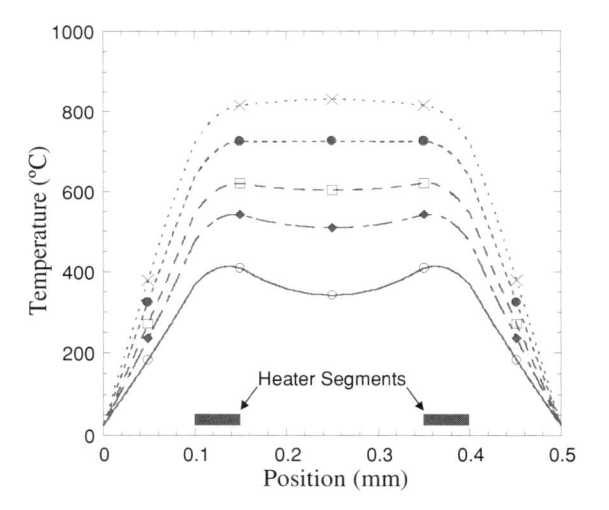

Figure 12. Temperature profile across the membrane for the two heater designs. —O— *Meandering heater design;* --◆-- *split heater design.*

Figure 13. Effect of reaction heat generation on the thermal uniformity of the new heater design for ammonia oxidation. —O— *5% NH_3 in air (19% conversion);* --◆-- *7.5% NH_3 (58%);* —□— *10% NH_3 (63%);* --●-- *12.5% NH_3 (66%);* ·····×····· *15% NH_3 (68%).*

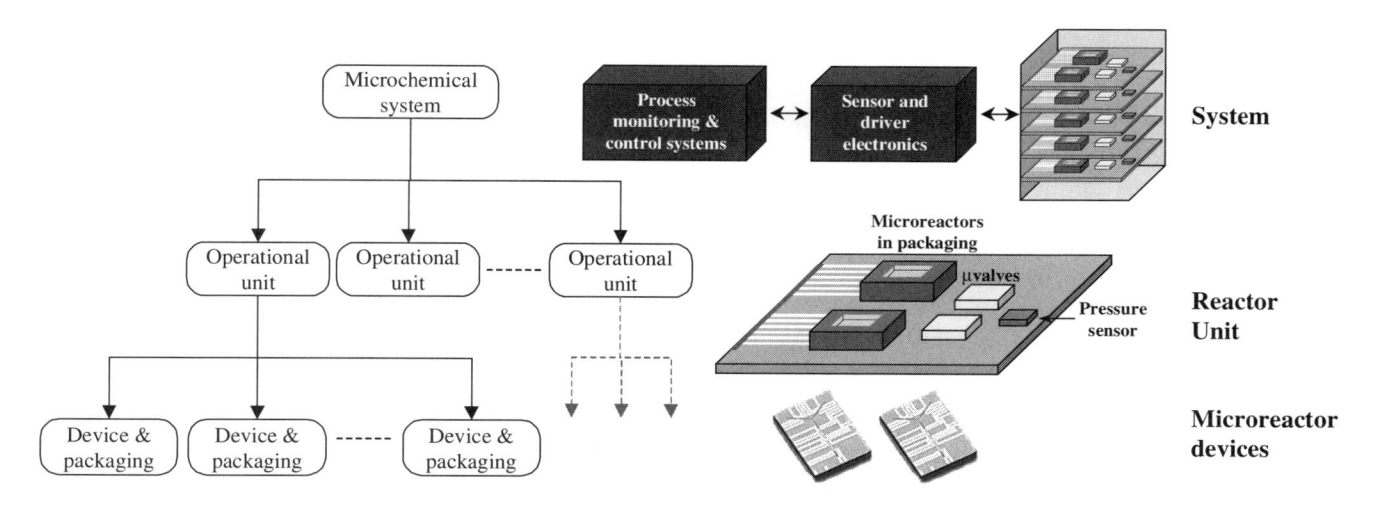

Figure 14. Proposed architecture for a microchemical system.

Simulations in Minichemical Plant Design

Unlike traditional plant simulators that focus on flow sheets and unit operations, minichemical plant design emphasizes detailed understanding of individual components. The reactor simulators already available such as Fluent and CFD-ACE+ can be used for detailed microreactor modeling, and additional tools are also available for simulation of other units, such as microvalves and microheat exchangers. These tools will need to be supported by experimental results that can be used to fit unknown parameters such as material properties in the device simulation. Experimental study will also be necessary to understand complex dynamic behavior that is beyond current modeling capabilities. These results can be used to generate transient models needed to simulate the system dynamics.

For example, detailed heat transfer simulations have been used in the MIT-DuPont microreactor scale-up project to examine the temperature profiles in the microreactor die packaging and the product transfer lines. The purpose was to ensure that the temperature limitations of the packaging were not exceeded while providing adequate heating of the transfer lines. This type of broader simulation showed the applicability of modeling to package design

Although detailed modeling results in improved understanding of the individual devices, further analysis is needed to construct an optimal design for a minichemical plant. A simplistic model would assume that each device performs the same and a deterministic prediction of the system performance is thus possible. This approach is valid for first pass analysis of the process economics, but a more realistic model is needed to produce an optimized system. The tolerances in manufacturing, catalyst performance, and process control are needed to predict overall system performance.

The device performance is related to the system performance through an understanding of the distribution functions of the important device operating parameters, such as catalyst activity and flow control. Stochastic modeling can be used to predict the probability density function of the system performance from detailed knowledge of the device and unit behavior. This requires knowing the sensitivity of the product output as a function of, among other variables, catalyst activity, flow rate, and feed composition. For initial design evaluations, this can be performed using simulations alone, but experimental data is needed to refine these predictions. Experiments are also critical in understanding variations in catalyst activity and other measures of system performance.

Optimization algorithms can then be used in combination with stochastic simulations and device models to determine the optimum unit and system layouts. For example, it is easier to construct an array of reactors that have common feeds and outlets with a single mass flow controller, but the sensitivity of the reactor to flow rate may limit the size of the array. Alternatively, the reactor failure rate may require the ability to shut-off flows to individual reactors.

Thus, the optimization problem requires a high level of understanding of the effect of individual devices and their integration on the unit performance. The situation becomes more difficult when the complexity of the computational models is considered. The choice of algorithm for analyzing the non-deterministic problems is critical to reduce computational time. Furthermore, the integration of an optimization routine poses an additional challenge.

Simulations in Minichemical Plant Operation

Model development will also be important in the advanced control algorithms required for minichemical plant operation. Beyond the level of individual reactor control, control schemes are required for unit and system operation. These controllers will need to evaluate unit behavior as well as set the operating conditions to meet production requirements. The diagnosis and shutdown of failed units will be critical functions since operator intervention must be minimized. The development of expert systems for minichemical plant operation will require models that can be used to identify corrective actions for process disturbances and to identify unit performance degradation.

Furthermore, these models will be critical in optimizing individual unit performance. For example, a degraded catalyst may require increased reactor temperature to maintain the required level of conversion. Alternatively, feed composition and flow rate can also be used to maintain performance if these parameters are adjustable on the unit level. Even more advanced adaptive control schemes can be used to refine the model as the unit is running.

The use of hundreds of units running in parallel, also allows diagnostic testing to be performed on-line by diverting the product stream from a unit to an exhaust while the test is running. Thus, the individual units can be retuned during operation. These potential operational advantages give compelling reasons for further system study.

Conclusions

The maturing of microfluidic and microreaction technology creates an enormous opportunity for sectors of the chemical processing industry. MEMS devices already have the potential for revolutionizing portable power generation devices and laboratory chemical analysis. Similarly, the production of minichemical plants based on microreactors represents a paradigm shift in the design methodology for chemical plants. The disruptive nature of the technology along with the conservative viewpoint of the chemical process industry presents a significant challenge in technology adoption.

The outcome of microfluidics in other areas, such as biological assay devices, will provide key information for developing processes based on MEMS components. However, the results of on-going research in the field of microreactors and their systems will be critical. To address the system aspects of microchemical production, MIT and DuPont have begun the construction of a parallel reactor system.

Previous efforts in using simulations to understand the operating behavior of the MIT-DuPont microreactor have been highly successful. These modeling results gave important information on the temperature and species profiles inside the operating reactor. The simulation tools have also been valuable in analyzing design changes, as in the case of the split heater design. In addition, the simulations made it possible to produce a flow anemometer with excellent performance without any experimental design iterations.

In the future, it is likely that models will be coupled with experimental results to replicate the behavior of the reactors and their units for system design. These models will then be used in stochastic simulations to predict system performance. Integrating these tools with optimization routines will decrease the time required to develop a final system design. Modeling will also be critical in the operation of the minichemical plant. Expert control systems will replace highly trained technicians and allow almost autonomous plant operation. Clearly, significant technical challenges remain in the design of microchemical systems, but research in other microfluidic application areas as well as in the microreaction community is progressing.

Acknowledgements

The authors thank the DARPA MicroFlumes Program for financial support under contract F30602-97-2-0100. DJQ thanks the National Science Foundation for his graduate fellowship.

References

Amato, I. (1998). Fomenting a revolution, in miniature. *Science*, **282**, 402-405.

Anderson, R. C., G. J. Bogdan, A. Puski, and X. Su (1998). Advances in integrated genetic analysis. In D. J. Harrison and A. v. d. Berg (Eds.), *μTAS '98 Workshop*, Kluwer Academic Publishers, Banff, Canada. 11-16.

Ayon, A. A., R. Braff, C. C. Lin, H. H. Sawin, and M. A. Schmidt (1999). Characterization of a time multiplexed inductively coupled plasma etcher. *J. Electrochem. Soc.*, **146**, 339-349.

Baselt, J. P., A. Förster, J. Herrmann, and D. Tiebes (1998). Microreactor Technology: Focusing the German activities in this novel and promising field of chemical process engineering. In *Process Miniaturization: 2nd International Conference on Microreaction Technology*, AIChE, New Orleans, 13-17.

Benson, R. S., and J. W. Ponton (1993). Process Miniaturization—A route to total environmental acceptability. *T. I. Chem. Eng.-Lond.*, **71 A**, 160-168.

Berg, A. v. d., and D. J. Harrison (Eds.) (1998). *Micro Total Analysis Systems '98*, Kluwer Academic Publishers, Banff, Canada.

Borman, S. (1999). Reducing time to drug discovery. *Chem. Eng. News*, March 8, 33-48.

Brenchley, D. L., and R. S. Wegeng (1998). Status of microchemical systems development in the United States of America. In *Process Miniaturization: 2nd International Conference on Microreaction Technology*, AIChE, New Orleans, LA. 18-23.

Brody, J. P., P. Yager, R. E. Goldstein, and R. H. Austin (1996). Biotechnology at low Reynolds numbers. *Biophys. J.*, **71**, 3430-3441.

Burns, J. R., C. Ramshaw, A. J. Bull, and P. Harston (1997). Development of a microreactor for chemical production. In W. Ehrfeld (Ed.), *Microreaction Technology: Proceedings of the First International Conference on Microreaction Technology,* Springer, Berlin, pp. 127-133.

Burns, J. R., C. Ramshaw, and P. Harston (1998). Developmnt of a microreactor for chemical production. In *Process Miniaturization: 2nd International Conference on Microreaction Technology,* AIChE, New Orleans, LA. pp. 39-44.

Dagani, R. (1998). A faster route to new materials. *Chemical & Engineering News*, 51-60.

DeWitt, S. H. (1999). Microreactors for Chemical Synthesis, *Curr. Opin. Chem. Biol.*, **3**, 350.

Ehrfeld, W., C. Gärtner, K. Golbig, V. Hessel, R. Konrad, H. Löwe, T. Richter, and C. Shulz (1997). Fabrication of components and systems for chemical and biological microreactors. In W. Ehrfeld (Ed.), *Microreaction Technology: Proceedings of the First International Conference on Microreaction Technology,* Springer, Berlin, 72-90.

Ehrfeld, W., K. Golbig, V. Hessel, H. Lowe, and T. Richter (1999). Characterization of mixing in micromixers by a test reaction: Single mixing units and mixer arrays. *Ind. Eng. Chem. Res.*, **38**, 1075-1082.

Ehrfeld, W., V. Hessel, H. Möbius, T. Richter, and K. Russow (1996). Potentials and realization of microreactors. In *Microsystem Technology for Chemical and Biological Microreactors: Papers of the Workshop on Microsystem Technology, Mainz, 20-21 February, 1995,* Vol. 132, DECHEMA, Frankfurt, 1-28.

Floyd, T. M., M. W. Losey, S. L. Firebaugh, K. F. Jensen, and M. A. Schmidt (1999). Novel liquid phase microreactors for safe production of hazardous specialty chemicals. In *3rd International Conference on Microreaction Technology,* Springer, Frankfurt.

Franz, A., K. F. Jensen, and M. A. Schmidt (1999). Palladium based micromembranes for hydrogen separation and hydrogenation/dehydrogenation reactions. In *Technical Digest 12th International Conference on MicroElectroMechanical Systems,* IEEE, Orlando, Florida. 382-385.

Freemantle, M. (1999). Downsizing chemistry. *Chemical & Engineering News*, February 22, 27-36.

Frye-Mason, G. C., R. J. Kottenstette, E. J. Heller, C. M. Matzke, S. A. Casalnuovo, P. R. Lewis, R. P. Manginell, W. K. Schubert, V. M. Hietala, and R. J. Shul (1998). Integrated chemical analysis systems for gas phase CW agent detection. In D. J. Harrison and A. v. d. Berg (Eds.), *µTAS '98 Workshop,* Kluwer Academic Publishers, Banff, Canada. 477-481.

Gross, N., and O. Port (1998). The 21st Century Economy: The coming molecular revolution. *Business Week*, August 31, 80-83.

Hagendorf, U., M. Janicke, F. Schüth, K. Schubert, and M. Fichtner (1998). A Pt/Al$_2$O$_3$ coated microstructured reactor/heat exchanger for the controlled H$_2$/O$_2$-reaction in the explosion regime. In *Process Miniaturization: 2nd International Conference on Microreaction Technology,* AIChE, New Orleans, LA. 205-210.

Hicks, J. (1999). Genetics and drug discovery dominate microarray research. *R&D Magazine*, February, 28-33.

Hsing, I.-M., R. Srinivasan, M. P. Harold, K. F. Jensen, and M. A. Schmidt (1999). Simulation of micromachined chemical reactors for heterogeneous partial oxidation reactions. *Chem. Eng. Sci*, (in press).

Jäckel, K. P. (1996). Microtechnology: Application opportunities in the chemical industry. In *Microsystem Technology for Chemical and Biological Microreactors: Papers of the Workshop on Microsystem Technology, 20-21 February, 1995,* Vol. 132, DECHEMA, Frankfurt, pp. 29-50.

Jandeleit, B., H. W. Turner, U. Tetsuo, J. A. M. v. Beek, and W. H. Weinberg (1998). Combinatorial methods in catalysis. *Cat.Tech.*, **2**, 101-123.

Jensen, K. F., I.-M. Hsing, R. Srinivasan, M. A. Schmidt, M. P. Harold, J. J. Lerou, and J. F. Ryley (1997). Reaction Engineering for Microreactor Systems. In W. Ehrfeld (Ed.), *Microreaction Technology: Proceedings of the First International Conference on Microreaction Technology,* Springer, Berlin, 2-9.

Jones, A., K. Flamm, K. Gabriel, R. v. Atta, and L. Mareth (1995). *Microelectromechanical Systems Opportunities: A Department of Defense Dual-Use Technology Industrial Assessment,* U.S. Department of Defense, Washington, D. C.

Konrad, R., W. Ehrfeld, H. Freimuth, R. Pommersheim, R. Schenk, and L. Weber (1997). Miniaturization as a demand of high-throughput synthesis and analysis of biomolecules. In W. Ehrfeld (Ed.), *Microreaction Technology: Proceedings of the First International Conference on Microreaction Technology,* Springer, Berlin, 348-356.

Lerou, J. J., M. P. Harold, J. Ryley, J. Ashmead, T. C. O'Brien, M. Johnson, J. Perrotto, C. T. Blaisdell, T. A. Rensi, and J. Nyquist (1996). Microfabricated minichemical systems: Technical feasibility. In *Microsystem Technology for Chemical and Biological Microreactors: Papers of the Workshop on Microsystem Technology, Mainz, 20-21 February, 1995,* Vol. 132, DECHEMA, Frankfurt, 51-69.

Losey, M. W., M. A. Schmidt, and K. F. Jensen (1999). A micro packed-bed reactor for chemical synthesis. In *3rd International Conference on Microreaction Technology,* Springer, Frankfurt.

Mandel, M. J. (1998). The 21st Century Economy: Innovation's great leap. *Business Week*, August 31, 60-63.

Marshall, S. (1999). Fundamental changes ahead for lab instrumentation. *R&D Magazine*, February, 18-25.

Marsili, R. (1999). Lab-on-a-chip poised to revolutionize sample prep. *R&D Magazine*, February, 34-40.

Martin, P. M., D. W. Matson, and W. D. Bennett (1998). Micofabrication methods for microchannel reactors and separations systems. In *Process Miniaturization: 2nd International Conference on Microreaction Technology,* AIChE, New Orleans, 75-80.

Mobius, H., W. Ehrfeld, V. Hessel, and T. Richter (1995). Sensor controlled processes in chemical microreactors. In *8th International Conference on Solid-State Sensors and Actuators and Eurosensors,* Vol. 1, IEEE, Stockholm, 775-778.

Paul, P. H., D. W. Arnold, and D. J. Rakestraw (1998). Electrokinetic generation of high pressures using porous microstructures. In D. J. Harrison and A. v. d. Berg (Eds.), *μTAS '98 Workshop,* Kluwer Academic Publishers, Banff, Canada. 49-52.

Petersen, K. E. (1982). Silicon as a mechanical material. *P. IEEE,* **70,** 420-457.

Quiram, D. J., I.-M. Hsing, A. J. Franz, R. Srinivasan, K. F. Jensen, and M. A. Schmidt (1998). Characterization of microchemical systems using simulations. In *Process Miniaturization: 2nd International Conference on Microreaction Technology,* AIChE, New Orleans, LA. 205-210.

Richter, T., W. Ehrfeld, K. Gebauer, K. Golbig, V. Hessel, H. Löwe, and A. Wolf (1998). Metallic microreactors: Components and integrated systems. In *Process Miniaturization: 2nd International Conference on Microreaction Technology,* AIChE, New Orleans, LA. 146-151.

Robins, I., J. Shaw, B. Miller, C. Turner, and M. Harper (1997). Solute transfer by liquid/liquid exchange without mixing in micro-contactor devices. In W. Ehrfeld (Ed.), *Microreaction Technology: Proceedings of the First International Conference on Microreaction Technology,* Springer, Berlin, 35-46.

Salimi-Moosavi, H., R. Szarka, P. Andersson, R. Smith, and D. J. Harrison (1998). Biology lab-on-a-chip for drug screening. In D. J. Harrison and A. v. d. Berg (Eds.), *μTAS '98 Workshop,* Kluwer Academic Publishers, Banff, Canada, 69-72.

Service, R. F. (1998). Coming soon: The pocket DNA sequencer. *Science,* **282,** 399-401.

Srinivasan, R., I.-M. Hsing, P. E. Berger, K. F. Jensen, S. L. Firebaugh, M. A. Schmidt, M. P. Harold, J. J. Lerou, and J. F. Ryley (1997). Micromachined reactors for catalytic partial oxidation reactions. *AIChE J.,* **43,** 3059-3069.

Stoldt, J., H. Wurziger, H. Kummeradt, U. Kopp, M. Hohmann, and N. Schwesinger (1999). Experiences with the use of microreactors in organic synthesis. In W. Ehrfeld (Ed.), *3rd International Conference on Microreaction Technology,* Springer, Berlin, Frankfurt, Germany.

TeGrotenhuis, W. E., R. J. Cameron, M. G. Butcher, P. M. Martin, and R. S. Wegeng (1998). Microchannel devices for efficient contacting of liquids in solvent extraction. In *Process Miniaturization: 2nd International Conference on Microreaction Technology,* AIChE, New Orleans, 329-334.

Terry, S. C., J. H. Jerman, and J. B. Angell (1979). A gas chromatographic air analyzer fabricated on a silicon wafer. *IEEE T. Electron Dev.,* **ED-26,** 1880-1886.

Wegeng, R. S., C. J. Call, and M. K. Drost (1996). Chemical system miniaturization. In *AIChE 1996 Spring National Meeting,* New Orleans, LA.

Wegeng, R. S., and M. K. Drost (1998). Opportunities for distributed processing using micro chemical systems. In *Process Miniaturization: 2nd International Conference on Microreaction Technology,* AIChE, New Orleans, 3-9.

Weinberg, W. H., B. Jandeleit, K. Self, and H. W. Turner (1998). Combinatorial methods in homogeneous and heterogeneous catalysis. *Current Opinion in Solid State & Materials Science,* **3,** 104-110.

Wießmeier, G., and D. Hönicke (1996). Microfabricated components for heterogeneously catalysed reactions. *J. Micromech. Microeng.,* **6,** 285-289.

Wörz, O., K. P. Jäckel, T. Richter, and A. Wolf (1998). Microreactors, a new efficient tool for optimum reactor design. In *Process Miniaturization: 2nd International Conference on Microreaction Technology,* AIChE, New Orleans, 183-185.

Zech, T., D. Honicke, A. Lohk, K. Golbig, and T. Richter (1999). Simultaneous screening of catalysts in microchannels: Methodology and experimental setup. In W. Ehrfeld (Ed.), *3rd International Conference on Microreaction Technology,* Springer, Frankfurt, Germany.

DECISION-MAKING BY DESIGN: EXPERIENCE WITH COMPUTER-AIDED ACTIVE LEARNING

Michael F. Doherty, Michael F. Malone and Robert S. Huss
Department of Chemical Engineering
University of Massachusetts
Amherst, MA 01003-3110

Montgomery M. Alger and Brian A. Watson
GE Plastics
Pittsfield, MA 01210

Abstract

Advances in computer aided design and simulation tools and reduced computing costs allow new uses for computing in engineering education. Models of sufficient fidelity and speed allow solutions of realistic problems as an integral part of the course experience. This allows a new emphasis on the importance of making, justifying and evaluating decisions in process design. This paper describes our experience over several years and courses in implementing and testing this approach to design education. We find that use of simulation and design tools as an integral part of lectures has a major impact on student evaluations, if roughly 30% of class time is spent using these tools for active learning.

Keywords

Conceptual design, Process synthesis, Decision-making, Education.

Introduction and Motivation

In our view, the role of the engineer in industry is to make, evaluate, and justify decisions in support of business. There is a growing need in design education to provide students with the skills needed to make and justify engineering decisions in support of business in a competitive global marketplace. We are developing and implementing a comprehensive plan to address these concerns in a new undergraduate curriculum model that emphasizes interactive learning and realistic applications in process engineering throughout the course of study. The goal is to develop a curriculum, which provides students early and ongoing exposure to a framework for decision-making and understanding of the impact of decisions on economics, environment, safety and operability. There is an emphasis on invention of a better process across all curriculum areas. To implement change on this scale we use a new teaching infrastructure in a "classroom of the future" and strategies to insure ongoing and rapid integration of industrially relevant research and manufacturing trends in the classroom.

This rather old and compelling idea of "learning by doing" goes beyond the particular application to design education. However, methods and the associated tools for design and process engineering are most advanced thanks in large part to advances detailed in previous FOCPAD Conferences and similar related meetings, e.g., Mah and Seider (1981); Westerberg and Chien (1984); Siirola, Grossmann and Stephanopoulos (1990); Biegler, L. T. and Doherty (1995).

The chemical process industries remain one of the strongest segments of the worldwide economy. This is due to the cost-effectiveness of well-designed chemical processes as well as to inventive chemistry. However, as we enter the next century, the industry faces major new challenges through increased global competition, greater regulatory pressures, and uncertain prices for energy, raw materials, and products. These competitive concerns increase focus on processes with subsystems that have tighter integration and coordination, on computing tools that communicate with each other, and on consideration of multiple design criteria including profitability, safety, operability, quality, and the environment.

Traditional chemical engineering curriculum models do not address most of these issues. For example, much of the typical engineering curriculum is focused on closed-end analysis problems where a system is well defined and sufficient information is given to completely solve a problem. For example, a flowsheet is often the starting point in a senior "capstone" course on design, which is then focused on computer simulation of the system. This focus is an excellent training in analysis, but does not necessarily develop the skills to treat open-ended problems. The unique and essential factor in solving such problems is the invention and ranking of alternatives, which in essence reduces to decision-making.

In the last two decades, many new ideas and tools have been developed to complement simulation with synthesis, e.g., Douglas (1988), Biegler, Grossmann and Westerberg (1997), Seider, Seader and Lewin (1999). Conceptual design combines synthesis and analysis and forces an examination of the decisions needed to invent a process. Systematic approaches articulate the economic objectives, trade-offs and constraints from operability and control, safety, the environment, and product quality. Any or all of these factors can be critical in selecting among the potentially profitable process alternatives.

We stress that the approach outlined here is not intended to replace analysis based on fundamental principles, which is typically well covered in the curriculum. It is the intent, however, to use analysis as one ingredient in exploring and comparing alternatives that maximize economic objectives and meet the constraints.

Thus, the program addresses the following primary needs:

1. The need for early and continued exposure to design and related decision making in the undergraduate curriculum.
2. The need for a process to integrate recent and relevant research ideas into the chemical engineering curriculum.
3. The need to educate students in synthesis as well as analysis and to develop the attitude and skills needed to make and justify engineering decisions based on economic objectives and realistic constraints.
4. The need to increase understanding and appreciation for the role of technology and technology decisions in a global context and in relation to markets, transportation, exchange rates, etc.

Specifically, this project addresses change in six of fourteen required "core" undergraduate courses as well as and introductory course for the freshman year. These are: Introduction to Chemical Engineering (freshman), Material Balances (sophomore), Staged Operations (junior), Process Design I and II (senior) and Process Control (senior). The approach was also used in a graduate course on Process Control. We focus primarily on the junior and senior courses in this paper.

Infrastructure

Computer design aids and new technologies afford the real opportunity to significantly alter the way in which chemical engineering is taught. Indeed, the chemical engineering curriculum of the twenty-first century can make major use of a "classroom of the future."

Construction and furnishing of such a classroom was completed in time for use for in spring 1997. A schematic of the classroom is shown in Figure 1. The facility contains 28 computer stations networked to two central servers that provide software tools and eventually access to real-time experiments in an adjacent undergraduate laboratory[1].

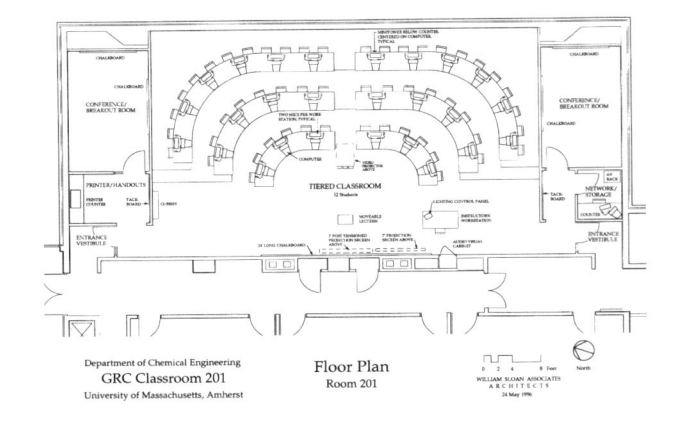

Figure 1. Schematic of the classroom.

This enables hands-on, interactive use of the methods, tools, control algorithms and experiments. Each class is divided into pairs of two students per station that

---

[1] We use 28 Pentium Pro 200's with CDROM and 4GB hard drives. This network is supported with a HP NetServer LH as the primary domain controller and a Dell Poweredge 4100 as the backup domain controller. All machines have Fast Ethernet cards to use the 100-mbps Ethernet network within the classroom. The network operates under Windows NT 4.0, with individual user accounts and security. General-purpose software (Microsoft Office 97, Netscape) was useful along with modeling and simulation software (Mathematica, MathCAD and Matlab). Specialist tools for chemical process design and simulation (HYCON and HYSYS) as well as computational chemistry and chemical education (Gaussian 94, Web Lab Viewer, Chemland, and several others).

actively participate in the classroom and interact with the instructor and teaching assistants, in place of passive listening in traditional lecture formats. Video projection equipment provides real-time, interactive classroom display of results and experiments from any station in the facility.

This classroom is the basic, required, implementing tool, which is the foundation for the capacity to affect widespread curriculum change. It enables integral use of computer-aided design and analysis software such as Matlab, HYSYS, or HYCON for use in the courses. HYCON is a commercial package for designing nonideal distillation systems developed by Hyprotech Ltd. (now AEA software technology) based on ideas and a prototype developed in research at UMass. HYCON is one of the more specialized tools we use in the curriculum, but offers effective high fidelity models for complex mixtures. It is also an effective means to integrate selected research results with undergraduate teaching.

In fact, the particular choices of software always involve tradeoffs of fidelity and applicability to a particular subject vs. cost and ease of use. For instance, MathCAD is widely available and relatively inexpensive, but does not solve problems as effectively as Matlab in process control or HYCON in separations. New emerging standards for component software described elsewhere in this conference (Braunschweig et al. (1999)) offer the excellent potential for major advances in ease of use and impact. In fact, such standards and their use appear to be essential to long term success in this sort of approach to design education.

The cost of construction and furnishing the facility was approximately $425,000. If the facility is fully utilized, approximately 16 classes per year can be taught. For a lifetime of 3 years, the cost is just approximately $177 per student per course in classes of 50. In addition to the construction costs, operations bring the total to approximately $200 per student per course. The facility also serves as a computing lab in the evening hours, which is a further justification for the resources.

Industrial Advisory Board

An industrial outreach program fosters an ongoing dialogue with practicing engineers. This is accomplished through our existing research contacts in the UMass Process Design and Control Center. The most valuable input generally concerns the nature of the problem and the decisions encountered. Special emphasis was made on a balanced approach and intelligent use of the technology so that the students learn to make decisions, but also to keep a firm basis in the fundamentals and analysis. Input on sample problem definitions relating to business aspects of decision making is particularly helpful; one such problem is described below.

Curriculum Redevelopment

Although it was not envisioned at the outset, the effective use of technology for active learning permits and, in fact, requires a change in teaching approach. Traditional "stand-and-deliver" lectures and related assignments and exams (Fig. 2) are naturally replaced by tighter interactions between students and instructors (Fig. 3) and eventually teaching assistants (Fig. 4).

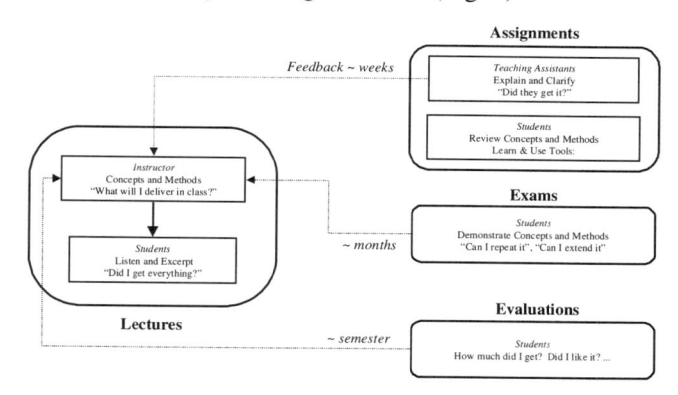

Figure 2. Staged operations in 1995.

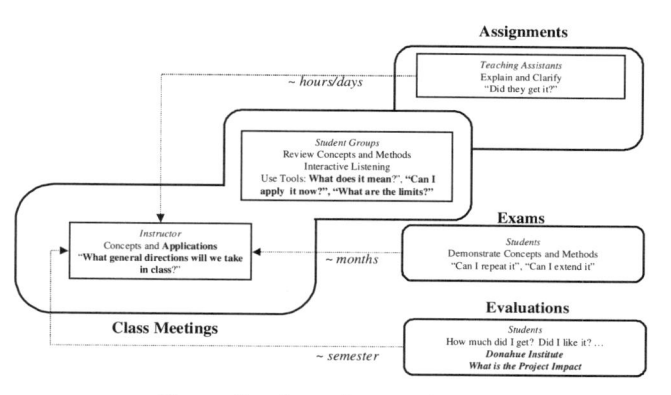

Figure 3. Staged operation in 1998.

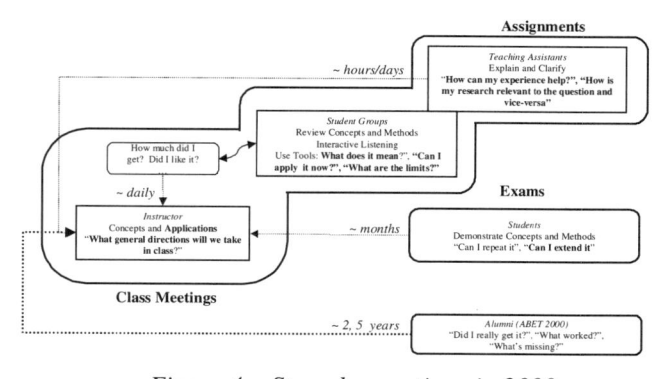

Figure 4. Staged operations in 2000.

It is interesting that the classroom processes are becoming much more integrated in a search for improved educational impact, in close parallel to trends in chemical process design.

In what follows, we describe in detail our experiences and results for three courses, one each at the freshman, junior, and senior levels.

Freshman Introduction to Chemical Engineering

The objectives of this course were to provide design experiences in chemical engineering, to encourage teamwork, and to develop communication and computational skills for freshmen. This module gave an introduction to chemical engineering by way of designing and implementing automatic controller schemes to chemical processes. Computational skills were developed and utilized in the analysis of automatic controller schemes. Presentation and discussion of controller dynamics encouraged communication skills. The structure of the course required students to be divided into groups of three or four to encourage and develop teamwork during solution of simulation examples and experimental work.

The major lecture topics included

1. Overview of chemical engineering processes and the importance of automatic control in practice.
2. Dynamics of simple systems and differential equations.
3. Solutions of simple differential equations.
4. Computer-aided simulations of dynamic systems using MathCAD.
5. Effects of feedback on stability and performance.
6. Applications in chemical engineering, using computer simulations.
7. Classroom experiment using CIMCAR (see below)

Students solved control examples using computational and analytical approaches. These examples were tailored to chemical engineering by adapting cases from the senior level chemical engineering control course, simplified so that freshmen could solve them. A typical example was the design of a controller that sets a product yield requirement in a stirred tank reactor, which puts concepts from chemistry and mathematics courses into an engineering context for students. The solution of differential equations pertinent to control examples is probably the most difficult aspect for students to master in this course. To avoid the course evolving into a math course, the differential equations were handled mostly using computer software such as MathCAD, emphasizing graphical formulation and presentation of simulation results. Simulations of more challenging control applications in chemical engineering were performed in class using Matlab.

To provide hands on experience with an actual automatic controller, students conducted in-class experiments. Typical control examples in chemical engineering are not practical in the classroom. However, the types of dynamics encountered in chemical engineering are common to many examples in electrical and mechanical engineering. The classroom experiments involving an electric car (CIMCAR) already designed by Professor T. Djaferis in ECE was a useful vehicle for students to observe first hand the effect of controller schemes. Students used CIMCAR to test the effectiveness and related dynamics of different controllers during collision avoidance problems. Experimental results using an intelligent car (approximately 6 inches long equipped with a camera) were compared to the simulated performance predicted by theoretical models.

The course was highly interactive in nature, with an emphasis on active learning and presentations via group projects in class. Students were divided into groups of three or four when working on simulation examples and CIMCAR experiments. Group participation involved analyzing the behavior of the controller and presenting solutions graphically, orally and in written form. This aspect of the course allowed students to develop collaborative and communication skills that are vital in engineering.

Junior Year Staged Operations

The curriculum for this course differs significantly from the traditional approach to teaching separations. The focus is on the systems approach to separating nonideal, multicomponent mixtures of commercial complexity. The main new features are:

1. Inclusion of process economics. It is not typical to introduce this subject into separations courses. However, we believe it is vital to give students an appreciation of both capital and operating costs associated with separation processes so that the major tradeoffs can be taught and process alternatives can be screened effectively.
2. Emphasis on process alternatives. There is never a single way to solve an engineering problem, and this is true in separation systems as well as many other areas of engineering. We begin by making a precise problem statement (e.g., separate a ternary mixture of acetaldehyde, methanol and water of a given composition into pure streams with certain specifications on the allowable impurity levels) that allows for several alternative separation strategies. Students are required to rank the alternatives according to an engineering figure of merit such as the total annual cost, total energy use, etc. These concepts are reinforced by repeated application of the ideas on different systems. Once the students are comfortable with the idea of multiple solutions to a fixed problem

statement they are introduced to...

3. Changing the problem statement to obtain a better solution. The problem statements that define process sub-systems (e.g., separation sub-systems) are often partial statements of the overall process system or the process goals. Better engineering solutions are frequently found by changing the sub-system definition or problem statement. For example, if two adjacent components are very difficult to separate it may be possible to eliminate one of them upstream before the separation system, e.g., by reaction. In other cases, it may be necessary to redefine the problem statement to satisfy quality criteria, e.g., by over-purifying a stream beyond what is called for in the problem statement in order to eliminate undesirable impurities. Students were challenged with a number of "systems issues" of this kind and did a surprisingly good job at asking the right kinds of questions as well as answering them.

4. HYCON design tool. The HYCON design tool was an important aid in the solution of class problems. The new design methods taught in the lectures are too sophisticated for the students to program efficiently themselves. This tool allows them to get access to relevant data and procedures such as phase diagrams, residue curves, azeotrope bifurcation diagrams, etc. However, the program will not calculate anything that is not requested by the user, so it is impossible to get "cookbook" answers with no user input. The structure of the program requires that the user make decisions and know how to interpret the answers.

5. Making decisions. The major emphasis in the course is getting the students to make and justify decisions. Even in the early stages of the course, students were required to perform their own selection and specification of phase equilibrium and physical property models for realistic mixtures, including multiple components and realistic, nonideal physical properties. They quickly developed enough intuition to pick a suitable model and to justify their choice. Similar emphasis was given to the choice of complete process models (How and when do you include heat effects? How many components are included in the initial design? etc.). By the end of the course, most students were confident enough to solve difficult problems of industrial complexity with minimal input from the instructor.

Students quickly developed the intuition and confidence to pick a suitable physical property model and justify their choice. This was then extended to process models, problem definitions, and finally to the most important and open-ended task of problem redefinition. By the end of the course, many students were confident enough to define and solve difficult problems of industrial complexity.

The outcomes include:

1. Students were exposed to more realistic engineering situations.
2. Students learned to use a commercial-grade design tool at the cutting edge of new tools available.
3. Selected recent research results were implemented and taught to students much faster than has been traditionally possible. The lag time was cut by several years in this new environment.
4. Students were better at making and justifying decisions than their predecessors.
5. A significant amount of time was spent on interactive learning, but very little was removed from the course. This made the course longer, and an extra 30-40 minute class was taught each week, but the need for this has diminished in subsequent course offerings.

Senior Process Design II

M. F. Malone and A. Nagurney (Department of Finance and Operations Management) taught this course in the spring, 1997 semester to 38 seniors. Dr. Peter D. Edwards from the DuPont Company also participated through visits with classroom lecturing and discussion groups focused on financial evaluation of chemical processes.

This is the second in a two-semester sequence of courses in chemical process design. The main goal is to complete a systematic approach to understanding chemical processes begun in the fall semester. The fall semester Process Design I course is focused on continuous systems involving vapor-liquid mixtures and single plants. In the spring, these ideas are used as a basis for understanding interconnected manufacturing facilities, including processes to produce polymers and other solids. This naturally brings out the need to understand the interactions of process design and control.

Approximately 1/3 of the course was spent becoming familiar with some modern optimization tools, their use in assessing the economic impact of engineering decisions and how this information is used in a business context. Towards this end, part of the course provided a survey of the fundamentals of quantitative techniques to improve decision-making and management performance. Emphasis is on the formulation of decision problems as mathematical models and on the selection of the appropriate techniques for analysis and solution. This component of the course used Excel, along with specialized add-on solvers, to compute solutions to the models. Many actual successful management science model implementations by well-known companies were used as illustrations of the methods. Typical cases included oil blending and process design to warehouse location and distribution as well as portfolio optimization problems. Hence, applications were drawn from all functional areas of business including finance, marketing, and operations. The management science tools that were covered included linear programming; network models and some algorithms including the transportation problem and variants; the assignment problem; project planning networks; integer programming models; nonlinear programming applications; and the basics of formal decision theory.

Example 1: The DeRosier Problem

This in-class exercise is introduced during the nonideal and azeotropic distillation module of our junior year Separations course. The first part of the exercise is designed to be a mechanical application of the principles taught in class on the analysis and design of ternary azeotropic distillations. The second part challenges the students to find a creative design to a typical engineering problem that, on face value, is impossible to achieve. Most students can't do this, but the entire class benefits from the discussion that erupts during the attempt. This problem is named after Robert Derosier who was the first student to develop a successful design.

The problem statement is as follows:

Part 1:
    Calculate the residue curve map for the mixture methanol-isopropanol-water at 1 atm pressure. Use the NRTL-ideal model to represent the VLE.
    Design a distillation column to separate a saturated liquid feed consisting of 40 mol% methanol, 40 mol% water, and 20 mol% isopropanol into saturated liquid products. The distillate is specified to contain 99 mol% methanol and 0.5 mol% water. The bottom product contains 0.5 mol% methanol. Use your engineering judgement to select a reflux ratio that trades off the vapor rate against the number of theoretical stages.

Part 2:
    After solving Part 1 you proudly show your design to the client who commissioned the job. She tells you that she wants as little water as possible in the methanol product because it kills the catalyst in her process. She will not accept methanol containing more than 50 ppm water. Since water is the heaviest component, this seems like an easy constraint to meet. Is it? What is the smallest composition of water that *you* can get in the methanol product? What is your design for the separation scheme?

The solution to Part 1 is straightforward and most students solve it rapidly. In Figure 5 we show the calculated residue curve map. Figure 6 shows the column composition profiles at minimum reflux, which is determined by the presence of a node pinch in the stripping profile just below the feed stage. We find r_{min} = 5.0. As a first estimate, we let the operating reflux ratio be 50% larger than the minimum value, and obtain the design shown in Fig. 7. The column has 28 theoretical stages with a feed on stage 6 (numbering from the top of the column).

Students typically begin solving Part 2 by successively lowering the mole fraction of water in the distillate and repeating the design strategy evolved for Part 1. Whereupon they find that when the distillate contains less than 3400 ppm water, the distillate composition lies in the lower distillation region and the rectifying profile does not intersect the stripping profile at any reflux ratio! This is shown in Figures 8 and 9. At this point many students feel smug that their superior engineering know-how has proved that the client cannot get what she wants. But in fact, she can, and the question is, how?

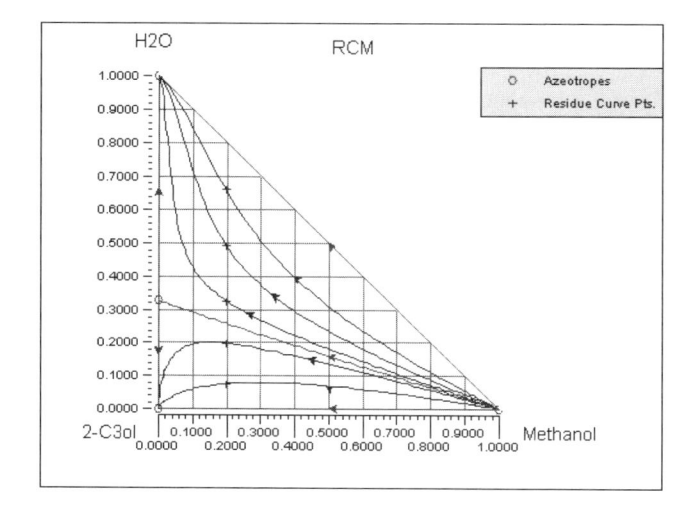

Figure 5. Residue curve map for the mixture methanol-water-isopropanol at 1 atm pressure.

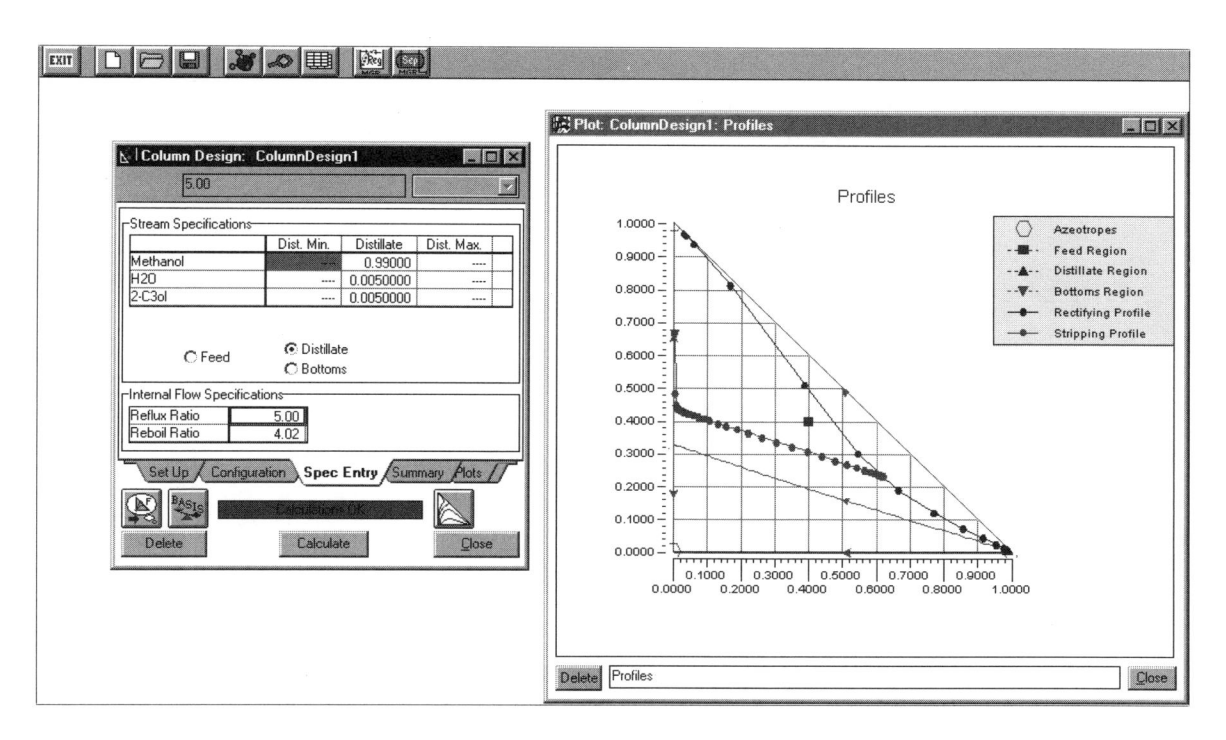

Figure 6. Minimum reflux profiles, r_{min}=5.0.

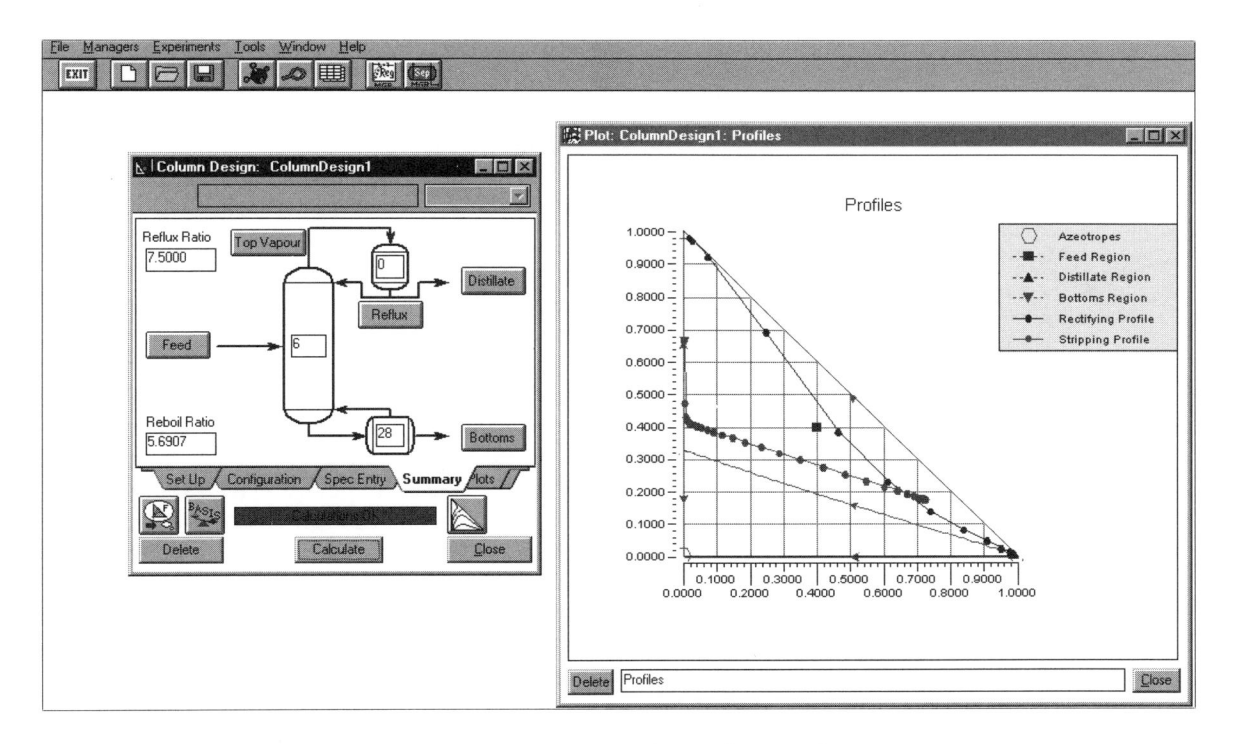

Figure 7. Base case design for Part 1.

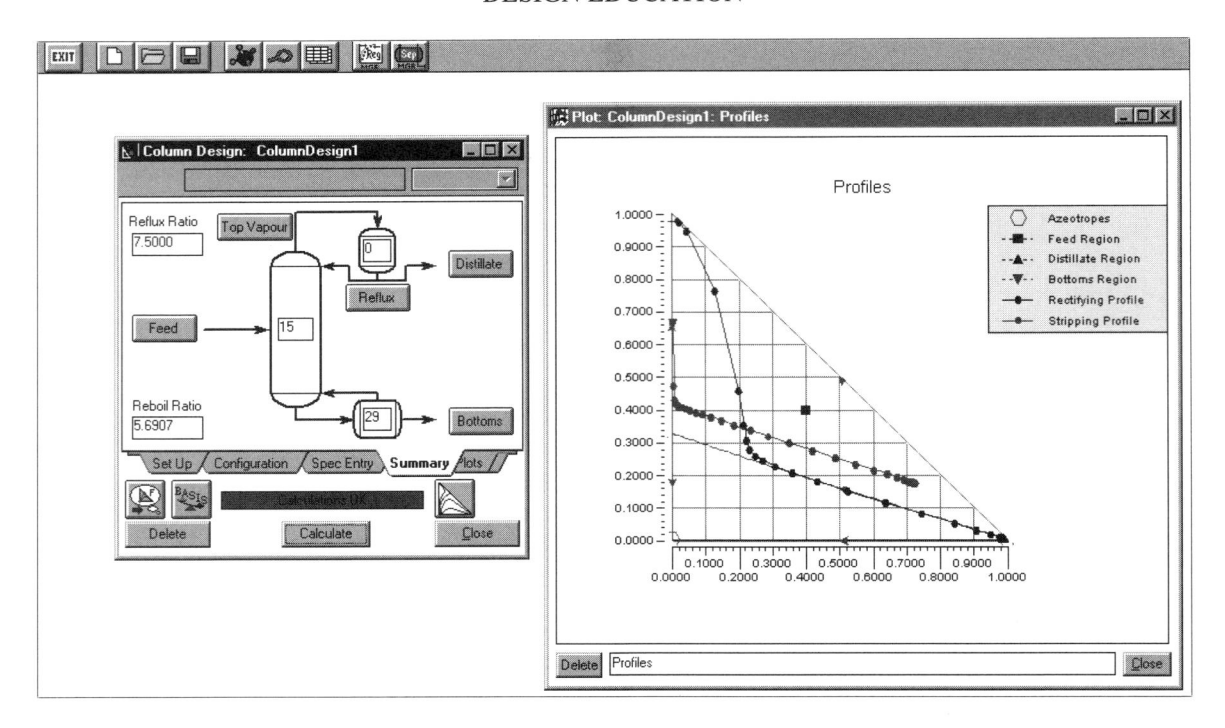

Figure 8. Minimum water composition in the distillate is 3400 ppm.

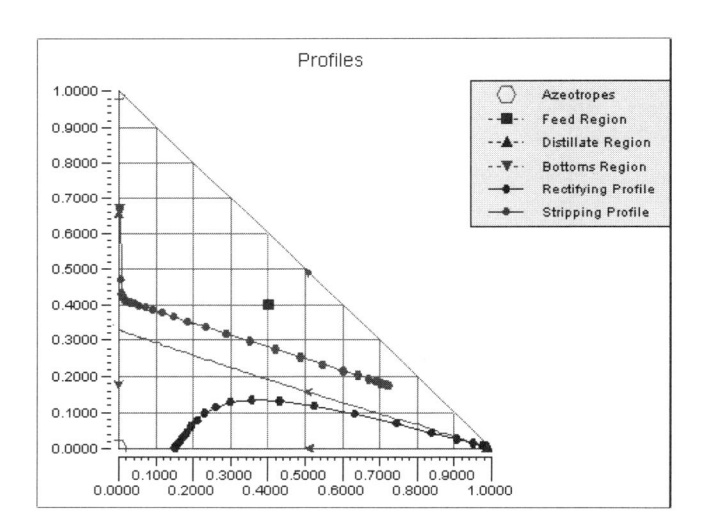

Figure 9. Water composition in the distillate is 3300 ppm, which is below the minimum value.

The solution was developed interactively with the audience at the conference. As was typical of our class experience, many alternative strategies were suggested at the conference within the exact constraints stated in the problem definition. Most alternatives involved adding more columns or other finishing separation steps.

However, the key to achieving our goal is to get the distillate below the distillation boundary into the 50 ppm water range. The best way to do this is to recognize that as the distillation boundary approaches the pure methanol vertex it becomes tangent to the base of the triangle, i.e., free of water. Therefore, we cannot get below the distillation boundary at 99 mol% methanol but we can drive the water content down by *increasing* the purity of the methanol product. The only remaining engineering issue is, how many extra stages are required to do this? Trying a methanol purity of 99.9 mol% does not do it, but 99.99 mol% does! The design is shown in Figure 10 where a 33 stage column operating at the same reflux ratio as the previous (unsuccessful) design achieves a methanol purity of 99.99 mol% and a water content of 50 ppm. Perhaps we can even charge a premium price for our methanol because it is of such high quality. The important lesson to be learned here is that we have achieved our manufacturing goal by relaxing some of the (softer) constraints in order to meet the hard constraint. That is, we have translated the most important part of the customer's order into a *redefinition* of the problem statement. Breaking out of the box is one of the most important aspects of engineering decision-making.

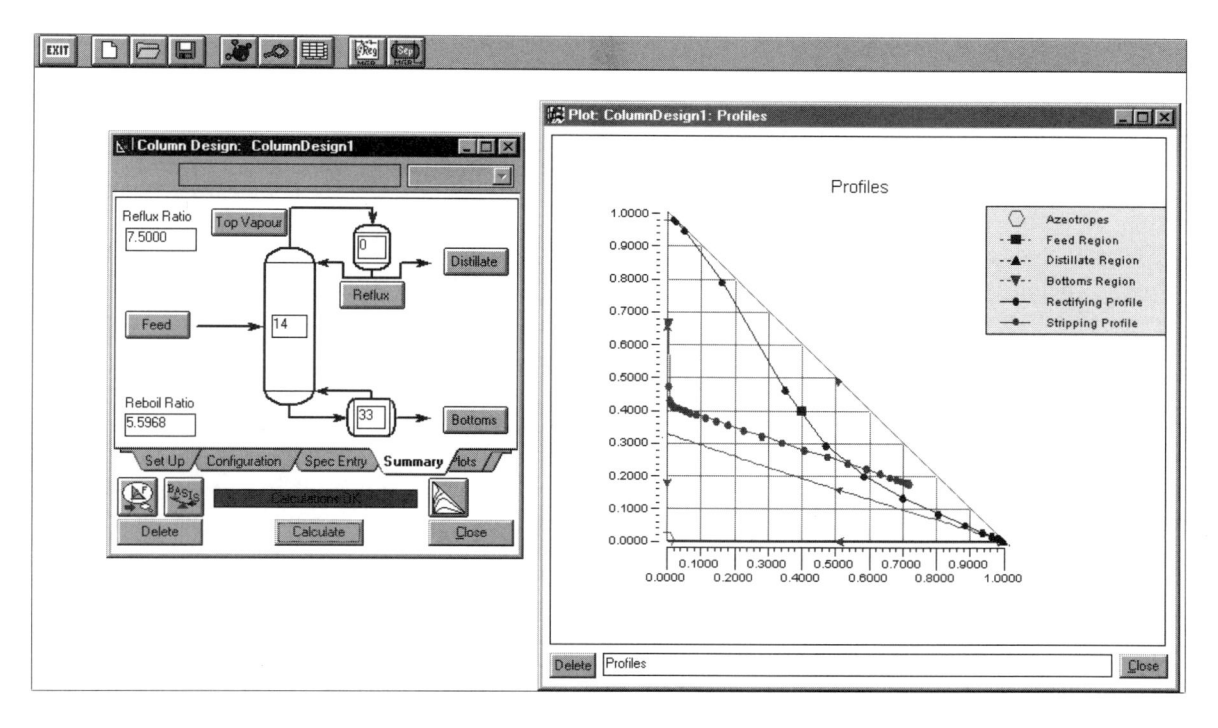

Figure 10. Column design for distillate stream containing 99.99 mole % methanol, 50 ppm water.

Example 2: Senior Design-Business Challenge Problem

As a two-week module in the Senior Design course, we worked together to provide a realistic business challenge problem, based on GE Plastics experience. We have now used this module two years in a row, with improvements implemented in the second year based on the problems encountered the first year. The students are given the task of taking a plant that is currently losing twenty million dollars a year and turn it into a business making at least five million dollars a year. The flowsheet for this plant has several "opportunities for improvement."

We start the module off with a lecture from our industry partner on business - wide economics. One of the key concepts is telling the students to change the traditional view of profitability from

$$\text{Profit} = \text{Revenue} - \text{Cost}$$

to

$$\text{Cost} = \text{Revenue} - \text{Profit}$$

That is, revenue and profit targets are pre-determined by external forces, i.e., Wall Street, corporate executives or, in this case by the instructors. The students can only manage the cost side of the equation through (1) identifying opportunities to reduce product costs through process analysis (2) funding business activities, such as marketing, to increase demand (3) funding R&D to develop lower cost manufacturing processes or new products with higher average selling prices and/or demand and (4) purchasing additional raw materials to enable the sale of more products. We also describe the basics of an income statement, including revenue, variable costs, base costs, and taxes. We demonstrate the classic basic break-even analysis, shown in Figure 11. If the slope of the total revenue curve is greater than the slope of the variable cost curve then there is some break-even point where the process becomes profitable with increased volume.

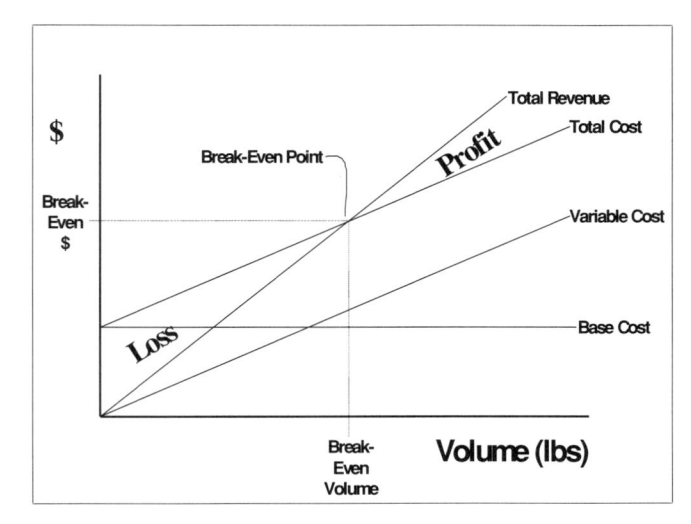

Figure 11. Break-even analysis.

For a problem statement, they receive the following

1. Background information with references to potential investment choices. This contains

many details on human interaction, and only careful reading of this will give them instructions on what investments are possible and what investments are wise.
2. A flowsheet of the plant, with incomplete flow and energy information, but enough to allow them to do some mass and energy balances.
3. Last year's budget and income statement.
4. Process chemistry and rate model (as known).
5. An Excel worksheet that simulates the experimental reactor with built-in random error.
6. A budget form with several defined areas for investment, and space for other programs

The assignment is to return a budget statement for the first cycle (1-year) with descriptions of what they expect to achieve with their investments and any instructions they have for modifying the flowsheet. The budget form contained the following investment areas:

1. New business development - look for new business opportunities for the product.
2. Marketing /Advertising - attract new business, maintain, possibly increase price.
3. Manufacturing
   - Process Control and Optimization: Investment above a critical value gives a significant reduction in the noise level of measurements around the reactor. This enables a significantly higher throughput.
   - Base Production Improvements - Reduce general operating costs or increase operating factor
4. R&D Budget
   - New Catalyst Development
   - Catalyst & Reaction Analysis: investment of $1 million in a kinetic model. This enables an accurate calculation of the amount of waste byproduct so that it can be reduced. The reduction in waste costs and ingredient savings exceeds the $1 million investment. Lower levels of investment do not provide the detailed model, but advice that this is a beneficial direction for future investment. However, for this year, this is just a cost with no benefit.
   - New Process Design: specify reactor, process chemistry, and separation systems. One major factor here is in design changes to recycle unreacted ingredients at the optimal level. This generally brings cost benefits in reduced ingredients use well in excess of the cost of implementing the recycle. It is surprisingly difficult for students to identify these seemingly obvious recycle opportunities!
   - Other R&D development - if they put money here without explanation, they just lose it!
5. Other Program Items - for projects they should learn about by careful reading of the problem statement.

This year we added the requirement that they give us the flowrates of each component in the reactor feed to get the production rate they expect. We added this requirement because the first year we taught this module students paid little attention to the flowsheet, randomly guessing on where to invest money.

We used a Microsoft Access database to apply rules for investments and send the information to an Excel spreadsheet with a complete model of the process to determine the income for that cycle. The rules included hidden investments like paying for an environmental evaluation of the plant, which is mentioned in the background information, but not explicitly listed on the budget sheet. If they invested at least $0.5 million in an environmental analysis, they avoided a $5 million fine in the second cycle. This grading could be done in class so the students could prepare a new budget by the next class. We had two breakout rooms with two teaching assistants, who functioned as "corporate experts," in each room to go over the budgets and answer some questions. Then the processed budgets were entered into the computer and the income statement was computed using the database. We had discussions at the beginning of each class talking about the type of decisions made and answering general questions.

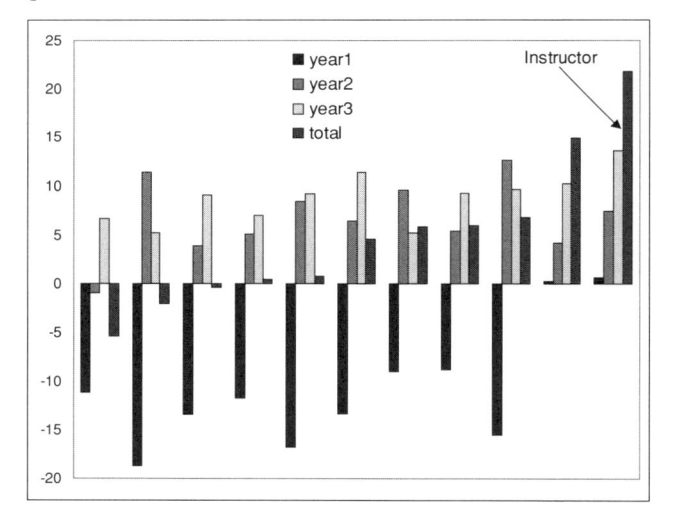

Figure 12. Economic performance for business challenge problem over three yearly cycles.

Figure 12 shows the income for each group over a three-cycle period and the total income for three cycles. The groups are ordered from left to right by the total income, with results from our own budgets getting the highest ranking. The instructor budgets made all good choices, but not the best possible. Also, it was possible to make more money in a particular year, but that does not necessarily lead to the best long-term results.

The best performing student groups combined elements of chance and common sense (just like the "real world") with engineering judgement. For example, if a group invested sufficiently (perhaps even on the first cycle) to get an optimization and control system on their reactor flowrates, they would produce as closely as possible to the demand. If they realized that some of the raw materials were not being recycled (the smart groups made this observation the first cycle), they would significantly reduce their materials costs.

Most groups lost money the first year because they did not meet demand, because they could not meet it with the current plant, or because they underspecified the flowrates to the reactor. Most groups used the initial values in the Excel spreadsheet we gave them to simulate the test reactor, which was no where near the capacity of the real process reactor.

The poorest performing groups did not carefully read the problem statement and failed to invest in important areas. Another common problem was for a group to continue investing heavily in an area that has already given a great benefit, like new catalyst development, or process control and optimization.

After our first year of teaching this module, we learned that we needed somehow to remind the students to think about the process. The requirement for them to give us the flowrates to the reactor did not help them make more money, since most groups specified flowrates well below the capacity of the reactor, but it did increase the number of groups implementing recycles. After the second year of teaching this module, we have decided to provide the problem statement as hypertext rather than a printed sheet, so we can provide them with extra information. This way we can explicitly tell them what all the possible investments are without making these choices obvious.

Evaluation

A detailed evaluation of the program impact for the 1997-1998 academic year was done by a third-party sub-contract.[2] The primary evaluation goals were to provide formative feedback to the faculty in support of ongoing program improvement, and to assess the extent to which program goals are attained.

---

[2] The Maurice A. Donahue Institute for Governmental Services is the public service and outreach unit of the University of Massachusetts President's Office, offering services involving economic and organizational development.

There are three primary evaluation questions:

To what extent do faculty integrate the technology in the classroom and recent research within the framework of their curricula and teaching styles?

What is the affective impact on students of classroom use of the technology? (e.g., attitudes re: the course, subject matter, and discipline)

What is the cognitive impact on students? (e.g., mastery of subject matter, problem-solving performance, course grades)

To address these questions, the evaluation used both quantitative and qualitative data sources, using a variety of data collection methods for each area. These include classroom observation, utilization logs (instructor and system), faculty interviews, student focus groups, end-of-semester course evaluations, focused student surveys, and course grades.

During the first year, the amount of class time devoted to student use of workstations differed considerably among courses. In courses with high utilization, students worked at stations approx. 20-25% of the time. In other courses, workstations were only used approximately 5% of the time. The major factors affecting first year utilization include the availability of relevant and reliable software, faculty familiarity with tools, and the time required adapting curricula and teaching styles.

Concerning student attitudes, we found that standard course evaluations, which would enable longitudinal tracking of courses, are unreliable measures of change. However, student interviews and focused surveys indicate positive perceptions of the classroom as a teaching tool, cooperative work with classmates and the benefits of work with software at workstations. Students consistently report a positive impact on their interest in the subject matter and in chemical engineering.

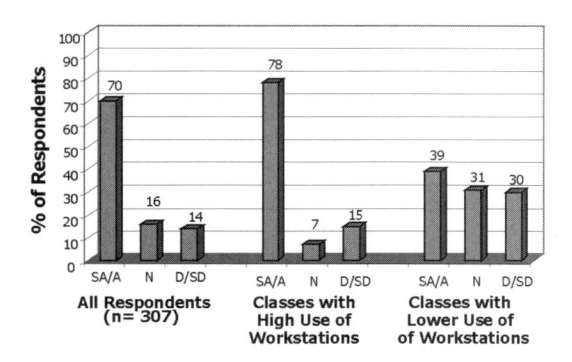

Figure 13. Student responses to "Use of the technology helped me learn the material more than I would have otherwise." The scale: Strongly Disagree, Disagree, Neutral, Agree, Strongly Agree; also applies to subsequent figures.

Specifically, 307 student responses were collected which could agree or disagree with the statement "The Alumni Classroom's technical capabilities helped me to learn more than I would have otherwise." The scale used was Strongly Disagree, Disagree, Neutral, Agree, Strongly Agree with the results shown in Figure 13. The responses of the same group to a similar question statement concerning the quality of learning were essentially identical.

Perhaps the most striking results from student surveys were in the reported effects of the approach on motivation. Student responses concerning motivation through use of technology is shown in Figure 14.

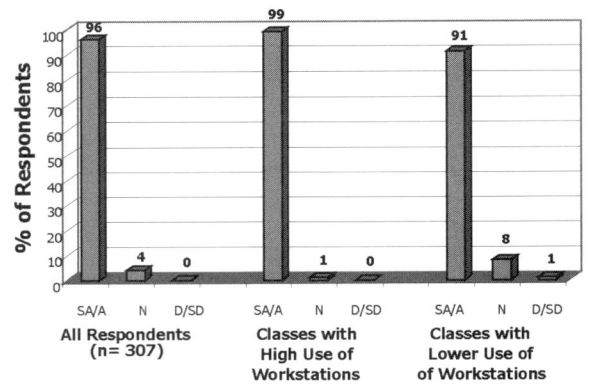

Figure 14. Student responses to "The Alumni Classroom's technical capabilities motivated me to learn better than in other classes"

Significant changes are found between the first and subsequent offerings of a course using this approach. In fact, the differences correlate more strongly with the experience of the instructor than with the particular course.

The more interesting results concern the percentage of time spent using the technology (Fig. 15) and the correlation with student evaluations of amount of learning (Fig. 16), motivation (Fig. 17) and impact on learning how to solve open ended problems (Fig. 18).

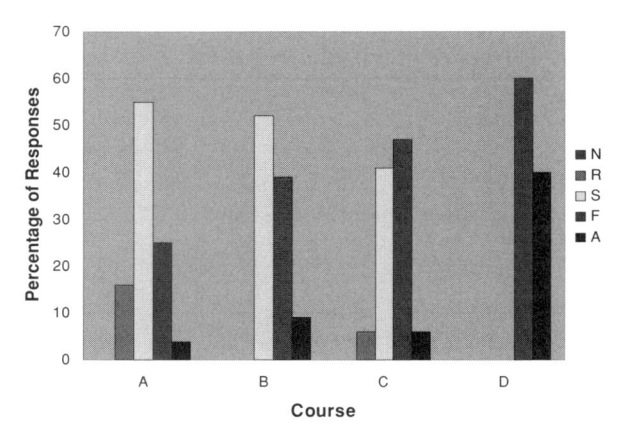

*Figure 15. Student Response to "How often did you work on a computer during class time?" Scale: **N**ever, **R**arely, **S**ometimes, **F**requently, **A**lways.*

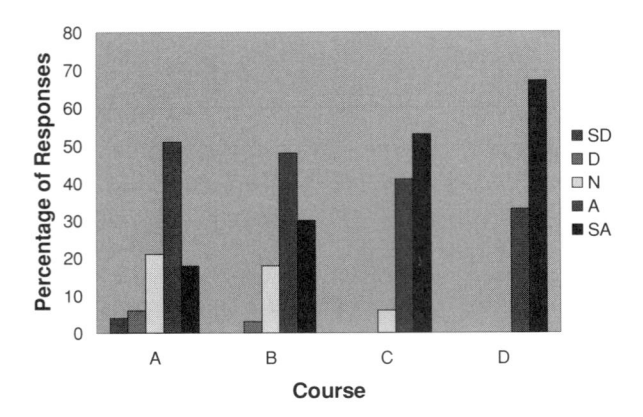

*Figure 16. Student response to "Use of the classroom's technical capabilities helped me learn the material in this course better than I would have otherwise." Scale: **S**trongly **D**isagree, **D**isagree, **N**eutral, **A**gree, **S**trongly **A**gree; also applies to subsequent figures.*

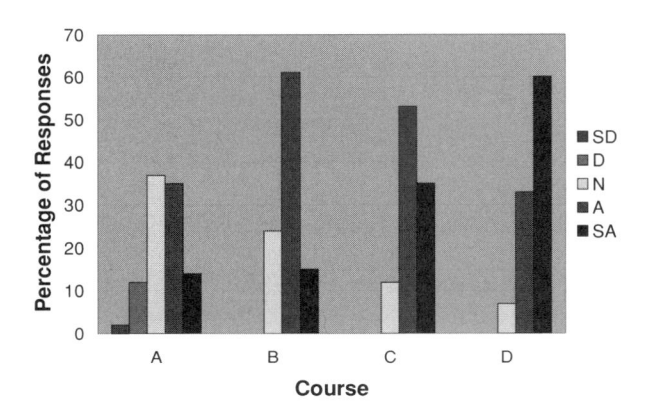

Figure 17. Student response to "Because of the classroom's technical capabilities, I was more motivated to learn in this class than in others."

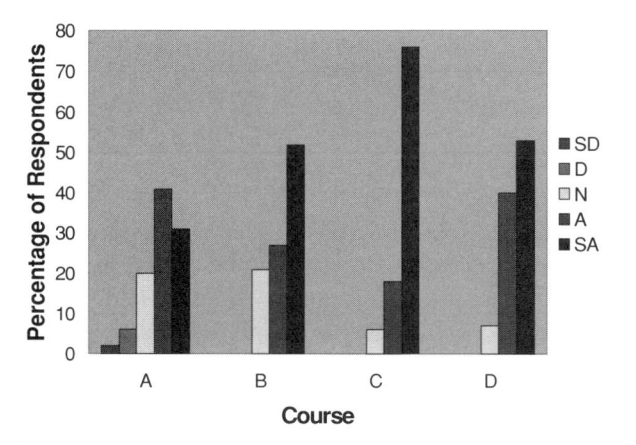

Figure 18. Response to "Use of the facilities contributes more to learning open ended problem solving skills than a traditional classroom."

Summary and Lessons Learned

We have found several important factors for implementing an interactive curriculum and technology.

Essential items:

The faculty must have the enthusiasm and release time to change from the traditional lecture format. They must be willing to adapt during class to student discoveries and interests.

1. Software tools must be familiar to the faculty and teaching staff, *before* the class starts.
2. Examples must be developed and tested in the working environment before classes.
3. A minimum amount of professional staff time is needed to maintain the computer systems, and for faculty/TA training and support.

Strong Preference:

1. Commercial software unless critical functionality is not available.
2. Team-teaching is advantageous, especially if one of the faculty is experienced with the technology and the other(s) not.
3. The first offering of a course in the new format should be taught by faculty with research experience in the subject and with experience in the software tools used.
4. Industrial involvement is important to let the students know they are learning about real world problems, and to demonstrate the kinds of interactions which occur in industry.

Needs

1. Component software and standards to develop, test and distribute tools with new functionality. Preferably Web-based.
2. Easy to use cost models
3. Faster transport models, e.g., for heat transfer and CFD

Acknowledgements

Financial support from the GE Fund and from alumni for construction of the classroom facility is gratefully acknowledged. We are also grateful to P. R. Westmoreland, S. C. Roberts, Z. Q. Zheng, M. Sutherland, K. M. Ng, R. L. Laurence, A Nagurney, and all of our students for their work in the classroom. E. Heller, H. Gibson and S. Liebowitz from the Donahue Institute collected and analyzed data for the evaluation. We are also grateful to the members of our Industrial Advisory Board from GE Plastics, BASF, Dow Corning Corp, DuPont Company, Eastman Chemical Company, Mitsubishi Chemical, Rohm & Haas, Shell International Chemical, Union Carbide Corporation, Unilever Research, Searle, Hyprotech Ltd., and UOP for their extensive comments and suggestions on the content and approach. We are also grateful to AEA Software Technology for HYSYS and HYCON licenses.

References

Biegler, L. T. and M. F. Doherty (Eds.) (1995). *Foundations of Computer-Aided Process Design*, Proceedings of the Fourth International Conference on Computer Aided Process Design, AIChE Symposium Series, 91

Biegler, L. T, I. E. Grossmann and A. W. Westerberg (1997). *Systematic Methods of Chemical Process Design*. Prentice-Hall, NJ.

Braunschweig, B. L., C. C. Pantelides, H. Britt and S. Sama (1999). Open software architectures for process modeling: current status and future perspective. These Proceedings.

Douglas, J. M. (1988). *Conceptual Design of Chemical Processes,* McGraw Hill, NY.

Mah, R. S. H and W. D. Seider (Eds.) (1981). *Foundations of Computer Aided Process Design*, Proceedings of the First International Conference on Computer Aided Process Design. Engineering Foundation, NY.

Seider, W. D., J. D. Seader and D. R. Lewin (1999). *Process Design Principles.* Wiley, NY.

Siirola, J. J., I. E. Grossmann and G. Stephanopoulos (Eds.) (1990*). Foundations of Computer-Aided Process Design.* Proceedings of the Third International Conference on Computer Aided Process Design. Elsevier, NY.

Westerberg, A. W. and H. H. Chien (Eds.) (1984). *Proceedings of the Second International Conference on Foundation of Computer-Aided Process Design.* Elsevier, NY.

REACTOR/TRANSPORT MODELS FOR DESIGN: HOW TO TEACH STUDENTS AND PRACTITIONERS TO USE THE COMPUTER WISELY

Bruce A. Finlayson
University of Washington
Seattle, WA 98195-1750

Brigette M. Rosendall
Bechtel Technology and Consulting
San Francisco, CA 94119-3965

Abstract

As computer-aided design becomes more prevalent in the process industry, it is essential that graduating engineers have the breadth to know the capabilities of CAD as well as the skepticism to interpret computer results wisely. Current day students are caught between an educational system that focuses on analytical solutions to simplified problems and CAD programs that provide generality and power that can be overwhelming. Students are only too willing to accept computer results at face value, which could be disastrous in a commercial environment. An important aspect of current day education should be to instill in students and practitioners a critical examination of results from their CAD program. This paper presents experience with CAD tools in an educational environment as well as an industrial environment.

Keywords

Transport, Chemical reactor, Bifurcations, MATLAB, FIDAP, CRDT, Industrial radiant furnace.

Introduction

The question is broader than educational, however. There is a growing body of applied mathematicians, too, who are trying to quantify or analyze the risk of using models (Wheeler, 1999). They are mainly looking at the risk of models (like global warming) where there may be no way to test them, but the models are used for public policy decisions. There are several questions that must be asked of models:

1. What is the risk of leaving out some phenomena that need to be included? Chemical engineers minimize this by comparing their predictions to experimental results.

2. Supposing the model includes the appropriate phenomena, there are four basic questions to answer:

   a. Is the discretization (or method) correct, so that solutions converge to the right answer? There can be false bifurcations and the shocks can be in the wrong place, as shown below.

   b. How sensitive are the results to the data used? (i.e., do a sensitivity analysis)

   c. How accurate are the results? (Error assessment)

   d. Did you make any mistakes? This is a hard one to get students to worry about; some of them have a tendency to throw things together, say they are done, and move on to something else.

This paper gives examples of how these questions are answered primarily for applications to reactor and transport models. Examples will be taken from problems solved mostly by undergraduates using (i) MATLAB or

Excel, as the 'numerical analysis engine'; (ii) a special reactor program, the Chemical Reactor Design Tool, CRDT; (iii) application of CRDT to an industrial design problem of a radiant furnace; and (iv) computational fluid dynamics (CFD). In order to keep the paper within reasonable bounds, all questions will not be answered for each example.

Undergraduate Transport Problems Solved with EXCEL.

The first example is for heat transfer in a cylinder; the thermal conductivity depends on temperature, and there is internal heat generation. The problem statement is:

$$\frac{1}{r}\frac{d}{dr}\left[rk(T)\frac{dT}{dr}\right] = -2G(1-r^2)$$

$$k = k_0 + a(T - T_0)$$

$$\left.\frac{dT}{dr}\right|_{r=R} = 0, \quad T = T_0 \text{ at } r = R$$

We pose two questions: what is the heat flux at the boundary of the cylinder? What is the peak temperature? The first question is answered with a little thought: at steady state all the energy generated inside has to be coming out. In this case, the heat generation rate varies with position, so a simple integral must be evaluated. These questions illustrate an important lesson: think first! The second question can't be answered as easily. We solve it here with the finite difference method, but we do the programming in steps to make it easier to check our results.

The non-dimensional version of the problem is

$$\frac{1}{r'}\frac{d}{dr'}\left[r'(1+a'\theta)\frac{d\theta}{dr'}\right] = -2G'(1-r'^2)$$

$$\left.\frac{d\theta}{dr'}\right|_{r'=0} = 0, \quad \theta(1) = 1$$

Next the primes are dropped for convenience. There are various ways this equation can be solved; here we differentiate it and collect terms.

$$(1+a\theta)\left[\frac{d^2\theta}{dr^2} + \frac{1}{r}\frac{d\theta}{dr}\right] + a\left[\frac{d\theta}{dr}\right]^2 = -2G(1-r^2) \quad (1)$$

A special equation is needed at the origin, since the radial position goes to zero there. We use l'Hospital's Rule

$$\lim_{r \to 0} \frac{1}{r}\frac{d\theta}{dr} = \frac{d^2\theta}{dr^2}(0)$$

and get a different equation for the first node.

$$\left.(1+a\theta)2\frac{d^2\theta}{dr^2}\right|_{r=0} = -2G$$

If the finite difference method is applied to Eqn. (1) we get

$$(1+a\theta_i)\left[\frac{\theta_{i+1} - 2\theta_i + \theta_{i-1}}{\Delta r^2} + \frac{1}{r_i}\frac{\theta_{i+1} - \theta_{i-1}}{2\Delta r}\right] + $$

$$a\left[\frac{\theta_{i+1} - \theta_{i-1}}{2\Delta r}\right]^2 = -2G(1-r_i^2), \quad r_i = (i-1)\Delta r \quad (2)$$

Eqn. (2) (plus appropriate equations at the first and last node) provide a set of nonlinear equations to be solved. These are solved using the iteration feature in Excel spreadsheets. The equation is rearranged into a formula for the i-th node

$$\theta_i = \frac{1}{2}\left\{\theta_{i+1} + \theta_{i-1} + \frac{\Delta r}{2r_i}(\theta_{i+1} - \theta_{i-1}) + \right.$$

$$\left. + [2G(1-r_i^2)\Delta r^2 + 0.25 a (\theta_{i+1} - \theta_{i-1})^2]/(1+a\theta_i)\right\}$$

and this equation is placed into one cell.

B1:= 0.5*{A1 + C1 + 0.5*B4*(C1 – A1)/B2 +

+ [2.*B5*B8 + 0.25*B7*(C1 – A1)^2]}/B3

where B2 is r_i (calculated for each node), B3 is $1 + a\theta_i$,

B4 is Δr, B5 is Δr^2, B6 is G, B7 is a, and B8 is $2*G*(1-r^2)$.

B1 is θ_i, A1 is θ_{i-1}, C1 is θ_{i+1}

The equation is copied into surrounding cells, and the iteration capability is turned on. After a few seconds the answer for θ at each node is shown in each cell. If a busy undergraduate with several classes, a part-time job, and a social life did that, would you accept the results as likely correct? I wouldn't, so let's see how to approach the problem in a way to improve the chances that the answers will be right. Solve a simpler, but related, problem first. Solve the problem with a = 0, G = 0, first. We know the answer is θ(r) = 1, and we make sure the program gives that answer. Then include the constant G, i.e. with the right-hand side = G. Now the exact answer is a quadratic function of r, and you can derive it and see that the computer gives the same result. (For this case, since the solution is a quadratic function of position, and the finite difference method is second order, the exact answer is obtained at the nodes; that won't be true in general; see the information on error assessment below.) Next put in the right-hand side = – 2 G (1 – r²). Put this in one cell, and put specific values in the cells on either side. Do hand calculations to see that the value of θ_i is correct. Putting $\theta_{i+1} = \theta_{i-1} = 0.5$ makes the result

$$\theta_i = \frac{1}{2} \left\{ 1 + 2 G (1 - r_i^2) \Delta r^2 \right\}$$

and it should give that answer to many (say 9) significant digits. However, even this isn't a good enough test. Since the values of the two adjacent nodes were the same, this check would not catch an error in which the formula used θ_{i+1} in place of θ_{i-1}. Thus, different values need to be used. In this step, one has to be sure that all terms are big enough to affect the result. Adding something that takes the value zero doesn't really check the formula, for example. Using $\Delta r = 10^{-6}$, $G = 1$ doesn't test the right-hand side.

Next, add in the variable thermal conductivity. Do this in one cell, put values in adjacent cells, and do the hand calculation to check. Once that check is satisfied, copy the formulas to all the cells, turn on the iteration feature, and wait for convergence. You can make one final check, using the results and putting them into the equation for one node, but if there is an error it will be hard to find at this stage; it is much better to have made all the prior steps correctly.

These are the steps you use to prevent mistakes in your work, which is one source of error. However, finite difference equations are approximate; the discretization step will introduce some error. How does one test that?

First test: do the truncation error analysis. Substitute the Taylor series

$$\theta_{i+1} = \theta_i + \left.\frac{d\theta}{dr}\right|_i \Delta r + \left.\frac{d^2\theta}{dr^2}\right|_i \frac{\Delta r^2}{2!} + \left.\frac{d^3\theta}{dr^3}\right|_i \frac{\Delta r^3}{3!} + \left.\frac{d^4\theta}{dr^4}\right|_i \frac{\Delta r^4}{4!} + ..$$

$$\theta_{i-1} = \theta_i - \left.\frac{d\theta}{dr}\right|_i \Delta r + \left.\frac{d^2\theta}{dr^2}\right|_i \frac{\Delta r^2}{2!} - \left.\frac{d^3\theta}{dr^3}\right|_i \frac{\Delta r^3}{3!} + \left.\frac{d^4\theta}{dr^4}\right|_i \frac{\Delta r^4}{4!} + ..$$

into the finite difference form of the equation. What you find is

$$2\theta_i = 2\theta_i + \left.\frac{d^2\theta}{dr^2}\right|_i + O(\Delta r^2) + \frac{1}{r} \left.\frac{d\theta}{dr}\right|_i +$$

$$\frac{1}{(1+a\theta)} 2G(1-r^2) + \frac{1}{(1+a\theta)} a \left(\left.\frac{d\theta}{dr}\right|_i\right)^2 + O(\Delta r^2)$$

Thus, the error term is proportional to Δr^2. As $\Delta r \to 0$, the equation we solve is closer and closer to the exact equation. Second, is our application of the method correct? We think it is – due to the checks above, but if it is, we expect errors to go as Δr^2, too (Keller, 1972).

Second test: We know the exact heat flux, so we could check that and see that our values approach the exact value with an error Δr^2. Consider first the case with a = 0 (i.e. making the problem linear). We know the exact solution for the flux, and in this case we can derive an exact solution for the temperature at the center. The flux is calculated with a one-sided formula that is second order in Δr (Finlayson, 1980, p. 68). Table 1 shows the results, and the errors clearly are proportional to Δr^2.

Table 1. Solution for Center Temperature and Heat Flux, Compared with the Analytical Solution; Generation Rate = 0.5 G (1 - r²), G = 0.5, a = 0 (linear problem).

Δr	T(0)	Error in T(0)	q(1)	Error in q(1)
0.2	1.191161	0.00366	0.274476	0.0195
0.1	1.188431	0.00093	0.256498	0.0065
0.05	1.187734	0.00023	0.251672	0.0017
Exact	1.187500		0.250000	
Extrapol.	1.187502		0.250063	

Third test: We next solve the problem with a = 1 and G = 0.5. Now we have a problem for which we don't know the exact solution (at least we've not tried to find it). The problem is nonlinear with variable heat generation. However, we can still do the same tests as before. The truncation error analysis was done above. Next we solve the problem on one grid and do a hand calculation to insure that the correct equation is being solved. Then, we solve the problem on finer grids and make sure the results are proportional to Δr^2. Table 2 and Fig. 1 shows this clearly for both center temperature and heat flux at the boundary. Extrapolation of the last two points using a straight line

☐

gives the value _1= 1.091651. We use this as the answer and calculate the approximate error shown in Table 2. Clearly the error decreases by about a factor of 4 as the Δr is halved, as required. The extrapolated heat flux is 0.250078, which is about 0.03% in error from the value of 0.25.

Now we have the answer. We claim the temperature is correct to ± 0.0001 (0.01%), and the heat flux is accurate within 0.0016 (0.64%). What does our claim depend upon?

(1) The finite difference equation is correct – the discretization error analysis showed that.
(2) The equations have been put into Excel correctly, based on a careful organization of our work and comparison with calculations.
(3) The discretization error has been estimated, and we can assign a number to it.

(4) In the case of heat flux, we can compare the result to an overall energy balance, which gives us the exact solution.

That is a good analysis, and it applies to problems without an analytical or 'exact' solutions, and to non-linear problems. For comparison, the extrapolated values are also given in Table 1, for the case when we do know the exact solution. The extrapolated values are very close to the exact answer.

Table 2. Solution for Center Temperature and Heat Flux; Generation Rate = 0.5 G (1 - r2), G = 0.5, a = 1 (nonlinear problem).

Δr	T(0)	Approx. Error in T(0)	q(1)	Approx. Error in q(1)
0.2	1.093282	0.00160	0.274123	0.0240
0.1	1.092066	0.00042	0.256499	0.0064
0.05	1.091754	0.00010	0.251683	0.00016
Extrapol.	1.091651		0.250078	

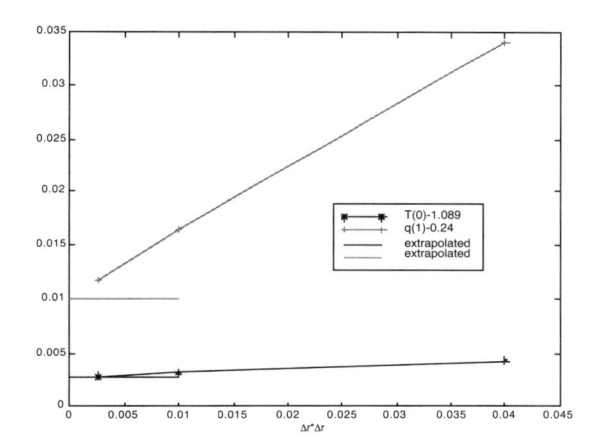

Figure 1. Center temperature and boundary heat flux as a function of mesh size; Generation rate = 0.5 G (1 - r2), G = 0.5, a = 1 (nonlinear problem).

Undergraduate Transport Problems Solved with MATLAB.

The next problem illustrates how one solves for sensitivity of the model to input data. Consider the two-dimensional heat transfer problem shown below. This is solved with the PDE toolbox in MATLAB (PDE Toolbox, 1997) which uses the finite element method. One must set the domain, identify boundary conditions, decide what terms need to be in the equation, let the computer generate a mesh, and solve the problem. The first mesh tried gave the following results.

$$\Delta^2 T = \frac{\partial^2 T}{\partial x^2} + \frac{\partial^2 T}{\partial y^2} = 0$$

T=1 on x=0, for all y; y=0, for all x

T=0 on y=1 for all x

$\frac{\partial T}{\partial n} = 0$ on x=1, for all y

T=0.5 in the center block

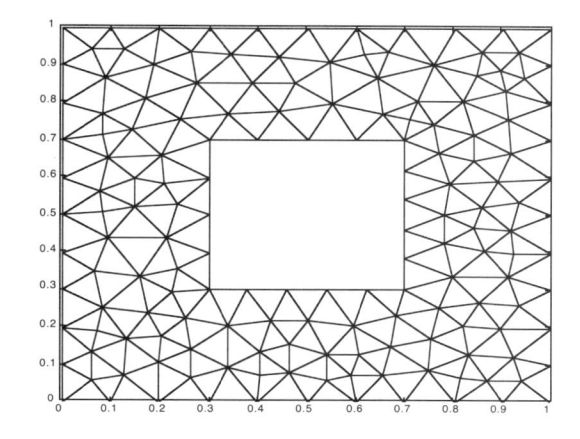

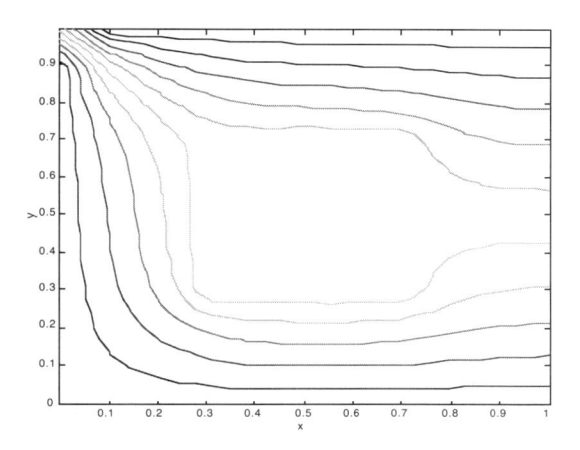

Figure 2. Heat transfer problem solved with finite element method; (a) Mesh, (b) Temperature contours.

The mesh needs to be refined (this is done by the click of a button in the PDE toolbox), and the new results are given below. These two cases indicate that the discretization errors are small enough and the finite element results are reliable. We could make this more precise, if desired, by looking at the solution at specific points obtained for the two solutions. It is harder to show the rate of convergence with mesh size, though, when using an automatic mesh generator. To check whether you have solved the right equation, the only real check you have is to look at the graphical user interface (gui) and see what are the coefficient values. In this case, one assumes that the program solves the equation it says it does, and even that can be checked by solving problems that others have solved reliably.

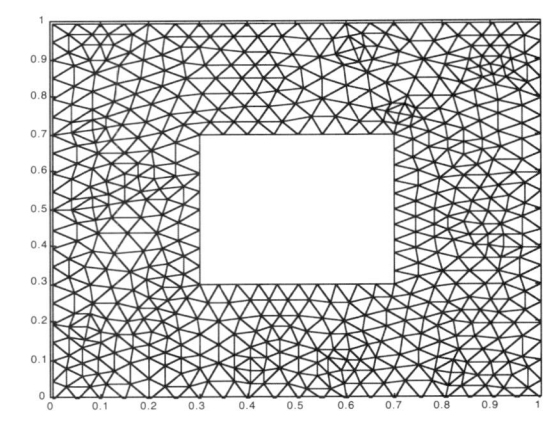

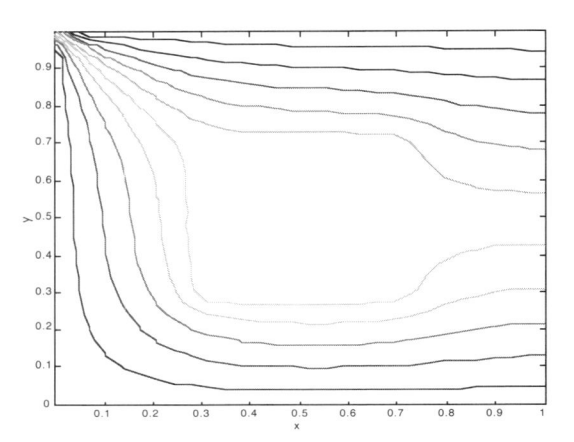

Figure 3. Heat transfer problem solved with finite element method; refined mesh; (a) Mesh, (b) Temperature contours.

Here, though, look at the sensitivity to the temperature in the center block, using a sensitivity equation since that approach can also be used for parameter estimation. Differentiate the equation and all boundary conditions with respect to T_0.

$$\frac{\partial}{\partial T_0}[\nabla^2 T = 0] \Rightarrow \nabla^2 \frac{\partial T}{\partial T_0} = 0$$

$$\frac{\partial}{\partial T_0}[T = 1] \Rightarrow \frac{\partial T}{\partial T_0} = 0 \text{ on } x = 0, \text{ for all } y$$

$$\text{and } y = 0, \text{ for all } x;$$

$$\frac{\partial}{\partial T_0}[T = 0] \Rightarrow \frac{\partial T}{\partial T_0} = 0 \text{ on } y = 1 \text{ for all } x$$

$$\frac{\partial}{\partial T_0}\left[\frac{\partial T}{\partial n} = 0\right] \Rightarrow \frac{\partial}{\partial n}\left[\frac{\partial T}{\partial T_0}\right] = 0 \text{ on } x = 1, \text{ for all } y$$

$$\frac{\partial}{\partial T_0}[T = 0.5] \Rightarrow \frac{\partial T}{\partial T_0} = 1 \text{ in the center block}$$

So, now solve for $z = \partial T / \partial T_{center}$.

$$\nabla^2 z = 0$$

$z = 0$ on $x = 0$, for all y; $y = 0$, for all x; $y = 1$ for all x

$$\frac{\partial z}{\partial n} = 0 \text{ on } x = 1, \text{ for all } y$$

$z = 1$ in the center block

The result is in Fig. 4.

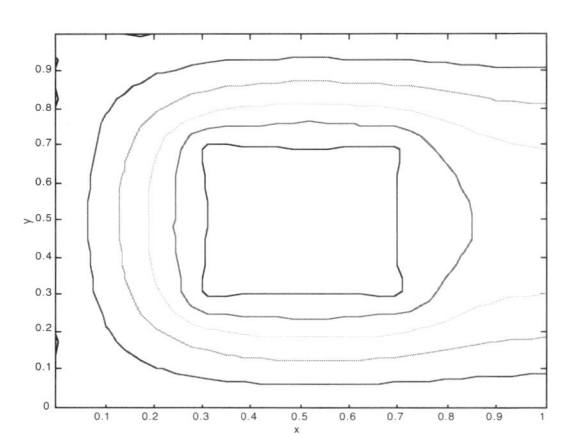

Figure 4. Contours of sensitivity to center-block temperature.

Now we know the temperature at every point, and how sensitive it is to the center temperature. The largest value of z along the line x = 1 is for y = 0.5. It is clear that the maximum sensitivity is at the middle of the right-hand side, so that is where we would place thermocouples. If we had heat transfer data we could determine the middle temperature by evaluating the model, comparing with data, forming an objective function (the absolute difference between calculational and experimental results), and use an optimization program to find the minimum. Since we have the derivative of the solution with respect to the optimization parameter, we can use an optimization routine that requires derivative evaluation. The same technique works with ordinary differential equations. So, now we have a technique for seeing how sensitive the result is to some part of our model.

False Bifurcations Obtained with the Method of Lines

The next example illustrates the importance of understanding the method one uses when discretizing in space. Many computer programs for modeling transient phenomena provide an ODE solver. The user then writes a routine that is essentially the method of lines. The example chosen here shows how that user-supplied routine can give results which look like bifurcations and chaos, but in reality are the result of not using a small enough time-step for the type of discretization used. This is particularly troublesome because the program can be absolutely correct, yet the results can look like nonsense. Thus, the testing procedure described above would not uncover an error, because there isn't one; yet, the results can look extremely weird.

Consider the diffusion equation with reaction, called the Fisher equation.

$$\frac{\partial u}{\partial t} = D \frac{\partial^2 u}{\partial x^2} + \alpha u(1-u)$$

If a finite difference method is applied to this equation using an explicit, first-order method, the result is

$$\frac{u_i^{n+1} - u_i^n}{\Delta t} = D \frac{u_{i-1}^n - 2 u_i^n + u_{i+1}^n}{\Delta x^2} + \alpha u_i^n (1 - u_i^n)$$

Following Mitchell and Bruch (1985) we transform the equation using

$$v_i^n = \frac{\alpha \Delta t}{1 + \alpha \Delta t} u_i^n$$

The result is

$$v_i^{n+1} = b(v_{i-1}^n - 2 v_i^n + v_{i+1}^n) + a u_i^n (1 - v_i^n)$$

where

$$a = 1 + \alpha \Delta t, \, b = D \Delta t / \Delta x^2$$

First solve this problem without diffusion, i.e., b = 0; the equation is then called the logistic equation (Mitchell and Brunch, 1985). It is solved for a variety of a or time steps, and the solution as t approaches infinity is plotted in Fig. 5. These solutions are obtained by starting from different initial conditions. Clearly for high values of a, or $\alpha\Delta t$, it is possible to get more than one solution. Next add diffusion; solutions for one value of a, and many values of b. The boundary conditions are no flux at x = 0, the value of v(1) = 0, and the initial condition is everywhere zero except at x = 0, where it is 0.6, 0.3, and 0.8 in successive runs to obtain all the solutions.

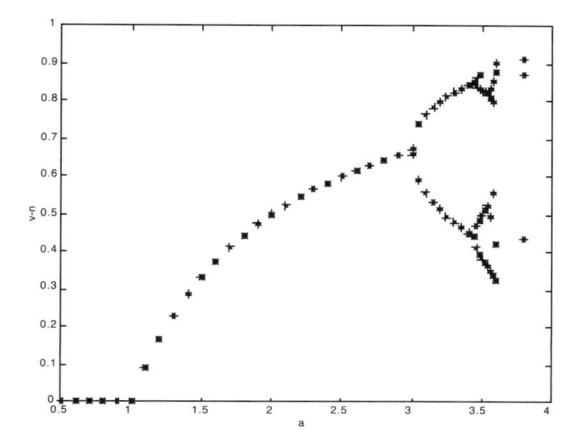

Figure 5. Solutions of logistic equation obtained from different initial conditions.

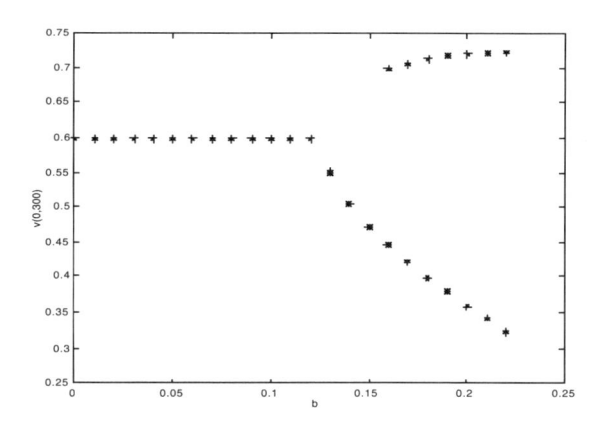

Figure 6. Solutions of the Fisher equation obtained from different initial conditions; finite difference method; a = 2.5, $\alpha \Delta t$ = 1.5, Δx = 0.5, $0 \le x \le 1$.

Results are shown in Fig. 6 for a = 2.50 (thus $\alpha\Delta t$ = 1.50), Δx = 0.5, and various b, i.e. various D. There is a clear bifurcation at b = 0.125. If one plotted the solution at infinite time versus both a and b, one can imagine Fig. 5 projecting out from the paper at the origin, b = 0, and a very complicated geometric pattern can be imagined. However, let us use a smaller Δt by a factor of 4, corresponding to a = 1.375, and a smaller Δx by half, corresponding to Δx = 0.25; the results are in Fig. 7. The value of $\Delta t/\Delta x^2$ is the same in the Figures 6 and 7, so that situations with the same b are for the same physical situation. Fig. 7 shows that the bifurcation has disappeared! Thus, it was a result of the discretization. If one reduces Δt without changing Δx (a = 1.375), then the bifurcations also disappear. Thus, the bifurcations in this case were a result of using the method of lines and a discretization (in Δx and Δt) that was too big.

Now Mickens (1989) argues that such discretizations as used here can lead to false bifurcations (as demonstrated above), and that a better form of the equation is to evaluate the reaction term at different times, or at different spatial positions, e.g.

$$\frac{u_i^{n+1} - u_i^n}{\Delta t} = D \frac{u_{i-1}^n - 2u_i^n + u_{i+1}^n}{\Delta x^2} + \alpha u_i^n (1 - u_{i+1}^{n+1})$$

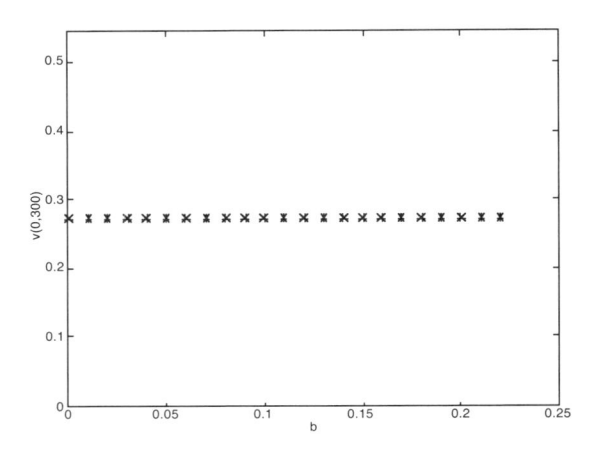

Figure 7. Solutions of the Fisher equation obtained from different initial conditions; Finite difference method; a = 1.375, $\alpha \Delta t$ = 0.375, Δx = 0.25, 0 ≤ x ≤ 1.

This is similar to what the Galerkin finite element method does. The Galerkin method gives (Finlayson, 1980)

$$\frac{1}{6} \frac{u_{i-1}^{n+1} - u_{i-1}^n}{\Delta t} + \frac{4}{6} \frac{u_i^{n+1} - u_i^n}{\Delta t} + \frac{1}{6} \frac{u_{i+1}^{n+1} - u_{i+1}^n}{\Delta t} =$$

$$D \frac{u_{i-1}^n - 2 u_i^n + u_{i+1}^n}{\Delta x^2} + \frac{\alpha}{6} (u_{i-1}^n + 4 u_i^n + u_{i+1}^n)$$

$$- \frac{\alpha}{12} [(u_{i-1}^n)^2 + 2 u_{i-1}^n u_i^n + 6 (u_i^n)^2 + 2 u_i^n u_{i+1}^n + (u_{i+1}^n)^2]$$

There is a variant of the Galerkin method, which makes the term multiplying the time derivative diagonal, too. It is called a lumped, Galerkin method. (The term lumped means that the left-hand side has been made diagonal by adding all coefficients and putting them on the diagonal). This form of the problem, plus the transformation to v, gives

$$v_i^{n+1} = b (v_{i-1}^n - 2 v_i^n + v_{i+1}^n) + \frac{a}{6} (v_{i-1}^n + 4 v_i^n + v_{i+1}^n) +$$

$$- \frac{a}{12} [(v_{i-1}^n)^2 + 2 v_{i-1}^n v_i^n + 6 (v_i^n)^2 + 2 v_i^n v_{i+1}^n + (v_{i+1}^n)^2]$$

Calculations with the lumped Galerkin method are shown in Fig. 8 for the same parameters used in Fig. 6. The bifurcations now occur at a much larger value of b, b = 0.2.

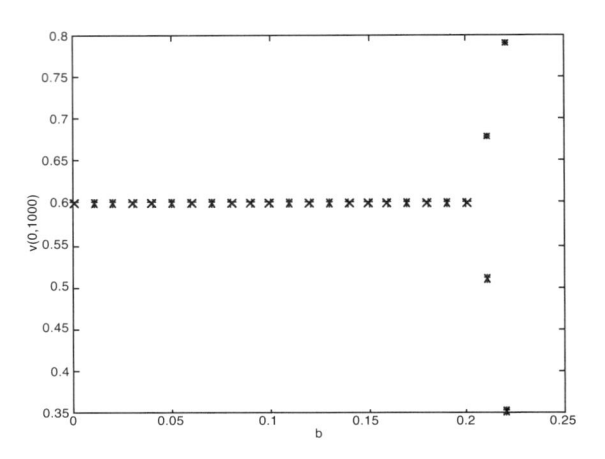

Figure 8. Solutions of the Fisher equation obtained from different initial conditions; Galerkin method, lumped; a = 2.5, $\alpha \Delta t$ = 1.5, Δx = 0.5, 0 ≤ x ≤ 1.

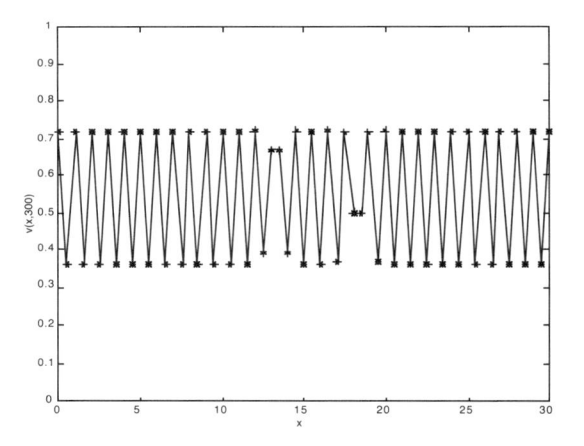

Figure 9. Spatial Variation of the Solution of the Fisher Equation Obtained from the Finite Difference Method; a = 2.5, $\alpha \Delta t$ = 1.5, b = 0.20, Δx = 0.5, 0 ≤ x ≤ 1

Lest one think this is a trivial matter, the solutions for one of these cases with bifurcations is shown in Fig. 9. The first response to seeing such a result is to seek to find an error in the computer code. Yet there isn't one. Thus, one must be very careful when using the method of lines; it is possible that more advanced integration methods will eliminate these false bifurcations, too. Of course, there are many situations for which bifurcations are real (Varma, et al., 1998). One must do careful analysis to distinguish the difference.

Incorrect Shock Movement with Method of Lines

Another area where careful numerical analysis is needed is a problem whose solution exhibits shocks. Consider adsorption in a packed bed with rapid mass transfer so that the fluid and solid adsorbate are in equilibrium. The equations are

$$\phi \frac{\partial c}{\partial t} + \phi V \frac{\partial c}{\partial x} + (1 - \phi) \frac{\partial n}{\partial t} = 0,$$

$$n = f(c) = \frac{\alpha c}{1 + Kc}$$

Let

$$\sigma(c) = \frac{d}{dc}\left(c + \frac{1-\phi}{\phi} n\right) = 1 + \frac{1-\phi}{\phi} \frac{df}{dc} =$$

$$= 1 + \frac{1-\phi}{\phi} \frac{\alpha}{(1+Kc)^2}$$

and transform the equations to

$$\sigma(c) \frac{\partial c}{\partial t} + V \frac{\partial c}{\partial x} = 0$$

This equation underlies models like those used by Strube and Schmidt-Traub (1998) after the equilibrium assumption has been made. Usually, special techniques are used to solve this equation (Finlayson, 1992; Poulain and Finlayson, 1993; Anklam, et al., 1997) in order to maintain a shock if one occurs. However, even those methods can be led astray if the equation is not solved in the correct form (Poulain and Finlayson, 1993).

If one solves the equations in the form

$$\frac{\partial c}{\partial t} + \frac{V}{\sigma(c)} \frac{\partial c}{\partial x} = 0$$

or

$$\frac{\partial c}{\partial t} + \frac{\partial M(c)}{\partial x} = 0, \quad M(c) = \int_0^c \frac{du}{\sigma(u)}$$

then an explicit, upwind method is

$$c_i^{n+1} = c_i^n - \frac{V \Delta t}{\Delta x}[M(c_i^n) - M(c_{i-1}^n)]$$

No matter how this equation is solved, the wave speed of a shock is (Poulain and Finlayson, 1993)

$$\frac{\text{shock speed}}{V} = \frac{M(c^l) - M(c^r)}{c^l - c^r}$$

If one solves the equation in the form

$$\frac{\partial g}{\partial t} + V \frac{\partial c}{\partial x} = 0, \quad g(c) = c\left[1 + \frac{1-\phi}{\phi} \frac{\alpha c}{1 + Kc}\right]$$

then the solution method is more complicated (given g(c), one must find c), and the wave speed is

$$\frac{\text{shock speed}}{V} = \frac{c^l - c^r}{g(c^l) - g(c^r)}$$

These differ by 9 % (0.547 vs. 0.500).

One can show (Poulain and Finlayson, 1993) that as one reduces Δt and Δx, the solution based on method 1 converges to its wave speed, and those based on method 2 converge to its wave speed, but the two wave speeds are different. As shown by Leveque (Leveque, 1992) it is necessary that the Rankine-Hugoniot shock speed be consistent with a conservation law. This is satisfied for method 2 but not for method 1. The conservation law for method 2 is just an overall mass balance including the fluid and adsorbed fluid. However, the more likely method of solution, method 1, leads to the incorrect solution, even when Δt and Δx approach zero. Thus, even when one has an ODE solver available, one must be careful how the partial differential equations are treated.

Reactor Problems Solved with the Chemical Reactor Design Tool (CRDT).

The next examples are for modeling chemical reactors with a computer code that the user accesses only through a graphical user interface (gui). Thus, it relates most closely to the present day use of computers by engineers who have not programmed the computer themselves. The CRDT has been described in detail (Rosendall and Finlayson, 1994) and applied to three industrial chemical reactors (Rosendall and Finlayson, 1994, 1995) Briefly, it permits a user to design chemical reactors including the realistic transport effects that are frequently present. When attempting to solve real problems, students are faced with difficulties which are primarily bookkeeping and manipulation rather than conceptual. Phenomena that might be important, and might be hard to include in a student-written program are:

• multiple reactions (leads to lots of bookkeeping, OK if reactor is not too complicated)
• the temperatures of the catalyst and the fluid may be different (requires solving sets of nonlinear algebraic equations along with the reactor model)
• internal mass transfer (requires solving two-point boundary value problems at every node)

- there may be cooling at the wall (leading to radial dispersion and partial differential equations).

Textbooks usually treat only simple systems such as batch, CSTR, or plug flow reactors, usually for only a few components. The inclusion of the above effects are time-consuming to include, and hence are seldom included even though the phenomena are sometimes important. This leads students to think that it is acceptable to leave out important phenomena just because they can't easily do the computation when it is included.

The CRDT can solve reactor equations with up to 20 components plus temperature when the reactors are CSTR, batch, plug flow, axial dispersion, or radial dispersion reactors. Phenomena included are:

- intraparticle heat and mass transfer is important
- significant mole changes occur
- significant pressure changes occur

These effects are especially important for selectivity, especially in non-isothermal cases, and this makes the models useful for pollution prevention by not making unwanted products. The user supplies a FORTRAN routine that evaluates the reaction rate, chooses the type of reactor and phenomena to include, and provides the parameters needed for that application. CRDT solves the problem and provides output files for plotting. Batch and plug flow reactors involves solving ordinary differential equations as initial value problems, and the capability to do that is in most numerical analysis packages today. One feature added by CRDT is the addition of intraparticle heat and mass transfer resistance: the user includes this by pressing a button, and coupled nonlinear differential-algebraic equations are solved. A sophisticated program will be invoked to identify if there are multiple solutions to any of the intraparticle problems. It uses linear programming techniques to get a good initial guess for the iterations, and to test for multiple solutions. This same capability is in all the reactor modules. Axial dispersion reactors can, of course, have multiple steady state solutions without any intraparticle resistance, and these are predicted as well. Finally, radial dispersion reactors require solving multiple partial differential equations. This is clearly beyond the capabilities of most undergraduates, at least for complicated problems. So, the program is very powerful. How do you know you've solved the problem correctly?

The first test is the reaction rate. The interface with the main CRDT must be correct. Because of the generality of the program (especially due to intraparticle resistance), the reaction rate subroutine has to be correct for every possible set of concentrations and temperature, even those unexpected in the eventual solution (they arise occasionally in iterative methods). So what did we do? We designed a program, test_rate, that uses the user's rate subroutine but interacts in the same way the program does. The user can run interactively with various inputs and obtain values of the rates of reaction. These can be checked with hand calculations to insure that the subroutine is correct under all conditions. The second test is to run sample cases (with a simpler reaction rate expression) to compare with analytical results. This is a most important step, since it assures the student that they are using the code as intended.

The third test is empirical. Run the CRDT with different sets of numerical parameters and make sure that the results do not depend on arbitrary choices of the numerical parameters. For example, when solving ordinary differential equations using either RKF45 or LSODE, the user sets an error criterion. Thus, the user needs to run the CRDT with at least two different error criteria to make sure that the results do not depend on the numerical analysis parameters. For partial differential equations, since the program uses the method of lines, either finite difference or orthogonal collocation is used to reduce the partial differential equations to sets of ordinary differential equations. We have the same tests for the error criterion for the ODE solver. However, we also have discretization error associated with the spatial discretization. For example, you can fix the number of finite difference grid points, solve with an ODE solver, reduce the error criterion, and solve again. The answer will be a more accurate solution of the ordinary differential equations, but not necessarily a more accurate solution of the partial differential equations. There may still be discretization error in space. So, one then finds solutions with N = 4, 8, 16, etc., and for each of these uses several error criteria in the ODE solver. One finds that as the number of finite difference points increases, the error criteria must become smaller to get any solution at all. Sometimes the discretization analysis is such that the error term has a positive term times Δt plus a negative term times Δx^2. Thus, by choosing a 'magic' mesh and step size, one can even make the error zero; just don't change anything or look at any other value! Thus, figuring out the solution to partial differential equations and the errors in the numerical method when both Δt and Δr are decreasing can be difficult.

As an example of the kind of things CRDT can do, consider a reactor to oxidize SO_2. The reaction is

$$SO_2 + \frac{1}{2} O_2 <=> SO_3$$

and the reaction rate is written in terms of partial pressures.

$$\text{Rate of oxidation} = \frac{k_1 p_1 p_2 - k_2 p_3 p_2^{1/2}}{p_1^{1/2}}$$

$$\ln k_1 = 12.07 - \frac{31000}{RT}, \quad \ln k_2 = 22.757 - \frac{53600}{RT}$$

The first index is for SO_2, then O_2, SO_3, and N_2. After writing a 10-line FORTRAN program and testing it with test_rate, the user is ready to use CRDT. In this paper the focus is on whether radial dispersion is important and how that is determined. Once the 10-line reaction rate program is written, the only thing the student has to do to handle radial dispersion is click a button, provide radial dispersion parameters, and choose either the finite difference or orthogonal collocation method. For a typical case (Hill, 1977; Rosendall and Finlayson, 1994) the radially-averaged concentration versus length is shown in Fig. 10.

The 2D profiles of SO_3 and temperature are shown in Fig. 11. One feature of the CRDT, since it is an educational program, is to let the user see the magnitude of various terms in the equations. Fig. 12 shows the radially-averaged diffusion and convection terms. The convection term is naturally negative for the reactants and positive for the products. Note particularly that the diffusion term is small compared with the convection term (thus suggesting a 1D model is OK) except for the last part of the reactor where the reaction is stronger. Other results (Rosendall and Finlayson, 1994) show that the reactor model including radial dispersion can be 15% shorter than the 1D model, to achieve the same conversion. Thus, the CRDT allows one to quantify the results of the assumptions made. With students it is important to keep asking for numerical values, rather than just vague generalities, so that they learn to figure out what phenomena is really important based on quantitative criteria.

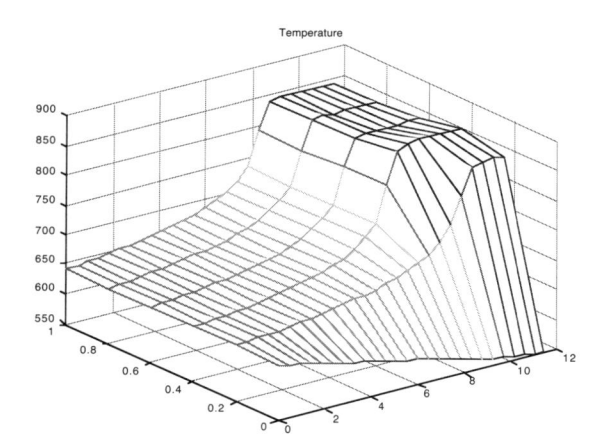

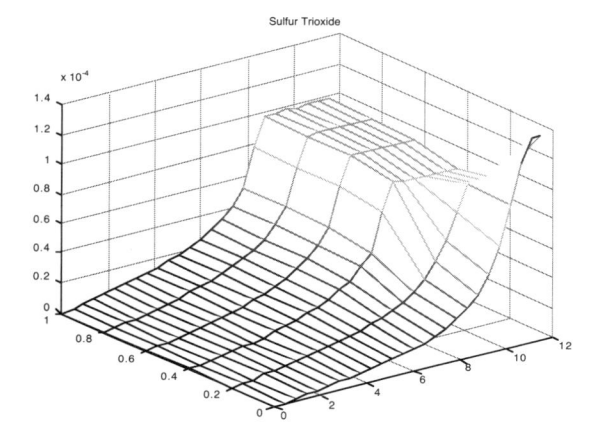

Figure 11. Reactor to oxidize sulfur dioxide; Temperature; (b) Sulfur trioxide.

The effects of radial gradients, heat and mass transfer limitations, and total molar and pressure changes would be most important in cases involving multiple reactions where selectivity is important, either for economic reasons or because one of the products is a pollutant. Thus, it is of interest to examine the effect of radial heat transfer for a case involving several reactions. The problem selected is the reaction of propylene and chlorine to form allyl chloride (Smith, 1970; Carberry, 1976).

The reactions are simplified here to include the three main ones, involving the formation of allyl chloride, 1,3-dichloropropene, and 1,2-dichloropropane.

$$Cl_2 + C_3H_6 \rightarrow CH_2=CHCH_2Cl + HCl$$
allyl chloride

$$Cl_2 + CH_2=CHCH_2Cl \rightarrow CHCl=CHCH_2Cl + HCl$$
allyl chloride   1,3–dichloropropene

$$Cl_2 + C_3H_6 \rightarrow CH_2ClCHClCH_3$$
1,2–dichloropropane

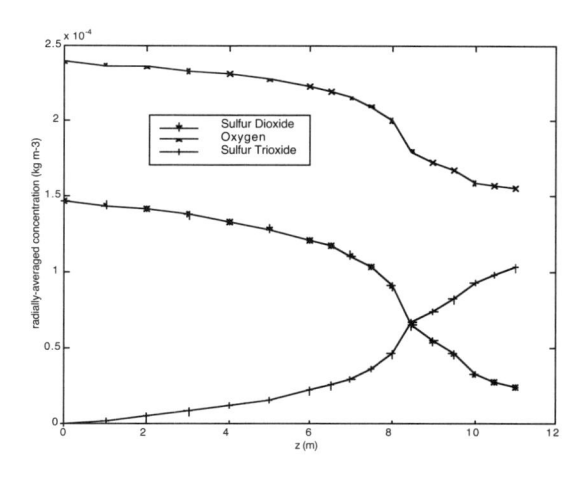

Figure 10. Reactor to oxidize sulfur dioxide, finite difference method radially, N = 5, 4th-5th order Runge-Kutta method axially.

small. Of course, the turbulent velocity profile is not exactly flat, and a small temperature difference radially can cause a reaction rate difference, which causes a concentration difference, which in turn affects the temperature. Thus, it is of interest to find out how strong those effects are.

The problem is solved in CRDT. The feed rate of propylene to chlorine is taken as 2.5 (industrial ranges are 1.7 to 3.8), and the parameters are chosen to agree with those in Carberry (1977). The equations for a 1D model are

$$\frac{dF_i}{dV} = Da_I RA_i, F_i = F_{j0} \text{ at } V = 0$$

$$\sum_{j=1}^{NC} F_j C_{pj} \frac{dT}{dV} = Da_{III} RT - St \ (T - T_c)$$

$$T = T_{in} \text{ at } V = 0$$

$$St = \frac{UA}{F_s C_{ps}}, \text{ where } F_s \text{ and } C_{ps} \text{ are standard}$$

molar flow rates and heat capacities

while those for a 2D model are

$$\frac{\partial F_i}{\partial V} = \alpha_i \nabla^2 C_i + Da_I RA_i$$

$$\sum_{j=1}^{NC} F_j C_{pj} \frac{\partial T}{\partial V} = \alpha_T \nabla^2 T + Da_{III} RT$$

$$-\frac{\partial T}{\partial r} = Bi_w \ (T - T_c) \text{ at } r = 1$$

The additional parameters needed for the 2D model are taken as

$$\alpha_i = \frac{V D C_s}{R^2 F_s}, \ \alpha_T = \frac{V k}{R^2 F_s C_{ps}}, \ Bi_w = \frac{h_w R}{k}$$

and here $\alpha = 0.022$ and $\alpha_T = 1$. The Biot number at the wall is chosen so that the heat losses at the wall are equivalent in the 1D and 2D models, by taking (Finlayson, 1980)

$$\frac{1}{St} = \frac{1}{2 \alpha_T} \left(\frac{1}{Bi_w} + \frac{1}{3} \right)$$

Here, $B_w = 0.27$. For the 2D model, the velocity is taken as the same average value as in the 1D model, but with a one-seventh power with respect to radius, which is a reasonable representation for turbulent flow in a pipe.

The results are shown in Figures 13-15. Figure 13 shows the radially-averaged concentration of chlorine versus the length of the reactor (chlorine is the limiting

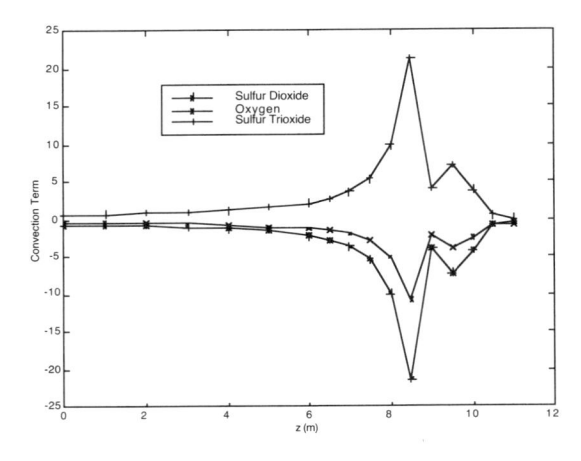

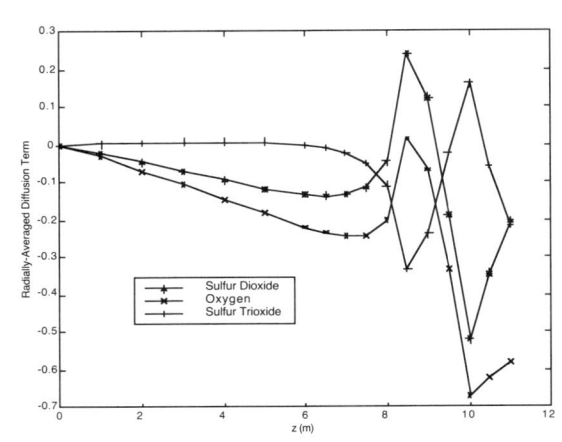

Figure 12. Reactor to oxidize sulfur dioxide; radially averaged convection and diffusion terms; (a) Convection; (b) Radial dispersion.

The total molar change will be small. However, the reaction rates depend on temperature and concentration as follows.

$$r_1 = A_1 \exp\left(-\frac{15,840}{RT}\right) C_{Cl_2} C_{C_3H_6}$$

$$r_2 = A_2 \exp\left(-\frac{23,760}{RT}\right) C_{Cl_2} C_{allyl \ chloride}$$

$$r_3 = A_3 \exp\left(-\frac{7920}{RT}\right) C_{Cl_2} C_{C_3H_6}$$

Thus, temperature will have a big effect on selectivity. At high temperatures, allyl chloride is favored, but at lower temperatures more 1,2-dichloropropane is formed. Typical operating conditions are 500 °C and 40 psia (Fairbairn, *et al.* 1947), and the reaction is carried out in the gas phase which is flowing in an empty tube. Most published models are one-dimensional; they ignore radial gradients of concentration and temperature on the grounds that they will be small in turbulent flow in a pipe because the velocity is flat. Using the Colburn analogy, it is assumed that the radial gradients of temperature are also reagent). The adiabatic result shows almost complete conversion of the chlorine, but the 1D and 2D models have about 5% of the initial chlorine unreacted. There is little difference between the 1D and 2D models when comparing the radially-averaged values. Thus, the radial effects are not large for this case. The radially-averaged temperature is shown in Figure 14. The adiabatic case approaches the adiabatic temperature rise (which in itself depends on selectivity, hence the transport conditions in the reactor). The 1D and 2D models give about the same radially-averaged temperature profile versus length.

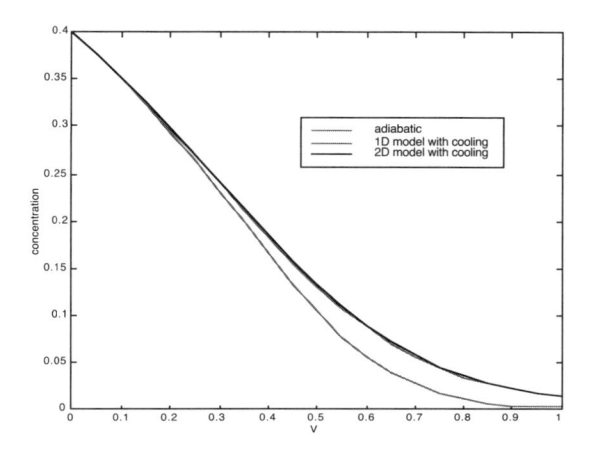

Figure 13. Average concentration of chlorine.

Despite this agreement of the radially-averaged values, there do exist profiles of the components in the radial direction. The outlet profile of chlorine is shown in Figure 15 for the 1D and 2D models. (The small dip near the centerline is due to the interpolation process.) While there is about a 40% change in the concentration of chlorine from the center to the wall, the average values are close (within 6%). In some cases that may be significant enough difference when trying to react the chemical completely.

This example shows how a 2D model can be done relatively easily once a 1D model has been constructed, and the effects of temperature and velocity profile are easily included. The CRDT is constructed to make this easy to do; the hard part is getting the numbers for the first model – the other models are within a click of a button.

Using the Chemical Reactor Design Tool (CRDT) with CFD programs.

The next example illustrates how one can use the CRDT in conjunction with CFD programs. The CFD model is of an industrial radiant furnace (Berkoe, *et al.*, 1998). The commercial CFD code, CFX version 4, was coupled to a one-dimensional model for the cracking reactions occurring inside the process tube passing through the fire-

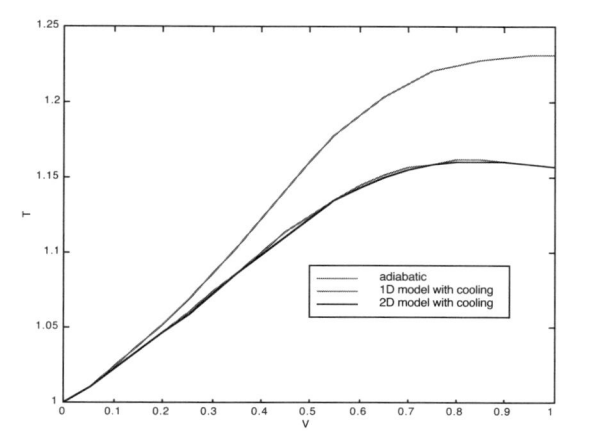

Figure 14. Radially-averaged temperature.

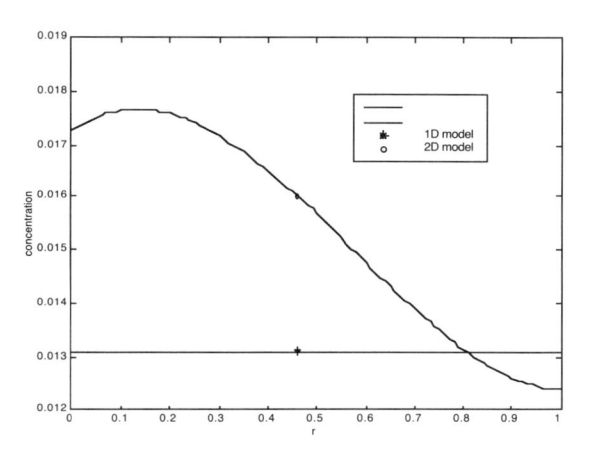

Figure 15. Chlorine concentration at exit.

Box. This model allows for the prediction and evaluation of detailed heat flux, temperature profiles, and flow distribution within the furnace. The CFX model accounts for

- 3D heat transfer due to radiation, convection, and conduction
- Combustion
- 3D fluid flow, including the effects of compressibility.

The one-dimensional, tube-side model includes

- Momentum balance for the process fluid
- Cracking reaction kinetics leading to vaporization
- Temperature dependent physical properties
- Heat transfer from the furnace side.

A typical situation is shown in Fig. 16.

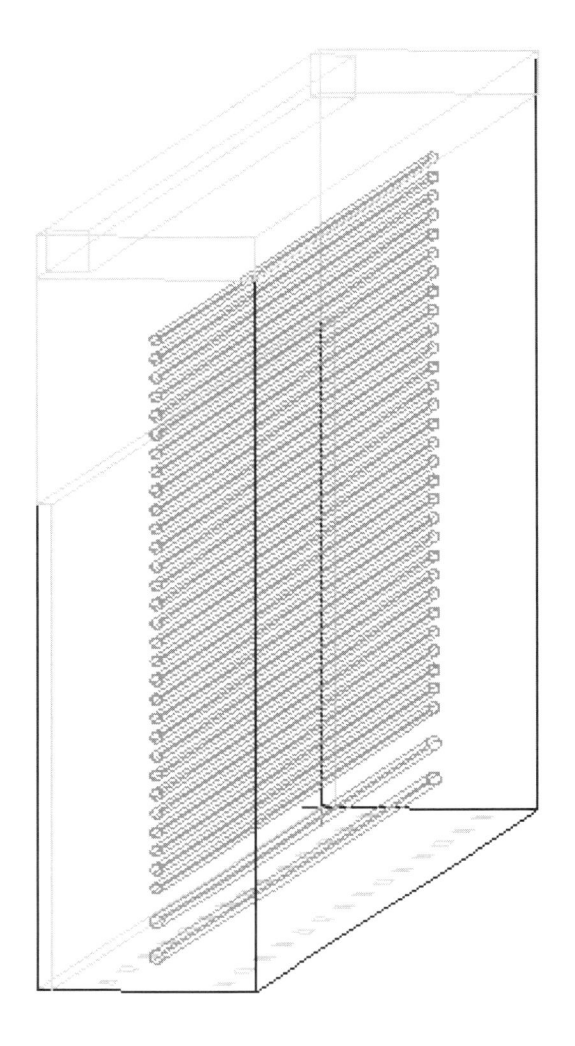

Figure 16. Industrial radiant furnace geometry.

The 3D furnace model and the 1D process tube model are coupled through the heat flux. The equations governing the process tube model are:

$$\frac{\dot{m} C_p}{A_{cs}} \frac{dT}{dz} = h a (T_o - T) + (-\Delta H_{rxn}) R$$

$$\frac{1}{A_{cs}} \frac{dF_i}{dz} = v_i R, \quad \frac{dp_t}{dz} = \rho \left(g_y \frac{dy}{dz} + u^2 Fr + u \frac{du}{dz} \right)$$

$$Fr = \frac{f}{2 d_t} + \frac{1.4}{\pi R_b} \left(0.051 + 0.19 \frac{d_t}{R_b} \right)$$

The reaction is

Resid(l) –> 0.742 HCGO(l) + 0.742 LCGO(g) + 0.725 Naphtha(g) + 1.82 LG(g)

Since some of the products are vapor, physical properties for the two phase mixture were taken as an average

$$\frac{1}{\mu} = \frac{\text{fraction vapor}}{\mu_v} + \frac{\text{fraction liquid}}{\mu_l}$$

The process tube model is coupled with the CFD model through a FORTRAN routine that the CFD model calls. The same principles discussed above apply here: one must check the FORTRAN program, and then one must test the use of the FORTRAN program in the CFD model to insure that it is being used correctly. But, how good is the 1D model? One can do all the testing and find the computer work is satisfactory, but if a truly 2D process tube model is necessary incorrect results will be given by the 1D model. This is an ideal application for CRDT, since once the 1D model is done, the 2D model is easily solved. Unfortunately, the CRDT is limited to one phase, so a special purpose program was written to compare 1D and 2D models, using the same approach taken in CRDT, namely using the orthogonal collocation method (Finlayson, 1980) to model radial dispersion of energy and mass. For the purpose of this test, only one tube was modeled, and the pressure was taken as constant in that tube. The boundary condition at the tube wall was that an applied flux was specified (i.e. the 1D and 2D models were tested outside of the CFX program).

For the test done here, the term h (T_o - T) is replaced by q, the average heat flux in the furnace. For the 2D model the equations are changed to

$$\frac{\dot{m} C_p}{A_{cs}} \frac{\partial T}{\partial z} = \frac{1}{r} \frac{\partial}{\partial r} \left(k r \frac{\partial T}{\partial r} \right) + (-\Delta H_{rxn}) R$$

$$\frac{1}{A_{cs}} \frac{\partial F_i}{\partial z} = \frac{1}{r} \frac{\partial}{\partial r} \left(D r \frac{\partial C_i}{\partial r} \right) + v_i R$$

$$\left.\frac{\partial T}{\partial r}\right|_{r=0} = 0, \left.\frac{\partial c}{\partial r}\right|_{r=0} = 0, -k \left.\frac{\partial T}{\partial r}\right|_{r=R} = q, \left.\frac{\partial c}{\partial r}\right|_{r=R} = 0$$

This 2D model is not the complete story, since the wall temperature is excluded, but it is sufficient to test the need for a 2D model.

First consider the standard tests for the importance of radial dispersion. Mears (1971) gives a test (derived for packed bed reactors) that is based on the principle that radial variation of temperature and concentration is unimportant when the radially averaged reaction rate differs from the local reaction rate at the wall by less than 1%. This is definitely not satisfied here, since the difference between the wall temperature and fluid temperature can up as large as 100 K. Another rule of thumb is that radial variations are important if the Biot number, $Bi_w = h R_t / k$, is greater than 10 (Rosendall and

Finlayson, 1995) Here it is typically 500. Thus, both standard *a priori* tests (admittedly derived for packed beds) say radial variations are important. However, it is the author's experience in modeling chemical reactors that if the temperature profile is approximately quadratic, then a one-term orthogonal collocation solution gives results as good as a six-term collocation solution (including all radial variations), and the one-term orthogonal collocation solution is the same as the 1D model. The only difference is that one is solving for the temperature at the collocation point rather than the average temperature. That is indeed what happened.. 17 shows the temperature profile for the 1D model and the temperature profile at two radial positions for the 2D orthogonal collocation model. While there is a radial variation in temperature, the average properties are very close to each other. Thus, the 2D model is quite appropriate for this industrial radiant furnace model.

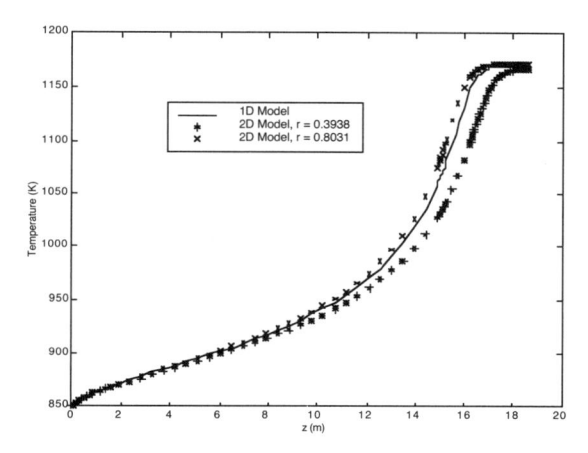

Figure17. Temperature in one tube, Comparing a 1D and 2D model.

Use of CFD Programs by Undergraduates

When undergraduates use a CFD program, the same principles discussed here apply. One needs to verify your use of the code by solving problems with known solutions. The first author usually has them do problems where we can use a mesh from an example problem. At the University of Washington we use the finite element program FIDAP, from Fluent Corporation, since that is available for a research study. In CFD, however, there are two important questions that the students have no experience with: upwinding and turbulence modeling.

Shown in Fig. 18 are two cases, one an accurate one [derived using methods from the book by Finlayson (1992)] and one using upwinding. In the simplest case, an upwinding finite difference derivative would use:

$$\left.\frac{du}{dx}\right|_i \approx \frac{u_{i+1} - u_{i-1}}{2\Delta x} \approx \frac{u_i - u_{i-1}}{\Delta x}$$

The net effect of this change is to minimize unwanted oscillations in the solution and allow the solution to be obtained much faster. Yet, to the student, it is simply a word entered onto a data set when using a CFD gui. If the upwinding is used for a conserved quantity (as in Fig. 18) then the effect of upwinding is to spread the material out, lowering the peak value. If the phenomenon depends on the concentration in a non-linear way, then obviously one will get incorrect results. For example, in chemical flooding of an oil field, the phenomenon depends on reaching a certain concentration level. The use of upwinding could cause the model to never reach that level. Every user of CFD needs to know this.

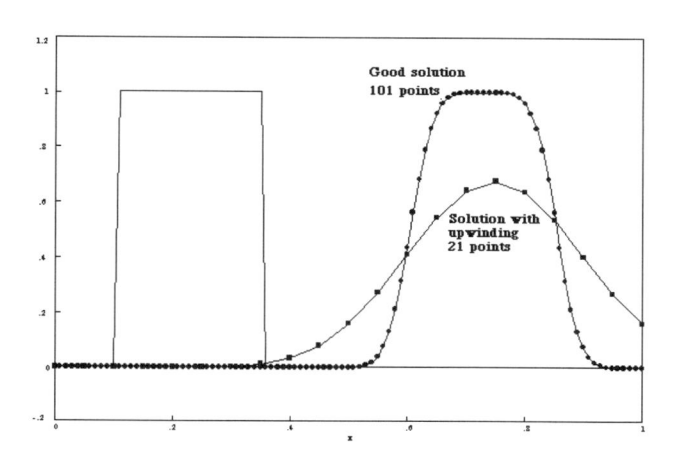

Figure 18. Concentration profile for flow through a packed bed. One curve uses upwind differencing (and is faster to calculate); the other uses the random choice method.

At the University of Washington we have an Undergraduate Fuel Cell project, whose goal is to design and build an amusement-park sized fuel-cell powered locomotive. This is an interdisciplinary, multi-campus project. The design of the fuel cell plate is complicated, since it is necessary for the fuel to go along very small channels, looping back and forth. One design question is: how small should the cross section of those channels be? If they are smaller, then the velocity is larger, and the mass transfer is better, which is desirable. However, if they are smaller, the pressure drop is larger, which is undesirable. The pressure drop at the corners of the loops is also important. One example, provided by Karen Fukuda, is shown in Fig. 19. In this example the flow is laminar, but the flow around the ends of the loops, and the pressure drop, can be obtained

Another case where the CFD capabilities exceed the knowledge base of students is when turbulent flow is modeled. Many programs, including FIDAP, use a k-ε model of turbulence. The parameters needed for the model have been measured in special situations, such as wall shear, free shear, etc., and must be used carefully in

2D and 3D situations that have more than one type of shear flow.

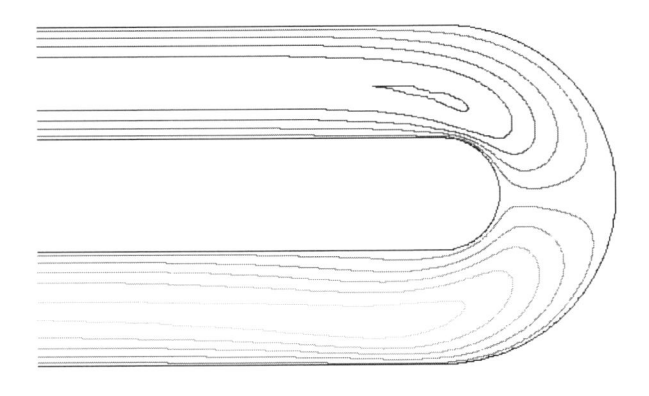

Figure 19. U_x velocity for one-phase flow in fuel cell plate (due to Karen Fukuda).

Use of Natural Boundary Conditions in the FEM

One advantage of the finite element method is the use of natural boundary conditions, and these allow students to solve problems in semi-infinite geometries without having to extend the mesh to infinity. The example chosen here is the solution to the nonlinear Poisson-Boltzmann equation that arises in colloidal chemistry.

$$\nabla^2 \psi = -\frac{1}{A} \cdot \sinh(A\psi)$$

For the case of a charged sphere inside a cylinder with zero potential, the solution using natural boundary conditions is shown in Fig. 20. The boundary condition along the top surface is just no applied flux (the natural boundary condition for the Galerkin method). Notice that the contours need not be perpendicular to the boundary, as they would be if we enforced a condition $\partial \psi/\partial n = 0$ or $\psi = 0$. While this is true far from the sphere, a much larger domain would be necessary to reach the position where this is true. The same principle applies to Stokes flow around a sphere or cylinder - one doesn't have to use a mesh covering the entire space to eliminate the effect of the wall; one just uses natural boundary conditions and specifies no applied mass flux on those outer boundaries. For additional information about the use of natural boundary conditions, see Finlayson (1992).

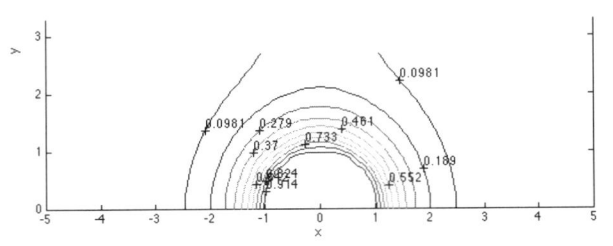

Figure 20. Potential around sphere in a cylinder; governed by the Poisson-Boltzmann equation, A = 2.3.

Conclusions

The power of computers and computer software has made it possible for students to solve problems that could only be imagined a few years ago. This paper has shown that this requires educators to change their emphasis from the mathematics of solving the problem to the tools needed to derive a nonlinear model and assess the numerical errors. Blind acceptance of computer results benefits no one except your competitors.

References

Anklam, M.R., R. K. Prud'homme, B. A. Finlayson (1997). Ion exchange chromatography laboratory. *Chem. Eng. Ed.*, **31**, 26-31.
Carberry, J. J. (1976). *Chemical and Catalytic Reaction Engineering*, McGraw-Hill, New York.
Fairbairn, A. W., H. A. Cheney, A. J. Chernaivsky (1947). Commercial scale manufacture of allyl chloride and allyl alcohol from propylene. *Chem. Eng. Prog.* **43** 280-290.
Finlayson, B. A. (1992). *Numerical Methods for Problems with Moving Fronts*. Ravenna Park, Seattle.
Finlayson, B. A. (1980). *Nonlinear Analysis in Chemical Engineering*. McGraw-Hill, New York.
Hill, C. G., Jr. (1977). *Introduction to Chemical Engineering Kinetics & Reactor Design*. Wiley, New York.
Keller, H. B. (1972). *Numerical Methods for Two-Point Boundary-Value Problems*, Blaisdell, New York.
LeVeque, R. J. (1992). *Numerical Methods for Conservation Laws*. Birkhäuser Verlag, Basel.
Mears, D. E. (1971). Diagnostic criteria for heat transfport limitations in fixed bed reactors. *J. Cat.*, **20**, 127.
Mickens, R. E. (1989). Exact solutions to a finite-difference model of a nonlinear reaction-advection equation: implications for numerical analysis. *Num. Meth. Part. Diff. Eqn.*, **5**, 313-325.
Mitchell, A. R. and J. C. Bruch, Jr. (1985). A numerical study of chaos in a reaction-diffusion equation. *Num. Meth. Part. Diff. Eqn.*, **1**, 13-23.
PDE Toolbox (1997). *MATLAB*. Mathworks, Inc.
Poulain, C. and B. A. Finlayson (1993). A comparison of numerical methods applied to nonlinear adsorption columns. *Int. J. Num. Meth. Fluids*, **17**, 839-859.
Rosendall, B. M. and B. A. Finlayson (1994). The chemical reactor design tool. *ASEE Proceedings*, 2219-2222, Edmonton, Canada.
Rosendall, B. and B. A. Finlayson (1995). Transport effects in packed-bed oxidation reactors. *Comp. Chem. Eng.*, **19**, 1207-1218.

Rosendall, B. M., Berkoe, J. M., and A. C. Heath (1998). Computational fluid dynamics model of an industrial cracking furnace. AIChE Annual Meeting, Miami Beach.

Smith, J. M. (1970) *Chemical Engineering Kinetics.* McGraw-Hill, New York.

Strube, J. and H. Schmidt-Traub (1998). Dynamic simulation of simulated-moving-bed chromatographic processes. *Comp. Chem. Eng.*, **22**, 1309-1317.

Varma, A., M. Morbidelli, and H. Wu (1998). *Parametric Sensitivity in Chemical Systems.* Cambridge University Press.

Wheeler, D. L. (1999). To improve their models, mathematicians seek a 'science of uncertainty', *Chronicle High. Ed.*, April 16.

PERSPECTIVES ON LIFECYCLE PROCESS MODELING

Wolfgang Marquardt, Lars von Wedel, and Birgit Bayer
Lehrstuhl für Prozesstechnik
RWTH Aachen
D-52056 Aachen, Germany

Abstract

The lifecycle concept has been stressed recently in many different ways in particular by the service and software industry. There is still a vague and fuzzy notion of what lifecycle (process) modeling really is and how it should be supported. This contribution tries to clarify the concept of the process lifecycle and the associated lifecycle process model. Different dimensions of integration across the lifecycle are defined. The state of the art regarding these dimensions is reviewed and future challenges are highlighted. A first step towards a comprehensive lifecycle model with emphasis on data rather than work processes is sketched. An advanced (informal) architecture of an integrated modeling tool based on our experience from previous tool development is suggested. Finally, promising areas of research as well as a novel understanding of collaboration between the major contributors towards advanced commercial products are suggested.

Keywords

Chemical process design and development, Chemical process modeling, simulation, and optimization, Computer supported collaborative work, Workflow management, Integrated design environments, Information modeling.

Introduction

Process modeling and simulation has become an established and widely used technology in industrial chemical process development, design, and operation. The various areas of application have led to some blurring of the terminology over the years. The following classical definition of process simulation has been given by Motard et al. (1975):

> Process simulation is the representation of a chemical process by a mathematical model which is then solved to obtain information about the performance of the chemical process.

The abstraction of the process behavior into a mathematical model requires a profound understanding of both, the physical, chemical or biological phenomena occurring *and* the objectives pursued with the application of the model. There is always a variety of different process models which describe the same chemical process from different perspectives to suit the requirements of a certain application. The evaluation of the model by means of some numerical technique produces an often vast amount of data which needs to be interpreted to extract information for the solution of the engineering problem.

This evaluation usually comes in two different flavors depending on the model specifications (see Fig. 1). The specification of the equipment design parameters, the operating conditions, and the feed conditions results in a *simulation problem* which is typically stated and solved during process analysis. In contrast to this *direct problem* a number of *inverse problems* are commonly formulated and solved. They typically arise if the causes (e.g. process structure, operating conditions, process disturbances) of some desired behavior (e.g. product specifications) or of some observed effect (e.g. plant measurement data) need to be determined. Inverse problems are naturally formulated as an optimization problem and solved by dedicated numerical algorithms. However, a sequence of simulation problems is often solved instead in industrial practice, where the specifications related to the desired or observed behavior are adjusted manually until the objective is met to a sufficient degree. Subsequently, we will not explicitly distinguish between simulation and optimization. Rather, process modeling, simulation and optimization will be subsumed by process modeling for short.

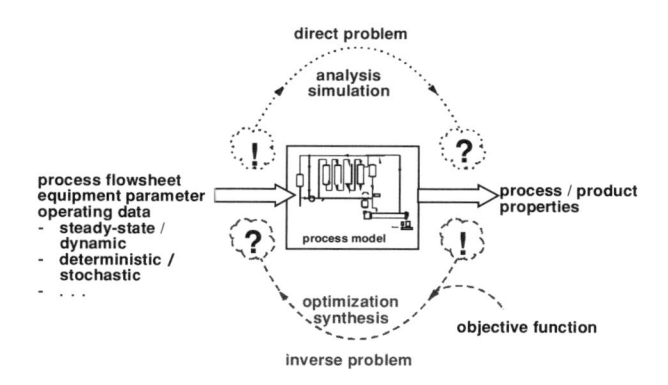

Figure 1. Direct and inverse problems: simulation and optimization for analysis and synthesis.

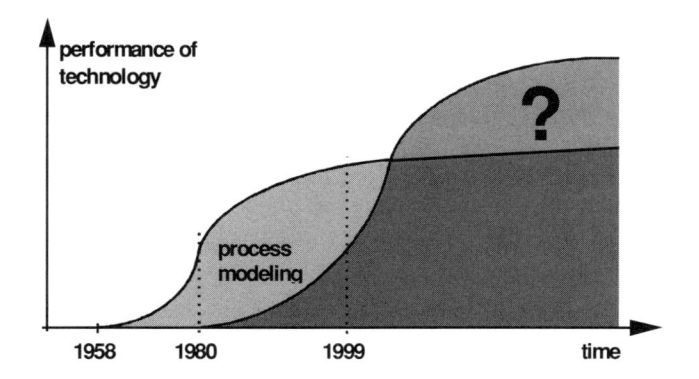

The Evolution of Process Modeling

Fig. 2 shows schematically the historical evolution of the field of process simulation (see Marquardt, 1999, for a more detailed summary) which has been initiated in the fifties by the process industries. Kesler and Kessler (1958) presented the first flowsheeting system and reported industrial applications. The major boost of development of process simulation concepts, methods, and software occurred in the seventies and eighties. In the eighties commercial products from a number of suppliers have become widely available, were successfully employed by the operating companies and integrated in their existing information technology infrastructure.

The nineties are largely characterized by consolidation of existing simulation methods and commercial software. Major improvement in the usability and robustness of the software largely enabled by the rapid developments in computing technology led to an even wider distribution of process modeling technology. In addition, numerical algorithms have been improved and the model libraries have been upgraded. Robustness and efficiency of the algorithms for large-scale problems as well as the lack of nonstandard (dynamic or detailed steady-state) models and computer-based support for their development and maintenance have been identified as major and still existing bottlenecks. Results of academic research in these areas are only slowly penetrating commercial products due to the high cost of software re-engineering and upgrading.

In recent years, the process modeling market has changed tremendously. There has been a concentration process among vendors which results in an undesirable dependency of the operating companies on the functionality of current products and future innovation capability of the few remaining vendors. Naturally, this market situation in conjunction with the high development cost favors stagnation of further technology development.

What Next?

In summary, process modeling has reached a quite mature state of development today which leads to the saturation in the performance-time curve in Fig. 2. Clever application of available technology is definitely of greater interest than perfecting it further. In face of the obvious saturation, the question naturally arises, which technology will follow and ultimately supersede current process modeling technology to produce a quantum leap in performance. We formulate the following prognosis:

> Process modeling will remain a key technology for model-based development, design and operation of chemical processes. It will be an integrated part of a more comprehensible technology which aims at the integration across the lifecycle. This objective can only be achieved by means of new business relationships among the major contributors.

This prognosis as well as the technological and business implications will be discussed in the remainder.

Figure 2. Evolution of process modeling.

Overview

We will start with a brief account of modeling and simulation in the lifecycle. Next we will discuss integration of models as well as modeling and design processes across the lifecycle. The resulting analysis can be interpreted as conceptual requirements for advanced computer-based support systems. Next, a sketch of a conceptual information model for lifecycle integration as a basis for software development is presented. This information model is directly employed to derive the architecture of an advanced system facilitating model integration, process simulation and optimization across the lifecycle. Last but not least major research directions and the need for a novel understanding of the business process leading to advanced commercial products are pointed out.

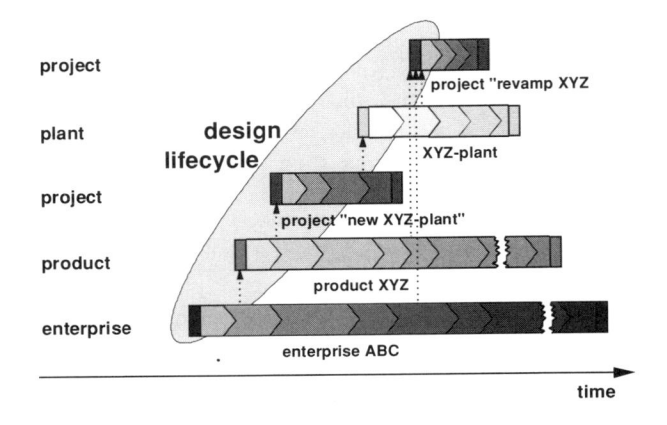

Figure 3. Different lifecycles.

Modeling in the Lifecycle

Lifecycle concepts have been developed and employed in software engineering for a long time (e.g. Alhir, 1998). This concept has been adopted recently by many service-oriented process engineering companies including the process modeling software vendors. Their objective is to provide services and tools for all the lifecycle phases of an artefact spanning its design, creation, use, and decommissioning. The lifecycle we are most interested in interferes with a number of other lifecycles as depicted in Fig. 3. There are the lifecycles of the enterprise, of a certain chemical product, of a project to design and build the manufacturing plant, of the manufacturing plant itself or of one or more projects to revamp and debottleneck the existing plant. Process modeling is carried out in one way or another in all these related lifecycles. By concatenation of these activities, we can define the *design lifecycle* (called lifecycle for short subsequently), which encompasses all the business processes carried out by different people from product design to operation support systems design (Schuler, 1998).

Modeling on Multiple Scales

So far, we have implicitly assumed that the granularity of the simulation model is orientated at the process units on the flowsheet level to analyse their overall behavior and their interaction in the plant. The phenomena driving the process unit behavior are typically captured only in a coarse manner. There are, however, many other kinds of models employed in the various phases of the lifecycle.

As shown in Fig. 4, these models may be organized on different scales. On the finest *length scale* there are quantum physical or molecular mechanics models to represent single atoms or molecules. On the following length scales, we find molecular dynamics or molecular simulation models of interacting molecules in a cluster, molecular dynamics or continuum models to describe phase interfaces, micro-structured materials including (multi-grain) particles, membranes or thin films, continuum models of single- or multi-phase compartments in a process unit with various degrees of resolution, models of the apparatus, the plant or the site, or even models capturing the whole enterprise. Roughly speaking, the models on a particular length scale are also associated with a characteristic time scale as shown in Fig. 4 (though a significant degree of variation is commonly encountered). In addition to the usually cited *temporal* and *spatial resolution* (e.g. Sapre and Katzer, 1995), we emphasize in addition the so-called *chemical resolution* of a model which captures the number and type of chemical (pseudo-) components. The chemical resolution is at least as detrimental for the prediction capabilities of a model as the spatial and the temporal resolution since the major phenomena crucially depend on the chemical species considered. The models on a certain scale are employed for specific tasks in the lifecycle. A brief summary is given next, a more detailed presentation is provided elsewhere (Marquardt, 1999).

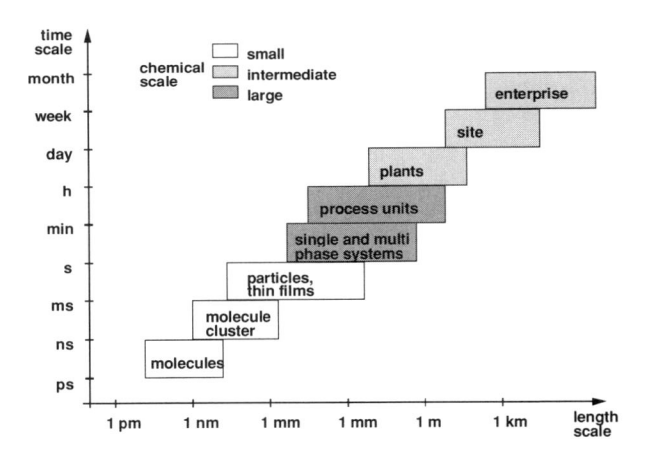

Figure 4. Scales in the design lifecycle.

Modeling for product development is increasingly carried out by means of computational chemistry techniques on small time and length scales ranging from individual molecules to micro-structured material entities considering one or at most a few chemical species. These methods are used to explore those material properties which are required for better process design (i.e. phase equilibria) by the design team, for its application (i.e. pharmaceuticals or agrochemicals) by the consumer, or for its processing (i.e. polymers or nanoparticles) by the client. If a product model needs to be incorporated into a larger model, a coarse scale description, ideally backed up by a mechanistic model on the molecular scale, is required for computational reasons.

There are numerous applications of *modeling for process development* including flowsheeting to assist labscale experiments, simulation or optimization to synthesize process structure or to determine best operating points on the level of the plant. The detailed *design and analysis of process units* is increasingly supported by means of detailed models capturing the transport and reaction processes in detail. Computational fluid dynamics (CFD) models have shown to be valuable tools to support novel reactor or separator design. Dynamic simulation and optimization is used for process analysis as well as for the planning of recipes for batch or grade transitions for continuous processes. Different types of models are required to assist for *engineering design* such as material strength calculations, plant layout; and pipe routing optimization which are just a few examples.

Modeling for process operations has gained increasing interest in particular for real-time applications such as real-time optimization, or model predictive control, as well as planning and scheduling. These techniques are expected to merge in the future by integrating different dynamic models on short and long time scales for the various tasks of an operation support system.

Critical Assessment

As the examples illustrate, the models employed during the lifecycle are widely differing in the modeling formalism adopted, in the facets of the physical and chemical phenomena captured, and in the adapted numerical methods used in order to match the requirements on a model's expressiveness and level of resolution on a certain scale at reasonable computational effort. Hence, there is always an intimate relation between the various models given by the underlying physical and chemical phenomena.

Despite this inherent relation, modeling is carried out largely independently in the different lifecycle phases. Models and solution methods are tailored to the objectives of a particular lifecycle phase and reflect the specific knowledge and experience of the developer. Usually, only local objectives are considered which are relevant to a particular task. Often, only the results obtained in one lifecycle phase are passed to the next phase by means of some written report. Typically, the rationale and the work processes which have led to the result as well as the knowledge accumulated and (at least in part) implicitly encoded in a model does not reach the subsequent lifecycle phases (because it is either not included in or not extracted from the report due to time constraints at either end). An integrated view of the objectives across the lifecycle is largely missing despite its importance for capturing the true potential in the lifecycle. Reuse of knowledge (facts and experience) in different phases of one lifecycle or even in different lifecycles is difficult and therefore limited or in many cases impossible. This is particularly true if the lifecycle spans widely differing scales (cf. Fig. 4) and if it extends over a long time period as often necessitated by the lifetime of a plant.

A lack of methodologies and software for bridging the gap between different lifecycle phases and the associated scales can be identified as a major cause for this general shortcoming of current simulation technologies.

Integration Across the Lifecycle

The potential of integration across the lifecycle has been widely acknowledged in the process industries (e.g. Ramage, 1998). A number of large companies have therefore launched strategical projects to improve lifecycle integration by an upgrade of their technology base but also by re-engineering the business processes in the lifecycle. The global objectives are the reduction of product and process development time and cost and the operation of a profitable plant producing high quality product at the demand of its customers at competitive prices.

Yet, there seems to be a lack of scientific principles to guide the integration effort across the lifecycle. These principles need to be established in future research focussing on the following issues of integration:

- Integration of models.
- Integration of solvers.
- Integration of design problem formulations.
- Integration of modeling work processes.
- Integration of lifecycle business processes.

These dimensions of integration are identified to be *the major perspectives of lifecycle process modeling*. A more detailed exposition will be presented next.

Model Integration

Different models describing different parts of a chemical process on the same or on differerent scales (Fig. 4) can be integrated to a single model. Ideally, the knowledge captured in the submodels should be different and not overlap. Examples include the combination of process unit models to a flowsheet model or the introduction of a physical property model into a process unit model.

These examples gives rise to the distinction of *vertical* and *horizontal model integration* (Schuler, 1998). In horizontal integration, models of process parts of comparable granularity but of probably different kind are combined to a model of the whole process. For example, we may integrate a CFD model of a reactor and a neural net model of a melt degassing unit with lumped mass and energy balance models of the remaining process units to form the model of a complete polymerisation process. In contrast, vertical model integration refers to the case where submodels are introduced in a supermodel in order to increase the degree of detail of the description of the physical and chemical phenomena occurring.

The technical implementation of model integration is largely determined by the way models are represented. We distinguish *procedural* and *declarative representations* (or *closed* and *open form models*, respectively). Declaratively represented models are coded in a generic modeling language. They are completely independent of the solvers applied. With the term solver we refer to any numerical algorithm applied to a model such as simulation or optimization. In contrast, procedural representations are intimately intertwined with a solver. They are coded in some programming language. Integration of declarative models requires aggregation of submodels on the level of the modeling language and is therefore similar to the concept of *data integration* in software engineering (Wasserman, 1990). In case of the integration of procedural models some concept of *control integration* must be applied to execute the partial models according to their aggregation in the complete model.

Model integration can be accomplished either *a-priori* or *a-posteriori*. In a-priori model integration the partial models are designed and implemented with a particular way of integration in mind. The decomposition of the complete model into partial models is only applied to reduce complexity. In this case, model integration can easily be accomplished, but the reuse of the partial models in other modeling projects is often impossible without their modification. This mode of modeling is often applied in custom modeling of a dedicated process using generic modeling tools such as PSE's *gPROMS*.

In contrast, a-posteriori model integration refers to the case where partial models can be integrated in an almost arbitrary manner. Consequently, the resulting models are not known when the partial models are developed. A-posteriori integration requires at least the definition of standards for the logical interfaces of the partial models to be integrated. This is sufficient for models which are represented by means of a declarative language. In case of procedural representations, software interfaces must be standardized in addition to allow inter-operation of the software on one platform. A typical example is process modeling by means of flowsheeting systems such as AspenTech's *AspenPlus* or Hyprotech's *HYSYS*. If the models are supposed to be distributed across different hardware or software platforms, some matching *middleware* (Adler, 1995) must finally be selected to facilitate *platform integration* (Wasserman, 1990).

Solver Integration

If we want to integrate procedurally represented models, model integration is obviously tightly linked to solver integration. Biegler et al. (1999) give a review with emphasis on steady-state simulation and optimization whereas multi-solver dynamic simulation has been reviewed elsewhere (Helget, 1997). Any multi-solver approach attempts to integrate the solvers of the partial models such that numerical stability can be guaranteed, user given accuracy bounds can be met, and robust and efficient solutions can be achieved. In those cases where a single simultaneous algorithm exists as an alternative, performance degradation of the multi-solver solution as compared to the single-solver solution must be minimized.

There are two major motivations for solver integration as an alternative to a single solver applied to a declaratively integrated model. First, only in this case numerical algorithms tailored to the solution characteristics (like largely varying time constants in differential-algebraic equation (DAE) or spatio-temporal patterns in partial differential equation (PDE) models) and to the particular structures (like the Jacobian patterns) of the partial models can be exploited. If different model types need to be integrated - i.e. models typically used on different scales such as molecular dynamics, CFD or lumped mass and energy balance models - multi-solvers seem to be the only feasible approach (e.g. Vlachos, 1997). Second, legacy software implementing validated models of some special unit or plant can be used with little or no recoding. In principle, solver integration can also be considered for declaratively represented models in the future, if information on the characteristics of partial models is available as part of the declarative representation to drive solver selection and integration (Joshi et al., 1998).

Integration of Design Problem Formulations

A number of case studies show, that the integration of traditionally separately treated design tasks into one coherent problem formulation can lead to significant benefits. A number of examples including the integration of process design, safety, control, and operation are given by Biegler et al. (1999) and van Schijndel and Pistikopoulos (1999). The resulting optimization problems are large-scale and of mixed type and are therefore challenging to solve. Biegler et al. (1999) refer to two major issues, global optimization and model fidelity. The need for global optimization must get more attention with integrated formulations due to the increasing likelihood of local (non-optimal) extrema. Realistic and hence fairly detailed models are required for reliable results, but inevitably increase numerical effort. In addition, integrated design problems across the lifecycle require consistency between the partial models on potentially different scales.

Modeling Process Integration

This consistency requirement naturally leads to a quite different dimension of integration, which we call *modeling process integration*. In contrast to the integration of model representations and their solution discussed so far, modeling process integration refers to the work processes which are associated with the development of models during the lifecycle phases.

Despite the differences between the models employed in the lifecyle on different scales, there is an inherent connection between them since they reflect different perspectives of the same physical and chemical phenomena contributing to the behavior of the chemical process modeled. The connections must be exploited in order to achieve consistency among the models and to facilitate the systematic reuse of the knowledge accumulated previously in later phases of the lifecycle.

To illustrate connections between models in the lifecycle we may take the example of a multiphase reactor. Fig. 5 shows schematically models on various scales. On the finest scale, there are quantum mechanical models to determine the force field of a single molecule to be used in the molecular simulations which result in the trajectories of the interacting molecules in the bulk phases or at the gas-liquid interface. Similarly, the activation energies and the reaction kinetic constants of the occuring elementary chemical reactions may be computed employing quantum mechanical models. The results of these computational chemistry computations will be used in the macroscopic correlations for bulk physical properties, for reaction rates, as well as intra- and interface heat, mass, and momentum fluxes. This information enters the differential balance equations of a single phase which combined in a certain way describe the reacting fluid in the multiphase reactor in a detailed manner. Obviously, the model resulting from the modeling process on one scale is either directly integrated into the model of the next scale (which would be a matter of vertical model integration according to our classification) or the models together with the results generated are used for lumping to bridge the multiple scales across the lifecycle.

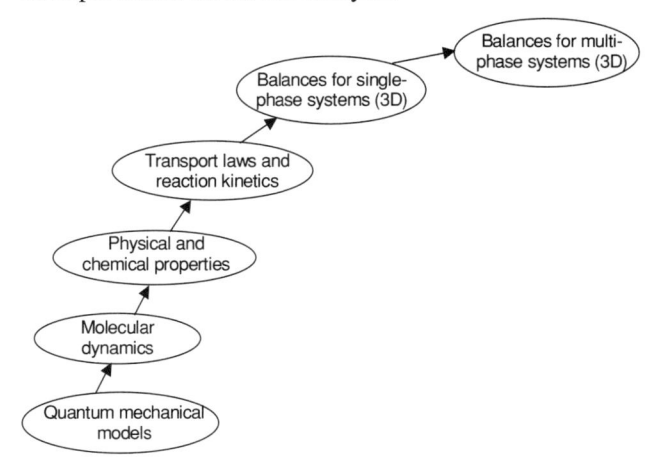

Figure 5. Connections between models on various scales in a bottom-up modeling process.

Such an integrated *bottom-up modeling process,* still visionary at this point, will not address all the requirements during the lifecycle where crude approximate models are mandatory to successfully deal with the computational and modeling complexity if problems on the scale of the plant, the site or even the enterprise are considered in the lifecycle. Approximate modeling and model reduction procedures are crucial for the success of this so-called *top-down modeling process.* This term is chosen, because we usually start on a coarse scale, suitably abstract the real process on the basis of the available understanding, and fill in the essential details employing models on the fine scales.

A detailed model stemming from a bottom-up approach will not be available in the future for industrial processes. This is not only because of the lacking theories, methods and tools to implement the bottom-up modeling process at reasonable cost. Rather, such an approach is often not practical because of the increasingly competitive business situation process industries are facing (Smith, 1997). Typically, a new manufacturing process is designed, built, and commissioned as quickly as possible with a minimum of deep knowledge to reduce time to market to the extent possible. If the product sells, one goes back and tries to better understand the process by more detailed modeling in order to revamp the design and the operation of the process to maximize profit.

Therefore, due to lacking or at least incomplete knowledge on the finer scales, we have to rely on some abstraction and simplification on the scale of interest in the context of a certain task in the lifecycle. Inevitably, there will be an overlap in the knowledge encoded in the various models. Hence, the systematic reuse of models, their parts, the results produced with them, or at least the knowledge encoded in them across the lifecycle is difficult to achieve or almost impossible if modeling process integration is not addressed systematically. Modeling process integration may also contribute to reach a certain level of consistency between the various models, if they ought to be integrated in a later lifecycle phase as discussed before.

Lifecycle Business Process Integration

Not all relevant information produced and processed during the lifecycle is appropriately captured by mathematical models and the associated modeling processes (Subrahmanian et al., 1993). Complete integration across the lifecycle therefore also requires to cover other aspects besides mathematical modeling. Examples include among others (i) the formulation and decision on requirements, the plant must ultimately meet, (ii) patent or literature search and its consequences, (iii) experimental investigations related to process development, (iv) detailed engineering design and its documentation, or (v) the formulation of operating procedures and quality assurance strategies. Consequently, the mathematical models need to be integrated with these and other types of information also arising during the lifecycle. This leads to an extended understanding of a *product data model* as compared to that typcially expressed in the product data integration literature (e.g. Book et al., 1994, Book, 1999). In particular, a family of mathematical models on multiple scales required for various activities in the lifecycle are viewed as a major and integral part of the product data model. Product data modeling and integration is hence generalized here and viewed as part of lifecycle business process integration.

The *lifecycle business process* includes all the activities carried out during the lifecycle by all people involved. This work process can be decomposed in a number of tasks which are ordered sequentially or in parallel. Each of these tasks is associated with a significant amount of information which is either needed during the task or which is produced by it. This information coincides with a certain view on the generalized product data in the sense introduced above. Information modeling can hence not be restricted to product data, even if understood in a generalized manner. Rather, the work processes in the lifecycle, providing the context of the product data, must be subject to information modeling on a resolution of sufficient granularity to match that of the product data. This *process data* must complement the product data to facilitate complete transparency of the lifecycle business process. Such a product/process information model is a prerequisite of the advocated business process lifecycle integration.

State of the Art, Trends, and Challenges

After having introduced the major dimensions of integration across the lifecycle, the state of the art will briefly be summarized in this section emphasizing available commercial technology. Trends and challenges are elucidated by comparing the requirements of the last section with the state of available technology.

Model Integration

Some kind of model integration is supported by a number of commercial process modeling system for a restricted part of the lifecycle typically associated with a particular scale of resolution. Integration is accomplished either with procedural (closed form) or declarative (open form) models. As an example, integration for steady-state flowsheeting, equipment rating and dynamic simulation as pursued by all the major vendors may be cited.

Vertical and horizontal model integration is most commonly employed. Closed form or open form process unit models are integrated horizontally on the flowsheet level. These models can be of different kinds such as classical flowsheeting models, CFD models, or neural nets. In contrast, closed form or open form physical property models are vertically integrated with the unit models. More advanced vertical integration is becoming available in most modeling systems, allowing the integration of any kind of (simple) procedure to customize phenomenological correlations in process unit models.

Model integration is facilitated tremendously if it can be based on declarative representations by generic equation-oriented modeling languages as available in a number of commercial systems such as AspenTech's *SpeedUp* and *ACM*, PSE's *gPROMS* or SimSci's *Romeo*. Despite the different expressiveness of these languages, all of them support horizontal and vertical model integration. The most powerful language of *gPROMS* supports an integrated structured representation of hybrid discrete-continuous partial integro-differential-algebraic equation (PIDAE) models (Barton and Pantelides, 1994, Oh and Pantelides, 1996) as well as their solution after a method of lines approximation. The language definition allows for almost arbitrary decompositions of a complex model into partial models.

Hence, within this paradigm all technical presuppositions are fulfilled to support model integration in a comprehensive manner. However, despite the maturity of equation-oriented modeling languages, model integration is a-priori in most cases. The developer of a dedicated complex custom model decomposes the model into partial models during the development process and codes them by means of the language constructs. Decisions on the model decomposition are often made on an ad-hoc basis instead of trying to achieve a more general solution. This way, reuse of partial models across the lifecycle or even in the lifecycle of different plants is severely restricted. A-posteriori integration, at least within the restrictions of a specific simulation system requires agreement among the modelers of an organization regarding model structuring concepts such as those reviewed by Marquardt (1996). These model structuring concepts must then be implemented on top of the modeling language of the target simulator to extend the language by domain specific concepts. Our own investigations (Cos, 1995) have shown that such an approach has been technically feasible on the basis of commerical software for some time. However, industrial realization has supposedly not been seriously attempted despite the economical potential of the approach.

True a-posteriori integration of models across different systems is possible if translators are available which translates a declaratively represented model coded in one modeling language into target modeling language. In addition to semantic differences between the constructs available in source and target languages, which may seriously impede model translation, there is the known n^2 converter problem occuring in any data integration problem (Book et al., 1994). As suggested by Barker et al. (1993), the integration problem can be significantly relaxed if a standard declarative model representation language would be available which could serve as a neutral model exchange format in complete analogy to product data exchange standards like *pdXi-STEP* (Book et al., 1994). The *Modelica* consortium aims at establishing such a language standard for general systems modeling (Mattsson et al. 1998). A similar problem occurs in the case of procedurally coded models. Here, a set of mediators (or wrappers) is required to transform between different interface specifications via a neutral format to access the model implementation.

Hence, integration practiced in commercial systems is only partly a-posteriori regardless whether the simulator is based on declaratively or procedurally represented models. The partial models offered in a model library have uniformly defined interfaces to allow for an almost arbitrary integration within the *same system on a single platform*.

The request of the simulator users for more flexible and powerful capabilities for a-posteriori model integration led to the EC funded *CAPE-OPEN (CO) project*, the second phase of which *(Global CAPE-OPEN)* has just been started. Simulator vendors and users as well as a number of academic experts in modeling technology have formed a consortium to work towards full a-posteriori integration of process unit and thermodynamic property models (Braunschweig et al., 1999, Köller et al. 1999). Standard interfaces are defined and prototypical component software implementations are developed for two major middleware platforms, Microsoft's *COM* (Box, 1998) and *CORBA* (Object Management Group, 1998) to demonstrate the feasibility of interoperability of process models among different commercial systems. If the *CO* interface specifications will be accepted as a de-facto-standard and actively supported by the major vendors, full a-posteriori integration of process unit and physical property models will be accomplished in the near future.

Solver Integration

There are a number of activities in pursuing the idea of solver integration (Biegler et al., 1999). The most notable system is Bayer's *Simulation Manager*, which enables heterogeneous multi-solver simulation by integrating different generic commercial and dedicated in-house simulators on a distributed platform. The software has been used in a number of in-house projects by Bayer, Germany, and is now commercially available as *SimSight* from SimSci. Other attempts towards solver integration include the foreign object interface in *gPROMS* (Kakhu et al., 1998), which currently facilitates the integration of steady-state models into a *gPROMS* model, and CHEOPS (von Wedel and Marquardt, 1999), a fully component based prototypical simulation system which aims at the a-posteriori integration of existing software - implementing models of different kinds on different scales - by means of configurable numerical strategies. It should be noted that the *CO* approach also provides a certain kind of solver integration, since *CO* process unit and physical property models are implemented as components which usually include the model together with a (tailored or generic) *CO* compliant solver.

Integration of Design Problem Formulations

For a review of the state of the art and the potential of optimization based integration of design problems we refer to Biegler et al. (1999) and van Schijndel and Pistikopoulos (1999).

Modeling Process Integration

To the authors' knowledge, there is virtually neither an academic nor a commercial system supporting modeling process integration as introduced above. This is not surprising since modeling methodologies which systematically address bridging of multiple scales of resolution are still scarce (Glimm and Sharp, 1997) and are definitely not worked out to an extent necessary to start general purpose software development.

In fact, only little attention has been paid to the modeling process as such. Marquardt and co-workers seem to be the first emphasizing the importance of the work process for chemical process modeling. Jarke and Marquardt (1995) present the prototypical (work) process-centered modeling support system *PRO-ART/CE* which has been adopted from *PRO-ART*, originally developed to support requirements engineering (Pohl et al., 1999). These ideas have been further detailed in the *ModKit* system (Bogusch et al., 1997, 1999).

Lifecycle Business Process Integration

The problem of lifecycle business process integration is closely related to enterprise and business process modeling (Mannarino et al., 1997, Fox and Gruninger, 1998). In contrast to these approaches, which have largely been adopted in the STEP project (ISO 10303, 1994) in order to guide product data modeling, it is anticipated that the lifecycle business process must be captured at a finer level of granularity to achieve a sufficient degree of integration, which is required to attain a noticable improvement.

Marquardt and Nagl (1998) argue that mere product data and control integration as currently employed by available technology is not sufficient to achieve true integration across the lifecycle. However, the pdXi and PISTEP product data integration initiatives (Burkett and Yang, 1995, Book et al., 1994, ISO 10303, 1997,1998) as well as process design databases like Akzo's *FET*, Fluor Daniel's MasterPlant (Nagy et al., 1999), Bayer's *PDW*, Aspen's *ZYQAD* or Innotec's *LogoCAD* support the persistent storage of all major design data such as process performance specifications, physical properties, simulation model specifications, simulation results, or equipment design parameters. These systems only capture part of the lifecycle (in particular conceptual design, frontend and to some extent basic engineering) and do not capture work processes which is essential for recapitulating and reusing procedural (how-to) design knowledge. Despite those shortcomings these developments point in the right directions and contribute to lifecycle business process integration.

Quantisci's *DRAMA* seems to be the only commercial attempt to address lifecycle business process integration at least in the area of conceptual process design. This product is based on the University of Edinburgh's *KBDS* (Banares-Alcantara and Lababidi, 1995). Other academic initiatives targeting the problem from a broader perspective are Carnegie-Mellon's *n-dim* project (Westerberg et al., 1997, Subrahmanian et al., 1997), the joint concurrent process engineering project at University of Leeds and at Tokyo Institute of Technology (Lu et al., 1997, Batres et al., 1998), the sharable engineering knowledge database project at Ohio State University (Miller et al., 1997), and RWTH Aachen's *IMPROVE* project (Marquardt and Nagl, 1998, Nagl and Westfechtel, 1999).

Lifecycle Process Modeling – A Major Challenge

Model integration, solver integration, integration of design problem formulations, and the integration of the modeling processes and ultimately all the lifecycle business processes have been identified as the major issues in lifecycle process modeling. We want to propose the following definition which is intentionally somewhat analogous to Motard's definition of process simulation given in the introduction:

> *Lifecycle process modeling* is the formal representation of the lifecycle business processes by an information model which is interpreted by means of computer-based support systems to achieve an integrated design of chemical products and their manufacturing processes.

This definition sets the stage for an outreaching and visionary research objective. It will take a concerted and long term interdisciplinary effort to make progress in this direction. Even if this objective will not be reachable in the next future, it hopefully helps to focus efforts and to guide research.

Conceptual information modeling is considered a major cornerstone to successfully deal with the complexity of the suggested research objective. The next section will therefore discuss this issue in detail.

Conceptual Information Modeling

Information modeling has become a standard task during the software design process. It comes in a number of different flavors depending on the objectives of the modeling activity. Fowler and Scott (1997) for example distinguish three different perspectives of information modeling with increasing level of detail:

a. *Conceptualization* targets at a proper understanding of the domain an information system is built for, with little or no regard of the actual software,
b. *specification* refers to the modeling of the software's functionality by precisely defining the interfaces of the individual modules, and
c. *implementation* finally covers all the details of software implementation by means of some programming language.

We are interested here in conceptual modeling only to contribute to a better understanding of the lifecycle.

What is a Conceptual Information Model?

A conceptual information model incorporates a common vocabulary and a common understanding of the lifecycle between all experts involved. It materializes the *shared memory* (Konda et al., 1992) of the members of the interdisciplinary lifecycle team. Obviously, the definition of vocabulary requires a conceptualization of the domain at its outset, i.e. the identification of the entities and activities together with their relationships which are assumed to be required for a representation of the business processes (*domain process model*) and their results (*domain product model*) in the lifecycle with sufficient detail and coverage. The conceptual lifecycle model is usually decomposed in a set of connected partial models to deal with the inherent complexity (cf. the decomposition of complex chemical process models into partial models). This conceptual model is closely related to a set of ontologies which represent factual as well as problem-solving knowledge (Gruber, 1993, Chandrasekaran et al., 1999) on the topics involved in the lifecycle (Batres and Naka, 1999).

The shared meaning reached by the construction of a conceptual model is crucial for any integration across the lifecycle, whether it will be supported by computer-aided tools or not. The conceptual model qualifies as the medium for information transfer and knowledge sharing across the lifecycle. As already mentioned above, it also forms the basis for building information systems to support the lifecycle business processes. In a first step, it may be refined to a formal, for example logic or frame (object) based domain specific representation language (Bench-Capon, 1990). Such formal languages are not only indispensible for sharing knowledge without ambiguity but also for the construction of persistent information repositories (knowledge bases, design databases etc.) which are at the heart of any computer-based design support system (e.g. Miller et al., 1997).

Further, the conceptual model is also used as a starting point for specification and implementation models for any other part of the information system to be build (Fowler and Scott, 1997). Finally, the conceptual model can guide the integration of existing applications (legacy software) to an integrated software environment to support the lifecycle business processes. In this context, it contributes to the conceptual basis of the middleware services connecting data sources and application programs (Wiederhold and Genesereth, 1997).

How to Build a Conceptual Information Model?

A conceptual information model is time-consuming and expensive to build. Hence, it should be useful for a number of different tasks. Ideally, one would hope that the information model is completely independent of the tasks it will be applied to. However, we have to recall the fundamental limitation of any modeling exercise: any model will reflect only those aspects of reality, which are both fully understood and considered revelant by the modelers. Though the conceptual model will most likely not cover arbitrarily diverging tasks, its design should aim to capture the breadth and depth of the lifecycle to the extent possible.

General guidelines for conceptual information modeling are available in numerous facets (i.e. Gruber, 1993, Fox and Gruninger, 1998, Chandrasekaran et al. 1999). Major issues are *completeness, generality* and *extensibility* (gradually represent all relevant information across the lifecycle), *perpicuity* (easy to understand to facilitate application across the lifecycle), *precision granularity* (primitives with no overlap in meaning on various levels of abstraction), *transparency* (similar solutions for similar problems), and *minimality* (a minimum number of concepts). Obviously, these requirements can not easily be met and usually require trial and error iterations. Our preferred approach is as follows.

A small development team is formed. The members should have a good knowledge of all the facets of the domain and of information modeling techniques. Lacking domain knowledge can be acquired in field studies for example by interviewing additional experts employing a case study approach (Yin, 1984) as practiced in the context of process modeling by Foss et al. (1998).

In the first modeling attempt, emphasis should be rather on a sekeleton of the conceptual model than on all its details. The model skeleton can be viewed as the equivalent to the architecture of an information system. It should include only the key concepts and their relations on a coarse level to reflect the basic model structure. An architecture must be designed to manage complexity and evolutionary development. In particular, information required in different places should be factored out into common packages to facilitate largely segregated and focussed development and improvement. Possible extensions must be planned during the design phase. Such an approach has been successfully employed by the *STEP* initiative (Burkett and Yang, 1995). The success of this project convincingly demonstrates how the collaborative and distributed development of a huge information model spanning widely differing domains can be simplified if an architecture of the information model is agreed upfront.

Typically, a graphical notation like *g-Express* in *STEP* (Schenk, 1994) or *UML* (e.g. Booch et al., 1999) is employed first to define the major concepts with only few details to reflect the particular knowledge in the different areas of the domain. The level of conceptual detail is gradually increased as the confidence in the overall structure is built up in the team. Agreement on the conceptualization in the team as well as a sound foundation can usually only be achieved if a formal and hence unambigious representation of the conceptual model is attempted subsequently. For this purpose, we have developed *VEDA*, a frame-based conceptual modeling language which focusses on the domain of mathematical models of chemical processes to be employed in process design and operations (Marquardt, 1992, 1996, Marquardt et al., 1993, Bogusch and Marquardt, 1997, Lohmann, 1997, Baumeister, 1996, 1999, Bogusch, 1999) as well as the associated work processes. A series of technical reports document the current status of the language and the conceptualization of mathematical process modeling (Marquardt et al., 1998). A redesign and an extension of the language is currently under way in order to capture the lifecycle more broadly. *ConceptBase*, a meta data management system (Jeusfeld et al., 1998) is used to support conceptual information modeling. It combines meta-modeling features with hybrid knowledge representation comprising deductive rules, constraints, and frames.

Westerberg and coworkers (Westerberg et al., 1997, Subrahmanian et al., 1997) have been developing *n-dim* to support business process modeling in different domains of engineering design. However, their approach is different in emphasis. They are looking primarily at actual design processes in some organization and identify factual and problem-solving concepts which characterize the particular design process by observing the design team. Rather than coming up with a *generally agreed* conceptual information model which is indispensible for lifecycle integration, they aim at the development of a *tailored* information model which can be used to implement a design support system for the design process studied. Their approach can be characterized rather as inductive in contrast to our deductive approach. Their work emphasizes on design as a social process whereas we focus more on the representation of domain knowledge across the lifecycle.

Subsequently, we will sketch some basic ideas underlying work in progress towards the development of a conceptual information model which ultimately aims at capturing the whole lifecycle. In contrast to related work (Miller et al., 1997, Lu et al., 1997, Batres et al., 1998), our modeling effort attempts to seamlessly integrate mathematical modeling on various scales and the associated modeling processes into the lifecycle business process as requested by the increasing use of multi-scale models in design. Model presentation is based on the *UML* notation (Fowler and Scott, 1997). Extensive use will be made of package diagrams to group related concepts in a partial model and of class diagrams to detail some of the concepts and their relations in the packages. A more detailed exposition will be available elsewhere (Bayer et al., 1999).

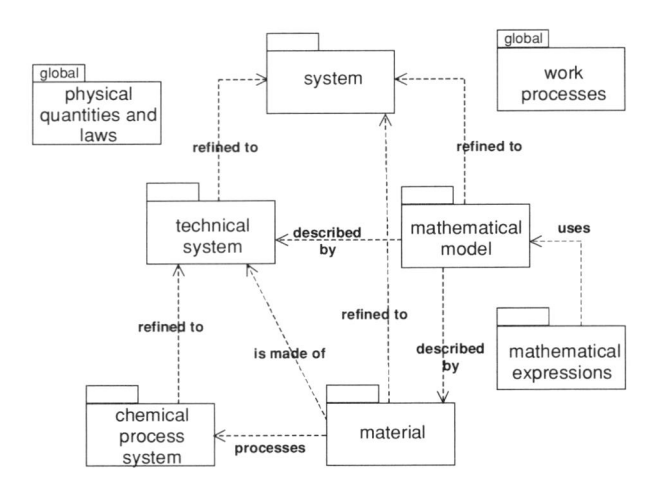

Figure 6. Conceptual lifecycle process model, overview.

Towards a Conceptual Lifecycle Process Model

A diagram depicting the major packages of a lifecycle process model is shown in Fig. 6. The *system package* captures general systems. It incorporates *perspectives* that are introduced as some kind of a filter which make only selected concepts belonging to the package of interest visible. One possible, formal definition of perspectives is given by Mariño et al. (1990). The perspectives considered during the lifecycle of a system are: the *requirements* which specify a system, the *specification* which fulfils these requirements, the *realization* which describes the reality the system abstracts, and finally the *behavior* which describes the observed behavior of the system. These perspectives are shown in Fig. 7.

The systems package is refined to a *technical systems package*, which itself is further refined to the *(chemical) process systems package*. Both packages use the *materials package* which captures concepts to describe the materials and their thermodynamics.

Technical systems and chemical process systems can be described by *mathematical models*, which are systems themselves. Models contain mathematical expressions such as equations, inequalities, etc.. Therefore, the model package uses the *mathematical expressions package* as depicted in Fig. 6.

Finally, there are two global packages (Fowler and Scott, 1997). One package captures taxonomies of *physical quantities, physical laws and their relations* (Bogusch and Marquardt, 1997). This package is used in different places. For example, it refines the concepts of a mathematical variable and an equation in the equation systems package. In contrast to variables, the physical quantities relate to a physical interpretation and are therefore associated with dimensions and a system of units. Similarly, physical laws render a physical meaning to equations. The second global package is related to *work processes* which are typically carried out during the lifecycle.

Obviously, the design of the information model structure follows a gradual refinement going from general systems to chemical process systems. The major packages are detailed subsequently regarding the knowledge they capture and the underlying rationale. Major emphasis will be on *product data models* in the lifecycle. *Work process models* will only be briefly touched.

General Systems

The notion of a general system is introduced in the sense of general systems theory (Bunge, 1979, Klir, 1985) and all the related concepts are grouped in the systems package as shown in Fig. 7.

The *requirement, specification, realization,* and *behavior* of a system are captured by perspectives. A system is elementary if it is not decomposable, or composite if it consists of other systems. This decomposition refers to systems of the same kind; systems of different nature can also be combined via the *refers-to* association as shown in Fig. 7. An *aggregated system* is an accumulation of parts with interrelation but without any structure. In contrast, a structured system is characterized by its internal structure and by its interface which connects the system with its environment (which can be another system) in a well defined manner.

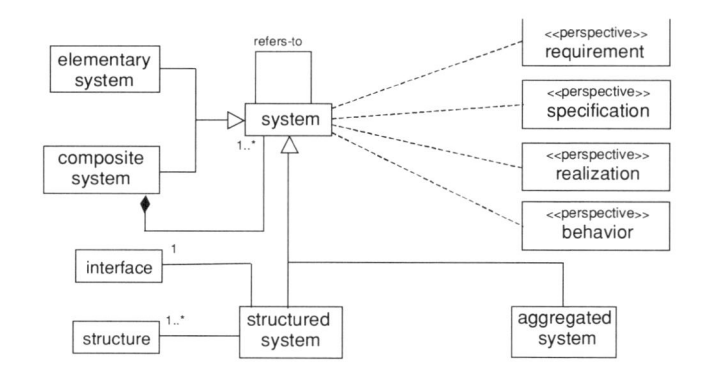

Figure 7. General system concepts and relations.

System Requirements and System Specifications

System requirements are an initial collection of information on the system under consideration and mainly deal with *what* the system will do. Requirements are not detailed here, but techniques from requirements engineering in software engineering (Gause, 1989) may be a useful starting point. In contrast to the requirements, the *system specification* describes *how* the system will fulfil its requirements, i.e. the intended behavior (the function) of the system and its intended realization to obtain this behavior.

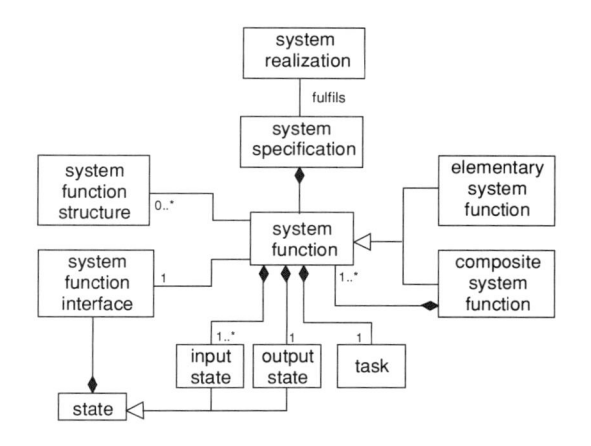

Figure 8. Class diagram of system specification with focus on the system function.

To specify a system's function, a state-task network (Miller et al., 1997, Enste and Epple, 1997, Sargent, 1998) can be employed. A state-task network is a directed bipartite graph with states and tasks as its nodes and links as its edges. Hence, states and tasks are the major concepts to define system functions. The representation of the intended realization is strongly dependent on the system under consideration. Different formalisms such as drawings, graphs, or algorithms are frequently employed.

The system function is depicted in the class diagram of Fig. 8. Functions are either elementary or composite. An *elementary system function* is comprised of a single *task* which transforms one or more *states* at its input to one state at its output, whereas a *composite system function* links a number of functions. The function sequence of a composite function is defined by a set of constraints which unify the *input state* of a function with the *output state* of its predecessor. The *interface* of a function can be defined as the set of all input and output states. Finally, a system function is associated with its *implementation* which corresponds to the system realization.

System Realization and Behavior

The realization of a system is called an *entity*. In contrast to a system, an entity is always associated with the *material* it is made of. In addition to its *interface* and its *structure*, an *elementary* or *composite entity* is characterized by its *behavior*, which is usually expressed by an ordered set of data associated with selected process quantities. The entity behavior concept typically holds all historical data during e.g. the operation of a (real) plant.

The behavior matches the intended specification of the system, if properly designed and operated.

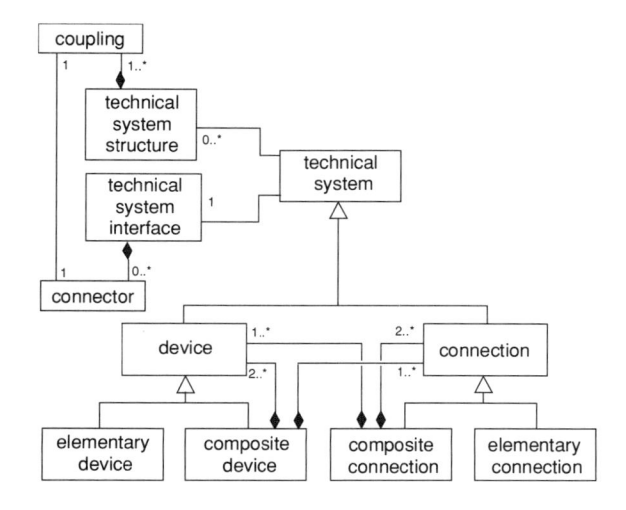

Figure 9. Class diagram of technical systems.

Technical Systems

The concepts in the systems package should be applicable to the description of any kind of system. It needs to be refined to address more specific systems. Here, the systems package is first refined to a *technical* and then to a *process systems package* (Fig. 6). The latter is related to the plant we want to build and operate. The major classes of the technical systems package are shown in Fig. 9. The technical system is a subclass of a structured system and thus inherits all its properties. In particular, it is associated with requirements, specification, realization, and behavior.

Technical systems are either *devices* or *connections*. Devices abstract those entities with major functionality, whereas connections link devices to complex technical systems (see Marquardt, 1996 for details in the context of chemical process systems). Devices and connections can occur as either *elementary* or *composite*. The interface of a technical system is specified by *connectors*. By their linkage with *couplings* technicals systems can be interconnected to form composite systems. With this mechanism the couplings describe the *technical system structure*.

Chemical Process Systems

A *chemical process system* extends the definition of a technical system (cf. Fig. 6) by adding additional structure. In particular, the chemical process system contains the *processing, operating, and managing subsystems* as its parts as introduced by Backx et al. (1998) (see also Batres et al., 1998), each with associated requirements, specification, realization, and behavior (Fig. 10).

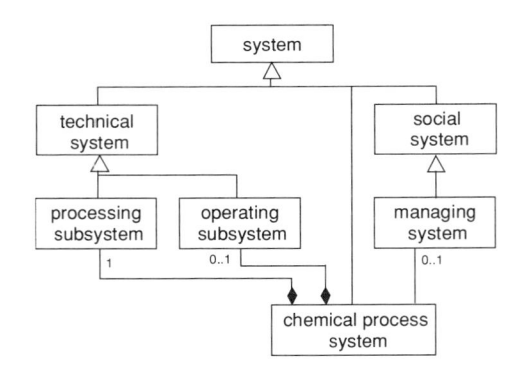

Figure 10. Class diagram of chemical process systems.

The processing system abstracts all entities related to the physical, chemical, or biological processing of materials. The operating subsystem operates (i.e. observes and controls) the processing subsystem and therefore captures all the entities related to information processing. The managing subsystem includes all the human decision makers. They monitor and set criteria for the processing subsystem through the operating subsystem. The perspectives of these subsystems form the perspectives of the chemical process system after aggregation. The concepts representing the subsystems of a chemical process system can be derived based on those introduced for the representation of structured and technical systems. We restrict the presentation here to a brief sketch of selected packages and refer to Bayer et al. (1999).

Chemical Process System Requirements, Specifications, and Realizations

The decomposition into processing, operating, and managing subsystem nicely distinguishes the three major contributors to the overall function of a chemical process system (Schuler, 1998). The allocation of specific requirements to one of these subsystems, the derivation of the matching specifications, their realization either as part of the processing, the operating, or the managing subsystems, and the description of their behavior as the sum of the behavior of the subsystems and their interrelation constitute the major tasks in the integration of design, control/operation, and operation/management of chemical process systems.

Hence, during the early phases of the lifecycle, the overall function of a chemical process system may be specified first using its associated specification concept. Later, this overall function is decomposed into the functions of the three subsystems which will subsequently planned and specified. These plans will be implemented by their realizations, e.g. either by a materials processing entity, an information processing entity or a human decision-maker.

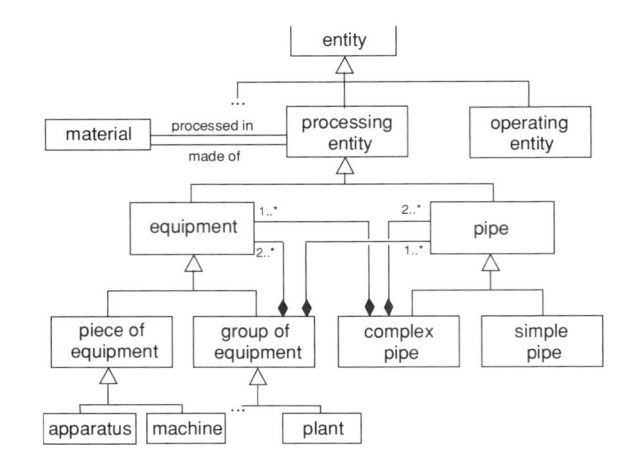

Figure 11. Class diagram of a processing subsystem realization.

Concepts for representing requirements, specifications, and realizations of the three chemical process subsystems follow directly from those introduced for technical and general systems above. These concepts are related to the *STEP* models AP 221 (ISO 10303, 1997) and to AP 231, *pdXi* (ISO 10303, 1998), describing the functional and realization perspective. A further refinement must consider the philosophy of the *STEP* models to the extent possible.

Exemplarily, Fig. 11 shows a class diagram for the realization of a processing subsystem. An entity is refined to a *processing entity*. The material any entity is associated with, is the material the processing entity is *made of* and the material it is *processing*. There are two major subclasses, the *equipment* and the *pipe*. They correspond to the device and the connection entities of the technical system. Equipment and pipe can be *elementary*, i.e. a piece of equipment or a simple pipe, or *composite*, i.e. a group of equipment or a complex pipe. A *plant* is a special composite processing entity, whereas apparatuses and machines are pieces of equipment.

Mathematical Expressions

The language of mathematics forms the foundation of any model developed and employed during the lifecycle. The *mathematical expressions package* includes therefore a fundamental set of mathematical concepts. Roughly speaking, there are variables and mathematical operators as the most elementary concepts. From these, relations of different types such as equations or inequality relations can be formed. They are combined to structured sets which among others comprise vector or matrix equations but also automata of different kind to provide various formalisms for mathematical modeling. The concepts allow the formulation of general continuous models comprised of PIDAE (Oh and Pantelides, 1996), purely discrete models (Zeigler, 1984) or hybrid discrete-continuous models (Barton and Pantelides, 1994,

Branicky et al., 1998) among others. Equation systems can be decomposed and aggregated to manage complexity. Couplings between subsystems can be expressed by identity constraints on the variables.

Class definitions of this kind (differing in expressiveness and in syntactic and semantic details) are available for example in *ASCEND* (Piela et al., 1991) *gPROMS* (Barton and Pantelides, 1994, Oh and Pantelides, 1996) or *Modelica* (Mattsson et al., 1998). A reconciled merge of these and other languages may be used to define the equation systems package. The concepts of the equation systems package are deliberately independent from any specific domain of application to acknowledge the universality of mathematics.

Mathematical Models

Concepts to represent a *mathematical model of a system* are shown in Fig. 12. A model is a specific kind of a structured system. Note that a model of a system can be associated to it via the *refers-to* link introduced in Fig. 7. As an example, a technical system is associated to its model, and the specification perspective allows access to this model (besides other specification concepts). Mathematical models consist of a set of expressions and variables defined in the mathematical expressions package as shown in Fig. 6. The model package includes *elementary* or *composite models* with an *interface*, a *structure*, a set of *expressions*, and a *behavior*. In contrast to the behavior of a technical system, the model behavior captures the results generated by the model in a simulation experiment. The interface of the system model hides some of its internals and explicitly enables well-defined access to it. For this purpose, variables occuring in the model behavior are published at the model *ports* which mirror the system connectors. For dynamical systems, these variables can be classified as either inputs, outputs or parameters, whereas the remaining variables are called states. The model expression concept holds all the mathematical expressions of the model. Control engineering block diagrams or flowsheeting models can be subsumed under the concept of system model.

So far, the model has been introduced as a declarative concept. In order to include a procedural view, a *model implementation* concept is added, which corresponds to the system realization. It associates a model with a *solver*. The solver itself is associated with an *algorithm* and a *programming language*. Often, not the initial system model, but rather a *derived* or an *approximated model* is linked to the solver. The derived system model may result from manipulation of the expressions for a more efficient or more robust numerical evaluation or from adding derived information such as sparsity patterns or Jacobian and Hessian matrices to the system model. The approximate system model results from a physically or mathematically motivated approximation such as a method of lines discretization of a PDE or a quasi-steady-state assumption introduced into a balance equation.

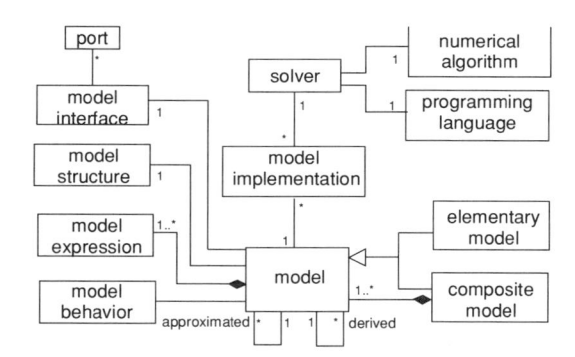

Figure 12. Class diagram of general models.

Chemical Process Models

There are three packages of models associated with the three parts of a chemical process system. The *model of the managing subsystem* is identical to models describing general systems, but may be refined by concepts used in organization modeling (Prietula et al., 1998). The *processing* and *operating subsystem models packages* group models describing elementary and composite devices and connections of processing and operating subsystems, respectively. In most cases, the models of the operating subsystem are captured with sufficient detail by the concepts in the models package. In particular, hybrid-continuous equation systems are used in the model equations concept to represent the behavior of the information processing entities.

Models are associated with all the parts of the materials processing subsystem. Hence, we distinguish *device and connection models*. A class diagram of the *elementary processing device model* is shown exemplarily in Fig. 13, which refines the elementary device model of a technical system. If an empirical (black box) modeling approach is preferred, the processing device model is the same as the (general) system model (Fig. 12). However, if a first principles (white box) approach is taken, a more specific model concept must be introduced. We adopt the notion of *a generalized phase model* as defined in previous work (Marquardt, 1996). It generalizes the model of a phase in classical thermodynamics by allowing a geometrical shape, heterogeneity, spatial nonuniformity and consequently non-equilibrium processes.

Besides *interface, equations,* and *behavior* inherited from the technical system model, the generalized phase is associated with concepts to capture the *balances* of selected extensive process quantities at a certain resolution (or scale, cf. Fig. 4), a *material* and *geometry model*. These concepts include a qualitative characterization and a set of physical equations (derived from the phyiscal quantities and physical laws package) and at the same time refining an equation system. All the equations are concatenated in the model equations concept of the generalized phase.

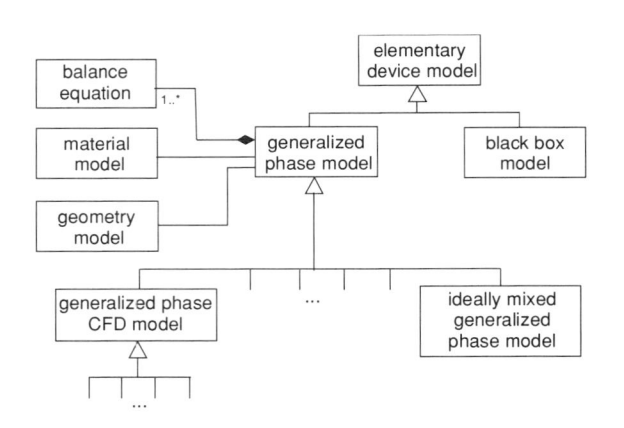

Figure 13. Class diagram of an elementary processing device model.

To accomodate the requirements on modeling in the lifecycle, refinements of a generalized phase need to be introduced. These concepts classify different models according to their level of resolution (or scale, cf. Fig. 4), which is closely related to their application in a lifecycle phase. As two examples, we have included an ideally mixed phase and a 3D computational fluid dynamics phase in Fig. 13. These models represent the same process systems device, namely the fluid contained in a reactor vessel, at different scales and for different purposes.

There is an analogous class diagram for the models of an *elementary processing connection*, namely the phase boundaries of generalized phases. This partial model and other details can be found in Bayer et al. (1999).

Material Systems

The concepts grouped in the material systems package describe the materials a plant is made of and the materials a plant produces from a macroscopic point of view. An extension to cover material representations also on microscales at different resolution is subject to future work.

Material is an instance of an aggregated system as depicted in Fig. 14, it can either occur as a *single-phase* or as a *multi-phase* system, which constitutes of single-phase materials. The single-phase material itself is formed by either *single species* or a *mixture of species* with an obvious composition relation. Any material is associated with a set of *macroscopic properties* which themselves can be represented either by (experimental) *property data* or by *property models*. Any material has a *state* and may be subject to *equilibrium constraints*. These concepts make use of the *physical quantities and physical laws*

package. The material property models and the equilibrium constraints refine an equation system; therefore they are part of the material model package which is a specialization of an aggregated model.

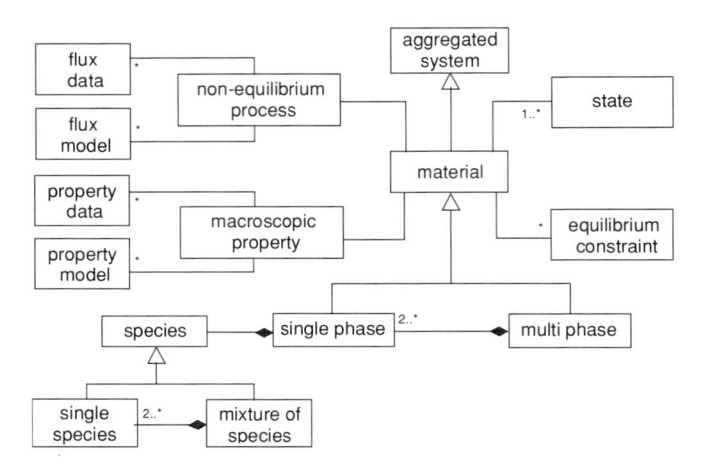

Figure 14. Class diagram of the material system and its model.

Work Processes

In the previous subsections concepts for capturing the product data during the lifecycle have been discussed. This information is used, manipulated, and extended by the activities carried out during the lifecycle. However, regardless of the detail the product data model tries to achieve, the history of an entity created is never covered to its full extent. In order to support the reuse and tracing of modeling process knowledge and to provide guidance during model development a proper representation of these activities must be included into the model (Pohl, 1996).

Therefore product data models have to be enriched with work process models. The definition of pre- and post-conditions of single activities can be used to deduce activities that can be applied on a certain entity and to guarantee the correct execution of activities. Further, given a certain activity, products required as input to it can be inferred. The conditions are mainly characterized by the product data that is used and created during that activity (Lohmann and Marquardt, 1996, Lohmann, 1997). From a conceptual point of view, activities can also be seen as methods that work on the product data. An extended treatment of this matter is given in (Bayer et al., 1999).

An Advanced Architecture for Integrated Modeling

This section illustrates how the conceptual information model of the previous section can guide the design of an advanced software architecture for integrated modeling during the lifecycle. Such an environment constitutes a major part of the integrated information system supporting the overall lifecycle business process.

The (informal) architecture of the information system presented in the following does not address all the requirements of lifecycle process modeling but rather focusses on the development and use of simulation (and optimization) models as a part of the lifecycle. The suggested architecture extends the design of the *ModKit* system (Bogusch et al. 1997, 1999, Lohmann, 1997, Bogusch, 1999), which has been under development for a number of years. It is related to the multigraph architecture introduced by Karsai et al. (1995).

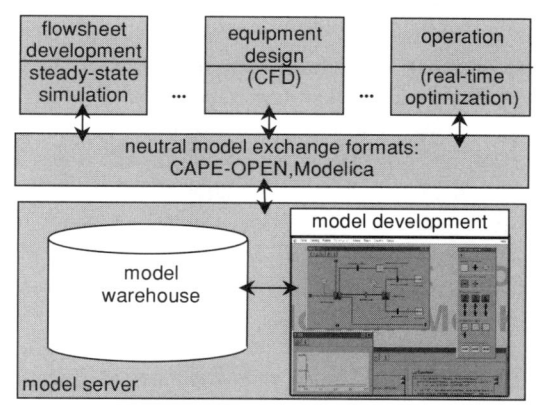

Figure 15. Coarse conceptual structure of an advanced integrated modeling system.

The integrating factors of the integrated modeling system (Fig. 15) are the concepts of the mathematical model package and its refinements as well as those in the modeling processes package as part of the work processes package. Consequently, all functionality related to model development, reuse, and maintenance is incorporated in a single software module, the *model server*. It interacts with a set of clients, comprising existing modeling and simulation tools or dedicated solvers, which are specific to the tasks on different scales in various phases of the lifecycle. Data and control integration of the model server with its clients is accomplished by means of neutral formats exploiting for example the forthcoming *CO* or *Modelica* standards.

The model server can conceptually be separated into two parts, one to support the model development, reuse and maintenance, and the other to provide persistent storage of process models in different representations by a model warehouse (Fig. 16).

Though existing modeling tools can be used for model development, the features of the model server are leveraged by a tailored model development environment as discussed subsequently. It can be viewed from three different perspectives: (i) User-machine interaction must be facilitated by a suitable graphical user interface comprising several task-specific *tools*. (ii) Well understood domain knowledge is incorporated in a set of intelligent *agents* in order to assist the modeler and in a model library to facilitate reuse of predefined models. (iii) A so-called *process enactment service* provides context-sensitive modeling support based on modeling process knowledge. Finally, an *event channel* enables asynchronous control integration within the environment.

The *model warehouse* is based on an object-oriented database management system (OODBMS) that is encapsulated by an *application programming interface* (API). The OODBMS contains a semantically rich, declarative representation of models based on the concepts described above as well as references to simulator-specific implementations. Various *code generators* provide import and export functionality to and from neutral formats. A *schema management* module controls the extension of the model warehouse schema.

Model Development

The *knowledge based graphical user interface* in the style of construction kits (Fischer and Lemke, 1988) supports the modeler during all lifecycle phases. Such a construction kit consists of a set of interactive tools (Fig. 16). The interaction between the modeler and a tool is based on the model concepts introduced above. These concepts are most intuitive (if reasonably) chosen and accelerate the learning curve for any novice or occasional user. According to its preferences, the user may either specifiy models on the level of an equation system, a general system or a chemical process system. Appropriate graphical presentations and menues for the specification of the concepts are provided for each of these views of a simulation model. Computations with the model using some solver as well as the visualization of the results can be integrated in the user interface to avoid unnecessary and time consuming switching between tools. A prototypical realization is available as part of *ModKit* using the infrastructure of Gensym's *G2*, which has been chosen as a rapid prototyping platform. In the future, developments based on Java can provide platform independence in a heterogeneous corporate network.

From a technical point of view the tools do not interoperate directly, but are decoupled by means of asynchronous events in order to enable new tools to be brought into the environment as needed (Brown, 1994). Using an event channel, events are reported to a central service which captures a model of the overall modeling process, i.e. the logic of the application. This *process enactment service* requests actions from tools or the user as encoded in the modeling process model according to the current context (see also Jarke and Marquardt, 1995) and thus implements the modeling process. Note, that the process enactment logic is not necessarily hardcoded. It could rather rely on a configurable implementation of the modeling process. Such an approach has been explored in *ModKit* to support modeling processes on a fine level of granularity (Lohmann, 1997, Bogusch et al. 1999). Here, the underlying representation of the modeling process was based on a model of the work process as introduced above.

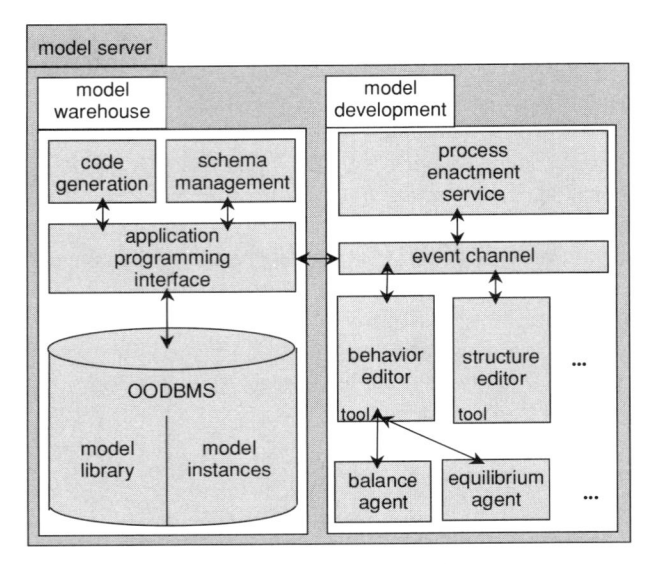

Figure 16. Architecture of a model server environment.

In a future design, the functionality of the user interface should be restricted only to the functionality of human-software interaction. Self-contained software components implementing intelligent agents (Russel and Norvig, 1995, Linninger et al., 1998) for specific tasks should assist a user with the selection, specification, and aggregation of predefined partial models contained in the model library and provide decision and documentation support functionality. A particular agent covers only a very limited domain of discourse coinciding with one of the concepts introduced in the conceptual model. For example, a phase equilibrium agent would provide the knowledge in the equilibrium constraints concept (Fig. 14) to assist a user in formulating phase equilibrium equations in a certain material including the selection of appropriate property models, a balance agent would implement the knowledge of the balances concept (Fig. 13) to assist a user in deriving appropriate balances for a generalized phase, or a solver selection agent would implement the knowledge how to select a solver best suited to the current model to solve (e.g. as in Casanova and Dongarra, 1998). The implementation of these agents obviously intertwins product and process data as captured in the conceptual information model on the fine granular level of a specific task. Interactive tools as described above are communicating with appropriate agents (Fig. 16).

Obviously, the type and granularity of the processes encoded in the agents themselves and in the process enactment service is an important design decision. Well-understood processes on a fine granular level that can be performed more or less automatically (in the sense of an algorithm) are suited for an agent-based approach. Processes on a modeling (or even business) level usually require a lot of human interaction via a tool and should therefore be supported by the process enactment service. In principle, existing software applications can be used and adapted to interact with the process enactment service if they fulfill the requirements stated by Pohl and Weidenhaupt (1997). Obviously, any newly developed tool should fit those by design.

The construction of a complete model is supported by a *library of predefined modeling objects* on various levels of granularity (e.g. Marquardt, 1996, for a detailed exposition) to facilitate the reuse of domain knowledge by selection, specification, and aggregation. These modeling objects are defined according to the major concepts of the conceptual data model. Their proper definition guarantees independent development, arbitrary horizontal and vertical integration as well as unconstrained reuse in different applications. The logical relations between the modeling objects identified during conceptual information modeling are implemented as part of the library. Ideally, the modeling objects should cover all the scales introduced in Fig. 4. A further major requirement on the library is its extensibility. Such a prototypical model library has been built in *ModKit* for a (very) restricted range of scales and applications. The model objects are coded by means of the knowledge representation system of *G2*, trading a reduced implementation effort against an unfavorably tight integration with the user interface.

Persistent Model Archivation

The *model library* is stored in the database (Fig. 17) in the sense of reusable class definitions together with model instances in a homogeneous schema - defined by a meta model - to capture *specific process models*.

Model instances are purely declarative symbolical representations of the model. However, they do not only include equations, but also a summarizing characterization, and a documentation of the modeling process they result of, including a justification of the decisions taken. The relations between the models describing a process from different perspectives are captured explicitly (for example by classification and generalization or referential relations). These relations are mandatory to keep track of earlier use and origin of particular modeling objects which is a presupposition for modeling process integration as introduced above. This way, process knowledge can be gradually compiled and assembled in mathematical models in a structured manner. In addition, reuse of partial models or complete models in later lifecycle phases or even in different projects is facilitated.

The database does *not only* store the *symbolic representation* of the model family. Rather, it may also store *any other format*, such as an executable open or closed form model coded as a software component with *CO* compliant interfaces (Fig. 17). These components could be generated from the symbolical representation by some code generator or checked into the database from an external source. Other tool specific model representations which may have been developed without using the model server can also be imported for archiving purposes.

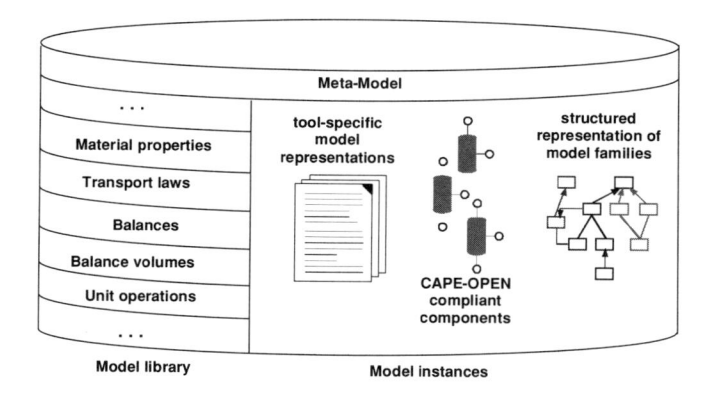

Figure 17. Structure of the model warehouse.

Importing and exporting models into and out of the database must be accomplished by means of a standard format. We suggest to use the *CO* standard for exchanging executable (closed or open form) models and *Modelica* for exchanging declaratively represented mathematical models. Hence, any exported model is converted from the internal highly structured and semantically rich representation into *Modelica* and then, if required, into a *CO* compatible component. Any imported model neither compliant with the *CO* or the *Modelica* format is translated to the internal symbolical representation to the extent possible. Only those models, which cannot be converted are stored in their native state.

Any partial model residing in the database in whatever format should be subsumed into the class hierarchy of the information model to document the semantic links between the partial models which are essential for lifecycle integration. Obviously, the granularity of these links is limited by the extent the structure of foreign models can be exploited during the translation process.

The database core must be encapsulated by a well defined application programming interface (API). It abstracts from the actual database schema by implementing a minimal set of operations to manipulate the database contents and decouples other parts of the proposed environment from the schema. The API is also linked to the event channel mentioned above in order to synchronize the graphical representation within interactive tools. Ideally, the API would be accessible by some component-based platform such as *CORBA* in order to allow access the model server via a network, probably from machines using different operating systems.

In contrast to *ModKit* where a flat class hierarchy and file-based storage is currently employed, the database should be built in the style of a *data warehouse* (Gray and Watson, 1998) employing *repository technology* (Bernstein and Dayal, 1994), to integrate different kinds of data from different sources in different representations such as the instances of the lifecycle model, *CO* compliant components or any other foreign model representation in its native format. For that purpose, a metamodel is defined which integrates the models of various sources. The schema of the model library and the model instances introduced above forms the most important part of the metamodel. Since the information model includes (work) process and product data in an integrated manner, the envisioned approach is extending current data warehouse architectures significantly (Jarke, 1999). The metamodel captures also the semantic links between the instances of the symbolic representation and the tool specific representations which may either hold complementary or redundant information in different formats. These links are not only essential for documentation but also for debugging since explicit links between the semantically rich symbolic model and the equation system solved treated by the solver can be established.

An *object-oriented database management system* (OODBMS) has been suggested earlier (Marquardt, 1992) to implement the persistent storage. It will limit or at least minimize the impedance mismatch between the database schema and the application information model. Further, a number of useful services such as fine-grained object locking and versioning or transaction and notification management are provided (Kemper and Moerkotte, 1994). Moreover, collaborative engineering processes are supported by long transactions and private workspaces. Our own technology evaluation in 1991 showed that the databases of that time were not sufficiently mature and versatile to be used as a core of an integrated modeling system. Later, Maffezoni et al. (1994) report on an early application of this technology in this context. Baumeister (1996, 1999) uses a relational database with a tailored model layer implementing the domain model by *C++* classes to bridge the mismatch between the application model and the database schema. The deductive database *ConceptBase* (Jeusfeld et al., 1998) is used to manage schema evolution. More recent experiments using current OODBMS such as *Versant* together with available component technology such as *COM* or *CORBA* have shown promising results regarding the realization of persistent and distributed object-oriented model archivation. One may conclude that a sufficiently mature technology platform is available today to build a database of the complexity we have to deal with.

A major issue we have experienced in our previous work is the evolution of the database schema. An evolutionary process must be incorporated by design since the conceptual information model and its implementation in the database schema (*schema management*, Fig. 16) will never be complete. An additional tool is required which checks the manual extensions of the schema regarding consistency, reorganizes the class hierarchy after an extension and extracts classes from the instances occuring in a particular chemical process by generalization. An approach based on a complete and decidable representation employing description logics has been explored by Sattler (1998) using *ModKit* class taxonomies as a testbed. An alternative technique to learn from instances and to generalize similar instances to a sensible class has been studied by Baumeister (1999). These attempts ultimately aim at the combination of the advantages of pure class-based and prototype-based object systems (Baumeister, 1996, 1999).

Heterogeneous Simulation and Optimization

Models of different scale and for different applications can be developed and stored by the functionalities described above. These models usually require different numerical methods to solve them efficiently. Therefore, the models are exported to dedicated existing target platforms to efficiently do the computations. This way existing software can be retained to leverage investment. Three different cases can be distinguished.

In the simplest case, a certain *model* will be *exported to a single tool* by transforming it into the tool specific format, which can then be read by the target tool to do the computations. Candidate tools are for example flowsheeting systems, general purpose process simulators with an equation-oriented model interface, CFD systems, or real-time optimizers (cf. Fig. 15). One may store the tool specific representations of the simulation or optimization experiment together with the computational results in the warehouse. Such an approach is implemented in *ModKit* at the moment. Models can be exported to *gPROMS* and *SpeedUp*, which are used as solvers.

If the *model is exported as a CO compliant component* representing either an equation system (Fig. 6) or a process systems model (Fig. 13), the model can be integrated with any *CO* compliant simulator in a more direct way (cf. Kakhu et al., 1998, for a first example). This way model libraries can be easily extended by means of custom built models. Alternatively, these components can be integrated to the complete model by fully component based simulators like CHEOPS (von Wedel and Marquardt, 1999) to facilitate integration of very different solvers typically used on various scales by means of flexible and configurable simulation or optimization strategies.

Finally, *export of parts of an integrated model to different simulation or optimization systems* can be envisioned. These systems are integrated in a coarse-grained manner to form a multi-solver for simulation and optimization (Biegler et al., 1999) as exemplified by the *Simulation Manager/SimSight* system.

Concluding Remarks

The definition of lifecycle process modeling as a unifying technology finally superseding current disciplinary process simulation and optimization constitutes a tremendous research and development effort. A suite of environments, such as the sketched modeling system, as integrated parts of a future information system can only be developed and integrated during a long term effort with carefully chosen milestones. The milestones can be defined either horizontally reflecting the tasks in the lifecycle or vertically to capture increasing level of detail. Obviously, one would focus first on those lifecycle phases and tasks which, on the one hand, are reasonably well-understood and, on the other hand, promise significant economical potential if they are integrated and efficiently supported by information technology. Software development must be accompanied by fundamental research in the following key areas.

Major Research Areas

Advanced modeling methodologies and techniques to bridge the scales are required for inter-scale model and modeling process integration. Two different routes may be pursued: (i) lumping and model reduction techniques applied prior to numerical solution such as ab-initio calculation of physical properties or meso-scale turbulence models and (ii) numerical algorithms applied to a high resolution model which automatically adapt to the required model resolution during the computations according to a user given error tolerance such as domain decomposition and multi-grid methods for PDE models or adaptive reduction of the chemical resolution in multi-component mixture processes (von Watzdorf and Marquardt, 1997).

Optimization-based integration of design problems spanning increasingly large parts of the lifecycle and solution algorithms which particularly address the increasing number of local extrema (potentially far from optimum). Besides the single-method, single-platform approaches employed for this purpose today (e.g van Schijndel and Pistikopoulos, 1999, and references cited by them), heterogeneous multi-method, multi-implementation and multi-platform problem-solving environments (Biegler et al., 1999) need to be investigated to accomodate the integration of legacy software as well as of widely differing model types and solution techniques specific to a certain scale.

Conceptual information modeling in the sense of ontology development (Gruber, 1993, Chandrasekaran et al. 1999) in knowledge engineering or domain modeling in object-oriented design (Fowler and Scott, 1997) aiming at a *commonly agreed view of the process lifecycle* is crucial for the creation of integrated information systems which can only be achieved in a long term effort with a large number of contributors. Such a conceptual model provides guidelines for the definition of the interfaces and the functionality covered by a certain software component. Besides more technical issues, an insufficient modularization of the functionality and its implementation in self-contained fine-grained or coarse-grained software components is a major hurdle for the integration of *existing* software. This problem experienced in our own integration efforts is (partly) due to a lack of a commonly agreed conceptual information model. Such an information model must capture product and process data in an integrated manner on matching levels of granularity to suit the requirements of the integration of the lifecycle business processes.

Ultimately, the conceptual information model must evolve in a commonly *agreed reference architecture* to support the lifecycle business processes. *Work process-centered information systems* as investigated in software engineering for a number of years (Finkelstein et al., 1994) and adopted to some extent in process engineering in recent years (Banares-Alcantara and Labibidi, 1995, Jarke and Marquardt, 1995, Westerberg et al., 1997, Nagl and Westfechtel, 1999) are considered an appropriate approach to effectively support collaborative activities during the lifecycle business process carried out in geographically distributed and interdisciplinary teams. Major constituents of such systems are a warehouse to persistently store the lifecycle process model, advanced tools for developing and maintaining lifecycle models and heterogeneous problem-solving environments for model execution. Existing legacy software needs to be integrated to the extent possible to protect the already huge investment.

A Promising Collaboration Paradigm

The given complexity of lifecycle process modeling and of the development of advanced commercial software to support lifecycle business processes in an integrated manner suggests a new understanding of collaboration between users, vendors, and academics. Long term and strategic partnership as well as clearly defined complementary roles must form the core of such a partnership. The client-vendor relationship practiced today in the area of process engineering software, where the end users license commercial technology and integrate them in their work processes, will be inappropriate if lifecycle process modeling is seriously addressed.

End users must get actively involved in the development effort spending significant human and financial resources. They must drive the development, which also includes to establish consortia of vendors, end users, and academics. They should formulate their requirements based on their problem understanding and continually monitor the evolving products during development with their own benchmark problems. A significant level of competence must be established in information technology not only to drive this process but also to adapt and upgrade commercial technology by in-house first-in-class methods and tools. Only then, they are able to guarantee effective application of generally available software employing enterprise specific knowhow to enable competitive advantage.

Academic research should focus on conceptual issues of such an advanced information system and develop the scientific basis in the areas discussed above. Close interaction with vendors and end users is essential to help targeting at relevant and economically interesting problem areas. New results must be implemented prototypically to show proof of concept and to enable benchmarking by end users and vendors on (at least semi-)realistic problems. These benchmarks must not only cover technical but also economical benefits of integrated modeling technology.

The *role of the vendors* is the development, marketing and maintenance of software products to enable lifecycle process modeling. Their development should aim at a highly reactive integration of research results from academia into their systems. This is only feasible, if their products are based on truly open software architectures. It seems to be impossible for one single company to establish and maintain competence in all the diverse areas to be covered for comprehensive lifecycle process modeling. Such competence is though essential for the provision of best-in-class and high performant integrated products. A specialization of the different vendors seems to be more appropriate focussing on a certain problem area (specialist) or on the integration of individual components to the integrated information system (generalist). A business culture combining cooperation and competition in the sense of co-opetition (Brandenburger and Nalebuff, 1996) is essential to facilitate economical success for all parties involved. The technical precondition is the commonly agreed conceptual information model and information system reference architecture as introduced above.

The suggested model requests a significant engagement of the end users. It will only be achievable if the added value in a purely economical rather than a technical sense can be demonstrated. Convincing success stories providing detailed and precise data are rare at this point. There seems to be no such case study emphasizing a business perspective for the benefits of an integration of microscale and macroscale modeling. End users themselves must make any effort to document (and even publish to the extent possible) how integrated modeling technology can reduce development cost and time to market. Such data will be crucial to justify the effort already spent and to get corporate management approval for the necessary significant investment in an area which is traditionally not regarded as core business. Credible data can only by produced by the end user community itself and neither by vendors nor academics. However, strategic partnerships between all end users, academics and vendors may also help to produce such data in pilot projects prior to an investment decision.

Despite the scarce data, extrapolation of the economical benefits captured with existing technology should indicate a sufficient economical incentive to seriously invest in lifecycle modeling and the associated information technology. This investment will allow *design process engineering* (Westerberg et al., 1997) complementing widely employed business process re-engineering (Hammer and Champy, 1993) to contribute to the development, manufacturing, and marketing of innovative and higly profitable products.

Acknowledgement

This work has been supported in part by Deutsche Forschungsgemeinschaft (Ma 1188/5, SFB 476) and the European Union (Brite-Euram 3512). Fruitful discussions with a number of colleagues helped to shape the ideas presented. In particular we acknowledge the contributions of M. Baumeister, L. T. Biegler, R. Bogusch, W. Geffers, M. Jarke, B. Lohmann, M. Nagl, and M. Pons. This paper has been written during the stay of the first author as a visiting professor at the Department of Chemical Engineering, University of Wisconsin, Madison. The generous support is gratefully acknowledged.

References

Adler, R. M. (1995). Emerging standards for component software. *IEEE Comp.* **28**, 3, 68-77.

Alhir, A.S. (1998). *UML in a Nutshell: A desktop Quick Reference Guide*. O'Reilly, Cambridge.

Backx, T., O. Bosgra, and W. Marquardt (1998). Towards intentional dynamics in supply chain conscious process operations. FOCAPO'98, Snowbird, Utah, 5.

Banares-Alcantara, R., and H.M.S. Lababidi (1995). Design support systems for process engineering. Part 1 and 2. *Comp. Chem. Eng.* **19**, 3, 267-301.

Barker, H.A., M. Chen, P.W. Grant, C.P. Jobling, and P. Townsend (1993). Open architecture for computer-aided control engineering. *IEEE Control Systems*, April, 17-27.

Barton, P.I., and C.C. Pantelides (1994). Modeling of combined discrete-continuous processes. *AIChE J.* **40**, 6, 966-979.

Batres, R., Y. Naka, and M.L. Lu (1998). A multidimensional design framework and its implementation in an engineering design environment. *5th ISPE Int. Conf. Concurrent Engineering*, Tokyo, July 15-17, 1998.

Bartres, R., and Y. Naka (1999). Process plant ontologies based on a multi-dimensional framework. This proceedings.

Baumeister, M. (1996). Ein objektorientiertes Datenmodell zur Beschreibung verfahrenstechnischer Prozesse. In Jahresbericht 1996 des Graduiertenkollegs *Informatik und Technik,* RWTH Aachen.

Baumeister, M. (1999). Dissertation, RWTH Aachen. In preparation.

Bayer, B., M. Eggersmann, J. Hackenberg, C. Krobb, W. Marquardt, R. Schneider, and L. von Wedel (1999). *Towards a Conceptual Lifecycle Process Model*. Technical Report, LPT, RWTH Aachen. In preparation.

Bench-Capon, T.J.M. (1990). *Knowledge Representation – An Approach to Artificial Intelligence*. Academic Press. London.

Bernstein, P.A., and U. Dayal (1994). An overview of repository technology. *Proceedings of the 20th VLDB Conference*, Santiago, Chile, 705-713.

Biegler, L.T., D. Alkaya, and K.J. Anselmo (1999). Multi-solver modeling for process simulation and optimization. This proceedings.

Bogusch, R., B. Lohmann, and W. Marquardt (1997). Ein System zur rechnergestützten Modellierung in der Verfahrenstechnik. In *Jahrbuch der GVC*. VDI-Verlag, Düsseldorf. 22-53.

Bogusch, R., and W. Marquardt (1997). A formal representation of process model equations. *Comput. Chem. Engng.* **21**, 10, 1105-1115.

Bogusch, R., B. Lohmann, and W. Marquardt (1999). Computer-aided process modeling with ModKit. Submitted to *Comp. Chem. Eng.*

Bogusch. R. (1999). Dissertation. RWTH Aachen. In preparation.

Booch, G., J. Rumbaugh, and I. Jacobson (1999). *The Unified Modeling Language User Guide*. Addison Wesley, Reading, MA.

Book, N. (1999). Information models for the electronic staorage and exchange of process engineering data. This proceedings.

Book, N., O. Sitton, R. Motard, M. Blaha, B. Maia-Goldstein, J. Hedrick, and J. Fielding (1994). The road to a common byte. *Chem. Eng*, September, 98-111.

Box, D. (1998). *Essential COM*. Addison-Wesley, Reading, MA.

Brandenburger, A.M., and B.J. Nalebuff (1996). *Co-opetition*. Currency Doubleday, New York.

Branicky, M.S., V.S. Borkar, and S.K. Mitter (1998). A unified framework for hybrid control: Model and optimal control theory. *IEEE Trans. Auto. Control,* **43**, 1, 31-45.

Braunschweig, B., H. Britt, C.C. Pantelides, and S. Sama (1999). Open software architectures for process modelling: current status and future perspectives. This proceedings.

Brown, A.W. (1994). Control integration through message passing in a software development environment. *Software Eng. J.,* **5**, 121-131.

Bunge, M.(1979). *Treatise on Basic Philosophy. Vol. 4: Ontolgy II: A World of Systems*. D. Riedel, Dordrecht.

Burkett, W., and Y. Yang (1995). The STEP integration information architecture. *Eng. Comp,.* **11**, 3.

Casanova, H., and Dongarra, J. (1998). Applying NetSolve's network-enabled server. *IEEE Comp. Sci. Eng.,* **3**, 57-67.

Chandrasekaran, B., J.R. Josephson, and V.R. Benjamins (1999). What are ontologies, and why do we need them? *IEEE Intelligent Sys.,* January/Feburary, 20-26.

Cos, R. (1995). Erstellung von Modellbausteinen zur rechnergestützten Reaktormodellierung. Diploma Thesis, LPT, RWTH Aachen.

Enste, U., and U. Epple (1997). Object-oriented concepts in process control. *Proc. IFAC '97*, Belfort, 251-256.

Finkelstein, A., J. Kramer, and B. Neuseibeh (1994). *Software Process Modeling and Technology*. Wiley, New York.

Fischer, G., and A. Lemke, (1988). Construction kits and design environments: Steps towards human problem-domain communication. *Human-Computer Interact.,* **3**, 179-222.

Foss, B., B. Lohmann, and W. Marquardt (1998). A field study of chemical process modeling. *J. Proc. Control*, **8**, 325-337.

Fowler, M., and K. Scott (1997). *UML Distilled*. Addison Wesley Longman Inc., Reading, MA.

Fox, M.S., and M. Gruninger (1998). Enterprise modeling. *AI Magazine* **19**, 3, 109-121.

Gause, D.C. (1989). *Exploring Requirements: Quality Before Design*. Dorset, New York.

Glimm, J., and D.H. Sharp (1997). Multiscale science: a challenge for the twenty-first century. *SIAM News*, **4**, 17-18.

Gray, P., and H.J. Watson (1998). Present and future directions in data warehousing. *The DATA BASE for Advances in Information System*, **29**, 3, 83-90.

Gruber, T.R. (1993). A translation approach to portable ontology specifications. *Knowledge Acquisition*, **5**, 3, 199-220.

Hammer, M., and J. Champy (1993). *Reengineering the Corporation: A Manifesto for Business Revolution*. Harper, New York.

Helget, A. (1997). Modulare Simulation verfahrenstechnischer Anlagen. *Fortschritt-Berichte VDI*, Reihe 20, Nr. 251. VDI-Verlag, Düsseldorf.

ISO 10303: Part 1 (1994). *Overview and Fundamental Principles*. ISO TC184/SC4.

ISO 10303: Part 221 (1997). *Functional Data and Their Schematic Representation for Process Plants*. ISO TC184/SC4/WG3 N600.

ISO 10303: Part 231 (1998). *Process Engineering Data: Process Design and Process Specifications of Major Equipment*. ISO TC184/SC4/WG3 N740.

Jarke, M., and W. Marquardt (1995). Design and evaluation of computer-aided modeling tools. In J. F. Davis, G. Stephanopoulos, V. Venkatasubrahmanian (Eds.), *Intelligent Systems in Process Engineering*. AIChE Symp. Ser. 312, **92**. 97-109.

Jarke, M., M.A. Jeusfeld, C. Quix, and P. Vassiliadis (1999). Architecture and quality in data warehouses: An extended repository approach. *Information Systems*, **24** (3), 229-253.

Jeusfeld, M.A., M. Jarke, H.W. Nissen, and M. Staudt (1998). ConceptBase. In P. Bernus, K. Mertins, G. Schmidt (Eds.), *Handbook on Architectures of Information Systems*. Springer-Verlag, Berlin, 265-285.

Joshi, A., N. Ramakrishnan, and E.N. Houstis (1998). Multiagent system support of networked scientific computing. *IEEE Internet Comp.*, May-June, 69-83.

Kakhu, A., B. Keeping, Y. Lu, and C.C. Pantelides (1998). An open software architecture for process modelling and model application. *Proc. FOCAPO'98*, 518-524.

Karsai, G., J. Sztipanovits, H. Franke, S. Padalkar, and F. DeCaria (1995). Model-embedded on-line problem solving environment for chemical engineering. *Proc. Int. Conf. on Engineering of Complex Computer Systems*, 227-233.

Kemper, A., and G. Moerkotte (1994). *Object-Oriented Database Management – Applications in Engineering and Computer Science*. Prentice Hall, London.

Kesler, M.G., and M.M. Kessler (1958). Engineering a process with a computer. *World Petrol*, **29**, 8, 60-63.

Klir, G.J. (1985). *Architecture of Systems Problem Solving*. Plenum Press, New York.

Köller, J., L. von Wedel, B. Braunschweig, M. Jarke, and W. Marquardt (1999). CAPE-OPEN: Experiences from a standardization effort in chemical industries. 1st IEEE Conf. Standardization and Innovation in Information Technology, Aachen, Germany.

Konda, S., I. Monarch, P. Sargent, and E. Subrahmanian (1992). Shared memory in design: A unifying theme for research and practice. *Res. Engng. Des.*, **4**, 23-42.

Kuipers, J.A.M., W.P.M. Swaaij (1998). Computational fluid dynamics applied to chemical reaction engineering. *Adv. Chem. Eng.*, **24**, 227-328.

Linninger, A., H. Krendl, and H. Pinger (1998). An initiative for integrated computer-aided process engineering. *Proc. Of FOCAPO'98*, 494-500.

Lohmann, B., and W. Marquardt (1996). On the systematization of the process of model development. *Comp. Chem. Eng.*, **21**, Suppl., S213-S218.

Lohmann, B. (1997). Ansätze zur Unterstützung von Arbeitsabläufen bei der rechnerbasierten Modellierung verfahrenstechnischer Prozesse. *Fortschritt-Berichte VDI*, Reihe **3**, Nr. 531. VDI-Verlag, Düsseldorf.

Lu, M.L., R. Batres, H.S. Li, and Y. Naka (1997). A G2 based MDOOM testbed for concurrent process engineering. *Comp. Chem. Eng.*, **21**, Suppl., S11-S16.

Maffezoni, C., R. Girelli, and P. Lluka (1994). Object-oriented database support in modeling and simulation. European Simulation and Modeling Conference, Barcelona.

Mannarino, G.S., G.P. Henning, and H.P. Leone (1997). Process industry information systems: modeling support tools. *Comp. Chem. Eng.*, **21**, Suppl., S667-S672.

Mariño, O., F. Rechenmann, and P. Uvietta (1990). Multiple perspectives and classification mechanism in object-oriented representation. ECAI-90.

Marquardt, W. (1992). An object-oriented representation of structured process models. *Comp. Chem. Eng.*, **16**, S329-S326.

Marquardt, W., Gerstlauer, A., and E.D. Gilles (1993). Modeling and representation of complex objects: A chemical engineering perspective. *Proc. Conf. Intelligent Engineering Applications / Artificial Intelligence in Engineering '93*. Edinburgh, Scotland, 219-228.

Marquardt, W. (1996). Trends in computer-aided process modeling. *Comp. Chem. Eng.*, **20**, 6-7, 591-609.

Marquardt, W., and M. Nagl (1998). Tool integration via interface standardization? *Dechema-Monographie*, **135**. VCH Weinheim. 95-128.

Marquardt, W., and the VEDA-group (1998). *The Chemical Engineering Data Model VEDA. Parts 1 to 6*. Technical Reports, LPT, RWTH Aachen.

Marquardt, W. (1999). Von der Prozeßsimulation zur Lebenszyklusmodellierung. *Chem. Ing. Tech.* To appear.

Mattsson, S.E., H. Elmqvist, and M. Otter (1998). Physical system modeling with Modelica. *Control Eng. Practice*, **6**, 4, 501-510.

Miller, D.C., J.R. Josephson, M.J. Elsass, J.F. Davis, and B. Chandrasekaran (1997). Sharable engineering knowledge databases for intelligent system applications. *Comp. Chem. Eng.*, **21**, S77-S82.

Motard, R.L., M. Shacham, and E.M. Rosen (1975). Steady state chemical process simulation. *AIChE J.,* **21**(3), 417-435.

Nagl, M., and B. Westfechtel (1999). *Integration von Entwicklungsumgebungen in Ingenieuranwendungen.* Springer-Verlag, Berlin.

Nagy, B., A. Shah, and K.D. Ganesan (1999). Streamlining projects through integrated front end Automation solutions. This proceedings.

Object Management Group (1998). *The common object request broker: Architecture and specification.* Available at ftp://ftp.omg.org/pub/docs/formal/98-12-01.pdf.

Oh, M., and C.C. Pantelides (1996). A modelling and simulation language for combined lumped and distributed parameter systems. *Comp. Chem. Eng.,* **20**, 6-7, 611-633.

Piela, P., T. G. Epperly, K. M. Westerberg, and A.W. Westerberg (1991). ASCEND – An object-oriented computer environment for modeling and analysis – The modeling language. *Comp. Chem. Eng.,* **15**, 1, 53-72.

Pohl, K. (1996). *Process-centered Requirements Engineering.* Wiley, New York.

Pohl, K., and K. Weidenhaupt (1997). A contextual approach for process-integrated tools. European Software Engineering Conference, ESEC'97. Garmisch-Partenkirchen, Germany.

Pohl, K., R. Klamma, K. Weidenhaupt, R. Dömges, P. Haumer, and M. Jarke (1999). Process-integrated (modeling) environments (PRIME): Foundations and implementation framework. ACM–TOSEM. To appear.

Prietula, M.J., K.M. Carley, and L. Gasser (1998). *Simulating Organizations.* AAAI Press, MIT Press, Menlo Park.

Ramage, M. (1998). Computation and competitiveness: Managing technology in the information age. *Proc. FOCAPO'98,* 126.

Russel, S.J., and P. Norvig (1995). *Artificial Intelligence: A Modern Approach.* Prentice-Hall.

Sapre, A.V., and J.R. Katzer (1995). Core of chemical engineering: one industrial view. *Ind. Eng. Chem. Res.,* **34**, 2202-2225.

Sargent, R.W.H. (1998). A functional approach to process synthesis and its application to distillation systems. *Comp. Chem. Eng.,* **22**, 1-2, 31-45.

Sattler, U. (1998). *Terminological Knowledge Representation Systems in a Process Engineering Application.* Dissertation. RWTH Aachen. Verlag Mainz, Aachen.

Schenk, D. (1994). *Information Modeling the EXPRESS Way.* Oxford University Press. New York.

Schuler, H. (1998). Prozeßführung. *Chem. Ing. Tech.,* **70**, 10, 1249-1264.

Smith, D. (1997). Personal communication.

Subrahmanian, E., S.L. Konda, Y. Reich, A. W. Westerberg, and I. Monarch (1993). Equations are not enough – Informal modeling in design. *AI EDAM,* **7**, 4, 257-274.

Subrahmanian, E., Y. Reich, S.L. Konda, A. Dutoit, D. Cunningham, R. Patrick, M. Thomas, and A. W. Westerberg (1997). The n-dim approach to creating design support systems. *Proc. DETC'97, 1997 ASME Design Engineering Technical Conference,* September 1997, Sacramento, CA.

van Schijndel, J., and S. Pistikopoulos (1999). Towards the integration of process design, process control and process operability – current status and future trends. This proceedings.

Vlachos, D. G. (1997). Multiscale integration hybrid algorithms for homogeneous-heterogeneous reactors. *AIChE J.,* **43**, 11, 3031-3041.

von Watzdorf, R., and W. Marquardt (1997). Fully adaptive model size reduction for multicomponent separation problems. *Comp. Chem. Eng.,* **21**, S811-S816.

von Wedel, L., and W. Marquardt (1999). CHEOPS: A case study in component-based process simulation. This proceedings.

Wasserman, A.I. (1990). Tool integration in software engineering environments. In F. Long (Ed.), *Software Environments,* Lecture Notes in Computer Science. Springer-Verlag, Berlin. 138-150.

Westerberg, A.W., E. Subrahmanian, and Y. Reich (1997). Designing the design process. *Comp. Chem. Eng.,* **21**, S1-S9.

Wiederhold, G., and M. Genesereth (1997). The conceptual basis for mediation services. *IEEE Expert.* September/October, 38-47.

Yin, R.K. (1984). *Case Study Research – Design and Methods.* Sage Publishers, London.

Zeigler, B.P. (1984). *Multifacetted Modelling and Discrete Event Simulation.* Academic Press, London.

STREAMLINING PROJECTS THROUGH INTEGRATED FRONT-END AUTOMATION SOLUTIONS

Bert Nagy, K. D. Ganesan, and Ashish Shah
Fluor Daniel
Greenville, SC 29607

Abstract

This paper describes how a large globally diversified Engineering company is automating and integrating front-end engineering design.

Keywords

Design engineering, Integration, Front-end integration, Design automation, High quality, Lower cost, and Shorter schedule.

Summary

The highly competitive business environment and pressure on business profits has forced engineering contractors to look for ways and means to achieve a quantum leap in efficiency. Development of an automated integrated engineering model is a solution which has emerged from the above. All engineering data generated in the project engineering process is stored in databases and shared across applications completely eliminating the need for redundant data entry. Since the development of such a solution model requires large resources and time, it is considered more prudent to purchase commercially available software for individual applications and configure and integrate these applications o make the integrated solution. Fluor Daniel has achieved considerable progress in setting up an integrated solution for project engineering development. With the emphasis shifted from documents to data we feel some of the conventional deliverables produced in the engineering life span of a project will eventually disappear.

Need for Automation

Unlike other fields in the chemical industry, no dramatic transformation has taken place recently in the way chemical projects are engineered. The only exciting thing that has happened in an otherwise routine business is the widespread use of computer applications. These computer applications have taken over most of the day-to-day design calculations. The engineering business has become highly competitive and continuously seeks out cheaper locations for execution. This pressure on business profits has forced the engineering industry to have a thorough look at the work processes, procedures and in general, the way we do our business seeking a quantum leap.

Front End Engineering

Front-end engineering aims to establish a detailed definition of the project scope in order to satisfy the owner's objectives for the capital investment. The front-end engineering package is more than just a set of engineering documents, it is a process of drawing together resources, facilities, people and organizations to translate a marketing and/or technology based opportunity into a capital project.

Most chemical companies understand the importance of a fit-for-purpose front-end engineering package as a means of controlling the project scope and cost. In most cases, the front-end engineering package also forms the basis for an owner/operator management approval for project funding.

Statistical project data shows that projects that are better defined at authorization cost less to execute than do poorly defined projects.

This not only results in enormous effort-hours for production of these documents but also leads to inconsistency of data and additional effort in checking

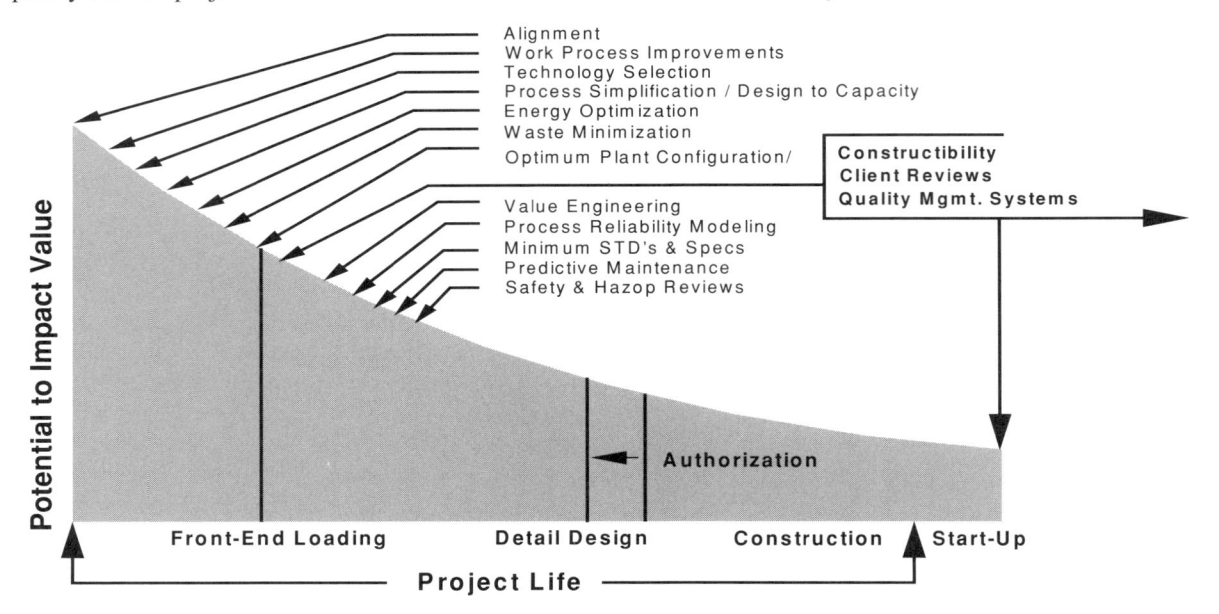

Figure 1. Cost focus to betterlife-cycle cost.

Figure 1 shows how project engineering work process elements potentially impact the "value" within the span of the project implementation time frame. Needless to say that the ability to influence value during the early part of project engineering (i.e., front end engineering) is significantly high. Often, during this period, several alternate processing options are investigated and several plant configurations are studied to improve the profitability and add value to the project. Ability to do techno-economic evaluation of these options on real time basis will allow several such options to be studied and the most economic plant configurations to be sought. This cannot be achieved without a high degree of automation to generate the simulation models, expand them to plant configurations, estimate a rough order of the magnitude of the investment cost and roll them into an economic model of the project (ROI calculations) – all in a matter of few hours/days.

Emphasis Away from Deliverables to Data

One major drawback to the traditional work process is that the entire design process is centered around the deliverables (documents) such as process flow diagrams, data sheets (equipment/instrument/electrical), Piping and Instrumentation Diagrams (P&IDs), piping drawings, building drawings, electrical drawings, etc. rather than the most important aspect, i.e. the "engineering data". This leads to a multiplicity of deliverables (documents) where the required engineering data is duplicated leading to an enormous effort to repeatedly key in the same information over and over again in different documents.

and correction. Identification of this inherent inefficiency in the traditional work process had prompted the search for a new work process where data generated in the design process could be automatically "moved" from one step to another as and when needed rather than keyed into documents repeatedly. This data integration throughout the project engineering process is expected to result in a significant increase in efficiency.

The Benefits

Development of an integrated engineering model, which will automatically share data, has the following benefits:

1. better consistency and hence accuracy of the data resulting in enhanced quality of deliverables
2. enable study of more technical options ("what if" scenarios) in shorter time period improving the quality of the solution
3. shorten the overall engineering process which will enable faster and more suitable decisions at every phase which in turn reduces the overall project schedule
4. make globally collaborative designs a reality
5. reduce design and plant life cycle costs
6. enable more accurate and early control level estimates
7. make repetitive designs easier and more cost effective

In summary, this results in increased quality and reduced project cost and schedule.

The Principle behind an Integrated Model

The first step in the development of an integrated model is an understanding of the workflow and deliverables. The traditional project development is normally split into the following three distinct phases:

Conceptual Engineering
Front End Engineering
Engineering, Procurement and Construction (EPC)

However, this paper focuses on the impact of the integrated solution only on the front-end engineering. The workflow and the corresponding deliverables are described in Figure 2.

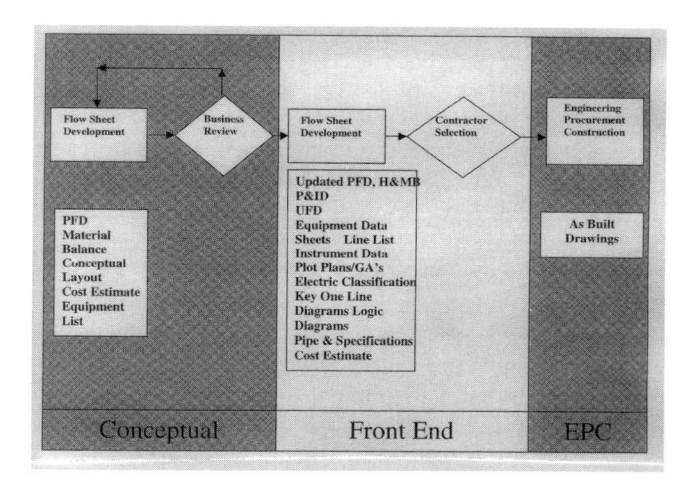

Figure 2. Conventional work flow.

In order to better understand the information flow in an integrated (front-end) solution, the traditional project engineering work process needs to be subdivided into the following typical work modules which would identify themselves with traditional deliverables:

1. Process Development (i.e. Process flow diagram with material and energy balance)
2. Preliminary/Conceptual Plant Layout
3. Conceptual Investment Estimate for the project.
4. Equipment Design Development (Data Sheets)
5. Engineering Data for Process, Piping and Control System (P&IDs, Line List, Instrument Index/Data Sheets))
6. Electrical Engineering
7. Preliminary Civil, Structural and Architectural (CSA) Engineering
8. Preliminary Plant Layout (Equipment Arrangement Drawings)
9. Control level Investment Estimates for the Project *
10. Material Management *
11. Construction Drawings * (Piping Plans, Sections and Isometrics)

* Normally not part of front-end engineering

A work process and information flow for a typical Integrated Model for Chemical Project Engineering is provided in Figure 3 for purposes of discussion.

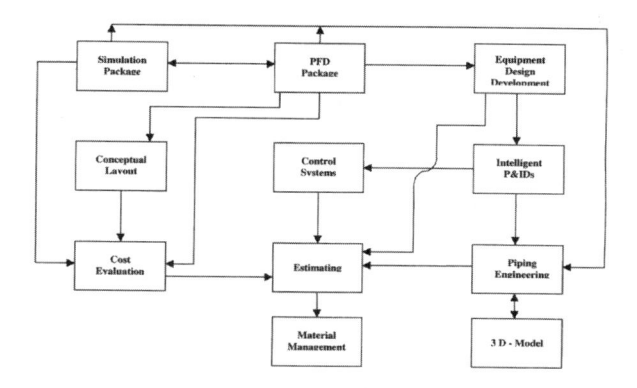

Figure 3. Integrated solution model – Information flow.

Conventional Front-end Work Process

In a conventional work process, a simulation of the process was run using a simulation program. A material balance and PFD's were prepared based on the simulation run. Process data sheets of equipment were developed and released for mechanical design. Mechanical engineers then designed and developed mechanical data sheets or mechanical design drawings, as the case may be. Using a cost estimation program and manually entering the data required for the estimate into the system, an estimate of cost for the project was then made. Any study of alternate configurations or process schemes was made by preparing the required design information manually and preparing a cost estimate for each case and then comparing the cases. P&IDs were drafted without any intelligence added to it. Process data for control system engineering will be provided in the form of process data sheets. Any transfer of information from process engineering to mechanical systems, piping and control systems will be done manually (meaning by manually keying in the information from process data sheets and forms).

Integrated Solution

In an integrated solution model, a simulation of the process is still being done using a simulation package. However, process data from a simulation package is directly transferred into a project database. Additional design data is added to this database to make the information complete. Data sheets, either process or mechanical are reports extracted out of this database. Any transfer of information between the project database and the cost estimation program is direct without the need for keying in information. In addition, it is possible to directly upload the information from a simulation model to a conceptual cost estimation program. This link facilitates techno-economic evaluation of alternate cases quick and easier. The P&IDs are intelligent (with a database behind it).

Transfer of information between the process simulation module, equipment design module, control system database and piping engineering database is direct (without the need for keying in data manually). The 3D modeling application to produce the construction drawings accesses all piping information available in Intelligent P&ID or other discipline specific databases.

Accurate project investment estimate is required at various stages of project development. This is achieved by automatically accessing the most recent engineering information from the estimation system. Once the material requirements become available from design development, this information is loaded to a material management system.

What is provided in Figure 3 is a schematic representation of the information flow. Due to obvious reasons, it is not prudent to transfer the data directly from application to application. This should be done through a central data warehouse. The configuration of the system will then look like what is provided in Figure 4.

Application Development

The degree of application development required for an Integrated Automation System of the type explained above is enormous and beyond the normal capability of an engineering organization. Some of the engineering organizations that have tried this development work have realized this rather too late. A more prudent approach would be to buy the individual application software from reputed vendors and configure them to suit specific needs. In this case, only the configuration work and integration link will need to be developed by an engineering contractor.

Development of an integrated solution for project engineering design is an active subject in the application development market and most application vendors have started talking to each other and are developing direct links between applications. These direct links are expected to further facilitate the development of the Integrated Solution.

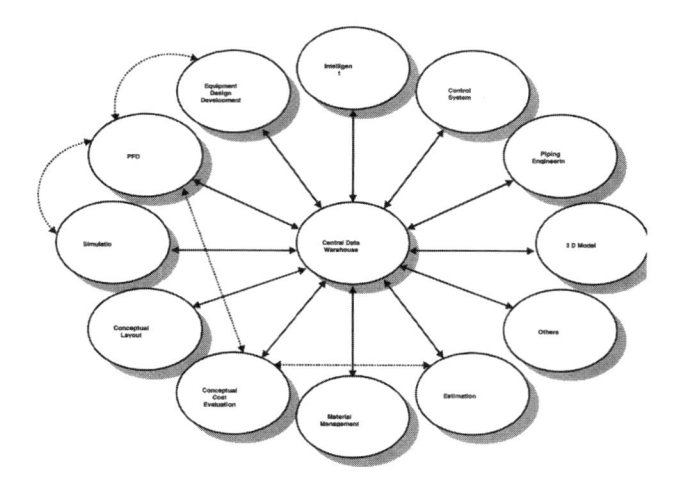

Figure 4. Integrated solution model.

Development at Fluor Daniel

As a pioneering engineering organization, Fluor Daniel has always been on the forefront of the development of the integrated design model. Faced with the challenge, the Fluor Daniel approach was to go back to the drafting board and think fresh. As a first step we reviewed the entire work process shifting the focus away from deliverables to "data". Our aim was identifying opportunities for integration. We reviewed all existing tools, forms, procedures and practices. We refined our work process to suit the integrated automation. We modified our tools to fit the integration and developed in-house tools and purchased new application tools. The aim was to select the best tool for each application. The emphasis was shifted from in-house development to collaboration with outside development vendors. Fluor Daniel has collaborated with several outside vendors to influence the development to match Fluor Daniel expectations. The customization and integration of tools were done in-house so that we would not be required to share our proprietary information with any vendor. Such customization included incorporation of Fluor Daniel methods, standards, practices, knowledge base and historical data into these application programs. The Fluor Daniel suite of integrated automation solutions is named MasterPlant[SM].

FrontRunner[SM] is the automation tool used to produce Process Flow Diagrams (PFD's) and Equipment Data Sheets (process and mechanical data sheets) and is an Aspen Zyqad[TM] based solution. It is an example of how Fluor Daniel has customized an out-of-the-box vendor program to suit specific Fluor Daniel requirements.

The conceptual layout development from PFD's is done using QuickPlant[SM] and OptimEyes[SM] - a combination of in-house and outsourced development incorporating a knowledge base developed in-house. Most of the disciplines' engineering databases are commercially available products configured for Fluor Daniel use.

Bottlenecks

There are several bottlenecks in the widespread use of the integrated solution. The most significant seems to be the effort required in training a large workforce for effective utilization of the new system. A work process improvement of this nature requires a concerted effort within the organization to upgrade workforce to be in step with technology development. This is a major task.

The other major bottleneck appears to be the constant upgrade of the application software, which requires an upgrade to the model on an almost continuous basis.

Conclusion

As indicated earlier, more vendors are developing links between their automation tools. This trend seems to slowly shift the responsibility for the integrated approach onto vendors rather than the end users.

So far, handling of data in an electronic format has been limited to the engineering contractor. Majority of data received from vendors is currently required to be keyed into the integrated model reducing certain degree of efficiency in the system. We see a trend in the future where more and more vendors will comply with our request to supply data via the extranets in electronic format compatible for extraction into the integrated data model.

As the engineering process becomes more and more automated and integrated, it is our vision that the focus will be more on management of data and **not** how to produce documents. Management of data in an automated, integrated manner is the best way to get there in a fast, accurate and repetitive. With this shift in emphasis, some of the conventional deliverables produced in the engineering life span of a project will become unnecessary and will eventually disappear.

OPEN SOFTWARE ARCHITECTURES FOR PROCESS MODELING: CURRENT STATUS AND FUTURE PERSPECTIVES

Bertrand L. Braunschweig
Computer Science and Applied Mathematics Department
Institut Français du Pétrole
92500 Rueil-Malmaison, France

Constantinos C. Pantelides
Centre for Process Systems Engineering
Imperial College of Science, Technology and Medicine
London SW7 2BY, United Kingdom

Herbert I. Britt
Aspen Technology, Inc.
Cambridge, MA U.S.A. 02141

Sergi Sama
AEA Technology Engineering Software-Hyprotech
08007 Barcelona, Spain

Abstract

Open software architectures are the way forward for the next generation of CAPE tools. They allow previously incompatible environments to interoperate through commonly agreed interface standards. They support life-cycle modeling and collaborative engineering across places and organizations. In this paper, we present two major initiatives in open software architecture for CAPE: the EU-sponsored project CAPE-OPEN, its follow-up Global CAPE-OPEN[1]. We also mention their links with other initiatives such as pdXi, STEP and OPC. We look at the benefits gained using such architectures, and at possible future directions for research and development in this area, with respect to European Commission's Fifth Framework Program covering years 1999-2002 and to the Vision 2020 documents.

Keywords

Open architectures, Interface standards, Middleware, Interoperability, Next generation computer-aided process engineering, CAPE-OPEN.

---

[1] CAPE-OPEN and Global CAPE-OPEN are funded by the European Community under the Industrial and Materials Technologies Programme (Brite-EuRam III), under contracts BRPR-CT96-0293 and BPR-CT98-9005. In addition, Global CAPE-OPEN follows the Intelligent Manufacturing Systems initiative promoting collaboration between six international regions.

Introduction

The ever-increasing usage of mathematical models in all aspects of process development, design and operation has been well documented in several recent studies (see, for instance, Foss et al., 1998). This increase in demand is matched by a significant increase in the supply of increasingly sophisticated modeling software from a variety of sources including process engineering software companies, automation system vendors and academic institutions as well as in-house developments within operating companies.

Some of the above software is intended to carry out a narrow, well-defined function such as the computation of physical properties, the simulation of a particular unit operation, or the numerical solution of certain types of mathematical problems arising in process simulation or optimization. On the other hand, other software tools are essentially environments that support the construction of a process model either from first-principles or from libraries of existing models, or both. They then allow the user to perform a variety of different tasks, such as process simulation or optimization, using this single model of the process (see Pantelides and Britt, 1995). To achieve their latter function, the second category of process tools incorporate or make use of several software tools of the first category. The distinction between these two kinds of software, albeit in practice not always as clear as described above, is particularly important for the purposes of this paper. We shall henceforth call them *Process Modeling Components* (PMCs) and *Process Modeling Environments* (PMEs) respectively.

The wealth of supply of process engineering software is clearly a positive factor for the process industry and has already resulted in major benefits. The recent industrial experience with the use of this software has revealed some interesting technical and commercial factors, which are worth noting:

1. *The need of integrated process modeling.*

    There is an increasing recognition of the need for adopting a more integrated view of the process that takes account of all-important interactions. This, in turn, implies that individual PMCs are of relatively limited applicability in isolation. For example, consider a specialized reactor that forms the heart of a certain plant. Although the availability of a software package implementing an accurate model for the reactor would certainly be very useful, it would be even more desirable to be able to couple this model with models of the separation part of the plant, thereby permitting an analysis of the trade-offs between conversion, selectivity and cost of separation. This indicates that the reactor PMC should ideally be incorporated within an appropriate PME that facilitates the necessary coupling.

2. *The adoption of generic PMEs.*

    In the 1960s and 1970s, almost every major process engineering company had developed internally its own PME, usually in the form of a steady-state modular flowsheeting package[2]. This trend continued well into the 1980s and, in some cases, the 1990s. However, it has increasingly become difficult for operating companies to justify the large cost of maintaining and supporting these tools, let alone to continue their development while keeping up with the rapid progress in process modeling as well as the increasing sophistication of the underlying software engineering and mathematical solution techniques. The current trend is, therefore, to cease the internal developments and, instead, to adopt PMEs provided by external software vendors. However, a small but often quite significant part of the internally developed technology (e.g. specialized unit operation models or physical property data) often represents a real competitive advantage for the individual companies. Consequently, in many cases it has been necessary to retrofit this technology within the generic PMEs brought in from external sources.

3. *The limited supply of PMEs.*

    The developments outlined above, together with the substantial increase in the sets of skills and resources required for developing, maintaining and supporting PMEs, has naturally limited the number of process engineering software companies that have been able to undertake such activities. In fact, the products of just three such companies currently cover most of the market.

4. *The increasing range of model-based applications.*

    Traditionally, process models have formed the basis for a relatively small number of different types of activities such as steady-state and dynamic simulation and optimization. However, it is increasingly recognized that the range of possible applications of such models is potentially much wider, covering, for example, such diverse activities as plant data reconciliation, controllability analysis, process safety validation, fault detection and so on.

---

[2] A similar trend was reflected in academic research, where a large number of steady-state and, later, dynamic simulation tools were being developed.

Each of the factors mentioned above is, on its own, quite understandable, reflecting the natural evolution and maturing of this sector of technology. However, all of them considered together point to certain serious problems that need to be addressed if the process industries are to reap the full benefit of technical progress in this area:

- There is an abundance of good-quality PMCs; however, these will be used widely by industry only if they are somehow incorporated within the relatively few PMEs that dominate the market.
- There is a strong incentive to develop and use new types of model-based applications. In most cases, the main cost of this activity, in terms of both money and time, is the development and validation of the process models on which these applications are based. Although many of these models already exist, they are usually embedded within PMEs and access to them is often severely restricted.

In other words, we require efficient and effective mechanisms (a) for incorporating new PMCs within existing PMEs, and (b) for making the models embedded within PMEs accessible to other applications.

The conventional approach to addressing need (a) has been for each software vendor to provide its own mechanisms for incorporating customer and third-party PMCs within its own PME(s), with each vendor defining its own Application Programming Interfaces (APIs), data formats, and protocols—as well as the scope of interfaces supported. Traditionally, these mechanisms have been based on procedure calls and file transfers. More recently, vendors have adopted industry standard "middleware" such as the Object Management Group's (OMG) CORBA (Object Management Group, 1997) and Microsoft's COM. While this approach, which might be termed "open proprietary", has proven effective for meeting need (a) when dealing with only one vendor or PME, it suffers from the serious drawback that a new PMC implementation or interface must be developed for each target PME. Since the cost of developing and maintaining multiple implementations of a PMC is considerable, and since vendors may not make the required interface documentation and tools openly available, this may lead to serious bottlenecks to the exploitation of rapidly evolving technologies.

The second need identified above is also inadequately addressed today. While most software vendors do provide mechanisms for integrating simulation models developed using their PMEs with other applications, for example using Microsoft Excel or Visual Basic, it is generally not possible to extract individual PMCs embedded within existing PMEs for use in other PMEs or applications. Furthermore, the available mechanisms suffer from the same drawbacks as those identified above—a separate implementation is required for each PME. The overall process is subject to much the same bottlenecks as those identified above, often magnified significantly in view of the much larger technical and commercial issues involved. This paper advocates an alternative "open standards" approach to addressing the above concerns. More specifically, this approach identifies several important classes of PMCs and defines standard software interfaces for each such class. These steps are seen as an essential pre-requisite for a long-term vision according to which:

- new PMCs will be constructed to adhere to these standards while existing ones will be modified ("wrapped") to do so;
- PMEs will make use of PMCs adhering to the standards;

thus allowing process engineers to use software from heterogeneous sources operating together to carry out complex model-based tasks.

The next section of this paper presents a description of the structure and scope of CAPE-OPEN, a major collaborative effort that has been spearheading developments in the direction mentioned above over the past couple of years. This is followed by a description of the underlying concepts and characteristics of the main PMC interfaces defined by CAPE-OPEN.

Inevitably, all non-trivial software initiatives are crucially dependent on the enabling state-of-the-art computer software and hardware technologies. Consequently, a section of this paper is devoted to the use of "middleware", such as CORBA and COM, for the formal definition and implementation of CAPE-OPEN compliant interfaces, as well as the work methodology adopted by CAPE-OPEN.

CAPE-OPEN is by no means the only major standardization effort in the area of process engineering. We, therefore, also consider its relation to other initiatives such as OPC and pdXi.

Finally, we consider some developments that are currently under way. Particular mention is made of the Global CAPE-OPEN project starting in July 1999. We conclude with some remarks on the current state, prospects and benefits of open architectures for process engineering software.

The CAPE-OPEN Project

The CAPE-OPEN ("*Computer Aided Process Engineering. Open Simulation Environment*") project involves the collaboration of a number of major operating companies[3], academic institutions[4] and process

---

[3] BP plc, BASF AG, BAYER AG, DuPont Iberica SA, Elf, ICI plc, IFP.

[4] RWTH Aachen, ICSTM London, ENSIGC Toulouse.

engineering software vendors[5]. The project had an overall budget of 3.25 MECU and was partly funded by the European Community under the Brite-EuRam III programme.

The CAPE-OPEN project started in January 1997 and was concluded in June 1999. Its main aims have included the following:

- the identification of major classes of PMCs and the formal definition of general interface(s) for each class;
- the construction and testing of appropriate prototype software demonstrating the use of the above interfaces and the benefits that may arise from it;
- the dissemination of the results of the project leading to the understanding, acceptance and adoption of open software architectures by the process engineering community.

Scope of the CAPE-OPEN Project

Open architectures can be beneficial for many different types of process engineering software. The specific focus of the CAPE-OPEN project has been on general tools for process modeling[6] and, in particular, their use for steady-state and dynamic simulation. Moreover, the project has recognized explicitly the *de facto* existence and widespread practical usage of two different types of such tools, namely the "modular" and "equation-orientated" ones[7].

In order to identify the key classes of PMCs that a standardization effort, such as CAPE-OPEN, needs to address, it is instructive to consider "typical" architectures for process modeling tools. A common, albeit somewhat simplified, architecture for modular tools is shown in Fig.1 below.

---

[5] AspenTech, Hyprotech, QuantiSci Ltd., SIMSCI.

[6] These tools are often known as "process simulators" although, as has already mentioned in this paper and elsewhere, they usually support several activities other than process simulation, all based around a common process model.

[7] The distinction between these two types is not always entirely clear. For the purposes of this paper, we shall use the term "modular" to refer to process modelling tools in which the models of unit operations have a certain responsibility for solving their own equations. On the other hand, in "equation-orientated" tools, models of unit operations simply provide their equations to a simulation executive that has overall responsibility for their solution.

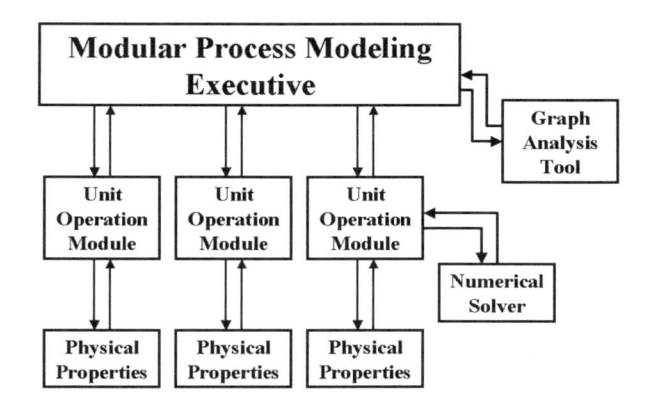

Figure 1. Typical architecture for modular process modeling tools.

As shown in the above figure, overall responsibility for constructing a model of a process and carrying out various computations with it is vested in a "process modeling executive". The latter interacts with modules describing individual unit operations. Typically, these will need to interact with packages used to compute the physical properties that occur within the unit operation models. As has already been mentioned, in the modular case each unit operation module will also need to solve the equations of the corresponding mathematical model; the solution may be performed by specialized algorithms (often coded within the unit operation modules themselves), or by making use of external numerical solvers (as shown in the case of the third unit operation module in Fig. 1).

An important characteristic of modular process modeling tools is the need for the executive to organise and co-ordinate the computations carried out by the individual unit operation modules. This often involves the analysis of the process flowsheet to identify "partitions" that may be solved independently, or sets of streams that have to be "torn" to remove cyclic dependencies (see Westerberg et al., 1979). Appropriate flowsheet analysis tools, usually based on graph-theoretical concepts, are used for this purpose.

A typical architecture for an equation-orientated package is shown in Fig. 2. Although the basic structure is not too different from that shown in Fig. 1, a fundamental difference is that the unit operation modules no longer have responsibility for solving their own equations. Instead, they pass information on them to the executive, which assembles them into a (typically large) set of equations, which it then solves by interacting with one or more appropriate numerical solvers.

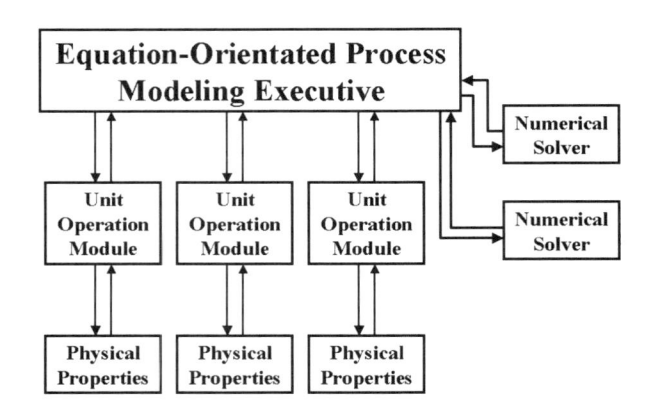

Figure 2. Typical architecture for equation-orientated process modeling tools.

The above analysis leads naturally to the identification of the following important classes of PMCs that are prime candidates for standardization:

- Physical properties
- Unit operation modules
- Numerical solvers
- Flowsheet analysis tools.

Moreover, the analysis suggests that the unit operation modules class has to be sub-divided into two distinct sub-classes exhibiting different behaviors, corresponding to use within modular and equation-orientated packages respectively. The different behaviors may be combined in a single module—so called "dual-mode" models.

The ultimate vision of CAPE-OPEN is to allow complex process modeling tasks and model-based applications to be performed successfully and cost-effectively via the collaborative use of software components coming from a wide variety of sources and possibly being executed on different computer hardware. An example of this is shown in Fig. 3. Here, one vendor supplies the PME (the simulator executive and the user interface), whereas the PMCs (e.g. one or more unit operations, the physical property calculations, and the solution algorithm) come from different suppliers. Extending this principle, the individual components will be able to communicate in a standard fashion with other environments such as process control and monitoring systems, or costing applications.

PMC Interfaces Defined by CAPE-OPEN

The previous section has identified some of the key classes of PMCs that are prime candidates for standardization. In view of the very wide range of materials and unit operations employed by the process industries as well as the range of solution techniques used for dealing with different model-based applications, it is not surprising that each of these PMC types can be sub-divided further into several more sub-classes. Clearly, not all of these can be handled within a project of limited duration and resource. This section considers in more detail the actual scope of each type of interface defined by the CAPE-OPEN project. It also describes the key concepts underpinning each interface and its main characteristics.

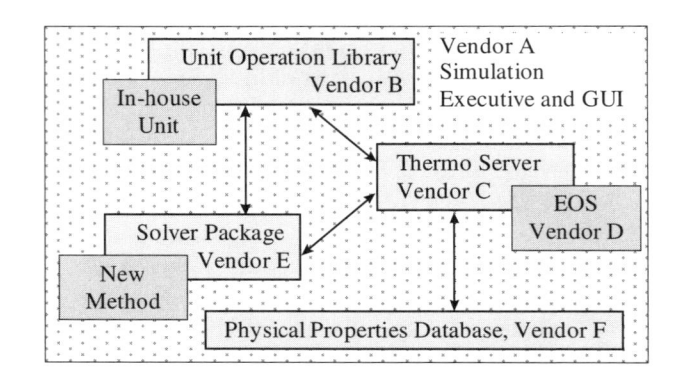

Figure 3. Vision of a typical CAPE-OPEN modeling tool

CAPE-OPEN Physical Property Interfaces

In the area of physical properties, CAPE-OPEN has focused on uniform fluids that are mixtures of pure components or pseudo-components, and whose quality can be described in terms of molar composition. The physical properties methods that have been provided with standardized interfaces are those required for the calculation of vapor-liquid-liquid-solid equilibria or subsets thereof, as well as other commonly used thermodynamic and transport properties.

A key concept in CAPE-OPEN is that of a *Material Object*. Typically, each distinct material appearing in a process[8] will be characterised by one such object.

Each unit operation module may interact with one or more material objects. For instance, a module modeling the separation of an input stream into vapor and liquid output streams using a non-equilibrium model may interact with 4 material objects, representing the material in the input and output streams as well as at the vapor-liquid interface respectively. This interaction may take a number of different forms including setting the values of the material object's independent ("state") variables, asking the object to compute a set of pure component or mixture properties for one or more phases at the current values of the state variables or to carry out phase equilibrium calculations, and requesting the current values of some of these properties. Partial derivatives of physical properties with respect to the independent variables can also be computed.

---

[8] This may include material in streams flowing between unit operations as well as material within individual unit operations.

In practice, material objects will compute physical properties or perform phase equilibrium calculations by reference to thermodynamic property packages which, in turn, carry out these tasks by making use of thermodynamic property calculation routines and equilibrium servers.

In order to support the implementation of the above framework, CAPE-OPEN has defined standard interfaces for material objects as well as thermodynamic property packages, calculation routines and equilibrium servers. The design of all these interfaces has paid particular attention to important issues such as extendibility and efficiency. For instance, the list of properties that any thermodynamic property package is capable of computing is not fixed but can be obtained by client software via a method provided by the corresponding object. Moreover, the interfaces support "batching" of the computation of two or more properties, thereby permitting the exploitation of any common computations shared by these properties.

CAPE-OPEN Unit Operation Module Interfaces

CAPE-OPEN has defined a comprehensive set of standard interfaces for unit operation modules being used within modular PMEs. We review some of the key concepts underpinning these interfaces below.

A unit operation module may have a number of *ports*, which allow it to be connected to other modules and to exchange material, energy or information with them. In the material case (which is also the most common), the port will be associated with a material object (see above). Ports also have directions (input, output, or input-output).

Unit operation modules also have sets of *parameters*. These represent information, which is not associated with the ports but which, nevertheless, the modules wish to expose to their clients. Typical examples include equipment design parameters (e.g. the geometry of a reactor) and important quantities computed by the module (e.g. the capital and operating cost of a reactor).

A unit operation module may have its own *user interface* that allows the user to configure each instance of this module in an appropriate fashion. Typically this configuration will take place at the time when the instance is inserted in a flowsheet and may involve a specification of the precise mode of operation of the unit and the provision of values for the associated degrees of freedom.

Finally, a unit operation module may be capable of producing one or more *reports* on the results of its computations. Other facilities include the ability of a unit to perform (potentially complex) *initialization* computations, to *save* its current state and to *restore* it at a subsequent point in time.

The unit operation interfaces defined by CAPE-OPEN allow PMEs and other clients to take advantage of the flexibility afforded by the above features. The computation of a unit operation module is triggered explicitly by its clients via the invocation of a method provided by the unit operation object.

Much of the above considerations also apply to equation-orientated unit operation objects. A key difference is that, instead of carrying out any computations, the main responsibility of the object is to form and expose a set of mathematical equations. Typically, an equation-orientated PME will assemble the sets of equations from the various units in the process into one large set; it will extend this set with equations reflecting unit connectivity and, potentially, other specifications; and it will solve the resulting square system of equations to produce the solution. A key prerequisite for this mode of operation to be feasible is the introduction of formal ways of specifying sets of equations of various kinds. This problem has also been addressed by CAPE-OPEN in the manner described below.

CAPE-OPEN Numerical Solver Interfaces

In the area of numerical solvers, CAPE-OPEN has focused on the solution algorithms that are necessary for carrying out steady-state and dynamic simulation of lumped systems. In particular, this includes algorithms for the solution of large, sparse systems of nonlinear algebraic equations (NLEs) and mixed (ordinary) differential and algebraic equations (DAEs). Algorithms for the solution of the large sparse systems of linear algebraic equations (LAEs) that often arise as sub-problems in the solution of NLEs and DAEs have also been considered.

A technical difficulty encountered in this context is the large amount of information that is necessary for the definition of a system of nonlinear equations. In fact, this amount increases as more and more sophisticated solution algorithms are being developed. For instance, most modern codes for the solution of large DAE systems require information on the sparsity structure of the system, as well as the ability to compute both the residuals and the partial derivatives of the equations. Even more sophisticated codes need further information on any discontinuities that may occur in the DAE system, the logical conditions that trigger these discontinuities and so on.

To overcome the above problem in a systematic manner, CAPE-OPEN has introduced a new concept, the *Equation Set Object* (ESO) as a software abstraction of a set of nonlinear algebraic or mixed (ordinary) differential and algebraic equations. The standard ESO interface allows access to the structure of the system (i.e. the number of variables and equations in it, and its sparsity pattern), as well as to information on the variables involved (i.e. their names, current values and lower and upper bounds). It also allows the ESO's clients to modify the current values of the variables and their time derivatives, and to request the corresponding values of the residuals and partial derivatives (Jacobian matrix) of a subset or all of the equations in the system.

The equations in an ESO may involve discontinuities (e.g. arising from transitions of flow regime from laminar to turbulent and vice versa, appearance and/or disappearance of thermodynamic phases, equipment failure and so on). Discontinuous equations in ESOs are represented as State-Transition Networks (STNs, see Pantelides, 1995, Avraam et al., 1998). At any particular time, the system is assumed to be in one of the states in the STN and its transient behavior is described by a set of DAEs which is itself an ESO. Transitions from one state to another occur when certain defined logical conditions become true; the ESO interface provides complete access to the structure of these logical conditions as well as allowing their evaluation. Such information is essential for the implementation of state-of-the-art algorithms for handling of discontinuities in dynamic simulation.

Any CAPE-OPEN compliant code for the solution of systems of NLEs or DAEs provides a "system factory" interface. Typically, client software[9] starts by creating an ESO that contains a complete mathematical description of the problem being solved. It then passes this ESO to the appropriate system factory to create a "system" object that combines an instance of the solver with the ESO to which the solver will be applied. The system object then provides appropriate methods for solving the problem completely or merely taking a single iteration (in the case of an NLE system) or advancing the solution over time (in the case of DAEs).

As explained above, the primary aim of the introduction of the ESO concept is to support the operation of CAPE-OPEN compliant nonlinear solvers. However, an important side benefit is that ESOs also provide a general mechanism for PMEs to expose the mathematical structure of models defined within these PMEs. Thus, ESOs may fulfil the rôle of "model servers" in the manner envisaged by Pantelides and Britt (1995) providing the basis for the development of new types of model-based applications beyond those that are supported by the PMEs themselves.

CAPE-OPEN Graph Analysis Tool Interfaces

A key part of the operation of sequential modular simulation systems is the analysis of the process flowsheet in order to determine a suitable calculation sequence for the unit operation modules (cf. Fig. 1). Thus, typically the set of units in the flowsheet is partitioned into one or more disjoint subsets (*maximal cyclic networks*, MCNs) which may then be solved in sequence rather than simultaneously ("ordering"). The units within each MCN are linked with one or more recycle loops which can be converged in an iterative manner via the identification of appropriate "tear streams" which allow the unit operation calculations to be sequenced. A more detailed description of these well-established operations may be found in Westerberg et al. (1979).

The above tasks are typically carried out using a set of tools that operate on the directed graph representation of the flowsheet. CAPE-OPEN has defined standard interfaces for the construction of these directed graphs, and for carrying out partitioning, ordering, tearing and sequencing operations on them.

CAPE-OPEN Implementation and Work Methodology

This section considers some of the issues involved in the implementation of CAPE-OPEN interfaces, and the work process itself adopted by the CAPE-OPEN project.

The Use of Middleware by CAPE-OPEN

The interfaces described in the previous section could, in principle, be implemented in a number of different ways including, for instance, as simple "subroutine" or "procedure" calls in standard procedural languages such as FORTRAN or C. However, CAPE-OPEN has chosen to adopt a component software and object-orientated approach which views each PMC as a separate object. All communication between objects is handled by "middleware" such as the Object Management Group's (OMG) CORBA (Object Management Group, 1997) and Microsoft's COM. These technologies provide standard mechanisms for one software object to interact with another based on a formal interface definition expressed in standard languages also provided by them. The communicating objects can be running as part of the same process, or in different processes on the same or different computer hardware connected in a network, thus providing "local/remote transparency". Issues such as differences in the computer languages in which the various objects are actually implemented, or in the representation of fundamental data types (e.g. real numbers) between different machines are handled automatically. All of these aspects are particularly important in view of the primary aim of CAPE-OPEN to support the interaction of process modeling software components from heterogeneous sources.

All CAPE-OPEN interfaces have been expressed in both CORBA and COM in order to be applicable to a wide variety of hardware platforms and operating systems, and to ensure that they are as "future-proof" as possible in a rapidly evolving environment.

Each CAPE-OPEN interface involves lists of interfaces, methods and arguments expressed in the CORBA IDL and the COM MIDL Interface Definition Languages. Developers of CAPE-OPEN compliant components will need to incorporate the same declarations in their applications and to use IDL compilers of either or

---

[9] These clients could be either unit operation modules (in the case of modular PMEs) or the simulation executive (in the case of equation-orientated PMEs).

both kinds to generate the corresponding instructions[10] in source language such as C, C++, Java, Smalltalk etc. The "wrapping code" generated in this manner can then be linked with the rest of the component. Legacy code, such as FORTRAN models, can also be used by encapsulation within CAPE-OPEN compliant wrappers.

As an example, Fig. 4 shows the simple standard interface specification for accessing some of the ports of a Unit Operation. The method belongs to the ICAPEUNIT interface, takes as input the type and directions of ports required, and returns a pointer to a portsInterface, which can then be used to communicate with the selected ports.

The CAPE-OPEN Work Process

The definition of interfaces throughout the project was done following a development process based on the UML object-orientated notation (Fowler and Scott, 1997) for all formal models of the interfaces, including the user requirements, producing use cases, sequence diagrams, state transition diagrams, class diagrams and, finally, interface diagrams which accompany the corresponding middleware implementation (see Table 1). In practice, an iterative approach where the different models and implementations were subject to progressive refinements had to be adopted. Overall, this work process proved to be both an efficient and an effective mechanism for developing commonly agreed standard interface specifications and prototypes meeting those specifications, in a project involving a relatively large number of people with widely different backgrounds.

The Global CAPE-OPEN Project

The CAPE-OPEN project was concluded on the 30th June 1999. Global CAPE-OPEN (GCO) is a follow-up project which started on 1st July 1999 with a planned duration of 30 months. GCO is an EC-funded project, under the Brite-EuRam framework, with a contribution of 2.5 Meuros from the European Commission and 3.2 Meuros from its partnership. It is also proposing to operate under the umbrella of the Intelligent Manufacturing Systems (IMS) initiative. IMS an industry-led, international Research & Development Program established in 1995 to develop the next generation of manufacturing and processing technologies. Recognizing that these technologies will be expensive to produce, and that no single organization has all the expertise needed, companies and research institutions from Australia, Canada, the European Union, Japan, Switzerland, and the United States have undertaken co-operative technology development to share costs, risks and expertise. Properly

---

[10] Mainly "header" files containing various declarations for stubs, and skeletons.

managed international co-operation in advanced manufacturing R&D, through IMS, can help improve manufacturing operations, enhance international competitiveness, and lead to technology breakthroughs via market-driven R&D.

The European part of GCO brings together an already unprecedented set of 19 European organizations from industry, academia, and software companies. The International GCO proposal adds to this three other consortia of various magnitudes in USA, Canada and Japan respectively. These partners will work together towards the GCO strategic objectives:

INTERFACE NAME	ICAPEUNIT
Method Name	GetPorts
Returns	CapeError
Return an interface to a collection containing a list of ports on the unit of a specific type and direction e.g. all input material ports.		
Arguments		

[in] streamType	CapeStreamType	the type of stream required: material, energy, information
[in] direction	CapeDirection	the direction of flow: input, output or any
[return] portsInterface	CapeInterface	a reference to the interface on the collection containing the specified ports

Figure 1. Typical specification of a CAPE-OPEN method.

1. Development of standards in new subfields of process modeling and simulation beyond those addressed by the initial CAPE-OPEN project (see above). Thus, GCO will address complex physical properties, kinetic models, new numerical algorithms and distributed models.
2. Support of the development of versions of simulation software conforming to the standard; several partners in GCO will develop CAPE-OPEN compliant components for internal use, research, or business.
3. Research on the integration of open process simulation technology in the work process; GCO will take real industrial cases and assess the use of CAPE-OPEN technology, including supporting software and user training.
4. Definition of open standards for new technologies beyond process modeling and simulation. GCO will develop prototypes for online systems, for discrete and mixed batch-continuous processes, for finer granularity

interfaces and for scheduling and planning systems.
5. Further dissemination of the technical results of CAPE-OPEN using both traditional and more modern mechanisms.
6. Definition of standards for modeling domains integration.
7. Assessment of the use and benefits of CAPE-OPEN standards for education and training;
8. Establishment of links with parallel efforts such as pdXi, OPC and STEP.
9. Establishment of an international standards body and integration laboratories network on process simulation. GCO will launch the "CAPE-OPEN Laboratories Network" (C.O.LaN), a non-profit institution aimed at maintaining the standards and providing services to CAPE-OPEN developers and users.

More generally, the major expected result of GCO will be the global acceptance of CAPE-OPEN as a standard for communication between components in process engineering simulation software, leading to the availability of software components offered by leading vendors, research institutes and specialized suppliers. This will enable the process industries to reach new quality and productivity levels in designing and operating their plants while opening new markets for suppliers of CAPE components. This will be a major technical breakthrough compared to the current state-of-the-art.

Relation of CAPE-OPEN to Other Standardization Activities for Process Engineering Software

The CAPE-OPEN initiative aims at minimizing the development time needed for interfacing process models, or elements of process models, with each other and with third-party elements. It specifically addresses the functional exchange of data and computation results pertaining to process simulation, optimisation and other types of model-based applications during run-time. In this section, we consider the relation between CAPE-OPEN and two other major standardization activities.

pdXi

The AIChE Process Data Exchange Institute (pdXi), is an initiative sponsored and funded by more than a dozen (mostly US) operating companies, E&C firms, and software vendors to promote the electronic exchange of process engineering data among diverse engineering applications and organizations. pdXi has addressed the development of a generalized data model and the development of a software toolkit for accessing data that conforms to the defined model. Here, the term "process data" refers to stream and equipment data and other data used to support the process engineering activity over the entire life cycle of processing facilities. Membership to pdXi is open to operating companies within the process industry, software vendors, engineering contracting firms and equipment design/manufacturing firms.

Table 1. The CAPE-OPEN Development Process.

Phase	Step	Goal
ANALYSIS	User requirements, text	Requirements in textual format
ANALYSIS	User requirements, Use Cases	Use Case models
DESIGN	Design Models	Sequence, state transition, and interface models using UML
DESIGN	UML Repository	UML models in project repository
SPECS	CAPE-OPEN/COM Specification	Draft interface specifications in Microsoft IDL
SPECS	CAPE-OPEN/CORBA Specification	Draft interface specifications in CORBA's IDL
IMPLEMENT	CAPE-OPEN/COM Implementation	Prototype MIDL implementation
IMPLEMENT	CO/CORBA Implementation	Prototype IDL implementation
VALIDATION	Standalone Testing	Tested component
VALIDATION	Integration testing	Tested specification
SPECS	CAPE-OPEN/COM final specifications	Approved specification
SPECS	CAPE-OPEN/CORBA final specifications	Approved specification

pdXi is sponsoring the ISO/STEP Application Protocol AP231 - *Process Engineering Data: Process Design and Process Specifications of Major Equipment*). The goal is for the pdXi data modeling work to result in a formal international standard for process data.

STEP (Standard for the Exchange of Product Model Data/ISO 10303) is an international activity for establishing standard data models and exchange protocols for a variety of manufacturing industries, including the process industries. In addition to AP231, other STEP activities for process industries include application protocols for detailed engineering: AP221 - *Functional Data and their Schematic Representation for Process Plants* and AP227 - *Plant Spatial Configuration*. All three process plant protocols will be "harmonized" in the future.

pdXi has also sponsored the development of an Application Programming Interface (API) and software toolkit. This toolkit would allow any application to communicate with pdXi formatted files through standard calls. pdXi is also considering an XML specification for the data model. We believe that pdXi and CAPE-OPEN are largely complementary activities with very little, if any, overlap between them. Consider, for example, a CAPE-OPEN compliant unit operation object used to model a heat exchanger. A pdXi file could be used to configure this object by defining the characteristics of a particular heat exchanger unit. The precise mechanism via which this configuration is done is of no concern to the CAPE-OPEN object interface, which is concerned, primarily with being able to compute the heat exchanger behavior.

OLE for Process Control (OPC)

The OPC Foundation is a non-profit organization promoting the establishment and use of standard interfaces for process control applications, based on Microsoft's OLE/COM middleware. OPC defines standard methods for accessing control systems functionalities and attributes of servers of real-time information including distributed process control (DCS) systems, programmable logic controllers (PLCs), smart field devices and analyzers. The OPC initiative is supported by a group of more than 150 providers and users of control systems hardware and software. Prototype OPC servers have been developed and demonstrated at various occasions.

The choice of OLE/COM as unique middleware makes the OPC standard rely on Microsoft technology. Hence, Unix or VMS compatibility will only be provided through software bridges like Iona's ORBIX for WINDOWS or through foreign implementations of COM on these platforms. Such bridges are now being made available on specific platforms, like the ENTIREX software (Software AG, 1997).

OPC is primarily concerned with real-time process data. As process models are increasingly used in a real-time operations support (e.g. in on-line optimization or as decision support tools), the availability of standard mechanisms for accessing plant data is particularly valuable. Once these data are acquired, then they may be used within process simulation and optimization activities – which are precisely those that CAPE-OPEN and its GCO continuation aim to support. Overall, then, OPC and CAPE-OPEN are highly complementary initiatives.

Possible Future Developments and Opportunities

The widespread use of *"plug-and-play"* PMCs is, of course, the main benefit expected from the introduction of open software architectures. It applies to all the components for which a standard interface is available, provided that implementors follow the interface specifications faithfully. It concerns commercial software (e.g. using X's physical properties package in Y's simulation executive environment); legacy software (e.g. using X's reactor model in Y's environment); niche software (e.g. using specific unit operations or physical properties in a standard simulation package).

Realistically, in view of the current environment, which is dominated largely by proprietary software packages, such "plug-and-play" usage will be initially limited to introducing a small number of external components within existing simulation packages. However, it may soon develop into a whole industry of pluggable PMCs, with semi-automatic facilities for selecting components based on meta-level descriptions of their capabilities and their cost, followed by electronic payment and downloading directly from the World Wide Web (see Edwards et al., 1996).

The achievements of the CAPE-OPEN project together with the further developments planned under GCO open new opportunities for the process industries. Once these ideas gain wide acceptance by the process engineering community, we may find ourselves facing some very major changes in the ways process engineering software is designed, developed, marketed, distributed and used.

One likely scenario is that emphasis will shift away from comprehensive, single-source process modeling systems towards a combination of PMEs and PMCs from diverse sources. Although vendors will still provide integrating process modeling environments (PMEs) within which these components will fit, we believe that most competitive advantage will be derived from two distinct sources:

- better software components (e.g. more accurate physical properties; high fidelity process models; more reliable and efficient numerical solvers);
- new model-based applications making use of the above components (e.g. real-time dynamic optimization; model-based fault detection; automated emergency handling systems).

The fundamental change brought by CAPE-OPEN and related activities as far as software suppliers are concerned is a lowering of the threshold involved in the technical development and commercial exploitation of new software. In particular, individual suppliers will be able to concentrate their activities on the areas where their know-how and expertise provide them with a distinct competitive advantage, relying on third-party PMCs for everything else.

On the other hand, users of process engineering software will be able to purchase from a variety of sources only those PMCs that are of immediate use to them. The purchase cost of a PMC is likely to be much less than that of an entire integrated process modeling system; moreover, the use of standard interfaces will facilitate the introduction of a new PMC or the replacement of one PMC by another. The overall effect as far as the process industry is concerned will hopefully be a dramatic lowering of the threshold of exploiting new technology.

Let us emphasize that it is not yet proven that open systems will lower overall costs or improve overall reliability. There is a possibility that the opposite will occur. Without a primary supplier, support costs may go up. Responsibility for correctness of results may become diffused and impossible to pin down. Data obtained from experience with a broad range of actual implementations is required to determine the tradeoffs. What is certain is that open systems and standards will allow the user the flexibility to use the PMCs that best meet his or her needs, and will open the door and lower the cost of entry for creating PMCs from a wide variety of sources. It is also certain that e-commerce will be heavily involved in making these PMCs accessible. For example, designers will probably shop for equipment such as pumps, packing, etc. over the Web, with the equipment vendors providing PMCs for simulation and design as a means of promoting use of their products. (e.g., an engineer on a column retrofit project evaluates packing from several different vendors by accessing them on the Web and running rating mode simulations to choose the one providing the best performance and overall cost, all in a few hours). This is the type of re-engineering of work processes that open systems and standards can enable.

Emerging Enabling Technologies and their Impact on Open Architectures for Process Engineering Software

We have already seen how the current state of enabling computer software, and in particular, the availability of middleware technologies, has influenced the outcome of the CAPE-OPEN project. It is, therefore, reasonable to expect that further developments in enabling technology may have an important impact on the future evolution and use of open standards and interfaces for process engineering software. In this section, we consider some emerging technologies that may have such an impact. Although some of this may appear futuristic given the current state of the technology, the rate of evolution in this area is so rapid that even ostensibly far-fetched propositions may become practically feasible within only a few years.

Wizards

Companions, or "wizards", have become familiar tools in modern office software environments. While this article is being typed, an invisible or iconized component monitors what is being typed, sometimes correcting spelling mistakes, sometimes suggesting better use of the word processing software, sometimes saving the work without necessarily informing the user. The user may also ask for advice on advanced features that he or she uses only infrequently.

In fact, companions for process simulation software already exist in some proprietary environments. These tools help to enter data, to select models, to check the solvability of a flowsheet, to determine the number of unknowns etc.

With the advent of open architectures for process engineering software, we will need intelligent software companions that are able to monitor, diagnose and propose remedies to process simulation models across commercial, technical and geographical barriers. We may also need "interface traffic managers", i.e. components that are able to observe functional exchanges through interfaces, therefore being in a position to propose solutions to problems such as the choice of the best properties model for a specific material, the choice of a solving mechanism, or the debugging of a convergence problem.

Java technology

Java is rapidly becoming the language of choice for web-based applications. The use of Java in other software development projects is still not very widespread because of performance problems with the currently available Java Virtual Machines. When these performance issues are addressed, Java will become a serious candidate for the development of new applications in scientific and technical domains as well as for major IT projects. Java is generally seen as a very clean object-oriented language and also provides the facilities that are needed for component-based development, including the increasingly popular Enterprise Java Beans middleware architecture. The CAPE community should not be the last to take advantage of the benefits of Java.

Collaborative CAPE environments

Computer-aided process design is a collaborative activity as the knowledge about chemical processes and their simulation is usually shared between several coworkers, and as the mergers in the engineering and in the process industries created global companies whose human resources are distributed on the planet. Whereas collaboration has taken place in the past through face-to-face meetings, telephone calls, exchange of reports and of files, current technology supports a vision of collaborative development process as a joint effort of several actors working in distant places. The CAPE-OPEN project has already made extensive use of the "*Basic Support for Collaborative Work*" (BSCW) system which has allowed more than 60 users to share standards documentation and the associate project information over the Internet (see Fig. 5). Although there are still some outstanding technical questions, the major challenge will be for workers to get used to sharing information by electronic means. We are not there yet, but the methods and tools developed in collaborative IT will definitely change the way we work together.

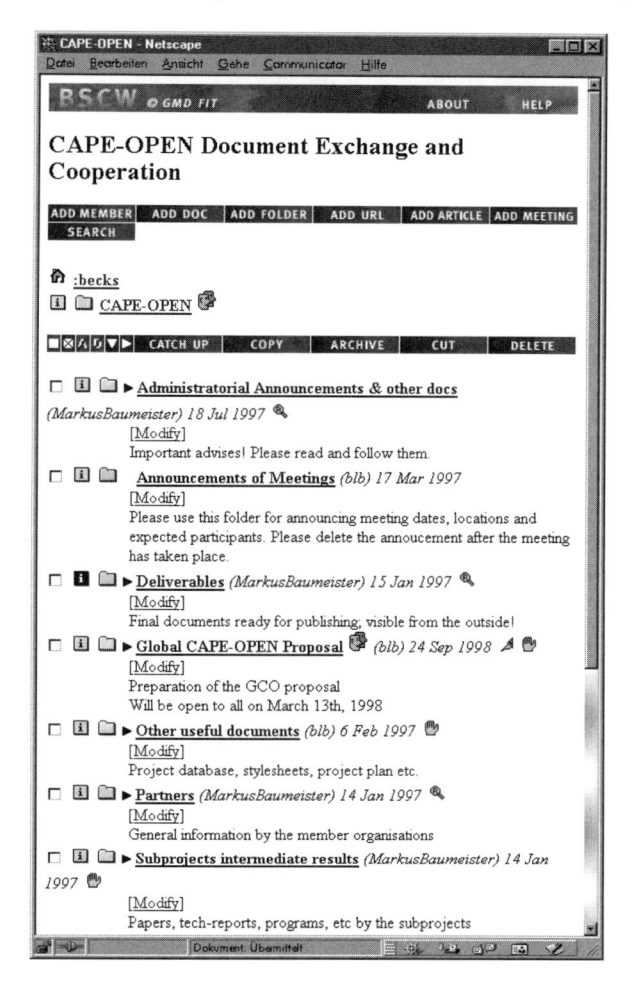

Figure 5. The BSCW environment for collaborative work on the WWW.

XML and DTD

Undoubtedly, Internet and web technologies now dominate information systems and networks. The technologies involved include, among other things, file formats (such as HTML), languages (cgi scripts, Java), communication protocols (HTTP, FTP etc.).

More recently, a new standard, the *Extended Markup Language* (XML), defined by the World Wide Web Consortium (W3C), has been emerging for the distribution of documents over the web (see http://www.xml.com). XML is based on the SGML language developed by the documentation community[11]. In contrast to the flat structure of HTML files, XML provides a structured but also extensible way of storing information. As noted on www.xml.com/xml/pub/98/10/guide1.html by Norman Walsh:

What XML brings to the web (and to the intranets and extranets) is the ability to express structured queries that exploit the structured form of XML files. This is much more advanced that current query mechanisms which rely on simple character string matching with little knowledge of the content of the documents they search and retrieve[12]. In contrast, with XML, a network containing many documents (such as the web) becomes a huge structured database, which can be queried with an efficiency similar to that achieved with SQL or object-orientated query mechanisms.

« In HTML, both the tag semantics and the tag set are fixed. An <h1> is always a first level heading and the tag <ati.product.code> is meaningless. The W3C, in conjunction with browser vendors and the WWW community, is constantly working to extend the definition of HTML to allow new tags to keep pace with changing technology and to bring variations in presentation (stylesheets) to the Web. However, these changes are always rigidly confined by what the browser vendors have implemented and by the fact that backward compatibility is paramount. And for people who want to disseminate information widely, features supported by only the latest releases of Netscape and Internet Explorer are not useful. XML specifies neither semantics nor a tag set. In fact XML is really a meta-language for describing markup languages. In other words, XML provides a facility to define tags and the structural relationships between them. Since there's no predefined tag set, there can't be any preconceived semantics. All of the semantics of an XML document will either be defined by the applications that process them or by stylesheets »

What XML brings to the web (and to the intranets and extranets) is the ability to express structured queries that exploit the structured form of XML files. This is much more advanced that current query mechanisms which rely on simple character string matching with little knowledge of the content of the documents they search and retrieve[13]. In contrast, with XML, a network containing many documents (such as the web) becomes a huge structured database, which can be queried with an efficiency similar to that achieved with SQL or object-orientated query mechanisms.

---

[11] XML involves a restricted set of SGML features.

[12] For example, the query "*how to calculate the fugacity of methane*" could well return a document containing the text "*I don't know how to calculate the fugacity of methane, please tell me*", which obviously misses the point.

[13] For example, the query "*how to calculate the fugacity of methane*" could well return a document containing the text "*I don't know how to calculate the fugacity of methane, please tell me*", which obviously misses the point.

An important aspect of XML is the *Document Type Declaration* (DTD). A DTD provides meta-information about XML documents by further defining the tags to be present in a document, their sequences, and their references to other elements[14]. Thus, DTDs allow the definition of a standard structure for complex documents, so that XML-enabled software (including the latest versions of standard web browsers such as Microsoft Internet Explorer 5 or Netscape Communicator 5) can understand information in the documents from the semantics of the DTD.

For all the above reasons, it is expected that XML will eventually replace HTML as a standard for the web; moreover, it is already being adopted as a standard for native file storage for applications, including the popular Office suite from Microsoft.

The widespread adoption of XML may lead to significant benefits in process engineering software. For example, the CAPE community has been considering the idea of defining a standard process simulation database interface[15]. In contrast to the CAPE-OPEN functional interfaces which allow process simulation components to communicate during the initialization and execution of a particular computation, a simulation database interface would allow the transfer of simulation data between applications. The simulation results, and in particular stream data and values of key variables within units (e.g. distillation column profiles) are often used for restarting the simulation at a later time. A standardized simulation interface would permit the use of the results of one simulator as a starting point for another. It would also be very useful for ensuring that results can be used even by the same simulator several years (and software versions) later. The pdXi initiative addresses the problem of data transfer between CAPE software, but more work is needed before it is ready to be used as an operational standard by process simulation vendors and by the user community. Therefore, today there is no practical way of storing and retrieving simulation data independently of the proprietary file formats of the existing simulation packages. An XML-based DTD for simulation results, e.g. based on the pdXi data model, could be a way of resolving this problem in a fairly short time.

Beyond simulation results, we could also store flowsheets and their associated data in XML documents. These could then be made available on a company's intranet, therefore allowing users to issue structured queries regarding the flowsheets from a web browser. We could also build re-use mechanisms for whole flowsheets or parts of flowsheets by querying the XML database of flowsheets. We could use commercial plug-ins for the browsers, communicating with the XML models base, thereby running the simulation from within the browser. We could publish flowsheets and results on the web for scientific purposes (or advertising). We could re-use the CAPE-OPEN interfaces, along with the pdXi data model, as the basis for defining an appropriate DTD.

Major National and International Initiatives and their Impact on Open Architectures for Process Engineering Software

Beyond the purely technical developments considered in the previous section, a number of national and international initiatives that are currently under way may have a significant effect on the move towards open architectures for process modeling tools. Below we consider the European Commission's Fifth Framework Programme and USA's Vision 2020 and PITAC reports in this context.

The Fifth Framework Programme for EU RTD

The Fifth Framework Programme (FP5) defines the European Union's research and development objectives for the period 1999-2002. With a budget of approximately 14 Billion Euros of funding, FP5 is a predominant research initiative, probably significantly bigger than any other such initiative in the world.

FP5 is implemented through four main "vertical programmes" and three "horizontal supporting actions". Two of the main vertical programmes, namely *Information Society Technologies (IST)* and *Competitive and Sustainable Growth (GROWTH)* directly incorporate key actions relating to computer-aided process engineering.

IST is the most important programme within FP5, with a budget of 3.6 Billion Euros over four years. Two of its four key actions, "2. *new methods of work and electronic commerce*" and "4. *essential technologies and infrastructures*" are of relevance to the new software architectures for CAPE. Action 2 addresses the use of new IS technologies for better working practices within the enterprise including, among other items, teamworking, knowledge management and sharing, security, digital object transfer, and electronic commerce. Action 4 is in charge of developing the future IS technologies that will support these practices, namely real-time systems, networks, component-based software engineering, and high-performance simulation, among others.

The GROWTH programme has a budget of up to 2.7 Billion Euros over four years. Its first key action "*Innovative Products, Processes and Organizations*" supports research and development towards building the competitive industry of the future through reduction of material content of products whilst increasing their service value, and through innovative, safer, cleaner and

---

[14] More formally, a DTD includes declarations for element types, attribute, internal and external entities, and parameters.

[15] Possible developments in this direction have already been considered by both pdXi and one of the work packages of CAPE-OPEN during an early stage of the project.

low natural resource intensity processes and products, including new organizational methods and tools for reaching those targets. This key action is part of a larger scope addressing also land, marine, and air transportation.

It is not our goal to present the full work programmes of IST and GROWTH. The interested reader can find detailed descriptions in European Commission documents available on www.cordis.lu. However, we can draw some perspectives concerning component-based software. Especially, IST key action 4.3 proposes research directives in component-based software engineering (European Commission, 1999):

«*Objective: To develop and validate the innovative processes, methods and tools necessary to design, implement and manage software-intensive systems using a component-based approach. The focus is on re-use, the incorporation of new technology COTS components and evolutionary re-configuration. The work should result in the definition of processes, methods and their supporting technologies that enable the smooth and auditable integration of components from multiple independent sources into complex systems and services, possibly taking advantage of the "system families" concepts. This work is to be complemented by technology-transfer and best-practice initiatives to stimulate both real-life practice improvement and the take-up of the associated technologies*»

Report of USA President's Information Technology Advisory Committee

On the other side of the Atlantic, the President's Information Technology Advisory Committee, in their recent report to Bill Clinton (PITAC, 1999), outlines component technologies as a high-level priority for IT research in the next five years:

«*Recommendation: Fund more fundamental research in software development methods and component technologies.*

The Committee recommends that research in software methods be aggressively pursued, especially in the area of automated support for software development and maintenance. Such research should:

- *Explore and create component-based software design and production techniques, and the scientific and technological foundations needed for a software component industry.*
- *Explore and create theories, languages and tools that support automated analysis, simulation, and testing of components and their aggregation into systems.*
- *Create a national library of certified domain-specific software components that can be reused by others.*
- *Explore and create techniques for using measurably reliable and secure components and their aggregation into predictably reliable and secure systems.*
- *Explore and create protocols, languages, and data structures to promote interoperability of applications running concurrently across wide-area networks.* »

Vision 2020

Technology Vision 2020 is a "visioning process" to develop a strategic plan to keep US chemical companies competitive in the global business environment. It is a joint effort of several organizations including the American Chemical Society, the American Institute of Chemical Engineers, and the Chemical Manufacturers Association. Technology Vision 2020 focuses on four areas of technology R&D:

(i) New Chemical Science and Engineering Technology
(ii) Supply Chain Management
(iii) Information Systems
(iv) Manufacturing and Operations

The participants of the visioning process concluded that "the growth and competitive advantage of our industry depend upon individual and collaborative efforts of industry, government, and academia to improve the nations R&D enterprise." This need for collaboration is the recurring theme of Vision 2020. The principal focus is to develop technology roadmaps that chart specific R&D needs for collaborative research to achieve the vision. This is done primarily through a series of topical workshops that bring together experts in the field.

The report "*Technology Vision 2020: The U.S. Chemical Industry*" describes the high level vision and needs. This reports places considerable emphasis on the need for open software systems, standard interfaces and data exchange formats, and integrated computer applications, and generally encourages efforts by other organizations and vendors to meet these needs. To our knowledge, none of the Vision 2020 workshop or roadmap activities so far have directly addressed this area. Thus CO/GCO can be seen as supportive to achieving Vision 2020.

For more information on Technology Vision 2020, and access to the complete report mentioned above, see http://www.chem.purdue.edu/ccr/v2020 .

In conclusion, it is clear that all the above initiatives aim at facilitating the widespread adoption of component software as *the* software technology of the next decade. The small steps that we are taking now (e.g. with CAPE-OPEN) form the foundations of a new wave which will turn the software industry into an industry of reusable components that can be mixed and matched in infinitely various ways. After generations of software technology, from machine code and assembly language through FORTRAN and structured programming to objects, it is now time to embark into this new generation, triggered by the growing demands of millions of users demanding and by the communication needs through the Internet.

Conclusion

It is the authors' belief that open software architectures are the way forward for the next generation of CAPE tools. They allow previously incompatible environments to interoperate through commonly agreed interface standards established on top of middleware platforms such as COM and CORBA. They contribute to the life-cycle approach for CAPE models that is promoted by commercial simulation suppliers and advocated in W. Marquardt's paper presented in this conference. Moreover, if process development for the next millenium is to be shared, as argued by A. Fowler in his keynote address for FOCAPD'99, open software architectures will be the technical basis allowing the sharing between organizations, departments within organizations, and individuals within departments.

Although much has been achieved by the recently completed CAPE-OPEN project, it represents just the beginning of the move of process engineering software towards truly open architectures. The recently started Global CAPE-OPEN project intends to cover the steady state and dynamic modeling of a much wider range of continuous, batch and hybrid processes. The material forms that fall into the ultimate scope of the standard includes electrolyte mixtures, reacting fluids and polymer solutions and particulates. In terms of model-based applications, it is intended that off-line simulation, optimization, parameter estimation and data reconciliation will be covered.

The simulation software provider companies consider CAPE-OPEN as a key aspect within their overall strategies of product development. Moreover, CAPE-OPEN is not just perceived as a standard for interoperability with third-party products but is also envisioned as an architecture for use by software components within the same company.

The adoption of the CAPE-OPEN standards by a large number of specialized software component providers is perceived as a way of extending the use of modeling software, which, in turn, should provide a competitive advantage to end users, PMC providers and also PME providers. From this perspective, CAPE-OPEN is considered as the basis for a successful collaboration among all these parties.

Acknowledgements

This paper uses updated parts of CAPE-OPEN's Conceptual Design Document, a collective writing by more than thirty authors from fifteen organizations, edited by Dr Tom Malik for the CAPE-OPEN Conceptual work package. It also uses elements from the European Global CAPE-OPEN work programme, a contractual document collectively authored by nineteen collaborating organizations, edited by Dr Bertrand Braunschweig. Sample diagrams and interfaces specifications are taken from the CAPE-OPEN interface specifications documents. For the latest versions of these documents, please consult the CAPE-OPEN and GCO web site.

The presentation of IMS was borrowed from IMS's web site at www.ims.org while the technical presentation of XML from www.xml.com

Excerpts of EC's FP5 programme and of PITAC report by their respective authors.

The authors are grateful to their colleagues in operating and process engineering software companies, and universities, for their contributions to the project and to many of the ideas presented here. However, the authors are solely responsible for the views expressed in this paper.

References

Avraam, M., Shah N., and C.C. Pantelides (1998). Modelling and optimisation of general hybrid systems in the continuous time domain. *Comp. Chem. Eng.*, **22**, S221-S228.

Baldwin, J., and J. de Almeida (1995). Information and data exchanger breakthrough - the new pdXi interface, AIChE 1995 Summer National Meeting, Boston.

Baldwin, J. (1996). Update of the status of the pdxi project, AIChE National Meeting, New Orleans.

Book N., et al. (1994). The road to a common cyte. Chem. Eng., **101**(9), 98.

Braunschweig, B. (1998). in Dhurjati P., and S. Cauvin, *IFAC Workshop on On-Line Fault Detection and Supervision in the Chemical Process Industries*, Solaize.

Brockschmidt K. (1996). *What OLE is Really About*, Microsoft Corporation

CAPE-OPEN Consortium (1997). *Conceptual Design Document*, Adobe Acrobat PDF document obtainable from http://www.quantisci.co.uk/CAPE-OPEN

Edwards, P.D., Hall, T.A., Merkel, G.A. and N.V. Patel (1997). New paradigms for conceptual process design. *AIChE Conference Proceeding*. Spring meeting.

European Commission (1999). *Information Society Technologies. A programme of Research, Technology Development & Demonstration under the 5 th Framework Programme.* Workprogramme downloadable from www.cordis.lu

European Commission (1998). *The Fifth Framework Programme* focuses on Community activities in the field of research, technological development and demonstration (RTD) for the period 1998 to 2002. Work Programme, *Competitive And Sustainable Growth*, available at www.cordis.lu

Foss, B.A., B. Lohmann and W. Marquardt (1998). A field study of the industrial modeling process. *J. Proc. Cont.*, **8**, 325-338.

Fowler M., and K. Scott (1997). *UML Distilled: Applying the Standard Object Modeling Language*. Addison-Wesley Object Technologies Series.

GMD (1999). http://bscw.gmd.de, Gesellschaft für Mathematik und DataVerarbeitung, BSCW web site

Microsoft (1999). http://www.microsoft.com/com/default.asp , web site for COM.

Object Management Group (1997). *The Common Object Request Broker: Architecture and Specification.*

OPC Foundation (1997). *OLE for Process Control , Data Access Standard, Version 1.0A*.

Pantelides, C.C. (1995). Modelling, simulation and optimisation of hybrid processes. Intl. Workshop on Analysis and Design of Event-Driven Operations in Process Systems, London, April.

Pantelides, C.C., and H. I. Britt (1995). Multipurpose process modeling environments. In L.T. Biegler. and M.F. Doherty (Eds.), *Proc. Conf. on Foundations of Computer-Aided Process Design '94*. CACHE Publications, Austin, 128-141.

pdXi: www.aiche.org/pdxi

President's Information Technology Advisory Committee (1999). *Report to the President: Information Technology Research: Investing in Our Future*. Downloadable from http://www.ccic.gov/ac/report

Rational (1999) UML Web Site: http://www.rational.com/uml

Software AG (1997). EntireX DCOM White Paper, Software AG and Microsoft

Walsh, N. (1998). *A Technical Introduction to XML*. Browsable at http://www.xml.com/xml/pub

Westerberg, A.W, H.P. Hutchinson, R.L. Motard and P. Winter (1979). *Process Flowsheeting*. Cambridge University Press, Cambridge, U.K.

TOWARDS TIGHTER INTEGRATION OF MOLECULAR DYNAMICS WITHIN PROCESS AND PRODUCT DESIGN COMPUTATIONS

Jelena Stefanović and Constantinos C. Pantelides
Centre for Process Systems Engineering
Imperial College of Science Technology and Medicine
London SW7 2BY, UK

Abstract

Molecular dynamics involves the application of classical Newtonian mechanics to a set of interacting molecules. In comparison with other computational chemistry techniques, it often provides a suitable trade-off between predictive power and computational cost.

Much of the current use of molecular dynamics is in replacing or complementing laboratory experiments for the generation of data points, which then form the basis for fitting macroscopic correlations (e.g. equations of state) used for product and process design. However, there may be significant advantages to be derived from the tighter integration of molecular dynamics within process and product design computations.

This paper seeks to establish a formal view of molecular dynamics as a deterministic mathematical mapping. One particular difficulty with the mapping constructed via conventional molecular dynamics calculations arises from the discontinuities caused by the use of periodic boundary conditions and the minimum image convention. An alternative molecular dynamics framework that eliminates the above discontinuities is proposed. This is based on the introduction of modified force functions that are almost everywhere continuous and differentiable, and exhibit a natural periodicity. These characteristics obviate the need for both the periodic boundary conditions and the minimum image convention.

The modified molecular dynamics technique defines a continuous and differentiable mathematical mapping. Consequently, it is now possible to compute the partial derivatives of this mapping via the integration of the adjoint equations of the classical Newtonian equations of motion. This development permits the use of molecular dynamics for the simultaneous and direct computation of both thermodynamic properties and their partial derivatives. It also allows the evaluation of the sensitivities of the predicted properties with respect to the intermolecular potential parameters.

Keywords

Molecular dynamics, Minimum image convention, Periodic boundary conditions, Adjoints.

Introduction

Computational chemistry permits the computation of macroscopic properties of materials based on models of behaviour either at the electronic level (employing the principles of quantum mechanics) or at the molecular level (based on the principles of statistical and classical mechanics). Within these two broad areas, a number of techniques have been developed. These differ greatly from each other in terms of their physical basis, computational cost and predictive accuracy.

Figure 1 considers the possible use of computational chemistry techniques in the context of their potential utilization in process and product design applications.

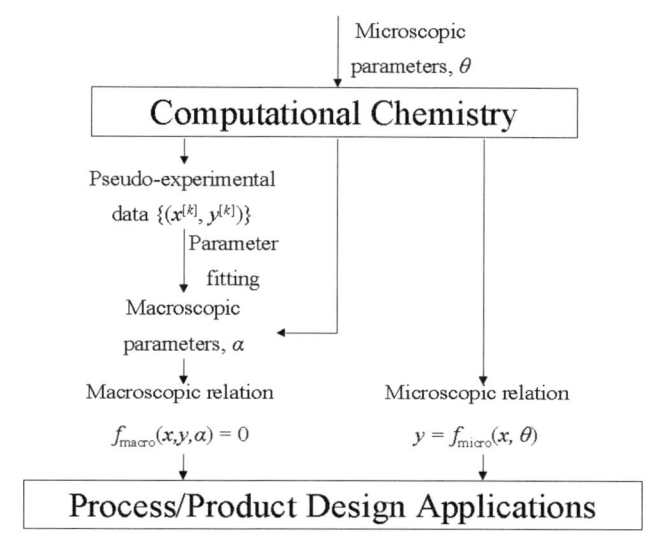

Figure 1. Possible routes for utilization of computational chemistry techniques in process design applications.

Starting from the bottom left of Fig. 1, x and y represent sets of *macroscopic system properties* (e.g. temperature, pressure, density, specific internal energy *etc*.) while the functions $f_{macro}(\cdot)$ are *macroscopic relations* typically required for process/product design applications. For instance, for a pure component, x could represent density and temperature, y could be pressure and $f_{macro}(\cdot)$ could be an equation of state.

The *macroscopic parameters* α are quantities that appear directly in the macroscopic relations $f_{macro}(\cdot)$, such as, for instance, the various binary interaction parameters in an equation of state for a mixture. These are normally obtained by fitting sets of experimental data $\{(x^{[k]}, y^{[k]}), k = 1, 2, ...\}$. The latter can be replaced (or complemented) by values computed using one of the techniques mentioned above. Such computations make use of a (usually small) number of *microscopic parameters*, θ, that characterize the properties of molecules, atoms and smaller entities as well as their interactions. For instance, molecular dynamics computations make use of intermolecular potential functions; here, then, θ would represent the parameters appearing in these potentials.

On the other hand, some computational chemistry techniques allow the direct computation of macroscopic parameters, α, from the microscopic parameters, θ. An example is provided by the recent work of Jónsdóttir *et al.* (1994, 1996, 1997) for the prediction of UNIQUAC interaction parameters using molecular mechanics.

The two possible routes mentioned above for exploiting computational chemistry in process/product design applications are shown in the left and middle parts of Fig. 1. Both of these result in a macroscopic relation $f_{macro}(\cdot)$, and it is the latter that is actually used by the end application. This is advantageous for at least two reasons:

- Process design applications typically require hundreds or thousands of references to the relations between macroscopic variables x and y; in view of the cost of currently available computational chemistry techniques, it is much more efficient to use a relatively simple macroscopic relation that fits complex microscopic behaviour at the molecular, atomic, or finer scales.

- The macroscopic relations $f_{macro}(\cdot)$ are usually identical to those *already* being used by existing process design applications; hence, no special adaptation of the latter is necessary.

However, there is, at least in principle, a third route (shown in the right part of Fig. 1). This is based on the recognition that most computational chemistry techniques can be viewed as mappings between the macroscopic variables x and the microscopic parameters θ on one hand, and the macroscopic variables y on the other; that is, these techniques are essentially procedures for computing y for given values of x and θ, thereby establishing functions of the form:

$$y = f_{micro}(x, \theta) \qquad (1)$$

This approach by-passes the macroscopic relations $f_{macro}(\cdot)$ and the parameters α appearing in them. This could be an important advantage in some cases; for instance, fitting molecular simulation results for a fluid obeying the Lennard-Jones intermolecular potential (which can be expressed in terms of only 2 microscopic parameters θ) to an equation of state has resulted in a modified Benedict-Webb-Rubin (MBWR) equation which requires no fewer than 33 macroscopic parameters α (see Sun and Teja (1998) and Johnson *et al.* (1993)). Thus, the microscopic description is much more compact than the macroscopic one.

The direct nature of the third route could, in principle, also offer certain advantages in product design applications seeking to establish directly the microscopic properties θ that impart a certain macroscopic behaviour to a particular substance.

However, for this direct route to become practically feasible for non-trivial applications, significant improvements in computational efficiency must be achieved. Moreover, other mathematical issues (e.g. the differentiability of the mapping $f_{micro}(\cdot)$) and their numerical counterparts (e.g. the computation of the partial derivatives $\partial f_{micro}(\cdot)/\partial x$ and $\partial f_{micro}(\cdot)/\partial \theta$) must be addressed.

The resolution of the above issues may also produce immediate benefits to the other two routes considered earlier. For instance, intermolecular potential parameters θ are often derived by fitting the pseudo-experimental data $\{(x^{[k]}, y^{[k]})\}$ to real experimental ones. However, the differentiability problems mentioned above prevent the use of systematic regression techniques; instead, trial-and-error procedures are often being used for this purpose (see, for instance, Errington and Panagiotopoulos, 1998).

Aims of This Paper

This work is based on a view of computational chemistry techniques as formal mathematical mappings relating *macroscopic* quantities of direct relevance to engineering applications. We have chosen to concentrate our initial efforts in the area of *molecular dynamics* for three main reasons:

- Molecular dynamics offers a wide scope for the prediction of thermodynamic, transport and equilibrium properties of engineering interest.
- Albeit significant, the computational cost associated with molecular dynamics computations, even in their current form, is not as high as that of other computational techniques such as quantum chemistry.
- The fully deterministic nature of this technique makes it more amenable to the type of approach that we intend to pursue than stochastic techniques such as those based on Monte-Carlo simulations.

In the rest of this paper, we first examine the molecular dynamics technique as a mathematical mapping between certain input and output quantities. We identify some of the improvements required to produce an algorithm that is well suited for process design purposes. Particular emphasis is placed on the discontinuities that are introduced by some elements of the conventional molecular dynamics technique.

We then propose a modification to the calculation of the forces exerted by one particle on another. The main aim of this modification is the elimination of the discontinuities mentioned above. This leads to a continuous and differentiable mapping being defined by a molecular dynamics computation.

Finally, we consider the computation of the partial derivatives of the molecular dynamics mapping. This is achieved via the formulation and solution of the adjoint equations of the classical Newtonian equations of motion.

The Conventional Molecular Dynamics Technique

In this section, we briefly review the structure of conventional molecular dynamics in the microcanonical ensemble of N particles with a given density ρ and energy E. We then consider this type of computation as a mathematical mapping between, on one hand, the values of the "input" variables (i.e. density and energy) and the parameters θ of the intermolecular potential, and, on the other, the computed "output" variables (e.g. temperature, pressure etc.). Finally, we consider the origins of discontinuities in the computation of this mapping.

Algorithm Structure

The basic algorithm for molecular dynamics can be outlined as follows:

1. Specify the macroscopic condition of the system in terms of density ρ and specific internal energy E.
2. Translate the above conditions for the simulation at the molecular level in terms of a number of particles N contained within a cube of size L (the "reference box") consistent with the specified density ρ.
3. Initialise the system by assigning the vector of initial particle positions $\mathbf{r}(0) = \mathbf{r}_0$ and velocities $\mathbf{v}(0) = \mathbf{v}_0$ consistent with the specified energy E.
4. Integrate Newton's equations of motion:

$$\dot{\mathbf{r}}_i = \mathbf{v}_i;\ \dot{\mathbf{v}}_i = \mathbf{F}_i/m_i;\quad i = 1,\ldots,N \quad (2)$$

from $t = 0$ to $t = t_f$. Here, the forces $\mathbf{F}_i$ are given by:

$$\mathbf{F}_i = -\nabla_{\mathbf{r}_i} U;\quad i = 1,\ldots,N \quad (3)$$

where U is the intermolecular potential.

5. Once the system has reached equilibrium (at a time denoted as $t_f - \tau$), the instantaneous system properties are calculated. For any instantaneous property, $p(t)$, the observable property p_{obs} can be determined as follows:

$$p_{obs} = \lim_{\tau \to \infty} \frac{1}{\tau} \int_{t_f-\tau}^{t_f} p(t)dt. \quad (4)$$

Some typical results of a molecular dynamics simulation are shown in Fig. 2. The simulation involved 256 argon atoms at a density of 39,960 mol/m^3 and energy of -3397 J/mol. The quantities plotted are the instantaneous temperature and pressure (cf. Eqns. (58) and (59) later in this paper). Here $t_f = 22$ ps; the time τ over which the averaging is performed is taken as $t_f/2$.

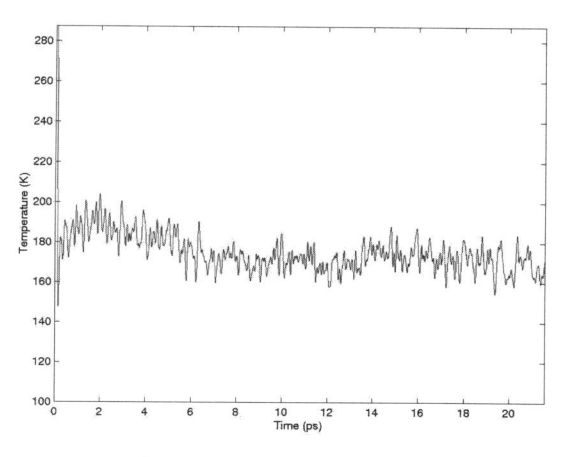

(a) Instantaneous temperature

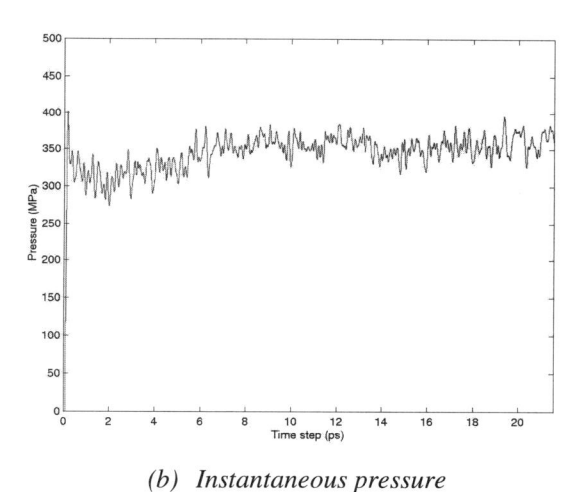

(b) Instantaneous pressure

Figure 2. Typical results of molecular dynamics simulation.

Molecular Dynamics as a Mathematical Mapping

The molecular dynamics algorithm described above can be viewed as the computation of a function of the form:

$$y = f(x, \theta) \qquad (5)$$

Here x denotes the vector of macroscopic input quantities (i.e., in the case of pure components, the density and energy), θ is a vector of microscopic parameters (e.g. the Lennard-Jones constants ε, σ) and y denotes one or more macroscopic properties (e.g. temperature, pressure, diffusion coefficients).

Since the molecular dynamics computation involves the solution of an initial value problem, the above function is well-defined in principle, provided the same deterministic procedure for generating the initial condition of the system is always used at step 3.

The partial derivatives $\partial f/\partial x$ and $\partial f/\partial \theta$ of a mapping established by the solution of a system of ordinary differential equations (ODEs) can normally be evaluated by integrating the corresponding *adjoint system* (Bryson and Ho, 1975). However, in this case, this computation is complicated by the occurrence of a large number of discontinuities. These are caused by two different elements of the conventional implementation of the molecular dynamics technique. We consider these below.

Discontinuities in Particle Positions: Periodic Boundary Conditions

The system of differential equations (2) is subject to the *periodic boundary conditions* (Born and von Karman, 1912). Thus, whenever a particle reaches a face of the reference box and is moving outwards, it is instantaneously transposed to the same position on the *opposite* face, moving inwards with the same velocity vector.

The periodic boundary conditions reflect the periodic nature of the spatial domain. As shown in Fig. 3, the reference box is surrounded by an infinite number of similar boxes, each containing N particles at identical relative position to those in the reference box. Whenever a particle leaves the reference box, an identical particle enters it from a neighbouring box (see particle 1 in Fig. 3). In practical terms, the aim of these conditions is to maintain a constant number of particles within the reference box. However, periodic boundary conditions introduce discontinuities in the positions of these particles.

Discontinuities in Particle Velocities: Minimum Image Convention

The computation of the forces $-\nabla_r U$ in Eqn. (3) is based on the so-called "minimum image convention" (Metropolis *et al.*, 1953). According to this convention, a particle i interacts with that image of another particle j which is nearest to it. Often this nearest image will not be within the reference box but in a neighbouring one (see the dashed box in Fig. 3).

The minimum image convention was introduced primarily as a way of reducing the computational cost of molecular dynamics. On the other hand, it gives rise to a discontinuity in the force exerted on a particle i by a particle j. The discontinuity occurs at the point at which the identity of the image of j that is nearest to i changes. This occurs when the particles are at a distance $L/2$ (as measured along any coordinate axis) apart and moving away from each other.

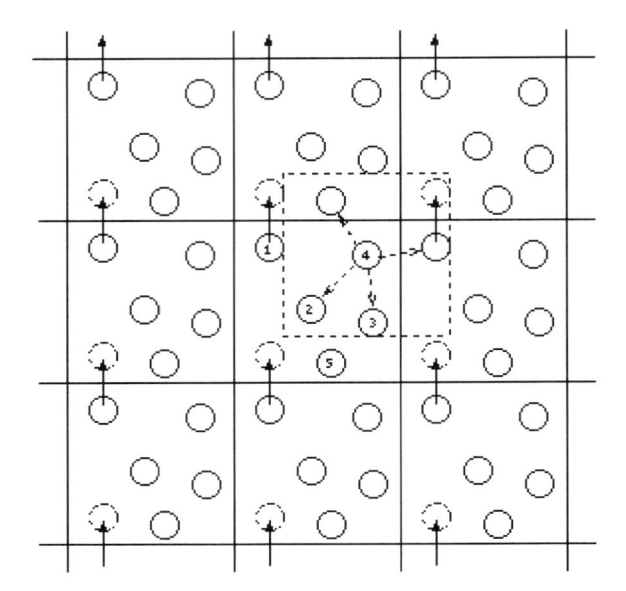

Figure 3. Periodic boundary conditions and minimum image convention.

As a result of this discontinuity, the force changes sign and, in most cases[16], direction instantaneously. Of course, this change is completely fictitious: in reality, the force is a continuous function of time. In order to mitigate this effect, the size L of the box must be chosen to be large enough for the force between two particles at a distance $L/2$ apart to be practically zero.

A Discontinuity-Free Molecular Dynamics Technique

In this section, we develop an alternative framework for molecular dynamics which eliminates the discontinuities inherent in the conventional simulation techniques. In particular, we shall demonstrate that the definition of appropriate force functions that exhibit certain spatial continuity and periodicity properties obviates the need for both the periodic boundary conditions and the minimum image convention in molecular dynamics.

Molecular Dynamics in 1-Dimensional Space

The basic ideas of our approach can be most easily understood by considering a one-dimensional line segment of length L containing two particles, denoted here as i and j, as shown in Fig. 4.

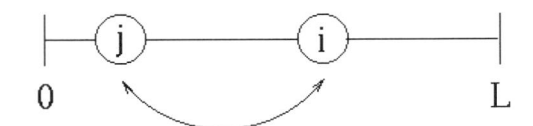

Figure 4. Reference 1-dimensional 2-particle system.

Suppose that the force F_i exerted on particle i by particle j ($\neq i$) is given by the gradient of the reduced Lennard-Jones potential and is of the form:

$$F_i = \frac{48}{(x_i - x_j)^{13}} - \frac{24}{(x_i - x_j)^7} \qquad (6)$$

Now if the *reference segment* shown in Fig. 4 is infinitely replicated in both directions in space, we obtain the system shown in Fig. 5.

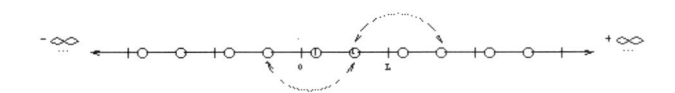

Figure 5. Infinite line representation of a 2-particle system.

In this case, the force acting on particle i is exerted by particle j ($\neq i$) within the same line segment, and also by *all* images of particle i and particle j in *all* other line segments:

$$F_i = \sum_{j \neq i} \sum_{k=-\infty}^{+\infty} \left[\frac{48}{(x_i - x_j + kL)^{13}} - \frac{24}{(x_i - x_j + kL)^7} \right] \\ + \sum_{\substack{k=-\infty \\ k \neq 0}}^{+\infty} \left[\frac{48}{(kL)^{13}} - \frac{24}{(kL)^7} \right] \qquad (7)$$

The last term in (7) corresponds to the summation of forces exerted on particle i by the images of itself in the surrounding line sections. It can easily be shown that this term is, in fact, equal to zero. Physically, this is because the forces exerted on i by its images in any two segments symmetric with respect to the reference segment (see the dotted lines in Fig. 4) cancel each other out.

Eqn. (7) can be more conveniently expressed using sums involving only non-negative k terms:

---

[16] Unless particle i happens to be on the straight line between two images of j involved in this discontinuity.

$$F_i = \sum_{j \neq i} \left\{ \frac{48}{L^{13}} \sum_{k=0}^{+\infty} \frac{1}{\left(\frac{x_i - x_j}{L} + k\right)^{13}} \right.$$
$$+ \frac{48(-1)^{13}}{L^{13}} \sum_{k=0}^{+\infty} \frac{1}{\left(1 - \frac{x_i - x_j}{L} + k\right)^{13}}$$
$$- \frac{24}{L^7} \sum_{k=0}^{+\infty} \frac{1}{\left(\frac{x_i - x_j}{L} + k\right)^{7}}$$
$$\left. - \frac{24(-1)^7}{L^7} \sum_{k=0}^{+\infty} \frac{1}{\left(1 - \frac{x_i - x_j}{L} + k\right)^{7}} \right\} \quad (8)$$

which can then be written as:

$$F_i = \sum_{j \neq i} \left\{ \frac{48}{L^{13}} \left[\zeta^{[13]}\left(\frac{x_i - x_j}{L}\right) - \zeta^{[13]}\left(1 - \frac{x_i - x_j}{L}\right) \right] \right.$$
$$\left. - \frac{24}{L^7} \left[\zeta^{[7]}\left(\frac{x_i - x_j}{L}\right) - \zeta^{[7]}\left(1 - \frac{x_i - x_j}{L}\right) \right] \right\} \quad (9)$$

where $\zeta^{[n]}(s)$ represents the generalised Riemann (or Hurwitz) zeta function defined as:

$$\zeta^{[n]}(s) \equiv \sum_{k=0}^{+\infty} \frac{1}{(k+s)^n}. \quad (10)$$

The one-dimensional case considered in this section, albeit physically unrealistic, illustrates certain interesting points. First, for the purposes of molecular dynamics computations, it is possible to define a "correct" Lennard-Jones force between two particles at a distance r apart given by:

$$F(r) \equiv \frac{48}{L^{13}} \left[\zeta^{[13]}(\hat{r}) - \zeta^{[13]}(1-\hat{r}) \right]$$
$$- \frac{24}{L^7} \left[\zeta^{[7]}(\hat{r}) - \zeta^{[7]}(1-\hat{r}) \right] \quad (11)$$

rather than the more conventional:

$$\tilde{F}(r) \equiv \frac{48}{L^{13}} \left(\frac{1}{\hat{r}^{13}}\right) - \frac{24}{L^7} \left(\frac{1}{\hat{r}^7}\right) \quad (12)$$

Here $\hat{r}$ is the *normalised* distance defined as:

$$\hat{r} \equiv \frac{r}{L} \quad (13)$$

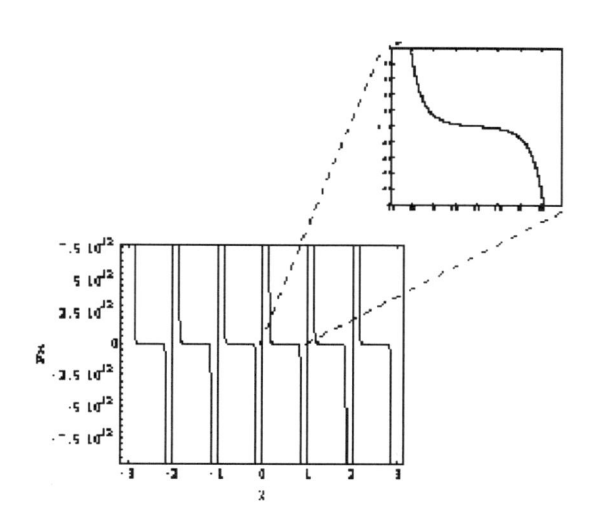

Figure 6. Modified one-dimensional interparticle force function.

The modified force function, shown graphically in Fig. 6, takes account of the interactions of the particles in the reference segment with those in all other segments. Hence, the minimum image convention is not invoked and no discontinuities arise when $\hat{r} = 0.5$. In fact, from equation (11) and also from Fig. 5, we can see that $F(0.5) = 0$, and that $F(\hat{r})$ is continuous and differentiable at $\hat{r} = 0.5$.

The function $F(\hat{r})$ is naturally periodic with a period of 1, i.e.:

$$F(\hat{r}) = F(\hat{r} \pm k), \quad \forall k \in \mathbb{N} \quad (14)$$

where $\mathbb{N}$ is the set of natural numbers. This periodicity can clearly be seen in Fig. 6. Later we will prove this property formally for a 2-dimensional generalisation of the force function considered here.

The periodic nature of the function $F(\cdot)$ has the important implication that it is no longer necessary to impose periodic boundary conditions in an explicit manner as is done in conventional molecular simulation: for instance, the force exerted on each other by two particles that are a distance $2.3L$ apart will be the same as the force that would be exerted were they only $0.3L$ apart. The *physical* reason for this property is quite obvious: since the function $F(\cdot)$ takes account of the interactions of one particle with *all* images of the other, it does not really matter which one of these images we consider to be the "real" particle.

Of course, the periodicity of $F(\cdot)$ is only of mathematical interest: in one-dimensional molecular dynamics, it is impossible for the distance between two particles to exceed L (i.e. for $\hat{r} \geq 1$) as this would imply that, at some time instant, the position of one particle would coincide exactly with that of (the image of) another. In the conventional molecular dynamics approach, this eventuality is avoided by the use of the minimum image convention together with the singularity in the force function $\tilde{F}(\cdot)$ (cf. Eqn. (12)) for $\hat{r} = 0$. Interestingly, in our modified framework, the same effect is ensured automatically by the singularity in the value of $F(\cdot)$ at $\hat{r} = 1$. Again, there is no need for the minimum image convention.

Molecular Dynamics in 2-Dimensional Space

We now consider an extension of the ideas presented in the previous section to the 2-dimensional case. This is qualitatively different to (and more interesting than) the 1-dimensional case while still allowing easy visualisation of the various concepts.

We start by considering a reference square of side L containing two particles i and j at positions (x_i, y_i) and (x_j, y_j) respectively. We now produce infinite replications of this square in all directions as shown in Fig. 7.

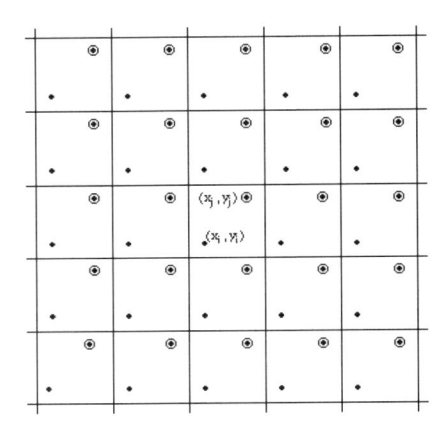

Figure 7. Infinite 2-dimensional space.

Assuming a reduced Lennard-Jones potential, the force F_i^x exerted on particle i by particle j in the x-direction is given by:

$$F_i^x = 48 \frac{x_i - x_j}{((x_i - x_j)^2 + (y_i - y_j)^2)^7} - 24 \frac{x_i - x_j}{((x_i - x_j)^2 + (y_i - y_j)^2)^4} \quad (15)$$

We now consider the force exerted on particle i in the reference box by (a) particle j in the reference box and all its images; and (b) all images of particle i. This is given by:

$$F_i^x = \sum_{j \neq i} \left\{ 48 \sum_{k=-\infty}^{+\infty} \sum_{k'=-\infty}^{+\infty} \frac{x_i - x_j + kL}{((x_i - x_j + kL)^2 + (y_i - y_j + k'L)^2)^7} - 24 \sum_{k=-\infty}^{+\infty} \sum_{k'=-\infty}^{+\infty} \frac{x_i - x_j + kL}{((x_i - x_j + kL)^2 + (y_i - y_j + k'L)^2)^4} \right\} + \sum_{k=-\infty}^{+\infty} \sum_{k'=-\infty}^{+\infty} \left\{ \frac{48kL}{((kL)^2 + (k'L)^2)^7} - \frac{24kL}{((kL)^2 + (k'L)^2)^4} \right\} \quad (16)$$

As in the 1-dimensional case, the last term on the right hand side, representing the interaction of particle with its own images, is zero. By defining the normalised distances X_{ij}, Y_{ij} as:

$$X_{ij} \equiv \frac{x_i - x_j}{L}, \quad Y_{ij} \equiv \frac{y_i - y_j}{L} \quad (17)$$

the above can be written as:

$$F_i^x = \sum_{j \neq i} \left\{ \frac{48}{L^{13}} \sum_{k=-\infty}^{+\infty} \sum_{k'=-\infty}^{+\infty} \frac{X_{ij} + k}{((X_{ij} + k)^2 + (Y_{ij} + k')^2)^7} - \frac{24}{L^7} \sum_{k=-\infty}^{+\infty} \sum_{k'=-\infty}^{+\infty} \frac{X_{ij} + k}{((X_{ij} + k)^2 + (Y_{ij} + k')^2)^4} \right\}$$
$$\forall (X_{ij}, Y_{ij}) \in \mathrm{R}^2 \setminus \mathrm{N}^{[2]} \quad (18)$$

We note that the above function is defined for all real values of the variables X_{ij}, Y_{ij} *except* those for which the pair (X_{ij}, Y_{ij}) belongs to a set $\mathrm{N}^{[2]}$ defined as:

$$\mathrm{N}^{[2]} \equiv \{(n, n') \mid n, n' \in \mathrm{Z}\} \quad (19)$$

where Z is the set of integers. In other words, X_{ij} and Y_{ij} cannot *both* take integer values at the same time. It can be seen that the function F_i^x would be infinite in such a case as the denominator of the term with $k = -n$ and $k' = -n'$ would be zero. Physically, this kind of singularity expresses the fact that if *both* X_{ij} and Y_{ij} were to take integer values, then the position of particle i in the reference box would coincide with that of an image of particle j in a box located X_{ij} places to the right and Y_{ij} places above the reference one. Hence, particles i and j in the reference box would overlap with each other.

The form of Eqn. (18) suggests that the correct expression for the force in the x-direction exerted by two particles on each other at a normalised distance (X, Y) apart is given by:

$$F^x(X,Y) = \frac{48}{L^{13}} \sum_{k=-\infty}^{+\infty} \sum_{k'=-\infty}^{+\infty} \frac{X+k}{((X+k)^2+(Y+k')^2)^7}$$
$$- \frac{24}{L^7} \sum_{k=-\infty}^{+\infty} \sum_{k'=-\infty}^{+\infty} \frac{X+k}{((X+k)^2+(Y+k')^2)^4}$$
$$\forall (X,Y) \in R^2 \setminus N^{[2]} \quad (20)$$

rather than the more conventional:

$$\tilde{F}^x(X,Y) = \frac{48}{L^{13}} \frac{X}{(X^2+Y^2)^7} - \frac{24}{L^7} \frac{X}{(X^2+Y^2)^4} \quad (21)$$

The following properties can be shown for the force function F^x:

Property I: Invariance Under Integral Shifts

$$F^x(X+k_x, Y+k_y) = F^x(X,Y), \quad \forall k_x, k_y \in Z \quad (22)$$

Proof: By definition,

$$F^x(X+k_x, Y+k_y)$$
$$= \frac{48}{L^{13}} \sum_{k=-\infty}^{+\infty} \sum_{k'=-\infty}^{+\infty} \frac{X+k+k_x}{((X+k+k_x)^2+(Y+k'+k_y)^2)^7}$$
$$- \frac{24}{L^7} \sum_{k=-\infty}^{+\infty} \sum_{k'=-\infty}^{+\infty} \frac{X+k+k_x}{((X+k+k_x)^2+(Y+k'+k_y)^2)^4}$$
(23)

By defining the shifted summation indices:

$$\hat{k} \equiv k+k_x; \quad \hat{k}' \equiv k+k_y \quad (24)$$

we can rewrite the above as:

$$F^x(X+k_x, Y+k_y)$$
$$= \frac{48}{L^{13}} \sum_{k=-\infty}^{+\infty} \sum_{k'=-\infty}^{+\infty} \frac{X+\hat{k}}{((X+\hat{k})^2+(Y+\hat{k}')^2)^7}$$
$$- \frac{24}{L^7} \sum_{k=-\infty}^{+\infty} \sum_{k'=-\infty}^{+\infty} \frac{X+\hat{k}}{((X+\hat{k})^2+(Y+\hat{k}')^2)^4}$$
$$= F^x(X,Y)$$
(25)

Q.E.D.

One implication of property I is that the force function $F^x(X,Y)$ is naturally periodic in both the X and the Y directions with a period of 1. There is, therefore, no need for explicitly enforced periodicity conditions designed to keep particles i and j within the reference box.

Property II: Computation Over a Limited Domain

$$F^x(X,Y) = F^x(\Phi(X), \Phi(Y)) \quad (26)$$

where the operator $\Phi(X)$ denotes the fractional part of a real number X (e.g. $\Phi(5.23) = 0.23$).

Proof: This property is a direct consequence of property I since $\Phi(X) = X - \lfloor X \rfloor$ and $\Phi(Y) = Y - \lfloor Y \rfloor$, where $\lfloor X \rfloor$ and $\lfloor Y \rfloor$ are the largest integer numbers that do not exceed X and Y respectively.

Q.E.D.

The implication of property II is that it is sufficient to be able to compute the function $F^x(X,Y)$ efficiently over the limited domain $X \in (-1,1)$, $Y \in (-1,1)$.

Property III: Computation Over a Limited Domain

$$F^x(1-X, Y) = -F^x(X,Y) \quad (27)$$

$$F^x(X, 1-Y) = F^x(X,Y) \quad (28)$$

Proof: Using the definition of the modified force function (Eqn. (20)), we have:

$$F^x(1-X,Y) = \frac{48}{L^{13}} \sum_{k=-\infty}^{+\infty} \sum_{k'=-\infty}^{+\infty} \frac{1-X+k}{((1-X+k)^2+(Y+k')^2)^7}$$
$$- \frac{24}{L^7} \sum_{k=-\infty}^{+\infty} \sum_{k'=-\infty}^{+\infty} \frac{1-X+k}{((1-X+k)^2+(Y+k')^2)^4}$$
$$\forall (X,Y) \in R^2 \setminus N^{[2]} \quad (29)$$

Using the same procedure as in the proof of property I, we define the following shifted summation indices:

$$\hat{k} \equiv -k-1, \quad \hat{k}' \equiv k' \quad (30)$$

and we rewrite the above as:

$$F^x(1-X,Y) = -\frac{48}{L^{13}} \sum_{\hat{k}=+\infty-1}^{-\infty-1} \sum_{\hat{k}'=-\infty}^{+\infty} \frac{X+\hat{k}}{((X+\hat{k})^2+(Y+\hat{k}')^2)^7}$$
$$+ \frac{24}{L^7} \sum_{\hat{k}=+\infty-1}^{-\infty-1} \sum_{\hat{k}'=-\infty}^{+\infty} \frac{X+\hat{k}}{((X+\hat{k})^2+(Y+\hat{k}')^2)^4}$$
$$= -F^x(X,Y)$$
(31)

From Eqn. (20), we also obtain:

$$F^x(X, 1-Y) = \frac{48}{L^{13}} \sum_{k=-\infty}^{+\infty} \sum_{k'=-\infty}^{+\infty} \frac{X+k}{((X+k)^2+(1-Y+k')^2)^7}$$
$$- \frac{24}{L^7} \sum_{k=-\infty}^{+\infty} \sum_{k'=-\infty}^{+\infty} \frac{X+k}{((X+k)^2+(1-Y+k')^2)^4}$$
$$\forall (X,Y) \in R^2 \setminus N^{[2]} \quad (32)$$

We define the following shifted summation indices:

$$\hat{k} \equiv k, \quad \hat{k}' \equiv -k'-1 \quad (33)$$

and rewrite the above as:

$$F^x(X, 1-Y) = \frac{48}{L^{13}} \sum_{\hat{k}=-\infty}^{+\infty} \sum_{\hat{k}'=+\infty}^{-\infty-1} \frac{X+\hat{k}}{((X+\hat{k})^2+(Y+\hat{k}')^2)^7}$$
$$- \frac{24}{L^7} \sum_{\hat{k}=-\infty}^{+\infty} \sum_{\hat{k}'=+\infty}^{-\infty-1} \frac{X+\hat{k}}{((X+\hat{k})^2+(Y+\hat{k}')^2)^4}$$
$$= F^x(X,Y) \quad (34)$$

Q.E.D.

Property III, in conjuction with properties I and II, implies that the domain over which the function $F^x(X,Y)$ needs to be evaluated can now be further reduced to $X \in [0, 0.5]$, $Y \in [0, 0.5]$. For instance, $F^x(2.9,-3.4) = F^x(0.9,-0.4) = -F^x(0.1,-0.4) = -F^x(0.1,0.6) = -F^x(0.1,0.4)$, where we have invoked successively properties II, III, I, and III.

So far in this section, we have concentrated on the force in the x-direction and its properties. The corresponding force function in the y-direction is given by:

$$F^y(X,Y) = \frac{48}{L^{13}} \sum_{k=-\infty}^{+\infty} \sum_{k'=-\infty}^{+\infty} \frac{Y+k'}{((X+k)^2+(Y+k')^2)^7}$$
$$- \frac{24}{L^7} \sum_{k=-\infty}^{+\infty} \sum_{k'=-\infty}^{+\infty} \frac{Y+k'}{((X+k)^2+(Y+k')^2)^4}$$
$$\forall (X,Y) \in R^2 \setminus N^{[2]} \quad (35)$$

and it has analogous properties to the function $F^x(X,Y)$ defined in Eqn. (20).

Property IV: Relation between $F^x(\cdot)$ and $F^y(\cdot)$

$$F^y(X,Y) = F^x(Y,X) \quad (36)$$

Proof: From Eqn. (19), we have:

$$F^x(Y,X) = \frac{48}{L^{13}} \sum_{k=-\infty}^{+\infty} \sum_{k'=-\infty}^{+\infty} \frac{Y+k}{((X+k')^2+(Y+k)^2)^7}$$
$$- \frac{24}{L^7} \sum_{k=-\infty}^{+\infty} \sum_{k'=-\infty}^{+\infty} \frac{Y+k}{((X+k')^2+(Y+k)^2)^4}$$
$$\forall (X,Y) \in R^2 \setminus N^{[2]} \quad (37)$$

As k and k' indices both cover the $-\infty$ to $+\infty$ range, expressions (37) and (35) are equivalent, hence $F^y(X,Y) \equiv F^x(Y,X)$. Q.E.D.

Molecular Dynamics in 3-Dimensional Space

The 3-dimensional case is a straightforward extension of the 2-dimensional one. Thus, the force extended on particle i by all other particles is given by:

$$F_i^x = \sum_{j \neq i} F^x(X_{ij}, Y_{ij}, Z_{ij}) \quad (38)$$

$$F_i^y = \sum_{j \neq i} F^y(X_{ij}, Y_{ij}, Z_{ij}) \quad (39)$$

$$F_i^z = \sum_{j \neq i} F^z(X_{ij}, Y_{ij}, Z_{ij}) \quad (40)$$

where we have defined the normalised distances:

$$X_{ij} \equiv \frac{x_i - x_j}{L}, \quad Y_{ij} \equiv \frac{y_i - y_j}{L}, \quad Z_{ij} \equiv \frac{z_i - z_j}{L} \quad (41)$$

The force functions $F^x(\cdot)$, $F^y(\cdot)$ and $F^z(\cdot)$ are defined as:

$$F^x(X,Y,Z)$$
$$= \frac{48}{L^{13}} \sum_{k=-\infty}^{+\infty} \sum_{k'=-\infty}^{+\infty} \sum_{k''=-\infty}^{+\infty} \frac{X+k}{((X+k)^2+(Y+k')^2+(Z+k'')^2)^7}$$
$$- \frac{24}{L^7} \sum_{k=-\infty}^{+\infty} \sum_{k'=-\infty}^{+\infty} \sum_{k''=-\infty}^{+\infty} \frac{X+k}{((X+k)^2+(Y+k')^2+(Z+k'')^2)^4}$$
$$\forall (X,Y,Z) \in R^3 \setminus N^{[3]} \quad (42)$$

$$F^y(X,Y,Z)$$
$$= \frac{48}{L^{13}} \sum_{k=-\infty}^{+\infty} \sum_{k'=-\infty}^{+\infty} \sum_{k''=-\infty}^{+\infty} \frac{Y+k'}{((X+k)^2+(Y+k')^2+(Z+k'')^2)^7}$$
$$- \frac{24}{L^7} \sum_{k=-\infty}^{+\infty} \sum_{k'=-\infty}^{+\infty} \sum_{k''=-\infty}^{+\infty} \frac{Y+k'}{((X+k)^2+(Y+k')^2+(Z+k'')^2)^4}$$
$$\forall (X,Y,Z) \in R^3 \setminus N^{[3]} \quad (43)$$

$$F^z(X,Y,Z)$$
$$= \frac{48}{L^{13}} \sum_{k=-\infty}^{+\infty} \sum_{k'=-\infty}^{+\infty} \sum_{k''=-\infty}^{+\infty} \frac{Z+k''}{((X+k)^2+(Y+k')^2+(Z+k'')^2)^7}$$
$$- \frac{24}{L^7} \sum_{k=-\infty}^{+\infty} \sum_{k'=-\infty}^{+\infty} \sum_{k''=-\infty}^{+\infty} \frac{Z+k''}{((X+k)^2+(Y+k')^2+(Z+k'')^2)^4}$$
$$\forall (X,Y,Z) \in \mathrm{R}^3 \setminus \mathrm{N}^{[3]} \tag{44}$$

where $\mathrm{N}^{[3]}$ is the set of all integer triplets:

$$\mathrm{N}^{[3]} \equiv \{(n,n',n'') \mid n,n',n'' \in \mathrm{Z}\} \tag{45}$$

Moreover, properties I-IV proven for the 2-dimensional case have straightforward extensions to the 3-dimensional one.

Comments

The molecular dynamics framework presented in this section takes account of interactions of each particle in a reference box with *all* the particles in *all* the boxes surrounding it. In this sense, it is similar to the Ewald (1921) summation which has been used to deal with long-range electrostatic interactions.

The key point is that, given any two particles, the forces between them are well-defined mathematical functions that, just like the forces in conventional molecular dynamics, depend only on their relative position. Moreover, these force functions are continuous and differentiable for all inter-particle distances except a set of values corresponding to physically impossible situations (i.e. the two particles essentially occupying the same position in space). Finally, the force functions are naturally periodic. Therefore, the force between two particles is identical to that between any two images of these particles; thus, there is no need to confine the N particles under consideration to remain within any given spatial boundary via the imposition of artificial periodic boundary conditions.

The Newtonian equations of motion (Eqns. (2) and (3)) making use of these modified force functions define a continuous system which is well behaved from both the mathematical and the numerical point of view. For instance, it is now feasible to consider rigorous ways for the computation of the gradients of the mapping $f(x,\theta)$ (cf. Eqn. (5)) with respect to the variables x and the parameters θ. Also, efficient multistep integration algorithms that are based on assumptions of polynomial continuity in the system variables can now be used for the solution of the underlying ordinary differential equations.

Partial Derivatives of Molecular Dynamics Mappings in the Microcanonical Ensemble

The framework for molecular dynamics simulation presented in the previous section establishes a continuous and differentiable mapping of the form:

$$y = f(x, \theta) \tag{46}$$

We wish to compute the partial derivatives of the observable system properties y (e.g. T and P) with respect to (a) the intermolecular potential parameters θ, and (b) the input macroscopic system quantities x (e.g. energy E and density ρ).

From the description of the molecular dynamics algorithm presented earlier in this paper, it is clear that x and θ affect directly *only* the initial conditions $\mathbf{r}_0$ and $\mathbf{v}_0$ of the ordinary differential equations (2). We will, therefore, follow a 3-stage strategy:

1. Compute the partial derivatives of the system properties y with respect to the initial system conditions $\mathbf{r}_0$, $\mathbf{v}_0$.
2. Compute the partial derivatives of the initial conditions with respect to the system potential parameters θ and input quantities x.
3. Combine the above using the chain rule:

$$\frac{\partial y}{\partial x} = \frac{\partial y}{\partial \mathbf{r}_0} \cdot \frac{\partial \mathbf{r}_0}{\partial x} + \frac{\partial y}{\partial \mathbf{v}_0} \cdot \frac{\partial \mathbf{v}_0}{\partial x} \tag{47}$$

$$\frac{\partial y}{\partial \theta} = \frac{\partial y}{\partial \mathbf{r}_0} \cdot \frac{\partial \mathbf{r}_0}{\partial \theta} + \frac{\partial y}{\partial \mathbf{v}_0} \cdot \frac{\partial \mathbf{v}_0}{\partial \theta} \tag{48}$$

Partial Derivatives With Respect To Initial System State

We start by considering a quantity y given by a simple time-integral expression:

$$y = \int_0^{t_f} p(\mathbf{r}, \mathbf{v}) dt \tag{49}$$

where t_f is some final time and p is a continuous differentiable function of the positions $\mathbf{r}$ and velocities $\mathbf{v}$ of all the particles under consideration. The behaviour of the system is subject to the Newtonian equations of motion:

$$\dot{\mathbf{r}}_i = \mathbf{v}_i; \quad \dot{\mathbf{v}}_i = \mathbf{F}_i(\mathbf{r}) \tag{50}$$

with the initial conditions:

$$\mathbf{r}(0) = \mathbf{r}_0; \quad \mathbf{v}(0) = \mathbf{v}_0. \tag{51}$$

The Adjoint System

We note that particle positions and velocities are functions not only of time t but also of the initial positions $\mathbf{r}_0$ and velocities $\mathbf{v}_0$. Therefore, y is also a function of $\mathbf{r}_0$ and $\mathbf{v}_0$. The partial derivatives $\partial y/\partial \mathbf{r}_0$ and $\partial y/\partial \mathbf{v}_0$ can be computed via the solution of the so-called *adjoint system* (Bryson and Ho, 1975):

$$\dot{\boldsymbol{\lambda}}_i^r = -\sum_{j=1}^{N} \mathbf{M}^{(j,i)} \boldsymbol{\lambda}_j^v - \frac{\partial p}{\partial \mathbf{r}_i} \quad , i = 1,..,N \quad (52)$$

$$\dot{\boldsymbol{\lambda}}_i^v = -\boldsymbol{\lambda}_i^r - \frac{\partial p}{\partial \mathbf{v}_i} \quad , i = 1,..,N \quad (53)$$

subject to the *final* time conditions:

$$\boldsymbol{\lambda}_i^r(t_f) = 0; \quad \boldsymbol{\lambda}_i^v(t_f) = 0 \quad , i = 1,..,N \quad (54)$$

where $\boldsymbol{\lambda}_i^r, \boldsymbol{\lambda}_i^v \in \mathbb{R}^3$ are vectors of adjoint variables, and $\mathbf{M}^{(j,i)}$ is a 3×3 matrix of the form:

$$\mathbf{M}^{(j,i)} = \begin{pmatrix} \frac{\partial F_i^x}{\partial x_j} & \frac{\partial F_i^y}{\partial x_j} & \frac{\partial F_i^z}{\partial x_j} \\ \frac{\partial F_i^x}{\partial y_j} & \frac{\partial F_i^y}{\partial y_j} & \frac{\partial F_i^z}{\partial y_j} \\ \frac{\partial F_i^x}{\partial z_j} & \frac{\partial F_i^y}{\partial z_j} & \frac{\partial F_i^z}{\partial z_j} \end{pmatrix} \quad (55)$$

where the partial derivatives of F_i^x, F_i^y, F_i^z can be obtained by differentiating equations (38)-(40) respectively.

The adjoint equations (52)-(53) form a set of ordinary differential equations in $\boldsymbol{\lambda}^r(t)$ and $\boldsymbol{\lambda}^v(t)$ subject to *final* (rather than initial) conditions (54). The system is integrated from $t = t_f$ to $t = 0$, and the required partial derivatives of y may be obtained from:

$$\frac{\partial y}{\partial \mathbf{r}_i(0)} = \boldsymbol{\lambda}_i^r(0); \quad \frac{\partial y}{\partial \mathbf{v}_i(0)} = \boldsymbol{\lambda}_i^v(0) \quad , i = 1,..,N \quad (56)$$

Solution of the Adjoint System

We note that the quantities $\partial p/\partial \mathbf{r}_i$, $\partial p/\partial \mathbf{v}_i$ and $\mathbf{M}^{(j,i)}$ appearing in the adjoint system (52)-(53) are generally functions of particle positions $\mathbf{r}(t)$ and velocities $\mathbf{v}(t)$. We will therefore need to know the latter for the integration of the adjoint system. Hence, we determine the gradients $\partial y/\partial \mathbf{r}(0)$ and $\partial y/\partial \mathbf{v}(0)$ using the following algorithm:

1. Perform a molecular dynamics simulation from time $t = 0$ to time $t = t_f$.
2. Starting from the end system configuration at time t_f, integrate the adjoint system of equations (52) and (53) together with the original Newton's equations of motion (50) from time $t = t_f$ to $t = 0$.
3. The values of $\boldsymbol{\lambda}^r$ and $\boldsymbol{\lambda}^v$ at $t = 0$ are the required gradients: $\partial y/\partial \mathbf{r}(0) = \boldsymbol{\lambda}^r(0)$ and $\partial y/\partial \mathbf{v}(0) = \boldsymbol{\lambda}^v(0)$.

Partial Derivatives of Time-Averaged Quantities

In practice, we are interested in the time-averaged values of the instantaneous system properties which are calculated *after* an initial equilibration period has been completed (cf. Fig. 2). Thus, time averaging is performed only during a final part of length τ of the total time horizon t_f.

The property of interest y can be expressed as (cf. Eqn. (4)):

$$y = \frac{1}{\tau} \int_{t_f-\tau}^{t_f} p(\mathbf{r},\mathbf{v}) dt. \quad (57)$$

For example, the temperature and pressure of the system can be expressed as (see, for instance, Allen and Tildesley, 1987):

$$T = \frac{1}{\tau} \int_{t_f-\tau}^{t_f} \frac{\sum_i \mathbf{v}_i^T(t) \cdot \mathbf{v}_i(t)}{3N} dt \quad (58)$$

$$P = \frac{1}{\tau} \int_{t_f-\tau}^{t_f} \left(\rho \frac{\sum_i \mathbf{v}_i^T(t) \cdot \mathbf{v}_i(t)}{3N} + \frac{\sum_i \mathbf{r}_i^T(t) \cdot \mathbf{F}_i(t)}{V} \right) dt \quad (59)$$

To compute the partial derivatives of the quantity y using the approach described in previous section, we express Eqn. (57) as:

$$y = \frac{1}{\tau}(\tilde{\tilde{y}} - \tilde{y}) \quad (60)$$

where:

$$\tilde{\tilde{y}} = \int_0^{t_f} p(\mathbf{r},\mathbf{v}) dt; \quad \tilde{y} = \int_0^{t_f-\tau} p(\mathbf{r},\mathbf{v}) dt. \quad (61)$$

We can then compute the partial derivatives of $\tilde{\tilde{y}}$ and $\tilde{y}$ with respect to $\mathbf{r}_0$ and $\mathbf{v}_0$ by solving the corresponding adjoint systems. Of course, both of these adjoint systems are of exactly the same form (i.e. Eqns. (52)-(53)), but the first is subject to the final conditions:

$$\tilde{\lambda}^r(t_f) = \tilde{\lambda}^v(t_f) = 0 \qquad (62)$$

while the second to:

$$\tilde{\lambda}^r(t_f - \tau) = \tilde{\lambda}^v(t_f - \tau) = 0. \qquad (63)$$

Hence we need *two* backward integrations, one starting from t_f and another from $t = t_f - \tau$, as illustrated in Fig. 8.

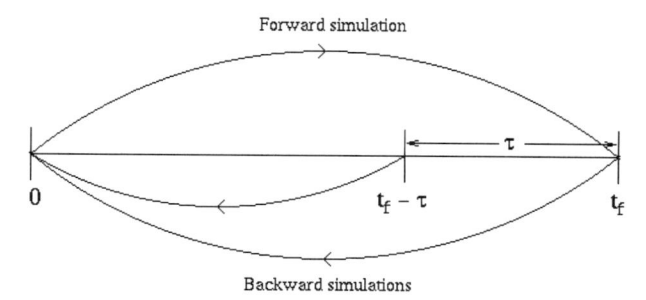

Figure 8. Calculation of partial derivatives of time-averaged quantities.

The required partial derivatives of y with respect to the initial particle positions and velocities are then given by:

$$\frac{\partial y}{\partial \mathbf{r}_i(0)} = \frac{1}{\tau}\left(\tilde{\tilde{\lambda}}_i^r(0) - \tilde{\lambda}_i^r(0)\right), \; i = 1,..,N \qquad (64)$$

$$\frac{\partial y}{\partial \mathbf{v}_i(0)} = \frac{1}{\tau}\left(\tilde{\tilde{\lambda}}_i^v(0) - \tilde{\lambda}_i^v(0)\right), \; i = 1,..,N \qquad (65)$$

Partial Derivatives of Initial System State

We have mentioned previously that the "input" macroscopic conditions of the system, energy E and density ρ, are translated into the microscopic system conditions. The reduced quantities of these variables are obtained using the interparticle potential parameters θ (e.g., the Lennard-Jones parameters ε and σ) as follows:

$$\rho^* = \rho\frac{N_A \sigma^3}{M_r}; \quad E^* = E\frac{N}{\varepsilon N_A} \qquad (66)$$

where N_A is Avogadro's number and M_r is the molecular weight. From the above expressions, we can obtain the partial derivatives of ρ^* and E^* with respect to the macroscopic system conditions ρ and E:

$$\frac{\partial \rho^*}{\partial \rho} = \frac{N_A \sigma^3}{M_r}; \quad \frac{\partial E^*}{\partial E} = \frac{N}{\varepsilon N_A} \qquad (67)$$

and, similarly, partial derivatives with respect to the system potential parameters ε and σ:

$$\frac{\partial \rho^*}{\partial \sigma} = 3\rho\frac{N_A \sigma^2}{M_r}; \quad \frac{\partial E^*}{\partial \varepsilon} = -E\frac{N}{\varepsilon^2 N_A} \qquad (68)$$

The procedure employed to obtain the initial vector of particle positions and velocities in our molecular dynamics simulations is fully deterministic:

1. Typically, the particles are placed at given (normalised) positions $\hat{\mathbf{r}}_i^*$ ($i = 1, \ldots, N$) corresponding to nodes on a regular grid defined on a unit cube. The initial vector of particle positions within a cube of side length L is therefore given by:

$$\mathbf{r}_i(0) = L\hat{\mathbf{r}}_i^* = \left(\frac{N}{\rho^*}\right)^{1/3}\hat{\mathbf{r}}_i^* \qquad (69)$$

The above expression determines the vector of initial particle positions, and this in turn determines the initial system potential energy $U^*(\mathbf{r}(0))$.

2. A random set of particle velocity vectors $\mathbf{V}_i$ ($i = 1,..,N$) is generated and then adjusted to give an overall momentum of zero. To ensure that the mapping $f(x,\theta)$ is fully deterministic, we need to use the *same* randomly generated vectors for all our molecular dynamics computations

The initial particle velocities are obtained by scaling $\mathbf{V}_i$ so that the initial value of the kinetic energy is equal to $E^* - U^*(\mathbf{r}(0))$. Thus:

$$\mathbf{v}_i(0) = \sqrt{\frac{2(E^* - U^*(\mathbf{r}(0)))}{\sum_j \mathbf{V}_j^T \mathbf{V}_j}}\,\mathbf{V}_i \qquad (70)$$

From (69) and (70), we obtain the following set of partial derivatives:

$$\frac{\partial \mathbf{r}_i(0)}{\partial \rho^*} = -\frac{\mathbf{r}_i(0)}{3\rho^*}; \quad \frac{\partial \mathbf{r}_i(0)}{\partial E^*} = 0 \qquad (71)$$

$$\frac{\partial \mathbf{v}_i(0)}{\partial \rho^*} = \frac{\sum_j\left(\frac{\partial U^*(\mathbf{r}(0))}{\partial \mathbf{r}_j(0)} \cdot \mathbf{r}_j(0)\right)}{3\rho^*\sqrt{2(E^* - U^*(\mathbf{r}(0)))\sum_j \mathbf{V}_j^T \mathbf{V}_j}}\,\mathbf{V}_i \qquad (72)$$

$$\frac{\partial \mathbf{v}_i(0)}{\partial E^*} = \frac{1}{\sqrt{2(E^* - U^*(\mathbf{r}(0)))\sum_j \mathbf{V}_j^T \mathbf{V}_j}} \mathbf{V}_i \quad (73)$$

Combining these with (67)-(68), we obtain the desired partial derivatives:

$$\frac{\partial \mathbf{r}_i(0)}{\partial \rho} = -\frac{\mathbf{r}_i(0)}{3\rho}; \quad \frac{\partial \mathbf{r}_i(0)}{\partial E} = 0 \quad (74)$$

$$\frac{\partial \mathbf{v}_i(0)}{\partial \rho} = \frac{\sum_j \left(\frac{\partial U^*(\mathbf{r}(0))}{\partial \mathbf{r}_j(0)} \cdot \mathbf{r}_j(0)\right)}{3\rho \sqrt{2(E^* - U^*(\mathbf{r}(0)))\sum_j \mathbf{V}_j^T \mathbf{V}_j}} \mathbf{V}_i \quad (75)$$

$$\frac{\partial \mathbf{v}_i(0)}{\partial E} = \frac{N}{\varepsilon N_A \sqrt{2(E^* - U^*(\mathbf{r}(0)))\sum_j \mathbf{V}_j^T \mathbf{V}_j}} \mathbf{V}_i \quad (76)$$

and:

$$\frac{\partial \mathbf{r}_i(0)}{\partial \sigma} = -\frac{\mathbf{r}_i(0)}{\sigma}; \quad \frac{\partial \mathbf{r}_i(0)}{\partial \varepsilon} = 0 \quad (77)$$

$$\frac{\partial \mathbf{v}_i(0)}{\partial \sigma} = \frac{\sum_j \left(\frac{\partial U^*(\mathbf{r}(0))}{\partial \mathbf{r}_j(0)} \cdot \mathbf{r}_j(0)\right)}{\sigma \sqrt{2(E^* - U^*(\mathbf{r}(0)))\sum_j \mathbf{V}_j^T \mathbf{V}_j}} \mathbf{V}_i \quad (78)$$

$$\frac{\partial \mathbf{v}_i(0)}{\partial \varepsilon} = -\frac{E^*}{\varepsilon \sqrt{2(E^* - U^*(\mathbf{r}(0)))\sum_j \mathbf{V}_j^T \mathbf{V}_j}} \mathbf{V}_i \quad (79)$$

The partial derivatives given by expressions (74)-(79) may be combined with those in expressions (64)-(65) as indicated by the chain rule Eqns. (47)-(48) to yield the desired partial derivatives of the system properties y with respect to the input variables ρ and E, and the intermolecular potential parameters ε and σ.

Concluding Remarks

Molecular dynamics computations have traditionally been viewed as simulation experiments that, given the values of certain input macroscopic variables and intermolecular potential parameters, compute the values of certain output macroscopic variables. In this paper, we have considered the mathematical mapping that such simulations define between the various input and output quantities, and have attempted to improve its properties by removing the discontinuities inherent in conventional molecular dynamics algorithms. Beyond their mathematical and numerical advantages, it could be argued that the modified force functions introduced for this purpose also have a firmer physical basis than force computations based on arbitrary rules such as the minimum image convention.

Of course, the modified force functions (42)-(44) are more expensive to compute than their simpler counterparts which effectively correspond to just the central term (i.e. for $k = k' = k'' = 0$) of the triple summations in these expressions. We are currently investigating a variety of techniques for addressing these important efficiency issues. The properties of the modified force functions are particularly relevant in this context. They imply that efficient and accurate evaluation procedures for these functions are necessary *only* over a cube with side of length 0.5 irrespective of the actual values that their arguments X, Y, Z may take during a simulation.

A key advantage of the modified molecular dynamics technique is that it allows the direct computation of all partial derivative quantities of the corresponding mathematical mapping. This information can be used in a number of different ways such as the estimation of intermolecular potential parameters from experimental data, parametric studies, or, in principle, the direct use of molecular dynamics within process and product design applications which involve computations that need partial derivative information.

Although the current paper has concentrated exclusively on the use of simple Lennard-Jones potentials between mono-atomic particles, the techniques developed are also applicable to other intermolecular potential functions. The existence of bond stretching, bending and torsional forces in poly-atomic molecules does not introduce further conceptual complications as these are exerted between atoms in the same molecule rather than across different molecules.

Finally, this paper has considered molecular dynamics in the microcanonical ensemble, i.e. one that maintains constant density and energy in the system under consideration. Extensions to other ensembles, such as the canonical ensemble (i.e. constant density and temperature) or the isobaric-isothermal ensemble (i.e. constant temperature and pressure) are also possible. Interestingly, the addition of constraints of constant temperature and/or pressure to the Newtonian equations of motion (2) leads to a constrained mechanical system which is described mathematically by a mixed set of differential and algebraic equations (DAEs) of index 2 (Brenan, Campbell and Petzold, 1989). The adjoint systems for such index 2 DAEs were formulated and studied by Gritsis, Pantelides and Sargent (1995) and this provides the necessary mathematical basis for the derivation of partial derivatives of molecular dynamics mappings in alternative ensembles (see Stefanović and Pantelides, 1999).

References

Allen, M. P., and D. J. Tildesley (1987). *Computer Simulation of Liquids*. Clarendon Press, Oxford.

Born, M. and T. V. Karman (1912). Über schwingungen in raumgittern. *Physic Z.*, **13**, 297-309.

Brenan, K. E., S. L Campbell, and L. R. Petzold (1989). *Numerical Solution of Initial-Value Problems in Differential-Algebraic Equations*. Elsevier, New York.

Bryson, A. E., and Y.–C. Ho (1975). *Applied Optimal Control*. Wiley, New York.

Errington, J. R., and A. Z. Panagiotopoulos (1998). Phase equilibria of the modified Buckingham exponential-6 potential from Hamiltonian scaling grand canonical Monte Carlo. *J. Chem. Phys.*, **109**, 1093-1100.

Ewald, P. P. (1921). Die Berechung optischer und elektrostatischer Gitterpotentiale. *Ann. Phys.*, **64**, 253-287.

Gritsis, D. M., C. C. Pantelides and R. W. H. Sargent (1995). Optimal control of systems described by index two differential-algebraic equations. *SIAM J. Sci. Comp.*, **16**, 1349-1366.

Johnson, J. K., J. A. Zollweg and K. E. Gubbins (1993). The Lennard-Jones equation of state revisited. *Mol. Phys.*, **78**, 591-618.

Jónsdóttir, S. and R. A. Klein (1997). UNIQUAC interaction parameters for molecules with –OH groups on adjacent carbon atoms in aqueous solution determined by molecular mechanics – glycols, glycerol and glucose. *Fluid Phase Equilibria*, **132**, 117-137.

Jónsdóttir, S., R. A. Klein and K. Rasmussen (1996). UNIQUAC interaction parameters for alkane/amine systems determined by molecular mechanics. *Fluid Phase Equilibria*, **115**, 59-72.

Jónsdóttir, S., K. Rasmussen and A. Fredenslund (1994). UNIQUAC parameters determined by molecular mechanics. *Fluid Phase Equilibria*, **100**, 121-138.

Metropolis, N., A. W. Rosenbluth, M. N. Rosenbluth, A. H. Teller and E. Teller (1953). Equation of state calculations by fast computing machines. *J. Chem Phys.*, **21**, 1087-1092.

Stefanović, J. and C. C. Pantelides (1999). *On the Mathematics of the Molecular Dynamics Technique*. Research Report, Centre for Process Systems Engineering, Imperial College of Science, Technology and Medicine, London SW7 2BY, United Kingdom.

Sun, T. and A. S. Teja (1998). Vapor-liquid and solid-fluid equilibrium calculations using a Lennard-Jones equation of state. *Ind. Eng. Chem. Res.*, **37**, 3151-3158.

A DESIGN PROCEDURE AND PREDICTIVE MODELS FOR SOLUTION CRYSTALLISATION PROCESSES

Sean K. Bermingham, Andreas M. Neumann, Herman J.M. Kramer,
Peter J.T. Verheijen, Gerda M. van Rosmalen and Johan Grievink
Process Systems Engineering Group and Laboratory for Process Equipment
Delft University of Technology
2628 BL Delft, The Netherlands

Abstract

Useful components and deficiencies of conventional crystallisation design procedures and models are identified. A systematic procedure consisting of a hierarchical decomposition into design of the product, task, flowsheet and individual crystalliser is presented. At each level the relevant specifications, variables and necessary knowledge are identified. For the flowsheet and crystalliser level a predictive crystallisation model is presented to design experiments, estimate kinetic parameters, and to analyse and optimise the behaviour of design alternatives. Currently, the procedure is applicable to cooling, flash-cooling and evaporative crystallisation. The developed model is restricted to systems with little agglomeration. Validation of the proposed design procedure and models requires industrial case studies.

Keywords

Hierarchical design, Scale-up, Compartmental models, Draft tube baffle crystalliser.

Introduction

Crystallisation involves the formation of one or more solid phases from a fluid phase or an amorphous solid phase. It is applied extensively in the chemical industry, both as a purification process and a separation process. The main advantage of crystallisation over distillation is the production of substances with a very high purity, at a low level of energy consumption, and at relatively mild process conditions.

Although crystallisation is one of the older unit operations in the chemical industry, the design and operation of crystallisation processes still pose many problems.

Many crystallisation plants frequently produce crystals, which do not satisfy the defined quality specifications. For instance, an excess of fine particles will typically result in poor filterability characteristics. Consequently increasing the cost of the downstream solid/liquid processing. Another example is the inclusion of mother liquor. After solid/liquid separation and drying of the crystalline product, e.g. during transportation or storage, mother liquor may seep from broken crystals. Subsequent re-crystallisation may cement the crystals together; a process referred to as the caking of crystals.

Operational problems also constitute a large portion of the problems encountered in crystallisation processes. Firstly, scale growth or crystal deposition on heat exchanger surfaces often limits plant availability significantly. Secondly, plant availability may also be reduced by pipe blockages as a result of scale growth or high solids concentrations. Finally, many crystallisation processes suffer from open loop unstable behaviour.

Despite the importance of crystallisation, there is a relative lack of systematic design procedures and predictive models to help avoid or overcome the before-mentioned problems. This relative lack in comparison with vapour/liquid processes is a major omission for the chemical engineering profession (Villadsen, 1997). It is however not a surprising omission, as the understanding of crystallisation processes, and of solids processes in general, is typically a degree more complex than that of vapour/liquid processes. This added complexity mainly results from:

- The fact that the product quality specifications cannot be solely defined in terms of chemical and phase composition. A crystalline product is also characterised by its size distribution, morphology, polymorphism and the amount of strain in the crystal lattice.
- The problems in dealing with the thermodynamic models of solid/liquid/vapour systems and even of liquid/vapour systems when electrolytes are involved.
- The difficulties in predicting the hydrodynamics of a multi-phase flow as a function of crystalliser and impeller geometry, operating conditions, crystal properties and crystal concentrations.
- The fact that the rates with which crystals are born, grow, dissolve, are attrited, break, agglomerate, etc. are not only a function of liquid phase process conditions but also of distributed crystal properties such as size, surface structure and internal energy.

Knowledge of crystal properties, thermodynamic properties, hydrodynamic conditions and particle mechanics as well as understanding of their interactions is essential to predict the spatially distributed, size dependent and time dependent crystallisation kinetics. After all, it is the kinetics that ultimately determines the properties of the crystalline product. At present it is impossible to derive this knowledge and understanding from first principle models only. Consequently, heuristics, tabulated data, laboratory and pilot plant scale experiments still constitute a major part of the domain knowledge necessary for the design and optimisation of crystallisation processes.

However, improved design and operation of crystallisation systems does not only require developments in various fields of fundamental knowledge. Equally important is the availability of a systematic design procedure, which can simplify the design problem, can help organise the design tasks and will consequently improve the quality of designs and speed up the design process. An additional benefit of a systematic design procedure is reproducibility of the design process, which is essential to identify any remaining errors in the applied domain knowledge and/or design procedure.

It is our opinion that such a design procedure is not available yet, although some very relevant work has been done in both the fields of process design and crystallisation.

Based on the pioneering work of Douglas (1985) concerning the conceptual design of vapour/liquid processes, Rajagopal et al. (1992) developed a hierarchical design procedure for solids processes in general. More specific design procedures have been developed for fractional crystallisation (Dye and Ng, 1995), reactive crystallisation (Berry and Ng, 1997) and the interactions between the crystallisation step and the downstream processing (Rossiter and Douglas, 1986). These procedures mainly focus upon the synthesis and (economic) evaluation tasks of the design process. The lack of predictive models for analysis and optimisation of design alternatives limits widespread application of systematic design procedures for crystallisation processes.

Valuable work with respect to predictive models includes the development of kinetic models based upon first principles (Gahn et al., 1997), hydrodynamic studies of crystallisers (de Jong et al., 1998) and compartmental modeling to account for the interactions between kinetics and hydrodynamics (Kramer et al., 1999).

Scope of this Paper

In the first part of this paper we will present a hierarchical procedure for the conceptual design of solution crystallisation processes. The proposed hierarchy consists of four design levels. The first two design levels have a product engineering character, whereas the last two design levels have a process engineering character. At each level of the design procedure we will (re-)consider the design specifications, design variables and the domain knowledge necessary to synthesise, analyse and optimise design alternatives. The design procedure is intended to cover various scenarios as regards the destination of the crystalline product (main product, by-product or waste product) and the role of the crystallisation process (conversion, separation or purification).

The second part of this paper focuses on predictive models that are necessary for the design of crystallisation processes, but are not readily available. This is especially the case for the third and fourth level of the design procedure, where the performance of respectively flowsheet alternatives and individual crystalliser alternatives need to be analysed and optimised. For this purpose two model structures have been developed. The kinetic models required by these model structures cannot be determined purely from first principles and can only rarely be found in literature. The newly developed model structures are therefore intended for both the domain knowledge acquisition task, i.e. experimental design and estimation of kinetic parameters, as well as the analysis/optimisation tasks of the design process. We will illustrate their use in the acquisition of domain knowledge (parameter estimation) and for analysis purposes (explorative simulations). The application of predictive models at levels one and two of the design procedure is not treated in this paper.

In its present form, the design procedure is applicable to cooling, flash-cooling and evaporative crystallisation processes. However, the developed model structures are currently still restricted to crystallisation systems, which exhibit negligible agglomeration.

A Conventional Crystallisation Design Procedure

In this section we will present a generalised, concise picture of the conventional procedures for the design of a crystallisation process.

Synthesis

The starting point is the specification of the available feed composition, required product purity, yield, production capacity and a rough measure for the crystal size distribution (CSD), such as the median crystal size or the volume fraction crystals below or above a certain size. The next step involves collecting thermodynamic data and physical properties for the crystallisation system. Subsequently the synthesis phase is entered, which involves selection and sizing issues. The crystallisation method is mainly determined by the thermodynamics of the solid/liquid equilibrium. A heuristic scheme for this selection process, based upon melt temperatures, T_{melt}, and equilibrium concentrations, C_{eq}, is given in Fig. 1.

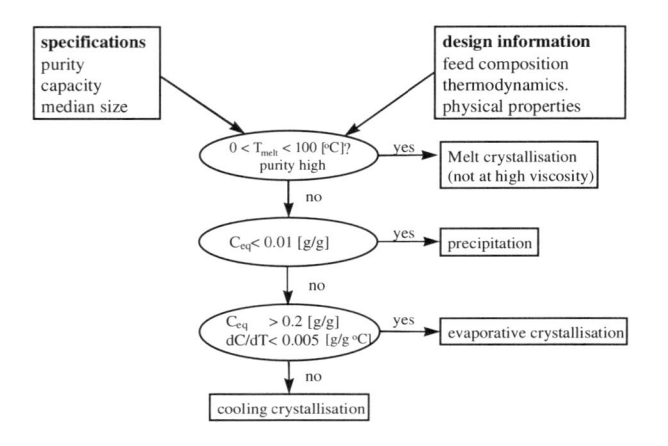

Figure 1. Crystallisation method selection (after Kramer et al. (1999)).

Another thermodynamically determined choice is the use of a certain solvent or additives to obtain a desired morphology. Subsequent selections involve the number of crystallisation steps, operation mode and number of stages. The first choice, the need for one or more recrystallisation steps is mainly determined by the amount of impurities in the feed and the desired final product purity. The operation mode, batch or continuous, is typically dictated by the production capacity and the desired shape of the CSD. Multiple stage operation is of interest for evaporative crystallisation processes with a high energy consumption. The next stage of the design process involves selection of crystalliser type and sizing the crystalliser. Crystalliser type selection is often influenced by company preferences and experiences. The crystalliser volume, V, is a simple function of the desired production capacity, P, the maximum solids concentration that the equipment can handle, ϕ, the desired median crystal size, L_{50}, and an estimate for the crystal growth rate, G:

$$V = \frac{P}{\phi \cdot \rho_c} \frac{L_{50}}{3.67 \cdot G} \quad (1)$$

The area required for heat exchange is a function of the production capacity, the selected crystallisation method, the selected heat transfer mechanism and the system's thermodynamics.

Analysis and Optimisation

At this point a so-called base case design has been realised. Now the importance assigned to the CSD determines whether any effort needs to be put into the analysis and optimisation of the base case design with respect to the CSD. These design tasks require relations between nucleation and growth rates of crystals on the one hand and crystalliser geometry and operating conditions on the other hand.

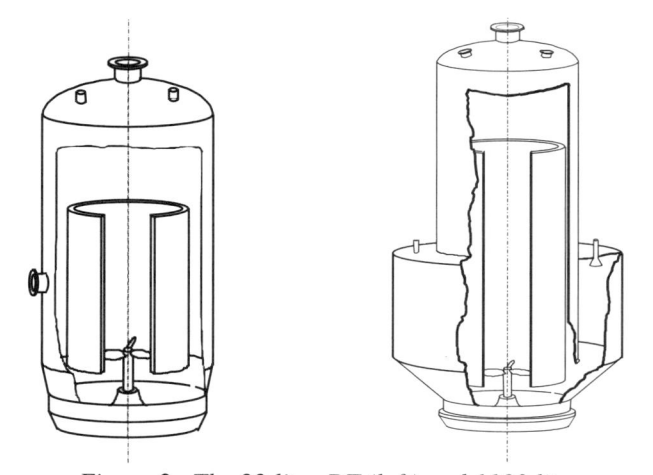

Figure 2. The 22 litre DT (left) and 1100 litre DTB evaporative crystalliser (right) at the Laboratory for Process Equipment.

Over the years many kinetic models have been proposed for this purpose, e.g. Ottens et al. (1972), Ploβ et al. (1989) and Ó Meadhra et al. (1996). All these models are power law models, ranging from purely empirical to partly fundamental. As a result, these models have a limited predictive value for scale-up and design, i.e. when applied to another scale of operation, crystalliser type or operation mode than the configuration used to estimate the model parameters. We will illustrate the limited predictive value below using experimental results obtained on a 22 litre draft tube (DT) and 1100 litre draft tube baffle (DTB) crystalliser and the model framework of Ó Meadhra.

On both crystallisers experiments were performed with a residence time of 75 minutes, a specific heat input of 120 kW·m^{-3} and similar impeller tip speeds. Both crystallisers were operated continuously in an evaporative mode at a constant temperature of 50°C using ammonium sulphate/water as the model system. Differences in process conditions between the two experiments are summarised in Table 1. A more comprehensive coverage of these experiments is given by Bermingham et al. (1998).

Table 1. Differences in Process Conditions between Eexperiments DT22 and DTB1100.

	DT22	DTB1100
Impeller frequency [s^{-1}]	16.7	5.33
Tip speed impeller [m.s^{-1}]	7.3	8.1
Circulation flow [m^3.s^{-1}]	0.015	0.200
Turnover time [s]	1.5	5.6
Mean spec. power input [W.kg^{-1}]	4.5	1.4
Fines withdrawal flow [m^3.s^{-1}]	-	0.002

Neumann et al. (1998a) used experiment DT22 to estimate the attrition parameters of the Ó Meadhra model. The measured trend of the median crystal size and the trend simulated with the estimated parameters are shown in Fig. 3. The steady-state values are in good agreement, but the dynamics of the fitted trend are more pronounced than observed experimentally.

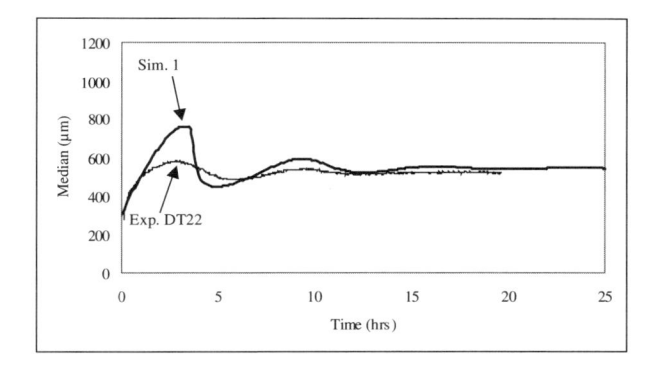

Figure 3. Measured and fitted trend of the median crystal size for experiment DT22.

The predictive capabilities of the Ó Meadhra kinetic model are tested by performing a simulation using the conditions for experiment DTB1100 and the parameters estimated from experiment DT22. For this simulation (sim. 2) the model used for the simulation of DT22 (sim. 1) was extended with a fines classification model and an ideal fines dissolver. More simulation features are given in the appendix.

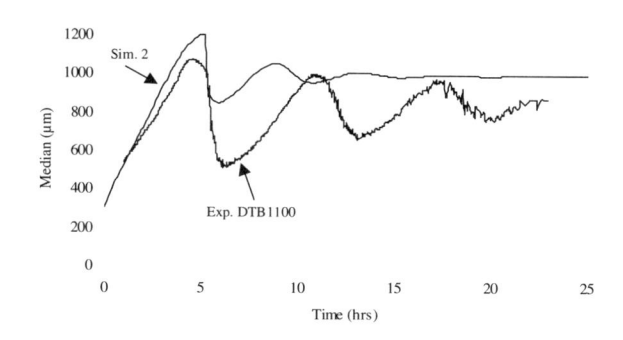

Figure 4. Measured and predicted trend of the median crystal size for experiment DTB1100.

Comparison of the median crystal size trend from simulation 2 and from experiment DTB1100 reveal a difference in dynamic behaviour and a difference of approximately 160 µm in expected steady state values (see Fig. 4). This difference may result from a number of factors, which are especially related to scale-up and crystalliser type:

- The spatial distribution of process conditions which is not accounted for, i.e. ideal mixing assumption.
- Incomplete dissolution of the fines removed from a DTB crystalliser. Conventional models assume complete dissolution.
- The kinetic model does not contain the right mechanisms to account for changes in nucleation rates due to differences in scale, geometry or operating conditions.

Evaluation of the Conventional Design Procedures

Numerous heuristics exist for the development of a base case design. For mainly two reasons, heuristics for selection issues are also useful in a systematic design procedure. First of all, many heuristics cannot yet be replaced by fundamental knowledge. Secondly, the application of heuristics is usually relatively simple and rapid.

Reliable tools for the analysis and optimisation of the base case design with respect to the CSD are not readily available. Due to the absence of such tools, full scale industrial processes cannot be designed on the basis of laboratory scale kinetic data only. In practice, additional experiments are therefore performed at one or two intermediate scales. Unless similar crystallisation processes exist, in which case new plants are often copied from existing ones. The development of reliable models for scale up purposes is thus essential in the development of a systematic design procedure for crystallisation processes.

Finally, when optimisation of the base case design fails to produce an adequate design, a design alternative with a different structure needs to be synthesised. To our knowledge none of the conventional methodologies cover this part of the design process.

A Systematic Design Procedure

The main incentive for the development of a systematic design procedure is the need for crystallisation process designs of a consistently high quality (*design effectivity*). A second incentive is the available time-to-market, which is decreasing continuously. A systematic design procedure can help speed up the design process, e.g. by removing the need for pilot scale experiments for scale-up purposes (*design efficiency*). An additional benefit of a systematic design procedure was already stated in the introduction, namely the reproducibility of the design process (*traceability of design decisions and rationales*). This is essential to improve on existing designs and identify any remaining errors in the applied domain knowledge and/or design procedure. A final objective is to create mutual awareness of the tasks that people have in a design team.

To achieve the above-mentioned aims, we propose a design hierarchy (see Table 2) in order to simplify the design problem and help organise the design tasks. A hierarchical decomposition was selected to reduce the number of design specifications, number of design variables and amount of knowledge that need to be considered simultaneously. The hierarchy consists of one level at which the initial design specifications are formulated and four design levels.

Table 2. Proposed Level Decomposition.

0	Initial design specifications
I	Design of the crystalline product
II	Physical/chemical design of the crystallisation task
III	Flowsheet design of the crystallisation process
IV	Design of a crystallisation stage

Table 3 shows the types of requirements considered at level 0. Depending on the destination of the crystalline product (main product, by product or waste product) and the role of the crystallisation process (conversion, separation or purification) some design specifications will be deemed necessary or desirable and others will be denoted irrelevant. Design process requirements are not treated in the remainder of this paper.

Design levels I through IV are aimed at finding design alternatives that meet the initial design specifications. As the designer progresses from one level to the next the emphasis shifts from product design (levels I and II) to process design (levels II and III) and ultimately to crystalliser design (level IV). In contrast with methodologies as proposed by Douglas (1985) the degree of detail does not automatically increase with each level. For product related issues it actually decreases, and for process and equipment related topics it increases.

Table 3. Level 0 - Initial Design Specifications.

Product performance requirements
In the crystalliser: no flotation, suspendability
Downstream handling: filterability, washability, dryability, dissolution rate, pneumatic handling, freedom from dust, flowability, mechanical strength
Customer application: no caking in storage, dissolution rate, mechanical strength, freedom from dust, bulk density or porosity, aesthetic appearance
Process requirements
production capacity, feed composition, yield, energy consumption, controllability, resiliency, availability, SHE considerations, battery limits and conditions
Design process requirements
design budget, time to market, in-house or licensed technology, available skilled design staff

At all four design levels the same tasks are performed, i.e. definition of design space and specifications, assessing domain knowledge, synthesis, analysis, evaluation and optimisation. (see Fig. 5). These tasks will be discussed briefly below and further on they will be exemplified at each level.

The first step at each level is to make an inventory of applicable design specifications and to identify the design space. The specifications consist of relevant initial design specifications from level 0 and eventual design specifications propagated from a previous level. Propagated design specifications are in fact design variables which are fixed after completion of a previous level. The design space is defined by the available design variables and operational variables.

The second step is to gather domain knowledge which relates the design variables and operational variables to the behaviour of a design alternative. This knowledge may consist of heuristics, experimental data and behavioural models. If parts of the domain knowledge are considered inadequate, additional experimental and modeling efforts may be required.

The third step, the synthesis task involves the creation of design alternatives. A design alternative is characterised by its structure and its scale. In this design procedure we will classify design and operational variables which determine the structure as discrete, and those that define the scale as continuous design and operational variables.

The fourth step is the analysis of the design alternatives. Subsequently, the set of behavioural results is evaluated (step five). The evaluation first of all concerns the compliance of the design alternatives' behaviour with the design specifications. If this evaluation is positive, the design alternative is propagated to a next design level, accompanied by new design specifications. These propagated specifications are in fact design variables and operational variables set at a previous level. If the design alternatives fail the evaluation step, their performance is optimised by modifying their structure and scale, i.e. by returning to the synthesis step. Normally, a screening of the economic potential will be part of any evaluation step. However, such screening is not covered in this paper. Secondly, the evaluation phase may be used to judge the quality of the applied domain knowledge. If a part of the knowledge is considered inadequate, additional experimental and modeling efforts may again be called for.

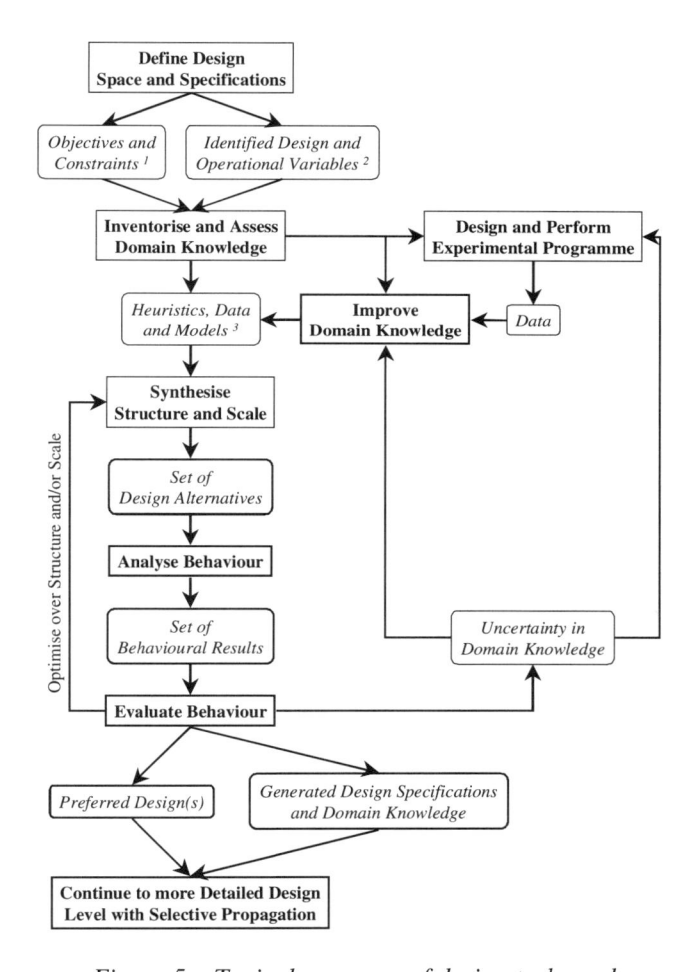

Figure 5. Typical sequence of design tasks and the outcome of these tasks. The superscripts refer to the column numbers of Tables 4 through 7.

A comprehensive execution of all the above-mentioned tasks at all four design levels, while considering all the design specifications mentioned in this paper, will obviously lead to very lengthy design trajectories. Assigning importance to the various initial design specifications is thus crucial, as the specifications taken into consideration largely determine the amount of effort put into each design level.

Design of the Crystalline Product

At this design level (see Table 4) the aim is to determine which product composition is required to meet the product performance criteria, i.e. the product related initial design specifications. For instance, filterability and freedom of dust are strongly related to the content of fine particles, aesthetic appearance is usually related to particle size uniformity, the bulk density is determined by the polymorphism and morphology of the crystals and the caking tendency in storage is related to the liquor inclusion content.

Most of the relevant domain knowledge for this level belongs to the field of particle technology, and has a strong empirical character. This is mainly due to the fact that product performance criteria are often equipment specific and hence cannot be defined generically. The development of fundamental knowledge in this field is further complicated by the need to account for distributed properties. Because a collection of particles can rarely be described by one size, one morphology, one purity, etc. Depending on the importance of the product performance criteria, much experimental work may be needed for this design level.

The design alternatives, i.e. sets of design variables, are propagated to design level II, where they are treated as design specifications.

Physical/Chemical Design of the Crystallisation Task

The design specifications of level II, the physical/-chemical design of the crystallisation task, are composed of process requirements from level 0 and propagated product composition characteristics from level I. Consult Table 5 for a list of specifications.

The polymorphism and morphology of the crystalline product are influenced by the choice of solvent and additives. The domain knowledge required for this selection issue was traditionally obtained from experimental work but is increasingly being replaced by molecular modeling. These tools can perform first principles calculations for the adsorption energy of a component on a specific crystal face.

The same knowledge is also essential to determine the sensitivity of the crystal purity, morphology and polymorphism for impurities in the feed, and hence determine the need for feed purification.

Table 4. Design Level I - Design of the Crystalline Product.

Design specifications Objectives and Constraints		Design Variables	Domain knowledge
Product: • Filterability • No caking in storage • No flotation • Suspendability • Washability • Dryability • Dissolution rate	• SHE considerations • Pneumatic handling • Freedom from dust • Flowability • Mechanical strength • Abrasion resistance • Bulk density or porosity • Aesthetic appearance	*Discrete:* • Polymorphism *Continuous:* • Morphology • Crystal size distribution • Purity • Maximum inclusion content	• Filterability tests and models: permeability and compressibility • Shear tests • Indentation tests • Caking tests • Flowability tests • Safety aspects

Table 5. Design Level II - Physical/Chemical Design of the Crystallisation Task.

Design specifications Objectives and Constraints	Design and Operational Variables	Domain knowledge
Process: • Production capacity • Feed composition • Yield • Energy consumption • Availability • SHE considerations *Product:* • Polymorphism • Morphology • Crystal size distribution • Purity • Maximum inclusion content	*Discrete:* • Crystallisation method • Feed purification • Recrystallisation step • Solvent(s) • Additive(s) • Material of construction *Continuous:* • Pressure range • Temperature range • Concentration solvent(s) • Concentration additive(s)	• Thermodynamic activity of species/components in solid, liquid and vapour phase • Adsorption (energy) of components/species on the various crystal faces • Scaling or encrustation tendency of components/species • Metastable zone with respect to homogeneous and heterogeneous primary nucleation • Physical properties, e.g. material densities and specific heats • Safety aspects

Table 6. Design Level III - Flowsheet Design of the Crystallisation Process.

Design specifications Objectives and Constraints	Design and Operational Variables	Domain knowledge
Process: • Production capacity • Feed composition • Crystallisation method • Yield • Pressure range • Temperature range • Supersaturation range • Energy consumption • Availability, controllability and Resiliency • SHE considerations *Product:* • Crystal size distribution	*Discrete:* • Operation mode • Number of stages • Feed configuration • Recycle structure • Location purge stream(s) *Continuous:* • Residence time in each stage or batch time • Recycle flow rates • Purge flow rate • Pressure and/or temperature in each stage • Heating/cooling duty or trajectory • Heat exchange rates	• Thermodynamic activity of species/components in solid, liquid and vapour phase • Physical properties, e.g. material densities, specific heats and viscosities • Crystallisation kinetics, i.e. rate expressions for the nucleation, growth, attrition, agglomeration and breakage of crystals • Fouling kinetics • Shape factors of the crystalline components • Fire and explosion index

Table 7. Design Level IV - Design of a Crystallisation Stage.

Design specifications Objectives and Constraints	Design and Operational Variables	Domain knowledge
Process: • Production capacity • Feed composition • Crystallisation method • Yield • Operation mode • Pressure and temperature range • Supersaturation range • Residence time (distribution) • Heat exchange rates • No boiling in heat exchanger • No entrainment of droplets by vapour • Suspension criterion • Availability, controllability and resiliency • SHE considerations *Product:* • Crystal size distribution	*Discrete:* • Crystalliser type • Fines classification and dissolution/clear liquor advance • Product classification • Heat exchanger type • Circulation device *Continuous:* • Equipment dimensions • Feed location • Product removal location • Solids concentration • Circulation flow rate • Operating conditions of classification devices • Flow rate through heat exchanger	• Equipment characteristics • Hydrodynamics • Thermodynamic activity of species/components in solid, liquid and vapour phase • Physical properties, e.g. material densities, specific heats and viscosities • Crystallisation kinetics, i.e. rate expressions for the nucleation, growth, attrition, agglomeration and breakage of crystals • Fouling kinetics • Shape factors of the crystalline components

Similarly to the conventional design procedure, the crystallisation method is mainly selected on the basis of the thermodynamics of the solid/liquid equilibrium (see Fig. 1). For many systems, e.g. those involving electrolytes, these equilibria still need to be determined experimentally. Other factors influencing this selection are the scaling tendencies of components present in the solution and the production capacity.

For instance, the availability of a direct cooling crystallisation process can be reduced significantly by scaling on the cooling surface. Depending on the added value of the product this loss in availability may or may not be considered a problem. Another example, a crystallisation method requiring a vacuum system is very inconvenient for low capacity processes. Operating conditions such as pressure and temperature are chosen such to obtain the highest possible yield, while obeying SHE constraints. Purity considerations determine the necessity for recrystallisation steps and the maximum crystal growth rate. With increasing growth rates the tendency of components to co-crystallise and entrapment of mother liquor increase. This introduces an upper limit for the supersaturation. The maximum allowable supersaturation also depends on the metastable zone. When the concentration exceeds this zone primary nucleation occurs, which is usually unwanted as it decreases the average crystal size.

Flowsheet Design of the Crystallisation Process

The realisations of the design and operational variables from design level II are propagated to design level III, the flowsheet design of the crystallisation process. Together with relevant initial design specifications from level 0, they constitute the objectives and constraints for level III. An overview of the design specifications, design and operational variables and domain knowledge for this level is given in Table 6. Note that at this level all product related specifications except the CSD have disappeared. If relevant, they are now present as operating windows for pressure, temperature and supersaturation. These operating windows are used, as it is currently unfeasible to incorporate calculations of product properties, such as morphology and polymorphism, into dynamic process models including population balances

Criteria to select batch or semi-batch operation include a low production capacity, a short time-to-market, a short product lifetime, high value products and a narrow CSD. A narrow CSD is also a reason to opt for multiple stage over single stage operation, as the shape of the CSD is influenced by the residence time distributions of the liquid and the particles. The feed temperature and the required CSD govern the feed configuration of a multiple stage process. The heat exchange rates are determined by the production capacity.

To analyse the consequences of the above-mentioned choices on the CSD a predictive process model consisting of thermodynamics, kinetics, mass, energy and population balances is required. Such a model should also be applicable to experimental design and estimation of kinetic parameters. As mentioned before, kinetic models cannot yet be derived from first principles only. Experimental design, experimentation and parameter estimation are hence an intrinsic step in crystallisation design. In the second part of this paper we will present a general crystallisation process model for analysis and optimisation with respect to the CSD as well as domain knowledge acquisition concerning the kinetics.

Another analysis, especially important for multiple stage evaporative crystallisation processes, involves the energy consumption and heat exchange surface area for the required heat exchange rates. Of an evaporative process consisting of N stages, the steam consumption and the total heat exchange surface area are proportional to respectively $1/N$ and N. Large deviations of the relationship for the surface area occur when a system exhibits significant boiling point elevations (BPE). When components with a high fouling tendency are present, time variant heat transfer coefficients need to be taken into account when sizing heat exchangers. For this purpose, another predictive model has been developed (Bermingham et al., 1999).

The resulting flowsheet design alternatives are propagated to level IV. If the suggested flowsheets contain multiple crystallisation stages, the relevant information for each stage is propagated to different instances of design level IV.

Design of a Crystalliser Stage

Practically all the design specifications from the previous level are present at this design level. The set is extended with equipment related specifications and with design and operational variables propagated from the previous level (see Table 7).

The first selection at this level involves the crystalliser type, e.g. fluidised bed crystalliser, DTB crystalliser, DT crystalliser, forced circulation crystalliser or simply a mixed tank. The order in which these crystallisers are mentioned here, usually coincides with decreasing crystal size for a certain crystallisation system. To increase the average crystal size and/or stabilise the crystalliser with respect to the CSD, certain crystallisers can be equipped with classification devices. The type of circulation device, impeller or pump, is determined by the crystalliser type and the mechanical properties of the crystals. For brittle materials a device with a high pumping number must be selected to prevent excessive attrition.

At this level, hydrodynamics is added to the fields of domain knowledge. Applications of this knowledge include the sizing and operation of classification devices, determination of minimum circulation rates for adequate particle suspension and optimisation of the product removal location. The use of computational fluid dynamics (CFD) packages to obtain hydrodynamic information is on the increase for crystallisation processes. However, it is not yet possible to combine CFD techniques with population balance modelling. To analyse the influence of kinetic-hydrodynamic interactions on the product CSD, the general crystallisation model mentioned at the previous design level can be used in a compartmental manner to account for spatially distributed process conditions.

The final design alternative can be propagated to the detailed engineering phase. This design level does not lie within the scope of this paper.

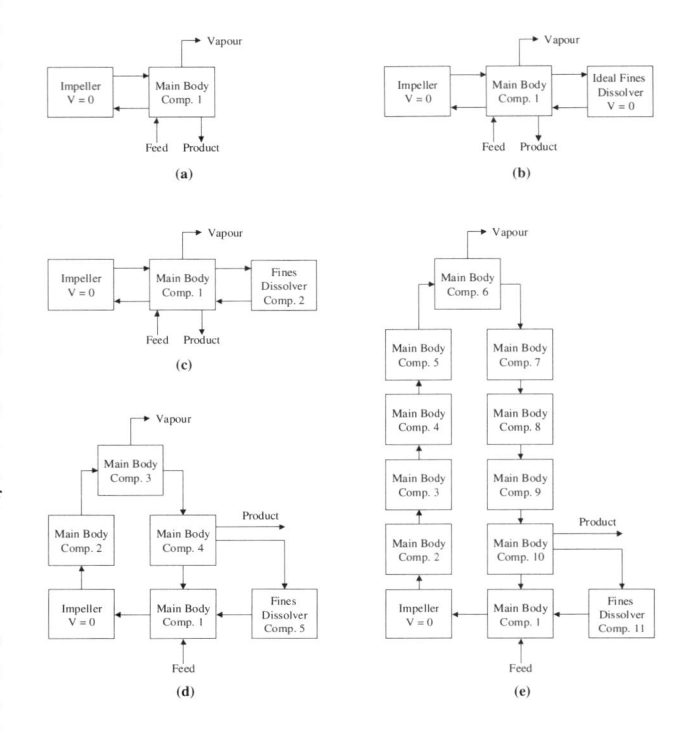

Figure 6. Compartment structures employed for simulations with the compartmental model.

A Predictive Crystallisation Model

A predictive crystallisation model with respect to the CSD has been developed for design levels III and IV. The model is intended for experimental design and parameter estimation (*domain knowledge acquisition*) as well as analysis and optimisation purposes (*predictive modelling*). The model consists of material balances, an energy balance, a population balance and a relation for the temperature-dependent equilibrium solute concentration. The main features of this model are:

- Kinetic model from Gahn et al. (1997). Nucleation as a function of crystalliser/-impeller geometry, impeller frequency and the CSD. Crystal growth rates determined by liquid phase composition and internal energy of crystals
- A fines dissolution model based on the kinetic parameters from the Gahn model.

- A conventional ideal fines dissolver for comparison reasons.
- A compartmental nature to describe the spatial distribution of the CSD, energy dissipation, supersaturation, etc.

Fig. 6 shows the compartment structures used for the simulations throughout this paper: *(a)* ideally mixed DT, *(b)* ideally mixed DTB with ideal fines dissolution, *(c)* ideally mixed DTB with real fines dissolution, *(d)* and *(e)* DTB with real fines dissolution and compartmentalised main body.

The subdivision of a crystalliser main body into multiple compartments, as done in Fig. 6(d) and 6(e), is performed on the basis of hydrodynamic analysis and characteristic times of the crystallisation phenomena (Kramer et al. (1999)).

The compartmental model is applied in the following sections for parameter estimation and for analysis purposes related to design levels III and IV.

Modelling of Single Crystallisers.

Estimation of Kinetic Parameters

Before the crystallisation model can be applied to a specific crystallisation system, the parameters of the Gahn kinetic model need to be estimated. For ammonium sulphate this was done by Neumann et al. (1998b) using experiment DT22 (see Table 1).

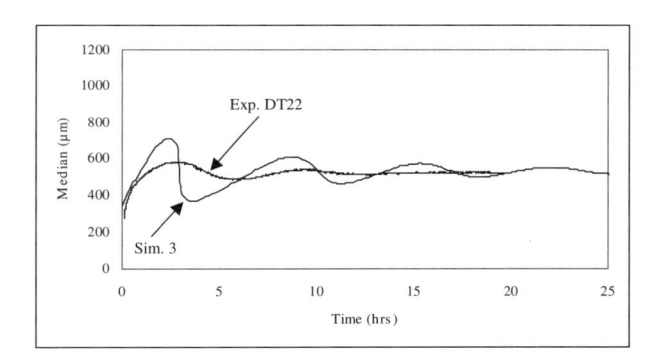

Figure 7. Measured and fitted trend of the median crystal size for experiment DT22.

The quality of fit (Fig. 7) is comparable to the one obtained with the Ó Meadhra kinetic model (Fig. 3), but the number of parameters involved is now only two as opposed to twelve.

Evaluation of the Model's Predictive Value

Using the same kinetic parameters, but a different compartment structure to account for differences in geometry and scale, experiment DTB1100 was simulated (sim. 6). The expected steady-state median crystal sizes only differ approximately 100 µm and the periods of the oscillations are in better agreement than when using the Ó Meadhra model (Fig. 4). The results are especially better when considering the six fold reduction in number of parameters going from the Ó Meadhra model to the Gahn model. Moreover, the Gahn model is presently the only model containing first principles relations between crystalliser geometry and product CSD.

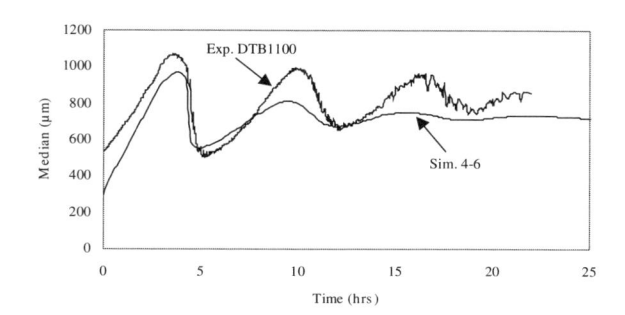

Figure 8. Measured and predicted trend of the median crystal size for experiment DTB1100.

No difference is seen between simulations 4, 5 and 6. By using three different compartment structures, these simulations show that incomplete fines dissolution and/or a spatial distribution of process conditions are not significant in the case of experiment DTB1100.

Influence of Crystalliser Scale on the Crystal Size Distribution

Because the volume specific surface area of both the crystalliser and the impeller decrease with increasing scale, the nucleation rate per unit volume should decrease with size, thus leading to larger crystals.

Table 8. Scale-Dependent Circulation Quantities.

Simulation nr.	7	8	9
Crystalliser volume [m^3]	0.022	1.1	360
Impeller frequency [s^{-1}]	19.52	5.38	0.756
Spec. power input [W·kg^{-1}]	7.27	1.52	0.21
Axial velocity [m·s^{-1}]	1	1	1
Internal circulation [m^3·s^{-1}]	0.018	0.20	9.04
Turnover time [s]	1.2	5.6	40

The capability of the model framework to predict this trend is shown in Fig. 9 for different scale crystallisers operating at identical conditions, i.e. same residence time, specific heat input and axial velocity. As for all simulations in this paper, more details can be found in the appendix.

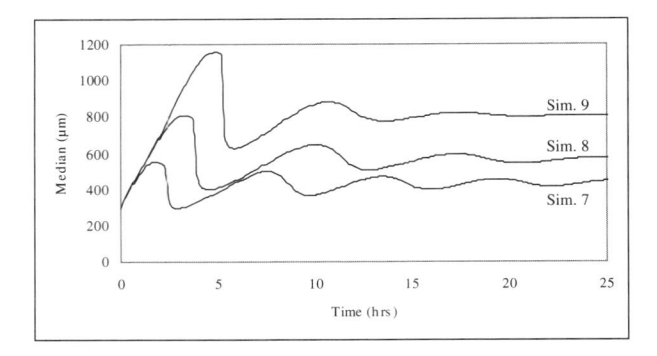

Figure 9. Median crystal size trends for DT crystallisers of 0.022, 1.1 and 360 m³.

Influence of Fines Removal Design and Operation on the Crystal Size Distribution

The model framework will now be applied to investigate sizing issues related to the fines removal system a fictive 360 m³ DTB crystalliser. First three fines removal characteristics are given:

- Fines residence time in the main body. This is the ratio of the main body volume and fines withdrawal rate.
- Vertical velocity in the annular zone. This is where the classification occurs
- Residence time in the dilution loop. This is the ratio of the dissolution loop volume and fines withdrawal rate.

Table 9. Different configurations of the fines removal system of a 360 m³ DTB crystalliser.

Simulation nr.	10	11	12	13
Fines withdrawal rate [m³·s⁻¹]	0.655	2.62	2.62	2.62
Velocity in annular zone [cm·s⁻¹]	0.263	0.263	2.63	2.63
Temperature increase fines [K]	18.9	4.72	4.72	4.72
Fines res. time in dilution loop [s]	100	100	100	10
Diss. rate [kg·s⁻¹]	0.130	0.188	2.32	1.27
Diss. rate [#·s⁻¹]	2.2E9	5.0E9	11E9	3.7E9
Mass-based diss. efficiency [kg·kg⁻¹]	0.97	0.84	0.38	0.06
Number-based diss. efficiency [#·#⁻¹]	0.97	0.89	0.79	0.30

The first configuration (sim. 10) is a scale-up of experiment DTB1100, i.e. same fines removal characteristics. As the temperature increase of the fines flow is rather high (18.9 K) in this set-up, simulation 11 was performed for a lower (factor 4) residence time in the main body, while keeping other fines removal characteristics constant. This new configuration has a very large annular zone: a cross sectional area of approximately 10^3 m². Therefore simulation 12 was performed with a higher velocity (factor 10) in the annular zone. The difference in classification behaviour is depicted in Fig. 10. For simulation 14, the residence time in the dilution loop was reduced tenfold to yield an industrial value.

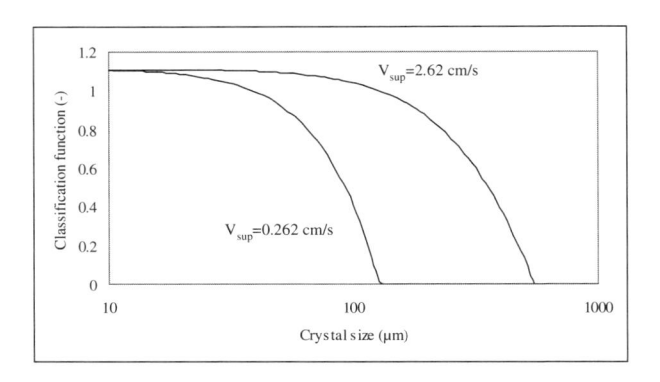

Figure 10. Calculated fines classification curves for velocities of 0.263 and 2.63 cm·s⁻¹.

The effect of the different fines removal configurations on the expected median crystal size can be seen in Fig. 11 and 12. Simulation 9 was included in Fig. 11 for comparison with a 360 m³ DT crystalliser.

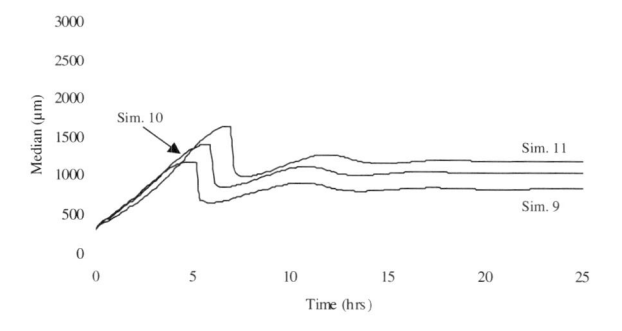

Figure 11. Influence of fines dissolution and withdrawal rate on the median crystal size.

The simulated effect of different fines removal configurations can be understood by investigating the mass-based and number-based dissolution rates given in Table 9. Because crystal dissolution is a size dependent phenomenon, these rates are not interchangeable. Clearly, the number based and not the mass based dissolution rate is of importance for the median size.

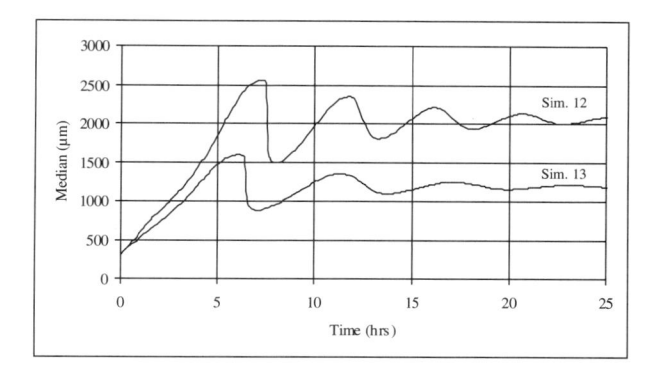

Figure 12. Influence of the dissolution loop volume on the median crystal size.

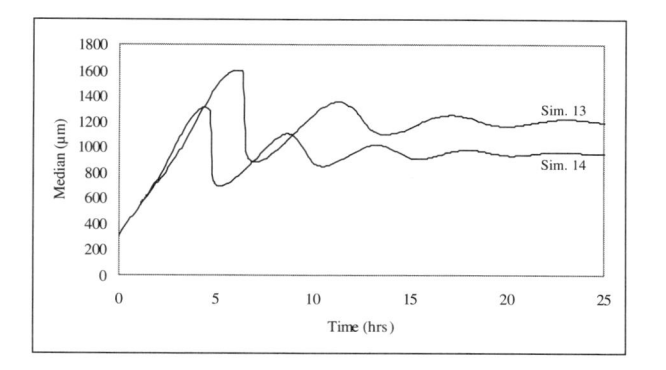

Figure 13. Influence of spatially distributed process conditions on the median crystal size.

Influence of Spatially Distributed Process Conditions on the Crystal Size Distribution

For the same fictive 360 m³ DTB crystalliser another simulation was performed to study whether the spatial distribution of process variables does play a role for this crystalliser. Recall that no effect was observed for experiment DTB1100. However, a 360 m³ crystalliser has significantly larger turnover times (see Table 8) and the characteristic times for the crystallisation phenomena are scale independent. Comparison of simulations 13 and 14 (main body modelled respectively with one and ten compartments) shows a large effect (see Fig. 13 and Table 10).

Table 10. Spatial Distribution of the Crystal Mass Production and Supersaturation.

	crys. mass production		Macroscopic supersaturation [kg·m⁻³]
	absolute [kg·s⁻¹]	specific [kg·m⁻³·s⁻¹]	
sim. 13			
MB-C1	15.64	0.0434	1.38
FD-C2	-1.252	-0.0478	-4.03
sim. 14			
MB-C1	0.015	0.0008	0.23
MB-C2	0.029	0.0007	0.24
MB-C3	0.030	0.0007	0.23
MB-C4	0.030	0.0007	0.23
MB-C5	0.030	0.0007	0.23
MB-C6	2.013	0.1118	2.66
MB-C7	3.905	0.0964	2.41
MB-C8	3.379	0.0834	2.18
MB-C9	2.935	0.0725	2.00
MB-C10	3.235	0.0799	1.78
FD-C11	-1.342	-0.0512	-3.60

Again, the effect can be best understood by a study of the number based and not the mass based dissolution rates. (see Table 11).

Table 11. Influence of Spatial Distribution on Fines Removal.

Simulation nr.	13	14
Dissolution rate [kg·s⁻¹]	1.27	1.35
Dissolution. rate [#·s⁻¹]	3.7E9	0.9E9
Mass-based diss. efficiency [kg·kg⁻¹]	0.063	0.046
Number-based diss. efficiency [#·#⁻¹]	0.30	0.14

Modelling Multiple Stage Crystallisation Systems

The influence of the number of stages on the crystal size distribution is investigated by comparing a single and triple stage configuration, both with a total volume of 360 m³. The triple stage configuration consists of identical 120 m³ DT crystallisers as opposed to one 360 m³ DT crystalliser. Operating conditions were identical for both configurations, i.e. same residence time, specific heat input and axial velocity.

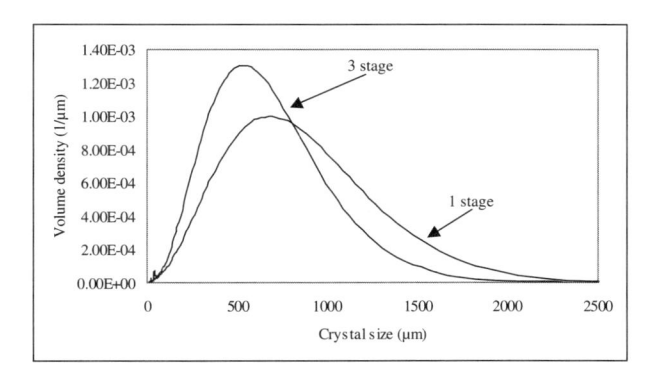

Figure 14. Simulated CSD's resulting from a single and triple stage process.

Fig. 14 shows a minor advantage of multiple stage operation for the CSD in case of ammonium sulphate.

Conclusions

Assessment of conventional design procedures reveals the value of heuristics for the rapid development of a base case design. Shortcoming are the lack of reliable analysis and optimisation tools and of a structuring of heuristics to provide design alternatives.

To systematically cope with the large number of initial design specifications, design and operational variables and the wide variety of domain knowledge involved, a hierarchical design procedure has been developed. Structuring of the relevant specifications, variables and knowledge at each design level simplifies the design problem and provides valuable insights for designers. It also highlights the major shortcomings in design knowledge, i.e. product performance/composition relations (level I) and predictive models for the concise analysis of flowsheets (level III) and comprehensive analysis of single crystallisers (level IV) with respect to the CSD and supersaturation. The hierarchical decomposition does not imply a once-through process with respect to the design levels.

To support the analysis step at levels III and IV, a crystallisation model has been developed. This model allows estimation of key kinetic parameters and prediction of the effect of crystalliser scale and type and of operating conditions on the local CSD and supersaturation without adjustment of kinetic parameters. However, the model's predictive capabilities need to be validated more decisively.

The ultimate aim of this design procedure, i.e. better designs in less time, cannot be guaranteed until the procedure as a whole is validated by means of industrial case studies. Main achievements so far are improved understanding of the design process and identification of bottlenecks in domain knowledge.

Finally, although the design procedure has been presented in the light of grassroots design, it can also be largely applied to retrofit and optimisation of process operation. One should think of synthesis at levels I and II and analysis tools at all levels.

Acknowledgements

STW, AKZO-Nobel, BASF, Bayer A.G., Dow Chemicals, DSM, DuPont de Nemours and Purac Biochem for supporting the UNIAK research program.

References

Bermingham, S.K., A.M. Neumann, H.J.M. Kramer, J. Grievink and G.M. van Rosmalen (1998). The influence of crystallizer scale and type on product quality. *Proceedings of the International Symposium on Industrial Crystallization, Tokyo*, 37-46.

Bermingham, S.K., M. Jonkers, G.J. Kruizinga, J.L.B. van Reisen and P.J.T. Verheijen (1999). Monitoring heat exchanger fouling for optimal operation of a multiple effect evaporator. *Comp. Chem. Eng.*, **23**, S771-S774.

Berry, D.A. and K.M. Ng (1997). Synthesis of reactive crystallization processes. *AIChE J.*, **43**, 1737-1750.

Douglas, J.M. (1985). A hierarchical decision procedure for process synthesis. *AIChE J.*, **31**, 353-362.

Dye, S.R. and K.M. Ng (1995). Fractional crystallisation: design alternatives and tradeoffs. *AIChE J.*, **41**, 2427.

Gahn, C. and A. Mersmann (1997). Theoretical prediction and experimental determination of attrition rates. *Trans IChemE.*, **75(A)**, 125-131.

Jong, M.D. de, R.E. Breslau, A.M. Neumann, H.J.M. Kramer, B. Scarlett (1998). Modelling the hydrodynamics of a 22 litre DT crystalliser. *Proceedings of the World Conference on Particle Technology III*, Brighton, UK.

Kramer, H.J.M., S.K. Bermingham and G.M. van Rosmalen (1999). Design of industrial crystallisers for a required product quality. *J. Crystal Growth.*, **198/199**, 729-737.

Mersmann, A. (1995). Fundamentals of crystallization. In A. Mersmann (Ed.), *Crystallization Technology Handbook*, Marcel Dekker, New York, 1-78.

Neumann, A.M., S.K. Bermingham, H.J.M. Kramer and G.M. van Rosmalen (1998a). Modelling the attrition pocess in a 22 litre DT crystallizer. Paper no. 36, International conference on mixing and crystallization, Malaysia.

Neumann, A.M., S.K. Bermingham, H.J.M. Kramer and G.M. van Rosmalen (1998b). Modelling the dynamic behaviour of a 22 litre evaporative draft tube crystallizer. *Proceedings of the International Symposium on Industrial Crystallization, Tianjin, China*, 222-226.

Ó Meadhra, R., H.J.M. Kramer and G.M. van Rosmalen (1996). A model for secondary nucleation in a suspension crystalliser. *AIChE J.*, **42**, 973-982.

Ottens, E.P.K., A.H. Janse and E.J. de Jong (1972). Secondary nucleation in a stirred vessel cooling crystalliser. *J. Crystal Growth*, **13/14**, 500-505.

Ploβ, R., T. Tengler, A. Mersmann (1989). A new model of the effect of stirring intensity on the rate of secondary nucleation. *Chem. Eng. Tech.*, **12**, 137-146.

Rajagopal, S., K.M. Ng and J.M. Douglas (1992). A hierarchical procedure for the conceptual design of solids processes. *Comp. Chem. Eng.*, **16**, 675-689.

Rossiter, A.P. and J.M. Douglas (1986). Design and optimisation of solids processes. I: A hierarchical decision procedure for process synthesis of solids systems. *Chem. Eng. Res. Des.*, **64**, 175-183.

Villadsen, J. (1997). Putting structure into chemical engineering. *Chem. Eng. Sci.*, **52**, 2857-2864.

Appendix: Crystalliser Specifications and Simulation Features

Table A.1. Crystalliser Specifications.

Crystalliser name	DT-1	DT-2	DTB-2	DT-3	DT-4	DTB-4
Volume main body [m^3]	0.022	1.1	1.1	120	360	360
Crystalliser diameter [m]	0.23	0.7	0.7	3.32	4.75	4.75
Draft tube diameter [m]	0.15	0.5	0.5	2.37	3.39	3.39
Impeller[‡] diameter [m]	0.14	0.485	0.485	2.32	3.34	3.34
Edge of impeller [m]	0.002	0.006	0.006	0.029	0.043	0.043
Breadth of impeller [m]	0.047	0.18	0.18	0.86	1.24	1.24
Power number [-] / Pumping number [-]	0.4 / 0.33	0.4 / 0.32	0.4 / 0.32	0.4 / 0.32	0.4 / 0.32	0.4 / 0.32
Volume dilution loop [m^3]	-	-	0.2	-	-	varied
Cross sectional area annular zone [m^2]	-	-	0.761	-	-	varied

[‡]All impellers are equipped with three blades; Angle of the blades is 0.436 rad.

Table A.2. General Simulation Features.

Crystalliser temperature [K]	323.15
Feed temperature [K]	323.15
Feed composition	saturated and crystal free
Heat input per unit volume [W·m^{-3}]	120·10^3
Product residence time [s]	4500
Solids concentration [vol. %]	10.2
Initial crystal size distribution	Normal distribution (L$_0$ = 300 µm, σ = 70 µm, 40 kg·m^3)
Discretisation crystal size domain	Logarithmic with a lower bound of 10 µm
Estimated parameters for Gahn kinetic model	k_r = 10^{-5} m^4·mole^{-1}·s^{-1} ; Γ_k = 1.9·10^{-4} kg·m^3·mole^{-1}·s^{-2}

Table A.3. Features per Simulation.

Simulation nr.	1	2	3	4	5	6	7
Crystalliser name (see Table A.I)	DT-1	DTB-2	DT-1	DTB-2	DTB-2	DTB-2	DT-1
Impeller frequency [s^{-1}]	16.67	5.33	16.67	5.33	5.33	5.33	19.52
Mean spec. power input [W·kg^{-1}]	4.53	1.48	4.53	1.48	1.48	1.48	7.27
Axial velocity [m·s^{-1}]	0.85	0.99	0.85	0.99	0.99	0.99	1
Internal circulation rate [m^3·s^{-1}]	0.015	0.19	0.015	0.19	0.19	0.19	0.018
Turnover time [s]	1.5	5.6	1.5	5.6	5.6	5.6	1.2
Product removal rate [dm^3·s^{-1}]	0.0049	0.244	0.0049	0.244	0.244	0.244	0.0049
Fines res. time in main body [s]	-	550	-	550	550	550	-
Fines withdrawal rate [m^3·s^{-1}]	-	0.002	-	0.002	0.002	0.002	-
Cross sect. area annular zone [m^2]	-	0.761	-	0.761	0.761	0.761	-
Velocity in annular zone [cm·s^{-1}]	-	0.263	-	0.263	0.263	0.263	-
Temperature increase fines flow [K]	-	18.9	-	18.9	18.9	18.9	-
Fines res. time in dilution loop [s]	-	-	-	-	100	100	-
Volume dissolution loop [m^3]	-	-	-	-	0.2	0.2	-
Compartmental model (see Fig. 6)	a	b	a	b	c	d	a
L$_{25}$ in final point of simulation	372	690	379	516	515	513	308
L$_{50}$ in final point of simulation	543	982	515	726	724	721	438
L$_{75}$ in final point of simulation	744	1344	644	949	949	945	543
CPU [hrs]	0.3	0.5	0.7	0.9	1.3	6.3	0.7

Table A.3 (cont'd). Features per Simulation.

Simulation nr.	8	9	10	11	12	13	14
Crystalliser name (see Table A.I)	DT-2	DT-4	DTB-4	DTB-4	DTB-4	DTB-4	DTB-4
Impeller frequency [s^{-1}]	5.38	0.756	0.756	0.756	0.756	0.756	0.756
Mean spec. power input [$W \cdot kg^{-1}$]	1.52	0.21	0.21	0.21	0.21	0.21	0.21
Axial velocity [$m \cdot s^{-1}$]	1	1	1	1	1	1	1
Internal circulation rate [$m^3 \cdot s^{-1}$]	0.20	9.04	9.04	9.04	9.04	9.04	9.04
Turnover time [s]	5.6	40	40	40	40	40	40
Product removal rate [$m^3 \cdot s^{-1}$]	0.244	80	80	80	80	80	80
Fines res. time in main body [s]	-	-	550	137.5	137.5	137.5	137.5
Fines withdrawal rate [$m^3 \cdot s^{-1}$]	-	-	0.655	2.62	2.62	2.62	2.62
Cross sect. area annular zone [m^2]	-	-	249	996	99.6	99.6	99.6
Velocity in annular zone [$cm \cdot s^{-1}$]	-	-	0.263	0.263	2.63	2.63	2.63
Temperature increase fines flow [K]	-	-	18.9	4.72	4.72	4.72	4.72
Fines res. time in dilution loop [s]	-	-	100	100	100	10	10
Volume dissolution loop [m^3]	-	-	65.5	262	262	26.2	26.2
Compartmental model (see Fig. 6)	a	a	c	c	c	c	e
L_{25} in final point of simulation	379	553	707	821	1433	822	664
L_{50} in final point of simulation	565	809	1016	1170	2044	1185	951
L_{75} in final point of simulation	771	1121	1390	1591	2760	1623	1293
CPU [hrs]	0.5	0.5	1.2	1.2	1.3	1.0	13.7

CONFERENCE SUMMARY AND CHALLENGES FOR FOCAPD-2004 AND BEYOND

Properties => Molecules => Reaction paths => Flowsheets => Plants => Supply chain

J. M. Douglas
Department of Chemical Engineering
University of Massachusetts
Amherst, MA 01003

J. A. Trainham
E. I. DuPont Co.
Wilmington, DE 19880

A.W. Westerberg
Department of Chemical Engineering
Carnegie Mellon University
Pittsburg, PA 15213

Abstract

This paper is broken down into three topical areas: 1) A summary of the conference proceedings, 2) a more detailed discussion of the omissions of existing research and challenges for the future, and 3) a discussion of how future technologies, and particularly increased computer speeds, may impact design. Rather than summarizing the conference by discussing the papers in the order in which they were presented, an attempt is made to put them into a perspective of the totality of process development problems with the hope of identifying topics that were not addressed in this conference and that have received little attention in the literature. This makes it apparent where additional research is needed and what are some of the challenges for the future. In addition, a broader challenge of extending design to include creating new value for a company is presented. Finally, the possible future of design research is discussed in terms of disruptive technologies and enabling technologies; consideration is given to the impact of far greater computer speeds.

Keywords

Conference summary.

PART 1 – SUMMARY OF THE CONFERENCE

Hierarchy of Design Detail

Design is a very large and complex problem, and therefore the industrial practice for many years has been to break the problem up into a hierarchy of levels of design detail, where each level adds more complexity to the flowsheet and allows more refined cost estimates. The purpose of this decomposition is to allow for an efficient abandonment of a design project, since experience indicates that only about 1% of ideas for new designs are ever commercialized. Different companies use different levels and the error bounds at each level are also different, but here we consider the levels in Table 1 (Douglas, 1988).

Table 1. Hierarchy of Design Detail.

1. Order of magnitude (best flowsheet) - error = 40%
2. Factored estimate (rigorous, optimum) - error = 25%
3. Budget authorization estimate (operability) - error = 12%
4. Project control estimate (drawings) - error = 6%
5. Contractor's estimate - error = 3%

Order of Magnitude Level

At the Order of Magnitude Level (also called a preliminary design or a conceptual design), we start with a new reaction chemistry discovered by a chemist (or a materials scientist) to make a specific product (or set of products), and the production rate is fixed by our Marketing Department. The initial goal is then to determine if any flowsheet can be created that will be profitable, and, if so, to find all of the process alternatives that could be considered. This effort requires a synthesis activity (given the inputs and the outputs, find all of the systems that can transform the inputs to the outputs) and an analysis effort (given the inputs and the system, find all of the outputs). It should be noted that this is a different definition of synthesis than is used in much of the design literature. The focus of the analysis is to estimate whether the profit is positive or negative and the order of magnitude of the profit, so that short-cut methods normally are used. The subsequent goal is to understand how the interactions in various process alternatives effect the costs, and how much detail (in physical property and unit models) needs to be considered in order to eventually find the best design.

Common practice used to be to allow a chemist a year, or more, to develop a complete chemistry database before an order of magnitude design was undertaken. However, the results often indicated that no flowsheet would be profitable (only about 1% of the ideas for new designs are ever commercialized), and if one was that the data were not taken in the most profitable operating region. By the time this was recognized, the chemist had dismantled the apparatus and had moved to a different project. Therefore, current practice in many companies is to start a conceptual design within a few weeks, or so, after a chemist discovers a new reaction (the physical properties of the new components, the reaction kinetics, the crystallization kinetics, the filtration characteristics, etc. are unknown), and to use economic design estimates to help set priorities for additional laboratory experiments as well as the region where experiments will be performed – this is better than factorial experimental designs. Sensitivity studies based on short-cut calculations are used to estimate how the various types of uncertainties propagate through both the reactor and the separation system recycle loops. With this approach the time-to-market can be reduced.

A systematic procedure for order of magnitude designs (starting with just the reaction chemistry and the production rate) was published by Douglas (1988). An updated version that is applicable to multistep reaction processes with V/L/L/S mixtures is given in Table 2. The means-end approach described by Siirola and Rudd (1971) or the means-end analysis with nonmonotonic planning described by Stephanopoulos, et al., (FOCAPD Paper No. I06, 1999) provide alternative approaches.

Table 2. Hierarchical Decision Procedure for Conceptual Design.

Level 0: Input Information
Level 1: Number of Plants and Plant Connections
Level 2: Input-output Structure for Each Plant
Level 3: Recycle Structure of the Flowsheet and Reactor Design
Level 4: Specification of the Separation System
    4a. General Structure - phase splits
    4b. Vapor Recovery System
    4c. Solid Recovery System
    4d. Liquid Separation Systems (organic & aqueous)
    4e. Integrate the separation systems for multiple plants
Level 5: Energy Integration
Level 6: Evaluate Alternatives

No physical property data other than that known by the chemist (the phase of the reaction, the molecular weights, the ambient phases of the components, etc.) is required up through Level 2 (in Table 2). Hence, the waste loads associated with the reaction chemistry can be assessed very quickly. Moreover, the economic potential at Level 2 (i.e. the product and byproduct values minus the raw materials and waste costs, as a function of the design variables, i.e., those that effect the selectivity) helps to narrow the range of the chemist's experimental program corresponding to profitable operation and provides a benchmark for making decisions. Thus, the required accuracy for both physical property and unit operation models can be assessed.

O'Connell and Neurock (FOCAPD Paper No. I02, 1999) provide a comprehensive review of the methods for estimating physical properties, which are needed to evaluate the reactor heat effects and for the synthesis and analysis of the separation system alternatives. With the significant advances in molecular modeling techniques that have been made, it appears as if the physical properties for new, small pharmaceutical molecules and the early steps (where small molecules are produced) in new ag. chemical processes, etc., can be estimated with sufficient accuracy for first designs. The very large molecules produced in many of these processes are beyond the current capability. Moreover, for most specialty chemical processes, and even for some petrochemical processes, the reaction chemistry is often partially unknown, i.e., in many petrochemical processes a "component" called "heavies", "other" or "impurities" is produced. The goal then becomes to determine how sensitive the flowsheet structure and the economic estimates are to these unknowns.

After at least one process alternative has been found to be sufficiently profitable to justify additional work and flowsheets for various alternatives generated have been sketched, the experts on safety, environmental impact, and control should be consulted. Often some alternatives might be inherently safer than others. The EPA is "pushing" industry to assess the environmental risk of all of the chemicals used in the reaction system before any work is done, but if all of the alternatives prove to be unprofitable this effort would be wasted. Moreover, the environmental impact of solvents can not be determined until the separation system synthesis step has been completed. Waste minimization is a natural component of the hierarchical procedure because costs must be assigned to every exit stream. The paper by Carberry, et al. (FOCAPD Paper No. I04, 1999) presents a nice overview of the environmental challenges faced by industry. Often some of the flowsheet alternatives are not resilient to trace components (trace components entering with the feeds, solvents, cake washing fluids, etc. or produced by the reaction system can not exit) so that the process is inoperable, but since different alternatives have exit points in different locations an initial screening of operability can be made on this basis. Other potentially, very difficult control problems might be noted for some of the other alternatives.

The apparent profitability of one or more flowsheet alternatives provides the economic justification for going back and measuring the physical properties of the new components, the reaction kinetics, the crystallization kinetics, the filtration characteristics, etc. The microchemical reactors discussed by Quiram, et al. (FOCAPD Paper No. I14, 1999) should be particularly useful for determining a kinetic model. Once a kinetic model is available it is possible to determine the best reactor configuration in the context of the complete process (i.e., the dominant trade-offs for processes with reactant recycles normally are the raw material losses because of selectivity effects and purge losses balanced against the reactor recycle costs, so that the best reactor configuration depends on the recycle system). The Chemical Reactor Design Tool discussed by Finlayson and Rosendall (FOCAPD Paper No. I16, 1999), or an equivalent, makes it possible to evaluate quickly the effect of an adiabatic temperature change, radial gradients, etc, on the selectivity, and then the impact of selectivity losses on the complete process can be re-evaluated. Hence, the incentive for including a sophisticated reactor optimization routine in a subsequent complete process optimization study can be established.

Similarly, the degree of complexity that needs to be considered for isolated crystallization systems, both the product and the flowsheet, is discussed by Bermingham, et al. (FOCAPD Paper No. I21, 1999). However, the interaction of the crystallization system with the complete flowsheet can have a significant impact on the best reactor configuration and the optimum design. The computational fluid dynamic codes such as described by Sinclair (FOCAPD Paper No. I13, 1999) can be used to assess the effect of the detail of a model on the complete process economics. The focus of these studies should be on the economics of the process interactions throughout the reactor and the separation system recycle loops, rather than specific process units. The goal of these conceptual design studies is to determine the problem definition for the subsequent optimization and control studies.

As an aside, the paper of Finlayson and Rosendall (FOCAPD Paper No. I16, 1999) presents some useful teaching examples to prevent students from misusing computers, and the paper by Doherty, et al. (FOCAPD Paper No. I15, 1999) emphasizes the need for exposing students to open-ended problems (generating alternative solutions and understanding the implications of these alternatives) throughout the undergraduate curriculum, since these are the types of problems that will be encountered in industry. Thus, they both emphasize the need to go beyond simply teaching students methods and to provide an introduction to "engineering judgement".

Factored Estimate Level

At the Factored Estimate Level we start with a few flowsheets that are profitable and we have decided on the degree of detail for the physical property and unit operation models, and then we use a simulator to develop a rigorous design for the "best few" alternatives. We retain more than one alternative because the cheapest steady state design might not be the cheapest when safety and control systems are added. In addition the design can be optimized for each alternative. Wright (Paper No. I07, 1999) presents a comprehensive overview the recent advances in both linear and nonlinear programming techniques, and provides references for codes for a variety of optimization algorithms that are available on web servers and can be used for a particular application. Grossmann and Hooker (FOCAPD Paper No. I08, 1999) discuss the advances in disjunctive programming to overcome the difficulties when MINLP (the most efficient way of screening a large number of process alternatives) is applied to very large problems. Floudas and Pardaleos (FOCAPD Paper No. I09, 1999) present the advances in global optimization methods (a problem that has plagued many), and they discuss applications to very wide selection of problems. Beigler (FOCAPD Paper No. I12, 1999) describes situations where it is necessary to retain some procedural models in an equation based simulator, and then describes hybrid strategies that use closed-form models within a simultaneous convergence and optimization strategy.

Stefanovic and Pantelides (FOCAPD Paper No. I20, 1999) note that equation based simulators suffer from not having derivative information available for thermodynamic quantities when conventional thermodynamic methods are used. Hence, they propose an alternative approach to molecular dynamics where modified force functions are introduced that are almost everywhere continuous and differentiable and exhibit a natural periodicity. With this approach the conventional assumptions for both the periodic boundary conditions and the minimum imaging convention are not needed, and derivative information becomes available. In addition, their approach allows the sensitivities of the predicted properties with respect to the intermolecular parameters to be estimated. Since their new method challenges the widely accepted basic assumptions of molecular simulations, it will be interesting to see the response of the thermodynamics community.

Budget Authorization Estimate Level

The optimum steady state design is not the process that should be built because the connections of the process with its environment have not been considered, i.e., disturbances entering the process often make the optimum steady state design inoperable. Hence, at the Budget Authorization Estimate Level, equipment overdesign to handle disturbances is considered, a control system is added, pressure relief valves and other safety systems are added, the plant layout and a piping diagram is developed, a PID diagram is prepared, etc. Often the optimum steady state designs for several process alternatives have about the same cost, and the apparent cheapest steady state design might not be the cheapest when control and safety systems have been added. Since more detail has been added to the flowsheet, a more accurate cost estimate can be made.

Tyreus and Luyben (FOCAPD Paper No. I11, 1999) present two examples where the optimum steady state design needs to be modified to make control possible, and they also discuss Shinnar's idea of partial control and the use of thermodynamic analysis (the rate of energy carriers in a process as these carriers undergo a change in their potentials) to identify the dominant control variables. The control of the dominant variables then significantly simplifies the plantwide control problem, and it makes it possible to shift the current ordering of the objectives (stability, coping with constraints, inventory control, economic objectives, and control of the recycle structure) to an ordering of economics, stability, constraints, recycle and inventory.

Project Control Estimate Level

At the Project Control Estimate Level, the large number of the other details of the design are completed, including drawings for all of the individual vessels. Many companies call this "Front End Engineering", although different companies draw the boundaries between the Levels at different places. A very nice discussion of Front End Engineering and its automation is presented by Nagy, et al. (FOCAPD Paper No. I18, 1999).

Issues that are Distributed Across the Various Levels of Detail

Safety and environmental impact must be considered at various levels of detail during the development of a design. Safety starts with the chemist's definition of the reaction path and the potential toxicity of the chemicals involved. It is also considered during the order of magnitude design in terms of explosive limits, the sensitivity of an adiabatic temperature rise, ensuring that toxic components are chemically destroyed before exiting the process, etc. Finally, a Hazops study is conducted once a PID diagram is available.

Waste loads can be estimated as the conceptual design is being developed, and can be compared for the various process alternatives and reaction paths. The environmental risk analysis and the fate of all of the chemicals can be estimated during a conceptual design, after at least one flowsheet has been shown to be profitable, using short-cut methods described in a forthcoming text by EPA on Green Engineering (Allen, et al., 1988). Fonyo, et al. (1994) refer to these as environmental problems intrinsic to the flowsheet. Environmental problems extrinsic to the flowsheet, i.e., tank cleaning, spills, fugitives, sampling, start-up, etc., need to be considered during the detailed stage of the design (the Budget Authorization Estimate Level).

Automation of the Levels of Detail

Fowler's (FOCAPD Paper No. I01, 1999) plenary address discusses in detail the new industrial paradigm of better/faster/cheaper (or at least 2 out of the 3) process development. One way this can be accomplished is by developing computer codes for the various levels of detail so that the drudgery associated with many of the details of a design can be minimized. Prototype codes for order of magnitude design have been discussed by Siirola and Powers (AIDES, 1971) using a means-end analysis, by Douglas and coworkers (PIP, Kirkwood, et al., 1988, and PIP II, Shultz and Douglas, 1999) for the hierarchical decision procedure and by Stephanopoulos and coworkers (Batch Design Kit, available from Hyprotech Corp.) for batch processes. Optimization codes were discussed in some detail at this FOCAPD Meeting, as noted above, and the automation of Front End Engineering was also presented, as noted above. Hence, the automation at each level has been demonstrated, although there is no compatibility among these codes.

The papers by Braunschweig, et al. (FOCAPD Paper No. I19, 1999) and Marquardt, et al. (FOCAPD Paper No.

I17, 1999) emphasize the need for component software to support life-cycle modeling and collaborative engineering across places and organizations. In particular, Marquardt's paper discusses the scales of modeling from molecules to molecular clusters to particles and thin films to single and multiphase systems to process units to plants to sites to enterprises, as well as from quantum mechanics models to molecular models to physical and chemical models to transport and kinetic models to balances for single phase systems to balances for multiphase systems. We would like the codes at all of these levels to be compatible because they are describing the same system at various levels of detail. The emphasis in Braunschweig's paper is more on the Global CAPE-OPEN project and the need for a capability to use software from a variety of sources in any particular application. The relationship between simulators/optimizers and downstream applications such as pdXi, STEP and OPC is also discussed. It is essential for academics to move in this direction in order to enable industry to evaluate their ideas.

Collapsing the Levels of Detail

Another aspect of better/faster/cheaper paradigm is to look for ways of collapsing the levels. In an earlier paper, Daichendt and Grossmann (1994) demonstrated that slight modifications of the hierarchical decision procedure of Douglas made it possible to use MINLP methods at each level of the hierarchy to accomplish a breadth first evaluation of all of the alternatives for cases where all of the required data are available (often the data required for adsorption, membranes, etc. are not). Beigler (FOCAPD Paper No. I12, 1999) demonstrated that little additional computational effort is required to simultaneous generate a rigorous design and its optimum operating conditions. van Schijndel and Pistikopoulos (FOCAPD Paper No. I12, 1999) considered the simultaneous design optimization and control optimization of some isolated subsystems. It seems to be apparent that this collapsing of the levels of detail will rapidly continue in the future (as computers become bigger and faster), and that there is a great need to develop component software for these problems.

Classifications of the Types of Chemical Processes

A taxonomy of chemical processes is given in Table 3. Most processes fall into more than one category, just as it is possible to discuss the papers presented in this meeting in more than one category. However, our goal here is to see if there are important problem areas that were not addressed at this meeting.

Table 3. Taxonomy of Chemical Processes.

A. Type of design problem - blend, separate (minerals), reactions
B. Continuous or batch
C. Product slate and plants
  1. Single product vs. multiple products
  2. Single plant vs. multiple plants
D. Type of product
  1. Structured products
  2. Pure chemicals
  3. Mixtures - gasoline
  4. Distribution functions - polymer or solids
E. Phases - vapor, organic liquid, aqueous liquid, solid
F. Industry (molecular weights of components and unit operations)
  1. Petrochemical
  2. Polymer
  3. Bio
  4. Specialty - chemical, monomer, ag., pharmaceutical
G. Value-added
  1. Therapeutic proteins 1 kg/yr, $1x10^6/kg
  2. Pharmaceuticals $0.1x10^6$ lb/yr, $25/lb
  3. Specialty $1x10^6$ lb/yr, $10/lb
  4. Commodity $100 x10^6$ lb/yr, $0.2/lb
  5. Super commodity $1000 x10^6$ lb/yr, $0.05/lb

Type of Process

Most of the problems considered in FOCAPD Meetings are reaction processes, or subsystems for these processes. At first thought, blending processes might be considered to be trivial because all of the feed components will be present in the product, but the formation of structured materials as discussed by Meeuse, et al. (FOCAPD Poster No. A13, 1999) illustrates the complexity encountered. Separation processes, such as mineral processing, where the product is contained in the feed, were not considered at this meeting.

Continuous or Batch

The selection between a continuous or a batch process (or some combination of both) has been briefly discussed by Douglas (1988). Normally, super commodities (C_2H_4, EtOH, etc. and commodities (production rates greater than $20x10^6$ lb/yr) are continuous, pharmaceuticals and many specialties with production rates less than $1x10^6$ lb/yr are batch, and ag. chemicals and specialty monomers with production rates between $1x10^6$ lb/yr and $20x10^6$ lb/yr can be either or some combination of both. Most of the papers in this FOCAPD Meeting were concerned with continuous processes, although the papers discussing pharmaceuticals were batch.

Product Slate and Plants

Single products can be produced in single plants or multiple plants, and multiple products can be produced in single plants or multiple plants. Understanding the plant structure early helps to define the structure of the design problem. For example, petrochemicals (e.g. the monomers for nylon 6, nylon 6,6, polyester, etc.) are often produced in multi-step reaction processes with recycles between plants and recycles within each plant. Agricultural chemicals usually are produced in multiple plants with a converging tree structure (no recycles between plants other than solvents), where the early plants that produce the small intermediates often have reactant recycles, while the later steps which produce solid intermediates seldom have reactant recycles (except for solvents). In contrast, the multi-step reaction processes that produce pharmaceuticals (see Stephanopoulos, et al., FOCAPD Paper No. I06, 1999) normally have serial structures with no internal or external recycles (except solvents), and the intermediates are all solids (by crystallization of the intermediates at each step, impurities are prevented from propagating downstream to the product).

Agricultural chemicals and specialty monomers are also characterized by single reactions that are stoichiometrically balanced, but the yields of many of the individual reaction steps are less than unity (the side reactions for many of the steps are unknown), while pharmaceuticals are characterized by multiple reaction steps where the dominant reactions are not stoichiometrically balanced and the side reactions are unknown. Obviously, if any of the components in the process are not known (i.e., the reaction stoichiometry is not complete), it is difficult to synthesize separation systems and any optimization study is not rigorous.

Because of the difference in the reaction chemistry, which leads to significant differences in the flowsheet structure (highly coupled, vs. converging trees vs. serial structures), the waste loads for the different types of processes are quite different. Shultz and Douglas, 1999, observed that the waste loads for hydrocarbons are normally 0 lb waste/lb product (the byproducts can be used to replace fuel), for petrochemical processes used to produce commodity monomers are in the range from 0.5 to about 5 lb waste/lb product, for specialty polymers about 12 lb waste/lb product, for agricultural chemicals about 20 lb waste/lb product, for one fermentation process 17 lb waste/lb product, and Stephanopoulos, et al., FOCAPD Paper No. I06, 1999) found that for pharmaceutical processes the loads were in the range from 20 to 100lb waste/lb product.

Type of Product

Most of the papers at FOCAPD Meetings have been concerned with the production of a single chemical, so that product purity is the only measure of product quality. Mixtures, such as obtained in the design of a refinery, have not been considered. Structured products were briefly discussed above as an example of a blending process, and the product quality depends on both purity and structure. Similarly the product from polymer processes is characterized by the molecular weight distribution as well as purity, and solids final products normally are characterized by a particle size distribution as well as purity (solids intermediates that are re-dissolved in the next reaction step normally do not have a size distribution specification).

Product quality issues were not discussed in this FOCAPD. An earlier study by McKenna and Malone (1990) on the conceptual design of polymer processes and their sensitivity to all of the process parameters (35) indicated that the parameters that had the greatest impact on the economics often had essentially no effect on the product quality (the polydispersity of the molecular weight distribution), and that the parameters that had the most effect on quality only had a small effect on the economic potential. A similar product quality study for processes that produce a final solid product does not seem to be available. In cases where the product quality is very sensitive to parameters that can not be measured very precisely, the process should be abandoned with the least design effort. Product quality problems for pure chemicals normally can be avoided by design modifications (i.e., if undesired components can be separated both before and after (to prevent their recycle) the products are removed.

Phases - Vapor, Organic Liquid, Aqueous Liquid, Solid

Most previous design research has been focused on V/L/L processes. Studies of solids systems usually have been limited to solids unit operations, such as described by Bermingham, et al. (FOCAPD Paper No. I21, 1999). However, the interaction among solids subsystems (crystallizer-filter-dryers) with the remainder of the process has received little attention. Similarly, processes that produce commodity monomers which contain solids processing operations (terephthalic acid, adipic acid, hexamethylene diamine, caprolactam, bishydroxyethyl terephthalate, etc.) are not mentioned in design texts nor have they been considered in either optimization or control studies. These problems are quite difficult, as evidenced by the industrial practice of always piloting the recycle parts of process containing solids processing steps (one unpublished survey indicated that there is a 100% guarantee of failure unless piloting is undertaken).

Type of Industry

This classification is useful because very different unit operations are encountered in different types of industries. For example, vapor-liquid separations in the polymer industry require wiped-film evaporators or counter rotating, devolatizing extruders instead of flash drums because of the high viscosities. As mentioned earlier, Mckenna and Malone (1990) have described a hierarchical decision procedure for the conceptual design of polymer processes. Similarly, small capacity bioprocesses often contain affinity chromatography units or electrophoresis separations, which are seldom encountered in commodity petrochemical processes. The differences encountered in the flowsheet structures and chemistry paths for ag. chemicals and pharmaceuticals were discussed above.

Kim's paper (FOCAPD Paper No. I03, 1999) presents a comprehensive overview of the information required at various stages of development of a pharmaceutical process, Stephanopoulos, et al. (FOCAPD Paper No. I06, 1999) present a systematic procedure for the conceptual design of pharmaceutical processes based on a means-end analysis with nonmonotonic planning, and Shah, et al. (FOCAPD Paper No. I05, 1999) describe the improvements that are possible for pharmaceutical processes by considering feed addition policies.

Value-added

Super-commodity and commodity processes have small profit margins, and so careful design and optimization is essential to stay in business. Specialty chemicals, such as ag. chemicals, with capacities in the range from 1×10^6 lb/yr to 20×10^6 lb/yr, have a much higher profit margin. This industry has been dominated by chemists, and usually the product was produced in existing batch facilities, with the various reaction steps carried out at different sites (i.e., equipment design was not considered). However, this production rate range is where a dedicated continuous or batch plant (or some combination) should be considered. A recent study by Seymour (1995) for a fenvalerate process indicated that a dedicated batch plant with numerous merged operations had the same cost as a continuous plant at about 3×10^6 lb/yr.

Pharmaceutical processes have much higher profit margins, they are batch processes run in existing equipment, and equipment design and optimization normally are not considered because it is believed to delay the time-to-market. Current engineering efforts, such as those described by Stephanopoulos, et al., (FOCAPD Paper No. I06, 1999) involve modifications of the chemist's recipe and solvent selection, along with an estimate of the raw materials costs and waste costs. Therapeutic proteins and similar substances have very high profit margins, and neither design nor optimization is considered to be important. It is interesting that the microprocesses discussed by Quiram, et al. (FOCAPD Paper No. I14, 1999) are estimated to have capacities of 4 kg/yr, which would satisfy the capacity requirements.

Conclusions from the Summary - Shifting the Research Emphasis

Fowler's (FOCAPD Paper No. I01, 1999) plenary address makes it apparent that the chemical industry is in the midst of significant changes. In contrast, most design research over the past 20 years has been (and still appears to be) focused on vapor-liquid, continuous processes (and subsystems for these processes). It is surprising that almost no research effort has been devoted the design and control of continuous processes producing solids (terephthalic acid, adipic acid, hexamethylene diamine, caprolactam, bishydroxyethyl terephthalate, etc.) Similarly, product quality issues for polymers and solids have received little attention, despite the large industrial focus on quality. The widely recognized trend in the chemical industries of de-emphasizing commodities and emphasizing specialties (with the added difficulties of dealing with incomplete chemistries and large molecules) has received little attention (except for a growing interest in pharmaceuticals). One of the major goals of FOCAPD Meetings is to help identify new agendas for research, and Fowler's presentation indicates that it is essential to develop a new agenda as soon as possible.

PART 2 – CHALLENGES FOR THE NEXT FOCAPD AND BEYOND

The "Grand Vision" that we have for chemical engineering is to develop tools to enable the design of products of the future and the processes from which they will be produced from their desired properties. These properties will be met by calculating the required molecular structure (and corresponding molecules). Industry will synthesize these molecules using the best possible chemistry, and they will manufacture using the best possible process. Such processes will be noteworthy as they will have industry leading cost and investment advantages, and the products and the processes will be sustainable from a business and environmental perspective.

The DuPont Company defines *Sustainable Growth* as: "Creating Shareholder and Societal Value while decreasing 'Footprint' along the *value chain*." (the supply chain with values added, see below). The footprint for a process is equivalent to the injuries, illnesses, waste and emissions it causes and the deplatable forms of energy and raw materials it uses. Companies will design/redesign the value chain for existing products and create new market opportunities for new products.

Omissions

In this FOCAPD meeting there were at least three major omissions, and we would like to see this remedied in the future. These omissions were: chemistry, solids and structural products, and life cycles. As one of us (JT) was a meeting organizer, he shoulders a share of the burden for these omissions.

In the area of chemistry, the meeting co-chairs spent considerable time strategizing how they could increase the chemistry content of the meeting and the participation of chemists. Unfortunately, they were unable to do so. Many people mentioned this subject, but they were short on action. Ultimately, all of us as engineers get paid because a chemist somewhere discovers new chemistry. At the very least, we engineers need to engage in the discovery process with the chemists, and this should be more than providing economics. Rather we should move in the direction of achieving the "Grand Vision." One of the real problems we as engineers have is that we just do not know enough chemistry (as stated most eloquently by George Stephanopolous in his talk at this meeting). In fact, in the U.S., chemical engineers that know chemistry are nearly a null set. We challenge all chemical engineering faculty to learn more chemistry and use it in their research. In industrial R&D today, more emphasis is being placed on chemical synthesis than on analysis because growth of the industry depends on creation of new compelling products.

We feel that everyone realizes the lack of research in the areas of solids and structural products processing. Our predictive capability for designing products and processes involving solids is minimal. Examples of processes that are crying out for action are: crystallization, filters, dryers, mills, reactors which use solids and ones that create solids, but particularly in the systems context. (The last paper presented at the meeting on crystallization is an excellent start in this area, but neglects system focus.) Among the products we include are agricultural, catalysts, films, fibers and particulate products. It has been estimated that well over 60% of the processes in the chemical industry contain solids.

In the area of life cycles, we heard from Wolfgang Manquardt who concentrated on process modeling software life cycles. As he also noted, there are life cycles, and there are life cycles. In general, we need much better tools for analyzing life cycles of plants, processes, and products.

Challenges

The challenge problems include the ultimate disposal of a molecule, and the life cycle of a chemical plant. For the chemical plant, we must contend with the "experience curve" of the current process versus "technology breakthroughs." Economic implications must include the cost of development versus plant duplication. The responses to "technology breakthroughs", range from developing alternative innovations or licensing the new technology and modifying the current process to simply shutting down. The ultimate fate of a molecule determines the sustainability of a product and a process. The challenge is to develop products and processes that improve the economics yet are sustainable. This is extraordinarily difficult because, in the market place, consumers are not willing to spend one penny more for a sustainable product (despite the hope and propaganda of various political groups). Legislation may force the issue, but think of the commercial opportunity if low cost, sustainable alternatives can be developed.

Other challenges for chemical engineers (and we have seen these before):

- How can we minimize the use of pilot plants (perhaps use design estimates to set priorities for experiments)?
- Can we ever get modeling to the state that we can avoid experiments?

Both questions have been mentioned at this conference. From an industrial perspective, we have a long way to go. Typically we just do not have enough fundamental understanding to design a robust process that will guarantee the financial goals of a project. Experience shows that piloting a process at commercial scale is financially unsound (because the critical experiments were not defined beforehand in terms of economics) and leads to non-optimal operations. Thus experiments will be with us a very long time.

A Grand Challenge Problem: Value Creation

Now we want to discuss a grand challenge problem that falls within our Grand Vision: How can we provide technology that allows us to redesign and collapse the value chain along the path to creating final products? We need first to define what we mean by a value chain. Most of us are very familiar with the supply chain for a product. It is the flow of material (kg's, if you like) from supplier to intermediate to monomer to polymer to fabricated product and finally to the customer. The *value chain* is the flow of money along the path from supplier to final product. Understanding it is to understand where and how value is created along the supply chain. The challenge is to maximize that share of the value produced at each and every step? An approach to consider is:

Understand where the dollars are and who is making them.

With such understanding, redesign the product and process that will enable one to collapse the value chain (e.g. maybe one could redesign in a way that allows one to go from supplier of raw materials (carbon source, energy, other ingredients) to polymer without going through a monomer. This requires new chemistry).

And typical of other value chain improvement steps, collapse the transition from chemical to mechanical processes.

Note here the emphasis is to turn from thinking about the amount of material flowing to thinking directly about the value. If a step adds no value, there is no advantage to improving how well it handles material. The challenge is to create more value. If a company can create the knowledge it needs to simplify the value chain, it can collect more profits. To have a major impact on this industry, we -- and that includes academics as well as engineers in industry -- need to understand value creation. As we move from the industrial age through the information age and on to the knowledge age, knowledge creation will be the ultimate deciding factor of the fate of our profession.

Summary of the Challenges

This conference has been about knowledge creation. As engineers, we need to focus our efforts on areas that will create more value for customers (like us), as well as to specific problem areas.

PART 3 – FUTURE TECHNOLOGIES AND DESIGN

In this section of the paper, we examine future technologies that we expect will impact how we carry out process design. These technologies can be disruptive and/or enabling. Disruptive technologies can cause a major reorganization or even the collapse of a business. Enabling technologies can lead to entirely new business activities. A technology may disruptive for one company and enabling for another.

Disruptive Technologies

Examples of a disruptive technology would be new chemistry or a new method to carry out a separation. It could be in the form of an innovative process simplification, such as happened for the manufacture of methyl acetate (Siirola, 1995; Siirola, 1996). It could be the collapse of the value chain wherein a company discovers how to bypass entire portions of the current process that are adding little value to the final product. As noted above, the response of the companies affected can be to become innovative, likely leading to the development of retrofits for their current processes. They could decide to license the new technology, or they could decide to shut down.

Disruption has several aspects worthy of study. First, can we anticipate it happening to a process? We get into the area of predicting what breakthroughs might occur and when they might occur. If we can make progress on such predictions, can we then rate processes according to how vulnerable they might be to predicted disruptions and use this measure as one of many by which we decide among processes to build? When hit with disruptive technologies, we often have to retrofit an existing process. As often stated in the past two decades, a second area on which we should concentrate much more effort is on developing design methods for retrofit. Part of that effort could be on developing counter measures for predicted technology breakthroughs before we need to employ them. This type of thinking takes us squarely into combining business considerations with technology. A third area then is for chemical engineers to work with business people. We really cannot ignore this interaction as business decisions typically have much more financial impact on a company than process optimization. For example, buying the right tanker of crude oil is much more important for the finances of an oil company than getting another 5% reduction in energy costs. Buying the right tanker of crude requires, among other things, that we know what we can process best in the next period of time, given the current state of our refineries and projected sales. Pushing this to the limit we mentioned earlier, we as engineers must learn a lot more about value chains both in our practice and our research.

Enabling Technologies

Enabling technologies are such things as the developing of standards. We already have some enabling technologies, but we fail to use them as well as we should. For example, we should train engineers to look at processes from a cost perspective using cost diagrams [Douglas, 1988]. This simple and much overlooked feedback can make a designer very aware that there are big cost "leaks" in a design that he has to reduce. We see in this conference another exciting technology in micro-manufacturing which will allow us to build multiple copies of entire chemical processes on a chip or new and much faster PCs for our desk.

Standards

Standards, when developed for control software, communication software or simulation software, allow each player to concentrate of his small part of the software technology and to make it work really well without having to worry about how it will "wire" into the rest of the world. The Global Cape Open project is about standards for simulation software. Researchers will be able to create new simulation, solver or physical property modules that plug easily into larger systems. Companies will benefit because they will be able to mix and match software from many vendors and from academia. The simulation companies have the interesting problem of developing a business plan that allows them to make a profit in this environment.

Standards have a downside to them also as they can prematurely slow down or stop important developments not anticipated by them. A company may see a significant improvement that it cannot fit into the current standards. It can either choose not to make the improvement or to create a nonstandard system. Thus standards must be "living things" that can improve over time. Additionally, it seems worthwhile that the community at large can have a very active role in that evolution. There should be ways to make changes at least for experimentation purposes.

An interesting thought, and one we shall discuss again momentarily, is to develop problem solving architectures that operate as independent agents that share a large, common information store -- reminiscent of the blackboard architectures that were popular in the early 1980s (Lien, et al, 1987). Today the blackboard can be a central database. There are now standards for querying such data stores. Each agent can watch what is happening in this information store and operate when it sees fit to make improvements - which can be to extract information, create a new result and place it back or to find bad results and simply remove them. Can we create standards for such agents and make them readily shared? If we can do this one right, we should often be able to add an agent or remove one without re-coding the software system in which they operate. Another approach is to place rules in such information stores that trigger agents into action.

Micromanufacturing

Quiram (FOCAPD Paper No. I14, 1999) very nicely described the area of micro-manufacturing and its implications for the future of our discipline. If we can readily build thousands of replicated processes on a chip and run them with slightly differing conditions, we can carry out massive screening experiments for materials, catalysts, and so forth. The trick is to find a way to report the results from these screenings, such as having different color appearing depending on the results - e.g., red is good, blue is bad. One can then expand the experiments where the previous batch produced a red color. Using these systems to manufacture new materials where some of the intermediates are dangerous is very intriguing. An explosion for one of these processes is only a small "poof" on the chip rather than a catastrophe, or it is the release of a very small amount of noxious substance rather than a large cloud that threatens for miles around. Developing design and control methods for these systems must be an opportunity waiting to happen.

Of course micro-manufacturing also is the technology that is giving us faster and faster computers, one of the most enabling of technologies for the process systems area. In the past few decades, we have experienced phenomenal increases in all aspects of computing. We have seen hardware get about 100 times faster per decade. In 1955 one of the first bridge design programs solved about 19 equations in about nine hours. By the mid-1970s we were proud of programs that were solving 700 equations. Today we talk almost glibly about 100,000 equations. Notably, there is already hardware today that is about 15,000 times faster than our current PCs. Supposedly we are to hit a barrier for manufacturing chips from silicon, but we still have over a decade of potential improvement. Algorithms have improved similarly, allowing us to do the same computations much faster on the same hardware. Readily purchased disk memory for PCs has gone from 10 megabytes in 1984 to about 10 gigabytes today, giving a factor of about a 75 fold increase per decade. Today networking is rapidly improving, from 10 megabit rates 15 years ago to gigabit rates today. Here it is the complete connectivity of everyone to everyone else that is making our world so different.

10,000 Times the Computer Power

We can easily imagine that we can have computing "power" in a decade that is about 10,000 times what we have now; what we solve today in a week will take only a minute. So what should we do with this projected computer power?

Faster Routine Computations

We can readily see the advantages of solving much larger problems when we examine the papers here by O'Connell and Neurock on physical property calculations (FOCAPD Paper No. I02, 1999), Stefanovic and Pantelides on quantum mechanical based distillation computations (FOCAPD Paper No. I20, 1999), Sinclair on computational fluids (FOCAPD Paper No. I13, 1999) and van Schijndel and Pistikopoulos (FOCAPD Paper No. I10, 1999) on collapsing levels of design detail. We will be able to solve what are large optimization problems today to global optimality (Floudas, FOCAPD Paper No. I09, 1999) in a few minutes.

Computers Also Handle Information

We do not use computers to do only large computations. We are also in the midst of an information revolution. It is routine today to capture organization-wide data in large databases, allowing the sharing of the information by people throughout the organization. The web literally exploded onto the scene in 1995. Today we act like it was always here, and we are rapidly using it as our primary source of all information.

The web makes clear one of the key issues about information. Given so much is available, how do we find the information we really want? Herb Simon noted in one of his seminars at Carnegie Mellon University that all this information has a cost – namely, it takes our time to look at it and time is our most limited and valuable resource. Search engines have remarkable algorithms to find and classify information, but we generally find that they come up with hundreds to thousands of irrelevant pages for each request.

We also would like to protect our valuable information. How can we create information systems that can produce responses that only we and not our competitors can see? How can we construct composite information in which we are allowed to see from detailed information we are not allowed to see – e.g., the total salaries paid by a company. More than encryption is involved in protecting data. How do we make information retrieval reliable? We would like the information to be not only safely delivered but reliably delivered when part of the system fails. These two issues are important and, with today's technology, very expensive to deliver. One can easily figure a factor of 100 times for each of the computing cycles that are required today to deliver a piece of unprotected information from a database. We appear to need that extra computer speed.

"Mean" Computational Problems

As we are able to solve larger and larger computational problems, we add more and more nonlinear detail to our models. Many times we formulate a problem, create the computer program to solve it and spend the next weeks to months trying to get a first solution. We may also find that changing one of the input parameters for the computation by as little as a tenth of one percent causes a previously successful computation to fail. Given such failure, either we have to find new ways to solve such problems, or engineers may be forced look for other design solutions to the problem.

An example of such a problem is to solve a distillation column model for an azeotropic mixture when the feed is in one distillation region and the products are in another. Most of the code in column solving routines is to get to a good enough first guess that the computations will then converge. This particular azeotropic case will defeat most approaches. For example, one may start the computation with a small number of trays, each having the feed composition on them. It is very unlikely that this computation will find its way into the other distillation region from such a starting point.

Interestingly, we find little of this experience with taking months to get to a first solution reported in papers and theses. It is easy to see why. A likely interpretation is that one was making only silly mistakes. However, another interpretation is that some of these problems are just plain difficult to solve, even when the program contains no errors. We might term such problems to be "mean" problems. We offer the conjecture that these problems not only defeat today's Newton-based algorithms, they will continue to defeat tomorrow's new and improve Newton-based algorithms.

One message we would like to convey here is that we as a community should be reporting these experiences. First of all, anyone who attempts to use the results where we do not will almost certainly experience the same difficulties and come to the conclusion that either the paper or thesis is wrong of that they are incompetent. Such experiences have to slow down the desire to pick up new technology. A second point is that we are almost certainly missing a research opportunity here. Only when we recognize enough of these types of problems will we be able to make an improved solving technology out of understanding them and improving our ability to solve them. This technology may include cataloguing mean problems. We may catalogue classes of problems we can solve robustly and learn how to build solutions to more complex ones from them (Tyner, 1999).

Asynchronous Teams of Agents – A-Teams

To return to our original theme for this section: we will have 10,000 times the computing power available in a decade. How might we use this power on mean problems?

We suggest that one approach is to form teams of agents who attack the problem in a cooperative manner. We already introduced the idea of using agents above when mentioning standards. Each agent can be one of the many possible methods we have available to solve problems in this class. There are several aspects of this approach worth considering (Quadrel, 1991; Murthy, 1992; de Souza, 1993; Talukdar and de Souza, 1994; Talukdar, 1998; Talukdar et al., 1998; Tyner, 1999).

First the agents cooperate. Each takes a partial or total solution from memory, works on it, and places its intermediate and/or final results back into memory. Another agent sees a solution in memory that is better than anything it has produced and switches to using this improved solution. The bandwidth for sharing is low – only through the sharing of partial and total solutions. Experience with A-teams is that they often solve a problem much faster than the linear speed-up one might expect by running N agents in parallel (Tsen, 1998 (train scheduling); Rachlin et al., 1996 (VLSI circuit routing)). To see why intuitively, imagine having three agents to do optimization: a genetic algorithm, a hill climber and a GRG algorithm. The genetic algorithm will place a few tens to hundreds of points all over the search space. The hill climber can pick any of these that look good and attempt to improve it. It will do acceptably well over a wide region and generally travel uphill. However, it will not rapidly converge when near a solution. The Newton algorithm can then pick up points produced by the other two and attempt rapid convergence. Failing it can put back any improved point and move on to another point in memory. The generic algorithm can use the improved points from the other two algorithms as parents in its generation of new points.

A second aspect of A-teams is that the agents operate autonomously. They decide for themselves when they can contribute to the problem solution. This is the idea behind the blackboard architecture proposed to create the first successful speech recognition software (Erman et al., 1980) and widely developed subsequently for many other problems (Wehe et al., 1987). Thus the development of the control structure for the cooperative behavior of the agents is actually not very difficult. One can add and remove agents without changing the code for the other agents.

Finally, the agents need not be complex nor do the whole task. Often the agents are intuitively designed to do something reasonable in producing a part of the solution. For example an agent might only attempt to simulate a column that is among those in one of the proposed flowsheets, each involving several columns. Another agent might use very approximate models to get good initial guesses for more rigorous ones only when another agent cannot make that model converge. Another might use an heuristic approach to make discrete assignments, which could quickly produce good first solutions that another more rigorous branch and bound search algorithm might use to establish bounds. Again, this aspect makes creating these systems much less difficult.

The secret is to put lots of agents operating autonomously and incrementally on the problem, rather like a colony of ants swarming all over the problem. The hope is that one of the ants will find its way to a solution. We can think to do this because we can imagine having 10,000 times the computer power. Interestingly, we may actually create algorithms that find solutions more reliably for mean problems and maybe even much more quickly.

Summary

In this paper, we looked at the invited papers to this conference and attempted to classify their contributions to improving the design process. We looked for omissions to suggest where we as a community should place more effort. We listed a number of challenge problems. We looked at what the future holds and where it could provide enabling technologies for future developments in design.

All of this material follows our grand vision of moving from the molecule to the chemistry, from the chemistry to the process, and from the process to the value chain.

Reference

Other than FOCAPD Papers

Allen, D.T., D. Shonnard and S. Weil, (1998). A new text on green engineering being developed by EPA.

Daichendt, M.M. and I.E. Grossmann (1994). Preliminary screening for the MINLP synthesis of process systems: Aggregation and decomposition techniques. *Comp. Chem. Eng.*, **18**, 663.

de Souza, P. (1993). *Asynchronous Organizations for Multi-Algorithm Problems*. Ph.D. Dissertation, Dept. of Electrical and Computer Engineering, Carnegie Mellon University, Pittsburgh, PA.

Douglas, J.M. (1988) *Conceptual Design of Chemical Processes*. McGraw-Hill, N.Y.

Erman, L.D., R. Hayes-Roth, V.R. Lesser, and D. Raj Reddy (1980). The Hearsay-II speech-understanding system: Integrating knowledge to resolve uncertainty. *Comp. Surveys*, **12** (2), 213-55.

Fonyo, Z., S. Kurlum and D.W.T. Rippen (1994) Process development for waste minimization: The retrofitting problem. *Comp. Chem. Eng.*, **18**, S591-95.

Kirkwood, R.L., Locke, M.H. and J.M. Douglas (1988). An expert system for synthesizing flowsheets and optimum designs. *Comp. Chem. Eng.*, **12**, 329.

Lien, K., G. Suzuki, and A.W. Westerberg (1987). The role of expert systems technology in design. *Chem. Eng. Sci.*, **42** (5), 1049-1071.

Mckenna, T.F. and M.F. Malone (1990). Polymer process design - I. Continuous production of chain growth homopolymers. *Comp. Chem. Eng.* **14**. 1127-1149.

Murthy, S., (1992). *Synergy in Cooperating Agents: Designing Manipulators from Task Specifications*. Ph.D. Dissertation, Dept. of Electrical and Computer Engineering, Carnegie Mellon University, Pittsburgh, PA.

Quadrel, R. W., (1991). *Asynchronous Design Environments: Architecture and Behavior*. Ph.D. Dissertation, Dept. of Architecture, Carnegie Mellon University, Pittsburgh, PA.

Rachlin, J., Wu, F., Murthy, S., Talukdar, S., Sturzenbecker, M., Akkiraju, R., Fuhrer, R., Aggarwal, A., Yeh, J., Henry, R., and R. Jayaraman (1996). *Forest View: A System for Integrated Scheduling in Complex Manufacturing Domains*. IBM report.

Seymour, C.B., (1995). *Conceptual Design of Dedicated Batch vs. Continuous Processing for Multistep Reaction Processes*. PhD Thesis, Univ. of Mass., Amherst, Ma.

Shultz, M.A. and J.M. Douglas, (1999). Stream costs - A first screening of reaction pathways. Accepted by *I&EC Res*.

Siirola, J.J. and D.F. Rudd, (1971). Computer-aided synthesis of chemical process designs. *Ind. Eng. Chem Fund.*, **10**, 353.

Siirola, J.J., (1995). An industrial perspective on process synthesis. *Foundations of Computer-Aided Process Design*, CACHE Corp., Austin, 222.

Siirola, J.J., (1996). Industrial applications of process synthesis. *Adv. in Chem. Eng.*, **23**, Academic Press, San Diego, CA, 1-62.

Talukdar, S.N., (1998). Autonomous cyber agents: rules for collaboration. *Proceedings of the Thirty-First Hawaii International Conference on Systems Science (HICSS-31)*, Hawaii.

Talukdar, S. N., Baerentzen, L., Gove, A., and P. S. de Souza (1998). Asynchronous teams: Cooperation schemes for autonomous agents. *J. Heuristics*, **4**, 295-321.

Talukdar, S. N. and P. S. de Souza, (1994). Insects, fish and computer-based super-agents. Systems and control theory for power systems. *Inst. of Math. App.* **64**, Springer-Verlag.

Tsen. C. K., (1995). *Solving Train Scheduling Problems Using A-Teams*. Ph.D. Dissertation, Dept. of Electrical and Computer Engineering, Carnegie Mellon University, Pittsburgh, PA.

Tyner, K.H., (1999). *Multiperiod Design of Azeotropic Separation Systems*. Ph.D. Thesis, Dept. of Chemical Engineering, Carnegie Mellon University, Pittsburgh, PA 15213.

Wehe, R.R., K. Lien, and A.W. Westerberg (1987). *Control Architecture Considerations for a Separation Systems Design Expert*. Keynote lecture, presented at Expert Systems Workshop, Columbia University, Mar. 9-10.

FOCAPD Invited Papers

IO1   A. E. Fowler, Dow Chemical. Process development in the new millenium - Hands on or modeled, in-sourced or out-sourced, solo or shared?

IO2   J. P. O'Connell and M. Neurock, University of Virginia. Trends in property estimation for process and product design.

IO3   Sangtae Kim, Parke-Davis Pharmaceutical Research. The web and flow of information in pharmaceutical R&D.

IO4	J. B. Carberry, R. W. Sylvester, and W. D. Smith, E. I. DuPont Co. Information and modeling for greener process design.	I15	M. F. Doherty, M. F. Malone and R. S. Huss, University of Massachusetts. Decision-making by design: Experience with computer-aided active learning.
IO5	N. Shah, N. J. Samsatli, and M. Sharif, Imperial College, London, J. N. Borland: The Britest project, Huddersfield, and L. G. Papageorgiou, University College, London, UK. Modeling and optimization for pharmaceutical and fine chemical process development.	I16	B. A. Finlayson, University of Washington, and B. M. Rosendall, Bechtel Corp. Reactor/transport models for design: How to teach students and practitioners to use the computer wisely.
IO6	G. Stephanopoulos, MIT; A. Linninger, University of Illinois, Chicago, and E. Salomone, INGAR, Argentina. Batch process development: Challenging traditional approaches.	I17	W. Marquardt, RWTH, Aachen. Perspectives on life cycle process modeling.
IO7	S. Wright, Argonne National Labs. Algorithms and software for linear and nonlinear programming.	I18	A. Shah, B. Nagy and K. D. Ganesan, Fluor-Daniel Energy and Chemicals. Streamlining projects through integrated front-end automation solutions.
IO8	I. E. Grossmann and J. Hooker, Carnegie Mellon University. Logic based approaches for mixed integer programming models and their application to process synthesis.	I19	B. Braunschweig, Institut Français du Pétrole, and C. C. Pantelides, Imperial College. Open software architectures for process modelling: Current status and future perspectives.
IO9	C. Floudas, Princeton University, and C. Pardalos, University of Florida. Recent Developments in global optimization methods and their relevance to process design.	I20	J. Stefanovic and C.C. Pantelides, Imperial College. Towards tighter integration of molecular dynamics within process and product design computations.
I10	J. van Schijndel, Shell International Chemicals BV, Amsterdam, and S. Pistikopoulos, Imperial College, London. Towards the integration of process design, process control, and process operability - Current status and future trends.	I21	S. K. Bermingham, H. J. M. Kramer, G. M. van Rosmalen, and Johan Grievink. A design procedure and predictive models for solution crystallisation processes.
I11	B. D. Tyreus and M. L. Luyben, E. I. DuPont Co. Industrial plantwide design for dynamic operability.	I22	J. M. Douglas, University of Massachusetts, J. A. Trainham, E. I. DuPont Co., and A. W. Westerberg, Carnegie Mellon University. Conference summary and challenges for FOCAPD 2004 and beyond.
I12	L. T. Biegler, Carnegie Mellon University. Multi-solver modeling for process simulation and optimization.		
I13	J. L. Sinclair, Purdue University. CFD for multiphase flow: Research codes and commercial software		
I14	K. F. Jensen, MIT. Integrated microchemical systems: Opportunities for process design.		

FOCAPD Contributed Paper Mentioned

A13	F. M. Meeuse, J. Grievink, J.; and P. J. T. Verheijen, P. J. T., Delft University of Technology, and M. L. M. van der Stappen, Unilever Research, the Netherlands. Conceptual design of processes for structured products.

SIMULTANEOUS DESIGN AND LAYOUT OF BATCH PROCESSING FACILITIES

Ana Paula Barbosa-Póvoa*
Unidade de Economia e Gestão Industrial
Instituto Superior Técnico
101 Lisboa, Portugal

Ricardo Mateus and Augusto Q. Novais
Dep. de Modelação e Simulação de Processos
Instituto Nacional de Engenharia e Tecnologia
1699 Lisboa, Portugal

Abstract

A mathematical formulation for the simultaneous design and layout of batch process facilities is proposed. The model determines *simultaneously* the optimal plant topology (*i.e.* the choice of the plant equipment and the associated connections), the layout (*i.e.* the arrangement of processing equipment storage vessels and their interconnecting pipe-work over a two-dimensional space) and the optimal plant operation.

The problem is formulated as a mixed integer linear problem; binary variables are introduced to characterise operational and topological choices. The applicability of the proposed model is illustrated via a representative example.

Keywords

Design, Layout, Batch facilities, Optimization, Mixed integer linear programming (MILP).

Introduction

The design of batch processing facilities involves a large number of interacting decisions. An important part of the latter is layout considerations concerning the spatial allocation of equipment items and their interconnections. Traditionally, such layout issues have been considered *a posteriori* once the main plant design task has been completed. However, the interactions of layout with the rest of the design decisions are often quite strong, and this renders a simultaneous approach more desirable.

Papageorgaki and Reklaitis (1991) addressed the general batch plant design problem where the main equipment was selected allowing a flexible unit-to-task allocation. Plant operation over several campaigns was studied and the optimal allocations of products to campaigns as well as the campaign lengths were determined. Later on, Barbosa-Póvoa and Macchietto (1994) studied a single campaign problem were different products could be produced. Plant topology was considered and the design problem led to the choice of both the main equipment items and the associated network of connections.

The layout problem within batch process facilities has recently been studied by several authors, mainly being treated as a stand-alone problem to be solved once the

---

* Author to whom correspondence should be addressed – email : apovoa@ist.utl.pt, tel : + 351 1 8417729, fax: + 351 1 841 79 79

main equipment items and their interconnections have been determined. Georgiadis et al. (1998) used a space discretization technique to consider the allocation of equipment items to floors and the detailed layout of each floor. Also, Papageorgiou and Rotstein (1998) proposed a mathematical programming model for determining the optimal process plant layout; their model employed a continuous representation of space and took account of many important features of the plant layout problem. Finally, a simultaneous approach to plant design, scheduling and layout was adopted by Realff et al. (1996) for the case of pipeless batch plants.

This paper presents a mathematical formulation for the simultaneous design and layout problem of batch process facilities. Based on the design model proposed by Barbosa-Póvoa and Macchietto (1994), detailed layout aspects are studied using a generalised approach of the work recently proposed by Papageorgiou and Rotstein (1998). The resulting model determines *simultaneously* the optimal plant topology (*i.e.* the choice of the plant equipment and the associated connections), the optimal layout (*i.e.* the arrangement of processing equipment storage vessels and their interconnecting pipe-work over a two-dimensional space) as well as the plant operation.

The complications arising from operational conditions, suitability and availability of equipment, the various layout constraints and the presence of equipment items of various sizes are taken into account.

The problem is formulated as a mixed integer linear problem where binary variables are introduced to characterise operational and topological choices.

The applicability of the proposed model is illustrated via a representative example.

Problem Representation

The problem representation uses *the maximal State-Task Network (mSTN)* as defined by Barbosa-Póvoa (1994).

The process recipes are defined through a *State-Task Representation (STNs,* Kondili et al (1988)*)* and the plant is characterised through a normal flowsheet - *equipment network* of vessels, processing units and possible connections.

Having the process recipes and the plant description the *mSTN* automatically combines the operations and equipment network by performing the mapping between the plant and process networks. This mapping is defined by:

- The suitability of each unit in the equipment network to carry out processing tasks, to store material states;
- The suitability of connections to transport material states;
- The resources - equipment, utilities, operators, etc. - required by each task.

The main advantage of this representation is that it unambiguously and explicitly represents all, and only, the location of material states and the allocations of processing, storage and transfer tasks, which are potentially necessary and structurally feasible within the problem in study.

Problem Definition

The simultaneous design and layout of batch facilities can be stated as follows.

Given:

- The STN descriptions of product recipes with associated parameters and resource requirements - equipment, utilities, etc.
- A plant superstructure - network of units and connections - with associated capacities and suitability.
- Equipment units space requirements.
- Production demands, a time horizon, operation mode and availability profiles of all resources.
- Capital costs for units and connections.
- Space and equipment allocation limitations within a two dimensional continuous space.

Determine:

- The plant configuration - equipment network and sizes.
- The plant layout – allocation of each equipment item (i.e. coordinates and orientation).
- An operations schedule - sizes, allocation and timing of all batches, storage and transfers.

So as to minimize the capital cost of the plant.

Problem Characteristics

The developed model allows the description of general processing networks described by multi-product and multipurpose plants.

A discretization of time into a number of intervals of equal duration is fixed a priori, with process events occurring at interval boundaries. Task processing times, operations horizon, etc. are integer multiples of the selected interval.

Design and operation decisions are represented by continuous variables (batch sizes, equipment capacities, amounts of materials, etc.) and discrete choices by binary variables (equipment existence, task allocations to equipment and time, etc.).

Equipment items (units, dedicated storage and connections) are selected optimally from the defined plant superstructure while operation is optimized so as to satisfy all constraints.

Product requirements are defined for each product as fixed or variable within ranges, in distributed (with arbitrary time profiles) or aggregate (over a cycle, horizon, etc.) form. Demands (and supplies) may be associated to specific equipment.

Equipment in discrete size range(s), mixed storage policies, shared intermediate material, material merging, splitting and recycling, in-phase and out-of-phase operation in any combination are allowed.

A single campaign structure with fixed product slate is assumed within a non-periodic operation. The plant is defined within a two dimensional continuous space. Equipment items to be allocated in the available space are described by rectangular shapes. Rectilinear distances are assumed providing a more realistic estimate of the piping costs as opposed to direct connections. Finally, multiple equipment connectivity inputs and outputs are considered as well as space limitations.

These characteristics were modeled considering the following sets of constraints:

- *Unit Allocation Constraints* – determine the assignment of units to processing tasks assuming that at any time each unit is idle or processing a single task and that a task cannot be pre-empted once started.
- *Unit Capacity and Batch Size Constraints* – define the equipment requirements as well as the necessary capacity.
- *Connectivity Constraints* – accounts for the plant topology, transfer capacity and connections suitability.
- *Dedicated Storage Constraints* – dedicated vessels are designed based on operational storage needs.
- *Mass Balances Constraints* – relate the amount in the state with the amount of material being produced, consumed and transferred by all incident tasks.
- *Production Requirement Constraints* – production levels of any product are allowed to float within given upper and lower bounds.
- *Equipment orientation constraints* - reflect the effect of the equipment orientation within the space availability.
- *Distance Restrictions* – define the location of the equipment accounting for distance restrictions.
- *Non-overlapping Constraints* – avoid the overlapping of equipment within the same area.
- *Equipment Input and Output Constraints* – define the connections input and output points for each equipment item.
- *Space Availability Constraints* – model the space availability accounting for the possible existence of restrictions.

Finally, the objective function is defined in terms of the capital cost of the plant. This accounts for the units and connections costs – a function of the selected equipment capacity, material and suitability. For the connections, the final piping length – rectilinear distance within the plant – is also considered.

All equations are expressed in linear form as well as the objective function, resulting in a MILP problem. This is solved by a branch and bound method using a standard package (CPLEX).

Example

A simple example is used to illustrate the applicability of the method presented. This is based on one of the examples solved by Barbosa-Póvoa and Macchietto (1994). A margin of optimality of 5% is assumed.

A plant must be designed to produce two different products through the following process recipe:

Step 1 (task T1): heat raw material S1 for 2 hours to produce the unstable intermediate S3.
Step 2 (task T2): process raw material S2 for 2 hours to form the intermediate S4.
Step 3 (task T3): mix intermediate material S3 with material S4 in the ratio of 60:40 and let them react for 4 hours to form product P1.
Step 4 (task T4): mix S3 with P1 in the ratio of 60:40 and let them react for 2 hours to form product P2.

The plant must be able to fulfill the production requirements of 80 tonne for materials P1 and P2 over a time horizon of 8 hr. The minimisation of the plant capital cost is considered as the main objective. The example is solved considering two different cases:

1. The stand-alone design problem, as proposed by Barbosa-Póvoa and Macchietto (1994), without layout considerations. The solution provides the final plant topology and associated schedule for an objective function defined as the cost of the units.
2. The design and layout problems are solved simultaneously using the model proposed in this paper. The solution provides the final plant topology, plant layout and schedule for an objective function defined as the total cost of units and connections. In this case a limited area availability is considered - 21 x 6 m^2 (X x Y).

The plant superstructure used is shown in Figure 1 and the equipment characteristics are given in Table 1.

For case 1, the final plant is characterised by the choice of units 1a, 1b and 2a and the vessels V1, V2, V5 and V6. This corresponds to a capital cost of 73000 c.u. In operational terms unit 1a is performing task T1, unit 1b task T2 and unit 2a is a multi-task unit performing tasks T3 and T4. No intermediate storage was chosen.

Table 1. Equipment Characteristics.

Unit	Suitability	Capacity [tonne]	Costs [10^3 c.u.]	Dimensions length/depth [m]
1a	T1/T2	70	14	8 / 8
1b	T1/T2	70	15	6 / 4
1c	T1	70	18	6 / 2
2a	T3/T4	120	40	5 / 5
V1	Store S1	Unlim.	1	6 / 3
V2	Store S2	Unlim.	1	6 / 3
V4	Store S4	50	3	5 / 1
V5	Store P1	Unlim.	1	4 / 3
V6	Store P2	Unlim.	1	4 / 3
Connections Costs : 150 c.u./m (currency units/m)					

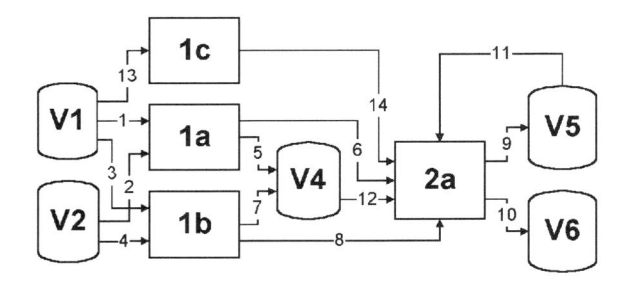

Figure 1. Plant superstructure topology.

For case 2 - where aspects of operation, design and layout are considered simultaneously - the results obtained are different. In this case the units chosen are respectively 1b, 1c and 2a apart from tanks V1, V2, V5 and V6. This corresponds to a capital cost of 77000 c.u. The operational schedule and layout are respectively illustrated in Figures 2 and 3.

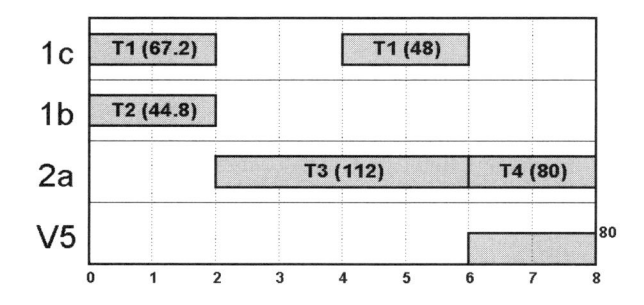

Figure 2. Plant scheduling for case 2.

The different plant topology is explained by the space availability restrictions. In this context unit 1a, which needs higher space requirements (8/8) when compared with unit 1c (6/2), can not be installed in the final plant due to the total space availability (21/6). Indeed, when trying to optimise the layout of the plant topology obtained from case 1, with the space restrictions defined in case 2, an infeasible problem is obtained. This shows how it may be important to consider design and layout aspects simultaneously when designing the plant.

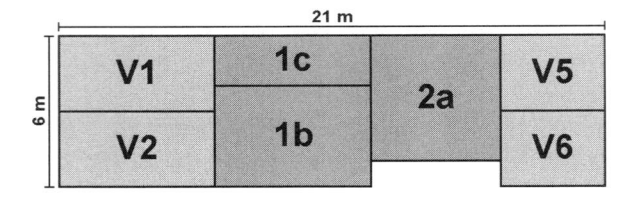

Figure 3. Plant layout for case 2.

In terms of computer statistics the final results are shown in Table 2.

Table 2. Problem Statistics (Pentium II, 300).

Case	Nodes	CPUs
1	6	negligible
2	3 896	30.8

Both cases are solved quite efficiently, although it is worth noting that case 1 does not cover layout. However, our yet limited experience with the model shows that improvements can be made if some of the already model intrinsic layout characteristics are further explored such as equipment orientation and pipe-work distance calculations. This is now under development.

Conclusions

The simultaneous design and layout of batch facilities has been studied. The optimal plant configuration and operation were obtained. Important aspects were considered such as plant operation, plant topology - the choice of the plant equipment and the associated connections – and plant detailed layout - arrangement of processing equipment storage vessels and their interconnecting pipe-work over a two-dimensional space.

The problem was formulated trough a Mixed Integer Linear Programming were binary variables define equipment choices, operability and space arrangement.

The present model seems promising and, as stated above, it is now under development by the authors in order to explore the modeling efficiency of the already embedded features as well as to consider more general features such as: cost of land, shop-floor sections and multi-floor plant layout, among others.

One of the main limitations of the model identified so far is related to the solution efficiency of large systems. In order to overcome this difficulty other solutions approaches are being addressed. These take advantage of the problem characteristics and will be published elsewhere.

Acknowledgments

The program PRAXIS XXI has supported this work - grant PRAXIS/2/2.1/TPAR/453/95.

References

Barbosa-Póvoa, A.P.F.D and S. Macchietto (1994). Detailed design of multipurpose batch plants. *Comp. Chem. Eng.,* **18**, 1013-1042.

Georgiadis, M.C., G.E. Rotstein and S. Macchietto (1997). Optimal layout design in multipurpose batch plants. *Ind. Eng. Chem. Res.,* **36**(11), 4852-4863.

Papageorgaki, S. and G. V. Reklaitis (1991). Optimal design of multipurpose batch plants – 1. Problem formulation, *Ind. Eng. Chem. Res.,* **20**(10), 4852-4856.

Papageorgiou, L. and G. E. Rotstein (1998). Continuous domain mathematical models for optimal process plant layout. *Ind. Eng. Chem. Res.*, **37**, 3631-3639.

Kondili, E, C.C. Pantelides and R.W.H. Sargent (1988), A general algorithm for scheduling batch operations. In *Proc. of 3rd Intl. Symp. on Process Systems Engineering*, Sydney, 253-274.

Realff, M. J., N. Shah and C. C. Pantelides (1996). Simultaneous design, layout and scheduling of pipeless batch plants. *Comp. Chem. Eng.,* **20**, 869-883.

INTEGRATED INTELLIGENT TOOLS FOR AUTOMATED BATCH OPERATING RECORD SYNTHESIS

Linas Mockus, Jonathan M. Vinson, and Prabir K. Basu
Monsanto, Pharma Sector
Skokie, IL 60077

Abstract

A batch operating record contains the detailed instructions an operator has to follow to manage a batch process safely and optimally. Manual construction of an operating record takes considerable time and is subject to human error because details from many different knowledge sources need to be considered. In this work, we describe a set of integrated intelligent tools for operating record synthesis. An automated system collects and retains all the requisite information and generates an operating record based on this information. This significantly reduces the time required to review the operating record. A tool for estimating the cost of required raw materials provides information about business side of the process. Waste stream and emission calculations allow engineer to estimate environmental impact. The simulation of unit operations demonstrates the feasibility of the process itself. A tool under development for process hazards analysis assists in evaluating the safety aspects.

Keywords

Operating instruction synthesis, Batch processing.

Introduction

Until recently, operating record synthesis was a manual process in the Searle pilot plant. A detailed review process has been developed to ensure that no inadvertent errors are carried into the final set of operating instructions (Basu et. al., 1997). For processes involving hazardous materials, the lack of adequate precautions may be disastrous. However, it does not make sense to use strong precautions where they are not warranted due to their nature and expense. Another important aspect is the scale up of a bench scale recipe to detailed plant floor instructions. In case of manual scale up it is subject to the experience of the people involved. An operator may receive quite different instructions, depending on the experience and motivation of the people performing scale-up. The review process is designed to reduce this variability.

Artificial intelligence techniques, such as expert systems, can help set precautions corresponding to the level of hazard in each operation based on objective information. Such information includes materials involved, temperatures and type of material handling (manual handling requires a higher level of precautions). Expert systems can also assist in the scale up process, ensuring that the correct parameters are examined for each type of operation, such as reaction, centrifugation or drying. To ensure consistent result for each new process the only sound alternative is to use an expert system to produce the same operating record given the same information.

In collaboration with several researchers, a suite of integrated tools has been developed for use at the Searle pilot plant and in research and development. These include the Batch Record Automation Tool (BRAT), the Process Assessment Tool (PAT), the Emission Evaluation Tool (EET), the Process Hazard Analysis Tool (PHAT), process simulation and several related tools.

BRAT assists in generating the detailed instructions for the operator. Although still under development, PHAT (Viswanathan et. al., 1999) analyzes the BRAT instructions for potential process hazards and assists in generating a hazards report. (A separate hazard analysis is conducted for equipment-specific hazards, as each piece of equipment might see several different processes.) Both of these tools have been developed in collaboration with Purdue University.

PAT conducts an initial cost screening, which accounts for the cost of raw materials, waste streams and general operating costs. PAT does not consider the cost of equipment since the program is used at the early stages of development. PAT has been developed in-house, based on other internal cost analysis software. EET calculates the vapor emissions for a process by using BRAT data. It has been developed in collaboration with Mitchell Scientific, Inc.

Hierarchical Representation of a Process Model

All of these tools rely on a hierarchical description of the batch process. The hierarchy contains five levels, similar to the ISA-S88 standard for batch automation: Process, Step, Unit Operation, Operation and Phase. Each of these is described below; they define increasing detail of a process from a general description of the product down to the exactly when to open valves or the settings for a control system.

This representation is useful in batch processing where parts of one process are frequently used in another. For example, two compounds might be produced via similar pathways. This representation permits one to easily copy the common elements.

Level 1: The Process

The highest level of the hierarchical description is the process itself. For our purposes, we take this as an entire chemical transformation from original raw materials to a salable product. Information at this stage might include how much material is to be produced per year and the overall cost of producing that material. Details about raw materials might be available at this level, but they are generally input at the Step or Unit Operation level.

Level 2: The Step

A Step describes an identifiable process in terms of manufacturing. Generally, a Step begins with raw materials and produces a dried intermediate (or the final product) which can be stored. One run of Step in a pilot or manufacturing plant is commonly called a batch and is frequently labeled with the Process and Step names (Step 3 of XYZ Process).

At this level, one specifies the desired product of the Step, the amount to be produced at the standard scale, the product yield, the throughput, the production time, and other generic information pertaining to the Step. The scale up factor may be needed to change an existing process to a different scale. As with the Process level, many details are left to the Unit Operation and Operation levels of detail. In fact, the throughput and yield information may be calculated based on detailed material balances from these levels. Essential raw materials may be specified at this level to begin estimating the cost of the Step and the overall Process. Equipment is not yet specified at this level, although general equipment size information can be deduced from the throughput and production quantity.

Level 3: The Unit Operation

The next level in hierarchy is a Unit Operation. As implied by its name, a Unit Operation represents a set of activities performed in a single unit (or a set of equipment), such as reaction, distillation or filtration. At this level, we specify general operating conditions, raw materials, chemistry, basic material balances and major equipment characteristics. Some operating conditions, like temperature or pressure, may vary during the execution of a Unit Operation; thus, the specification of those conditions is left for the next level of detail. General information about operating conditions like the adiabatic or exothermic character of a reaction may be specified here. Assignment of equipment to a Unit Operation need not be made until the Step is scheduled for production at which point the scheduler selects equipment based on availability and characteristics of the Unit Operation. An intelligent scheduler might be able to determine that corrosive materials exist in the mixture and recommend appropriate vessels.

It is possible to automatically synthesize Unit Operations from the Step description. One may supply characteristics of the reaction and separation in the Step definition that are then used to generate the correct sequence of Unit Operations to achieve the Step production needs (Douglas, 1988; Jaksland et. al., 1995). We are exploring the possibility of using some of these synthesis ideas (Vinson et. al., 1999). In the meantime, our systems require that the user specify the sequence of Unit Operations. The resulting description is a high-level flow sheet of the Step, indicating the operating sequence; react, distill, filter, dry.

Level 4: The Operation

Each Unit Operation consists of an ordered set of *Operations*, representing the next level of detail. The changes in the operating conditions of a process are captured in each Operation, such as charge, heat, hold and stir. Each Operation specifies in detail the materials to charge, the type of charge (solid or by pump), the temperature and time of holding, or agitator speed. Equipment used by each Operation is also set at this level.

As with Unit Operations, it is possible to synthesize Operations based on the information given in the levels above. For example, the reaction Unit Operation usually involves charging of raw materials, heating to some temperature, stirring and cooling. The specific materials and operating conditions involved would be extracted from a description of the reaction. Then the sequence of Operations that achieve the desired goals of the Unit Operation would be generated. (Crooks et. al., 1994; Fusillo and Powers, 1987; Lakshmanan and Stephanopoulos, 1988abc; Rotstein et. al., 1994). However, more sophisticated reactions involve charging of raw materials and solvents in a specific sequence with possible heating, holding, stirring, and cooling repeated several times.

Level 5: The Phase

The Phase represents the lowest level of detail. In a fully automated facility, a Phase might be the sequence of controller and valve settings required to for a charge or heat operation. For manually operated plants and pilot plants, phases are represented as standard operating procedures with which the operators are intimately familiar. So far, the tools discussed in this paper are not concerned with this level of detail.

Views of the Hierarchical Information

Once the information has been collected, the tools mentioned previously access and display the appropriate details. Most of the information is entered through the BRAT interface, although both and PHAT can take direct user input as well. One can think of each tool as providing a different view of this information.

The most important view of the data from an operations standpoint is the Batch Record, or sequence of operating instructions, used by the plant operators. This represents the Operations-level details of one Step in a Process. Phase-level details are captured as references to the appropriate SOP within the instruction text. Since the batch record is used in regulated activities, it must be safe and accurate. Instructions representing the same operation must look the same for every batch record generated. One of the main goals behind the development of BRAT is to provide this kind of consistency, both for the engineers developing the batch record and the operators who will use it in the plant. This consistency has an additional benefit during safety and quality reviews of the document.

As BRAT was designed for operations in our pilot plant, it must be flexible to meet the needs of new processes that come into the plant. As such, the engineers who use BRAT are able to modify some portions of the final instructions to meet special needs of the process and ever-changing regulations. As these needs are discovered, BRAT is modified to incorporate the new engineering practices. Since its full implementation in January 1998, BRAT has seen the addition of several such changes. One example is the addition of three new charge setups that are new standards in the pilot plant. These used to require special subset of instructions, but are now captured in SOP's and make the Batch Record easier to follow.

There are many other views of this hierarchical collection of information. An alternate view of the Operations in BRAT graphically displays each operation in columns representing pieces of equipment. The drawing is such that parallel operations are held in the same row. Each operation is labeled with its corresponding instruction number, the duration of the operation and the cumulative volume in the equipment during that operation. This view is useful for understanding the entire process in a quick view. Where one hundred instructions might occupy fifty pages, the flow sheet view of the same operations will only need four or five pages. BRAT has a similar view of the Unit Operations that contains the name of the Unit Operation and the connection to the others.

Another set of BRAT views at the Operation level represents specific information required by support groups at the pilot plant. The analytical group receives a list of those operations where samples are taken in order to schedule their activities. Similarly, the material management group receives a bill of materials and a list of operations that require chemicals or other operational materials. BRAT can also produce a material balance, based solely on user input, for use in waste stream handling, protective equipment selection, simulation and other standard engineering calculations.

PHAT, which conducts hazard analyses, uses the information stored within BRAT along with some additional user input for more Operation-level details that are not required by BRAT. When PHAT will be fully implemented, it will check the safety of the instructions as they appear in the Batch Record. This involves an automated process that examines a variety of potential upsets in the process and determines whether the upset might lead to a hazardous situation.

Also related to the hazard analysis are other views of the Operations where the pressure is very high or very low. This information could then be compared to the maximum and minimum allowable working pressure of the equipment to ensure that the equipment meets the needs of the process. Similarly, vent size calculations could be made for all operations at the highest pressure. Such analyses could be made for temperature and the materials involved in the Unit Operation. A material interaction matrix is a list of all the materials associated with a Step and the known or potential interactions between them. From a safety standpoint, one must know when two materials must be kept segregated.

While BRAT and PHAT focus mainly on the Operation level of detail and are unconcerned with the overall Process, PAT is interested in the Process, Step and Unit Operation levels. It uses information at these levels of detail (raw materials, material balance, and duration) to determine the cost of producing the active pharmaceutical ingredient. This view of the information entails retrieving data from other sources, such as a purchasing database. Once PAT collects and displays the information, the user might be interested in determining the affect of improvements to the process. As PAT is written in Excel, this task is a simple matter of a quick sensitivity analysis.

EET, the Emission Evaluation Tool, is a view of the Operations in terms of vapor emissions from the Step. It uses Environmental Protection Agency standards for vapor emission calculations for operations such as charge, heat and nitrogen sweep that allow vapors to escape to the atmosphere.

Simulation and other tools can also use the information stored in Operations and Unit Operations as a starting point for the creation of more detailed material balances and estimates of the time required for each Operation and the entire Step. Beyond simulation, a variety of tools can use material balance information at any level of detail. For example, a scheduler might be interested in how many kilograms of product a Step could provide given a maximum vessel size of 500 gallons.

Future Work

Currently, the information in PAT is stored separately from BRAT, but we have created routines that pass Unit Operation information from PAT to BRAT in order to speed the development of the BRAT model of a Step. Ideally, this will be two-way street where modifications in BRAT can be examined for their effect on the cost in PAT. The most sensible format for this is a common database, full of tables representing everything from the Process down to each Operation. This would permit the construction of tools that access and modify this data, based on their particular view of the hierarchy.

A step forward would be full automation of the process when operator instructions are fed to an automated plant. Another alternative is to feed operator instructions to an electronic batch record system (EBR) which displays the instructions on operator consoles in the plant. It sets equipment set points; reads appropriate data (from scales in the case of charge), and even perform in-process analyses.

Information necessary for an operating record synthesis may be used for other purposes as well. Currently this information is used to calculate emissions. We are planning to integrate other engineering calculations (vent sizing, simulation, cost estimation, etc.) by using an "engineering toolbox" concept.

The next stage would be process scheduling and control. The integration of all these activities is very important aspect in making facility safe and profitable and is really challenging task.

Conclusions

We presented some of the tools used to automate everyday activities in the Searle pilot plant. Our experience shows that it is quite cumbersome do develop a tool encompassing all activities, a "Swiss army knife" to solve all engineering problems. Instead, we want to use tools that already exist or do their specific job very well. A suite of tools for batch record generation, emission calculations, costing, simulation, vent sizing is much more flexible approach, assuming that these tools are able to talk with each other and to use a common database. Another requirement is ease of use, the open architecture of a tool and its ability to grow by incorporating new techniques and regulations (i.e., GMP requirements). Research and Development activities require a lot of intelligence. A suite of tools to automate repeating activities allows engineer to focus more on process development.

References

Arzen, K. E. (1994). Grafchart for intelligent supervisory control applications. *Automatica*, **30**, 1513-1525.

Basu, P. K., J. Quaadgrass, J. E. Holleman, R. A. Mack, and A. R. Noren (1997). Pharmaceutical pilot plants are different. *Chem. Eng. Prog.*, **93**, 66-75.

Crooks, C. A., S. F. Evans, and S. Macchietto (1994). An application of automated operating procedure synthesis in the nuclear industry. *Comp. Chem. Eng.*, **18**, S385-S389.

Douglas, J. M. (1988). *Conceptual Design of Chemical Processes*. McGraw Hill, New York.

Fusillo, R. H. and G. J. Powers (1987). A synthesis method for chemical pilot plant operating procedures. *Comp. Chem. Eng.*, **11**, 369-S382.

Jaksland, C. A., R. Gani and K. M. Lien (1995). Separation process design and synthesis based on thermodynamic insights. *Chem. Eng. Sci.*, **50**, No. 3.

Lakshmanan, R. and G. Stephanopoulos (1988a). Synthesis of operating procedures for complete chemical plants –I. Hierarchical structured modeling for nonlinear planning. *Comp. Chem. Eng.*, **12**, 985-1002.

Lakshmanan, R. and G. Stephanopoulos (1988b). Synthesis of operating procedures for complete chemical plants – II. A nonlinear planning methodology. *Comp. Chem. Eng.*, **12**, 1003-1021.

Lakshmanan, R. and G. Stephanopoulos (1988c). Synthesis of operating procedures for complete chemical plants – III. Planning in the presence of qualitative mixing constrains. *Comp. Chem. Eng.*, **14**, 301-317.

Rotstein, G. E., R. Lavie, and D. R. Lewin (1994). Automatic synthesis of batch plant procedures: process oriented approach. *AIChE J.*, **40**.

Vinson, J. M., P. K. Basu, R. A. Gani, M. Hostrup, and C. A. Jaksland (1999). Separation synthesis in the pharmaceutical industry. Process Systems Engineering Topical Conference, AIChE Spring Meeting 1999. Paper T002.

Viswanathan, S., J. Zhao, V. Venkatasubramanian, L. Mockus, J. M. Vinson, A. Noren, and P. K. Basu (1999). Integrating operating procedure synthesis and hazards analysis automation tools for batch processes. *Comp. Chem. Eng.*, (submitted).

Viswanathan, S., L. Mockus, V. Venkatasubramanian, P. K. Basu, R. Mack, P. Cherukat, and V. Iskos (1998). iTOPS: An intelligent tool for operating procedures synthesis. *Comp. Chem. Eng.*, **22**, S601-S608.

DESIGN OF MULTIPRODUCT BATCH PLANTS WITH SEMICONTINUOUS STAGES AND STORAGE TANKS

Jorge M. Montagna and Aldo R. Vecchietti
INGAR – Instituto de Desarrollo y Diseño
UTN – Facultad Regional Santa Fe
Santa Fe, Argentina

Abstract

In this work a detailed model for the design of multiproduct batch plant is presented. The impact of the intermediate storage tank allocation is analyzed considering different alternatives. In previous works presented in the literature, the relationship between the tank location and the semicontinuous stages was ignored. In the present model we analyzed several expressions related to the operation time for the stages of the batch plant. These expressions are included in the Mixed Integer Non Linear Model (MINLP) formulated for the plant design.

Keywords

Multiproduct batch plants, Batch plant design, Intermediate storage tanks.

Introduction

In the last years several works have treated the allocation of storage tanks for the design of multiproduct batch plants. Takamatsu et al (1982) presented a noncontinuous expression to size the storage tanks. Afterwards, Karimi and Reklaitis (1985) developed a more conservative continuous expression. As the constraint obtained was not a posynomial expression, the optimization model lost its geometric program structure, as was posed for Grossmann and Sargent (1979). Modi and Karimi (1989) worked with a posynomial expression to solve the sizing problem. Then, Ravemark and Rippin (1998) consider the optimal tank location. However, they do not treat the impact of the storage tank allocation when semicontinuous stages are involved. Finally, Vecchietti and Montagna (1998) presented a model, considering only batch stages, where the position of a tank can be different for each product.

This work analyzes the alternatives for sizing and allocating intermediate storage tanks for a batch plant which includes semicontinuous stages. The different alternatives are modeled with binary variables. The resulting MINLP model is solved using GAMS/DICOPT++. The results obtained solving several examples with this approach are showed.

Operation Time and Limiting Cycle Time

The operation time for the batch stage j for product i is given by the following expression:

$$T_{ij} = T_{ij}^0 + T_{ij}^1 B_{ij}^{\delta_{ij}} \qquad \forall i, \forall j \qquad (1)$$

where T_{ij}^0, T_{ij}^1 and δ_{ij} are coefficients for product i and stage j. B_{ij} is the batch size for product i at stage j. The operation time can be variable according to B_{ij}.

In plants operating with monoproduct campaigns, one batch of product i is obtained during a TL_i time interval. TL_i, the limiting cycle time of product i, is given by the maximum of the operating time of the stages needed to produce i. If semicontinuous stages located up and downstream of a batch stage are operating, the batch stage is not available for another batch cycle. Then

$$TL_i \geq \frac{1}{M_j}\left(\theta_{ij_d} + T_{ij}^0 + T_{ij}^1 B_{ij}^{\delta_{ij}} + \theta_{ij_u}\right) \quad \forall i, \forall j \quad (2)$$

θ_{ik} is the operating time for semicontinuous stage k, product i; k assumes the values j_d and j_u corresponding to units up and downstream of batch stage j. M_j is the number of units in parallel out of phase for the stage j.

The expression for θ_{ik} is the following:

$$\theta_{ik} = \frac{D_{ik} B_{ij}}{N_k R_k} \quad \forall i, \forall k \quad (3)$$

where D_{ik} is the size factor corresponding to the semicontinuous unit k for product i, R_k the rate of this unit and N_k the number of units in parallel in phase. TL_i must be greater than the operating time of the semicontinuous units:

$$TL_i \geq \theta_{ik} \quad \forall i, \forall k \quad (4)$$

In (2), (3) and (4) we did not have an expression for the batch size B_{ij}. We can have several situations. Let us consider the case of Fig. 1 where no tank is located next to the batch stage j. In this case the batch size for units j, $j+1$, k, $k-1$ and $k+1$ is the same. Particularly we have:

$$TL_i \geq \frac{1}{M_j}\left(\theta_{i,k-1} + T_{ij} + \theta_{ik}\right) \quad \forall i, \text{batch stage } j \quad (5)$$

$$TL_i \geq \frac{1}{M_j}\left(\theta_{ik} + T_{i,j+1} + \theta_{i,k+1}\right) \quad \forall i, \text{stage } j+1 \quad (6)$$

Both expressions (5) and (6) have the time of the unit k (θ_{ik}), which is located between unit j and $j+1$.

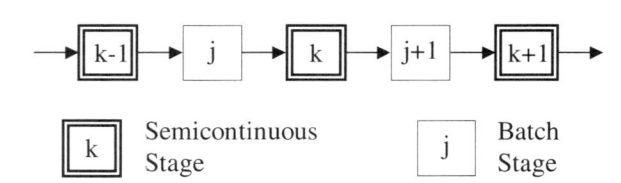

Figure 1. Batch stage j without storage tank.

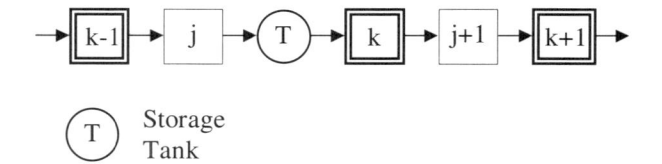

Figure 2. Storage tank between batch stage j and semicontinuous stage k.

With the inclusion of an intermediate storage tank the process is decoupled in two subprocesses s and s'. If S tanks are located S+1 subprocesses are defined. Each subprocess has its own limiting cycle time and batch size.

In the model of Ravemark et al. (1998) if j is the optimal location of an intermediate storage tank means that the tank is located after the batch stage j. We need to be more specific when semicontinuous stages are presented up and downstream the stage j. Figures 2 and 3 show two alternatives. In case of Fig. 2, the expressions for TL_i are:

$$TL_{is} \geq \frac{1}{M_j}\left(\theta_{i,k-1} + T_{ij}\right) \quad \forall i, \text{stage } j, \text{subprocess } s \quad (7)$$

$$TL_{is'} \geq \frac{\theta_{ik} + T_{i,j+1} + \theta_{i,k+1}}{M_{j+1}} \quad \forall i, \text{stage } j+1, \text{subprocess } s' \quad (8)$$

The time corresponding to the semicontinuous unit k has been included only on (8) for the batch stage $j+1$.

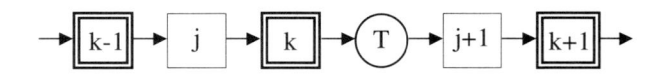

Figure 3. Storage tank between batch stage j+1 and semicontinuous stage k.

For the case of Fig. 3 we have:

$$TL_{is} \geq \frac{1}{M_j}\left(\theta_{i,k-1} + T_{ij} + \theta_{i,k}\right) \quad \forall i, \text{stage } j, \text{subprocess } s \quad (9)$$

$$TL_{is'} \geq \frac{1}{M_{j+1}}\left(T_{i,j+1} + \theta_{i,k+1}\right) \quad \forall i, \text{stage } j+1, \text{subprocess } s' \quad (10)$$

The same location j of an intermediate tank implies different expressions for the limiting cycle time. Another important detail is that the batch size for the semicontinuous unit k (Eqn. 3) is not the same for all the cases presented. For Fig. 2 the batch size is equal to the stage $j+1$, and for Fig. 3 is equal to the stage j.

In the previous analysis, the semicontinuous units are grouped in a train between two consecutive batch stages. For simplicity, in this presentation we will consider only one semicontinuous stage.

Allocation of Intermediate Storage Tanks

Two positions p will be considered depending on the tank location up and downstream the semicontinuous train between the batch stages j and $j+1$:

- before the semicontinuous train between the batch units j and $j+1$: $p=$**b**efore (Fig. 2).
- after the semicontinuous train between the batch units j and $j+1$: $p=$**a**fter (Fig. 3).

In the case when no semicontinuous stages exists between the batch stages j and j+1 we have only one location for the intermediate storage tank.

The binary variables y_{jp} are introduced, whose value are given by:

$$y_{jp} = \begin{cases} 1 & \text{if the tank is located at } j, \text{ position } p \\ 0 & \text{otherwise} \end{cases}$$

Taking into account that two tanks can not be included at the same position j, up and down the semicontinuous stage, the following constraint applies:

$$y_{ja} + y_{jb} \leq 1 \qquad \forall j \qquad (11)$$

Three alternatives are possible for the tank location:

1. No tank is located at positions $(j-1,a)$ and (j,b) (Fig. 1). The limiting cycle time for the stage j is given by:

$$TL_{is} \geq \frac{\theta_{ij_d} + T_{ij} + \theta_{ij_u}}{M_j} - F_{ij}(y_{j-1,a} + y_{jb}) \quad \forall i,j,s \qquad (12)$$

where F_{ij} is a constant large enough to apply the constraint like a Big-M constraint.

2. A tank is located after the stage j (Fig. 2):

$$TL_{is} \geq \frac{\theta_{ij_d} + T_{ij}}{M_j} - F_{ij}(y_{j-1,a} + y_{jb}) \quad \forall i,j,s \qquad (13)$$

3. A tank is located before the stage j (Fig. 3):

$$TL_{is} \geq \frac{T_{ij} + \theta_{ij_u}}{M_j} - F_{ij}(y_{j-1,a} + y_{jb}) \quad \forall i,j,s \qquad (14)$$

According to the value of the variables y_{jp} only one of the previous expressions is active for determining TL_{is}.

It is important to note that if no semicontinuous units exist in the plant, the constraints (12)-(14) are not required anymore, instead constraint (2) applies.

Model Formulation

The following assumptions have been made:

- The plant consists of N semicontinuous and M batch stages, to produce P different products.
- At each batch stage j there are $M_j\, G_j$ units in parallel, (M_j out of phase and G_j in phase) each one with a size V_j.
- Only monoproduct campaigns are admitted
- When a storage tank is not allocated, batches are transferred with the ZW (Zero Wait) policy.
- At each semicontinuous stage k there are N_k units in parallel in phase which have a processing rate Rk.
- The production requirements Q_i for products i in the time horizon H are known.

The objective function is minimize the capital cost of the plant:

$$Min\ C = \sum_{j=1}^{M} a_j M_j G_j V_j^{\alpha_j} + \sum_{k=1}^{N} b_k N_k R_k^{\beta_k} + \sum_{j=1}^{M-1} \sum_{p=a,b} c_{jp} VT_{jp}^{\gamma_{jp}} \qquad (15)$$

where $a_j,\ \alpha_j,\ b_k,\ \beta_k,\ c_{jp}$ and γ_{jp} are cost coefficients according to the unit type. VT_{jp} is the size of the storage tank located after the batch stage j, in position p.

The units of batch stage j must be large enough to operate a batch of every product. Taking into account that the size factor S_{ij} is known and that G_j units in parallel in phase are operating, then:

$$V_j \geq \frac{S_{ij}\, B_{ij}}{G_j} \qquad \forall i, \forall j \qquad (16)$$

If a tank is included between the batch stages j and $j+1$, the batch sizes B_{ij} and $B_{i,j+1}$ are different, otherwise they are equal. To consider this behavior, the following expressions should be satisfied (Ravemark, 1995):

$$1 + \left(\frac{1}{\Phi} - 1\right)(y_{ja} + y_{jb}) \leq \frac{B_{ij}}{B_{i,j+1}} \leq 1 + (\Phi-1)(y_{ja} + y_{jb}) \qquad (17)$$

$$\forall i, \forall j = 1,\dots,M-1$$

The productivity P_i for product i is given by B_{ij}/TL_{is}, $\forall i, j, s$, where unit j belongs to subprocess s. P_i must be the same for all the subprocesses of the plant to avoid accumulations in the intermediate storage tanks. Using P_i into (12)-(14), the variable TL_{is} is eliminated. When a tank is placed at position j, from (17) it is obvious that the batch sizes are different and, therefore, the limiting cycle time.

The TL_{is} must be the maximum of all stages, including the semicontinuous stages. Constraints (4) are added, after replacing θ_{ik} from the expression (3). Previously, we had not mentioned how to determine the batch size to be used in expression (3). Using the expression of P_i mentioned above, both TL_{is} and B_{ij} can be eliminated from (3) and (4). The same productivity in all the stages is attained for product i.

Q_i quantities of product i must be produced over the horizon time H such that:

$$\sum_{i=1}^{P} Q_i/P_i \leq H \qquad (18)$$

The expression for sizing of the storage tanks is:

$$VT_{jp} \geq ST_{ijp}(B_{ij} + B_{i,j+1}) - F_j(1 - y_{jp}) \quad \forall i, j \neq M, p \quad (19)$$

where ST_{ijp} is the size factor for the storage tank.

Finally the MINLP model consists of minimizing the objective function (15) subject to constraints (11)-(14), (16)-(19). It is a posynomial model, so the equations are convexified to guarantee the global optimum.

Results

Table 1 presents the data of an example. Table 2 shows the results of three problems solved with these data.

First the problem without storage tanks was solved. Second, we solved the model proposed in this paper. The solution showed a reduction in the cost of the plant of 14 %. Two tanks were allocated after stage 1 in position before and after stage 2, in position after. To assess the difference between considering the storage tank allocated before or after the semicontinuous stage, a last problem was solved, changing the solution obtained. Two tanks were allocated fixing their position: one after batch stage 1 in position after, and the other after batch stage 2, in position before. The objective function was increased 8 % from the solution of the problem 2.

Conclusions

This paper presents a model for the design of multiproduct batch plants when the allocation of intermediate storage tanks and semicontinuous stages are considered simultaneously. From the results obtained in the example, it is obvious that the position of the tank respect to the semicontinuous stages must be considered to attain the appropriate storage tank location and sizing. The allocations of the intermediate storage tanks change the expressions for the TL_i when semicontinuous stages are involved. New constraints must be included in the model.

Table 1. Data for Example.

P=2		M=3		N=4	
H=7200		Q_A = 90,000		Q_B=70,000	
Semicontinuous Stages: R_k^L = 100; R_k^U = 1,800					
	k=1	k=2	k=3	k=4	
b_k	250	200	210	210	
β_k	0.4	0.85	0.62	0.62	
D_{Ak}	2.74	2.74	1.44	1.44	
D_{Bk}	1.34	1.34	1.65	1.65	
Batch Stages: V_j^L = 100; V_j^U = 2,400					
	j=1	j=2	j=3		
a_j	592.	582.	1200.		
α_j	0.65	0.39	0.52		
S_{Aj}	2.74	1.44	1.2		
S_{Bj}	2.34	1.65	1.2		
$T0_{Aj}$	1.5	5.0	10.0		
$T0_{Bj}$	1.2	7.0	9.0		
$T1_{ij} \forall i$	0.0172	0.612E-5	0.0364		
$\delta_{ij} \forall i$	0.865	2.	0.823		
Storage Tank: VT_{jp}^L = 100; VT_{jp}^U = 2,400					
c_{jp}= 278. $\forall p$		γ_{jp}=0.49 $\forall p$		ST_{ijp}=S_{ij} $\forall p$	

Table 2. Results for Example.

Example 1: C = 101,734.				
Semicontinuous Stages $N_k = 1 \forall k$				
	k=1	k=2	k=3	k=4
R_k	882.97	147.73	196.47	100.00
Batch Stages $G_j = 1 \forall j$				
	j=1		j=2	j=3
V_j | M_j	275. | 1		166. | 2	120. | 3
Example 2: C = 87,856.				
Semicontinuous Stages $N_k = 1 \forall k$				
	k=1	k=2	k=3	k=4
R_k	634.53	100.	100.	246.72
Batch Stages $G_j = 1 \forall j$				
	j=1		j=2	j=3
V_j | M_j	131. | 1		202. | 2	441. | 1
Storage Tanks	j=1, p=b		j=2, p=a	
VT	409.71		808.65	
Example 3: C = 94,780.				
Semicontinuous Stages $N_k = 1 \forall k$				
	k=1	k=2	k=3	k=4
R_k	219.07	100.	174.83	174.83
Batch Stages $G_j = 1 \forall j$				
	j=1		j=2	j=3
V_j | M_j	100. | 2		100. | 1	199. | 2
Storage Tanks	j=1, p=a		j=2, p=b	
VT	297.82		128.76	

References

Grossmann I. E. and R. W. H. Sargent (1979). Optimum design of multipurpose chemical plants. *Ind. Eng. Chem. Process Des. Dev.*, **18**, 343-348.

Karimi I. A. and G. V. Reklaitis (1985). Intermediate storage in noncontinuous processes involving stages of parallel units. *AIChE J.*, **31**, 44-52

Modi A. K. and I. A. Karimi (1989). Design of multiproduct batch processes with finite intermediate storage. *Comp. Chem. Eng.*, **13**(1/2), 127-139.

Ravemark D.E. (1995). *Optimization Models for Design and Operation of Chemical Batch Processes*. Ph. D. Thesis, Swiss Federal Institute of Technology, Zurich.

Ravemark D. E. and D. W. T. Rippin (1998). Optimal design of a multiproduct batch plant. *Comp. Chem. Eng.*, **22**, 177-183.

Takamatsu, T., Hashimoto, I. and S. Hasebe (1982). Optimal design and operation of a batch process with intermediate storage tanks. *Ind. Eng. Chem. Process Des. Dev.*, **21**, 431-440.

Vecchietti A. R. and J.. M. Montagna (1998). Alternatives in the optimal allocation of intermediate storage tank in multiproduct batch plants. *Comp. Chem Eng.*, **22**, S801-S804.

COMPUTATIONAL EXPERIENCE SOLVING THE DESIGN OF MULTIPRODUCT BATCH PLANTS MODELED AS NON-LINEAR CONTINUOUS/DISCRETE PROBLEMS

Jorge M. Montagna and Aldo R. Vecchietti
INGAR – Instituto de Desarrollo y Diseño
UTN – Facultad Regional Santa Fe
Santa Fe, Argentina

Abstract

This note presents the computational experience in the solution of a set of selected optimization problems for the design of multiproduct batch plants. Two representations for the model are assessed: MINLP and disjunctive. The model is formulated with different degrees of difficulties: from weakly non-linear convex to non-convex form of the constraints. For the solution of the MINLP problems two algorithms are applied: Extended Cutting Plane (ECP) and Outer Approximation/Equality Relaxation/Augmented Penalty (OA/ER/AP). DICOPT[++] is the solver selected for the OA/ER/AP algorithm. A DICOPT[++] modification linked to GAMS has been developed for the ECP algorithm. LOGMIP is used for the solution of the problems with the disjunctive formulation. The objective is to analyze the performance of the different approaches solving several problems. The goal is to increase the knowledge about the formulation behavior and the algorithm to apply to the design of batch plants.

Keywords

MINLP algorithms, Multiproduct batch plants, Generalized disjunctive programming.

Introduction

Mathematical programming techniques for addressing problems in Process Systems Engineering are used extensively for industrial and academic problems. Many applications in the synthesis and process design require the treatment of both continuous and discrete variables. Due to the non-linear nature of the equations involved in the models, problems are modeled as mixed integer non-linear program (MINLP). The design and scheduling of batch processes is an area where that type of modeling is widespread used. These models assume an algebraic representation of the equations, and discrete variables are mainly restricted to 0-1 values. One reason for the increased use of MINLP models is the availability of different methods for solving this type of optimization problems. Branch and Bound, Generalized Benders Decomposition (GBD) and Outer Approximation (OA) are the more known algorithms for solving MINLP problems. New techniques and methodologies are investigated in order to find new and more robust algorithms to solve MINLP problems. Examples of recent research driven in this area are the Extended Cutting Plane by Westerlund an Pettersson (1995). MINLP models are difficult to solve. One important issue to reach the solution is to provide an efficient model for the discrete decisions. Raman and Grossmann (1994) have proposed a modeling framework in which the discrete-continuous problem is modeled as a generalized disjunctive program. In this approach the mixed-integer logic is represented through disjunctions and integer logic through propositions. Turkay and Grossmann (1996) have proposed an algorithm to solve the disjunctive formulation. In this work is shown that the main advantages of generalized disjunctive programs for structural flowsheet optimization are its robustness and computational efficiency when compared to algebraic MINLP models and algorithms.

The above modeling schemes provide several alternatives and solution methods for the same problem. Depending on the representation selected, the computational efficiency and robustness to achieve the solution can be greatly affected. There are several issues in this area that need to be investigated to increase the knowledge about the behavior of the models and algorithms to be applied to a particular process-engineering problem. In this work the idea is to explore that behavior for the structural design of batch plants.

Batch Design Model

The batch design model used includes the possibility of having batch units in parallel in phase and out of phase and also the allocation of intermediate storage tank. The problem formulation has the following assumptions:

- The plant consists of M batch stages, to produce P different products.
- At each batch stage j there are M_j units in parallel out of phase and N_j in phase with size V_j.
- Only monoproduct campaigns are admitted
- When a storage tank is not allocated, batches are transferred with the ZW (Zero Wait) policy.
- The production requirements Q_i for products i in the time horizon H are known.

Equations of the problem have been convexified through the following variable transformation:

$$b_{i,j} = \log B_{i,j}; \quad n_j = \log N_j; \quad m_j = \log M_j$$
$$e_i = \log E_i; \quad v_j = \log V_j; \quad vt_j = \log VT_j$$

where B_{ij} is the batch size for product i in stage j, VT_j is the intermediate storage tank size located between stage j and $j+1$, E_i is the inverse of the productivity rate. Then the MINLP problem formulation is as follows:

$$\text{Min } C = \sum_{j=1}^{M} a_j \exp(m_j + n_j + \alpha_j v_j) + \sum_{j=1}^{M-1} c_j \exp(\gamma_j vt_j) \quad (1)$$

subject to:

$$v_j \geq \log(S_{ij}) + b_{ij} - n_j \quad \forall i, \forall j \quad (2)$$

$$e_i \geq \log(T_{ij}) - b_{ij} - m_j \quad \forall i, \forall j \quad (3)$$

$$H \geq \sum_{i=1}^{P} Q_i \exp(e_i) \quad (4)$$

$$n_j = \sum_k coef_k \, yn_{kj} \quad \forall j \quad (5)$$

$$\sum_k yn_{kj} = 1 \quad \forall j \quad (6)$$

$$m_j = \sum_k coef_k \, ym_{kj} \quad \forall j \quad (7)$$

$$\sum_k ym_{kj} = 1 \quad \forall j \quad (8)$$

$$vt_j \geq \log(2S_{ij}^*) + b_{i,j+1} - K_{ij}(1 - y_j) \quad \forall i, \forall j = 1, M-1 \quad (9)$$

$$vt_j \geq \log(2S_{ij}^*) + b_{ij} - K_{ij}(1 - y_j) \quad \forall i, \forall j = 1, M-1 \quad (10)$$

$$b_{ij} - b_{i,j+1} \leq \phi' y_j \quad \forall i, \forall j = 1, M-1 \quad (11)$$

$$b_{ij} - b_{i,j+1} \geq -\phi' y_j \quad \forall i, \forall j = 1, M-1 \quad (12)$$

where C is the investment cost of the plant, a_j, α_j, c_j and γ_j are cost coefficients, S_{ij}^* are the size factors for intermediate storage tank, S_{ij} are the size factors for stage j product i, y_j are the binary variable whose value is 1 if a storage tank is allocated at position j, otherwise is 0; y_{nk}, y_{mk} are binary variables for selecting the number of units in parallel operating in and out of phase respectively. T_{ij} are constant values to express the batch time for product i at stage j. K_{ij} are constants large enough to formulate constraints (9) and (10) as Big-M type. ϕ' is equal to log ϕ, where ϕ is the maximum allowed for the ratio between the batch size up and down the storage tank (Ravemark, 1995).

Note that in the formulation presented all the constraints are linear with the exception of (4). In the following, this formulation will be Model I. Model I is a weakly non-linear convex model. In order to increase the non-linearity of the model the linear constraints (9) and (10) for sizing the storage tank were replaced with the following expression:

$$S_{ij}^*[\exp(b_{ij} - vt_j) + \exp(b_{i,j+1} - vt_j)] - K_{ij}(1 - y_j) \leq 1 \quad (13)$$
$$\forall i, \forall j = 1, M-1$$

With this replacement, Model II is generated, maintaining the convexity of the model.

The next step was to include in the formulation an expression for a variable batch time, so T_{ij} is defined by:

$$T_{ij} = \frac{T_{ij}^0 + T_{ij}^1 B_{ij}^{\delta_{ij}}}{M_j} \quad \forall i, \forall j \quad (14)$$

After convexifying, the final form of the constraint (3) is:

$$\frac{T_{ij}^0 + T_{ij}^1 \exp(\delta_{ij} b_{ij})}{\exp(e_i + m_j + b_{ij})} \leq 1 \quad \forall i, \forall j \quad (15)$$

Including (13) and (15) instead of (9), (10) and (3) we have the convex Model III.

Using expressions for estimating the size and time factors instead of the constant values of Model I, II, and III generated a non-convex formulation (Model IV). The constraints used in Model IV are: (2)-(8), (11)-(13) plus the equations (most of them non-convex) used for estimating the size and time factors as function of the process variables (Asenjo et al, 1999).

Solution Methods

Outer-Approximation (OA)

This method decomposes the original MINLP problem into two subproblems: the NLP subproblem where all binary variables are fixed to 0 or 1 values, and the master subproblem, where a linearization of the non-linear equations of the original MINLP is performed. These linearizations render a mixed-integer linear (MILP) subproblem. The objective of the MILP master subproblem is to predict the lower bound and the values of the binary variables for each major iteration. The objective of the NLP subproblem is to optimize the continuous variables and predict an upper bound. The search is terminated when the lower bound is equal or exceeds the upper bound. The starting point can be one of these two alternatives: provide an initial point for all binary variables, or execute the relaxed MINLP allowing the binary variables to take any value between 0 and 1. While the major iterations proceeds, the MILP master problems accumulate new constraints. These constraints consist of a set of linear approximations of the nonlinear constraints in the space of both binary and continuous variables. The original formulation of the OA has been improved. Viswanathan and Grossmann (1990) have proposed the Outer-Approximation/Equality-Relaxation/Augmented-Penalty Method (OA/ER/AP). In this method convexity conditions of the equations may not hold.

Extended Cutting Plane (ECP)

Westerlund and Peterson (1995) have proposed this method for solving convex MINLP. The authors claim this algorithm is suitable for large weakly non-linear convex MINLP, since the method is based only on solving a set of MILP problems. The algorithm starts giving an initial point for the continuous and binary variables, and then at each MILP solution the non-linear constraints are evaluated. The more violated constraint is linearized. This linearization is used as an approximation of the constraint. At the first iteration the linearization is added to the original set of linear constraints to generate the first MILP to be solved. While the iterations proceed linearization are added to the previous MILP. The convergence is reached when no more non-linear constraints are violated. An extension for non-convex problem was reported, but global convergence is no ensured. The method has the advantage that it does not solve NLP problems. In our case we have implemented this method linked to GAMS using DICOPT++ as a base.

Extension of the Logic-Based Outer Approximation

In the original version of the Logic-Based Outer Approximation, the algorithm was proposed mainly for synthesis of process networks. The disjunctive model was based on the modeling framework proposed by Raman and Grossmann (1994). Based on that work Vecchietti and Grossmann (1997) have proposed a hybrid modeling formulation for discrete-continuous non-linear problems for process system engineering. The model involved disjunctions, binary variables and integer or mixed-integer constraints. The model for the design of batch plants consists of the objective function (1) and the constraints (2)-(9). The constraints (10), (11), (12) and (13) of Model I have been replaced by the following set of disjunctions:

$$\begin{bmatrix} Y_j \\ vt_j \geq \log(2S_{ij}^*) + b_{ij} \\ vt_j \geq \log(2S_{ij}^*) + b_{i,j+1} \\ b_{ij} - b_{i,j+1} \geq \phi \\ b_{ij} - b_{i,j+1} \geq -\phi \end{bmatrix} \vee \begin{bmatrix} \neg Y_j \\ vt_j = 0 \\ b_{ij} = b_{i,j+1} \end{bmatrix} \quad (16)$$

In this case the Boolean variable Y_j replaces the binary variable y_j. If the Boolean variable is true (Y_j = True) it means that a storage tank will be allocated after stage j, then the left term of the disjunction apply. If the Boolean variable is false ($\neg Y_j$ = False) then no storage tank will be allocated after stage j. Therefore, the batch size of stage j is the same than stage $j+1$, and the size of the storage tank at that position is zero. The algorithm for solving this problem is described in Vecchietti and Grossmann (1997).

Results

Two batch design examples were selected, the first consists of six batch stages and five products. The data for this example were taken from Ravemark (1995). This example was formulated in two forms: Model I and II. The results obtained with this example are shown in table I. The second example corresponds to a batch plant design of a protein production plant. It consists of 8 batch stages and four products: Cryophilic Protease, Chymosin, Insuline and Vaccine for Hepatitis B. This example was formulated in two forms convex and non-convex. For the convex model constant time and size factors were used, the data can be extracted from Montagna et al. (1998). Model II and III were used to solve this problem, the results obtained are in Table II. For the non-convex model, the estimation of the time and size factors was made by non-convex equations of the process variables. The data for this example can be extracted from Asenjo et al. (1999). This problem was modeled as Model IV. The results for this example are also in table II. The examples have been executed on a Pentium II PC 266 MHz.

The solution of the protein plant for Model II is quite different of the others (Model III and IV) because we have guessed poor constant values for the time factors. The aim on the solution of Model II was to perform a comparison on the solution algorithms.

From tables 1 and 2 it is obvious that the performance of the disjunctive formulation and algorithm is the best in the execution time for all the cases solved. The performance is better when the complexity of the problem increases. It is important to note that the solution reached for the non-convex case (Table 2, Model IV) is worst than the one obtained with DICOPT++. One reason can be the starting point. DICOPT++ starts with the relaxed problem. This is not possible with the Logic-Based algorithm. The performance of the ECP algorithm is comparable the other solvers when the problem is weakly non-linear and the amount of binary variables is not large (see the execution time for Table 1 Model I and for Table 2, Model II). For the Model III and Model IV, after 20 iterations the execution was interrupted before reaching the solution, because the convergence of ECP was slow. At that point the execution time for ECP was high comparing to the other solvers. In general the performance of DICOPT++ remains in the middle between the other two methods in execution time.

References

Asenjo J., O. Irribarren, M. Montagna, A. Vecchietti, and J. Pinto (1999). Process performance models in the optimization of protein production plants. *Biotech. Progress* (submitted).

Montagna J. M., O. Irribarren, A. R. Vecchietti, J. A. Asenjo and J. M Pinto (1998). Optimal design of protein production plants with time and size factor process models. *Biotech. Progress* (submitted).

Raman R. and I.E. Grossmann (1994). Modeling and computational thecniques for logic based integer programming. *Comp. Chem. Eng.*, **18**, 7, 563-578.

Ravemark, E. (1995). *Optimization Models for Design and Operation of Chemical Batch Processes*. Ph. D. Thesis.

Turkay M. and I.E. Grossmann (1996). Logic-based algorithms for the optimal synthesis of process networks. *Comp. Chem. Eng.*, **20**, (8), 959-978.

Vecchietti A. and I.E. Grossmann (1997). LOGMIP: A disjunctive 0-1 nonlinear optimizer for process system models. *Comp. Chem. Eng.*, **21**, 427-432.

Viswanathan J.V. and I.E. Grossmann (1990). A combined penalty function and outer-approximation method for MINLP optimization. *Comp. Chem. Eng.*, **14**, (7), 769-778.

Westerlund and Pettersson (1995). An extended cutting plane method for solving convex MINLP problems. *Comp. Chem. Eng.*, **19**, S131-36.

Table 1. Results Obtained Solving the Example from Ravemark.

Model	Algorithm	# equations	# nonlinear equations	# variables	# binary variables	Objective function	Iterations	Time
I	OA/ER/AP	187	2	113	53	262483	7 major	34"
I	ECP	188	2	114	53	262483	28 MIP	35.2"
I	Logic	187	2	113	47	262483	3 major	21.9"
II	OA/ER/AP	162	31	113	53	261729	7 major	43"
II	ECP	163	31	114	53	261729	29 MIP	178"
II	Logic	162	31	113	47	261729	3 major	30.5"

Table 2. Results Obtained Solving the Batch Design of a Protein Production Plant.

Model	Algorithm	# equations	# nonlinear equations	# variables	# binary variables	Objective function	Iterations	Time
II	OA/ER/AP	183	33	180	103	585642	10 major	77"
II	ECP	184	33	181	103	585642	9 MIP	65.67"
II	Logic	183	33	180	97	585642	3 major	35.3"
III	OA/ER/AP	183	65	113	103	853726	11 major	258"
III	ECP	184	65	114	103	not reached	20 MIP	771"
III	Logic	183	65	113	97	853726	5 major	25.4"
IV	OA/ER/AP	285	126	297	87	809320	19 major	1532"
IV	ECP	286	126	298	87	not reached	20 MIP	2512"
IV	Logic	285	126	297	80	819726	3 major	53.4"

PRODUCTION SIMULATION FOR PIPELESS BATCH PROCESS DESIGN AND OPERATION

Hirokazu Nishitani
Nara Institute of Science and Technology
Nara 630-0101, Japan

Tadao Niwa
Asahi Engineering Co., Ltd.
Tokyo 108-0075, Japan

Abstract

The dynamics of the pipeless batch plant are very complex because they include many design problems which require solving such as determining the number of vessels, functional stations and AGVs. The layout of stations, vessel moving rules and job scheduling are also problems that need to be considered. In this paper, the systems approach using modeling and simulation is applied to system analysis aiming at optimal design of the plant. The dynamic behavior of the plant is modeled as a discrete event system, where the resources are mobile vessels, operational station, and AGVs. Modeling is achieved by describing the sequence of events with the corresponding actions for resource management. These models are embedded in a realtime application software for production simulation of the plant. A design support system composed of modeling, simulation and evaluation subsystems is also developed and its usefulness is assessed.

Keywords

Pipeless batch plant, Discrete event system, Production simulation, Design-by-analysis, Design support system.

Introduction

Batch processing using mobile vessels was developed to achieve a flexible and effective production system for the multi-product variable-quantity production of specialty chemicals. The concept was created by decomposing the functions of unit operations in the conventional batch processing plant from an operational viewpoint and reorganizing the batch cycle as a sequence of operational tasks. In this production system, each mobile vessel can produce a different product by moving successively to necessary stations, which achieve operational tasks according to the product recipe. This system not only eliminates the heavy chore of pipe cleaning, but also brings greater production flexibility. Consequently, this production scheme enables us to put new products on the market from the post R & D stage in the shortest amount of lead time (Niwa, 1991; 1993; 1994).

The pipeless batch process concept was proposed in the early 1980's. Soon, base technologies of hardware and software to realize the new concept were developed. The first plant was started up in 1986. So far more than 20 pipeless batch plants have been constructed in Japan. Most of them are for the production for paints, inks and industrial chemicals (Yamashita, 1997). The plants have been increasing in size and the largest one at present consists of approximately 30 stations and 20 mobile vessels. However, most of them were designed carefully using many heuristics. Optimizing all the design factors such as the number of vessels, functional stations and AGVs, the layout of these stations, vessel moving rules and job scheduling requires evaluation through appropriate modeling and simulation. Such simulation can give an accurate estimate of production time required and production rate for new products and the designer can recognize obstacles that need attention. System analysis through repeated simulation runs enables us to obtain an optimal choice for various design factors (Niwa, Kataoka & Nishitani, 1998a).

Outline of the Pipeless Batch Plant

Unit operations are a useful concept for the process design of continuous processes. For batch processes, however, unit operations should be rearranged according to actual operational tasks. For example, operational tasks such as the transporting, measuring and mixing of fluids can be lumped together into a raw material charging unit. Fluid transportation, heat transfer, and reaction shall be placed together into a reacting unit. The units based on the operational tasks are referred to as functional stations. Each functional station should be independent from the other functional stations. On the other hand, mobile vessels are inherently independent from each other. So each vessel can select stations based on the recipe of a product. Figure 1 illustrates a new structure of batch processes using mobile vessels. This scheme is called a pipeless batch plant because there is no pipeline between units. The pipeless batch plant is also called the registered trademark such as AIBOS and MILOX.

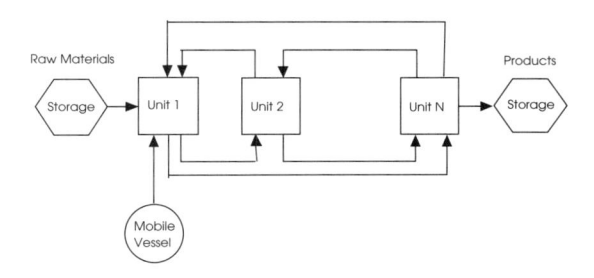

Figure 1 Batch process plant using mobile vessels.

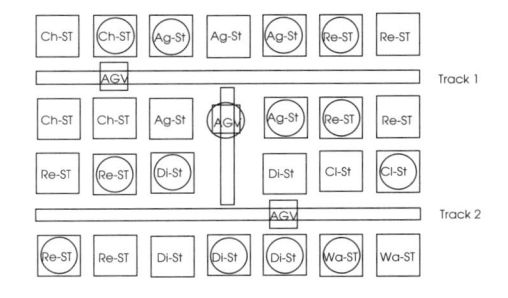

Figure 2. An example of the pipeless batch plant.

Figure 2 illustrates a typical plant configuration with two tracks for 20-30 stations. In this figure, a functional station is represented by a square and a mobile vessel by a circle. A mobile vessel is transported by the automated-guided-vehicle (AGV) to the destination station where the assigned functional tasks are implemented to the commodity in the vessel. Operation of the pipeless batch plant is achieved according to the production plan of the day. In other words, a set of batch production jobs is determined daily. The set of batch cycle processes, the sequence of these processes, and the required time for each process are assigned to a job. An example of the cycle time of a job is shown as follows:

Charging(30min)--Agitating(100min)--
Reacting(300min)--Discharging(160min)--
Cleaning(40 min)

Modeling and Simulation of the Pipeless Batch Plant

The system performances of this production system depend on both equipment design and operational strategies. This makes it difficult to optimize the pipeless batch plant analytically. For example, a set of batch production jobs is decided to achieve the production plan of the day. First, we must determine the production schedule of these jobs. Each job to produce a product requires a mobile vessel and a set of functional stations according to its recipe. We can assign equipment to each job either at a specified time or at a step when it is needed. The commodity in the mobile vessel receives treatment at each functional station. The mobile vessel is transported by an AGV on a track. In some cases, changes of AGVs are necessary. Thus, there are many design factors such as the plant configuration, the number of vessels, stations and AGVs, the layout of stations, the vessel moving rules, job scheduling and daily operation hours (Niwa, Kataoka & Nishitani, 1998b).

Basically, the pipeless batch plant is regarded as a discrete event system, of which resources are mobile vessels, functional stations and AGVs. So the dynamic behavior of a batch production system is described by a sequence of events, which means there are beginnings and endings of activities such as batch cycle processes, mobile vessel moving, and services at functional stations (Akatsuka, Furumatsu & Nishitani, 1997).

The following is the basis of modeling of the pipeless batch plant: A batch job needs to reserve necessary resources. An activity will start by registering the successive event information. In cases when resources cannot be reserved, the event is put into the corresponding queue and the event must wait until the resources become available. In the pipeless batch plant, there are three kinds of resources; mobile vessels, functional stations, and AGVs. A corresponding queue must be prepared for each resource. In our study, a job reserves a mobile vessel, the next station, and AGVs in this order. Actions such as reserve/catch/release resources are described to proceed the sequence of events. Modeling resembles programming of sequential control using the sequential function chart.

The outline of modeling of the pipeless batch plant is summarized as follows:

(1) Job generation: A set of daily jobs is generated and all the elements enter into vessel-queue.
(2) Vessel reservation (Job starting): A job is selected out of the vessel-queue and it reserves a vessel. The job with a reserved vessel enters into the queue for the next station.
(3) Station reservation: A job with a reserved vessel is selected out of the next station queue and it reserves the following station. The job enters into the AGV-queue.
(4) AGV reservation: A job, for which both a vessel and a next station are reserved, is selected out of the AGV-queue and it reserves an AGV. If the next station is on the other track, all necessary AGVs for transportation are reserved.
(5) AGV catch: The job catches the reserved AGV. The AGV moves to the present station. The vessel gets on the AGV. The job releases the present station.
(6) Vessel transportation: The AGV with the vessel moves to the next station reserved.
(7) AGV release: The vessel gets off the AGV and the job catches the reserved station. The job releases the AGV.
(8) Operation at station: Operational tasks at the station start and finish after a specified time.

In the same way, station reservation is repeated until the last operational task of a job is completed. When the entire batch cycle of a job is finished, the job releases the vessel. When a job reserves a resource, some practical solutions are used in the industrial plant. In our study, these heuristics were acquired and represented by a search algorithm for a mobile vessel, the next station and the AGVs. These control algorithms determine a sequence of discrete events.

Development of a Design Support System

The plant performances depend on both hardware of equipment and software of operation. A practical design method known as design-by-analysis was applied to solve such a complex design problem. Requirements for a design support system are summarized as follows: Design process includes changes of various design factors. So these changes must be easily embedded into the model. Also many heuristics are used for the operation of mobile vessels, functional stations and AGVs. These rules must be described easily in modeling. Effective representation of simulation results is also an important requirement for design-by-analysis. Animation of production simulation enhances evaluation availability of the designer. Taking these requirements into account, we used a realtime application tool, G2.

The design support system is composed of modeling, simulation and evaluation subsystems, and whose structure is shown in Fig.3. The minimum number of mobile vessels, and the kinds of functional stations and AGVs can be calculated under a simple assumption of full operation. Based on this information, an augmented plant configuration is made. Then, a proposed plant is displayed with a diagram, which is made by putting icons of functional stations in the augmented plant configuration. The results of simulation, i.e., the sequence of events occurred in the plant, are displayed as animation. This visual information is effective to show the changes of the plant state to the designer.

In general, a plant composed of equipment using the minimum number under the ideal condition cannot satisfy the specified production rate. From the simulation results, the designer must find bottlenecks in the proposed plant. For this purpose, we defined four states of each functional station; (1) A vessel with a job is achieving the task at the station. (2) A vessel with a job is waiting for the next station after completion of tasks at the station. (3) A vessel with no job is staying at the station. (4) No vessel is staying. The ratio of each state gives the designer hints to locate the bottlenecks by looking at the queue length of each type of stations. We can use both the historical data and averaged data of these indices. These results are displayed both graphically and as data values. Figure 4 is an example of display for design evaluation.

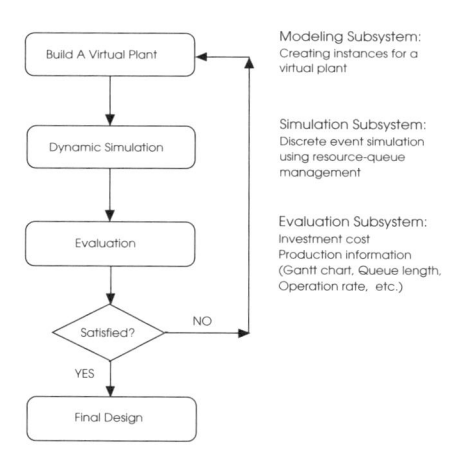

Figure 3. Structure of a design support system.

Illustrative Example

Design data for nine products is summarized in Table 1. The minimum number of mobile vessels and functional stations under full operation is calculated as follows:

Mobile vessels (22), Charging Station (2), Agitating Station (5), Reacting Station (10), Discharging Station (5) Cleaning Station (2). This data is used as an initial design. All the stations of the initial design are put on the augmented plant configuration and production simulation is achieved under a set of given conditions. An interactive design process using design-by-analysis is represented by the dendrogram as shown in Fig.5. As the final design, case-8 with additional one agitation station and one waiting station was selected from a cost-minimum viewpoint.

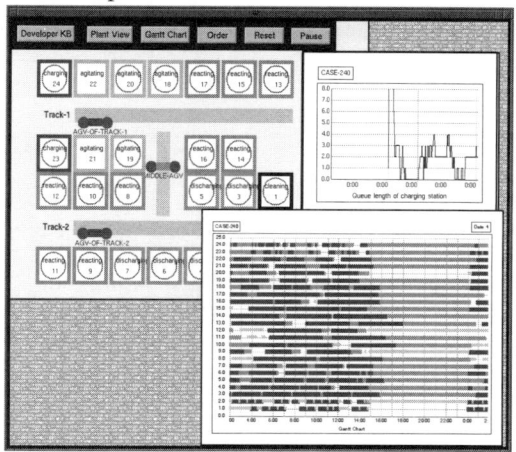

Figure 4. Display for design evaluation.

Table 1. Design Data for Nine Products.

Product (Seq. #)	Production Rate [m3/day]	Time Required at Each Station [min]					Cycle Time [min]
(Batch Sequence 1)		Ch-ST	Ag-ST	Re-ST	Di-ST	Cl-ST	
(Batch Sequence 2)		Ch-ST	Re-ST	Ag-ST	Di-ST	Cl-St	
A (1)	18	30	100	300	160	40	630
B (1)	12	20	200	200	120	40	580
C (2)	18	30	300	100	160	40	630
D (2)	12	10	350	150	150	50	710
E (2)	12	40	150	-	100	50	340
F (1)	6	50	170	-	70	30	320
G (1)	12	20	150	250	30	10	460
H (2)	12	20	250	150	30	30	480
I (1)	6	40	-	-	100	30	170

Daily Operation Time: 900 [min/day]

Moving Vessel Volume: 3.0 [m3]

Relative Cost: Vessel 40, Ch-ST 30, Ag-ST 30, Re-ST 30, Di-ST 30, Cl-ST 50, Wa-ST 10

Conclusion

The pipeless batch plant was developed to achieve an effective production system for multi-product variable-quantity production for specialty chemicals. The system performances of this production system depend on both process design and operational strategies. In this study, modeling and simulation methods were developed aiming at system analysis and optimal design of the complex system. The pipeless batch plant is regarded as a discrete event system, where the resources are mobile vessels, operational stations, and AGVs. Modeling is achieved by describing the sequence of events with corresponding actions and related resource management. These models are effectively used in a realtime application software with the object oriented techniques. A design support system with modeling, simulation, evaluation subsystems was developed based on a practical design-by-analysis method. This system helps make a rational decision through performance evaluation for a proposed design and ensure the optimal system structure for the pipeless batch plant. The usefulness of the system was examined by solving some industrial examples.

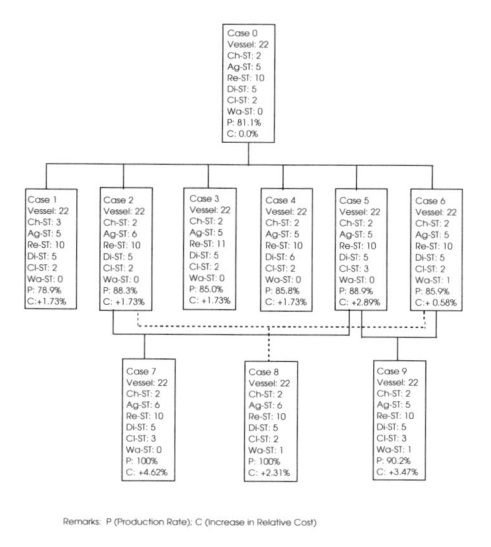

Figure 5. Dendrogram of a design process.

References

Akatsuka, T., N. Furumatsu, and H. Nishitani (1997). Modeling and simulation of combined continuous and discrete system. *J. Chem. Eng. Japan*, **30**, 2, 285-292.

Niwa, T. (1991). Transferable vessel-type multi-purpose batch process. *Proc. of 4th Symposium on Process Systems Engineering* (Montebello, Canada), **IV**, 2.1-2.15.

Niwa, T. (1993). Pipeless plants boost batch processing, *Chem. Eng.*, **100**, 6, 102-108.

Niwa, T. (1994). Evaluation of pipeless plant and recipe base operation. *Proc. of 5th Symposium on Process Systems Engineering* (Kyongju, Korea), 497-502.

Niwa, T., M. Kataoka, H. Nishitani (1998a). Modeling and simulation of pipeless batch process. *Kagaku Kogaku Ronbunshu*, **24**, 3, 437-444.

Niwa, T., M. Kataoka, and H. Nishitani (1998b). Development of design support system for pipeless batch plants. *Kagaku Kogaku Ronbunshu*, **24**, 4, 620-627.

Yamashita, S. (1997). Flexible production system for paints using mobile vessels. *Kagaku Kogaku*, **61**, 9, 653-655.

SEMI-CONTINUOUS OPERATION OF A MIDDLE-VESSEL DISTILLATION COLUMN

James R. Phimister and Warren D. Seider
Department of Chemical Engineering
University of Pennsylvania
Philadelphia, PA 19104-6393

Abstract

The semi-continuous operation of a middle-vessel column (MVC) is presented for the distillation of near-ideal and azeotropic ternary mixtures, extending semi-continuous distillation for the separation of pressure-sensitive homogeneous azeotropic pairs (Phimister and Seider, 1998). The MVC is a batch distillation column with an internal or external vessel that provides feed for and draws from the middle of the column.

Due to the two product streams, a MVC has shown superior performance to a typical batch process, where distillate cuts of the light and intermediate products are taken sequentially. Semi-continuous operation of the MVC retains this advantage, while not requiring that the column be discharged between batches. Compared to continuous processing, where two or three distillation towers are needed, investment costs are substantially lower, and downtime when changing between mixtures to be separated is often shorter. In this paper, three case studies, including two homogeneous azeotropic systems, are examined. In future work, comparisons of semi-continuous MVCs with batch and continuous distillations are planned.

Keywords

Distillation, Azeotropic distillation, Middle-vessel column, Distillation dynamics.

Introduction

Within the past decade there has been considerable interest in the operation of the middle-vessel column (Hasebe et al., 1996; Safrit and Westerberg, 1997; Farschman and Diwekar, 1998), with emphasis on batch operation. In this paper, the advantages of semi-continuous operation of the MVC are examined, including desirable attributes of continuous processing, that is, closed-loop control and the lack of column dumping, as well as advantages of batch processing, that is, process flexibility and lower investment costs. Simulation studies illustrate the flexibility of the MVC for the separation of near-ideal as well as azeotropic mixtures.

In a MVC, a large vessel is connected to trays in the vicinity of the middle of the column. The MVC simulated in this paper has a liquid sidedraw from the middle tray, which feeds the middle vessel, with liquid returned to the tray below the middle tray, as shown in Figure 1. The process feed is charged cyclically to the middle vessel, allowing for continuous operation of the column. Alternative MVC configurations have been proposed in which vapor from the middle tray is fed to the middle vessel, the entire liquid stream is drawn from the middle

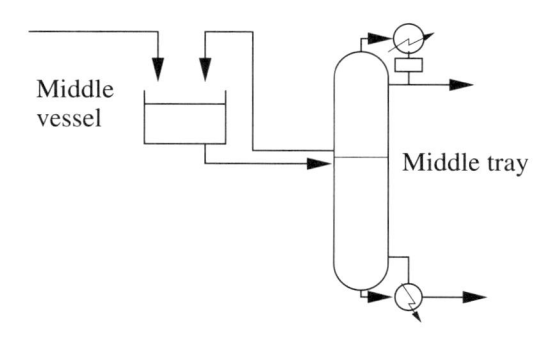

Figure 1. Middle-vessel column.

tray, and the middle vessel functions as a reboiler. The operating strategies presented in this paper appear to be generally applicable to these MVC configurations with similar results expected.

Three examples are presented that illustrate the MVC in semi-continuous operation. First, the ideal mixture of hexane, heptane and octane is separated into three species in high purity. Bottoms and distillate products of hexane and octane, respectively, are removed almost continuously. Heptane concentrates in the middle vessel and is discharged cyclically once the desired purity is achieved. Simulations for the semi-continuous distillation of two azeotropic mixtures are also presented. In the first, the acetone-methanol azeotrope is broken, using water as an extractive agent, by alternately concentrating acetone in the middle vessel and methanol in the distillate. In the second, a low-boiling azeotrope is separated using an intermediate-boiling entrainer. The light and heavy species are alternately discharged from the distillate and the bottoms, respectively, while the middle vessel is concentrated in the entrainer cyclically.

Figure 2 shows a configuration, involving a single column and five tanks, for the separation of all three mixtures, with streams S1, S3, S4, S5, S7, and S9, used in all of the separations, and the other streams used selectively. Tanks T2 – T5 are used in all of the separations with tank T1 employed when azeotropic mixtures are separated.

The column used in all of the examples consists of 25 trays, including the reboiler. A rigorous dynamic simulation using an implicit integrator and accounting for thermal and mass dynamics is performed according to the operating strategy prescribed. The column is loaded with an equimolar ternary mixture on each tray. T2 contains 100 m^3 of this mixture. The flow rate of stream S3 is set to 5 m^3/hr, and when the distillate or bottoms product is on specification, its flow rate is 2 m^3/hr. The flow rate of S4 is adjusted to maintain a constant molar holdup in the column. The column is fed on the 13th tray down, and the sidedraw, S4, is the liquid from the 12th tray. A campaign time of 200 hr is selected to show the cyclic behavior of all of the separations. Neither the campaign operating strategies nor the column parameters have been optimized, and hence, significant improvements in productivity are envisioned.

Middle-vessel Dynamics

The middle-vessel composition is affected both by thermodynamics and mass-transfer effects. The composition of the middle vessel moves towards the liquid composition of the middle tray (at vapor-liquid equilibrium), as the liquid sidedraw feeds the vessel. However, Cheong and Barton (1999a) observe that when negligible tray holdup is assumed relative to middle-vessel holdup, the composition of the middle vessel moves away from the resultant vector connecting the middle-vessel composition with the distillate and bottoms compositions, as shown in Figure 3. These two effects appear to result in paradoxical behavior because the middle-vessel composition moves both towards and away from the composition of the middle tray. To resolve this conflict, recognize that the middle-vessel composition must move towards the middle-tray composition, and hence, the tray and middle-vessel compositions must move away from the distillate and bottoms compositions.

Case Study A. Ideal Ternary Mixture

Mixtures containing heavy and light impurities can be purified in a MVC, with the heavy and light impurities removed in the bottoms and distillate, respectively, while the desired product concentrates in the middle vessel. For these mixtures, Hasebe and coworkers (1996) conclude that a batch MVC consistently provides superior performance compared to batch rectification followed by

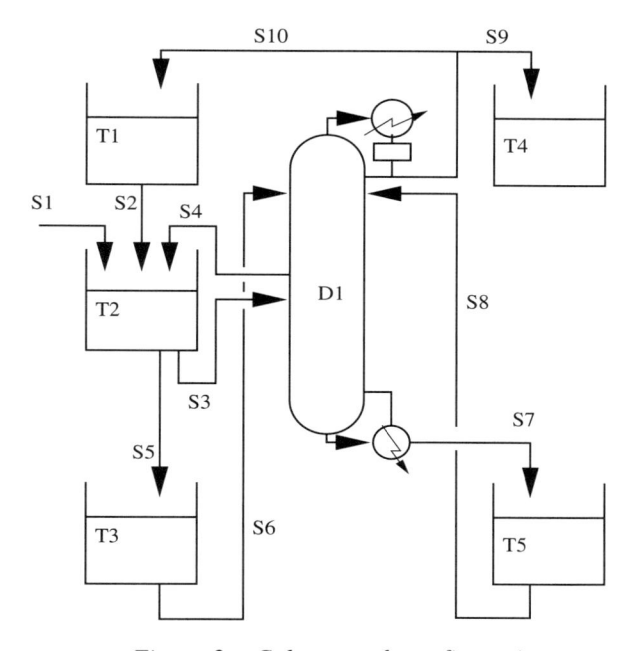

Figure 2. Column-tank configuration.

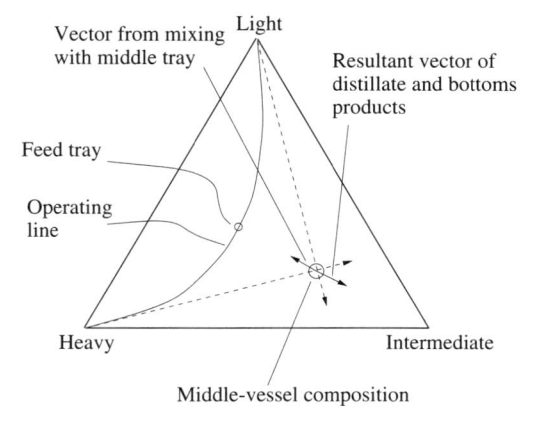

Figure 3. Composition dynamics in a middle-vessel.

batch stripping.

A similar ternary mixture consisting of equimolar n-hexane, n-heptane, and n-octane is examined. Its residue-curve map is shown in Figure 4A. In this process, the n-hexane and n-octane products are collected in tanks T4 and T5, respectively. Upon achieving 98 mol% n-heptane in tank T2, its contents are dumped into product vessel T3, and subsequently tank T2 is recharged using stream S1. This operating strategy is illustrated over the campaign by the profiles shown in Figures 5A1 and 5A2. Tanks T4 and T5 fill nearly continuously, though slowly as the middle vessel approaches the specified intermediate composition. At 35, 95, and 150 hr, the middle vessel is suitably concentrated in n-heptane, its contents are dumped into tank T3, and the vessel is recharged.

The distillate and bottoms product compositions do not vary significantly. Unlike batch processing, where manual control of the distillate and bottoms flow rates is often preferable, product compositions are controlled by PID controllers, by adjusting the reflux or reboil flow rates, respectively. Level controllers are used to maintain liquid inventories in the sump and reflux drum.

A *slop cut* is not removed in this operating strategy, though it would be attractive. During the first twenty hours, compared with the second twenty hours ($t = 20 - 40$ hr), the bulk of the n-hexane and n-octane products are removed. If, during the second twenty hours, a *slop cut* is removed with the distillate and bottoms product fed to tank T1, the middle vessel reaches its product specification sooner. After the contents of the middle vessel are dumped into tank T3, the contents of tank T1 and the process feed are added to tank T2 before this procedure is repeated.

Case Study B. Extractive System

Safrit and Westerberg (1997) separate a low-boiling azeotrope, acetone-methanol, using water as an extractive agent, in a batch MVC. Their operating strategy is adapted herein for semi-continuous operation. See Figure 4B for a residue-curve map for the acetone-methanol-water system.

The column and tank T2 are loaded with 100 m^3 of an equimolar mixture of acetone and methanol. Tank T5 is loaded with 1,000 kmol of water. The process operates in three modes: (1) Methanol concentrates in the middle-vessel as the distillate, which approaches the low-boiling azeotrope of methanol and acetone, is fed to tank T1, and water is collected in tank T5. When the methanol purity is achieved in tank T2, its contents are dumped into tank T3; (2) Having emptied tank T2, the contents of tank T1 are dumped into tank T2; hence, the feed to the column is near the azeotropic composition. Water, the extractive agent, is fed to the top of the column in stream S8 and acetone distillate is fed to tank T4. The middle vessel fills with a water-methanol mixture; (3) After sufficient acetone is removed, the cycling of water is discontinued and water is collected in tank T5. A near-azeotropic distillate is removed in stream S10, and tank T2 becomes concentrated in methanol, as in mode 1. When methanol is concentrated sufficiently, the contents of tank T2 are dumped, the tank is recharged with the equimolar feed, and the process returns to mode 1.

It is desirable to send the near-azeotropic distillate to tank T1, rather than operate the column at infinite reflux. While the near-azeotropic distillate is removed, the

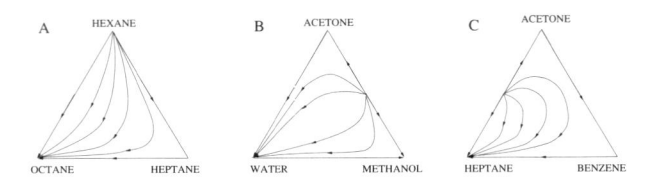

Figure 4. Residue-curve maps for case studies.

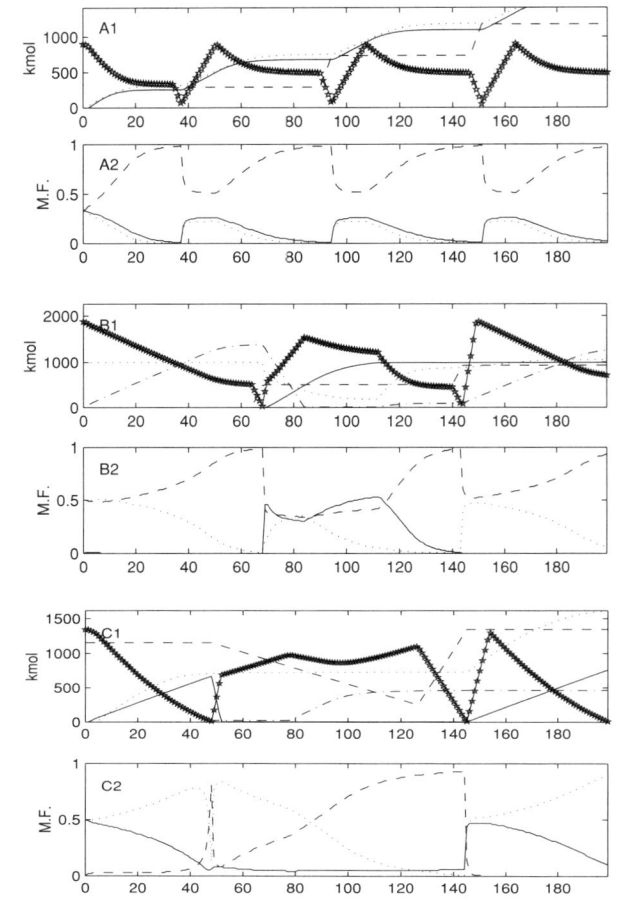

Figure 5. Case study profiles. A1, B1 and C1: tank molar holdups for case studies A, B, and C, respectively. Line styles: $-\cdot-$ T1, *** T2, $--$ T3, ____ T4, $\cdots$ T5. A2, B2, C2: middle-vessel mole fractions for case studies A, B, and C, respectively. A2: ____ hexane, $--$ heptane, $\cdots$ octane. B2: ____ water, $--$ methanol, $\cdots$ acetone. B3: $\cdots$ acetone, $--$ benzene, ____ heptane. Abscissa in hours.

middle-vessel composition is concentrated in methanol.

The profiles in Figures 5B1 and 5B2 illustrate that the process is successfully operated semi-continuously. Tank T2 is dumped upon achieving 98 mol% methanol at 60 and 120 hr. The extractive mode (2) and the charging mode (3) follow. Prior to recharging tank T2 with the process feed, observe that tanks T3 and T4 have obtained similar molar holdups of their respective products, and hence, the feed has been separated.

Provided that the liquid holdup within the column is small, the process quickly transfers from the concentration of methanol in the middle vessel (modes 2 and 3), to the removal of acetone in the distillate (mode 1). This is possible since the distillate mole fraction of acetone changes only from 0.78 to 0.98, while the mole fraction of water in the bottoms product remains greater than 0.98. Observe that, soon after the contents of tank T2 are dumped at $t = 65$ hr, tank T4 begins to fill with acetone.

Similar to Case Study A, a PID controller maintains a constant bottoms temperature (inferred composition), ensuring that the bottoms product remains concentrated in water. This set point can be maintained throughout modes 1-3. During the extraction mode (2), the middle vessel fills with a mixture of methanol and water. To avoid any unnecessary mixing, acetone is recovered from this mixture (in mode 3), prior to re-charging the middle-vessel with the process feed.

Case Study C. Intermediate-boiling Entrainer

An intermediate-boiling entrainer, benzene, can be used to continuously separate a low-boiling binary azeotrope of acetone and n-heptane (Laroche et al. (1992)). The residue-curve map is shown in Figure 4C. A similar system, using an intermediate-boiling entrainer to separate a minimum-boiling azeotrope in a MVC, is presented by Cheong and Barton (1999b). However, the operating strategy presented herein, differs considerably from the strategy advocated in their paper.

In the MVC configuration (Figure 2), all entries with the exception of stream S8 are employed. As in Case Study B, the process operates in three modes: (1) An equimolar charge in tank T2 is fed to the column and the bottoms product, n-heptane, is accumulated in tank T5. A near-azeotropic mixture of acetone and n-heptane collects in tank T1; (2) the contents of vessel T1 are dumped into tank T2, which feeds the column. Benzene, in tank T3, is fed to the top of the column, and is recovered in the middle vessel. The distillate is nearly-pure acetone, which is collected in product vessel, T4. Some n-heptane is removed in tank T5 while a benzene-heptane mixture is accumulated in middle vessel T2; (3) the middle vessel becomes concentrated in the entrainer. Its contents are dumped into tank T3, and operation returns to mode 1.

The simulation profiles are shown in Figures 5C1 and 5C2. During the first forty hours of the campaign, n-heptane is accumulated in tank T5 while the middle vessel is emptied. In mode 2, the middle vessel is charged with the near-azeotropic mixture and the entrainer, benzene, is fed near the top of the column. Observe that there is an undesirable delay in removing the acetone product, from $t = 50 - 80$ hr. Optimization of the entrainer flow rate and feed tray position should reduce this time. Also, rather than operate at infinite reflux until the distillate achieves the desired purity, removal of a slop cut with the distillate recycled to the middle vessel should reduce this time. At 125 hr, the distillate and bottoms flow rates are zero, and the middle vessel is concentrated in the entrainer. The contents of the middle vessel are discharged, after which it is recharged with the equimolar acetone–heptane feed.

As in Case Studies A and B, PID-control of the bottoms composition is desirable since it remains nearly constant throughout modes 1-3. Also, the mole fraction of acetone varies from 0.92 to 0.98, and hence, the composition profiles swing quickly as operation shifts from mode to mode. Lastly, since the entrainer remains within the system, a high purity of entrainer in the middle vessel is *not* required, though it is desirable to reduce the n-heptane composition in the middle vessel.

Conclusions

Three case studies are presented that introduce the semi-continuous operation of a middle-vessel column. Semi-continuous operating strategies provide for the complete separation of three types of ternary mixtures in a configuration that involves a single distillation column and five tanks. The strategies apply to the separation of azeotropic mixtures.

References

Cheong, W., and P.I. Barton (1999a). Azeotropic distillation in a middle vessel batch column. I. Model formulation and linear separation boundaries. *Ind. Eng. Chem. Res.*, **38**, 1505-1530.

Cheong, W., and P.I. Barton (1999b). Azeotropic distillation in a middle vessel batch column. II. Nonlinear separation boundaries. *Ind. Eng. Chem. Res.*, **38**, 1531-1548.

Farschman, C.A. and U. Diwekar (1998). Dual composition control in a novel batch distillation column. *Ind. Eng. Chem. Res.*, **37**, 89-96.

Hasebe, S.T., T. Kurooka, B.B.A. Aziz, I. Hashimoto, and T. Watanabe (1996). Simultaneous separation of light and heavy impurities by a complex batch distillation column. *J. Chem. Eng. Japan.* **29**, 1000-1006.

Laroche, L. N. Bekiaris, H.W. Andersen, and M. Morari (1992). The curious behavior of homogeneous azeotropic distillation – implications for entrainer selection, *AIChE J.*, **38**, 1309-1328.

Phimister J.R., and W.D. Seider (1999). Semi-continuous, pressure-swing distillation. Submitted to *Ind. Eng. Chem. Res.*

Safrit, B.T., and A.W. Westerberg (1997). Improved operational policies for batch extractive distillation columns. *Ind. Eng. Chem. Res.*, **36**, 436-443.

PROCESS PERFORMANCE MODELS IN THE OPTIMAL DESIGN OF PROTEIN PRODUCTION PLANTS

José M Pinto
Universidade do Sao Paulo
Sao Paulo SP 05508-900 Brazil

Juan A. Asenjo
Universidad de Chile
Santiago de Chile, Chile

Jorge M. Montagna, Aldo R. Vecchietti and Oscar A. Iribarren
INGAR- Instituto de Desarrollo y Diseño
3000 Santa Fe, Argentina

Abstract

In this work we propose a process performance model to simultaneously optimize the design and the operation of multiproduct batch plants. While the constant time and size factors model is the most widespread used to model multiproduct batch processes, the process performance models describe these time and size factors as functions of the process variables selected as optimization variables. Thus the same mathematical model for the plant is used in both approaches, with the process performance models as additional constraints for the second case. The process performance models are obtained from the mass balances and kinetic expressions that describe each unit operation. A plant for producing human Insulin, Vaccine for hepatitis B, Chymosin and a cryophilic Protease is used as an example. It is shown on that example that the additional degrees of freedom introduced by the process performance models render a superior design. The contribution of this approach lies in its inherent modularity, which removes the bound on the size of the problems that can be solved. Given a set of values for the process decision variables, the process performance models are able to predict a set of fixed size and time factors. This makes a great difference with respect to the global solution approach for which the problem size rapidly becomes bound, thus making the solution impractical.

Keywords

Multiproduct batch plants, Design, Optimization, Process performance models, Protein production.

Introduction

The design and structural optimization of multiproduct batch plants have been widely investigated in recent years. The aim is to determine the plant configuration and the equipment sizes that minimize the capital cost. The most common strategy for solving the problem is to consider constant values for the size and cycle time factors. These values are obtained from laboratory or pilot plant data. When obtaining these values only the product under consideration is evaluated, no interactions with the other products are analyzed. If these factors are used in the design of a multiproduct batch plant the performance of the optimal design is reduced. Another approach is to incorporate process information into the design by predicting the size and time factors through process performance models for the unit stages (Salomone and Iribarren, 1992). The process performance models define the size and time factors as functions of the process variables selected to optimize the plant. The use of process decision variables in the model allows a better representation and approximation for the size and time factors allowing a more accurate design of a multiproduct batch plant.

In this work a plant consisting of 8 stages for producing Human Insulin, Vaccine for Hepatitis B, Chymosin and Cryophilic Protease by genetically engineered *Saccharomyces cerevisiae* was modeled and optimized. Process performance models for size and time factors have been introduced. An example of how these models have been obtained is described in the paper. For the design and structural optimization of the multiproduct batch plant, a modular model that includes the capability of having parallel units in and out-of phase, semicontinuous units, and intermediate storage tanks has been considered. The results obtained with the process performance models have been compared with the traditional approach considering constant size and time factors. The model developed is a non-convex MINLP solved with DICOPT++ (Viswanathan and Grossmann, 1990). The design obtained with the process performance models renders a superior solution than the traditional approach.

Process Description

Figure 1 shows the flowsheet of a multiproduct batch plant intended for the production of Human Insulin, Vaccine for Hepatitis B, Chymosin and a cryophilic Protease, produced by *Saccharomyces cerevisiae*.

All these are proteins produced as the cells grow in the fermentor. The Vaccine and Protease are intracellular; hence, in these two cases the first microfilter is used to concentrate the cell suspension, which is then sent to the homogenizer for cell wall disruption to liberate the intracellular proteins. The second microfilter is used to remove the cell debris from the solution of proteins.

The ultrafiltration prior to the extractor is used for concentrating the solutions in order to minimize the extractor volume.

Ultrafiltration is used again for concentrating the solution (in case this is required before the chromatographic steps), and finally the last stage is a chromatography where selective binding is used to further separate the product of interest from other proteins.

Insulin and Chymosin are extracellular products. In both cases the product protein is separated from the cells in the first microfilter. Cells and some of the liquid supernatant stay behind. In order to reduce the amount of valuable product lost in the retentate, extra water is added to the cell suspension. The filtration operation with make up water is also called diafiltration and dilutes the solution of proteins.

The homogenizer and microfilter for cell debris removal are not used when the product is extracellular; however, the ultrafilter is necessary to concentrate the dilute solution prior to extraction. The final steps of extraction, ultrafiltration and chromatography are common.

Rather than using all the available detail for each product, first level process performance models will allow preliminary estimates on the economic viability of this multiproduct facility such as equipment size required, idle times and percentage usage of units by each product. More important than detail is the consistency of the data used concerning the demand of each resource by each product.

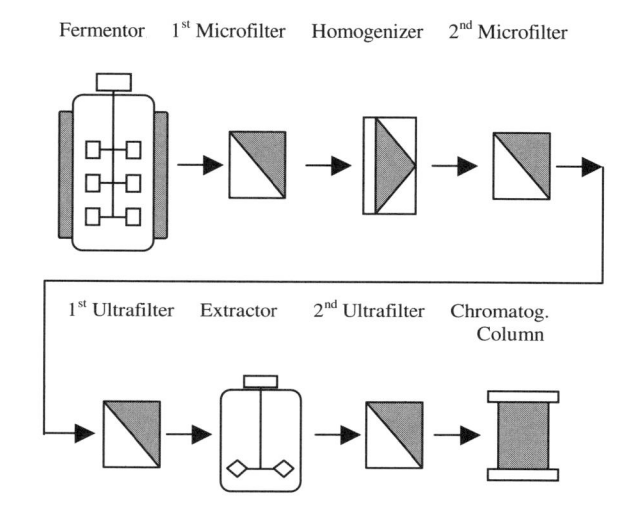

Figure 1. Batch plant flowsheet for protein production.

Process Performance Models

Using constant size and time factors gives rise to a posynomial model for the process, as the one presented in our previous paper (Montagna et al., 1998). To get these constant factors it is necessary to guess a value for every process variable so as to cover the degrees of freedom of the process mass balances. In the present paper we use process performance models retaining the influence of the process variables with the largest impact in the process economy, in line with Douglas (1988).

Once these variables have been selected, we write the mass balances and kinetic equations that describe each stage estimating constant values for every non-selected process variable, but not for the chosen process variables that will be further optimized. As a result, we get analytical expressions for the size and time factors that will be functions of these process variables.

In conceptual terms, the mathematical optimization model for the design of the multiproduct batch plant will be exactly the same as the one presented by Montagna et al (1998) if size and time factors were constant, *plus* the additional constraints describing these factors as functions of the process variables. As a result, it is expected that the introduction of these new degrees of freedom into the optimization model will produce a better design.

The process variables that have been selected as optimization variables are: the biomass concentration at the fermentor $(X_{i,fer})$ and the microfilter 1 $(X_{i,mf1})$ for all products, the volumetric ratio of diafiltration water to suspension feed at microfilter 1 $(W_{i,mf1})$ for extracellular Insulin and Chymosin and at microfilter 2 $(W_{i,mf2})$ for intracellular Vaccine and Protease, the number of passes through the homogenizer (NP_i) for intracellular Vaccine and Protease, and the volumetric ratio (R_i) of PEG to Phosphate phases at the extractor for all products. The subscript i means the product i

Due to the reduced space available for the paper, in the following section we will describe the steps involved on the generation of the process performance model only for the chromatographic column. The models for the rest of the stages and products are produced in a similar form.

Chromathographic Column

The batch volume in BV_{in} to this stage is the batch volume out BV_{out} from ultrafilter 2 that are given by the following equation for i=Insulin and Chymosin:

$$BV_{in} = \frac{X_p^{in}\,\eta_{i,ext} + X_c^{in}\,\dfrac{R_i}{(R_i+1)^2}}{50\,X_p^{in}\,\eta_{i,ext}\,\eta_{i,chr}} B_i \quad (1)$$

where $\eta_{i,ext}$ is the extractor yield for product i, $\eta_{i,chr}$ is the yield of the chromathographic column, X_p^{in} is the product concentration at the input and X_c^{in} is the contaminant concentration at the input, B_i is the batch size for product i after the chromathographic column. For i=Vaccine and Protease we have:

$$BV_{in} = \frac{X_p^{in}\,\eta_{i,ext} + X_c^{in}\,\dfrac{R_i}{(R_i+1)^2}}{50\,X_p^{in}\,\eta_{i,ext}\,\eta_{i,chr}\,\exp(-0.03\,NP_i)} B_i \quad (2)$$

These expressions divided by the respective B_i and multiplied by 1.25 to allow 80% occupancy are the size factors for the chromatographic column vessel for all products.

We assume that the chromatographic column works at a constant linear velocity of:

$$v = 4 m/h \quad (3)$$

and that the packing has a binding capacity:

$$bc = 20\ kg/m^3 \quad (4)$$

Just a percent of this maximum capacity is used, to avoid excessive product breakthrough. A 50% capacity usage was assumed, and that this % leads to a yield for this stage and product i:

$$\eta_{i,chr} = 0.95 \quad (5)$$

Then the size factor is given by:

$$S = \frac{1}{0.5 \cdot bc} = 0.1 \frac{m^3}{kg} \quad (6)$$

A column height of 0.5 m was assumed, which is large enough to allow high column resolution, still compatible with reasonable linear velocities. Then, the cross sectional area of the column (A) is:

$$A[m^2] = \frac{V_{chr}\,[m^3]}{0.5 \cdot m} = 2\,V_{chr} \quad (7)$$

The time required by the BV_{in} to pass through the column is:

$$T[h] = \frac{BV_{in}\,[m^3]}{A[m^2]\,v[m/h]} = \frac{BV_{in}}{8\,V_{chr}} \quad (8)$$

Elution plus washing-regeneration solution volumes were assumed to amount to 3 times the column volume, and the linear velocity for these processes to be the same *4m/h* as for loading. This gives a fixed amount of time:

$$T^0_{chr} = 0.375\ h \quad (9)$$

to be added to the time in (8). Replacing BV_{in} in (8) and adding the downtime in (9) gives the time expressions for the chromatographic column for all products.

Results

Assigning reasonable estimates for the process variables and replacing them into the equations of the process performance models, we have obtained values for the size and time factors that we fixed as constants. Then, we solved the models with these constant values. These estimated values are showed in Table 1. Afterwards we used the same values as initial points for optimizing the plant cost with the performance models. Both approaches have been solved with and without the allocation of intermediate storage tanks. It means that four mathematical models have been solved:

1. Fixed size and time factors with no intermediate storage tanks.
2. Process Performance models for size and time factors with no intermediate storage tanks.

3. Fixed size and time factors with the allocation of intermediate storage.
4. Process Performance models for size and time factors with the allocation of intermediate storage.

The process variable values obtained at the optimum with models II and IV are shown in Table 2.

From Table 3 it can be seen that the introduction of intermediate storage tanks leads to a considerable reduction in the cost of the plant. This reduction is obtained with both approaches (fixed factors or process performance models).

With the process performance models an additional saving in the cost of the plant is obtained comparing against the fixed factors. This reduction is around 15% ($ 1,505,326 vs. $ 1,770,418) with the model without intermediate storage, and 13,1% ($ 800,138 vs. $ 920,790) with tanks.

Table 1. Estimated Values of the Process Variables.

Product	$X_{i,fer}$	$X_{i,mf1}$	$W_{i,mf1}$	$W_{i,mf2}$	NP_i	R_i
Protease	50.	200.	-	1.50	3.0	1.0
Chimosin	50.	200.	1.25	-	-	1.0
Insulin	50.	200.	1.25	-	-	1.0
Vaccine	50.	200.	-	1.50	3.0	1.0

Table 2. Optimum Values for the Process Variables Obtained with Models II and IV.

		$X_{i,fer}$	$X_{i,mf1}$	$W_{i,mf1}$	$W_{i,mf2}$	NP_i	R_i
II	P	31.25	250.0	-	1.50	2.39	0.635
	C	39.00	248.7	0.10	-	-	0.635
	I	46.50	250.0	0.35	-	-	0.635
	V	38.50	250.0	-	1.50	2.36	0.475
IV	P	45.0	250.0	-	1.45	2.21	0.58
	C	49.1	250.0	0.20	-	-	0.63
	I	49.5	250.0	0.21	-	-	0.63
	V	45.0	250.0	-	1.45	2.21	0.58

P=Protease; C=Chymosin; I=Insulin; V=Vaccine

Table 3. Plant Cost Obtained with Four Models.

Model	I	II	III	IV
Plant Cost	1,770,418	1,505,326	920,790	800,138

The introduction of intermediate storage tanks reduces the number of fermentors in parallel (out-of-phase) from five to one in both approaches (constant factors and process performance), and the number of chromatographic columns in parallel (in-phase), from three to one for fixed factors, from two to one for process performance models. These two units are the most expensive of the plant. The allocation of intermediate storage tanks permits a reduction on the equipment volume, cost and idle times. The units operate with reduced size and larger cycle time, making a better utilization of their capacity.

The process performance models allow a further accommodation of the size and time factors allowing an extra reduction of the idle times and unit sizes.

Conclusions

A process performance optimization model for the design of a protein production batch plant has been developed. The model is a MINLP (Mixed Integer Non-linear Program). The model considers the inclusions of parallel units in-phase and out-of-phase, and the allocation of intermediate storage tanks. It is not trivial to estimate good constant time and size factors for the plant. This difficulty is overcome by using the proposed process performance models.

The performance model is non-linear and non-convex and so more difficult to solve than the fixed factors model which is a geometric program. A careful tuning of the initial points was necessary to reach the solutions. The optimal solutions obtained are local optima; it is not guaranteed to obtain the global optimum with these models.

We compared the results of the performance models against fixed size and time factors. The results obtained show that the process performance models give a better solution compared with the fixed factors approach, as expected because of the extra degrees of freedom that this model introduces. The improvement is in the order of a 15 % lower cost with reduced idle times of the equipment. As compared with a global solution approach (Bathia and Biegler, 1996) we expect that the modularity of the one presented here permits to solver much larger problems. That is because the global approach discretize the dynamics over finite time elements and then globally solve the resulting system of equations.

Acknowledgements

The authors would like to acknowledge financial support received from VITAE within the Cooperation Program Argentina-Brazil-Chile under grant B-11487/6B004.

References

Asenjo J. A (1990). *Separation Processes in Biotechnology*. Marcel Dekker, New York.

Bathia T. and L. Biegler (1996). Dynamic Optimization in the design and scheduling of multiproduct batch plants. *Ind. Eng. Chem. Res.*, **35**, 7, 2234-2246.

Douglas J. M. (1988). *Conceptual Design of Chemical Processes*. McGraw-Hill New York.

Montagna J. M., Vecchietti A. R., Iribarren O. A., Pinto J. M. and J. Asenjo (1998). Optimization of protein production plants with a time and size factors process model. Submitted to *Biotech. Progress*.

Salomone H. E. and O. A. Iribarren (1992). Posynomial modeling of batch plants. *Comp. Chem. Eng.*, **16**, 173-184.

Viswanathan J.V. and I. E. Grossmann (1990). A combined penalty function and outer approximation method for MINLP optimization. *Comp. and Chem. Eng.*, **14**, 769-782.

THERE IS MORE TO PROCESS SYNTHESIS THAN SYNTHESISING A PROCESS

David Glasser, Diane Hildebrandt, and Craig McGregor
Centre for Optimisation, Modelling and Process Synthesis
University of the Witwatersrand
Johannesburg, South Africa

Abstract

Traditionally new processes have been developed in a sequential way, that is an opportunity is identified, the process chemistry is investigated and finally the information is handed over to the engineers in order to design a process flowsheet. There are a number of problems with this approach as will be discussed in this paper. We will firstly discuss the sequential approach and then illustrated it by set of four case studies. These will show situations where firstly, by the choice of the experiments that had been done in the laboratory the process was effectively fixed. Secondly, because all the experiments were planned and completed before the modelling was done, inappropriate experiments, that did not cover the correct range of conditions, were performed. Thirdly, ideas for modifying the reaction scheme to allow for easier subsequent separation (often the most expensive part of the plant) could not be pursued. An approach for synthesising a process that is more iterative in nature will be presented. The advantages of this approach should be better processes that are developed more rapidly, with less time and money spent on irrelevant experiments or ones with inappropriate accuracy. What will, however, be needed are a new breed of engineers which for the want of a better name, we have called Process Synthesis Engineers. They will need to have a sufficiently broad background to be able to work with the laboratory personnel, to critically evaluate experimental techniques, model the results, as well as develop new flowsheets. If industry accepts this model, these people will clearly need to be trained and it will be up to the universities to provide courses and training programmes in order to supply the demand.

Keywords

Process synthesis, Design, Process development, Management.

Introduction

One of the main areas that chemical engineers have been involved in is the discipline of Process Design, which may involve using data measured in laboratory experiments to design and costing a chemical plant. This mechanism traditionally requires the practitioner to have a good deal of experience that is brought to bear in order to come up with a flowsheet. Once this flowsheet is produced there is some effort to optimise the operating parameters for the proposed plant. This is where the process simulation packages such as Aspen and Hysim have found a great deal of use. Much of the initial work of the designer is of such a nature that it is difficult to automate. With the advent of cheap and powerful computers there is now a need to develop computer algorithms to develop good flowsheets. Biegler, Grossmann and Westerberg (1997) review the current methods for systematically designing a chemical process. Two general approaches have been taken to try to achieve these objectives.

Table 1. Case Studies in Process Development.

Stage of Project	Problems Encountered
Case 1: • Called in after laboratory programme was complete.	• While doing modelling found evidence that reaction rates were mass transfer limited • New experiment showed tenfold increase in reaction rate. • Whole new series of experiments required.
Case 2: • Asked to do VLE measurements on a completely new system.	• With simplified, relatively accurate, technique could do complete measurements for three component quickly • Results showed two liquid phases and insolubility problems. • Could interpret results for design purposes although not originally asked to do it.

Case 3:
• Experimental programme complete and modelling of catalytic kinetics results already done.
	• Models were not satisfactory, as they were empirical and thus applicable only to a batch process.
	• Large recycle around reactor envisaged but experiments done only for pure feed.
	• If experiments with correct feeds were done:
	  • Able to predict reactor results directly from experimental data (model free)
	  • Able to directly compare CSTR and PFR measurements for experimental consistency

Case 4: • Called in after experimental programme complete and process flowsheet already developed.	• Model used for solid conversion kinetics only suitable for batch process. • Process not satisfactory as no account was taken of distribution of conversion of product particles.

The first is via expert systems. Here one tries to distil from good designers what they are doing and encapsulate this information in what essentially amounts to a programmed set of heuristics or rules by which one makes choices concerning the flowsheet. This approach is limited by the difficulty of getting the relevant information in a form suitable for the computer, but even more so by the limited current state of knowledge or development of heuristics.

The second (Mehta and Kokossis, 1997, Papalexandri and Pistikopoulos, 1996) is to use the power of the computer to try to evaluate a much larger set of possibilities than would normally be covered in a more conventional process design. The idea behind this method is to allow complex interactions between all the possible pieces of equipment and to choose the combination that minimises some objective function. This approach recognises that the way in which one organises the units in the flowsheet can have a significant impact on the profitability of a plant. In spite of the speed and storage of the modern computer, these latter calculations still present a formidable challenge and the size of problem that can be handled is somewhat limited. Furthermore, there is no guarantee that the superstructure that has been chosen is sufficiently general to represent all the possible cases. Interpreting the final answer, even supposing it is correct, in terms of equipment is also not straightforward.

Recently the Attainable Region (AR) method has been used to develop theory (Glasser and Hildebrandt, 1997) that allows for a more rational choice of the fundamental physical processes to be used and the way in which these can best be combined. Once this has been done the interpretation of the results in terms of equipment is reasonably straightforward. The AR method soon runs into difficulties as the size of the problem to be solved increases. The combination of the superstructure approach, where the superstructure is made to conform with the theoretical results from AR approach is a topic of current and ongoing research (Lakshmanan and Biegler, 1996).

Designing a New Approach to Process Synthesis

All the above discussion relates to the traditional sequential way in which new processes have been developed or improved. The use of complex process simulators only increases the speed of calculations and thus allows more cases to be studied but does not alter the

Table 2. Traditional Process Design and Some of the Problems that Can Arise.

Traditional Approach	Problems
Opportunity identified	Choice of chemical route decided before the engineers have had an input
Process chemistry identified	Experiments that have been performed limit the process that can be designed
Laboratory results handed over to engineers to design process	All the opportunities for the process design not fully investigated

sequential manner in which things are done. In our experience, problems with developing and designing new processes often arise because of this sequential approach used to manage the development process. A series of actual case studies in which problems where encountered are presented in Table 1. These case studies, which have involved a large range of companies in different countries and industries, have been typical of our experience and have occurred in almost every case we have been involved with. This suggests the problems are structural in nature rather than associated with poor management within a particular organisation. These problems result in wasted resources, longer times to develop the final design and inefficient processes.

With all the new tools and computing power that is available, one might ask whether there are better and more efficient ways to develop new processes. In order to attempt to answer this question let us try to characterise the steps that might take place in a typical traditional process design and then look at the problems and difficulties that may arise as a result of following this path. This is outlined in Table 2.

We note that even what is currently termed Process Synthesis, fits into this paradigm in that it is a technique that assumes we have the all the knowledge that we are going to get, albeit incomplete, and attempts to design the "best" process based on this information.

It can be seen that the traditional approach is sequential in nature and that many problems arise from this. It is also know in optimisation theory that a sequential approach will not in general find the global optimum. It could be postulated that the same is likely to hold true for the design of a new process. In order to overcome this difficulty a new approach is needed and the development of this approach is the subject of the remainder of the paper.

Let us now try to examine what might actually be required in trying to efficiently synthesise a new chemical process and attempt to devise a procedure for doing it. While not every process will go through the same steps it is suggested that most of them can be characterised or at least approximated by what follows.

A company decides it wishes to investigate the production of a new chemical. It might, of course, only be a new chemical for that company but not new for the market, but this does not really affect the issue. What is required at this stage is a quick and simple process design to enable an initial costing to be done to evaluate if the chemical can be produced and sold for a profit. This requires some degree of optimisation otherwise a potentially profitable product might be rejected because of a poor flowsheet. What is required is a very simple design based on a minimum amount of laboratory results. These results need not be of high accuracy as the design methods will generally be short cut methods that will themselves have limited accuracy. Thus it would be advantageous if the design engineer collaborates with the laboratory and helps to specify the experimental results that are required to proceed further. This clearly suggests that from the earliest stages in a project the design engineers need to be involved in the specification of the actual laboratory test programme. In order to do this adequately, the laboratory needs to have available a range of simplified techniques for rapidly measuring quantities such as reaction rates, heat effects, liquid-liquid and vapour-liquid equilibrium. In the sequential approach previously discussed the laboratory has very little idea for what purpose the data will be used, that is either for initial scoping or final design, and so they are forced to do high accuracy expensive and time consuming measurements. It is just because the laboratory does not know what the results are to be used for and therefore always produces high accuracy results, that the simplified techniques are not usually routinely available.

One may find at this stage that many projects or at least some of the alternatives are scrapped and so it is imperative not to spend too much time and money on this part. The current Process Synthesis techniques are probably not appropriate for this task as one would not like to have to produce detailed equipment designs in order to do costing and in any case one does not want to spend a long time with expensive flowsheeting packages to do this. What is required is some method so that the different designs can be compared on a simplified basis. The approach must be based on asking, "what is the simplest calculation that I can do that will still adequately differentiate between the alternatives?" There is thus a need for going back to the kind of short cut techniques that were used in the period just before the advent of the computer and further developing these methods in a form suitable for use with the computer. In particular one would also like to have variables available that result directly from these calculations that suggest the cost of the process. Thus for reactors we might like to use the mean residence time as a measure of the size in comparing different scenarios. Clearly different types of units such as plug flow reactors and CSTR's will have different costs per unit size but within the accuracy required at this stage this is likely to be of small account. A new variable that has similar properties for separators has recently been developed (Jobson *et. al*. 1996).

We believe that what is required in general is an approach which is more iterative in nature where these iterations involve the laboratory work as well quick costing techniques. It is suggested that a process synthesis engineer (possibly a new type of person with new skills) will need to be involved with the process from concept through to flowsheet. At the very earliest stage a putative flowsheet will need to be prepared by a team which involves both the laboratory personnel and the process engineer. This flowsheet will be used to determine the subsequent chemistry that will need to be studied. By this we mean not only the range of conditions that will need to be covered but also the required accuracy. Furthermore the knowledge of the presence of recycles in the flowsheet is also needed to help to determine the likely feeds to units. This is needed for instance to characterise the feed to chemical reactors to ensure the kinetics are measured covering the correct range of input conditions. At each appropriate stage, the experimental results will be interpreted, the flowsheet will be updated and this will help inform the laboratory of the next set of experiments that will need to be undertaken.

The process synthesis engineer referred to above will need skills that are currently not usually found in a single person, as there has been a tendency to train specialists rather than generalists. This is fine for the people who are developing new process synthesis techniques in universities but is probably not appropriate for the people who are doing the real process synthesis in industry. From the above description of what will be involved, it is clear that a process synthesis engineer would need to have a good grasp of chemistry and be able to link with laboratory people concerning experimental programmes and the methods used to generate data of appropriate accuracy. They would furthermore need to be able to model these results in a form suitable for use in the computer programmes that they have access to. Using these programmes they will produce optimised flowsheets which will be used to further direct the laboratory programme. The current MSc and Ph.D. graduates from the universities tend to only be skilled in some of these areas. In fact, there seems to be a particularly large divide between those that know about experimentation and those that know about theory. This is very negative for the whole area of process synthesis.

Not surprisingly the process synthesis engineer would have to have the range of skills that were needed by a process designer of the period before the advent of computers but these people would now have available a much wider range of theory and resources with which to practice their skills. It is suggested that at present it is not likely to be possible to teach this in an undergraduate programme and that it be done as a postgraduate course of some sort.

Conclusions

We believe that there are structural problems with the current methods by which processes are designed, mainly related to the sequential way in which the experiments and design are done. A more iterative method is suggested and we believe that is likely to be better. In order to make the best use of the new approach new short cut experimental, design and costing techniques will need to be developed. Furthermore a new breed of engineer called a Process Synthesis Engineer will need to be trained. One might even regard this as a resuscitation of an old breed!

References

Biegler, L.T., I.E. Grossman and A.W. Westerberg (1997). *Systematic Methods of Chemical Process Design*. Prentice Hall, New Jersey.

Jobson, M., D. Hildebrandt, and D. Glasser (1996). Variables indicating the cost of vapour-liquid equilibrium separation processes. *Chem. Eng. Sci.*, **51**, 3739-4757.

Glasser, D., and D. Hildebrandt (1997). Reactor and process synthesis. *Comp. Chem. Eng.*, **21**, S35-S41.

Lakshmanan, A. and L.T. Biegler (1996). Synthesis of optimal chemical reactor networks. *Ind. End. Chem. Res*, **35**, 1344-1353.

Mehta, V.L. and A. Kokossis (1997). Development of novel multiphase reactors using a systematic design framework. *Comp. Chem. Eng.*, **21**, S325-S330

Papalexandri, K.P. and E.N. Pistikopoulos (1996). Generalised modular representation framework for process synthesis. *AIChE J.*, **42**, 1010-1032.

OVERALL PROCESS DESIGN AND OPTIMIZATION - AN INDUSTRIAL PERSPECTIVE

J. S. Kussi, H. J. Leimkühler, and R. Perne
Bayer AG
51368 Leverkusen, Germany

Abstract

This paper describes the way in which processes are currently developed and optimized within a chemical firm, paying attention to all aspects of the production process. For well-understood processes the use of mathematical models and simulation alongside laboratory and plant experiments has proven successful. For new processes that are not as well understood, process synthesis tools based on assembled knowledge offer advantages over optimization based synthesis methods, which select particular designs from a predefined superstructure.

Keywords

Process development, Process synthesis, Material resource planning, Simulation.

Introduction

The life cycle of a process extends from the process development, through the planning and construction of the production process, plant commissioning, optimization of the running process, to the application of the acquired knowledge to the design and construction of a new plant. Creative ideas based on experience that exploit the technical possibilities are sought throughout the life of the process. Process synthesis methods provide a systematic approach, possessing advantages over the intuitive, experience-based approaches commonly used to solve problems that arise during the life cycle of a process.

In addition to the chemical and physical behavior, other factors influence and restrict the space of feasible process designs. For instance, the process life cycle is influenced by the supply chain (procurement and storage of raw material, production, product storage and distribution), customer expectations (quality, amount, delivery appointments), and competition from other companies. Therefore, from an industrial perspective, the successful definition of the process structure is constrained by many factors that include the process control and production logistics, which can only be evaluated by examining the complete production process.

The development and support of an entire production process requires systematic approaches carried out by interdisciplinary teams. On the one hand, this requires concentrating on each particular aspect (e.g., unit operations) of the process early in the process development. On the other hand, this demands that the team remains focused on the total production process and not on the detailed optimization of particular unit operations in isolation. This implies that portions of the production process which may be overlooked when focusing solely on the process unit operations, such as the material resource planning and the optimal design of raw material and product storage, play and integral role in the design of total production processes.

In the following sections, an example of how an existing, well understood production process was optimized using an intuitive procedure to incorporate all aspects of the production process during the optimization is presented. Another example shows how the methods of process synthesis can be applied in a heuristic, numeric approach to process development.

Sum of Suboptima ≠ Overall Optimum

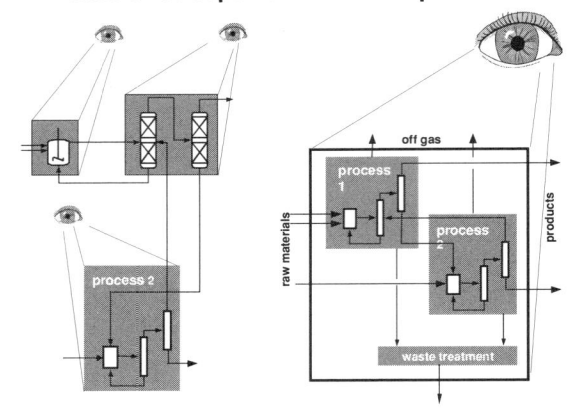

Figure 1. Overall process design and optimization.

Process Analysis, Design, and Control

The following example problem demonstrates how the development of a total production process is carried out within industry (Perne and Endesfelder, 1998). Our understanding of the process was systematically assembled and improved using process analysis and modeling techniques validated with pilot plant and laboratory experiments. Simulation studies were employed to scale-up the process and define the process structure and process control strategy. This example considers the optimization of an existing production process for which the assembled process understanding is high.

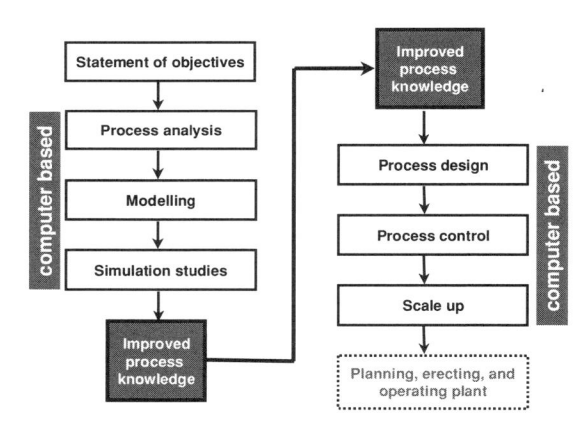

Figure 2. Methodical procedure of overall process design and optimization.

Process improvements concentrated on product quality and process economics. The existing production process is comprised of five steps: a batch reaction, solution formation, a continuous polymerization, solvent evaporation, and distillation. The primary objective focused on the first three process steps. The quality of the final product depends largely on the viscosity of the polymer solution before the evaporation step which must be kept within narrow limits. This criteria places restrictions on the properties of the intermediates produced by the batch reaction and solution formation steps. In addition, the concentration of side products in the reactor effluent must be minimized. Therefore, first objective concentrated on minimizing the side products created during the batch reaction, storage, and solution preparation steps, while improving process economics. The control of the polymerization reaction also had a decided influence on the ability of the process to achieve the desired quality, so it was chosen as the second process optimization objective.

The process steps were modeled based on the chemical and physical behavior of the system (figure 3). Some of the required process parameters were taken from the open literature, and the rest were determined from laboratory experiments. The polymerization reaction was based not only on end group kinetics, but also on the growth and structure of polymer chains produced. The latter determine the molecular weight distribution and the average molecular weight whose relation to the viscosity of the polymer solution was investigated. The models of each processing step included process control structure and control profiles. These models enabled a comparison of the model of the continuous polymerization reactor with the available data taken from the production process.

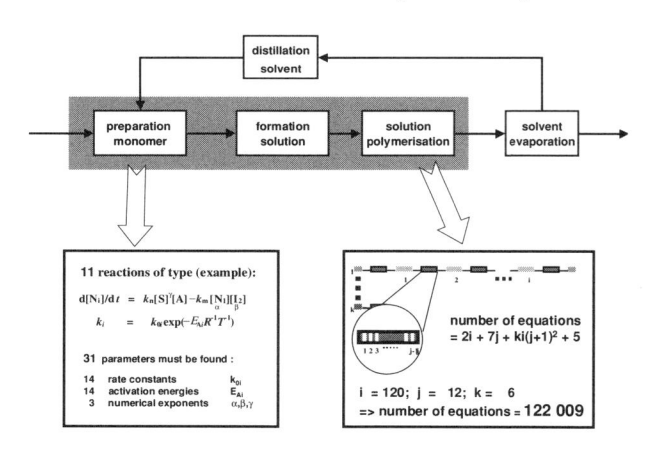

11 reactions of type (example):
$$d[N_i]/dt = k_m[S]^j[A] - k_m[N_i][I_2]$$
$$k_i = k_{0i}\exp(-E_{Ai}R^{-1}T^{-1})$$

31 parameters must be found:
- 14 rate constants k_{0i}
- 14 activation energies E_{Ai}
- 3 numerical exponents α, β, γ

number of equations $= 2i + 7j + ki(j+1)^2 + 5$

$i = 120$; $j = 12$; $k = 6$
=> number of equations = 122 009

Figure 3. Mathematical model of the process.

The first phase of the analysis provided a set of simulation modules, validated against experimental and the available production data, from which all of the interesting process structures such as batch-melt, batch-solvent, continuous melt, and continuous solvent for the first two process steps were investigated.

Computer simulations were employed to explore these process alternatives with respect to the side product formation. Figure 4 compares the results of these structures to the existing design. The new process alternatives substantially reduced the side product formation and improved the economics of the production process.

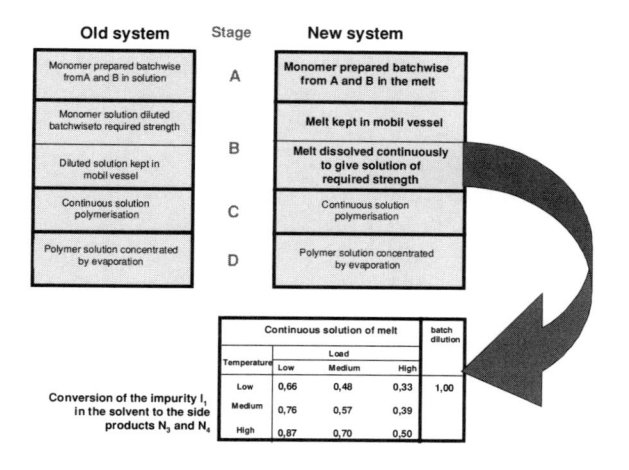

Figure 4. Comparison of old and new process structure.

Once the structure of the process was established, the operating profiles and equipment design for the individual processing units was optimized to achieve the required polymer solution viscosity. The model calculations provide the specifications for each unit, on which the company's unit operation specialists base their detailed design. For example, the solution preparation should homogenize the polymer melt and solvent as quickly as possible at minimum temperature. To operate at low temperature, the mixing equipment must possess adequate cooling capacity and the feed temperatures must not be too high. However, these constraints place constraints not only on the design of the mixing equipment, but also on the control profiles employed during reaction. In addition, the effect of operating the distillation column to increase the purity of the solvent needs to be evaluated with respect to side product formation. This requires optimization of the entire process, including not only the design of each unit operation, but also the interaction between the operating profiles of each. Modifications of the product recipe may be required. Through such an approach, we assure that the entire production process, and not just its parts, runs optimally.

We found that by applying advanced control methods, the distillation column can achieve a solvent purity that leads to improved product quality. The simulation studies also revealed that proper control of the polymer solution viscosity can only be achieved through the use of advanced process control concepts. A process control concept, capable of functioning under production conditions, was developed based on the mathematical models of the process and tested using simulation.

The process that was developed and optimized using process models scaled-up to the production facility based on the simulation results, which were validated using only laboratory experiments; no pilot plant studies were required, except to test the new concept for the preparation of the monomer solution. As a result, the time needed for the process development was cut in half. Since the new process has been in use, the interval between shutdowns for cleaning has been increased by a factor of ten. The economics of the process have been improved through the doubling of the capacity of the first two manufacturing steps.

Systematic Methods for Process Synthesis

Existing methods for process synthesis have concentrated on the development of processes for which relatively little information is available and for which a great deal of effort is required to investigate process alternatives through laboratory and pilot plant experiments. The systematic methods for process synthesis can be grouped into two categories (Han et. al., 1996):

- Creative methods which work from a rough definition of process goals expert systems to define a set of structural process alternatives.
- Optimization-based methods which determine the optimal process using mathematical optimization methods (MILP, MINLP) (Floudas, 1995) from a predefined superstructure of process alternatives.

Our experience indicates the neither of these methods is employed with regularity. However, in particular cases, these methods have proven their worth through successful applications. The following section relates our experience using a creative process synthesis method, PROSYN (Schembecker and Simmrock, 1996).

Application of PROSYN to an Industrial Example

The process synthesis tool PROSYN was applied in the early phase of the development of new process to produce solar grade silicon for the photovoltaic industry. The manufacturing process for high purity silicon requires the catalytic conversion of trichlorosilane ($SiHCl_3$) to lower-boiling silane (SiH_4) and higher-boiling silicontetrachloride ($SiCl_4$). Existing manufacturing processes for SiH_4 are comprised of two fixed bed reactors each requiring two distillation columns resulting in an energy intensive process with high equipment costs. With little detailed knowledge of the physical properties and chemical kinetics, PROSYN was employed to investigate the question, which type of process is best suited for the production of silane. PROSYN was provided with the following information: processing constraints, the vapor pressure of the pure components, the reaction stoichiometry, literature data providing rough information about the reaction equilibrium, and qualitative information about the reaction kinetics. This level of information is typical during the early stages of process development.

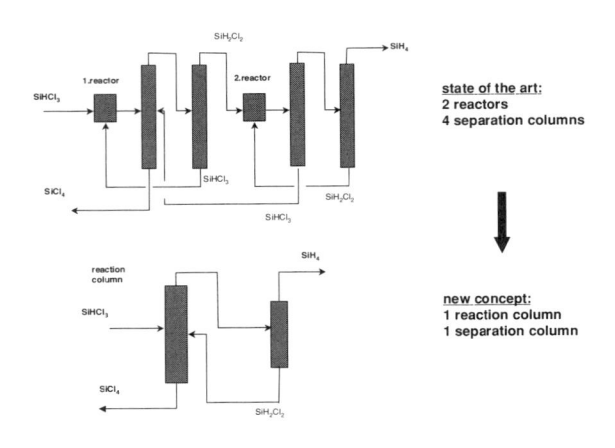

Figure 5. Conclusion of the PROSYN study.

The study with PROSYN provided several alternatives for the production process a) a cascade of CSTRs with gas separation in each and b) a simultaneous reaction and distillation carried out in a reactive distillation column. PROSYN provided several process alternatives with a minimum of information enabling an assessment of the processes. The reactive distillation process has both lower equipment and energy costs (see figure 5); this alternative was also proposed by a cross-functional development team comprised of experts using an intuitive development methodology. The result provided by the systematic approach within PROSYN agreed with the solution found by our conventional methods; a reactive distillation column has economic advantages for an equilibrium limited reaction with significant differences between the component vapor pressures. However, the following advantages obtained from the systematic development approach are not limited to this example:

- meaningful results provided from a limited amount of data, enabling such studies early in the development process
- an unbiased consideration of possible technical alternatives is assured
- the results are independent of the expertise and experience of the engineer
- the economic advantage of a particular alternative is quantifiable
- potential savings can be measured against technical risk (for new technologies)

Evaluating the Available Methods of Process Synthesis

In spite of this promising example, such methods are not yet employed by process engineers on a daily basis. This contrasts the position of simulation programs such as AspenPlus, Hysys, etc., which are commonly employed to compare creativ% concepts using simulation to quantify the merits of various process alternatives. However, an experienced process engineer is required to generate creative ideas, in what is often an unsystematic process.

Another factor that detracts from the application of process synthesis tools is the fact that entirely new process development activities for large continuous chemical process occur infrequently. Far more frequently, new processes are developed to manufacture active ingredients for specialty, agricultural, and pharmaceutical chemicals, in which the development process is subject to strict time constraints. Since the products must be brought to market quickly, they are typically manufactured in existing batch manufacturing facilities. For these systems we are unaware of applicable methods.

Mathematical programming methods have been suggested as a systematic method to choose between alternative process superstructures. On the one hand, these methods have been successfully applied to tasks such as the scheduling of multipurpose manufacturing facilities over several planning horizons. On the other hand, mathematical optimization methods have rarely been applied for the structural optimization of industrial processes because the generation of superstructures embedding a rich set of processing alternatives has not yet been appropriately resolved.

Conclusions

The development of total production processes, as the example demonstrates, is successfully carried out within industry. Good and open lines of communication between various parts of the company are known to be prerequisite for industrial development projects. Only through both good communication can both dimensions of the project scope be appropriately addressed: 1) technical support throughout the lifetime of the process and 2) attention to all aspects of the material flow through the production process, from the ordering of raw materials to the delivery of the product.

Currently, formal process synthesis methods address only small portions of the entire problem. However, for technology assessment, which requires economic comparisons of alternative processes for which little process information is often available, process synthesis tools can provide much needed assistance. Our opinion is that numeric-heuristic process synthesis methods provide advantages because they are capable of defining the process structure. The mathematical programming methods are appropriate for optimizing of production planning and material flow processes.

The creative process synthesis methods (e.g., PROSYN) must begin to address industrially relevant batch process development problems and must continue to extend their capabilities to address continuous processes. Moreover, process synthesis methods must improve the incorporation of process control and process logistic, particularly for batch processes in which the two are tightly coupled.

References

Floudas, Christodoulos A. (1995). *Nonlinear and Mixed-Integer Optimization*. Oxford University Press, New York.

Han, C., Stephanopoulos, G.; and Y.A. Liu, (1996). Knowledge-based approaches in process synthesis. *AIChE Symp. Ser.*, **92**, No. 312, 148-159.

Perne, R., and A. Endesfelder, (1998); Process analysis, design and control – a case study. *Comp. Chem. Eng.*, **22**, S1071.

Schembecker, G. and K.H. Simmrock, (1996). Heuristic-numeric process synthesis with PROSYN. *AIChE Symp. Ser.*, **92**, No. 312, 275-278.

A NEW CONCEPT IN PROCESS SYNTHESIS: MAXIMISING THE AVERAGE RATE

Craig McGregor and Diane Hildebrandt
School of Process and Materials Engineering
University of the Witwatersrand
Johannesburg, South Africa

Abstract

For any fundamental process (such as reaction, heat transfer or mass transfer) an instantaneous rate can be defined based on the dynamic behaviour of a batch system. Unfortunately, the instantaneous rate gives no indication of the global behaviour of the process, which is instead measured by the average process rate. The optimal design is simply the design that maximises the average rate that can be achieved over the entire process. A simple process synthesis procedure would then involve the search for regions of operation where high instantaneous rates can be maintained and that consequently result in a high average rate. This type of analysis is particularly suitable to graphical techniques and gives good insight into the system under consideration. For complex systems, such as reactive distillation systems, there are many fundamental processes each with an associated average rate. Here the objective is to maximise some overall average rate formed from the contributions of the average rates for the different processes. The concept is illustrated in this paper on some well-known reactor network synthesis problems. The advantage of this approach is that it can be used to synthesise near optimal designs without resorting to complicated process synthesis or optimisations techniques.

Keywords

Process synthesis, Optimisation, Attainable regions, Process rate.

Introduction

The problem with many process synthesis techniques is that they are difficult to implement and provide very little insight into the behaviour of the system under consideration. Consequently, industry has been slow to apply these techniques to their problems. A method that can, in principle, be easily applied to the complicated problems of industry would be more readily accepted. Such a method need not necessarily find the global optimum but should instead lead quickly to a good design that is near optimal.

The aim of this paper is to illustrate how the complement principle of Feinberg and Hildebrandt (1997), developed to prove a number of attainable region results, can be developed into a tool that satisfies these objectives. The type of analysis suggested is particularly suitable to graphical techniques that give good insight into the system under consideration

Reaction

To develop this simple process synthesis technique requires an understanding of the difference between the two different concepts of the instantaneous reaction rate and the average reaction rate.

The *instantaneous* reaction rate is defined such that the change of state **c** in a batch reactor is given by

$$\frac{d\mathbf{c}}{d\tau} = \mathbf{r}(\mathbf{c}) \qquad (1)$$

where **r** is the vector of instantaneous reaction rates r, that are defined according to the definition of c and τ.

The idea of an *average* reaction rate for a reactor network was first proposed in the *complement principle* of Feinberg and Hildebrandt (1997). The complement principle states that a reactor network is described by:

$$\bar{c} - \bar{c}_0 = \bar{r}\tau \tag{2}$$

where $\bar{r}$ is the average reaction rate vector. This average is a weighted sum of all the instantaneous reaction rates obtained because of the different operating conditions throughout the reactor network.

There are two different ways that the average rate can be evaluated. The first, given by Feinberg and Hildebrandt (1997), is the volume-averaged reaction rate:

$$\bar{r} = \int r(c)\,d\tau(c) \tag{3}$$

while the second, based on the equation for a PFR, is the state-averaged reaction rate:

$$\frac{c-c_0}{\bar{r}} = \int_{c_0}^{c} \frac{1}{r(c)}\,dc \tag{4}$$

Both averages give the same value but are evaluated in different ways, and may be useful for different situations. For example, it is obvious from Eqn. (4) that the average rate is heavily dependent on the lowest rates in the network, as these will give very large values of $1/r(c)$ and consequently a low $\bar{r}$.

It is easy to see from Eqn. (2) that in order to maximise the conversion to product for a given space times τ the average rate of formation of product $\bar{r}$ must be maximised. Conversely, the production of by-product is minimised by minimising the average rate of formation of by-product.

The formal mathematical optimisation of $\bar{r}$ is a very difficult problem to solve. The available techniques, such as Attainable Region analysis and superstructure optimisation, tend to be very complicated, both to understand and to implement.

Instead of trying to find the exact maximum $\bar{r}$ a simple process synthesis procedure would try to find a reaction network that resulted in a high *average* reaction rate. Such a procedure would involve the search for regions of operation where high *instantaneous* rates can be maintained, while avoiding regions of low *instantaneous* rates. There must, of course, also be a feasible path between the regions of high instantaneous rates. This feasible path can then be interpreted as a reactor network. This network could be subsequently optimised, if desired.

This principle of maximising the average rate will be illustrated using two examples.

Reaction Example 1

The first example is the adiabatic, exothermic-reversible reaction studied by Glasser *et al.* (1992), where the rate of change of conversion is given by:

$$r_X = k_1(1-X)\exp\left(\frac{-E_1}{RT}\right) - k_2 X \exp\left(\frac{-E_2}{RT}\right) \tag{5}$$

where $T = T_0 + T_{ad} X$.

Table 1. System Constants for Example 1.

k_1	k_2	E_1/R	E_2/R	T_{ad}
5×10^5	5×10^8	4 000	8 000	200

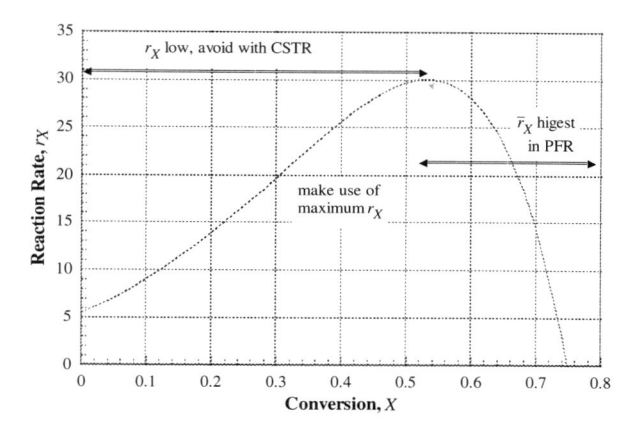

Figure 1. Instantaneous rate versus conversion for adiabatic, exothermic-reversible reaction.

The instantaneous rate, shown in Fig. 1, goes through a maximum. At low conversions the rate is low and should be avoided, which of course can be achieved by using a CSTR to jump to the maximum instantaneous rate. After this maximum rate the instantaneous rates drop again. However, in this case the lower rates cannot be avoided using a CSTR, as the reactor will itself operate at a low rate. Instead, a PFR is used so that at least the initial instantaneous rates in the reactor are high, so maximising the average rate in the reactor. This suggests a reactor network of a CSTR followed by a PFR, the well-known result. Fig. 1 is of course very similar to the plot suggested by Levenspiel (1965). The average rate, and consequently the required space-time, can be evaluated by:

$$\frac{X}{\bar{r}_X} = \int_0^X \frac{1}{r'_X(X)}\,dX \tag{6}$$

where r'_X is given by Eqn. 5 for X greater than the conversion of the maximum rate and by a constant, namely the maximum instantaneous rate, for lower conversions.

If the adiabatic constraint is removed then the temperature can be controlled either using external cooling or using cold shot cooling. In this case the principle of maximising the rate suggests that the system should be operated near the optimum temperature profile. Nicol (1998) has extensively studied this system.

Reaction Example 2

The second example uses the well-known van de Vusse kinetics, as studied by Glasser et al (1987):

$$A \leftrightarrow B \rightarrow C, \ 2A \rightarrow D$$

where the rates of formation of A and B are given by:

$$\begin{aligned} r_A &= -k_1 c_A + k_2 c_B - k_4 c_A^2 \\ r_B &= k_1 c_A - k_2 c_B - k_3 c_B \end{aligned} \quad (7)$$

The feed to the system is pure A of concentration c_{A0} of 1. In this example the objective is to maximise the production of B for a particular conversion of A. The instantaneous selectivity, s, is then defined by

$$s = \frac{r_B}{-r_A} \quad (8)$$

The average selectivity, $\bar{s}$, is the average "rate" that must be maximised in this problem, which is easily visualised since the slope of the line between the system product and feed in a plot of c_B versus c_A is just $-\bar{s}$.

Table 2. Rate Constants for Example 2.

k_1	k_2	k_3	k_4
1	1	1	10

A contour map of the selectivity is given in Fig. 2. A number of regions are of interest in this figure. The highest instantaneous selectivities are obtained near the origin, that is at low concentrations of A and B. This at first seems advantageous but a deeper look shows that the average selectivity here is low as the slope of the line between feed and operating region is small. Hence, the reactor network should not operate near this region. The highest average selectivity is given by a line with a slope of -1, and passing through the feed point. However, along this line the instantaneous selectivities are very low, indicating that no feasible path will lead to a selectivity of 1. Next, it should be noted that the instantaneous selectivities are low near the feed. This region should also be avoided, suggesting that a CSTR should be operated from the feed. In this figure a line of "optimum" s can be found by drawing tangents from the feed to the selectivity contours. This line is in many ways similar to the "optimum" temperature profile found for exothermic-reversible systems. The point on the optimum selectivity line where $\bar{s} = s$ is the operating point for the maximum-selectivity CSTR. Above the optimum selectivity line is a region where the instantaneous selectivities are

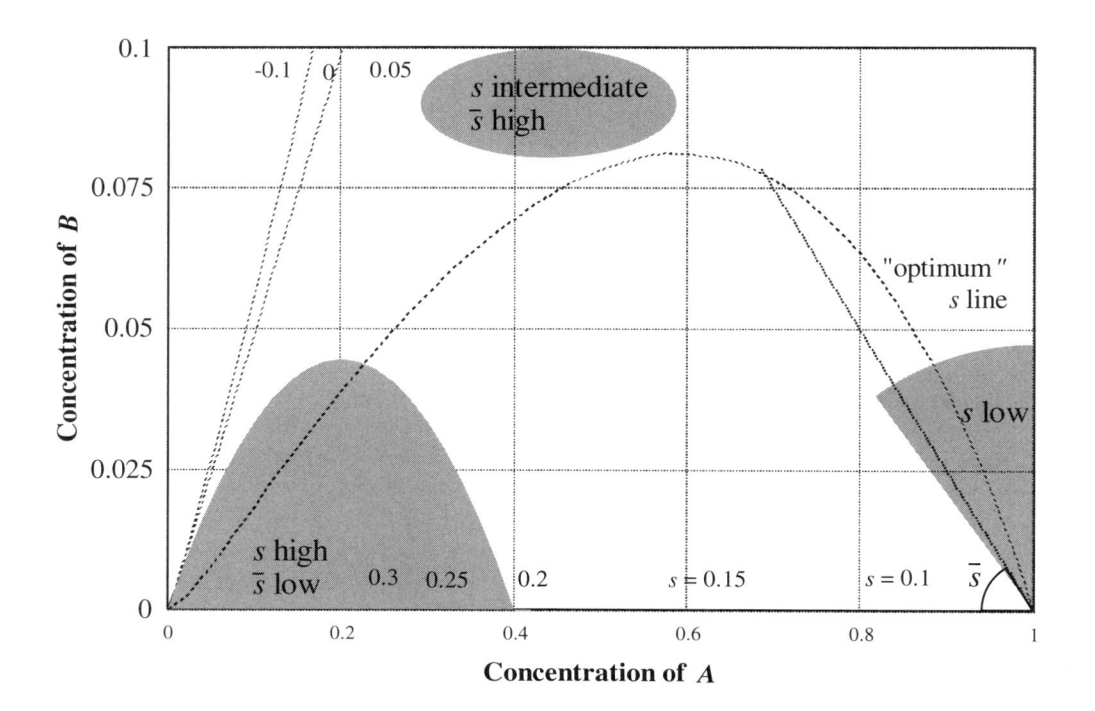

Figure 2. Selectivity contour plot for van de Vusse kinetics (Example 2).

moderately high, but decreasing. A PFR should be operated in this region in order to achieve the maximum average selectivity at higher conversions. Although the average selectivity is lower than in the optimum CSTR, both the conversion and concentration of B are higher, and these are obtained without lowering the average selectivity by much.

Thus, in order to achieve the highest conversion to B for a particular conversion of A the principle of maximising the average rate suggests the use of a CSTR followed by a PFR, and the maximum conversion is achieved by operating the PFR until the instantaneous selectivity is zero.

Other process synthesis results can also be achieved by applying the principle of maximum average rate to Fig. 2. For example, higher selectivities can be achieved by diluting the feed, but this is obtained at the expense of lower B concentrations and consequently the separation will be more expensive down-stream. This is, of course, a trade-off that can be optimised; however, the principle of maximum average rate highlights this and suggests a course of further investigation.

Other Fundamental Processes

For complex systems, such as reactive distillation systems, there are many fundamental processes each with an associated average rate. Here the objective is to maximise some overall average rate formed from the contributions of the average rates for the different processes. For example, the best economic design is that which maximises the average rate of generation of profit. In such complex systems it may not be possible to find the maximum rate using graphical techniques, but certainly regions of operations can be identified where the rates of the different processes are high. These regions of high rates can then be used to identify likely candidates for the optimal process for further study.

This requires that the complement principle be extended to consider other fundamental processes besides reaction, such as heat or mass transfer. If the instantaneous rate of heat transfer is expressed by a heat flux vector $\mathbf{q}$ and mass transfer by the mass flux vector $\mathbf{j}$, then the complement principle would require that

$$\bar{\mathbf{c}} - \bar{\mathbf{c}}_0 = \bar{\mathbf{r}}\tau + \bar{\mathbf{q}}\sigma_q + \bar{\mathbf{j}}\sigma_j \qquad (9)$$

where σ is the surface area for transport per unit feed flowrate. Alternatively, Eqn. (9) could be rewritten as:

$$\bar{\mathbf{c}} - \bar{\mathbf{c}}_0 = \bar{\mathbf{p}}\chi = \left(\bar{\beta}_r \bar{\mathbf{r}} + \bar{\beta}_q \bar{\mathbf{q}} + \bar{\beta}_j \bar{\mathbf{j}}\right)\chi \qquad (10)$$

where the vectors contain the instantaneous rates defined in terms of the generation of some cost function, χ. In this case contour plots of the instantaneous rates of reaction, heat transfer and mass transfer would indicate areas of operation for reaction, heat exchange and mass exchange and where there is a reasonable trade-off obtained by using the processes simultaneous. As discussed for Eqn. 4, the average processing rates p of the average process vector $\mathbf{p}$ are dominated by the terms with the smallest instantaneous rates and any synthesis procedure should concentrate on avoiding these.

Conclusion

The approach offers the major advantages in that it can be used to synthesise near optimal designs without resorting to complicated optimisations techniques such as attainable region analysis or superstructure optimisation and can be used to gain insight into the various trade-off that must be made within a process.

Nomenclature

$\mathbf{c}$	=	state vector
$\mathbf{c}_0$	=	feed state vector
c	=	concentration
E	=	activation energy
$\mathbf{j}$	=	mass flux vector
k	=	rate constant
$\mathbf{q}$	=	heat flux vector
$\mathbf{r}$	=	reaction vector
r	=	reaction rate
R	=	molar gas constant
s	=	selectivity
T	=	temperature
T_0	=	feed temperature
T_{ad}	=	reaction adiabatic temperature rise for $X = 1$
X	=	conversion
β	=	policy variable
χ	=	processing cost per unit feed flowrate
σ	=	surface area per unit feed flowrate
τ	=	space-time

References

Feinberg, M., and D. Hildebrandt (1997). Optimal reactor design from a geometric viewpoint – I. Universal properties of the attainable region. *Chem. Eng. Sci.*, **52**, 1637-1665.

Glasser, B., D. Hildebrandt, and D. Glasser (1992). Optimal Mixing for exothermic reversible reactions, *Ind. Eng. Chem. Res.*, **31**, 1541-1549.

Glasser, D., D. Hildebrandt, and C. Crowe (1987). A geometric approach to steady flow reactors: The attainable region and optimisation in concentration space, *Ind. Eng. Chem. Res.*, **26**, 1803-1810.

Levenspiel, O. (1962). *Chemical Reaction Engineering*, 2nd ed., Wiley, New York, 163-200.

Nicol, W. (1998). *Extending the Attainable Region Technique by Including Heat Exchange*, PhD Thesis, University of the Witwatersrand, South Africa.

CONCEPTUAL DESIGN OF PROCESSES FOR STRUCTURED PRODUCTS

F. Michiel Meeuse, Johan Grievink, and Peter J. T. Verheijen
Department of Chemical Technology
Delft University of Technology
2628 BL Delft, The Netherlands

Michel L. M. vander Stappen
Unilever Research
3133 AT Vlaardingen, The Netherlands

Abstract

Systematic approaches to the design of processes for structured products are still relatively underdeveloped, despite their gaining industrial importance. We propose a method for the conceptual design of processes for structured products, which is an expansion of Douglas' hierarchical decomposition. The novelty of the method is threefold: firstly, the internal structure of the product is considered explicitly rather than only composition, secondly the method is applicable to multiple product plants, this is almost always a necessity for the type of products considered here, and thirdly, the batch/continuous choice does not have to be made beforehand for the entire process, which means that hybrid batch/continuous processes can easily be considered. The hierarchical approach proved its potential by finding feasible designs for the example considered; mayonnaise and dressings. Additional effort has to be spent on the generation of more knowledge-based rules for flowsheet synthesis and the selection of equipment.

Keywords

Conceptual design, Hierarchical decomposition, Structured products, Emulsions, Mayonnaise and dressings.

Introduction

The performance properties of certain polymers, but also of non-chemical products like paint, liquid detergents, ice cream and mayonnaise depend critically on the internal microstructure of such products. These so-called structured products have an internal spatial structure on a nano or a micro scale. Such structures can manifest themselves in different ways, e.g. on a molecular scale in polymers, or on a nano or micro scale as phase domains with distributed sizes, e.g. droplets, embedded in other thermodynamic phases, as in emulsions. These products therefore cannot be characterised only by a composition. Take for instance mayonnaise. It consists of 80 vol% small oil droplets, dispersed in an aqueous matrix, stabilised by a protein network. The quality of the product depends critically on the droplet size distribution of the oil droplets.

Despite the industrial relevance of structured products, chemical engineering has paid only limited attention to these products (Villadsen, 1997). Traditionally, conceptual design in the food and cosmetics industries takes place in an evolutionary way, the main focus being on equipment design, rather than on the systematic exploration of alternative processing configurations. The purpose of this paper is to propose a method for the conceptual design of processes for structured products, based on the method developed by

Douglas (1988). Douglas and Stephanopoulos (1995) give a more recent description. Since most structured product plants must accommodate for multiple products, the method has to cover this. An optimisation-based method is not considered, since the required mathematical models are generally not available for this type of process.

Approach

A level decomposition is proposed, based on the level decomposition presented by Douglas and Stephanopoulos (1995):

0. Input information
1. Processing structure
2. Plant I/O structure
3. Task structure
4. Unit operations
5. Equipment design

The batch/continuous choice does not have to be made beforehand. Parts of the process can be batch, whereas other parts can be continuous. The choice will be made in level 4, for each part of the process separately.

Level 0

At this level the battery limits and conditions need to be defined. The input information to be given is divided into two classes: basis of design and physical properties. The basis of design consists of process targets and constraints, the desired microstructure, a description of the desired physical chemical transformations and cost data.

It is not necessary to specify all the input information in this level; input data related to a finer degree of detail can be given in the scope of design at subsequent levels.

Level 1

At this level the processing structure is determined, resulting in transformation blocks. All physical chemical transformations which change the internal physical structure of the product and which occur under the same conditions are grouped in one block. This is similar to the multiple plant description.

So transformations, like crystallising, emulsification and reaction, are grouped in separate blocks. Transformations like feed heating, followed by reaction can be grouped in the same block.

Level 2

Level 2 considers the input/output structure. First it is determined where each ingredient is added to the process. Therefore an Ingredients Table is created, describing the function and the place of each ingredient in the final structure of the product.

Overall mass balances are created for all products, resulting in capacity requirements for the different processing blocks determined in level 1. Furthermore, the split ratios of feed distributions over the blocks and of recycles are set and the processing capacities of the blocks are optimised.

Level 3

At level 3 each block is decomposed into sub-systems. Each sub-system is associated with a certain functional task. For structured product processes the following general sub-systems are proposed:

- feed preparation, to change the state of the ingredients to the required state for the next sub-system (e.g. heating, cooling, dissolving).
- reaction (including fermentation), to change the chemical identity of the product.
- micro-scale phase assembly or transition, to change the internal structure of the product.
- separation, to change the composition of the phases of the product.
- preservation, to prolong the shelf life of the product.

There is no strict order for these sections; the ingredients can, for example, be sterilised before the phase assembly, or the product can be sterilised after the phase assembly. An optimal structure needs to be determined for every process.

Level 4

At this level specific unit operations and main types of equipment for the tasks identified in level 3 are selected. Characteristic parameters that determine the performance of the unit operations, like shear rate and residence time, next to temperature and pressures, are determined.

Level 5

At this level the equipment is further specified and auxiliary equipment like pumps are selected. The parameters like shear rate and residence time, determined in level 4, are targets for this level.

Example: Mayonnaise and Dressings

Mayonnaise and dressings are essentially oil in water emulsions, stabilised by a protein network. The oil fractions vary from about 20 % for low-fat dressings, to about 80 % for mayonnaise. Ingredients for mayonnaise and dressings include water, oil, egg yolk, vinegar, salt, sugar, spices and pieces vegetables. Fig 1 shows schematically the structure of mayonnaise.

This example considers only the design up till level 4. The heuristics presented are based on prior knowledge and simulation results.

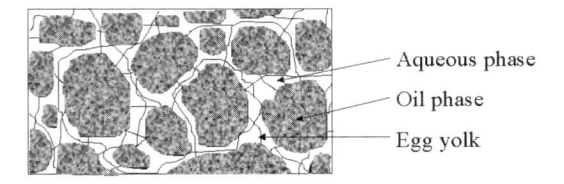

Figure 1. Structure of mayonnaise.

Level 0: Input information

An important quality specification of mayonnaise is the droplet size distribution. The oil droplets have to be small enough to get the correct consistency. A typical droplet size for the oil droplets is 2 – 8 μm (Ranken et al., 1997). This droplet size can be obtained by break-up of the oil droplets due to shear stresses in the production process.
Walstra (1993) and Ottino et al. (1999) give extensive overviews of possible break-up mechanisms.

We assume the following product portfolio for 30 h production time: 150 ton 80% oil product X, 100 ton 40% oil product Y and 50 ton 20% oil product Z.

Level 1: Processing structure

For products with an oil fraction below 80 % there are two processing options; first part of the continuous phase is added, resulting in a pre-emulsion with an internal phase fraction of 80 % which is subsequently diluted with additional continuous phase, or all continuous phase is added directly.

With the given product portfolio, the processing structure for this process consists of two blocks. In the emulsification block the emulsions are prepared. In the mixing block the emulsions can be diluted with additional continuous phase.

Level 2: Input/Output structure

Table 1 shows an ingredients table for mayonnaise. On the basis of this table the following heuristics about where the ingredients have to enter the process become clear:

The egg yolk has to be added before emulsification starts.
The vinegar has to be mixed with the water before emulsification starts.
The salt and sugar has to be dissolved in the water before emulsification starts.

Now mass balances can be generated. For products with a fraction dispersed phase smaller than 80%, it has to be determined which part of the water phase is fed to the emulsification block, and which part is added in the mixing block. The following heuristic was determined:

The minimum amount of water should be added to the emulsification block.

Table 1. Mayonnaise Ingredients Table.

Ingredient	Fraction	Function	Location
oil	80%		disp. phase
egg yolk	8%	increase stability	interface
water	7%		cont. phase
vinegar	3%	taste, preservation increase stability	cont. phase
salt	1%	taste, increase stability	cont. phase
sugar	1%	taste	cont. phase

Products with different oil fractions require different capacities of the blocks. The design capacity of the different blocks for the product portfolio given in level 0 is determined. The heuristic presented above will lead to the capacity requirements shown in Table 2.

Table 2. Required Capacities.

Product	Emulsification block [ton]	By-pass [ton]
X	150	0
Y	52	48
Z	14	36
total	216	84

Therefore the minimum capacity of the emulsification block should be 216 ton / 30 hr = 7.2 ton/hr. The following heuristic is proposed:

The design capacity of the emulsification block should be 130 % of the minimum capacity.

So the design capacity of the emulsification block is 9.4 ton/hr, and this will result in a production time for product X of 16.0 hr. To calculate the by-pass capacity the following heuristic is proposed:

The by-pass capacity should be the same for all products.

The resulting by-pass capacity is 84 ton / (30 -16) hr = 6 ton/hr. This leads to the results shown in Table 3.

Table 3. Results Capacity Calculations.

Product	Production [ton]	Production time [hr]	Production rate [ton/hr]
X	150	16.0	9.4
Y	100	8.0	12.5
Z	50	6.0	8.3
total	300	30.0	

Level 3

Here each block determined in level 2 is decomposed in subsystems. Since a preservation step is not considered, only a feed preparation step and an assembly step are present. This leads to the structure shown in Fig 2. Only the emulsification block will be discussed here.

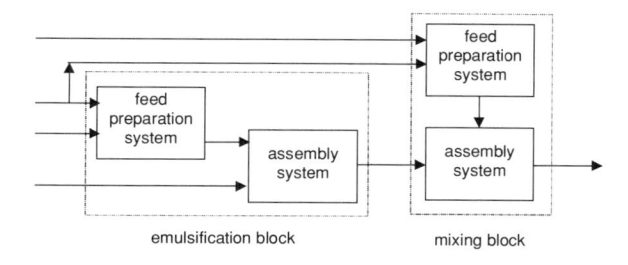

Figure 2. Structure of the mayonnaise/dressings process.

The tasks that have to be fulfilled in the feed preparation system of the emulsification block are to:

- dissolve salt and sugar in the water.
- mix vinegar and egg yolk in the water.

The targets that should be met are a 100 % dissolution of the salt and the sugar and complete mixing with the vinegar and the egg yolk.

The tasks that have to be done in the assembly system of the emulsification block are to:

- disperse the oil in the water.
- reduce the oil droplet size to the specification.

The target that should be met is the specified droplet size distribution

Level 4

Here the major equipment has to be selected and specified for the blocks determined in level 3. In this paper we will only focus on the assembly section of the emulsification block.

A preliminary equipment selection for the emulsification step is made based on Walstra (1993). Table 4 shows the selected equipment. Based on the criteria shown in Table 4, a colloid mill is selected.

Table 4. Criteria for Emulsification Equipment Selection.

Criterion	Stirred vessel	Colloid mill	Static mixer
droplet size range (μm)	5 – 100	1 – 20	10 – 100
high internal phase emulsions	±	+	-
cont. processing	±	+	+
batch processing	+	+	-
residence time	long	very short	short

The alternative designs of colloid mills have been evaluated numerically. A short-cut model was developed, based on a rigorous population balance based model by Wieringa et al. (1996).

Up till now, no choice batch/continuos has been made. We selected continuous operation, however heuristics for this choice are currently lacking.

The process thus designed is rather similar to processes for mayonnaise and dressings production which are nowadays common in industry (Lopez, 1987). The method results in a plant that is less overdesigned than a traditional designed plant.

Conclusions

It can be concluded that the use of hierarchical decomposition can certainly be extended to structured product processes. The method has the following main features: the internal structure of the product can be considered explicitly, the method is applicable to multiple product plants, and the batch/continuous choice does not have to be made beforehand for the entire process which means that hybrid batch/continuous processes can easily be considered.

Additional effort has to be put into generating more knowledge-based rules for flowsheet synthesis and the selection of equipment, for the products discussed in this paper, as well as for different types of products which are not yet considered.

References

Douglas, J. M. (1988). *Conceptual Design of Chemical Processes*, McGraw-Hill, New York.

Douglas, J. M. and Stephanopoulos, G. (1995). Hierarchical approaches in conceptual process design: framework and computer aided implementation. In L. T. Biegler, M. F. Doherty (Eds) *Foundations of Computer Aided Process Design*, AIChE Symposium series, No. 304, **91**, 183–197.

Lopez, A. (1987). *A Complete Course in Canning and Related Processes*, 12th edition, Book III. *Processing Procedures for Canned Food Products*. The Canning Trade Inc., Baltimore, 420–436.

Ottino, J. M., DeRoussel, P., Hansen, S., and Khakhar, D. V. (1999). Mixing and dispersion of viscous liquids and powdered solids. To be published in *Adv. Chem. Eng.*

Ranken, M. D., Kill, R. C. and C.G. Baker (1997). *Food Industries Manual*, 24th edition, Chapman & Hill, London, 358.

Villadsen, J. (1997). Putting structure into chemical engineering. *Chem. Eng. Sci.*, **52**, 17, 2857–2864

Walstra, P. (1993). Principles of emulsion formation. *Chem. Eng. Sci.,* **48,** 2, 333-349.

Wieringa, J. A., van Dieren, F., Janssen, J. J. M. and W.G.M. Agterof, (1996). Droplet break-up mechanisms during emulsification in colloid mills at high dispersed phase volume fraction. *Trans. I.Chem.E. part A*, **74**, 554-562.

INDUSTRIAL EXAMPLES OF PROCESS SYNTHESIS ANALYSIS DRIVING RESEARCH AND DEVELOPMENT

Daniel L. Terrill
Eastman Chemical Company
Kingsport, TN 37662

Abstract

In today's business environment of faster, better, and cheaper, analysis that reduces the amount of expensive, time-consuming piloting adds significant value to an R&D program. Three industrial examples are given which illustrate how process synthesis analysis enhanced commercialization. In these examples the fundamental issues behind the process synthesis are: screening alternative reaction systems and their respective flowsheets, merging processes with similar chemistry, and utilizing reversible reactions in waste minimization. Process synthesis tools, including hierarchical methods and superstructure methods, need to be able to accommodate these types of issues.

Research on a hydrogenation process found that the reaction is viable in both the liquid and vapor phases with homogeneous and heterogeneous catalysts. Flowsheet development and optimization were used to direct the R&D. The analysis improved the product cost by 19% and reduced the amount of piloting needed.

Laboratory reactor experiments of another process showed that a minor byproduct was formed. This byproduct was the desired product of another process also under development. Analysis indicated that value could be added by merging the two processes to co-produce the products. Pilot experiments demonstrated the merged process. The capital investment was reduced by 28%.

In a wastewater stream, it was determined that a reactant could be recovered from the stream by back-reacting some of the byproducts in it. Analysis indicated the yield and capital investment necessary to have a profitable process, which established a reactor performance criterion. This criterion was used to decide whether to continue or end the R&D efforts early on in the project.

Keywords

Process synthesis, Process analysis, Optimization, Pilot plant, Research, Development.

Introduction

Process synthesis involves the creative work of attempting to minimize the cost of producing a desired compound as quickly and inexpensively as possible. The process involves selecting possible reactor configurations, various separation techniques, and possible connectivity between these, while optimizing an objective function. Recent work utilizes a superstructure to formulate the problem with its many alternatives and then attempts to evaluate the cost for the various alternatives using mathematical programming (Gundersen and Grossmann, 1990; Bagajewicz and Maniousiuothakis, 1992; Friedler et al., 1993; Grossmann, 1996; Yeomans and Grossmann, 1998). Other work focuses on reducing the number of alternatives to be evaluated by using the known information about the process in a hierarchy of design decisions, before beginning to evaluate alternatives

(Douglas, 1985; 1988; Linnhoff et al., 1982, Daichendt and Grossmann, 1998).

Presented here are industrial examples of process synthesis problems. The purpose of this paper is to give industrial examples of process synthesis and to show how the analysis added valued to the R&D. They illustrate the questions and the analysis that this engineer faced in pursuing the commercialization of these processes. A hierarchical approach was taken to identify flowsheet alternatives. Process analysis was done during the research and development of these processes to reduce the effort and costs at the appropriate time during the project. It is difficult to determine the correct timing to discontinue development of the seemingly more-expensive alternatives. For example, one catalyst system may have better activity than another catalyst system, but upon further development, it may be found that the first system deactivates faster than the second, thereby eliminating any value added from higher catalyst activity. Providing analysis at the proper time in the life of a project provides useful information for making the difficult R&D decisions.

This work contains three examples. The first example involves a hydrogenation where there are several viable reaction systems. Each reaction system generates a different separation system. And optimization of the key design variables for the processes of these different reaction systems yield quite different results. The second example involves the simultaneous development of two processes with related chemistry. At a point in their development, it became obvious that both chemicals could be produced from one reactor, eliminating duplication and adding value. In the third example, in attempting to clean up a waste stream destined for the treatment plant, the most plentiful compounds in the stream reversibly react under acidic conditions to valuable starting material. Establishing a capital investment hurdle necessary for the project to be profitable enabled the appropriate level of R&D to be done. In each of these examples, doing the appropriate level of analysis at the appropriate time saved significant development costs and accelerated the project toward commercialization.

Example 1 – Reactor Synthesis

The reaction considered involves hydrogenating a raw material, A, into a product of higher value, B. The parallel byproduct reaction exists, producing product C.

$$A + H_2 \rightarrow B \quad (1)$$
$$A + H_2 \rightarrow C \quad (2)$$

Very early in the process development the product cost was determined. Product cost is defined as the cost to make the chemical and make a profit. It includes the manufacturing cost, the cost of capital, overhead, and profit. The product cost from the input-output structure indicated that the product was more valuable than its raw materials.

Chemists developed three viable reaction systems: a liquid-phase process with a solid-supported catalyst, a liquid-phase process with a homogeneous catalyst, and a vapor-phase, solid-supported catalyst process. A superstructure can be constructed which incorporates the three reaction systems. With each system, unique decisions arise regarding the separation and recycle of catalyst and hydrogen, making the superstructure very large. A hierarchical approach can lead to possible flowsheets, which need to be evaluated. Fig. 1-3 show one promising alternative for each of the reactor systems. In addition, for each alternative, an objective function needs

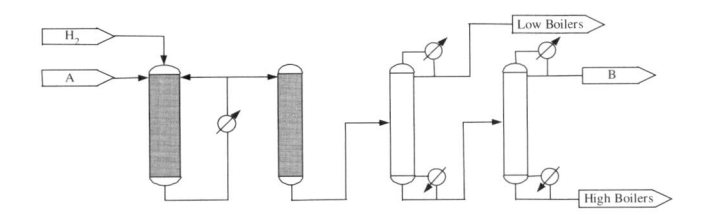

Figure 1. Liquid-phase, heterogeneous-catalyst reactor alternative.

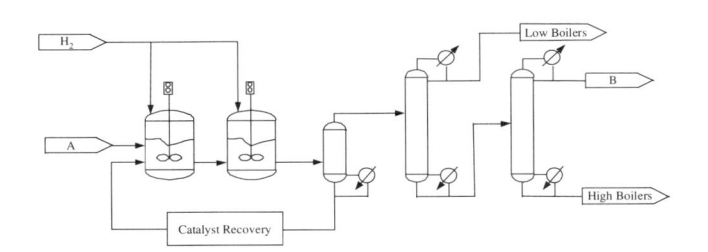

Figure 2. Liquid-phase homogeneous-catalyst reactor alternative.

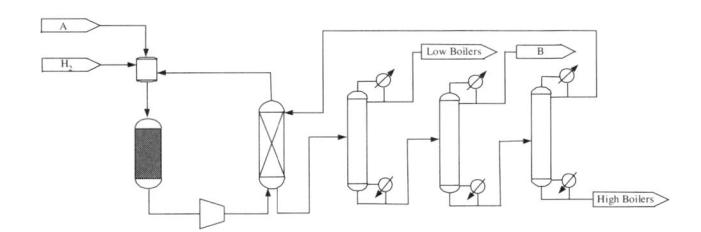

Figure 3. Vapor-phase heterogeneous-catalyst reactor alternative.

to be determined and optimized as a function of the key design variables. For this process, the key design variables include reactor temperature, reactor cooling utility, reactor pressure, catalyst loading, fractional recovery of hydrogen, fractional recovery of B, distillation column pressure, and distillation column reflux ratio. Product cost was selected as the objective function. As an example, Fig. 4 shows the product cost as a function of reactor temperature for these three reactor configuration alternatives. In this case, cooling tower water (CTW) or refrigeration can provide reactor cooling. The lowest product cost, as seen in Fig. 4, is a liquid-phase process with a heterogeneous catalyst (trickle bed) where the reactor cooling utility is refrigeration. Other key design variables were investigated similarly.

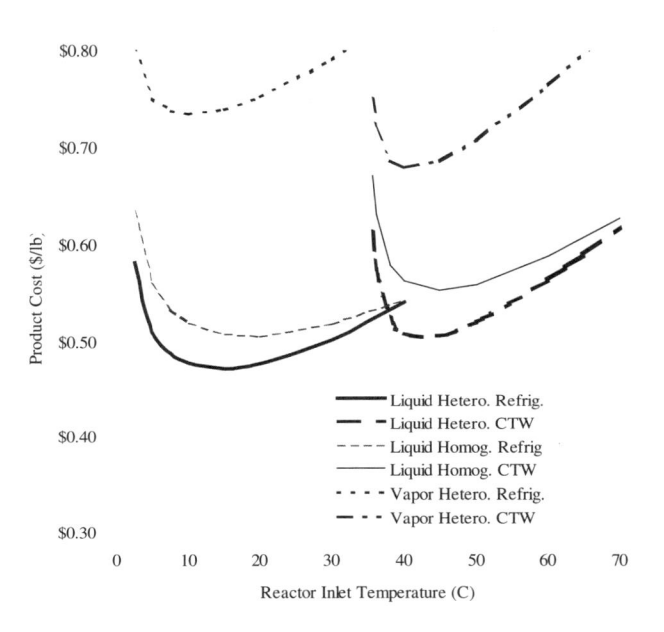

Figure 4. Optimization of reactor temperature for three process alternatives.

Analysis exemplified in Fig. 4 was used to make R&D decisions. It is obvious from Fig. 4 that the vapor-phase reactor system is a more-expensive alternative than the others. Research surrounding that process was ended. Work continued on the two liquid-phase reactor systems until it was found that the homogeneous catalyst had significant deactivation problems. The liquid-phase, solid-supported reactor system was piloted and a range of temperatures was investigated. These are the decisions that were made. It is always difficult to know in hindsight if the two disqualified reactor systems could have been improved by additional research.

Reactor temperature is a very important design variable in this case and warrants further discussion. As seen in Fig. 4, there is an optimum reactor temperature for each alternative due to process tradeoffs. As the temperature increases, there are greater selectivity losses. As the temperature decreases, the size of the reactor cooler increases. Of course, the refrigeration utility cost is much larger than the cost of cooling tower utility. But because of the unique kinetics, the optimal reactor temperature occurs for an alternative that uses refrigeration.

With a parallel side reaction, relative reaction rates can be varied by changing the temperature when the activation energies for the two reactions are different. In this particular case, the activation energy for the side reaction is larger than that of the primary reaction, so that decreasing the temperature increases the selectivity. Also, both activation energies are small so that the reaction rate of the primary reaction does not decrease substantially (and the reactor size and cost increase substantially) as the temperature is reduced. These characteristics lead to the result that the optimum reactor inlet temperature requires the use of refrigeration, a non-intuitive solution.

Example 2 – Merging Processes

During the reactor piloting for a potential new product, a byproduct was being produced. This byproduct is valuable and was the focus of a parallel research effort. Fig. 1 and Fig. 5 show possible flowsheets for the two processes. Further reactor testing demonstrated that both products could be produced in the same reactor with only minor alterations. Fig. 6 shows the combined process. The alterations included changing the type of support for the catalyst. Piloting efforts were modified to merge the two processes. The product cost for the two products decreased from 0.63$/lb and 0.87 $/lb to 0.56 $/lb and $0.81 $/lb, respectively. The capital investment decreased by 28%. The new process has the flexibility to vary the mixture of the two products over a large range.

Process synthesis problem formulation needs to be able to include the capability to merge processes that have similar chemistry. In a hierarchical approach, there needs to be a step which considers merging processes of similar chemistry. In a superstructure approach, there also needs to be a means to incorporate processes with similar chemistry.

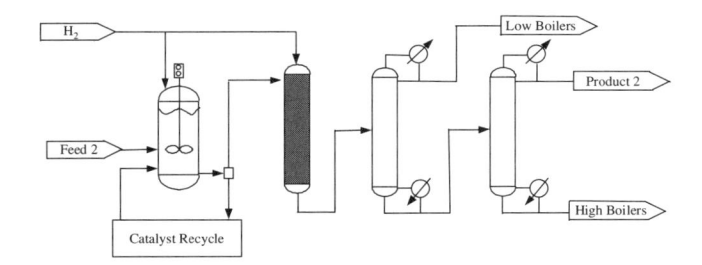

Figure 5. Second process.

Example 3 – Wastewater Process Improvement

Research in a waste minimization problem led to the observation that several compounds reversibly react from byproducts to a valuable raw material. It is known that the reaction will occur at a reasonable rate in acidic conditions. Preliminary analysis indicated that a capital investment of $2.3 M was justified from the recovery of the valuable raw material. Laboratory experimentation was justified on this basis. From the lab work, the proposed reactor type was demonstrated and a kinetic model was developed. Additionally, the operating conditions and materials of construction were validated. The final capital investment was determined to be $5.7 M. The project is not economically justified, until additional factors can contribute to the value of the project. The project was discontinued without incurring the expense of piloting.

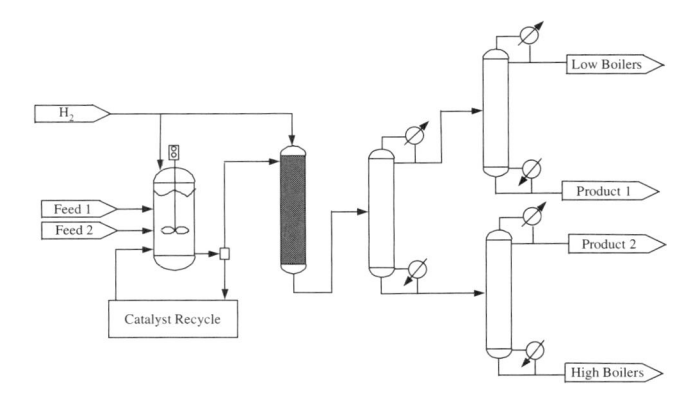

Figure 6. Combined process.

Conclusions

These three examples illustrate types of process synthesis problems encountered in industry. In all three cases, analysis was performed early in the life of the project which provided an expectation of the value of the project. This analysis was used to justify laboratory and pilot plant experimentation, and was used to direct the R&D efforts. Significant costs were eliminated by avoiding efforts that did not look to add the most value to the project. It is difficult to determine the effect of such significant issues as catalyst life, reactor selectivity, and separation methods for unknown species, so that several alternatives need to be investigated until these issues are resolved.

In these examples the fundamental issues behind the process synthesis are screening alternative reaction systems and their respective flowsheets, merging processes with similar chemistry, and utilizing reversible reactions. Process synthesis tools, including hierarchical methods and superstructure methods, need to be able to accommodate these types of issues.

References

Bagajewicz, M.J. and V. Manousiouthakis (1992). Mass/heat exchange network representation of distillation networks. *AIChE J.*, **38**, 1769.

Daichendt, M.M. and I.E. Grossmann (1998). Integration of hierarchical decomposition and mathematical programming for the synthesis of process flowsheets. *Comp. Chem. Eng,* **22**, 147-175.

Douglas, J.M. (1985). A hierarchical decision procedure for process synthesis. *AIChE J.*, **31**, 353.

Douglas, J.M. (1988). *Conceptual Design of Chemical Processes.* McGraw-Hill.

Friedler, F., K. Tarjan, Y.W. Huang, and L.T. Fan (1993). Graph-theoretic approach to process synthesis: polynomial algorithm for maximal structure generation. *Comp. Chem. Eng.*, **17**, 929.

Grossmann, I.E. (1996). Mixed-integer optimization techniques for algorithmic process synthesis. *Adv. Chem. Eng., Process Synthesis*, **23**, 171-246.

Gundersen, T., and I.E. Grossmann (1990). Improved optimization strategies for automated heat exchanger network synthesis through physical insights. *Comp. Chem. Eng.,* **14**, 925.

Linnhoff, B., D.W. Townsend, D. Boland, G.F. Hewitt, B.E.A. Thomas, A.R. Guy, and R.H. Marsland (1982). *User Guide on Process Integration for the Efficient Use of Energy.* IChemE, Rugby, UK.

Yeomans, H. and I.E. Grossmann (1998). A systematic modeling framework of superstructure optimization in process synthesis. Accepted for publication.

DRIVING FORCE COST AND ITS USE IN OPTIMAL PROCESS DESIGN

Jianguo Xu
Air Products and Chemicals, Inc.
Allentown, PA 18195

Abstract

The concept of driving force cost is introduced to evaluate the cost associated with providing the driving force in a unit process. It includes the cost of the energy resource consumed and the cost of the energy conversion equipment used to generate this driving force. Such a concept explains why in some cases an improvement in the process reduces energy cost and total equipment cost at the same time, and shows that the benefit of an energy efficiency improvement effort is not limited to energy cost reduction. The paper then relates the driving force consumption with that of exergy, which is often related with, and many times proportional to, the energy cost for the four elemental components of the non-energy conversion processes: mass transfer, heat transfer, fluid flow, and chemical reactions. Several heuristic rules are derived from this analysis. A more involved mathematical analysis using this concept yields very simple rules for optimal design of heat exchange processes.

Keywords

Process design, Process optimization, Conceptual process design, Driving force cost, Heat exchange.

Introduction

The objective function of optimization of processes, and very often that of innovation, is to minimize the cost of producing a certain product. That is, to minimize the sum of the initial capital equipment cost and the operating cost. When the recovery is fixed, it is often equivalent to minimizing the sum of equipment cost and energy cost. This is often how optimization problems are formulated, and how they are solved. The classical example used in undergraduate courses is the illustration using the cost - operating condition plot in which the capital cost, operating cost, and their sum are plotted against a design or an operating variable, such as the diameter of the pipe for fluid transport, see Edgar and Himmelblau (1988). It is based on the observation from such a classification that people often think that these two cost terms are rivals in optimization: an effort to reduce energy cost typically results in an increase in equipment cost since reduced energy cost means reduced driving force for the process to occur, which has to be made up by a larger equipment size. While such a notion is in many cases true, there are also cases in which an effort to reduce energy cost also reduces the cost of equipment. The reason is that the reduction in driving force, which is often related with energy cost, also results in a reduction in the cost of the equipment in generating that driving force. For example, when the pressure drop of a gas stream is the driving force for a certain process, and the higher pressure of the gas is obtained by gas compression, then the cost of the driving force includes the cost of electrical power and that of the compressor, the motor, the inter-coolers and aftercooler and the cooling water, if any, and the transformer etc. A reduction in the driving force also results in the reduction in the cost of the energy conversion equipment, which may be a major cost item in some applications. For example, in cryogenic air separation, the cost of the air compression related equipment is often higher than that of the distillations columns (the separation unit). The similar is true in many other gas separation processes.

The concept of driving force cost is introduced to address this and other issues. As will be seen from the text below, use of this concept allows us to relate the driving force, which is inversely proportional to the size of the non-energy conversion equipment, and therefore related with its cost, with the cost to create this driving force, and establish some rules that will be very useful in conceptual process design and in process optimization.

Driving Force and Driving Force Cost

The driving force cost includes all the cost items in generating that driving force. It is related with exergy consumption, along with some other operating conditions. For example, in many electrically driven processes, the energy cost portion of the driving force cost can be considered virtually proportional to the consumption of exergy used to create that driving force. The cost of the energy conversion equipment is also often a monotone function of the consumption of that type of exergy, although the operating condition may have a significant effect, too. As a consequence, the driving force cost can be directly related with the consumption of the exergy equivalent of the driving force. In some circumstances, it can be considered nearly proportional to the consumption of that type of exergy. Notice that I used "that type of exergy" since the unit costs of different forms of exergy, as related to the specific types of driving force, can be very different.

Table 1. Driving Force and the Corresponding Exergy Consumption of the Elemental Phenomena.

Phenomenon	Driving Force $\langle DF \rangle$	Exergy Consumption ΔEx ($d\Delta Ex/dl$ for fluid flows and mass transfer)
Incompressible Fluid Flow	$\sqrt{(dp/dl)}$	$F(dp/dl)/\rho$
Ideal Gases Flow	$\sqrt{(dp/dl)p}$	$-FRT_0 \ln(1-(dp/dl)/p)$
Mass Transfer	da/dl	$-FRT_0 \ln(1-\langle DF \rangle/a)$
Heat Transfer	$T_1 - T_2$	$T_0 q \langle DF \rangle / [T_1(T_1 - \langle DF \rangle)]$
Chemical Reaction $\sum_i \alpha_i A_i = \sum_j \beta_j B_j$	$\prod_i a_{A_i}^{\alpha_i} - \prod_j a_{B_j}^{\beta_j}/K$	$-FRT_0 \ln(1-\langle DF \rangle/\prod_i \alpha_{A_i}^{\alpha_i})$

Conventionally, driving force is defined as the differential part of the rate at which the non-energy conversion process occurs, such as pressure gradient, temperature gradient, activity gradient etc. In this paper, however, we will define it as the non-temperature variable part of the rate. The reason for so doing is that we are trying to relate the equipment size, which is inversely proportional to the driving force, and the exergy consumption. We want to include in the driving force term as many adjustable terms as possible (Although temperature is also adjustable in some cases, in many other cases it is more constrained. Besides, it typically does not affect the exergy consumption in a straightforward manner, or does not affect it at all. Therefore, we will exclude it from such a definition in this paper).

Driving force can be different for different processes. However, it is possible to extract some useful patterns. It may be said that there are basically four elemental components in the non-energy conversion processes in the process industry: mass transfer, heat transfer, fluid flow, and chemical reactions. I will first discuss the relationship between the driving force and its corresponding exergy consumption for these four elements to obtain some basic rules.

Fluid Flow

Under most process conditions, flows are turbulent. When a flow is turbulent, the mass transport rate is approximately proportional to the square root of pressure gradient. dp/dl, multiplied by its density, ρ. Therefore, its driving force is really the square root of the product of the pressure gradient and density. Since density does not change in incompressible fluids, this term drops.

The relation between driving force and exergy consumption is very different for liquids and gases. For incompressible fluids, which most liquids basically are, the exergy consumption per unit length is proportional to the pressure gradient (or the driving force squared). In the case of ideal gases, which is the model for most gases at low reduced pressures or high reduced temperatures, the change in gas volume causes the exergy consumption to deviate from this linear relationship with pressure gradient. See Table 1 for such relationships. Notice F is the flow rate, p the pressure, R the gas constant, and T_0 the ambient temperature.

Not all the flows in the process industry are turbulent ones. For example, the flows inside many hollow fiber membranes are more likely to be laminar ones. In such a case, the mass flow rate is proportional to the pressure gradient multiplied by the density. Therefore, the product of the pressure gradient and the density instead of its square root should be used to describe the driving force for mass transport. The exergy consumption formula is kept the same.

Mass Transfer

Most mass transfer processes follow Fick's law. In that case, the driving force for mass transfer is the activity gradient, *da/dl*. For ideal solutions, it is the gradient in concentrations. In the case of ideal gases, it is that in partial pressures. The exergy consumption, on the other hand, is proportional to the logarithm of the ratio of the activities. See the third row for the driving force and exergy consumption for mass transfer processes.

There are exceptions to such rules. For example, in the electro-chemical transport processes, as well as in the very slow mass transport processes such as that in solid phase, the mass transfer process my be better described to be proportional to the difference in Gibbs free energy. In that case, the exergy consumption is proportional to the driving force. These cases are, however, much less frequently encountered in the process industry.

Heat Transfer

Except in the cases in which radiation plays an important role, such as in very high temperature processes, heat transfer is typically considered to be proportional to the temperature difference between the heat releasing (at T_1) and heat receiving (at T_2) streams. The exergy consumption, however, is equal to the heat transferred, q, multiplied by the difference in Carnot efficiency (with the ambient as the heat sink). See the 4th column of Table 1.

In most heat exchange processes, it also holds that the heat transfer ΔT is much smaller than the absolute temperature, T, of the heat exchanging fluids. In that case, the exergy loss can be considered to be proportional to the heat transfer ΔT and inversely proportional to the absolute temperature squared.

Chemical Reactions

For a single phase chemical reaction of a solution $\sum_i \alpha_i A_i = \sum_j \beta_j B_j$, the driving force is equal to $\prod_i a_{A_i}^{\alpha_i} - \prod_j a_{B_j}^{\beta_j} / K$, in which K is the equilibrium constant. The exergy consumption is equal to $-FRT_0 \ln(1 - \langle DF \rangle / \prod_i \alpha_{A_i}^{\alpha_i})$. From this we know that the exergy consumption is not proportional to the driving force. For a heterogeneous reaction, the driving force still contains the $\prod_i a_{A_i}^{\alpha_i} - \prod_j a_{B_j}^{\beta_j} / K$ term, but the remaining part of the reaction rate that otherwise contains only a temperature dependent rate coefficient may include a case specific function of at least some of the activities of the reaction mixture. Therefore, the driving force can vary greatly from reaction to reaction. The exergy consumption, on the other hand, does not change.

From the basic relationships listed in Table 1, we can derive several heuristic rules about the optimal operating conditions for these processes to take place:

Fluid Flow

If the pressure exergy of the fluid at the product end can be fully used, it is desirable to transport gases at the highest possible pressure. That is partly because the exergy loss at the same driving force is smaller as pressure increases. That is of course the constant pipe volume (or rather the constant flow cross-section) situation. In many cases, the pressure will be let down when the gas arrives at the destination, constituting a parasitic loss of exergy. Even so, it is still a general practice to compress the gas to a high pressure before it is fed to a pipeline for a long distance transport.

On the other hand, since the density of an incompressible fluid is not adjustable, its exergy consumption is proportional to the pressure gradient only, there is no benefit in increasing the pressure to a higher value than what the end user needs. Therefore, the fluid is typically pumped to a pressure that is needed by the end user, plus a certain safety margin if applicable.

Such an analysis is also useful in more complex processes in which more than fluid flow occurs. It can be quantified, too.

Mass Transfer

As can be seen that for the same driving force if activity, a, is greater, the exergy loss is smaller. For ideal solutions, activity is equivalent to concentration. That means that if the concentration of the source is greater, the exergy loss for the same mass transfer driving force is smaller. In ideal gases, partial pressure can be used in the place of concentration, which is the product of the mole fraction and the total pressure. Therefore, at the same mole fraction conditions, the exergy loss at the same driving force is smaller when the pressure is greater. That is why higher pressures are preferred for many membrane gas separation processes if the pressure exergy contained in the product gas can be fully used (Xu and Agrawal, 1996).

Heat Transfer

As can be seen from Table 1 that the exergy loss for the same heat transfer driving force is inversely proportional to the absolute temperature squared. From this, one might want to conclude that a higher temperature is desirable to minimize exergy loss. However, typically the operating temperature of a heat exchanger is not arbitrarily chosen. The unit cost of heat transfer area may also vary with temperature. The optimal operating temperature for a heat exchanger is typically not that straightforward. In the next section, we will derive the optimal ΔT with some more sophisticated analysis.

Chemical Reactions

Different chemical reactions are very different. Therefore, there is no single rule that fits every reaction, or "most reactions". For example, for a combustion reaction and a neutralization reaction, the reaction rate is typically not as important as in slower reactions. Besides, in many reactions, conversion and product yield are often very important economic indicators, while the chemical exergy consumed in the reaction process is often not very expensive. Therefore, in such cases, less attention is paid to the "cost of driving force" than in other processes. Due to space limit, I will not be able to discuss in more detail about chemical reactions in this paper.

As mentioned before, the operating condition may have a significant effect on the cost of the energy conversion equipment. For example, since the size of a compressor is related with the volumetric flow, the compressor is larger, and therefore more expensive at a lower pressure. This further favors the use of higher pressures in fluid transport, mass transfer and related processes. Therefore, these are also important factors in determining the cost of driving force, and therefore the optimal design of processes.

The concept of driving force cost can be conveniently used to derive optimal design policies, as will be demonstrated in the next section.

Optimal Design of Heat Exchange Processes

In heat exchange process design, the most frequently asked questions are often: what is the optimal heat transfer ΔT, and what mass flow rate should be used? As simple as the heat exchange process may be (it does not involve chemical and composition change), there have been different schools of thoughts on this subject. One of the widely used method is to determine a certain minimum ΔT, and try to arrange the heat exchange processes such that the heat transfer ΔT is close to, but not smaller than, that minimum ΔT. Other schools include that advocating the use of Carnot efficiency differential in the place of ΔT since the loss of exergy is proportional to the difference in Carnot efficiency rather than ΔT. This would require ΔT to be proportional to T squared if $\Delta T/T$ is much smaller than one. Yet another school, represented by Bejan (1979), and Steinmeyer (1984, 1992), believe the optimal ΔT should be approximately proportional to the absolute temperature. We will try to find such an optimal ΔT through the use of the concept of "driving force cost".

Let's assume that the Fourier heat transfer equation takes the form

$$r_i \approx u_i \Delta T_i \quad (1)$$

in which r_i is the rate of heat transfer at i-th location, and u_i the heat transfer coefficient at location i. ΔT_i is the heat transfer temperature difference at location i. The heat transfer area of the heat exchanger, C, is thus:

$$C = \int_0^Q \frac{dq}{u \Delta T} \quad (2)$$

in which Q is the duty of the heat exchanger, u is the heat transfer coefficient, which can be considered independent of the driving force. Temperature T is a function of q when exchange of sensible heat is involved.

The rate of exergy loss follows

$$\Delta Ex = T_0 \int_0^Q \left(\frac{\Delta T}{T(T - \Delta T)}\right) dq \quad (3)$$

In order to optimize the driving force, we will have to set the derivative of the cost function with respect to the set of the driving forces $\Delta \mathbf{T}$ to zero. Notice that $\Delta \mathbf{T}$ is the set of driving force ΔT at any location i. At the optimum, the derivative of production cost, which is the sum of the cost of the non-energy conversion equipment and those of the driving forces, with respect to $\Delta \mathbf{T}$ is zero:

$$a \frac{\partial}{\partial \Delta \mathbf{T}} \int_0^Q \frac{dq}{u \Delta T} + eT_0 \frac{\partial}{\partial \Delta \mathbf{T}} \int_0^Q \frac{\Delta T}{T(T - \Delta T)} dq = 0 \quad (4)$$

$$\left(a = \frac{dA(C)}{dC}, \quad e = \frac{dE(\Delta Ex)}{d\Delta Ex}\right) \quad (5)$$

in which $A(C)$ is the cost of the heat exchanger as a function of the area, C, and $E(\Delta Ex)$ the capitalized cost of ΔT as a function of the rate of thermal exergy consumption in the heat exchanger. a and e can be considered to be the unit costs of the heat transfer area and thermal exergy, respectively. Notice that a and e are not necessarily equal to the total cost of the heat transfer area divided by the area and the total cost of ΔT divided by the total exergy consumption. For sub-ambient temperature processes, such a consideration is often a good approximation.

Since the policy at one point in the heat exchanger does not affect the objective function at a different position in this particular problem, and since T is much greater than ΔT, equation (4) is equivalent to equation (6) for all the locations in the heat exchanger:

$$a \frac{-dq}{u_i \Delta T_i^2} + eT_0 dq / T_i^2 = 0 \quad (6)$$

For equation (6) to hold, we should have

$$eT_0 / T_i^2 - \frac{a}{u_i \Delta T_i^2} = 0 \quad (7)$$

that is

$$\Delta T_i = T_i \sqrt{\frac{a}{eT_0 u_i}} \quad (8)$$

Equation (8) is exactly what Steinmeyer (1984, 1992) suggested for determining the optimal ΔT in reboilers. It gives the optimal driving force in a heat exchanger if this driving force can be manipulated without constraints. It also shows that the optimal ΔT should be proportional to the square root of the unit cost of the heat exchanger, and proportional to the square root of the ratio of the unit cost of the heat transfer area to that of the thermal exergy, and inversely proportional to the square root of the heat transfer coefficient.

From the above result, it is not difficult to further derive that at the optimal design conditions, the cost of heat transfer ΔT is equal to the cost of the heat transfer area (which is a portion of the cost of the heat exchanger).

The reader can also derive that if the head gradient is proportional to velocity squared, and the heat transfer coefficient is proportional to the mass velocity to 0.8^{th} power, the optimal mass velocity in a heat exchanger with a fixed length is such that the cost of the pressure drop in the heat exchanger is 40% of the heat transfer ΔT cost.

In order to show how this method results in savings, the reader can calculate the thermal exergy loss in the process of cooling a fluid with a constant heat capacity from 300K to 5 K in a heat exchanger with a fixed area and heat transfer coefficient when $\Delta T/T$ is much smaller than 1. The reader will then learn that the exergy loss when the ΔT is proportional to T is **less than 1/3** of that when ΔT is constant or proportional to T squared. Since the cost of heat transfer ΔT is a monotone function of the thermal exergy consumption in a heat exchanger, and that relationship is nearly linear in sub-ambient processes in which the source of the driving force is electricity, we know that the operating cost of that heat exchange process is much smaller when ΔT is proportional to T than using the other two policies.

Conclusion

By isolating the cost of generating the driving forces from that of the equipment in which the non-energy conversion process takes place, one can see how the operating parameters affect the different cost items with much less complexities and confusion. It also shows that the effort to increase driving force efficiency is not solely to reduce energy consumption. In many applications, such as in fluid transport and mass transfer processes, it often results in simplified rules for optimal design. In heat exchange processes, this paper showed how the application of such a concept results in the very simple rules for optimal design.

References

Bejan, A., (1979). A general variational principle for thermal insulation system design. *Int. J. Heat Mass Transfer*, **22**, 219-228.

Edgar, T. F. and D. M. Himmelblau (1988). *Optimization of Chemical Processes*, McGraw-Hill, New York. 503-506.

Steinmeyer, D. (1984). Process energy conservation. *Encyclopedia of Chemical Technology*, Supplement volume. Wiley, New York.

Steinmeyer, D. (1992). Optimum DP and DT in heat exchange. *Hydrocarbon Processing*, **4**, 53-56

Xu, J. and R. Agrawal (1996), Membrane separation process analysis and design strategies based on thermodynamic efficiency of permeation. *Chem. Eng. Sci.*, **51**, 3, 365-385.

INCORPORATION OF PRIOR PLANT KNOWLEDGE INTO NONLINEAR DYNAMIC BLACK BOX MODELS BASED ON INFORMATION ANALYSIS

Jun Hsien Lin, Shi-Shang Jang, and M. Subramaniam
National Tsing Hua University
Hsinchu, Taiwan

Shyan-Shu Shieh
Chang Jung University
Tainan County, Taiwan.

Abstract

To improve process modeling, the focus has shifted to nonlinear modeling such as artificial neural network (ANN) in the last decade. Very often, data used for modeling are not enough or inappropriate to catch the characteristics of processes. In this study, information free energy based experiment design is used to improve the quality of operational data set. For a complicated dynamic process, non-convex feasible regions are very likely to be encountered. Experiment design may suggest false points in the region where system can not reach. A novel non-convex triangulation is developed to handle non-convex feasible regions. Furthermore, we proposed a scheme to incorporate prior process knowledge to help obtain a reliable nonlinear dynamic model. We use Akaike Information Criterion to tune the penalty when different types of prior knowledge are used in modeling. ANN models acquired by the above approach are implemented in a simulated three component distillation tower. The results demonstrate that the proposed approach dramatically reduces prediction errors. And the acquired models are so trustworthy that model predictive control is very robust even in the noise contaminated operation environments.

Keywords

Dynamic modeling, Experiment design, Information theory, Non-convex hull.

Introduction

Almost all chemical processes are nonlinear systems and most of them are too complicated and too expensive to be modeled by first principles. Only empirical models, such as dynamic matrix model, can be used in the area of model predictive control (MPC) for the process industries. Unlike the linear case, there exists no systematic approach for the development of a nonlinear empirical model. Furthermore, partial plant knowledge usually exists in the process. The incorporation of prior plant knowledge into an empirical model is hence an important issue. One of the objectives of this paper is to derive a systematic approach to assess the effectiveness of prior plant knowledge, and thereafter incorporate prior plant knowledge into an empirical model.

Akaike Information was originally proposed as a regression index (Akaike 1970). The minimization of Akaike information criteria (AIC) has been widely implemented on model diagnosis and model structure selection (Tong 1990). The physical meaning of AIC is the compromise between regression errors and degree of freedom of a regression model. Conditional Akaike information criteria (CAIC) is used in this work to determine the optimal combination of experiment data and several types of prior knowledge. Prior plant knowledge can be equations derived from first principle models or operational behavior of a process plant, e.g. smoothness. We combine these partial knowledge data with on-line plant data together. A neural network model that trains these data together is derived. We also consider that the data obtained from different sources should be weighted differently. The weighting values can be determined by optimizing CAIC.

In this work, we term the "completeness" of a model as an empirical model always performs interpolatively as it predicts any new event. However, only limited works focus on this property previously. Our previous works (Chen et al., 1998, Lin and Jang, 1998) proposed that the model should determine new experiments to insure the quality of its prediction by investigating the total information free energy of the training set. Furthermore, we proposed to implement Delaunay triangulation to find the boundaries of feasible operation events. However, we assumed the convexity of the operating boundaries at that time. This is not always true in highly nonlinear systems such as high purity separation columns. In this work, we, one step further, developed the generalized Delaunay triangulation approach to find the non-convex operating region of a nonlinear system. The information analysis based on the non-convex operating region guarantees the "completeness" of an empirical model. This implies that the empirical model always performs interpolations.

Information entropy is derived by Shannon (Shannon, 1948) to evaluate the uncertainty of a random variable. Recently, many applications of this theory can be found in the literatures. However, most of these applications concentrated in the area of pattern recognition and image processing and only a few in the process engineering□Shieh, 1992□. In contrast to information entropy, the authors derived information enthalpy to evaluate the nonlinearity of the plant (Chen et al., 1998, Lin and Jang, 1998). We also argued that it is more appropriate to put new experiments according to the combination of information entropy and information enthalpy – information free energy. In this work, we determine the most efficient experiments based on non-convex feasible territory that is extended by the existed plant data and partial plant knowledge.

Model predictive control of high purity separation columns has been a challenging problem in the area of process control due to its nonlinearity and multivariable nature. In this work, an empirical multivariable model that incorporates steady state information of a high purity separation column is developed. The neural network model is "complete" and hence will not perform extrapolatively in plant operation. The MPC based on the empirical model works nicely for the dual temperature control of the high purity separation column.

Theory

Consider the following generalized lumped system:

$$\dot{\mathbf{x}} = \mathbf{f}(\mathbf{x}, \mathbf{m}) \qquad (1)$$
$$\mathbf{y} = \mathbf{g}(\mathbf{x}, \mathbf{m})$$

where $\mathbf{x} \in R^n$ are termed state variables, $\mathbf{m} \in R^m$ are plant inputs, and $\mathbf{y} \in R^p$ are system outputs or the on-line measurements of the plant. Assume that the above system can be equivalent to the following discrete time system under a determined sampling time T:

$$\mathbf{y}_{k+1} = \mathbf{h}(\mathbf{y}_k, \mathbf{y}_{k-1}, \cdots, \mathbf{y}_{k-s}, \mathbf{m}_k, \cdots, \mathbf{m}_{k-s'}) \qquad (2)$$

where k is current time. Denote an event happened to system (1) or (2):

$$\phi^i = [\mathbf{y}_k, \mathbf{y}_{k-1}, \cdots, \mathbf{y}_{k-s}, \mathbf{m}_k, \cdots, \mathbf{m}_{k-s'}] \qquad (3)$$

Denote the feasible region of system (1) or (2):

$$\Phi = \{\phi^i | \phi^i = \text{all possible events}, i = 1, 2, \cdots\} \qquad (4)$$

Problem Formulations

Consider a system (1) or (2), there exists an experimental event set

$$\Omega = \{\omega^i | \omega^i, i = 1, \cdots N\} \subset \Phi$$
$$\overline{y}_{k+1} = \mathbf{h}(\Omega) \qquad (5)$$

All events, Ω, are determined by experiments, and N is the size of the experimental event set. $\overline{y}_{k+1}$ is on-line measurement and contaminated by noise $N(0, \sigma^2_j)$.

Partial knowledge is given as

$$\Psi = \{\psi^i = [\hat{\mathbf{y}}_k, \hat{\mathbf{y}}_{k-1}, \cdots, \hat{\mathbf{y}}_{k-s}, \mathbf{m}_k, \cdots, \mathbf{m}_{k-s'}]\}$$
$$\hat{\mathbf{y}}_{k+1} = h'(\Psi) \qquad (6)$$

We here assume that the output for ψ is not noisy but the knowledge can be incorrect, i.e., h'≠h.

The first purpose of this work is to find the following neural model based on the training set, such that

$$\hat{y}_{k+1} = ANN(\Omega \cup \Psi) \cong h(\Phi) \qquad (7)$$

$\hat{y}_{k+1}$ is the output of neural model. To utilize prior knowledge as well as experimental data, we propose to obtain a gray-box ANN model by minimizing the following objective function:

$$J(\mathbf{w}) = \sum_{i=1}^{N} \left| \hat{\mathbf{y}}_{k+1}(\boldsymbol{\omega}^i, \mathbf{w}) - \overline{\mathbf{y}}_{k+1}(\boldsymbol{\omega}^i) \right| + \lambda \sum_{i=1}^{p} \left\| \frac{\partial^2 \hat{\mathbf{y}}_{k+1}(\boldsymbol{\omega}^i)}{\partial \boldsymbol{\omega}^i} \right\|^2 + \beta \sum_{i=1}^{R} \left\| \hat{\mathbf{y}}_{k+1} - \hat{\mathbf{y}}_{k+1}(\boldsymbol{\psi}_i) \right\| \quad (8)$$

where w is weightings of neural model.

Given total N pieces of data, which can be separated into two sets, one is a training set (K pieces of data), the other is a test set (N-K), we determine the penalty $\pounds a = (\lambda, \beta)$ by minimize following conditional AIC (Tong, 1990)

$$\underset{\zeta}{Min} V(\zeta) = \ln \left[\frac{1}{N-K} \sum_{i=1}^{N-K} \left| \hat{\mathbf{y}}_{k+1}(\boldsymbol{\omega}^i, \zeta) - \overline{\mathbf{y}}(\boldsymbol{\omega}^i, \zeta) \right|^2 \right] + \frac{2}{N-K} \dim[\zeta, w] \quad (9)$$

Development of Extended Experimental Data Set

Definition (Convex Hull):

Given Ω, then the convex hull of Ω is the smallest convex set containing Ω, denoted as $C(\Omega)$ (Lin and Jang, 1998).

Definition (Non-Convex Hull):

Given Ω, divide Ω into m subset. Non-convex hull, $NC(\Omega) = \bigcup_{i=1}^{m} C(\Omega_i)$, can be decided as follows:

$$\min_{i=1,...,m} \bigcup_{i=1}^{m} Area(C(\Omega_i)) \quad \text{s.t.} \quad \bigcup_{i=1}^{m} C_i \text{ is compact.} \quad (10)$$

The feasible region can be estimated as $C(\Omega)$ or $NC(\Omega)$. The following problem should be solved for the placement of a new data point:

$$\underset{\varpi_{NEW}}{Min} G = H - TS$$
$$s.t \quad y_{k+1} = ANN(\varpi_{NEW})$$
$$\Delta y_{k-j,\max} = NF(y_{k-j-1}) \quad (11)$$
$$m_{\min} \leq m_{k-l} \leq m_{\max}$$
$$\left| \Delta m_{k-i} \right| \leq \Delta m_{\max}$$

where NF: non-convex constraint on y_k
 H: information enthalpy
 T: information temperature.
 S: information entropy.

The solution of the above equation needs an input sequence to perform an experiment. (Lin and Jang, 1998). The algorithm is shown in Figure 1.

An Illustrated Example

A distillation column is identified as a 2x2 system

$$[y_{k+1}^1, y_{k+1}^2] = f(m_k^1, m_k^2, y_k^1, y_{k-1}^1, y_k^2, y_{k-1}^2) \quad (12)$$

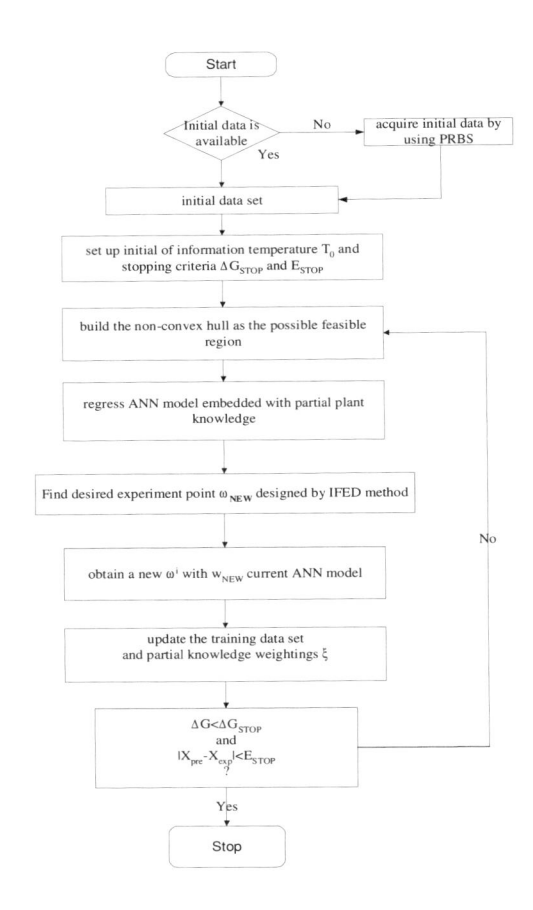

Figure 1. Flow chart of IFED.

where y_k^1 and y_k^2 are bottom and top temperature respectively; m_k^1, reboiler heat duty, m_k^2; reflux flow rate. The column is an 18-tray tower with 3 components feed, 50% benzene, 35% toluene, 15% xylene. The specifications are detailed in Lin's work (1999). Temperature measures are contaminated with white noise distributed as N(0,0.05). All process input satisfied the constraints, 2850 $\square$ m_k^1 $\square$ 3350 kBTU/hr, 65 $\square$ m_k^2 $\square$ 135 lb-mole/hr, Δm_k^1 $\square$ 50, Δm_k^2 $\square$ 6.

Three different types of training data set are obtained based on the following training data set.

(1) A training set with 600 data points using random amplitude signal, denoted as RAS (Pottman and Seborg, 1992).
(2) A training set with 600 data points using pseudo random binary signal, denoted as PRBS.
(3) ANN model based on the IFED training data set composed of 160 IFED points, 120 initial PRBS points and 36 steady state data, denoted as IFED.

The penalties of steady state and smooth factor are [2.9605 0.0025] according to minimization of CAIC.

In the case (3), all the points are located in the smallest region of non-convex hull. For the purpose of clarity, only the first 50 of 160 IFED points are shown in Fig. 2. The dotted line represents the non-convex hull of the distillation column.

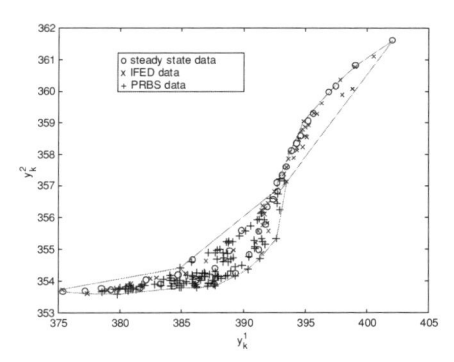

Figure 2. Non-convex hull for distillation column.

Three different types of ANN models are obtained and implemented for MPC. To demonstrate the performances of MPC, the distillation column, initially stable at [355 3900.2], top set point is changed from 355 to 354.5 at start and bottom set point is changed from 390.2 to 390.7 at 300 min.

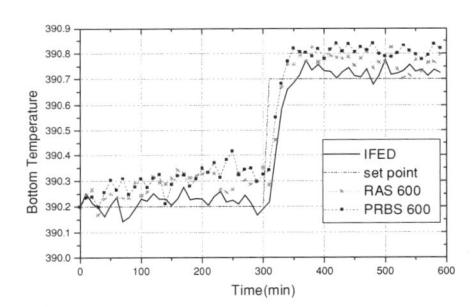

(a) Response of bottom temperature control.

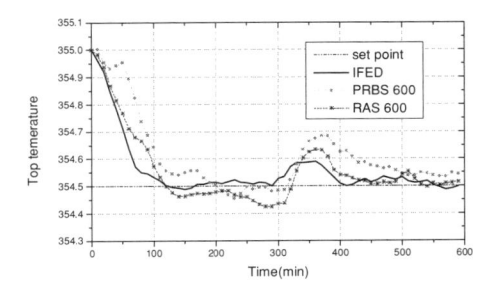

(b) Response of top temperature control.

Figure 3. MPC results of the three ANN models.

The performances of IFED approach are the best. IFED has a shorter settling time both in servo control and regulation control. There are no significant off-set in the both control loop only for IFED model.

Summarily, IFED case outperforms the other two in terms of offset and settling time.

Conclusions

A novel ANN training approach that can incorporate partial plant knowledge and noisy experimental data is derived. The ANN modeling scheme is applied to develop a dynamic model for a chemical process with a non-convex feasible region. Generalized non-convex hull is derived to determine the non-convex region.

A realistic multi-component distillation system is also simulated to show the superiority of this approach. The simulation results demonstrate that this approach is very useful for the development of gray box models (empirical model with partial plant knowledge) that can be implemented for multi-variable nonlinear model predictive control.

References

Akaike, H. (1970). Statistical predictor identification. *Ann. Inst. Statist. Math.*, **22**, 203-217.

Chen, J., Wong, D.S.H., Jang, S.S., and S.L. Yang (1998). Product and process development using artificial neural net model and information theory. *AICHE J.* **44**, 876-887.

Lin, J.S., and S.S. Jang (1998). Nonlinear dynamic artificial neural network modeling using an information theory based experiment design approach. *Ind. Eng. Chem., Res.,* **37**, 3640-3651.

Lin, J.S., and S.S. Jang (1999). Incorporation of partial plant knowledge into a multivariable dynamic artificial neural network model based on information analysis and experimental design. Paper submitted to *AICHE J.*

Pottman, M., and D.E. Seborg (1992). Identification of a non-linear process using a reciprocal multi-quadratic function. *J. Proc. Control*, **2**, 4,189-203.

Shannon, C.E. (1948). A mathematical theory of communication. *Bell Sys. Tech. J.*, **27**, 379-423.

Shieh, S. S. (1992). *Automatic Knowledge Acquisition from Routine Data for Process and Quality Control*. Doctoral dissertation, Washington University, St. Louis.

Tong, H. (1990). *Non-Linear Time Series: A Dynamical System Approach*. Oxford Univ. Press.

CONCEPTUAL DYNAMIC MODELS FOR THE DESIGN OF BATCH DISTILLATIONS

Enrique Salomone and José Espinosa
INGAR-CONICET
3000 Santa Fe, Argentina

Abstract

We present a methodology for modeling the operational behavior of batch distillation columns processing non-ideal and azeotropic mixtures. The methodology is based on a dynamic conceptual model for predicting the separation performance at the limiting condition of infinite number of stages. The level of detail is enough to preserve the non-ideal thermodynamic behavior and still simple enough to allow rapid simulation of dynamic runs suitable for the estimation of the minimum reflux needed for a final separation specification. The model also supports the estimation of the accumulated recoveries and operating time.

Keywords

Batch distillation, Azeotropic mixtures, Conceptual models, Separation performance, Design.

Introduction

Fractionation in batch distillation columns is one of the most common technologies used in pharmaceutical and chemical synthesis industries for separating solvent mixtures in waste-solvent streams. Since typical solvents are likely to form azeotropic mixtures, the design of recovery strategies based in distillation needs to be supported by conceptual operational models capable to represent the behavior of non-ideal and azeotropic mixtures.

The models we have developed consist in dynamic representations of limiting operation modes for batch columns: columns with infinite stages operating at a given reflux ratio and columns with a given number of stages operating at total reflux. These two extreme modes of operation allow the exploration of the whole space of possible separations and therefore we have found them very useful for being used at the conceptual design stage.

Given the latest significant advances in the thermodynamic analysis of azeotropic mixtures allowing the prediction of azeotropes, distillation regions and boundaries together with feasible sequences of distillate cuts and maximum attainable separations (Safrit and Westerberg, 1997; Rooks et al., 1998), the method in this paper incorporate this knowledge and extend the analysis to the operational issues in batch distillation.

For the infinite stage limiting condition, the method is based on the estimation of the instantaneous top composition without resorting to multistage calculations. By using linear approximations of the composition profile in the neighborhood of the pinch points, the model identifies the geometry of the profiles and the controlling pinch corresponding to the present still composition and reflux ratio. The method extends the ideas in Offers, Düssel and Stichlmair (1995) to variable distillate composition at a given reflux ratio and therefore suitable for column simulation.

The total reflux limiting condition can be adequately described by solving a sequence of equilibria and mass balance computations for a given still composition and assuming a differential rate of product withdrawal from the column at any time and is not discussed in this paper.

Instantaneous Product Composition

In order to introduce the concepts behind the method we will first focus our attention on the geometry of the instantaneous internal profiles in an infinite stages rectifying column and operating at different reflux ratios while inspecting the all-possible distillate compositions from a given instantaneous still composition.

Let's consider the given still composition x_B in Fig. 1, y^*_{xB} represents its vapor in equilibrium and must be located on a tangent to the residue curve that passes through the still composition. From x_B, different distillate compositions can be achieved for different values of the reflux ratio. Under total reflux operation, it can be expected that the internal liquid profile resemble the residue curve passing through the still composition. The top product will be the more volatile species Methanol and a pinch will be placed at the top of the column. For small values of the reflux above zero the distillates are aligned with the equilibrium vector y^*_{xB} - x_B starting from y^*_{xB} when the reflux equals zero up to the point x^*_D where the composition of the heaviest component becomes zero. We will refer as R^* to the reflux ratio corresponding to this condition. For all these separations a pinch is located at the end of the column immediately above the boiler. The pinch composition coincides with the composition of the mixture in the still. Offers, Düssel and Stichlmair (1995) called such instantaneous separations as "preferred separations". As Düssel (1996) shown, for a reflux ratio equals to R^* a binary saddle pinch x_p^{II} between the lighter components Methanol and Ethanol appears. Due to the saddle the profile becomes sharp, the internal profile between the instantaneous still composition and the saddle lies on a straight line containing both the saddle and the still compositions. For values of the reflux greater then R^* the distillate compositions are located in the binary edge corresponding to the lighter components between x^*_D and x_D^{max}. There is no more a pinch at the end of the column, but the binary saddle pinch remains with its composition x_p^{II} (invariant).

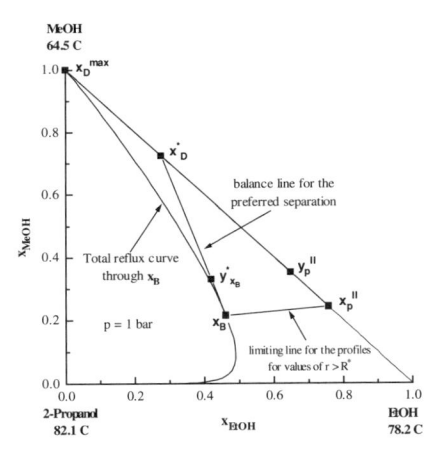

Figure 1. Instantaneous product composition region.

The internal profiles between the instantaneous still composition and the invariant saddle lie on a straight line. At x_D^{max} the composition of the second heaviest component becomes zero, we refer to this reflux condition as R^{**}. Finally, for values of the reflux greater than R^{**}, the profiles resemble the total reflux separation with a pinch at the column top and hence, there is not a saddle pinch.

Operation Regimes at Constant Reflux

In a column operating at constant reflux, the still composition is varying all the time and therefore, during a typical run, different profile patterns may occur for the same reflux ratio.

To illustrate the above ideas, let's imagine the operation of an infinite batch column under finite reflux operation policy for a ternary mixture. At the startup the column will operate at total reflux and hence, the top composition will be pure volatile component with a pinch at the top of the column. Then, assuming that a high reflux ratio (greater than the R^{**} corresponding to the initial feed) is established, the profiles will behave as in the case of total reflux. The column will be producing a pure distillate during certain period of time. Because of the depletion of the light component from the still, the column will not be able to sustain a pure top composition and the top will become a binary mixture containing the two lightest components. The situation now is such that the operating reflux ratio is between the R^* and R^{**} corresponding to the instantaneous still composition. Therefore, the separation proceeds with a profile having a saddle pinch between the lighter species. Afterwards, the column will operate with a pinch at the end of the column. The pinch composition will coincide with the instantaneous still composition and the instantaneous value of R^* is such that the operating reflux is between 0 and R^*. In a similar way, quaternary mixtures can traverse through four operating regimes while the column is operated at constant reflux; namely: total reflux, binary pinchs controlling the separation, ternary- and finally quaternary-controlling pinchs. The corresponding geometry of the profiles in the neighborhood of the still composition will be non-sharp profiles with a pinch at the column top, profiles lying on planes, linear profiles and non-sharp profiles with their pinch compositions coincident with the instantaneous still composition, respectively.

Instantaneous Separation Performance

The concepts discussed in the preceding sections can be used to develop an algorithm for the prediction of the instantaneous top composition of a batch column having infinite stages, given the still composition and the operating reflux ratio.

The estimation of the instantaneous top composition makes possible the integration of the mass balances of an infinite batch column assuming negligible holdup in the column and condenser.

For non-ideal mixtures, the characteristic pinch points are no longer invariant and in general the profiles between the still composition and the controlling pinch are no longer straight lines connecting them. However, in the neighborhood of the still composition, the linearization of the profile is still a good approximation. This situation is illustrated in Fig. 2 for a ternary system with a controlling binary saddle pinch. The profile leaves the point x_B in the direction given by the eigenvector v_I resulting from the solution of the eigenvalue problem of the Jacobian of the equilibrium, which correspond to a linerization of the profile at x_B (Pöllmann and Blass, 1994). Thus, it is possible to solve the overall mass balance around the column given the still composition, its vapor in equilibrium y^*_{xB} and the operation reflux. The composition of the liquid stream immediately above the boiler (x_N) resulting from the overall balance will be placed on the line that approximates the actual profile.

For small values of the reflux ratio ($r<R^*$) a pinch is located at the end of the column immediately above the boiler. The pinch composition coincides with the composition of the mixture in the still and therefore x_N is equal to x_B. The distillate is located on the preferred separation line. If the heavy species does not distribute in the distillate, a linearization of the profile must be done to calculate a distillate containing the lightest components. In this case, x_N is located on the line indicated by the eigenvector. Finally, if the intermediate component does not distribute in the distillate, only the lower boiling component appears at the top of the column. With the same basic ideas, an algorithm can be developed for multicomponent mixtures.

Examples

We illustrate the use of the model to analyze the separation of an azeotropic mixture by means of an entrainer. Figure 3 shows the still path departing from F_1 corresponding to a column operating at a reflux ratio of 5. The feed to the process F_1 consists of a mixture between the original feed F composed by 2-Propanol and water (azeotropic composition) and the Cyclohexane-rich phase E. From this mixture the ternary azeotrope is obtained at the top of the column until the still path reaches the edge of the triangle. At this moment, an intermediate cut consisting of the azeotropic mixture is removed as overhead product. An optimal feed M will avoid the undesirable intermediate cut.

Two unstable separatrixes limit the achievable top compositions for the given feed and hence, these distillation boundaries must be taken into account into the short-cut procedure. Given an instantaneous still composition, preferred separation is considered until intersection of that line with a linear approximation of the distillation boundary is achieved. If the actual reflux ratio is greater than R^*, a linearization of the liquid profile around the still composition and a search of the top composition lying on the linearized distillation separatrix must be done.

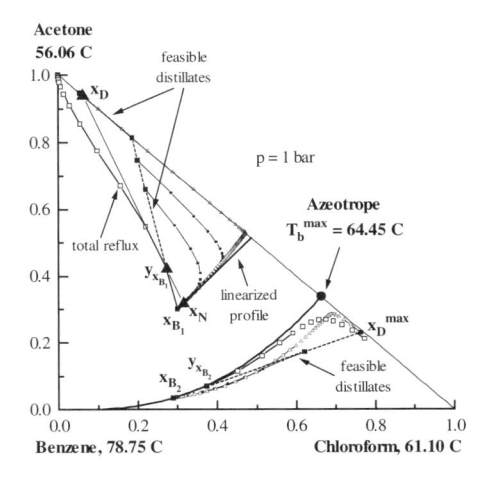

Figure 2. Linearization of the profiles in non-ideal systems.

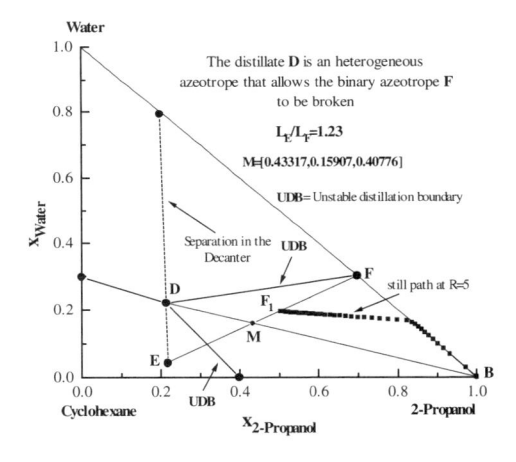

Figure 3. Still path at finite reflux of a system showing unstable distillation boundaries.

As a second example we analyze a system with four distillation regions. Figure 4 shows still paths calculated from our model at both low and high reflux ratios for six initial feeds. The system considered presents a stable distillation boundary that joins Methanol (stable node), the ternary saddle azeotrope and the maximum boiling azeotrope Chloroform-Acetone (stable node). Also, there exists an unstable boundary joining the minimum boiling azeotropes Acetone-Methanol (unstable node), Chloroform-Methanol (unstable node) and the ternary saddle azeotrope. As a result of these boundaries four distillation regions divide the concentration simplex as was noted by Van Dongen and Doherty (1985). Each one of the distillation regions has one unstable node, one or more saddle nodes and one stable node.

As in the previous example, the unstable distillation boundary limits the products that can be achieved at the top of the column. In the context of our short-cut procedure, two lines replaced the curved distillation boundary: one line joining the saddle with the azeotrope Acetone-Methanol, the other joining the saddle with the azeotrope Chloroform-Methanol.

Once a still path reaches a stable separatrix only top products lying on the tangent to the stable boundary at the still composition are permitted (Van Dongen and Doherty, 1985). That means that the only possible products are on the preferred separation line corresponding to an instantaneous still composition starting from its vapor in equilibrium until a point either on a unstable boundary or on a border of the composition simplex is reached. Figure 2 shows the feasible distillates for a still composition lying on the stable separatrix of the system Acetone-Chloroform-Benzene.

To reproduce the mentioned behavior a nonlinear approximation for the stable distillation boundary was included into the algorithm. At each step of the integration during a typical run the location of the still composition with respect to the boundary was subjected to verification. If the still composition pertains to the stable boundary only preferred separation is permitted starting from this point. Another way to check this is by obtaining the residue curve for each bottom. A swap in the unstable node indicates a crossing of the stable boundary. This approach does not require an equation to represent the stable boundary. We detect the change in the present unstable node by calculating the distillate composition at a very high reflux (i.e. $R = 1000$).

Returning to Fig. 4, note that at low values of the reflux ratio, the still paths resemble the residue curves through the corresponding initial composition. At high values of the reflux, each one of the paths moves along a straight line through the initial composition away from one of the unstable nodes.

As the still paths approach the stable distillation boundary (see points 1, 3, 4 and 5 in Fig. 4) they follow the curve toward one of the stable nodes. During this step of the distillation only top products along the tangent to the curve are possible.

Conclusions

We have developed a short-cut procedure for estimating the instantaneous separation performance for batch columns having infinite number of stages that it is applicable to non-ideal mixtures. The model assumes a negligible holdup for the column and condenser and allows the estimation of the instantaneous top composition without resorting to stage by stage computation. The key aspect of the method is the determination of the operating regimes at each instant of time. The regime is characterized by a controlling pinch, that in turn controls the geometry of the internal profiles. With this model, both a method for rapid simulation of the batch operation and a procedure for calculating the minimum energy demand can be developed as an extension of the method for ideal systems described in a previous work (Salomone, Chiotti and Iribarren, 1997).

Acknowledgements

The authors are grateful to CONICET and Soteica SRL for supporting this research.

References

Düssel, R. (1996). *Zerlegung azeotroper Gemische durch Batch-Rektifikation*. Ph. D. Thesis, Technische Universität München, Deutschland.

Offers, H., R. Düssel and J. Stichlmair (1995). Minimum energy requirement of distillation processes. *Comp. Chem. Eng.*, **19**, S247-S252.

Pöllmann, P. and E. Blass (1994). Best products of homogeneous azeotropic distillations. *Gas. Sep. Purif*, **8**, 4, 193-227.

Rooks, R. E., V. Julka, M. F. Doherty and M. F. Malone (1998). Structure of distillation regions for multicomponent azeotropic mixtures. *AIChE J.*, **44**, 6, 1382-1391.

Safrit, B. T. and A. W. Westerberg (1997). Algorithm for generating the distillation regions for azeotropic multicomponent mixtures. *Ind. Eng. Chem. Res.*, **36**, 1827-1840.

Salomone, H. E., O. J. Chiotti and O. A. Iribarren (1997). Short-cut design procedure for batch distillations. *Ind. Eng. Chem. Res.*, **36**, 1, 130-136.

van Dongen, D. B. and M. F. Doherty (1985). On the dynamics of distillation processes-VI. Batch distillation. *Chem. Eng. Sci.*, **40**, 11, 2087-2093.

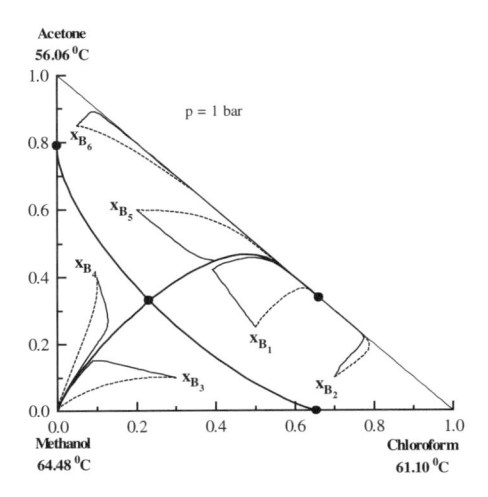

Figure 4. Typical still paths of a system with four distillation regions.

PLANT DESIGN BASED ON ECONOMIC AND STATIC CONTROLLABILITY CRITERIA

Panagiotis Seferlis and Johan Grievink
Department of Chemical Engineering
Delft University of Technology
2628 BL Delft, The Netherlands

Abstract

A methodology is developed that screens process designs and control system configurations, defined in terms of selected controlled and manipulated variables, that exhibit poor steady-state behavior under the presence of multiple simultaneous disturbances. The steady-state disturbance rejection characteristics of the nonlinear process models are considered in conjunction with economic criteria for the plant design. The steady-state control objectives for the system are incorporated within an optimization framework. Sensitivity information of the optimal solution is utilized to modify the design parameters for improved static controllability performance. The design approach allows the study of plants with unequal number of manipulated and controlled variables and the investigation of cases with saturation of manipulated variables and hard bounds on the controlled variables.

Keywords

Process design, Static controllability, Control structure selection, Parametric optimization, Sensitivity analysis, Continuation methods.

Introduction

In the most widely used procedures for chemical process design, controllability aspects of process systems are considered only in a very late stage (Douglas, 1988). Therefore, the achieved performance of the control system is irreversibly affected by the limitations imposed by the process flowsheet configuration and the equipment design. The present work incorporates the use of static controllability criteria in the early design of process systems and plantwide control systems. The level of control system design considered is defined as the selection of the potential set of controlled and manipulated variables that would ensure the satisfaction of the control objectives. The selection of the set of controlled variables affects the quality of the static behavior of the system due to the usually nonlinear relationship between the input and output variables. The manipulated variables are generally equal to the legitimate control valves in the system, which are primarily affected by the process design (flowsheet configuration and process equipment selection), and constitute the system's degrees of freedom. However, the full count of all possible combinations between potential manipulated and controlled variables for different process flowsheets may be very large. Thus, optimization based approaches with embedded dynamic models (Schweiger and Floudas, 1997) and the incorporation of matrix metrics (singular values, condition number) and interaction measures for the control quality (Luyben and Floudas, 1994) in the design of process and control systems may become prohibitive.

Yi and Luyben (1995) assessed the ability of different control structures to alleviate the steady-state effects of step changes in single variables. Lewin (1996) introduced the concept of disturbance cost to quantify the required control effort to keep the output variables on the desired setpoint level. This procedure allows the investigation of the effects of directionality in disturbances over a wide range of frequencies on the control quality utilizing linearized process models.

The current approach investigates the steady-state effort in terms of manipulated variables variation to alleviate the effects of multiple simultaneous disturbances of finite magnitude on the control objectives. Nonlinear process models for different flowsheet configurations and sets of controlled and manipulated variables are employed. The main objective is to perform an effective and rigorous screening and rank ordering for a set of candidate process design alternatives based on economic and static controllability criteria.

Problem Formulation

Initially, a set of alternative flowsheet configurations is determined. For each design the optimal operating conditions and equipment sizing around a nominal operating point are determined based on economic criteria. Thereafter, a set of control objectives is specified for the plant. The satisfaction of the control objectives is ensured by keeping the values of a proper set of controlled variables at a desired level. The manipulated variables are easily identified for each flowsheet while making use of the entire available input space for the system. On the other hand, a number of scenarios of disturbances that are expected to influence the operation of the plant are constructed. Currently, only disturbances of deterministic type are considered whose variation is much slower than the slowest time constant in the plant. Setpoint changes are also regarded as slow disturbances.

The present work aims to describe a large variety of control objectives that require the controlled variables to either remain at a constant steady-state value (setpoint) or allow variations around a target value of a specified magnitude. A quadratic objective function is introduced that penalizes deviations of the controlled variables from a target value in a least squares sense. Similarly, the sum of squared relative changes of the manipulated variables from some nominal steady-state optimal values is weighted in the objective function. The formulation is able to investigate situations where a smaller number of manipulated variables than control objectives are available. Furthermore, in cases where an excess of manipulated variables is available it would be possible to drive the process to the most profitable, from an economic point of view, operating point. Such an approach defines priorities in the satisfaction of the control objectives and sets preferences in the use of manipulated variables equivalent to those of a multivariable optimal control algorithm. The steady-state disturbance rejection problem is formulated within an optimization framework. Given a set of fixed structural design variables (e.g. number of stages in a distillation column, volume of a reactor), $\mathbf{d}$, a set of setpoints for the controlled variables, $\mathbf{y}_{sp}$, a set of optimal steady-state values for the manipulated variables, $\mathbf{u}_{ss}$, a set of model parameters and disturbances, ε, and symmetric weighting matrices for the deviations of the controlled and manipulated variables, $\mathbf{W}_y$ and $\mathbf{W}_u$, respectively, the following optimization problem is constructed:

$$\begin{aligned} \text{Min } f &= (\mathbf{y} - \mathbf{y}_{sp})^T \mathbf{W}_y (\mathbf{y} - \mathbf{y}_{sp}) + (\mathbf{u} - \mathbf{u}_{ss})^T \mathbf{W}_u (\mathbf{u} - \mathbf{u}_{ss}) \\ \text{s.t. } &\mathbf{h}(\mathbf{x},\mathbf{y},\mathbf{u},\mathbf{d},\varepsilon) = \mathbf{0}, \quad \mathbf{g}(\mathbf{x},\mathbf{y},\mathbf{u},\mathbf{d},\varepsilon) \le \mathbf{0} \\ &\mathbf{y}^l \le \mathbf{y} \le \mathbf{y}^u, \quad \mathbf{u}^l \le \mathbf{u} \le \mathbf{u}^u, \quad \mathbf{x}^l \le \mathbf{x} \le \mathbf{x}^u \end{aligned} \quad (P1)$$

The entries in the weighting matrices may also represent the cost related importance of the different control objectives and input resources of the system. Upper and lower hard bounds for the controlled, $\mathbf{y}$, manipulated, $\mathbf{u}$, and process, $\mathbf{x}$, variables are present which define the allowable ranges of variation for the different parameter values, ε. In case of saturation of a manipulated variable, one degree of freedom will be consumed. The individual entries of the weighting matrices, especially the diagonal elements, will then determine the way the system will react based on the relative importance of the control objectives and the cost associated with the use of each manipulated variable. For instance, a large weight on $\mathbf{W}_y$ would maintain the corresponding controlled variables closer to the setpoint level compared to a weight of a smaller magnitude. The integrity of a given control system can be easily explored with the removal of a manipulated variable from the input set.

The objective of the formulation is twofold. First, identify inadequate process designs and control structures that require large changes in the manipulated variables for small disturbance magnitudes. Second, determine the capacity requirements for the process equipment or the range for manipulated variables in order to compensate for the effects of disturbances, a result that actually acts as a feedback to the general process design problem.

Solution Method - Analysis

The parameterized set of the first-order Karush-Kuhn-Tucker optimality conditions for the nonlinear program (P1) is derived. The set is augmented with the relations that govern the variations of multiple disturbances (Seferlis and Hrymak, 1994), $\Delta\varepsilon$, as follows.

$$\begin{bmatrix} \nabla \mathbf{f} + \lambda^T \nabla \mathbf{h} + \mu^T \nabla \mathbf{g}_A \\ \mathbf{h} \\ \mathbf{g}_A \\ \Delta\varepsilon - \theta\zeta \end{bmatrix} = 0 \quad (1)$$

The first entry in Eqn. (1) represents the gradient of the Lagrangian function of the nonlinear program, with respect to vectors $\mathbf{x}$, $\mathbf{y}$, $\mathbf{u}$, $\mathbf{d}$, and ε. Vectors λ and μ denote the Lagrange multipliers associated with the equality, $\mathbf{h}$, and active inequality constraints, $\mathbf{g}_A$, respectively.

The trajectory of the optimal solution is calculated for load and model parameter changes along predefined directions, θ, in the multidimensional disturbance space using a predictor-corrector type of continuation method. Variable ζ acts as the independent continuation parameter, which represents the magnitude of perturbation. The technique can handle active set changes and hard bounds on all variables efficiently, while optimality is ensured by inspection of the sign of the Lagrange multipliers associated with the active inequalities at every continuation point.

The ability of the different control structures to compensate for the effects of disturbances is assessed using a performance index Ω defined as follows:

$$\Omega = \int_{\zeta_{initial}}^{\zeta_{final}} w(\zeta)\, f\big(\mathbf{u}^*(\zeta), \mathbf{y}^*(\zeta)\big) d\zeta \qquad (2)$$

$w(\zeta)$ is a weighting function that determines the significance of each calculated segment of the optimal solution path along the perturbation direction (e.g. larger weight may be used for small magnitudes that are more likely to occur during plant operation). Vector **y** may be augmented with implicit or unmeasured control objectives that are not directly controlled.

The method eliminates from further consideration, those control structures that have poor steady-state disturbance rejection characteristics (large value for Ω), a property, which is independent of the choice of a control algorithm and the pairing of the control loops. A large change in the manipulated variables would imply large error in the controlled variables during the dynamic transition from one steady-state operating point to another. However, this is not a sufficient condition and should not be the only criterion for selecting a proper control structure but it is rather a necessary condition for a good performance by the control system.

Design Parameter Modification

The influence of disturbances on the plant operation results in an increase of the associated operating costs. These costs are quantified in terms of additional resources required (e.g. raw material, steam, cooling water requirements) and deviations from the target set for the controlled variables. The performance index, Ω, can be improved by adaptively modifying the structural design variables, **d**, of the plant that are kept constant during the disturbance scenarios and solution of Eqn. (1).

Sensitivity analysis information of the operating conditions with respect to the structural design parameters is evaluated along the optimal solution path for Eqn. (1). The sensitivity matrix, **P**, shown in Eqn. (3), provides a measure of the relative change of the process variables for infinitesimal changes in the design parameters.

$$\mathbf{P} = \begin{bmatrix} \nabla_d \mathbf{x}^*(\varepsilon) & \nabla_d \mathbf{y}^*(\varepsilon) & \nabla_d \mathbf{u}^*(\varepsilon) \end{bmatrix} \qquad (3)$$

The sensitivity information results from the solution of Eqn. (1) at no additional computational cost. Matrix **P** is analyzed using singular value decomposition in order to identify the design parameters whose modification would cause the largest influence in the system. The eigenvector that corresponds to the largest in magnitude eigenvalue of $\mathbf{P}^T\mathbf{P}$, represents the direction of variation in the design parameters space with the dominant impact on the process variables (Seferlis and Grievink, 1998). Moreover, the eigenvector element entries reveal the contribution of each design parameter to the eigenvector direction.

Reactor/Separator/Recycle System

Four different designs for the reactor-separator-recycle system shown in Fig. 1 (Luyben and Luyben, 1995) are analyzed with the proposed methodology. An optimal economic design is obtained using isothermal models for the reactors and stage-by-stage calculations for the distillation columns, with or without reaction, using orthogonal collocation on finite elements techniques (Seferlis and Hrymak, 1994), so that a continuous representation for the number of stages is achieved.

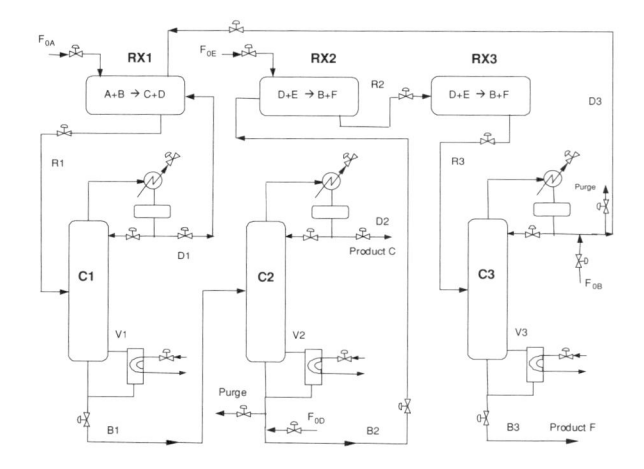

Figure 1. The Reactor/Separation/Recycle (R/S/R) flowsheet.

Selection of controlled and manipulated variables sets that lead to "snowball" effects due to positive recycle feedback have been eliminated quickly and attention has been given on more promising process designs. The total annual costs (investment and operating) for designs I-IV, shown in Table 1, suggest the use of a reactive distillation column (C3) for the second reaction step as an attractive choice. The control objectives are summarized as follows:

1. Maintain the desired product quality.
2. Achieve high conversion for the reactions.

3. Maintain plant operation close to economic optimum.

Generally, it is assumed that perfect control is achieved for all flow and level controllers in the plant, a requirement that can be however relaxed. The disturbance scenario that is simulated corresponds to a simultaneous decrease of the kinetic constants for both occurring reactions and an increase of composition for component "B" in the feed stream F_{0A}. The final value for the continuation parameter, ζ, corresponds to a 50% decrease in the kinetic constants and an increase from 0.0 to 0.15 for the mole fraction of "B". At the same time the reboiler's vapor flowrate in column C3 is kept constant, thus removing a degree of freedom from the system to examine an interesting case of saturation for a key manipulated variable.

Table 1. Alternative Designs for the R/S/R System.

Design	RX3	C3 with reaction	Purge/ Feed D	Costs 10^6\$/yr	Ω
I	Yes	No	Yes	1.306	121.6
II	Yes	No	No	1.306	170.0
III	No	Yes	Yes	1.133	78.3
IV	Yes	Yes	Yes	1.134	79.4
V	No	Yes	Yes	1.142	77.4

Figure 2 shows the behavior of the optimal solution path along the entire range of parametric variation. It can be easily seen that all designs perform reasonably well for small perturbations, but designs III and IV have a definite advantage for larger perturbations since they manage to keep the product quality closer to the desired specifications. Such behavior can be attributed to the ability of the reactive column to allow higher concentration for the reactants in every stage compared to the two CSTR in series. The values for the corresponding performance indices shown in Table 1 quantify the differences.

The sensitivity information along the solution path is utilized to identify the critical structural design parameters in order to improve the system's static controllability. This leads to design V with larger total volume for the reactor RX1 than design III. The static controllability performance index for design V, reported in Table 1, is slightly improved from that of design III due to the additional resources for reaction, at a small additional investment cost.

Conclusions

The illustrative example depicts the advantages of the proposed procedure for rigorously assessing and effectively screening alternative process flowsheets and control structures based on the economic potential and static controllability characteristics. Furthermore, sensitivity information is utilized to improve the steady-state disturbance rejection behavior of the process design by proper adjustment of the most important design parameters. Nonlinear effects and interactions are fully explored and critical constraints identified. The system's behavior under manipulated variable saturation is studied, and the impact of disturbance directionality on the static controllability is evaluated.

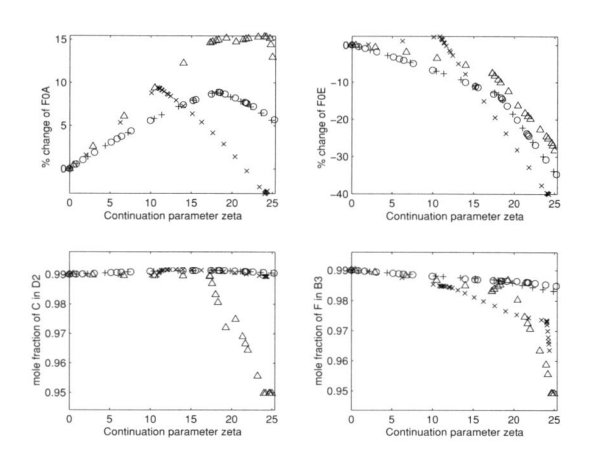

Figure 2. Optimal solution path for designs I-IV of Table 1 ('x'- I, 'Δ' – II, 'o' - III, '+' - IV).

However, a final decision on the plant's control system requires additional exploration of the dynamic behavior of the plant design and the dynamic performance of the control system.

References

Douglas, J. M. (1988). *Conceptual Design of Chemical Processes*. McGraw-Hill, New York.

Lewin, D. R. (1996). A simple tool for disturbance resiliency diagnosis and feedforward control design. *Comp. Chem. Eng.*, **20**, 13-25.

Luyben, M. L., and W. L. Luyben (1995). Design and control of a complex process involving two reaction steps, three distillation columns, and two recycle streams. *Ind. Eng. Chem. Res.*, **34**, 3885-3898.

Luyben, M. L., and C. A. Floudas (1994). A multi-objective optimization approach for analyzing the interaction of design and control. *Comp. Chem. Eng.*, **18**, 933-969.

Schweiger, C. A., and C. A. Floudas (1997). Interactions of design and control: Optimization with dynamic models. In W. W. Hager and P. M. Pardalos (Eds), *Optimal Control: Theory, Algorithms and Applications*. Kluwer Academic Pub. 388-435.

Seferlis, P., and A. N. Hrymak (1994). Optimization of distillation units using collocation models. *AIChE J.*, **40**, 813-825.

Seferlis, P., and A. N. Hrymak (1996). Sensitivity analysis for chemical process optimization. *Comp. Chem. Eng.*, **20**, 1177-1200.

Seferlis, P., and J. Grievink (1998). Optimal design of reactive distillation units under uncertainty using collocation models. AIChE Annual Meeting, Miami Beach.

Yi, C. K., and W. L. Luyben (1995). Evaluation of plant-wide control structures by steady-state disturbance sensitivity analysis. *Ind. Eng. Chem. Res.*, **34**, 2393-2405.

INFORMATION MODELS FOR CONTROL ENGINEERING DATA

Vernon A. Smith[1] and Neil L. Book
Department of Chemical Engineering
University of Missouri-Rolla
Rolla, MO 65409

Abstract

Information sharing is an important part of the daily design activities in modern industry. Unfortunately, the information to be shared has grown larger in quantity and more complicated causing the need to devote more time to managing the information and less time to using it. To combat this, the Process Data eXchange Institute (PDXI), sponsored by the American Institute of Chemical Engineers, developed information models to manage and exchange information specifically related to process engineering. A STEP (Standards for the Exchange of Product Model Data or ISO 10303) application protocol (AP) based on these information models is currently under international review (AP 231). The STEP standards define methods and protocols for the electronic storage and exchange of information captured by STEP models.

The purpose of the models presented here is to extend the existing process engineering information models to include control engineering information. Control engineering information is categorized into three models: the Signal Model, the Control System Equipment Model, and the Control Device Model. These models integrate with the existing process engineering models. The Signal Model describes the various types of signals and how they can be used. The Control System Equipment Model describes the different types of control system equipment which has as a principle function the generation, transmission, indication, and/or manipulation of signals. The Control Device Model describes several of the types of control devices that are the "wetted" portion of a control system (comes in contact with process materials) and that senses or manipulates process fluid variables. The Control Engineering Information Models extend the domain of the information that can be managed and exchanged with the PDXI models to include design data for control systems.

Keywords

Information models, Control engineering, Control equipment, STEP standards.

Introduction

This document introduces three new OMT object models (Rumbaugh, et al. 1991) for control engineering. They are the Signal Information Model, the Control System Equipment Information Model, and the Control Device Information Model. These models are harmonized with Revision 1 of the PDXI (1995) Planning Level Model which already contains references to the highest levels of the models introduced in this paper. In constructing the

---

[1] Vernon A. Smith is currently employed by Kellogg Brown & Root, Inc.

Control Device Information Model, specification forms for process measurement and control instruments developed by the ISA SP20 Committee (1993) were used to specify attributes of control devices.

Context for Control Engineering Models

The purpose of the control engineering models is to extend the existing process engineering information models (Blaha, et al. 1991, 1993; Book, et al. 1992, 1994a, 1994b; Fielding, et al. 1995; Motard, et al. 1995; NIST, 1995; PDXI, 1995). None of the current models contain an in depth representation of control engineering data. The figures in this paper show the upper level of detailed control engineering information models which expand existing models to include control engineering. Control engineering data is broken up into three models; the Signal Information Model, the Control System Equipment Information Model, and the Control Device Information Model.

Figure 1 is not a part of the control engineering models. It shows the class relationships from Revision 1 of the Planning Level Model to which external references are made from the control engineering information and provides a frame of reference for the control engineering models. The Signal Information Model emanates from the Signal_Port class, and the Control System Equipment Information Model and Control Device Information Model are extensions from the Control_Device and Control_System_Equipment classes, all of which are included in the Planning Level Model.

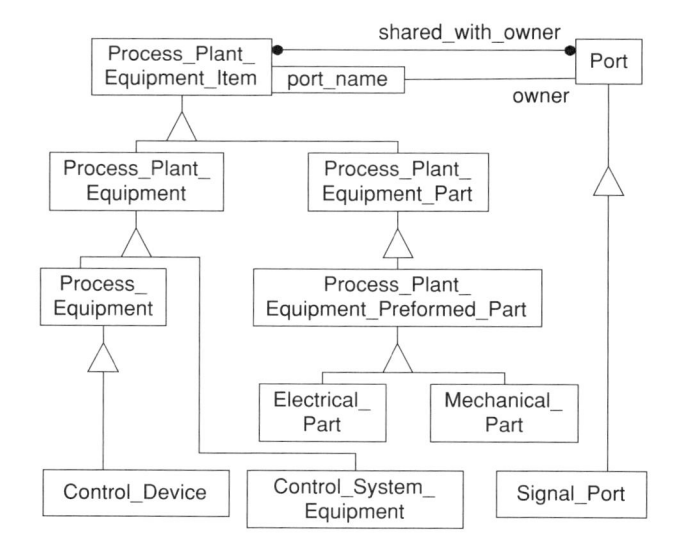

Figure 1. Overview model.

Signal Information Model

Figure 2 shows the Signal hierarchy and is the Signal OMT Information Model. A signal, which is defined as a message, energy, or force sent along some media for control purposes, can be associated to signal_port, allowing a signal to go from one piece of process_plant_equipment to another. A signal can be classified as electrical, pneumatic_hydraulic, or radio_frequency. Additionally, an electrical signal can be classified as an amperage, digital, or voltage signal, and a voltage signal can be either alternating current or direct current.

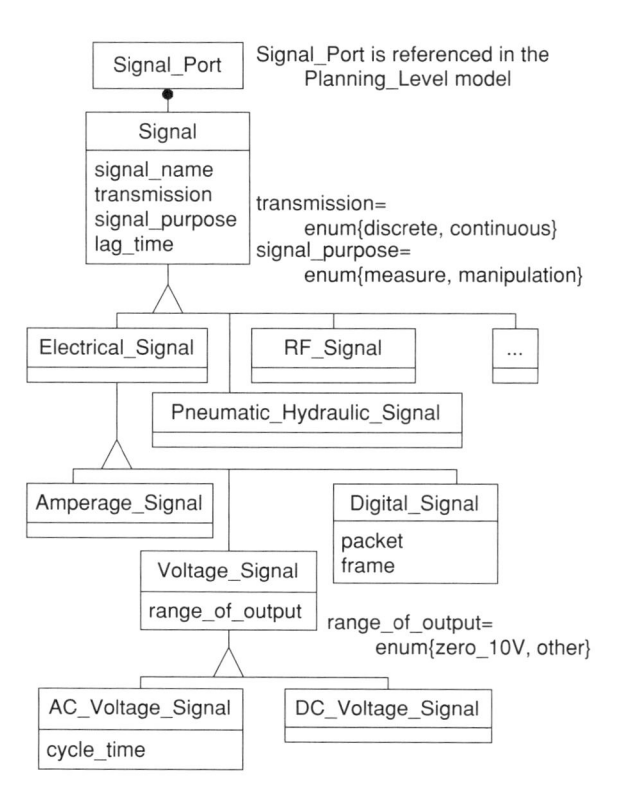

Figure 2. Signal hierarchy.

A signal can have transmission methods that are discrete or continuous. Also, a signal can have a lag time, the time between signal transmission and reception and can be used to either measure a process variable or manipulate one.

This captures most signals that are used in a control environment. Examples include a pheumatic_hydrulic signal used to manipulate a control valve and a voltage signal indicating a temperature from a thermocouple.

Control System Equipment Information Model

Control_System_Equipment is a type of process plant equipment that has as a principle function the generation, transmission, indication, and/or manipulation of signals and may be specified by a location. It can have signal ports but does not come in contact with process materials.

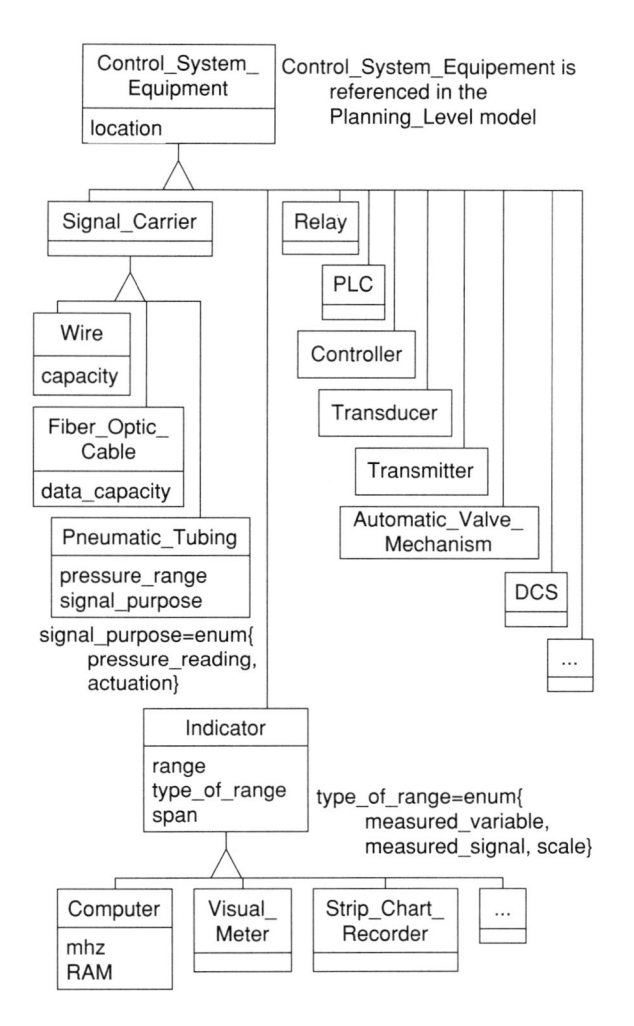

Figure 3 shows the Control_System_Equipment hierarchy. Control_System_Equipment can be classified as a Signal_Carrier, an Indicator, a Controller, a Transducer, a Transmitter, an Automatic_Valve_Mechanism, a Relay, a PLC, or a DCS.

Other items include a Transducer, which converts an input signal of one form to an output signal of another form; a Transmitter, which translates a low level signal to a higher level signal suitable for transmission; a Controller, which compares a process variable to a set point and signals a change to another device and; an Automatic_Valve_Mechanism, which can manipulate a Control_Valve when sent a signal to do so.

The control system equipment make up most of the items in control systems and the Control_System_Equipment information model makes an attempt to capture the equipment most commonly used.

Control Device Information Model

A Control_Device is the portion of a control system which comes in contact with process materials, physically or by radiation, and senses or varies process fluid variables. A Control_Device has a control type, purpose, and transmission.

Figure 4 shows the Control_Device hierarchy. A Control_Device can be classified as an Actuating_Device or a Sensing_Device. An Actuating_Device can be a Control_Valve or a Motor_Controller, and a Sensing_Device can be a Pressure_Device, a Temperature_Device, a Flowmeter_Device, or a Liquid_Level_Device.

Figure 3. Control_System_Equipment hierarchy.

A Signal_Carrier can be further classified as a Wire, a Fiber_Optic_Cable, or Pneumatic_Tubing, and an Indicator can be a Computer, a Visual_Meter, or a Strip_Chart_Recorder.

Signal carriers are used to transmit signals. Typically wire is used to carry a voltage or amperage signal, and pneumatic tubing can either be used to measure a pressure or for actuation.

An indicator is a source of data on process control status which may be heard and/or seen. A strip chart recorder chronologically records data, a visual meter displays process variable values visually but cannot record information, and a computer is an electronic device capable of data logging and process monitoring.

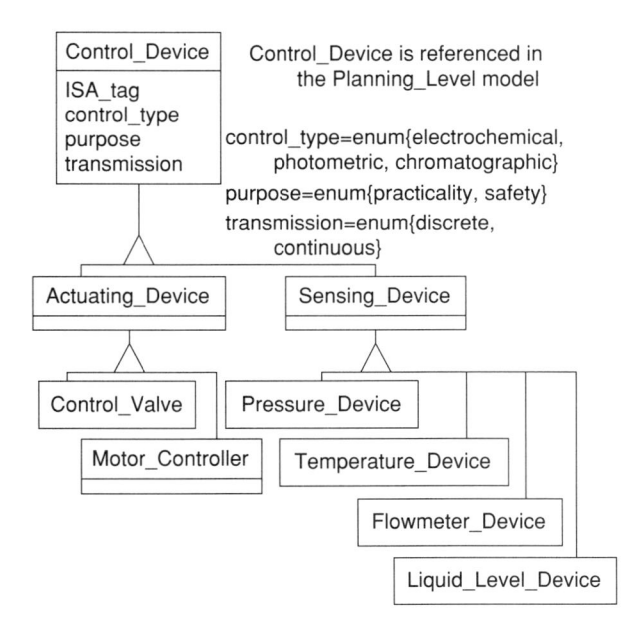

Figure 4. Control_Device hierarchy.

An actuating device can alter process fluid flow for control purposes, and a sensing device can collect process data.

Pressure_Device can be further classified as an Absolute_Pressure_Device, a Differential_Pressure_Device, or a Relative_Pressure_Device. An Absolute_Pressure_Device will return an absolute pressure measurement while a Relative_Pressure_Device will return the pressure relative to atmospheric pressure. A Differential_Pressure_Device will return the pressure difference between two points in a process.

A Temperature_Device can be classified as a Force_Temperature_Device, an Electrical_Temperature_Device, like a thermocouple or RTD, or a Radiation_Pyrometer_Temperature_Device.

Flowmeter_Device can be classified as an Inferential_Flowmeter, a Velocity_Flowmeter, a Volumetric_Flowmeter, and a Mass_Flowmeter. Each of these can be further classified, e.g. Orifice_Plate_Flowmeter or Coriolis_Mass_Flowmeter.

A Control_Valve is a valve which is used to alter or control a flow or pressure of a stream and can be an aggregation of an Automatic_Valve_Mechanism, Manual_ Valve_Mechanism, or Valve_Body.

Note: Automatic_Valve_Mechanism is included in the Control_System_Equipment_Model (Fig 3), and Valve_ Body is included in the Piping Model (NIST, 1995).

Results

The information models presented here were tested with an application programming interface by creating a STEP Physical File for an example scenario and then reading specific attributes from that STEP Physical File with the application programming interface. Data tested was taken from a simplified heat exchanger control system involving two control loops with several items from each of the three models.

The first simplified control loop is for the control of the process fluid flow rate to a heat exchanger. A flowmeter sends a signal to DCS which then sends a signal to a control valve mechanism. The second simplified control loop controls the outlet temperature of the process fluid exiting the heat exchanger. A thermocouple sends a signal to a PLC which sends a signal to a control valve mechanism.

Elements from the Signal information model that are in this example are an Amperage_Signal from a flowmeter, a Voltage_Signal from a thermocouple, and two Pneumatic_Hydraulic_Signals to two control valves. Each of these types of signals can be seen in the Signal information model (Fig 2).

The example includes the following pieces of Control System Equipment: a wire from the thermocouple, a wire from the flowmeter, and two sets of pneumatic tubing to the control valves, two control valve mechanisms, a PLC, and a DCS. These items are included in the Control System Equipment information model (Fig 3).

Control Device items from the example are two control valves, a flowmeter, and a thermocouple. These items are in the Control Device information model (Fig 4). The device Thermocouple is not shown in Fig 4, but is included within the hierarchy of Temperature_Device.

Conclusions

The information models presented in this paper have been shown to effectively capture control engineering information. The current need for control engineering information models is filled with these three separate models for Signal, Control System Equipment, and Control Devices which are already referenced by the existing PDXI Planning Level Model (Fig. 1).

The models are capable of capturing information for most equipment involved in control systems. Subsequent revisions of these models will serve to broaden the models to include less common or new items that are not specifically captured in detail.

References

Blaha, M. R., R. L. Motard, and J. Mehta (1991). Structure and methodology in engineering information management. AIChE Annual Meeting, Los Angeles.

Blaha, M. R., N. L. Book, and R. L. Motard (1993). Data modeling methodology and the PDXI model. AIChE National Meeting, Houston.

Book, N. L., O. C. Sitton, R. L. Motard, M. R. Blaha, B. L. Goldstein, J. L. Hedrick, and J. J. Fielding (1992). Data model based information management systems. AIChE National Meeting, Miami.

Book, N. L., O. C. Sitton, R. L. Motard, M. R. Blaha, B. L. Maia-Goldstein, J. L. Hedrick, and J. J. Fielding (1994a). The road to a common byte. *Chem. Eng.*, 98-110.

Book, N. L., O. C. Sitton, R. L. Motard, M. R. Blaha, B. L. Maia-Goldstein, J. L. Hedrick, and J. J. Fielding (1994b). *The PDXI Data File Interchange Format Project Deliverables*, Vol I-IV. AIChE.

Fielding, J. J., N. L. Book, O. C. Sitton, M. R. Blaha, B. L. Maia-Goldstein, J. L. Hedrick, and R. L. Motard (1995). Methodology for data modeling for the process industries. *Foundations of Computer-Aided Process Design, AIChE Symposium Series*, **91**, 304, CACHE/AIChE.

ISA SP20 (1993). *Specification Forms for Process Measurement and Control Instruments*, ISA dS20.3, Draft, 1993.

Motard, R. L., M. R. Blaha, N. L. Book, and J. J. Fielding (1995). Process engineering databases from the PDXI perspective. *Foundations of Computer-Aided Process Design, AIChE Symposium Series*, **91**, 304, CACHE/AIChE.

NIST (1995). *Piping Model*, Revision 0, Draft Document. The National Institute of Standards and Technology Deliverables.

PDXI (1995). *Planning Level Model*, Revision 1, Draft Document. The PDXI Data File Interchange Format Project Deliverables, AIChE.

Rumbaugh, J., M. Blaha, W. Premerlani, F. Eddy, and W. Lorensen (1991). *Object-Oriented Modeling and Design*. Prentice-Hall.

DESIGN AND ANALYSIS OF OPTIMAL WASTE-TREATMENT POLICIES

Andreas A. Linninger and Aninda Chakraborty
Department of Chemical Engineering
University of Illinois at Chicago
Chicago, IL 60607
e-mail: {linninge, achakr1}@uic.edu

Abstract

In pharmaceutical and specialty chemical production, stringent product purity and quality requirements tend to cause high waste to product ratios. Therefore a large number of different types and levels of waste are leaving pharmaceutical production campaigns and need material recovery and/or adequate treatment. This paper briefly discusses the results of the application of a hybrid methodology onto a case study from the pharmaceutical industry. The hybrid methodology consists of automatic synthesis of feasible alternatives, the superstructure generation, followed by rigorous combinatorial optimization. The methodology was also applied to find the best trade-off between conflicting objectives such as economic and ecological targets. An extension to consider uncertainty pertaining to different waste loads was also presented.

Keywords

Flowsheet synthesis, Pollution prevention, Multi-objective optimization, Uncertainty.

Introduction

Synthesis of optimal waste treatment policies requires knowledge on how to select technically feasible treatment technologies. For a given waste stream numerous alternatives may be available. With the inclusion of recycle and reuse options for pollution prevention, the problem space is increased even more. In this case physical separation techniques for material recovery must be considered alongside destructive treatment options.

Traditionally, process optimization in engineering used to be concerned with finding the optimal process parameters for a given process flowsheet. The underlying base case flowsheet may have emerged from earlier research and development work. Recently, algorithmic process synthesis methodologies are gaining more attention, e.g. Grossman, 1996. Following this more innovative paradigm, a large number of structural alternatives corresponding to different choices of raw materials, auxiliary chemicals such as solvents or utilities, needs to be considered. Unfortunately, even small problems may host a huge space of design alternatives. This combinatorial aspect makes computer-aided methods attractive for the solution of such synthesis problems, e.g. Friedler et al., 1994; Ali et al., 1998.

This paper will first review a hybrid technology for the automatic computer-aided synthesis of plant-wide material recovery and waste treatment policies. Results will be discussed with reference to a case study from a hypothetical pharmaceutical plant. It will be shown how multi-objective optimization of the resulting superstructure can help plant managers to implement feasible pollution prevention efforts and/or reach desired levels of compliance at minimal cost. A simplistic approach for the consideration of uncertainty related to different waste load scenarios is also derived.

Superstructure Generation

The first stage aims at synthesizing all feasible treatment options for a set of wastes of known amount and composition. The methodology for superstructure genera-

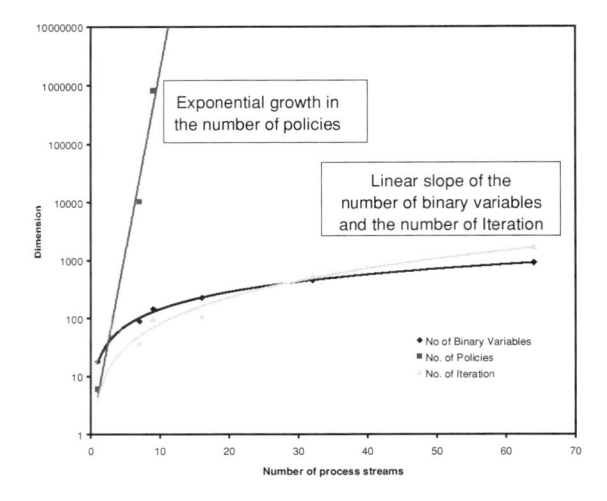

Figure 1. Relative growth of the policies and the binary variables.

tion is derived form earlier work for the assessment of pharmaceutical wastes, e.g. Linninger and Stephanopoulos, 1994. A detailed desciption is beyond the scope of this discussion. A more comprehensive treatment can be found in Linninger and Chakraborty, 1999a; Chakraborty and Linninger, 1999a and 1999b.

The reasoning mechanism involves three steps: (i) diagnosis (ii) preselection (iii) execution. The *diagnosis step* characterizes each waste w_{i0} and detects possible regulatory offences. For each offence a specific treatment goal is generated. *Preselection* searches the treatment database, T, which holds about 40 treatment options for destructive waste treatment and material recovery. A treatment, t, is *a-treatment-option* for a waste, w_{i0}, if it can satisfy at least one goal, g. In the *Execution step*, all treatment options found in steps two are "applied" on waste w_{i0}. This simulation produces j additional nodes, w_{ij}, that represent the residuals resulting from the treatment. These residual calculations rely on short-cut prediction methods using an operation-centered modeling paradigm as described in the BDK project, e.g. Linninger, et al, 1994. After the completion of *diagnosis, pre-selection, execution*, each non-compliant residual, w_{ij}, is submitted to another reasoning cycle. Repeated application leads to a tree of all feasible treatment alternatives.

An ordered sequence of options that treats a waste w_{i0} and all its residuals to compliance is called a treatment path p. The set of all feasible treatment paths p for a given waste w_{i0} is called a treatment tree. Two treatment paths are called *disjunct*, if they do not belong to the same tree. The superstructure S is the union of all treatment trees for all wastes W. A treatment policy, π_i is a set of *disjunct* treatment paths p with exactly one feasible path per waste. The number of embedded policies is equal to the cardinality of the cross product of all disjunct paths.

Optimal Waste Treatment Policy Management

Although the superstructure generated in phase one contains all feasible treatment policies, it does not answer to plant-specific constraints. In stage two of the proposed methodology, plant managers can find optimal waste management strategies using rigorous mathematical programming techniques. It is worth noting that the operating parameters of the different treatments like temperature, pressure etc. are not subject to optimization. This is due to fact that that treatment and recovery facilities, e.g. furnaces and solvent recovery, are operated at fixed conditions irrespective of the waste loads. Simultaneous optimization of operating parameter using MINLP techniques would in principle not pose any problem to the presented methodology.

Table 1. Total Cost of Different Policies and Associated Relative Cost for Pollution Prevention.

Waste Treatment Policy	Total Cost[$/yr]*	Cost Effort[$/yr]**
Unconstrained (Base case)	-1349087	–
Most Expensive	1526979	2876066
No Onsite Recycle	-633898	715188
80 % RRR Capacity	1451706	2800794
25 % Red. Incin. Capacity	-1171066	178020

The options for finding a plant-wide optimal policy include (i) exploring different levels of utilization of plant–specific infrastructure and resources (ii) locating the best trade-off between conflicting economical and ecological objectives and (iii) finding the globally optimal waste treatment policy subject to a deliberate set of capacity, environmental and logistic constraints. The details of the problem formulation as a constrained integer program are given in Linninger and Chakraborty, 1999b.

A case study with wastes from a pharmaceutical production process for the manufacturing of a maleate salt was used to test the methodology. The recipe Linninger et al., 1994 was simulated using the BDK environment [Hyprotech, 1998]. The resulting waste streams were submitted to computer-aided treatment analysis. The resulting waste streams were submitted to computer-aided treatment analysis. Technical aspects of the case study are described in more detail in Chakraborty and Linninger, 1998; optimal solutions were found below 1 CPU second on a 300 MHz Pentium II PC using the CONOPT algorithm of GAMS, 1996. All waste streams in the superstructure gave rise to treatment trees composed of 2 – 15 distinct treatment paths and up to 40 distinct treatment steps. The number of different policies on a plant-wide level is equal to the cross product of all treatment trees and therefore grows exponentially. The total number of policies in the case study exceeds 10^{10} alternative flowsheets.

The simulation runs also compare highest and lowest level of respective resource consumption. It is worth mentioning that this flexibility is attainable through fully compliant policies. Table 1 summarizes the cost associated with distinct pollution prevention efforts. Here, negative cost implies benefits due to recovery and reuse of valuable

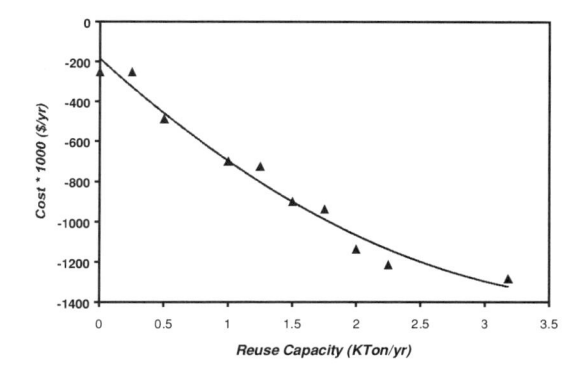

Figure 2. Pareto-optimal policies for different reuse capacities.

materials. It allows one to quantify the relation of distinct waste management strategies and their respective cost and environmental impact. In this case study, a reduction of incineration capacity by 25% leads to a cost increase of approximately 178,000 $/yr.

Figure 1 illustrates the exponential growth of treatment policies that comes with increasing number of waste streams. Despite this fact, execution time stayed small in all our experiments. It is safe to expect that the methodology can effectively solve problems involving hundreds of streams in reasonable CPU times. This feature may be attributed to beneficial exploitation of the underlying branch-and-bound algorithm that attains the integer optimum with a small number of cuts.

Multi-Objective Optimization.

Design with ecological considerations is a multi-objective problem, whose solution is given by the "optimal" trade-off among competing goals. Pollution prevention design balances non-commensurate objectives such as process economics alongside environmental impact. A solution, σ, to a multi-objective optimization problem is *Pareto-optimal* if no other feasible solution is at least as good as σ with respect to every objective, and strictly better than σ with respect to at least one goal.

The parabolic-shaped efficiency frontier or trade-off curve of Fig. 2 includes all *Pareto-optimal* policies. Each triangle represents the impact of structurally alternative treatment policy. Above it lies the space of inferior designs. Under the curve no feasible solution exists. The globally least expensive policy recovers 3,250 tons of solvents per year. Reduction of the recovery capacity would lead to a shift of the dominating solution towards the left of the abscissa. Hence smaller recovery benefits or higher operating cost would be the effect. Fig. 3 depicts the trade-off among the competing goals recycling versus CO_2 emissions. The locus of *Pareto-optima* delineates the policies with smallest CO_2 emissions for a specific level of recovery per year. Clearly, a reduction in the availability of solvent recovery capacities would lead to higher atmospheric emissions.

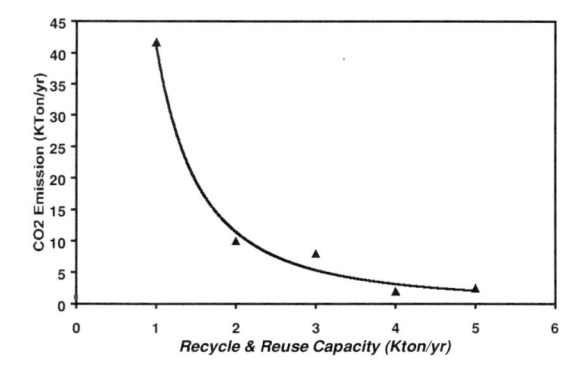

Figure 3. Pareto-optimal curve of material recovery vs. CO_2 emissions.

Table 2. Waste Loads of the Different Scenarios Evolving over Time.

Wastes [kg/yr]	Scenario 1	Scenario 2	Scenario 3
W1	453600	907200	226800
W2	453600	907200	226800
W3	453600	1360800	151200
W4	453600	1134000	181440
W5	453600	680400	302400
W6	453600	340200	604800
W7	453600	680400	302400
Total	3175200	6010200	1995840

Uncertainty in Waste Management

Uncertainty in waste management arises from the different waste loads generated from the pharmaceutical production plants. The wastes may change due to different campaigns reflecting seasonal variations in production. Clearly, practical waste management strategies need to exhibit robustness against changing operational conditions. Otherwise an "optimal" policy may never be effective due to varying conditions or even become infeasible if the variations are too large. For the subsequent discussion of uncertainty, we assume that the distribution of waste loads follows three distinct scenarios. Table 2 shows the different scenarios considered for wastes from the single stage of the case study. The compositions of the waste streams are assumed to be constant.

The waste loads evolve over time in a probabilistic manner and hence it can be described as a stochastic process. It is assumed that the probability of how the waste loads evolve depends only on the present state and is independent of events in the past. The one step stationary transition probabilities for this situation can be described by a 3×3 matrix with the different transition probabilities P_{ij}. $P_{ij}^{(1)}$ denotes the probability of going from state i to state j in one period of time. These transition probabilities are assumed as follows:

$$P^{(1)} = \begin{bmatrix} 0.7 & 0.1 & 0.2 \\ 0.5 & 0.15 & 0.35 \\ 0.4 & 0.25 & 0.25 \end{bmatrix}$$

The evolution of possible scenarios can be described as different states in a Markov chain. For our simple model, these transition probabilities are not expected to vary over time. Note that the row sum of the transition matrix adds up to unity. Over a long period of time, i.e. $n \to \infty$, the matrix P can also be used to compute the steady state probability of each scenario.

The steady state values, π_j, with $\pi_j = \lim_{n \to \infty} P_{ij}^{(n)}$ satisfy the following equations:

$$\pi_j = \sum_{i=0}^{M} \pi_i P_{ij}^{(1)}, \quad (1)$$

$$\sum_{j=0}^{M} \pi_j = 1. \quad (2)$$

The π_j are called the *steady state probabilities* of the Markov chain and the resulting values are shown below:

$$P^{(\infty)} = \begin{bmatrix} 0.59 & 0.15 & 0.26 \\ 0.59 & 0.15 & 0.26 \\ 0.59 & 0.15 & 0.26 \end{bmatrix}$$

For a fixed set of plant constraints, two distinct waste management policies can be defined: (i) the adjusting policy and (ii) the robust policy.

Adjusting Policy. For each scenario, we can find the optimal policy, P. In each time period, we adjust to the best available policy P'. Adoption of this approach over many time periods would give rise to a long term expected average treatment cost given by $\Sigma(C_j \times \pi_j)$, where C_j is the cost of the optimal policy of the scenario j. From Table 3 the long run expected average cost is found at –547616 $/yr. The negative cost implies a net benefit due to reuse and recovery of useful material.

Robust Policy. The robust policy asks for an invariably optimal design for all scenarios that minimizes the expected cost. Clearly this policy must be feasible in all scenarios and may not coincide with any of the independent optima. The average expected cost found for this robust policy is –428093 $/yr.

Although the *adjusting policy* leads to better performance than the *robust policy*, but may be impossible to implement, since it requires accurate information of the waste loads emanating from the production campaigns.

Conclusions

A hybrid methodology for the computer-aided synthesis and optimization of plant-wide optimal policies was applied to a case study from a batch pharmaceutical plant. The three stages of the case study gave rise to a total number of policies exceeding 10^{10}. This figure indicates an exponential growth in the number of policies for a linear growth in binary variable and iteration time. There was a great extent of flexibility in the policies and the optimal constrained policies differed from the base case policy by the cost required for the specific pollution prevention. Comparison of the cost and effort for recovery and treatment options associated with alternative production routes may help to select the chemical recipe with best trade-off between economic and ecological performance. Consideration of uncertainty due to varying waste loads and composition is of particular importance. Two simple approaches were presented. *The robust policy* produces an optimal design with minimal expected cost feasible in all variant scenarios of the time horizon. *The adjusting policy*, on the other hand, requires the design to adapt dynamically to the changing waste scenario. This would require measurability of the changing effluent streams, but may lead to higher maintenance and capital cost.

Table 3. Optimal Policy Cost for the Different Scenarios.

Scenario	Optimal Policy Cost, C_j ($/yr)	Steady state probability, π_j
1	-566930	0.59
2	-573720	0.15
3	-488728	0.26

Acknowledgements

The research was funded by the UIC – CRB Grant S98. Provision of the GAMS language and the optimization codes from GAMS Development Corp. is gratefully acknowledged.

References

Ali, S, Linninger, A. A. and G. Stephanopoulos (1998). Synthesis of batch processing schemes as synthesis of operating procedures: a means-ends analysis and non-monotonic planning approach. Paper 216c, AICHE Annual Meeting, Miami.

Chakraborty, A. and A. A. Linninger (1998). *Synthesis of Waste Treatment Alternatives for Batch Pharmaceutical Wastes: A Case Study.* UIC-LPPD Report, Chicago, Aug. 12.

Chakraborty, A. and A. Linninger (1999a). Systematic assessment of treatment policies for liquid wastes from pharmaceutical plants: A case study. Paper 24c, AIChE Meeting, Houston.

Chakraborty, A. and A. Linninger (1999b). Synthesis of treatment flowsheets for pharmaceutical wastes. Paper 98c, AIChE Meeting, Houston.

Friedler, F., Varga, J.B., and L.T. Fan (1994) Algorithmic approach to integration of total flowsheet synthesis and waste minimization: pollution prevention via process and product modifications. *AIChE Symposium Series*, **90**(303).

GAMS Development Corporation (1996). *GAMS, A Users Guide; Release 2.25.*

Grossman I. E. (1996). Mixed-integer optimization techniques for algorithmic process synthesis. In *Process Synthesis*, J. Anderson (Eds.), Academic Press.

Hyprotech (1998). *Batch Design Kit, Version 1.0.0., Alpha Build 17.*

Linninger, A. A., Ali, S. E., Stephanopoulos, E., Han C., and G. Stephanopoulos (1994). Synthesis and assessment of batch processes for pollution prevention. *AIChE*

Symposium Series, **90**(303), 46-53.

Linninger, A. A. and A. Chakraborty (1999a). Plant-wide optimal waste management. *Comp. Chem. Eng.*, **23**, S67-S70.

Linninger A. A. and A. Chakraborty (1999b). Synthesis and optimization of waste treatment flowsheets. *Comp. Chem. Eng.* (accepted).

INTEGRATED DESIGN AND OPERABILITY ANALYSIS FOR ECONOMICAL WASTE MINIMIZATION

Usha Gollapalli, Mauricio Dantus, and Karen High
Oklahoma State University
Stillwater, OK 74078

Abstract

The main focus of this research is to integrate environmental impact, profitability, and operability at the early stages of process design. A general methodology has been developed to evaluate process alternatives using a stochastic multiple objective optimization approach that identifies the best alternative that maximizes the economic performance and minimizes the environmental impact considering the uncertainties associated in the decision making process. Current work focuses on additional improvements to the methodology where operability is considered at the design phase by expanding the multiple objective optimization approach to include the criterion for operability. The operability criterion used in this research to evaluate the process is controllability of the process. A survey o the operability studies in industries was also conducted.

Keywords

Waste minimization, Multi-objective optimization, Stochastic optimization, Operability, Controllability.

Introduction

The main objectives of this paper are to develop a methodology to design a process to minimize waste, maximize profit and maximize operability. Previous research has focussed on waste minimization. A general methodology has been developed to evaluate process alternatives in terms of their profitability and environmental impact using a multiple objective optimization approach. The method takes into account process uncertainties in a decision theory framework and incorporates stochastic optimization techniques (Dantus, 1999).

Current work focuses on additional improvements to the methodology where operability is also considered at the design phase by expanding the multiple objective optimization approach to include one of the criterion for operability: controllability. Also, industrialists' opinion is sought on operability.

Measuring Environmental Impact

The analysis and final decision to implement a waste minimization project depends on its potential benefits, generally expressed in monetary terms. However, the identification of such benefits from a waste management and regulatory perspective is not so easily accomplished. Leading to an underestimation of environmental costs.

The environmental costs should not be the only factor considered in the evaluation of source reduction alternatives. With the same degree of importance, the overall environmental impact of the process θ

$$\theta = \frac{\sum_{i=1}^{n}\sum_{j=1}^{m} r_i f_i m_{j,i} \Phi_j}{P} \quad (1)$$

—generally difficult to quantify in monetary terms— should be considered as a complementary decision tool. Parameters in Eqn. (1) include:

f_i = flowrate of stream i (kg/hr)
$m_{i,j}$ = mass fraction of component j in stream I
r_i = release factor
P = total mass of product obtained (kg/hr)
Φ_j = environmental impact index of component j

This non-monetary valuation technique based on the work by Mallick *et al.* (1996), and combining the works by Davis *et al.* (1994), Bouwes and Hassur (1997), and the Hazard Ranking System Final Rule (Federal Register 1990) calculates the environmental impact of each chemical present in the waste stream and is a function of the specific chemical amount, its release and exposure potential, and its environmental and human health effects.

Since the "waste streams" might not be the only emission source in the process, Eqn. (1) includes a release factor r ($0 \leq r \leq 1$) that accounts for the release potential of a given stream.

Measuring Profit

The second objective used to evaluate alternatives seeks to maximize the amount of profit that can be obtained from a particular investment. Among the different profitability tools available in the literature the annual equivalent profit (AEP) was selected as the second objective's attribute.

The AEP includes all the process related costs that within the environmental accounting framework can be divided in five groups: usual costs, direct costs, hidden costs, liability costs, and less tangible benefits. A review of how these can be estimated is given by Dantus (1999).

Measuring Operability

The research in the previous sections dealt with the assessment of the impact of a process on the environment. Current work focuses on another important design objective, operability. Operability of an industrial process is defined in this research as the ability of a chemical process to be controllable, flexible, resilient and safe. The definition of operability varies from author to author in that each defines it according to a different set of criteria (Gollapalli, 1999). A qualitative and a quantitative analysis of controllability of a process has been conducted (Gollapalli, 1999).

Hashimoto (1995) said that there was an immediate need for bridging the gap between academia and industry, and later Schijndel (1999) showed that the gap has been substantially bridged. Thus, a lot of research nowadays is focussed more on the industrial needs. Hence, this research conducted a survey on the operability studies done in industries.

Controllability

Controllability is the ability of the process to track set points and reject disturbances. Controllability has been assessed by the method of Singular Value Analysis (SVA) (Seborg, 1989), which helps in the design and comparison of different control strategies using, a controllability index called condition number.

Industrial Survey

In the survey a questionnaire was sent to industries. The questionnaire was designed with the idea of improving operability studies and give a sharper focus to research in the operability area. Following are the questions posed to the industrialists and the responses to the questions.

> Does your company do methodical evaluations of process designs for operability/controllability to guide process design at the PFD stage?

There were as many positive responses as negative to this question.

> If Yes, what are the criteria that you evaluate it upon, and what methods do you use to analyze the criteria?

The criteria that the industries stressed upon are controllability, safety, environmental impact and profitability (already discussed by the current research), product quality, HAZOP studies, data validity and soft sensor predictability. Some of the tools that they have adopted for the operability studies are CONSYD, pilot plants thermodynamic models in ASPEN, and dynamic simulation to design alternatives.

> If No, why don't you think it is necessary to evaluate the designs based on these criteria?

The most common response was that operability study can not be done at the process flow diagram (PFD) stage. A process and instrumentation diagram (P&ID) is very essential to perform operability analysis of the process. Some were satisfied with their designs based on past experience, and consider it uneconomical to spend so much time and effort on an operability analysis.

> What criteria for operability/controllability do you think should be looked into to evaluate a process design?

The responses to this question were very informative, as they were drawn from different fields and many past experiences. Again, the most crucial criterion was profitability. Other criteria like environmental impact, product quality, controllability, good dynamics, ergonomics, intrinsic stability, instrumentation location, reliability, noiselessness, equipment response to upsets and vessel residence times and availability have also been suggested.

> What should academia do to provide you with an incentive to perform the operability/controllability analysis and evaluate process designs?

The major demand from the industries to academia is that they design a tool for the operability studies that is easy to apply by any process engineer and that which enhances profitability of their processes. A tool based on both steady state information, for specifying control points, and dynamic information should be developed. Also, the tool should be compatible to standard simulation packages like ASPEN PLUS™ and HYSYS.

Multiple Objective Stochastic Optimization for Evaluating Uncertainties

A multiple objective stochastic optimization problem is given by

$$\text{Min } z_1 = f_1(\mathbf{x}, \mathbf{y}, \Omega), \ldots z_n = f_n(\mathbf{x}, \mathbf{y}, \Omega)$$

subject to: (2)

$$g(\mathbf{x}, \mathbf{y}, \Omega) = 0; \ h(\mathbf{x}, \mathbf{y}, \Omega) \leq 0$$

where the optimum answer corresponds to the values of the continuous and discrete variables $\mathbf{x}^*$ and $\mathbf{y}^*$ respectively that maximize or minimize a set of objectives z_i over all possible values taken by the uncertain parameters Ω, subject to a set of equality $g()$ and inequality $h()$ constraints.

An approach usually taken for solving Eqn. (2) is to replace the stochastic problem by a suitable deterministic problem (Stancu-Minasian, 1990). In this case, the problem is solved by finding the solution vectors $\mathbf{y}^*$, $\mathbf{x}^*$ that minimize the expected value of the objective function, subject to some a priori distribution of Ω.

Once the stochastic problem has been reformulated in a deterministic form, Eqn. (2) can be solved using for example traditional mixed integer non linear programming (MINLP) techniques combined with multiple objective optimization approaches. In spite of their success, the traditional MINLP approach to synthesis may pose certain problems especially with sequential process simulators such as ASPEN PLUS (Chaudhuri and Diwekar, 1997). Furthermore, MINLP methods can get trapped into some neighborhood within the search region, leading to a local solution and failing to find the global optimum.

An alternative approach that circumvents the problems associated with MINLP algorithms is the use of random search methods. These methods have the feature of exploring more globally the feasibility region of a given problem, thus having a good possibility of finding the global optimum (Maffioli, 1987). Among the different random approaches used the method that has probably received the most attention is the simulated annealing algorithm, that is based on the analogy between the simulation of the annealing of solids and the solving of large combinatorial optimization problems (van Laarhoven and Aarts, 1987).

The algorithm's main criticism is its high computational requirements due to the large amount of trials that need to be evaluated. To reduce the number of configurations to be analyzed and to increase the algorithm's efficiency research directions have looked at either making a careful selection of the main parameters —also known as the "cooling schedule"— or modify the method's structure and design features.

An interesting approach is the one given by Painton and Diwekar (1995), who developed a modified algorithm: the "stochastic annealing algorithm." This algorithm is combined with multiple objective optimization approaches that evaluate competing objectives. For example, when evaluating pollution prevention alternatives, there is generally no investment option that maximizes the process' profit and minimizes its environmental impact. Hence, a sacrifice of the first objective is required to obtain a better performance of the second objective. As a consequence, the optimum solution obtained will be considered as the best compromise solution according to the decision maker's preference structure (Vanderpooten, 1990).

Among the various multiple objective optimization algorithms with prior articulation of preferences, the methodology incorporates the compromise programming (CP) approach (Zeleny, 1973) that identifies solutions that are closest to the ideal point $\mathbf{z}^*$

$$\mathbf{z}^* = \left(z_1^*, z_2^*, \ldots z_n^*\right) \quad (3)$$

The ideal solution is generally not feasible. However, it can be used to evaluate the set of attainable nondominated solutions. In this case, the compromise solution is such that it minimizes the closeness or distance L_j to the ideal point (Goicoechea et al., 1982).

$$L_j = \sum_{i=1}^{n} \gamma_i^j \left(z_i^* - z_i(\mathbf{x})\right)^j \quad (4)$$

The preference weight γ is used to represent the relative importance of each objective. The DM's preferences are also expressed in the compromise index j ($1 \leq j \leq \infty$), which represents the DM's concern with respect to the maximal deviation (Goicoechea et al., 1982).

The two-objective optimization problem is combined with stochastic optimization concepts to solve Eqn. (5).

$$\min E(L_j) = \gamma_{AEP}\left(\frac{E(AEP^*) - E(AEP)}{E(AEP^*) - E(AEP^{**})}\right)^j + \gamma_\theta\left(\frac{E(\theta) - E(\theta^*)}{E(\theta^{**}) - E(\theta^*)}\right) \quad (5)$$

Here, $AEP^{**} = \min AEP(\mathbf{x},\mathbf{y})$, $AEP^* = \max AEP(\mathbf{x},\mathbf{y})$, $\theta^{**} = \max \theta(\mathbf{x},\mathbf{y})$, and $\theta^* = \min \theta(\mathbf{x},\mathbf{y})$.

In summary, the evaluation of pollution prevention alternatives under uncertainty is applied combing the compromise programming approach and the stochastic annealing algorithm to solve Eqn. (8). Once the minimum and maximum points in Eqn. (8) have been determined using the stochastic annealing algorithm, the objective function is minimized for a value of $j = 1, 2$, and ∞, for the specific preference weights γ_i assigned to each objective using the ASPEN PLUS process simulator. The details are given in Dantus (1999).

Case Study

The proposed methodology was successfully applied to the production of methyl chloride (CH$_3$Cl) by the thermal chlorination of methane (Dantus, 1999). This process was optimized for two competing objectives considering three continuous variables, one binary variable, and four uncertain parameters. The multiple objective stochastic optimization approach identified the selection of an alternative reactor operated at 488 °C and a chlorine flowrate of 160 kgmol/h.

Conclusions

A realistic approach to design is to overcome the gamut of industrial challenges like environmental liabilities and uneconomic operating methods. This approach should be the ultimate goal of industry. This work presents a unique way to evaluate alternatives to process design in terms of environmental impact, profitability and controllability. The survey on the operability studies in industries has directed the focus of this research to develop tools that meet with the requirements of industry, thus bridging the gap between academia and industry.

Acknowledgements

The present work was made possible through the support of the *Consejo Nacional de Ciencia y Tecnología*, the Fulbright program, The Environmental Institute of Oklahoma State University and the Measurement and Control Engineering Center.

References

Bouwes, N. W. and S. M. Hassur (1997). *Toxic Release Inventory Relative Risk-Based Environmental Indicators Methodology*. US Environmental Protection Agency.

Chaudhuri, P. D. and U. M. Diwekar (1997). Synthesis under uncertainty with simulators. *Comp. Chem. Eng.*, **21**, 733-738.

Dantus, M. M. (1999). *Methodology for the Design of Economical and Environmental Friendly Processes: An Uncertainty Approach*. Ph.D. Dissertation. Oklahoma State University, Stillwater

Davis, G. A., L. Kincaid, M. Swanson, T. Schultz, J. Bartmess, B. Griffith, and S. Jones (1994). *Chemical Hazard Evaluation for Management Strategies: A Method for Ranking and Scoring Chemicals by Potential Human Health and Environmental Impacts*. US Environmental Protection Agency, EPA/600/R-94/177.

Federal Register (1990). Hazard ranking system final rule. *Federal Register,* **55** (241, December 14), 51532-51667.

Goicoechea, A., D. R. Hansen, and L. Duckstein (1982). *Multiobjective Decision Analysis with Engineering and Business Applications*. John Wiley & Sons, New York.

Gollapalli, U. (1999). *Feasibility and Operability Analysis of Chemical Process Alternatives*. MS Thesis. Oklahoma State University, Stillwater, OK.

Hashimoto, I. and E. Zafiriou (1995). Design for operations and control. *Foundations of Computer-Aided Process Design, AIChE Symposium Series,* **91**, 103-104.

Maffioli, F. (1987). Randomized heuristics for NP-hard problems. In *Stochastics in Combinatorial Optimization,* G. Andreatta, F. Mason, and P. Serafini (Eds.),. World Scientific, Singapore, 76-93.

Mallick, S. K., H. Cabezas, J. C. Bare, and S. K. Sikdar (1996). A pollution reduction methodology for chemical process simulators. *Ind. Eng. Chem. Res.*, **35**, 4128-4138.

Moore, C. (1986). Application of singular value decomposition to the design, analysis, and control of industrial processes. *Proceedings of American Control Conference,* 643-650.

Painton, L. and U. Diwekar (1995). Stochastic annealing for synthesis under uncertainty. *European J. Ops. Res.*, **83**, 489-502.

Schijndel, J. V. and E. N. Pistikopoulos (1999). Towards the integration of process design, process control and process operability - Current status and future trends. This proceedings.

Seborg, D. E., T. F. Edgar and D. A. Mellichamp. (1989). *Process Dynamics and Control*. Wiley, New York.

Stancu-Minasian, I. M. (1990). Overview of different approaches for solving stochastic programming problems with multiple objective functions. In *Stochastic Versus Fuzzy Approaches to Multiobjective Mathematical Programming Under Uncertainty,* R. Slowinski and J. Teghem (Eds.). Kluwer Academic, Dordrecht, 71-101.

van Laarhoven, P. J. M. and E. H. L. Aarts (1987). *Simulated Annealing: Theory and Applications*. D. Reidel Publishing Company, Dordrecht.

Vanderpooten, D. (1990). Multiobjective programming: basic concepts and approaches. In *Stochastic Versus Fuzzy Approaches to Multiobjective Mathematical Programming Under Uncertainty*. Kluwer Academic, Dordrecht, 7-22.

Zeleny, M. (1973). Compromise programming. In *Multiple Criteria Decision Making,* J. L. Cochrane and M. Zeleny (Eds.). University of South Carolina Press, Columbia, 262-301.

INDUSTRIALLY APPLIED PROCESS SYNTHESIS METHOD CREATES SYNERGY BETWEEN ECONOMY AND SUSTAINABILITY

Jan Harmsen
Shell International Chemicals B.V.
1031 CM Amsterdam, the Netherlands

Peter Hinderink, Jo Sijben, Axel Gottschalk, and Gerhard Schembecker
Process Design Center B.V.
NL 4811 XD Breda, the Netherlands

Abstract

The future chemical process industry has to be both cost-effective and sustainable. To satisfy these requirements, new chemical processes have to be designed from scratch, via a systematic method. This method comprises the mobilisation of all relevant knowledge, the use of state-of-the art tools like heuristic expert-systems and flowsheeting simulators, and activities like mathematical programming, exergy analysis, and costing. This method, which generally is referred to as 'process synthesis', has been successfully adopted by a multi-disciplinary team involving both people of Process Design Center and Shell International Chemicals. It resulted in a new conceptual design, requiring less equipment, less energy, and fewer raw materials.

Keywords

Process synthesis, Sustainability, Conceptual design, Design methods.

Introduction

More and more chemical companies want to contribute to the sustainable development of this world. According to Lemkowitz, et.al. (1998), this means amongst others that scarce resources are not depleted and that ecological systems are kept in tact.

Shell International Chemicals is a world-wide operating company (part of Royal Dutch/Shell Group) that is committed to contribute to sustainable development as well as being competitive. It requires therefore materials- and energy-efficient processes which are also cost-effective. By state-of-the-art design methods, these apparently conflicting targets can be harmonised. Consequently, the following tight objective, has been imposed to ourselves:

the synthesis of a new chemical process with considerably lower fixed and variable cost, and less energy requirements than the existing process,

and in addition, requiring little development effort.

Project Approach

A process synthesis project team has been established comprising various people from both Process Design Center (PDC) and Shell International Chemicals, all having different skills.

The contribution of Shell is by the supply of (proprietary) process-specific information, e.g., kinetics, and of generic knowledge, e.g. safety, environmental, and process engineering.

PDC has been founded in order to adopt all the expertise that exists within the companies Keuken & the Koning and Gesellschaft für heuristisch-numerische Beratungs-systeme (GHN) for large process synthesis projects. Under the umbrella of PDC, the following expertise is readily accessible:

- heuristic expert systems for a structured search to process alternatives (e.g., reactor selection, column sequencing);
- physical property services;
- process simulation;
- exergy analysis;
- mathematical programming (optimisation);
- economic evaluation

Because of the multi-discipline character of the project team and because of only a few hard-core constraints being imposed to the process to be synthesised, many approaches and ideas emerge during the project. This evolutionary character of a process synthesis study has to be managed carefully. Project management is also part of the PDC activities.

This paper refers to a process synthesis project which took about eight months to completion and which represented a total cost of around 350,000 US$ (client and consultant). Financial support from the Netherlands Agency for Energy and Environment (Novem) is acknowledged.

Heuristic Approach with PROSYN

PDC applies a heuristic-numeric approach in their process synthesis studies. It is a systematic and proven methodology in the search for promising process alternatives. This approach is embedded in a software package to facilitate and speed-up the development of new, technically and economic feasible process concepts. This software package, named PROSYN®, actually is a knowledge-based system, i.e., a set of databanks containing heuristic rules instead of numerical data. These databanks, denoted as 'expert-systems' are comparable with experienced process engineers having specific knowledge on e.g., reactor selection or complex distillation. Within PROSYN, these expert-systems are activated when needed by a managing-system, which also organises the flow of information supplied to the system and generated by the system. A schematic of the various expert-systems available within PROSYN is shown in Figure 1.

In our studies, PROSYN is frequently applied for reaction engineering issues and separation problems. In the module for *reaction engineering issues*, suitable reactor-types are indicated and directions to optimal operating conditions are given, based on either the wish to maximise yield or selectivity. For *separation issues*, modules for the most important unit operations are available. These modules give decisive answers to the suitability of a specific separation technique for a given separation problem. In addition they give hints for the arrangement of the separation units, i.e., column sequencing.

PROSYN assists chemical engineers to develop process concepts and to search for alternatives. It can be adopted both for the conceptual design of new processes and the redesign of existing processes. Because all heuristic rules essentially are founded on 'cost effectiveness', a process synthesis approach comprising PROSYN is cost driven.

A more comprehensive outline of the heuristic approach in process synthesis is given by Schembecker and Simmrock (1996).

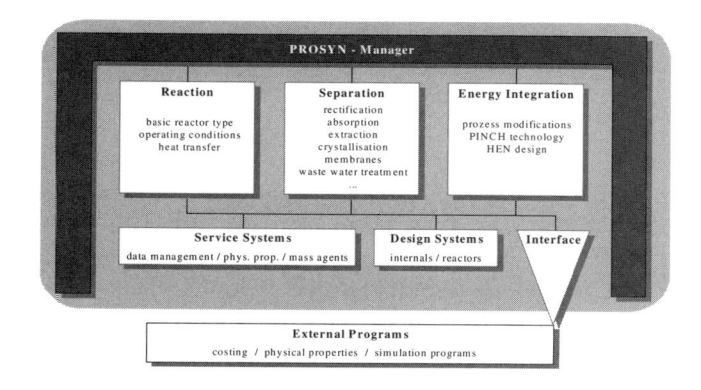

Figure 1. Structure of PROSYN.

A reactor selection method developed within Shell Chemicals (Harmsen, et. al., 1998) also uses the Reaction subsection of PROSYN.

Process Simulation, Optimisation, and Evaluation

Although the tool 'PROSYN' facilitates the process synthesis procedure, it is only part of the work. It helps to gather all required data and it sets the process synthesis team to systematically think about the process. PROSYN gives hints/directions and it needs to be used in combination with a rigorous flowsheeting simulator in order to determine whether the heuristics-based hints are technically and/or economically feasible. In addition, room is left for flowsheet optimisation by dedicated methods (mathematical programming) and for exergy analysis as described by Hinderink et al. (1996).

Process Simulation

In this study, a commercially available flowsheeting simulator has been applied for 'the building of conceptual flowsheet(s)' on the basis of the process alternatives which are generated by PROSYN or which are put-forward during brainstorm sessions. In this way, the technical feasibility of the various process alternatives is evaluated. It is often necessary to extend this simulator with user-defined subroutines, dedicated to e.g., special kinetics or physical property calculations.

Missing data, sensitivities of specific assumptions on process performance, limitations of the various process alternatives, and last but not least, new ideas, they all emerge during the phase of 'flowsheet building' by process simulation. This makes the whole trajectory of process synthesis rather evolving, driven by a tight combination of heuristics and process simulation.

Mathematical Programming (MINLP Optimisation)

Instead of carrying out numerous simulations to find a 'kind of' optimum process design, a more systematic approach to the optimisation-problem has been adopted in this study. The method in view is denoted as the Mixed Integer Non Linear Programming method (MINLP). It is a systematic approach to find optima in problems where numerous variables, both continuous and discrete, are closely related. Therefore, one of the strengths of this new optimisation method is its ability to handle superstructures.

In the MINLP optimisation approach, there is freedom to choose the level of abstraction. In other words, we can decide to use rigorous mathematical calculation procedures for each unit-operation of which the MINLP superstructure is build-up, or we can describe each unit-operation by non-linear mathematical relations catching only those variables and degrees of freedom that are relevant. Being lumped models, these relations are based on fundamental chemical and physical knowledge, and are derived from rigorous calculations. The latter approach requires some initial insight in the optimisation problem and its variables in order to be able to make adequate model reductions. However, if properly carried out, this approach greatly enhances finding the optimum. State of the art software facilitates optimisation. Moreover, optimisation-software generate lots of add-on information, e.g., Lagrange multipliers, that helps one to evaluate the consequences of constraints. Optimisation is based on a user-defined cost function suitable for the specific MINLP case.

Cost Evaluation

For a fair cost-judgement, both the new process design and the existing process design have to be evaluated in a similar manner. Furthermore, the processes to be compared have to be elaborated to the same level of optimisation. Assumptions with respect to choice of materials, equipment-type, and utilities are of importance and they are made by the project team after consulting experts in the field.

Results

The process synthesis approach described in the present paper has been successfully applied on a Shell process. Results are summarised in the table below.

The new process concept includes less unit-operations, which is the result of a new reactor configuration in strong relation with a new work-up section. Capital and variable cost have been reduced considerably and the new process is more energy-efficient as can be seen from the shift in utility requirement from high pressure steam to medium pressure steam.

Table 2 shows that our process synthesis methodology is fairly successful so far.

Table 1. Achievement of a Shell Process Synthesis Project.

	Existing process	New process
Cost		
capital cost	100 %	50 %
variable cost	100 %	84 %
Energy requirement		
electricity	9 MW	10 MW
high pressure steam	+570 kta	-840 kta
medium pressure steam	+120 kta	+1140 kta
low pressure steam	-180 kta	-450 kta

Table 2. Savings Achieved in Other PDC Process Synthesis Projects.

Project	Capital Cost	Variable Cost	Total Cost
1	21%	9%	20%
2	38%	41%	39%
3	20%	55%	36%
4		81%	61%
5	18%	62%	51%
6	-26%	64%	18%

Conclusions

A structured approach to process synthesis on the basis of state-of-the-art heuristic expert-systems, flowsheeting simulation, and mathematical programming methods, carried out by a creative multi-discipline team comprising both chemists and process engineers experienced in the field, and people without a direct relation to the process, has appeared to be effective.

It has lead to a new conceptual design, showing 50% less capital investment than the existing process, and requiring far less energy. So, our target of considerably improving both process economics and energy efficiency has been achieved. Moreover, the new design requires only a few years of development effort.

References

Harmsen G.J. and van Hulst H. (1998). Reactor selection methodology, Presentation at CAPE symposium: *Process Synthesis; Art and Application*, Amsterdam.

Hinderink A.P., et al., (1996). Exergy analysis with a flowsheeting simulator: Part 2: Application: synthesis gas production from natural gas, *Chem. Eng. Sci.*, **51**, 20, 4701-4715.

Lemkowitz S.M., Harmsen G.J. and H.N. Nugteren (1998). The challenge of 'sustainable development' to chemical engineering: A new paradigm for the 21st century?, M.C. Roco et al (Editors), *American Institute of Chemical Engineers (AIChE), Miami Beach*, Volume II, 599-610.

Schembecker, G. and Simmrock K.H. (1996). Heuristic-Numeric Process Synthesis with PROSYN. In *International Conference on Intelligent Systems in Process Engineering*, J. F. Davis et al. (Ed.). *AIChE Symposium Series*, **92**, 312, 275-278.

POLLUTION PREVENTION
VIA MASS INTEGRATION

H. Dennis Spriggs
Matrix Process Integration
Houston, TX 77257-0414

Mahmoud M. El-Halwagi
Chemical Engineering Department
Auburn University
Auburn, AL 36849

Abstract

Pollution prevention is one of the key objectives of a processing facility. Notwithstanding its importance, it must be reconciled with other process objectives such as cost effectiveness, quality assurance, yield enhancement, debottlenecking, safety, and energy conservation. Therefore, process integration can play a unique role in integrating environmental issues with the other techno-economic aspects of the plant. In particular, mass integration has emerged over the past decade as a powerful and attractive framework for providing a unique framework for systematically developing sustainable pollution prevention strategies. Mass integration is a holistic approach to the routing, generation, and separation of streams and species throughout the process. It emphasizes the unity of the process units and objectives and reconciles process objectives such as profitability, product quality, yield enhancement, debottlenecking and environmentally-benign manufacturing. This paper presents an overview of mass integration science and technology and its application in the area of pollution prevention. First, the driving forces for industrial waste reduction are discussed. Then, mass integration hierarchy is presented. Next, an overview of mass-integration science and methods is presented with special emphasis on their critical role in pollution prevention.

Keywords

Pollution prevention, Mass integration, Process synthesis, Optimization, Resource conservation.

Introduction

There are two primary driving forces for a plant to undertake pollution prevention activities; ecological and techno-economic. The ecological driver is typically in response to environmental regulation, growing public pressure, as well as self-induced industrial initiatives. The techno-economic driver is motivated by the realization that significant processing and economic benefits can accrue as a result of overall enhancement of process performance and a conservation of process resources. This enhancement translates into higher productivity, enhanced yield, debottlenecking of the process, higher recovery of valuable chemicals, and reduced usage of material utilities (e.g. water, solvents, etc.). For both categories, it is necessary to develop an insightful understanding of the issues associated with mass flows and interactions. In this regard, *mass integration* can play a significant role. Mass integration is a holistic approach to the generation, separation, and routing of species and streams throughout the process. It is a systematic methodology that provides a fundamental understanding of the global flow of mass within the process and employs it in identifying performance targets and optimizing the allocation and generation of streams and species. For recent literature on mass integration, the reader is referred to the textbook by El-Halwagi (1997) and the review article by El-Halwagi and Spriggs (1998). Mass-related objectives such as pollution prevention are at the heart of mass integration.

Elements of Mass Integration

There are three main factors used in screening candidate pollution prevention strategies; economics, impact, and acceptability. Economics can be assessed by a variety of criteria such as cost, return on investment, payback period, etc. In the environmental context, impact is a measure of the effectiveness of the proposed solution in reducing negative ecological and hazard consequences of the process such as reduction in emissions and effluents from the plant. Acceptability is a measure of how likely the proposed strategies to be accepted and implemented by the plant. This factor is plant based as it differs from one company to the other depending on their operating policies, corporate environment, and level of comfort within the plant to adopt process modifications (based on various objectives such as safety, controllability, reliability, etc). Based on these factors, one can classify candidate strategies via mass integration into three categories (Fig. 1):

- No/Low Cost Changes
- Moderate Cost Modifications, and
- New Technologies

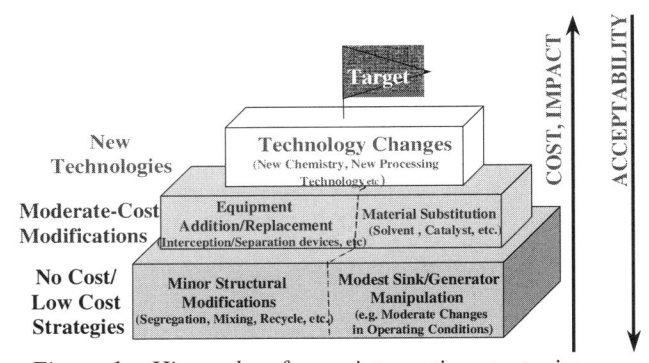

Figure 1. Hierarchy of mass integration strategies.

These strategies are typically in ascending order of cost and impact and in descending order of acceptability. The following sections provide more details on these strategies.

Low/No Cost Strategies

These strategies can be broadly classified into two categories: structural and conditional modifications. The structure-based changes pertain to low/now cost in process configuration such as stream rerouting (e.g. segregation and recycle) which involves piping and pumping primarily. Conditional changes include moderate adjustments in process variables (e.g. packing replacement, catalyst changes, reflux ratio adjustment) and operating conditions (e.g. temperature, pressure, etc.) which require no or modest capital expenditure (Noureldin and El-Halwagi; 1998, 1999). On the other hand, examples of low/no cost structural changes include segregation, mixing, and recycle. Recycle refers to the utilization of a pollutant-laden stream (a source) in a process unit (a sink). Each sink has a number of constraints on the characteristics (e.g. flowrate and composition) of feed that it can process. If a source satisfies these constraints it may be directly recycled to or reused in the sink. However, if the source violates these constraints segregation, mixing, and/or interception may be used to prepare the stream for recycle. A particularly useful tool for identifying recycle strategies is the source-sink mapping diagram. It is a visualization tool (which has a mathematical analogue) that can be used to determine direct-recycling opportunities for streams with single components (El-Halwagi and Spriggs, 1996), multiple components (Parthasarathy and El-Halwagi, 1997), and infinite components (Shelley et al., 1998).

Moderate/High Cost Strategies

In addition to the aforementioned low-cost strategies, waste reduction can be achieved by replacing or adding new equipment, species, or chemical pathways. The following sections illustrate these concepts.

Equipment Addition/Replacement

Interception denotes the utilization of new unit operations to adjust the composition, flowrate and other properties of the pollutant-laden streams to make them acceptable for existing process sinks. A particularly important class of interception devices is separation systems. These separations may be induced by the use of mass-separating agents (MSAs) and/or energy separating agents (ESAs). A systematic technique is needed to screen the multitude of separating agents and separation technologies to find the optimal separation system. Waste interception networks "WINs" can be employed to remove pollutants from in-plant and terminal streams and to integrate gaseous and liquid pollution (El-Halwagi et al., 1996). The synthesis of MSA-induced physical-separation systems is referred to as the synthesis of mass-exchange networks (MENs) (El-Halwagi and Manousiouthakis, 1989a). The subject of physical MENs has been addressed extensively in literature including MENs with a single transferable component (El-Halwagi and Manousiouthakis, 1989a, 1990a), those with multiple transferable components (El-Halwagi and Manousiouthakis, 1989b), those involving regeneration of the MSAs (El-Halwagi and Manousiouthakis, 1990b) and , mass exchange combined with heat exchange (Srinivas and El-Halwagi, 1994a), removal of fixed loads (Kiperstok and Sharratt, 1995), variable supply and target compositions (Garrison et al., 1995), fixed-cost targeting (Hallale and Fraser, 1997), MENs providing flexible performance (Zhu and El-Halwagi, 1995; Papalexandri and Pistikopoulos, 1994), and controllable MENs (Huang and Fan, 1995). Furthermore, mass-pinch diagrams have been developed for a single lean stream for resource conservation such as minimizing water use (Wang and Smith, 1994; Dhole et al., 1996; Kuo and Smith, 1998) and managing process hydrogen (Towler, 1996). Interception networks using reactive MSAs are termed reactive mass exchange networks (REAMEN) (Srinivas and El-Halwagi, 1994a,b; El-Halwagi and Srinivas, 1992). Network synthesis techniques have also been devised for other separation systems that can be used in intercepting pollutants. These systems include pressure-driven membrane separations (e.g., Crabtree et al., 1998; El-Halwagi, 1992, 1993), heat-induced separation networks (HISENs) (e.g., Dunn and Srinivas, 1997; Dunn et al., 1995; Dye et al., 1995; Richburg and El-Halwagi, 1995; El-Halwagi et al., 1995; Dunn and El-Halwagi, 1994) and distillation sequences (e.g., Malone and Doherty, 1995; Quesada and Grossmann, 1995; Kovacs et al., 1993; Wahnschafft et al., 1991).

Material Substition/Benign Chemistry

Material In many cases, it is possible to replace environmentally hazardous chemicals with more benign species without compromising the technical and economic performance of the process. Examples include alternative solvents, polymers and refrigerants. Two main frameworks have been employed to synthesize alternative materials: knowledge base and computer-aided optimization. Knowledge-based approaches depend on understanding the criteria of the materials to be replaced along with general rules and algorithms that link properties with structure. Examples of this approach can be found in literature (e.g., Joback and Stephanopoulos, 1990; Constantinou et al., 1994). Furthermore, software can be used to screen solvents based on their properties and performance. An example of this approach is the PARIS (Program for Assisting in the Replacement of Industrial Solvents) software (e.g. US EPA, 1994; Cabezas and Zhao, 1998). Computer-aided optimization approaches are based on formulating the molecular design problem as an optimization program which seeks to maximize a performance function or minimize deviation from desired properties subject to various constraints including structural feasibility, property-structure correlations and environmental criteria. Examples of this approach include synthesis of solvents (e.g., Brignole et al. , 1986; Odele and Macchietto, 1990; 1995; Hamad and El-Halwagi, 1998), polymers (e.g.,Vaidyanathan and El-Halwagi, 1996) and refrigerants (e.g., Achenie and Duvedi, 1996). Process synthesis techniques can also be integrated with product synthesis tools (e.g. Hamad and El-Halwagi; 1998). Environmentally-benign reactions can also be systematically generated. In this regard, preliminary screening ought to be conducted first to identify overall reaction alternatives that meet process requirements in terms of desired product, cost effectiveness, environmental acceptability, and thermodynamic feasibility. At this stage, minimum details are to be invoked. The problem of synthesizing environmentally-acceptable reactions "EAR's" has been introduced by Crabtree and El-Halwagi (1994) to generate cost-effective, environmentally-benign reactions using modest thermodynamic and economic data.

Conclusions

Mass integration is a comprehensive and insightful framework for pollution prevention. It systematically provides cost-effective strategies that address the root cause of the environmental problems. These strategies are typically simple, implementable but not intuitively obvious. They include stream allocation, species generation and separation, manipulation of process variables, and the use of environmentally-benign chemistry and species. Because of the fundamental nature of this approach, it is generally applicable to processing facilities throughout the chemical process industry. The comprehensive nature of this approach enables the reconciliation of the various process techno-economic process with environmental targets and constraints.

References

Achenie, L. E. K., and A.P. Duvedi (1996). Designing environmentally safe refrigerants using mathematical programming. *Chem. Eng. Sci.*, **51**(15), 3727-3739.

Brignole, E. A., Bottini, S., and R. Gani (1986). A strategy for the design and selection of solvents for separation processes. *Fluid Phase Equilibria*, **29**, 125-132.

Cabezas, H. and R. Zhao (1998). Designing environmentally benign solvents: Physical property considerations. AIChE Spring Meeing, New Orleans.

Constantinou, L., Jacksland, C., Bagherpour, K., Gani, R. and L. Bogle (1994). Application of group contribution approach to tackle environmentally-related problems. *AIChE Symp. Ser.*, **90**(303),105-116.

Crabtree, E. W., M. M. El-Halwagi and R. F. Dunn (1998). Synthesis of hybrid gas permeation membrane/condensation systems for pollution prevention. *J. Air Waste Mgt. Assoc.*, **48**, 616-626.

Crabtree, E. and M. El-Halwagi (1994). Synthesis of environmentally-acceptable reactions. *AIChE Symp. Ser.*, **90**(303), 117-127.

Dhole, V., Ramchandani, N., Tainsh, R., and M. Wasilewski (1996). Make your process water pay for itself. *Chem. Eng.*, Jan., 100-103.

Dunn, R. F., and M. M. El-Halwagi (1994). Optimal design of multicomponent VOC Condensation Systems. *J. Hazard. Mater.*, **38**, 187-206.

Dunn, R. F., and B. K. Srinivas (1997). Synthesis of heat-induced waste minimization networks (HIWAMINs). *Adv. Env. Res.*, **1**(3), 275-301.

Dunn, R. F., Zhu, M., Srinivas, B. K., and M. M. El-Halwagi (1995). Optimal design of energy induced separation networks for VOC recovery. *AIChE Symp. Ser.*, **90**(303), 74-85.

Dye, S. R., Berry, D. A. and K. M. Ng (1995). Synthesis of crytallization-based separation schemes, *AIChE Symp. Ser.*, **91**(304), 238-241.

El-Halwagi, M. M., *Pollution Prevention through Process Integration: Systematic Design Tools*. Academic Press, San Diego, 1997.

El-Halwagi, M. M. (1993). Optimal design of membrane hybrid systems for waste reduction. *Sep. Sci. Technol.*, **28**(1-

3), 283-307.

El-Halwagi, M. M. (1992). Synthesis of reverse osmosis networks for waste reduction. *AIChE J.*, **38**(8), 1185-1198.

El-Halwagi, M. M., Hamad, A. A. and G. W. Garrison (1996). Synthesis of waste interception and allocation networks. *AIChE J.*, **42**(11), 3087-3101.

El-Halwagi, M.M. and V. Manousiouthakis (1989a). Synthesis of mass-exchange networks. *AIChE J.*, **35**(8), 1233-1244.

El-Halwagi, M. M., and V. Manousiouthakis (1989b). Design and analysis of mass exchange networks with multicomponent targets. AIChE Meeting, San Francisco.

El-Halwagi, M. M., and V. Manousiouthakis (1990a). Automatic synthesis of mass exchange networks with single-component targets. *Chem. Eng. Sci.*, **45**(9), 2813-2831.

El-Halwagi, M. M., and V. Manousiouthakis (1990b). Simultaneous synthesis of mass exchange and regeneration networks. *AIChE J.*, **36**(8), 1209-1219.

El-Halwagi, M. M. and H. D. Spriggs (1998). Employ mass integration to achieve truly integrated process design. *Chem. Eng. Prog.*, August, 22-44.

El-Halwagi, M. M. and H. D. Spriggs (1996). An integrated approach to cost and energy-efficient pollution prevention. *Proceedings of Fifth World Congr. of Chem. Eng.*, Vol. III, 344-349.

El-Halwagi, M. M and B. K. Srinivas (1992). Synthesis of reactive mass exchange networks. *Chem. Eng. Sci.*, **47**(8), 2113-2119.

El-Halwagi, M.M., Srinivas, B. K., and R. F. Dunn (1995). Synthesis of optimal heat-induced separation networks. *Chem. Eng. Sci.*, **50**(1), 81-97.

Garrison, G. W., Cooley, B. L., and M. M. El-Halwagi (1995b). Synthesis of mass exchange networks with multiple target mass separating agents. *Dev. Chem. Eng. Miner. Proc.*, **3**(1), 31-49

Hallale, N., and D. M. Fraser (1997). Synthesis of cost optimum gas treating process using pinch analysis. *Proceedings, Top. Conf. On Sep. Sci. and Techs.*, W. S. Ho, and R. G. Luo, eds., Part II. AIChE, New York, 1708-1713.

Hamad, A. A., and M. M. El-Halwagi (1998). Simultaneous synthesis of mass separating agents and interception networks. *Trans. I. ChemE.*, **76**, Part A, 376–388.

Huang, Y. L., and L. T. Fan (1995). Intelligent process design and control for in-plant waste minimization. In *Waste Minimization Through Process Design*, A. P. Rossiter, ed. McGraw Hill, New York, 165-180.

Joback, K. G., and G. Stephanopoulos (1990). Designing molecules possessing desired physical property values. In *FOCAPD III* (J. J. Siirola, I. Grossmann, and G. Stephanopoulos, eds.), 363-387. CACHE/Elsevier, New York.

Kiperstok, A., and P. N. Sharratt (1995). On the optimization of mass exchange networks for removal of pollutants. *Trans. I. ChemE.*, **73**, Part B, 271-277.

Kovacs, Z., F. Friedler, and L. T. Fan. Recycling in a separation process structure. *AIChE J.*, **39**(6), 1087-1089.

Kuo, W. C. J. and R. Smith (1998). Designing for the interactions between water use and effluent treatment. *Trans. I. ChemE.*, **76**, Part A, 287-301.

Malone, M. and M. Doherty (1995). Separation system synthesis for nonideal liquid mixtures. *AIChE Symp. Ser.*, **91**(304), 9-18.

Noureldin, M. B. and M. M. El-Halwagi (1998). A shortcut approach to integrating process design and operation. AIChE Spring Meeting, New Orleans.

Noureldin, M. B. and M. M. El-Halwagi (1999). Interval-based targeting for pollution prevention through mass integration. *Comp. Chem. Eng.* (in press).

Odele, O. and S. Macchietto, (1993). Computer aided molecular design: A novel method for optimal solvent selection. *Fluid Phase Equilibria*, **82**, 47-54.

Papalexandri, K. P., and E. N. Pistikopoulos (1994). A multiperiod MINLP model for the synthesis of heat and mass exchange networks. *Comp. Chem. Eng.*, **18**(12), 1125-1139.

Parthasarathy, G., and M. M. El-Halwagi (1997). Mass integration for multicomponent nonideal systems. AIChE Annual Meeting, Los Angeles.

Quesada, I. and I. E. Grossmann (1995). Global optimization of bilinear process networks with multicomponent flows, *Comp. Chem. Eng.*, **19**(12), 1219-1242.

Richburg, A. and M. M. El-Halwagi (1995). A graphical approach to the optimal design of heat-induced separation networks for VOC recovery. *AIChE Symp. Ser.*, **91**(304), 256-259.

Shelley, M., Parthasarathy, G. and M. M. El-Halwagi (1998). Clustering techniques for mass integration for complex hydrocarbon mixtures. AIChE Annual Meeting, Miami.

Srinivas, B. K., and M. M. El-Halwagi (1994a). Synthesis of reactive mass-exchange networks with general nonlinear equilibrium functions. *AIChE J.*, **40**(3), 463-472.

Srinivas, B. K., and M. M. El-Halwagi, (1994b). Synthesis of combined heat reactive mass-exchange networks. *Chem. Eng. Sci.*, **49**(13), 2059-2074.

Towler, G. P. (1996). Refinery hydrogen management cost analysis of chemically integrated facilities. *Ind. Eng. Chem. Res.*, **35**(7), 2378-2388.

Vaidyanathan, R and M. M. El-Halwagi, (1996). Computer-aided synthesis of polymers and blends with target properties. *Ind. Eng. Chem. Res.*, **35**, 627-634.

Wahnschafft, O. M., Jurian, T. P., and A. W. Westerberg (1991). SPLIT: A separation process designer. *Comp. Chem. Eng.*, **15**, 565-581.

Wang, Y. P., and R. Smith, (1994). Wastewater minimization. *Chem. Eng. Sci.*, **49**(7), 981-1006.

Zhu, M., and M. M. El-Halwagi (1995). Synthesis of flexible mass exchange networks. *Chem. Eng. Comm.*, **138**, 193-211.

MODELING AND DESIGN OF AN ENVIRONMENTALLY BENIGN REACTION PROCESS

Benito A. Stradi, Gang Xu, Joan F. Brennecke, and Mark A. Stadtherr
Department of Chemical Engineering
University of Notre Dame
Notre Dame, IN 46556

Abstract

High-pressure carbon dioxide is attractive as an environmentally-benign replacement for organic solvents, such as benzene, in the synthesis of compounds of high chiral purity. Of particular interest here is the allylic epoxidation of *trans*-2-hexen-1-ol to (2R,3R)-(+)-3-propyloxiranemethanol. In the modeling and design of this reaction system, difficulties were encountered when standard tools were used to model the phase behavior of the compounds present. These difficulties are demonstrated by modeling the high-pressure phase behavior of carbon dioxide with *trans*-2-hexen-1-ol and with *tert*-butyl alcohol. Several examples are used to illustrate the problems encountered. The major problems occurred near the three-phase boundary and in the region with retrograde behavior. By using a technique based on interval mathematics, these difficulties were eliminated and correct results obtained. Final results of the phase behavior modeling lead to an improved design that employs a much lower pressure than originally proposed.

Keywords

Phase equilibrium, Environmentally benign processing, Solvent substitution, Supercritical fluids, Interval analysis, Reaction engineering.

Introduction

High-pressure gases and supercritical fluids are attractive substitutes for organic solvents in important chemical reactions. Carbon dioxide is particularly attractive as an environmentally benign solvent. Numerous reactions have been done successfully in liquid and supercritical CO_2, sometimes with rates and selectivities as good as or better than can be achieved in conventional liquids. The solvent properties of CO_2 can be tuned by selecting appropriate operating conditions. Carbon dioxide is also non-flammable, non-toxic and relatively inexpensive. In addition, there is a wealth of literature detailing its physical and chemical properties, and there are numerous studies of binary and ternary mixtures containing carbon dioxide.

Because the solubility of many compounds is rather low in CO_2, high pressures may be needed to achieve reasonable concentrations. Also, because CO_2 may not equally solubilize products and reactants, high pressures are used routinely to guarantee a homogeneous, single-phase reaction process. However, the use of lower pressures may be possible and is desirable to reduce costs and improve safety.

In order to select lower pressure conditions for the reaction, knowledge of the phase behavior of the reaction mixture in carbon dioxide is needed. In the absence of extensive experimental data, a modeling tool should be used to make educated predictions of the most promising conditions. These can then be explored further with a reduced number of experiments. Thus, there is a need for reliable modeling tools that can provide adequate predictions of high-pressure phase behavior.

The reaction process studied here is the allylic

epoxidation of *trans*-2-hexen-1-ol to (2R,3R)-(+)-3-propyloxiranemethanol. This was chosen because it is one of a class of industrially important syntheses for producing compounds with high chiral purity. Normally, this reaction has been performed in organic solvents (e.g., benzene), and results in two stereochemical centers formed with very high enantiomeric selectivity. Recently, however, Tumas (1996) has shown that it can also be done in liquid CO_2 at 30°C and 346 bar, also with very high enantiomeric selectivity. Thus, the subject of our study was to determine the phase behavior of the reaction mixture for this process (Fig. 1) in high-pressure CO_2. Titanium (IV) isopropoxide is present as a catalyst; diisopropyl L-tartrate is a ligand used to control chirality; *tert*-butyl hydroperoxide acts as an oxygen donor and renders *tert*-butyl alcohol as a byproduct. The reaction is not equilibrium limited and goes essentially to completion.

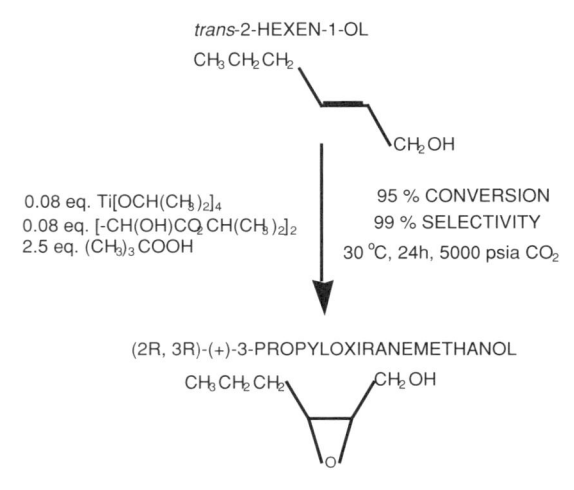

Figure 1. Allylic epoxidation (Tumas, 1996).

The phase behavior of this system is controlled by multiple interactions on the molecular level. These can be captured, at least partially, by using a cubic equation of state (EOS) model, and by then determining the binary interaction coefficients of each component with CO_2. The intrinsic assumption is that the interactions with the solvent are dominant in determining phase behavior. Since the CO_2 solvent is the major component of the mixture, this represents a reasonable initial assumption. Thus, first the binaries with CO_2 are modeled, and the resulting binary interaction coefficients are then used in modeling the complete multicomponent system.

The modeling of high-pressure phase behavior can be a very challenging computational problem. We present several examples below that illustrate the difficulties that arise in computing the phase equilibrium using standard tools, namely Aspen Plus (Aspen Technology, Inc.) and IVC-SEP (Hytoft and Gani, 1996). In addition, we introduce a new computational tool (INTFLASH), based on interval mathematics, that can handle these difficulties, and correctly determine the phase equilibrium in all cases. We concentrate on the high-pressure phase behavior of carbon dioxide with *trans*-2-hexen-1-ol and with *tert*-butyl alcohol. However, similar computational challenges were found with the other compounds.

Methodology

The Peng-Robinson EOS was used to model both the liquid and gas phases. Standard van der Waals mixing rules were used with a single temperature-independent binary interaction parameter (k_{ij}) for each binary with CO_2. The standard modeling tools were, from Aspen Plus, the FLASH3 module, which can be used for vapor/liquid and vapor/liquid/liquid equilibrium calculations, and the RGIBBS module, which can be used for phase equilibrium or combined phase and reaction equilibrium calculations, and, from IVC-SEP, the two-phase flash routine LNGFLASH, which employs Michelsen's well-known approach. In addition, a new multicomponent, multiphase phase equilibrium routine (INTFLASH), based on interval mathematics, is used.

INTFLASH combines local methods for doing phase split calculations with a global method for verifying phase stability. The key is the use of a technique based on interval analysis in performing the phase stability analysis. This involves the *global* optimization of a tangent plane distance function to verify that the global minimum is nonnegative (otherwise the phase being analyzed will split). By using an interval Newton/generalized bisection technique, the global optimum can be determined *with mathematical and computational certainty*. As applied to EOS models, this approach was first described by Hua *et al.* (1996), with later generalizations and improvements given by Hua *et al.* (1998). By incorporating this technique for global phase stability, INTFLASH provides a guarantee that correct phase equilibrium results are obtained. Complete details of INTFLASH will be provided elsewhere.

Results and Discussion

As noted above, we concentrate here only on the high-pressure phase behavior of CO_2 with *trans*-2-hexen-1-ol and with *tert*-butyl alcohol. The critical temperature T_c, critical pressure P_c, acentric factor ω, and binary interaction parameter k_{1j} with CO_2 (component 1) of these two compounds are given in Table 1.

Table 1. Properties of Compounds.

Compound	T_C (K)	P_C (bar)	ω	k_{1j}
trans-2-Hexen-1-ol	601.76	36.73	0.724	0.084
tert-Butyl alcohol	506.21	39.73	0.611	0.108

The binary interaction parameters were determined by parameter estimation using experimental VLE data. Details of the physical property determinations are provided in Stradi *et al.* (1998). Phase equilibrium calculations were then done with these model parameters in order to compare the model predictions to experimental measurements. In Table 2, we present the results of some of these computations, focusing on examples of cases in which difficulties were encountered using standard tools.

Referring to Table 2, the first five cases are for CO_2 and *trans*-2-hexen-1-ol at 303.15 K. In discussing these problems it is useful to refer to the P-x-y diagram shown in Fig. 2, which shows both the model prediction and our experimental data. At 303.15 K, the model predicts a three-phase line at 68.8 bar; above this pressure there are two two-phase envelopes, a very small vapor/liquid envelope (shown in the inset) near the pure CO_2 axis, and a larger high-pressure envelope in which the CO_2-rich phase has a liquid-like density. Below the three-phase line, there is a single large vapor/liquid envelope. The model curves were verified by direct examination of the Gibbs energy surface and application of tangent plane analysis. The experimental points follow the low-pressure envelope. This is expected since we used a static measurement apparatus. The high-pressure envelope may be accessible using a variable volume high-pressure cell.

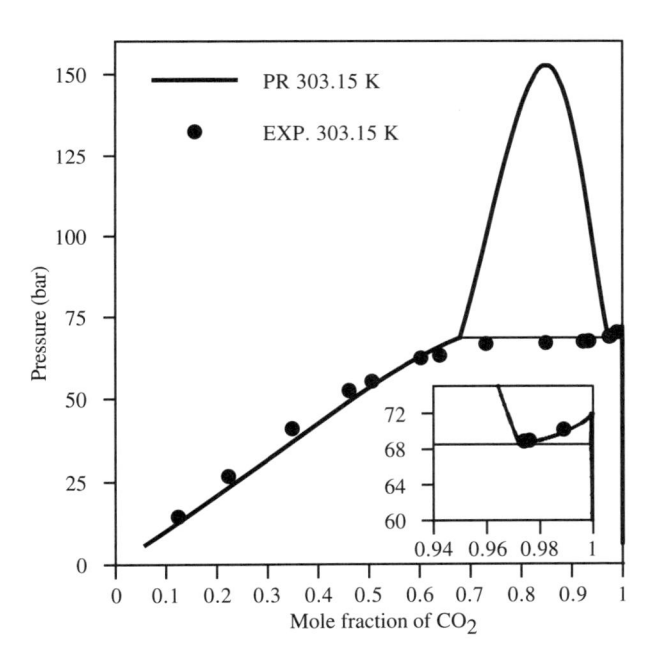

Figure 2. P-x-y plot for trans-2-hexen-1-ol/CO_2.

The first case is at a pressure well below the three-phase line, and all four tools for computing the phase split generate the correct answer. In case 2, at a pressure just below the three-phase line, FLASH3, RGIBBS and INTFLASH converge normally to the correct answer. LNGFLASH indicates correctly that the mixture will split, but it fails to converge to a result for the phase compositions in 1000 iterations. This is unusual, as this code generally converges in well under 10 iterations.

In the third case, the pressure is just above the three-phase line, and the conventional tools LNGFLASH, FLASH3 and RGIBBS all generate the same, but incorrect result. These tools essentially do not "see" the very small vapor/liquid envelope just above the three-phase line. Case 4 is at a pressure just slightly (0.00101 bar) higher than the previous case. FLASH3 and RGIBBS continue to give the wrong result; however, now LNGFLASH does give the correct answer. In general, the unpredictability with which reliable results can be generated was an issue in using all three conventional tools near the three-phase line. Only INTFLASH was consistently reliable.

In the fifth case, the pressure is lower than in the previous two cases, but still above the three-phase line. Here all three conventional tools predict incorrectly that there is no phase split, while INTFLASH gives the correct results. The difficulties encountered by the conventional tools here, and in the previous two cases, appear to be related to the presence of multiple real volume roots from the cubic equation of state near the correct composition of the CO_2-rich phase. In INTFLASH, the existence of multiple real volume roots, and selection of the right ones, is never an issue, since the method automatically uses the right volume roots.

The next three cases also involve *trans*-2-hexen-1-ol, now at a temperature of 323.15 K. At this temperature, we are just above the temperature range in which a three-phase-line is possible. The system has a single two-phase envelope. Looking at these three cases (cases 6, 7 and 8), each of the conventional tools fails once. RGIBBS fails to predict the phase split in case 7, and FLASH3 fails to predict the phase split in case 8. In case 6, LNGFLASH predicts that the mixture will split, but encounters a numerical computation error and does not converge to a result for the phase compositions. This is again indicative of the unpredictability with which reliable results can be obtained using the conventional tools. Again, only INTFLASH solves all the problems correctly.

The final case involves the mixture of CO_2 with *tert*-butyl alcohol at 305.95 K and 74.66 bar. This is in a region of retrograde condensation. Here both FLASH3 and INTFLASH give the correct result. LNGFLASH indicates correctly that the mixture will split, but it fails to converge after 1000 iterations to a result for the phase compositions. RGIBBS fails to predict the phase split.

Multicomponent Phase Behavior

After determining the binary interaction parameters of all reactants and products with CO_2 (Stradi *et al.*, 1998), these were used to make predictions of the multicomponent phase behavior, assuming that all other binary interaction parameters were zero. Results indicated that the desired single phase reaction mixture could be maintained at operating pressures considerably lower than the 346 bar originally used by Tumas (1996), in fact as low as about 125 bar. Based on this prediction, Tumas's group at Los Alamos repeated their experiments at a pressure of only 136 bar. Their results confirmed single phase behavior, and also indicated that very high enantiomeric selectivity was maintained. Thus, by using modeling tools to interactively guide experimental work, an improved design was achieved that uses a much lower pressure than originally proposed.

Concluding Remarks

Especially near three-phase regions and regions of retrograde condensation, conventional tools for modeling phase behavior may become unreliable. The unreliability may be manifested in several ways, including convergence failures, computing the wrong number of phases, and computing incorrect phase compositions. From point to point within these difficult regions the reliability of the conventional tools is very hard to predict. Only the new INTFLASH tool was completely reliable, always computing the correct number of phases and phase compositions.

INTFLASH is reliable since it uses interval methods, which eliminate the need for initial guesses and provide mathematical and computational guarantees of reliability, though these guarantees come at the expense of additional computation time (Hua et al., 1998). INTFLASH can also be applied to multicomponent and multiphase (more than two phases) problems. A complementary tool for reliably computing all critical points of mixtures would also be useful in this context. This would establish the limits of stability of mixtures without computation of the entire phase diagram.

The approach to modeling and design used here is simple and readily applicable to other reaction systems involving environmentally benign replacement solvents. Only binary interaction coefficients are needed, many of which can be found in or computed from published data. The results are insightful and allow targeting experimentation time to those conditions most likely to produce good designs.

Acknowledgement

This work was supported in part by the Environmental Protection Agency under grant R824731-01-0, and by the National Science Foundation grants CTS95-22835 and EEC97-00537-CRCD. We also greatly appreciate fellowship support by the Helen Kellogg Institute for International Studies and the Coca-Cola Company (B. A. S).

References

Hua, Z., Brennecke, J. F. and M. A. Stadtherr (1996). Reliable prediction of phase stability using an interval Newton method. *Fluid Phase Equil.*, **116**, 52-59.

Hua, J. Z., Brennecke, J. F. and M. A. Stadtherr (1998). Enhanced interval analysis for phase stability: Cubic equation of state models. *Ind. Eng. Chem. Res.*, **37** 1519-1527.

Hytoft, G. and R. Gani (1996). *IVC-SEP Program Package*, Danmarks Tekniske Universitet, Lyngby, Denmark.

Stradi, B., Kohn, J. P., Stadtherr, M. A. and J. F. Brennecke (1998). Phase behavior of the reactants, products and catalysts involved in the allylic epoxidation of *trans*-2-hexen-1-ol to (2R,3R)-(+)-3-propyloxiranemethanol in high pressure carbon dioxide. *J. Supercritical Fluids*, **12**, 109-122.

Tumas, W. (1996). Unpublished results. Los Alamos National Laboratory.

Table 2. Examples Showing Computational Difficulties.[1]

	Binary Mixture			LNGFLASH		FLASH3		RGIBBS		INTFLASH	
Case	Feed (z_{CO2})	Temp. (K)	Pressure (bar)	x_{CO2}	y_{CO2}	x_{CO2}	y_{CO2}	x_{CO2}	y_{CO2}	x_{CO2}	y_{CO2}
	trans-2-Hexen-1-ol/CO_2										
1	0.985	303.15	63.819	0.6133	0.9995	0.6134	0.9995	0.6134	0.9995	0.6131	0.9995
2			67.871	**NC1**		0.6686	0.9993	0.6686	0.9993	0.6683	0.9993
3	0.800	303.15	71.00725	**0.7315**	**0.9986**	**0.7310**	**0.9987**	**0.7309**	**0.9987**	0.6846	0.9690
4			71.00826	0.6842	0.9690	**0.7310**	**0.9987**	**0.7309**	**0.9987**	0.6846	0.9690
5	0.700	303.15	70.09	**NPS**		**NPS**		**NPS**		0.6828	0.9702
6	0.970	323.15	97.75	**NC2**		0.6267	0.9948	0.6267	0.9949	0.6281	0.9947
7	0.742	323.15	130	0.7234	0.9554	0.7232	0.9560	**NPS**		0.7240	0.9554
8[2]			135	0.7345	0.9490	**NPS**		0.7347	0.9515	0.7352	0.9489
	tert-Butyl Alcohol/CO_2										
9	0.995	305.95	74.66	**NC1**		0.9935	0.9962	**NPS**		0.9937	0.9963

[1] The mole fractions x_{CO2} and y_{CO2} in each phase are given. Entries in bold indicate incorrect results. The notation NPS indicates that no phase split was predicted. NC1 and NC2 indicate that the program predicted a phase split, but that the phase split calculation did not converge. With NC1 there was no convergence after 1000 iterations, and with NC2 numerical computation error occurred, the program generating the result NaN (not a number).

[2] RGIBBS gives an answer but with an error message.

SIMPLIFIED SYSTEM DISTURBANCE PROPAGATION MODELS AND MODEL-BASED PROCESS SYNTHESIS:
DEVELOPMENT OF HIGHLY CONTROLLABLE PROCESSES

Q. Z. Yan, Y. H. Yang, and Y. L. Huang
Department of Chemical Engineering and Materials Science
Wayne State University
Detroit, MI 48202

Abstract

Process integration techniques have been increasingly used in the process and allied industries in recent years, mainly because of the attractiveness in energy and material savings, and pollution prevention. This paper focuses on the improved process integration via integrated process design and control. As an integral part of the integration methodology, various first principles-based, simplified disturbance propagation (DP) models are developed to characterize DP through complex network systems. The models are linear in matrix form, and demonstrate excellent prediction capability for process synthesis. A novel and systematic model-based integration methodology is also introduced to design cost-effective and highly structurally controllable process network systems.

Keywords

Process integration, Disturbance propagation models, Heat exchanger networks, Mass exchanger networks.

Introduction

Process integration techniques have been increasingly employed in the process and allied industries to reduce energy and material costs and, more recently, to minimize waste. Industrial practice has shown, however, that improper integration can cause various operational problems, and may make economic and environmental goals unachievable (Morari, 1992; Luyben and Floudas, 1994). Consequently, to ensure successful process integration, an integrated process must be structurally highly controllable. Naturally, the integration of process design and control is becoming one of the most promising, but most difficult areas in the process systems engineering (Papalexandri and Pistikopoulos, 1994; El-Halwagi and Manousiouthakis, 1990; Yang *et al.*, 1996, 1999).

The integration of process design and control has been traditionally performed in the process control synthesis phase, where controllability is assessed. An uncontrollable process identified will be required for modification or even redesign (Morari, 1992). A structurally controllable process will be eventually derived through an iterative procedure of process design and process control design. This time-consuming and less efficient approach has been challenged by the approaches to process alternative screening or process design. These approaches intend to derive a highly controllable process before control design.

This paper centers on the earliest integration of process design and control in the overall process engineering activities. The main focus is on the development of a novel and systematic model-based integration methodology for synthesizing cost-effective and highly structurally controllable process network systems. In a network, heavy interactions among units are the main obstacle to achieving economic, environmental, and operational goals. In this text, process structural controllability is referred to the ability of structural disturbance rejection. As an integral part, first principles-based, simplified disturbance propagation (DP) models are developed to characterize DP through heat exchanger networks (HENs) and mass exchanger networks (MENs). The models establish the relationships among the network manipulated variables, disturbance variables, and controlled variables. They are linear in matrix form, and demonstrate excellent predication capability for process synthesis. By incorporating these DP models, a comprehensive synthesis strategy is developed targeting cost-effective and maximum disturbance rejection. Process flowsheet is developed successfully by the integration procedure with little computational effort. More importantly, the synthesis procedure will generate a process network with recommended pairing of manipulated variables and controlled variables. With this methodology, major structural problems can be prevented during process synthesis. A highly controllable process system should be identified in a systematic manner.

To demonstrate the attractiveness of this approach, process dynamic simulations are also performed to evaluate the resultant process networks. The disturbance rejection capabilities at the steady state by the DP models and the dynamic models are consistent in general. The efficacy of the methodology is demonstrated by solving practical MEN and HEN problems where optimal selection of manipulated variables are also derived during process synthesis. The methodology will show its great potential in advancing current process integration techniques to achieve both economic and operational goals in broad industrial applications.

Unit-Based Heat Exchanger and Mass Exchanger Model

Basic Equations. A heat exchanger (HE) with bypasses and mass exchanger (ME) with recycles as potential manipulated variables for disturbance rejection are sketched in Figure 1.

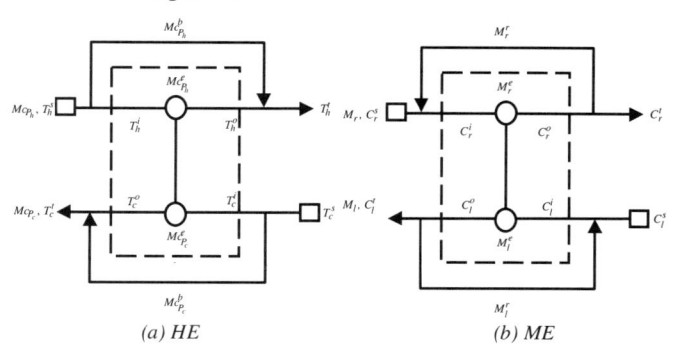

Figure 1. Sketch of an HE with bypass and an ME with recycles.

The energy balance and heat transfer equations for a HE can be readily derived as follows.

$$Mc_{p_h} T_h^t = Mc_{p_h}^b T_h^s + Mc_{p_h}^e T_h^o \tag{1}$$

$$Mc_{p_c} T_c^t = Mc_{p_c}^b T_c^s + Mc_{p_c}^e T_c^o \tag{2}$$

$$Q = UA \frac{(T_h^i - T_c^o) + (T_h^o - T_c^i)}{2}$$
$$= Mc_{p_h}^e (T_h^i - T_h^o) = Mc_{p_c}^e (T_c^o - T_c^i) \tag{3}$$

Basic Disturbance Model. Based on the above fundamental equations, we can derive a disturbance propagation and control model to characterize the response of the output of an HE to source disturbances and potential correction actions. The resultant model is of the following structure similar to a state space model, except that there is no time involved.

$$\delta T^t = B \, \delta f + D_t \delta T^s + D_m \delta Mc_P \tag{4}$$

where

$$\delta T^t = (\delta T_h^t \quad \delta T_c^t)^T$$

$$\delta T^s = (\delta T_h^s \quad \delta T_c^s)^T$$

$$\delta Mc_P = (\delta Mc_{P_h} \quad \delta Mc_{P_c})^T$$

$$\delta f = (\delta f_h \quad \delta f_c)^T$$

$$B = \begin{pmatrix} \dfrac{\alpha(T_h^s - T_h^t)}{2(1-f_h)^2} & \dfrac{\beta(T_h^s - T_h^t)}{2(1-f_c)^2} \\[2ex] -\dfrac{\alpha(T_c^t - T_c^s)}{2(1-f_h)^2} & -\dfrac{\beta(T_c^t - T_c^s)}{2(1-f_c)^2} \end{pmatrix}$$

$$D_t = \begin{pmatrix} 1-\alpha & \alpha \\ \beta & 1-\beta \end{pmatrix}$$

$$D_m = \begin{pmatrix} \alpha_h\left(2 - \dfrac{\alpha}{1-f_h}\right) & -\dfrac{\alpha\alpha_c}{1-f_c} \\[2ex] \dfrac{\beta\alpha_h}{1-f_h} & -\alpha_c\left(2 - \dfrac{\beta}{1-f_c}\right) \end{pmatrix}$$

and

$$\alpha = \frac{T_h^s - T_h^t}{T_h^s - T_c^s} \qquad \beta = \frac{T_c^t - T_c^s}{T_h^s - T_c^s}$$

$$\alpha_h = \frac{T_h^s - T_h^t}{2Mc_{P_h}} \qquad \alpha_c = \frac{T_c^t - T_c^s}{2Mc_{P_c}}$$

For an ME, the first principals are similar: the key component is transferred from the rich stream to the lean stream. The mass transfer results in the concentration decrease of the component in the rich stream from C_r^s to C_r^t and the increment of that in the lean stream from C_l^s to C_l^t. Based on experimental results, recycle of the rich stream and/or the lean stream is introduced as potential manipulated variables. Using basic mass balance, equilibrium and operating relationships, a basic disturbance propagation and control model for an ME can be:

$$\delta C^t = B\,\delta f + D_c\,\delta C^s + D_m\,\delta M \quad (5)$$

where

$$\delta C^t = \begin{pmatrix} \delta C_r^t & \delta C_l^t \end{pmatrix}^T$$
$$\delta f = \begin{pmatrix} \delta f_r & \delta f_l \end{pmatrix}^T$$
$$\delta C^s = \begin{pmatrix} \delta C_r^s & \delta C_l^s \end{pmatrix}^T$$
$$\delta M = \begin{pmatrix} \delta M_r & \delta M_l \end{pmatrix}^T$$

and B, D_c, D_m are corresponding coefficient matrices.

Structural Representation of HEN and MEN

To synthesize a highly controllable HEN and MEN, a key step is how to mathematically describe a network structure (or its topology). Structural matrix S is introduced which contains all information of stream connections and stream matching sequences. The matrix has the dimension of $2N_e \times 2N_e$; where N_e is the total number of exchangers. It can be decomposed into sub-matrices S_1 and S_2 that are, respectively, of the dimensions of $(N_h + N_c) \times 2N_e$ and $(2N_e - N_h - N_c) \times 2N_e$, where N_h and N_c are the number of hot streams and that of cold streams, respectively, if it is for a HEN.

In sub-matrix S_1, each row is designated for a hot or cold stream. The columns are divided into N_e pairs; each pair is designated for a specific HE. In each pair, the left column and the right column are, respectively, assigned to the hot and the cold stream going through the HE. Each element of S_1 represents a connection mode between a stream and the HE; it can be either 1 (representing a stream going through the HE) or 0 (representing a stream not going through it).

In sub-matrix S_2, each row is designated for an intermediate stream. The definition of columns is the same as that for sub-matrix S_1. Each element of S_2 represents one of the three connection modes of an intermediate stream with an HE. We assign [1], {1}, or 0 to an element to represent an intermediate stream entering, leaving, or not going through an HE.

Through structural matrix S, three kinds of conversion matrices, V_1, V_2, and V_3, can be derived for the process system construction.

System Disturbance Propagation and Control Model

The unit-based models in Eqs. (4), (5) form the basis for system models. For a HEN, the model has the following structure, which is identical to a unit model:

$$\delta \underline{T}^t = \underline{B}\delta\underline{f} + \underline{D}_t \delta\underline{T}^s + \underline{D}_m \delta \underline{Mc}_p \quad (6)$$

where

$$\delta \underline{T}^t = \begin{pmatrix} \delta T_{h_1}^t & \cdots & \delta T_{h_{N_h}}^t & \delta T_{c_1}^t & \cdots & \delta T_{c_{N_c}}^t \end{pmatrix}^T$$

$$\delta \underline{f} = \begin{pmatrix} \left(\delta f_{E_1}^{in}\right)^T & \left(\delta f_{E_2}^{in}\right)^T & \cdots & \left(\delta f_{E_{N_e}}^{in}\right)^T \end{pmatrix}^T$$

$$\delta \underline{T}^s = \begin{pmatrix} \delta T_{h_1}^s & \cdots & \delta T_{h_{N_h}}^s & \delta T_{c_1}^s & \cdots & \delta T_{c_{N_c}}^s \end{pmatrix}^T$$

$$\delta \underline{Mc}_P = \begin{pmatrix} \delta Mc_{P_{h_1}} & \cdots & \delta Mc_{P_{h_{N_h}}} & \delta Mc_{P_{c_1}} & \cdots & \delta Mc_{P_{c_{N_c}}} \end{pmatrix}^T$$

The system model establishes the cause-effect relationship at the steady state level. Quantitatively, we can evaluate the derivation of all system outputs when various disturbances enter the system and when relevant control actions take place. The significance of this model is that the resultant process evaluation for alternative screening will be much more reasonable, as compared to the model without control terms. Note that the system model for a MEN will be structurally the same as that for a HEN in Eq. (6), except that it has the variables of concentration, mass flowrate, and recycling, rather than temperature, heat capacity flowrate, and bypass.

Applications

Yee and Grossmmann (1990) employed an MINLP algorithm to solve the problem in Figure 2. This problem was first attempted by Linnhoff et al. (1982). For this problem, we introduce various disturbances specified below.

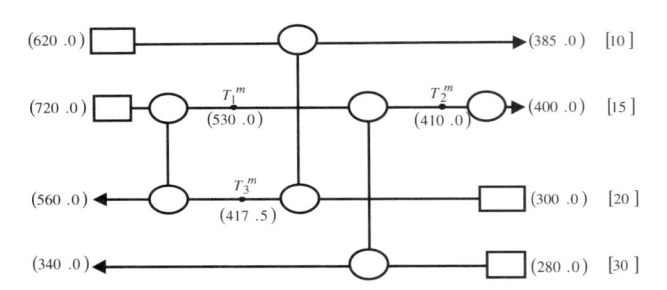

Figure 2. Yee et. al. solution (1990): temperature (K); [ ]: mass flowrate heat capacity (kW/K).

$$\delta T_s^+ = [5 \ 0 \ 0 \ 0]^T \qquad \delta T_s^- = [0 \ 0 \ -5 \ -5]^T$$
$$\delta Mc_p^+ = [0.5 \ 0 \ 0 \ -1.5]^T \qquad \delta Mc_p^- = [-0.5 \ 0 \ 0 \ 1.5]^T$$

By using our methodology, the system model with all bypass options can be identified below.

$$\begin{pmatrix} \delta T_{h_1}^t \\ \delta T_2^m \\ \delta T_{c_1}^t \\ \delta T_{c_2}^t \end{pmatrix} = \begin{pmatrix} 86.3 & 43.1 & 0 & 0 & 0 & 0 \\ -14.1 & -7.05 & 31.0 & 23.3 & 28.8 & 14.4 \\ -22.8 & -11.4 & -44.8 & -33.6 & 0 & 0 \\ -6.50 & -3.25 & 14.3 & 10.7 & -14.4 & -7.2 \end{pmatrix} \begin{pmatrix} \delta f_{h_{E_1}} \\ \delta f_{c_{E_1}} \\ \delta f_{h_{E_2}} \\ \delta f_{c_{E_2}} \\ \delta f_{h_{E_3}} \\ \delta f_{c_{E_3}} \end{pmatrix}$$

$$+ \begin{pmatrix} 0.27 & 0 & 0.73 & 0 \\ 0.12 & 0.19 & 0.21 & 0.48 \\ 0.19 & 0.47 & 0.34 & 0 \\ 0.06 & 0.09 & 0.10 & 0.76 \end{pmatrix} \begin{pmatrix} \delta T_{h_1}^s \\ \delta T_{h_2}^s \\ \delta T_{c_1}^s \\ \delta T_{c_2}^s \end{pmatrix} + \begin{pmatrix} 14.9 & 0 & -2.16 & 0 \\ 1.41 & 10.6 & -2.73 & -0.48 \\ 2.28 & 2.98 & -7.98 & 0 \\ 0.65 & 3.05 & -1.26 & -1.76 \end{pmatrix} \begin{pmatrix} \delta Mc_{P_{h_1}} \\ \delta Mc_{P_{h_2}} \\ \delta Mc_{P_{c_1}} \\ \delta Mc_{P_{c_2}} \end{pmatrix}$$

Based on this model, an extended relative gain array (RGA) and iterative approaches are developed to determine the optimal bypass placement and nominal fraction as manipulated variables and control strategy to keep the system outputs at their required control precision. The retrofit design for this process under the uncertain source disturbances is shown in Figure 3. The total area-increment of heat exchangers is 24.4% and cost increment is 14.6%. With the comparison with Uzturk's MINLP design (1997), this retrofit design has much lower cost increment and more robust system controllability.

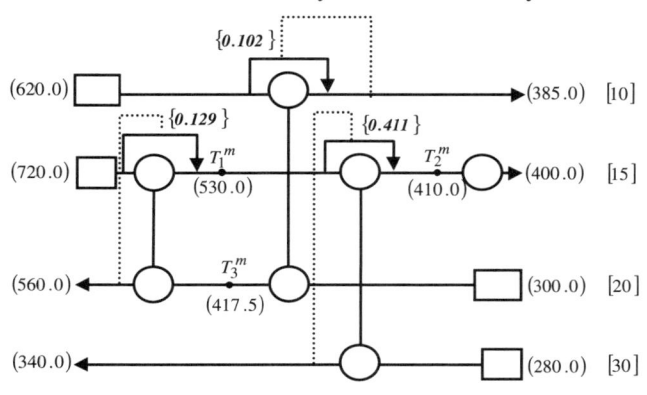

Figure 3. Control Strategy for Yee et. al. solution (1990) under uncertain disturbances. {}: nominal bypass fraction; dotted line is control loop scheme from manipulated variable to controlled target.

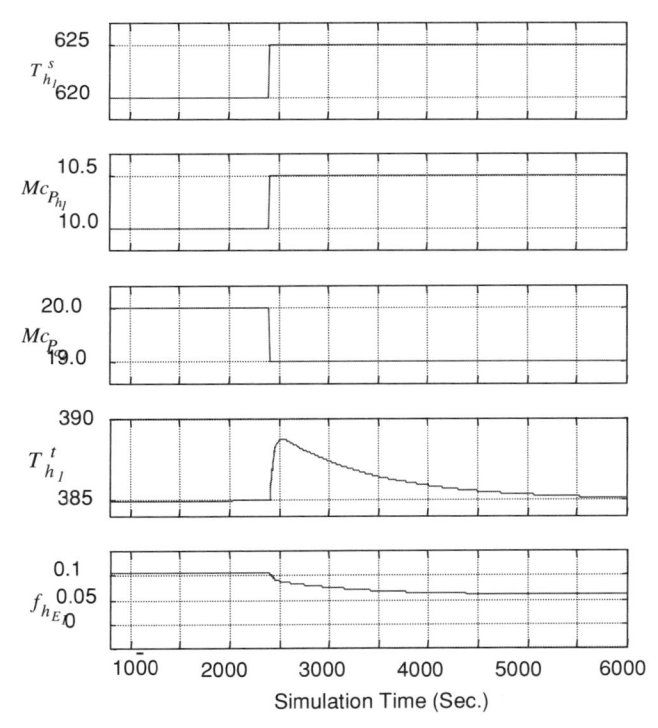

Figure 4. The dynamic performance of the first control loop in Figure 3 under source disturbance.

In order to verify the process controllability, dynamic simulation is conducted. The dynamic performance of the first control loop is illustrated in Figure 4, in which the first three curves show the source disturbances, the fourth one is the controlled target temperature, and the last curve is the bypass adjustment.

Conclusion

In this paper, a system disturbance rejection and control (DR&C) model is introduced to predict precisely the output fluctuations of process streams in any HEN and/or MEN. The model gives an explicit representation of structural DP, which allows us to predict systematically and comprehensively how each output is affected by various disturbance sources. Meanwhile, the synthesis procedure generates a process network with recommended pairing of manipulated variables and controlled variables. The disturbance rejection capabilities at the steady state by the DR&C models and the dynamic models are consistent. With this methodology, major structural problems should be prevented during process synthesis. A highly controllable process system should be identified in a systematic manner.

Acknowledgment

The support from the NSF through the grant CTS-9414494 is gratefully acknowledged.

Nomenclature

- A heat or mass exchanger area
- B control action process gain matrix
- D source disturbance gain matrix
- f manipulated variable vector
- Mc_P mass flowrate heat capacity
- S system structural matrix
- T temperature
- U overall heat transfer coefficient

Superscripts
- b bypass
- e heat exchanger
- s source temperature
- t target temperature

Subscripts
- c cold stream
- e heat exchanger
- f bypass fraction
- h hot stream
- s hot streams and cold streams
- l Lean stream

References

El-Halwagi, M. M. and V. Manousiouthakis (1990). *Chem. Eng. Sci.*, **45**.

Linnhoff, B. and E. Kotjabasakis (1986). *Chem. Eng. Prog.*, **82**, 23-28.

Luyben, M. L. and C. A. Floudas (1994). *Comp. Chem. Eng.*, **18**, 933-969.

McAvoy, T. J. (1987) in *Recent Development in Chemical Process and Plant Design*, Y. A. Liu, H. A. McGee, Jr., and W. R. Epperly (eds.). John Wiley & Sons, New York, NY, 289-325.

Morari, M. (1992). *Preprints of the IFAC Workshop on Interactions between Process Design and Process Control.* London, UK, 3-16.

Papalexandri K. P. and E. N. Pistikopoulos (1994). *Comp. Chem. Eng.*, **18**, 1125-1139.

Uzturk, D. and U. Akman (1997). *Comp. Chem. Eng.*, **21**, S373-S378.

Yang, Y. H., J. P. Gong, and Y. L. Huang (1996). *Ind. Eng. Chem. Res.*, **35**, 4550-4558.

Yang, Y. H., Yan, Q. Z. and Y. L. Huang (1999). *Trans. I. ChemE.*, **77**, May.

Yee, T. F. and Grossmann, I.E. (1990). *Comp. Chem. Eng.*, **14**, 1165-1184.

IMPROVING EFFICIENCY OF DISTILLATION WITH NEW THERMALLY-COUPLED CONFIGURATIONS OF COLUMNS

Rakesh Agrawal and Zbigniew T. Fidkowski
Air Products and Chemicals, Inc.
Allentown, PA 18195-1501

Abstract

New, simplified thermally coupled systems of columns for ternary separation, containing one one-way and one two-way connection between the columns, were recently proposed by Agrawal and Fidkowski (1999). The new thermally coupled systems are easier to control and often consume no more heat than the fully coupled systems. Thermodynamic efficiencies of the new systems distilling ideal saturated liquids into pure liquid product streams are calculated and compared with efficiencies of known distillation configurations. The new configurations significantly extend the zone of feed compositions where thermally coupled configurations are the most efficient. The new configurations appear to be particularly useful when separation of components is difficult. The new configurations can be instantly and effortlessly compared with other distillation schemes by using a computer program to calculate the efficiencies of various possible configurations for ternary distillation.

Keywords

Ternary distillation, Thermally coupled columns, Thermodynamic efficiency, Minimum energy requirements, Operability.

Introduction

Several configurations of distillation columns are known for ternary separations (King, 1980), including direct split (DS), indirect split (IS), side rectifier (SR), side stripper (SS) and fully coupled system (FC). Among them, the fully thermally coupled system of distillation columns (known also as the Petlyuk system, Petlyuk et al., 1965) requires the least amount of heat for the separation of an ideal ternary mixture into pure products (Fidkowski and Krolikowski, 1986). Because of their attractiveness, complex column configurations, including thermally coupled systems, were extensively studied in literature (for example Westerberg, 1985; Glinos and Malone, 1985). However, the thermodynamic efficiency of the fully thermally coupled system is not always attractive, due to the fact that all the heat is being supplied at the highest temperature and removed at the lowest temperature. A systematic study of the thermodynamic efficiency for various ternary systems (Agrawal and Fidkowski, 1998) shows that the region of feed compositions where the efficiency of the fully thermally coupled system is highest is quite limited. The efficiencies of the conventional direct split or indirect split systems (which do not have the lowest heat requirements) are often much higher than the efficiency of the fully thermally coupled system.

New, simplified thermally coupled systems of columns for ternary separations were recently proposed by Agrawal and Fidkowski (1999). All previous thermally coupled systems of distillation columns have had two-way material connections between the columns. The new systems contain one one-way and one two-way connection between the columns. One of the material streams connecting the columns in the fully coupled system has been replaced with a heat exchanger (reboiler or condenser). Depending on where the one-way connection is located and which phase stream (liquid or vapor) is used in the one-way connection, there are four different possibilities for these new systems. They have been named as follows: side Rectifier with Vapor connection (RV), side Rectifier with

Liquid connection (RL), side Stripper with Vapor connection (SV) and side Stripper with Liquid connection (SL). The systems RV and SL may have lower total vapor flow than RL and SV, respectively. Systems RV and SL are shown in Figure 1.

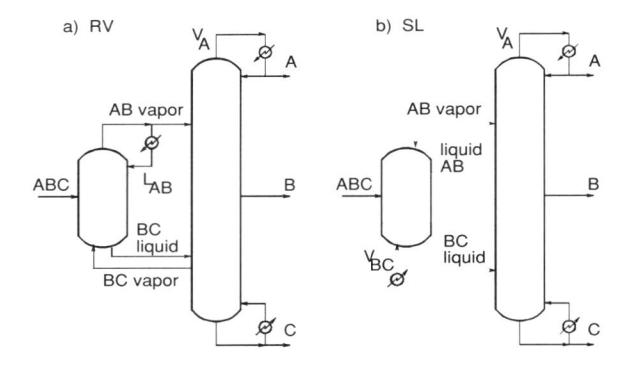

Figure 1. New thermally coupled configurations: (a) side rectifier with vapor connection - RV, (b) side stripper with liquid connection - SL.

It was shown by Agrawal and Fidkowski (1999) that the new thermally coupled systems are easier to control. For certain feed compositions and relative volatilities they consume no more heat than fully coupled systems, and frequently much less heat than the side stripper or side rectifier configurations, or a system with a prefractionator.

The objective of this paper is to calculate thermodynamic efficiencies for the two most attractive new systems - RV and SL. Unlike the fully coupled system, with RV and SL it is possible to remove or supply a portion of the heat duty at an intermediate temperature of (condensing or boiling) binary mixture. Therefore, we anticipate that these new systems will extend considerably the zone of feed compositions where thermally coupled configurations are the most efficient ones.

Efficiency Calculations

Calculations are performed for ideal mixtures with constant relative volatilities and constant molar overflow. Feed, a saturated liquid mixture of components A, B and C (where A is the most volatile component and C is the least volatile component) is separated into nearly pure components, which are withdrawn as saturated liquids. Minimum reflux conditions are assumed.

The following definition of distillation thermodynamic efficiency is used:

$$\eta = \frac{\text{minimum work of separation}}{\text{minimum work of separation} + \text{exergy loss}} \quad (1)$$

Exergy loss is calculated by balancing the exergies of all the material streams entering and leaving the distillation column system, excluding the reboilers and condensers. It was shown by Agrawal and Herron (1997) that it is possible to eliminate entirely the temperatures from the efficiency equations and express the efficiency only in terms of relative volatilities, feed composition and corresponding flow rates of material streams. The flow rates (at minimum reflux conditions) can then be expressed as functions of relative volatilities and feed composition. More details on calculating thermodynamic efficiencies of various column configurations are given in Agrawal and Fidkowski (1998). The final equations used to calculate thermodynamic efficiency of RV and SL are given below:

$$\eta_{RV} = \frac{-z_A \ln z_A - z_B \ln z_B - z_C \ln z_C}{V_A \ln \alpha_A + L_{AB}\left[\ln \alpha_B + \int_0^1 \ln \frac{1}{K_{B,AB}} dq \right]} \quad (2)$$

$$\eta_{SL} = \frac{-z_A \ln z_A - z_B \ln z_B - z_C \ln z_C}{V_A \ln \alpha_A - V_{BC} \int_0^1 \ln \frac{1}{K_{C,BC}} dq} \quad (3)$$

Although Eqs. 2 and 3 are valid both at and above minimum reflux, we chose to compare the efficiencies at minimum reflux. For a given mixture, the value of minimum vapor flows in RV and SL depends on the split of components in the first column (Agrawal and Fidkowski, 1999). Let us denote by β the fraction of component B leaving the first column in the distillate. An example of the dependence of the total vapor flow on β is shown in Fig. 2. A horizontal, flat section PR can be noted in Fig. 2. Throughout this zone, the total minimum vapor flow is at its minimum value (i.e., the vapor flow at pinched conditions is minimum). At $\beta = \beta_P$ the vapor flow in the first column is minimum; at $\beta = \beta_R$ the second column is pinched at both feed locations. The values of β_P and β_R depend on relative volatilities and feed composition. However, unlike for the fully coupled system (Fidkowski and Krolikowski, 1986), the value of β_R can lie beyond the (0,1) range. In that case, the only distinct minimum in the total vapor flow is located at $\beta = \beta_P$. Unlike the fully coupled system, it can be shown that if $\beta_R < \beta_P$ in RV configuration (or $\beta_R > \beta_P$ in SL configuration) the section between points P and R in Fig. 2 is not exactly horizontal, although its slope is small in comparison with the slope of the other sections. In this case, there is only one distinct minimum vapor flow at $\beta = \beta_P$.

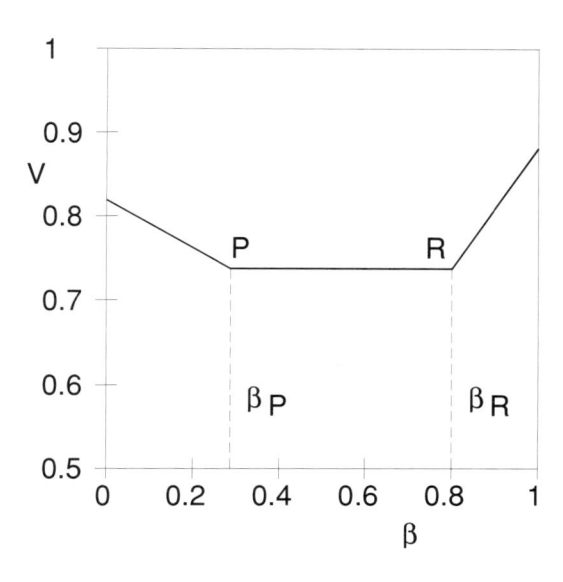

Figure 2. Example of minimum vapor flow in RV as a function of β.

Similar rules may be observed for the efficiency calculations:

1. If the value of β_R is beyond the (0,1) range, the maximum efficiency is always at $\beta = \beta_P$.
2. For $\beta_R \in (0,1)$, if $\beta_R > \beta_P$ in RV (or $\beta_R < \beta_P$ in SL), then the constant minimum vapor flow in the PR section is observed and the maximum efficiency is always at $\beta = \beta_R$
3. For $\beta_R \in (0,1)$, if $\beta_R < \beta_P$ in RV (or $\beta_R > \beta_P$ in SL), then the PR section is not horizontal and maximum efficiency is usually for $\beta = \beta_P$, but there are cases where the maximum efficiency is at $\beta = \beta_R$.

Therefore, when the maximum efficiency is sought, the values of β_P, β_R and the corresponding efficiencies need to be examined.

Results and Discussion

A computer program was written to calculate the thermodynamic efficiency of RV and SL, using the assumptions and equations described above. Additionally, the values of efficiencies for all the other ternary configurations (DS, IS, SR, SS and FC) were also calculated. This enabled comparison of the efficiencies of all the ternary configurations on the feed composition triangles. The comparison has been carried out for various type of splits, assuming that both the A/B split and, independently, the B/C split may be easy or difficult. Relative volatilities of 2.5 and 1.1 were arbitrarily chosen to represent the easy split and the difficult split. This gives four possibilities for the values of relative volatilities: $\alpha_A = 6.25$ and $\alpha_B = 2.5$ (easy, easy split), $\alpha_A = 2.75$ and $\alpha_B = 1.1$ (easy, difficult), $\alpha_A = 2.75$ and $\alpha_B = 2.5$ (diffi-cult, easy) and finally $\alpha_A = 1.21$ and $\alpha_B = 1.1$ (difficult, difficult). The difficulty of splits in ternary mixture may also be expressed in terms of the Ease of Separation Index (ESI), proposed by Tedder and Rudd (1978).

$$ESI = \frac{K_A / K_B}{K_B / K_C} = \frac{\alpha_A}{\alpha_B^2} \qquad (4)$$

For the easy - easy split (*ESI* = 1) the results of calculations are shown in Fig. 3.

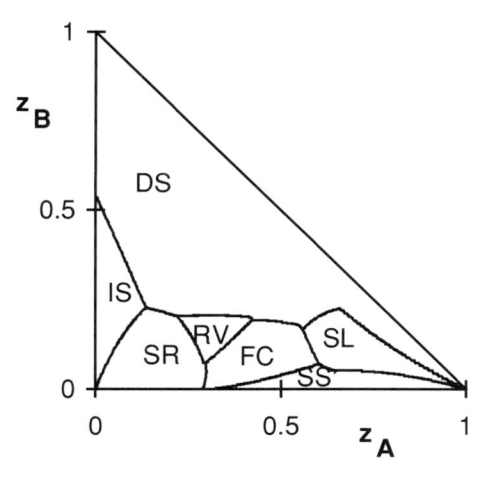

Figure 3. Optimum efficiency regions on a feed composition triangle for $\alpha_A = 6.25$ and $\alpha_B = 2.5$.

The optimal regions for RV and SL are adjacent to the FC region. They more than double the zone of feed compositions where the thermally linked systems are the most efficient. The improvement in efficiency is not very significant (about 2 or 3 %), but at the same time the reduction in the minimum vapor flow may be as big as 20% in comparison to DS or IS.

For dissimilar splits (easy - difficult or difficult - easy, where *ESI* is much lower or higher than 1) the thermodynamic efficiencies of RV and SL are not attractive; there is always another column configuration with a higher thermodynamic efficiency.

For a difficult-difficult split (*ESI* = 1), however, the regions where RV and SL are the most efficient grow considerably - Fig. 4. The new thermally coupled systems eliminate entirely the conventional direct split and indirect split configurations. The relative efficiency increase due to the use of the new configurations is about 10 - 20 % with the simultaneous reduction of minimum vapor flow by as much as 40 %.

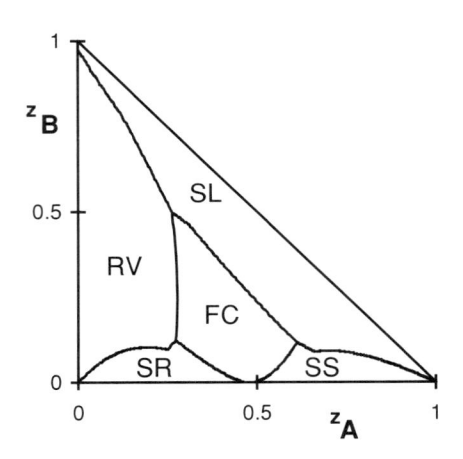

Figure 4. Optimum efficiency regions on a feed composition triangle for $\alpha_A = 1.21$ and $\alpha_B = 1.1$.

The optimality regions of RV and SL are always adjacent to the optimality region for FC. SL is optimal for feed compositions leaner in component C and RV - for liquid mixtures leaner in component A.

It should also be noted that the heat exchange in a binary reboiler or condenser (BC or AB) happens over a temperature range. This usually decreases the overall efficiency of the distillation system and the heat exchangers together, unless the heating or cooling medium mimics the temperature change of the mixture BC or AB.

Conclusions

The new thermally coupled systems (RV and SL) are simplifications of the fully coupled system created by replacing one of the interconnecting material streams with a reboiler or a condenser (Fig. 1). They offer improved thermodynamic efficiencies and simultaneously a reduction of the heat requirements for separation, especially for mixtures with *ESI* close to 1. The advantages are greater for mixtures that are difficult to separate. Improvement of thermodynamic efficiency of the distillation system will be offset by increased irreversibility in the added heat exchanger where binary mixture is boiled (or condensed) over a temperature range. System SL is optimal for feed compositions leaner in the heavy component (C). RV is optimal for liquid mixtures leaner in component A.

Nomenclature

DS = direct split
ESI = ease of separation index, Eqn. 4
FC = fully coupled system of columns
IS = indirect split
K_i = equilibrium coefficient for component i
$K_{B,AB}$ = equilibrium coefficient for component B in binary mixture AB
$K_{C,BC}$ = equilibrium coefficient for component C in binary mixture BC
L_{AB} = liquid flow rate from condenser AB (Fig. 1), kmol/s
q = mole fraction of liquid
RV = side rectifier with vapor connection, Fig. 1a
SL = side stripper with liquid connection, Fig. 1b
SR = side rectifier
SS = side stripper
V_A = vapor flow rate to condenser A (Fig. 1), kmol/s
V_{BC} = vapor flow rate from reboiler BC (Fig. 1), kmol/s
z_i = feed mole fraction of component i, ($i = A, B, C$)
α_i = relative volatility of component i, ($i = A, B$) with respect to component C
β = fraction of component B leaving the first column in the distillate
β_P = value of β at which the vapor flow in the first column of RV, SL or FC is minimum
β_R = value of β at which the second column of RV, SL or FC is pinched at both feed locations
η = thermodynamic efficiency

References

Agrawal, R., and Z. T. Fidkowski (1998). Are thermally coupled distillation columns always thermodynamically more efficient for ternary distillations? *Ind. Eng. Chem. Res.*, **37**, 3444-3454.

Agrawal, R., and Z. T. Fidkowski (1999). New thermally coupled schemes for ternary distillation. *AIChE J.*, **45**, 485-496.

Agrawal, R., and D. M. Herron (1997). Optimal thermodynamic feed conditions for distillation of ideal binary mixtures. *AIChE J.*, **43**, 2984-2996.

Fidkowski, Z. T., and L. Krolikowski (1986). Thermally coupled system of distillation columns: optimization procedure. *AIChE J.*, **32**, 537-546.

Glinos, K., and M. F. Malone (1985). Minimum vapor flows in distillation column with a sidestream stripper. *I&EC Process Des. Dev.*, **24**, 1087-1090.

King, C. J. (1980). *Separation Processes*, McGraw-Hill, New York, 2nd edition, Chapter 13, 702-712.

Petlyuk, F. B., V. M. Platonov, and D. M. Slavinskii (1965). Thermodynamically optimal method of separating multicomponent mixtures. *Int. Chem. Eng.*, **5** (3), 555-561.

Tedder, D. W., and D. F. Rudd (1978). Parametric studies in industrial distillation: Part 1, Design comparisons. *AIChE J.*, **24**, 303-315.

Westerberg, A. W. (1985). The Synthesis of distillation-based separation systems. *Comp. Chem. Eng.*, **9**, 421-429.

EXPERIMENTAL AND COMPUTATIONAL SCREENING METHODS FOR REACTIVE DISTILLATION

Bernd Bessling
BASF-Corporation
Wyandotte, MI 48192-3729

Abstract

Especially for complex hybrid operations like reactive distillation, a good screening method could save a lot of development time. Now a screening method is introduced, which is based on the analogies between distillation and reactive distillation. For distillation, the preferred distillation is well known. In the preferred distillation all products are at minimum reflux condition on a straight line through the feed concentration and its equilibrium vapor. In preferred reactive distillation the straight line goes through the feed and its vapor concentration using transformed composition variables developed by Barbosa and Doherty. This straight line also represents a material balance. Using this material balance a conversion for the reactive distillation can be calculated; this is called the preferred conversion. By comparing the preferred conversion with the equilibrium conversion of the chemical reaction, it can be determined if a reactive distillation should be used.

Keywords

Reactive distillation, Preferred reactive distillation, Preferred conversion, Screening methods.

Introduction

Screening methods are becoming more and more important in the chemical industry to compress the time between the first idea and the final plant concept. Until now such methods have not been available for reactive distillation despite the fact that, especially for this hybrid operation, a good screening method could save a lot of development time. In most cases intensive research on physical properties, computer simulations and miniplant experiments are necessary, in order to decide whether reactive distillation is the right technology. Herein a screening method is developed, which is based on the analogies between distillation and reactive distillation.

Fundamentals

One of the most important results of the reactive distillation research in the last fifteen years was the idea to use transformed concentration coordinates to simplify the description of reactive distillation processes with equilibrium reactions. The chemical reaction for a system with nc components, the species Ψ_i and one equilibrium reaction is presented as:

$$\sum_{i=1}^{nc} v_i \Psi_i = 0 \qquad (1)$$

The transformed liquid and vapor concentrations developed by Barbosa and Doherty (1987, 1988a-c) are given by (for i=1,....nc; and i≠p):

$$X_i = \frac{x_i - \frac{v_i x_P}{v_P}}{1 - \frac{v_T x_P}{v_P}} \qquad Y_i = \frac{y_i - \frac{v_i y_P}{v_P}}{1 - \frac{v_T y_P}{v_P}} \qquad (2)$$

Barbosa proposed (1988c) to use the transformed concentration coordinates to solve reactive distillation problems using a similar set of equations compared to the normal distillation. For example the operating line for the rectifying section of a reactive distillation is given by:

$$Y_{i,n-1} = \frac{v_n^T}{v_n^T + 1} X_{i,n} + \frac{1}{v_n^T + 1} X_{i,D} \quad (3)$$

For an infinite reflux ratio the operating line for a reactive distillation is given by:

$$v_n^T \to \infty : Y_{i,n-1} = X_{i,n} \quad (4)$$

The transformed liquid composition X on stage n is identical with the transformed vapor concentration Y arising from stage n-1. The vapor concentration is a function of the liquid concentration and the pressure. The concentration profiles described by equation 4 are called reactive distillation lines. The reactive distillation line concept was used by Bessling (1997, 1998a) to predict the product regions of reactive distillation processes. Top and bottom products of a reactive distillation column at infinite reflux must be located on the reactive distillation line and the material balance line.

The structure of the reactive distillation line diagrams depends on the vapor/liquid and chemical equilibria. Evaluating the structure of the reactive distillation line diagram it can be determined whether a specific reactive distillation is feasible or not. Due to the hybrid character of a reactive distillation, it is usually economically advantageous only in those cases in which the distillation enhances the reaction or vice versa.

The Preferred Reactive Distillation

A basic concept for the evaluation of a normal distillation is the preferred separation, see Stichlmair (1988a-b). In the case of the preferred distillation, the top and the bottom product compositions lie on the material balance line through the feed composition x_F, and its vapor y_F, compare Figure 1. The feed composition x_F is identical with the concentration on the feed stage. In the case of a sharp preferred distillation, the top product is free of high boiling components and the bottom product is free of low boiling components. The medium boiler is distributed between the top and the bottom products. The minimum reflux ratio for the preferred distillation is given by:

$$v_{min,pref} = \frac{x_{i,D} - y_{i,F}}{y_{i,F} - x_{i,F}} \quad (5)$$

The tangent to the feed distillation line at the feed concentration gives the balance of the preferred distillation.

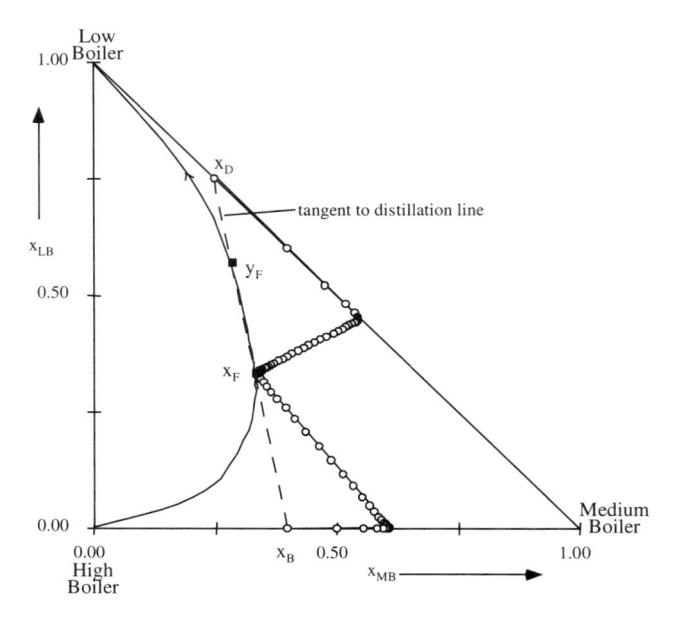

Figure 1. Concentration profile of a sharp preferred nonreactive distillation of an ideal three component mixture.

The preferred distillation concept was applied by Bessling (1998b) on reactive distillation problems with one equilibrium reaction. The transformed concentration coordinates have to fulfill the same requirements as the corresponding concentration variables did in case of the normal distillation. The minimum reflux ratio of the preferred reactive distillation is accordingly defined by:

$$v^T_{min,pref} = \frac{X_{i,D} - Y_{i,F}}{Y_{i,F} - X_{i,F}} \quad (6)$$

Figure 2 shows the concentration profile of a sharp reactive distillation with the reaction C+D⇌A+B.

The tangent to the reactive distillation line going through the feed composition X_F gives the balance of the preferred reactive distillation. The balance of the sharp preferred reactive distillation is given by the intersection of the tangent with the edges of the diagram. This balance line automatically determines the conversion of the sharp reactive distillation, also called the preferred conversion.

In Figure 2 the preferred conversion for a stoichiometric mixture of C and D is 52.5%. Due to the equilibrium constant of only 0.1, the equilibrium conversion is 24%. Operating the column with a reflux ratio which is higher than the preferred reflux ratio of 1.1 increases the conversion.

On the other side the same concentration profile also represents the preferred reactive separation for the reaction

A+B⇌C+D. The equilibrium conversion is 76% due to the high chemical equilibrium constant (K=1/0.1=10). Despite the high equilibrium constant, the preferred conversion is only 100%-52.5%=47.5%. Operating the column with the preferred reflux ratio has decreased the conversion. The reactive distillation is worse than a simple combination of reaction and distillation.

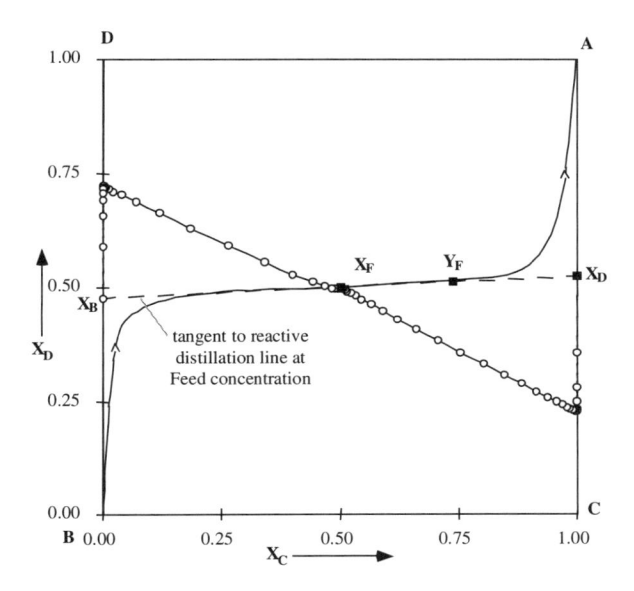

Figure 2. Sharp preferred reactive distillation for the reaction C+D ⇌ A+B. The Chemical equilibrium constant is K=0.1. The relative volatilities of the ideal system C/D/A/B are 4/2/8/1.

According to Stichlmair (1993), the preferred distillation can be interpreted as the natural separation between a high and a low boiling component. The medium boiling component is distributed between the top and the bottom product.

The natural influence of distillation on the reaction should be that the distillation increases the conversion.

Therefore the following heuristic rule, also called the tangent criterion, was defined by Bessling (1998b):

A system is suitable for reactive distillation, if the conversion for the sharp reactive distillation is higher than the equilibrium conversion.

The sharp preferred conversion can be determined by extending the tangent from the feed concentration to the edges of the diagram or to reactive distillation boundaries. Using this rule, it is possible for the first time, to determine the suitability of a system taking the chemical equilibrium and vapor liquid equilibrium simultaneously into account. This rule can be applied to reactive distillation problems, in which all products are elements of the reaction space. Bessling (1997) defined the reaction space as that part of the concentration space in which the conditions for the chemical equilibrium are fulfilled.

To apply the rule, it is only necessary to know the composition of the feed at chemical equilibrium and the vapor composition arising from that concentration. This can be done, by using a physical property model or simply by one simultaneous measurement of the chemical and the vapor liquid equilibria.

Application on the Methyl Formate Synthesis

Methyl formate can be produced by the esterification of formic acid with methanol.

$FA + MeOH \rightleftharpoons MF + H2O \quad K=5$

Figure 3 shows the boiling point surface of that system at 3.0 bar.

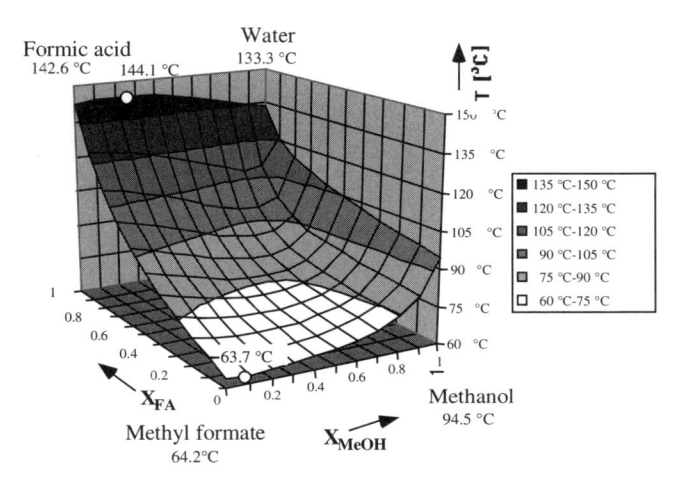

Figure 3. Boiling point surface for the system FA + MeOH ⇌ MF+H2O at 3 bar.

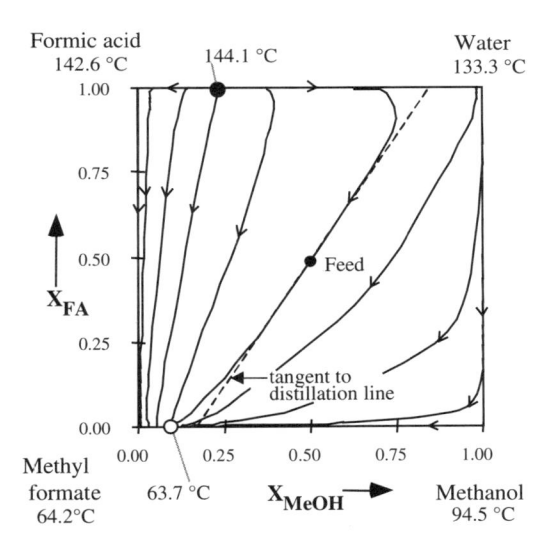

Figure 4. Reactive distillation lines for the system FA + MeOH ⇌ MF + H2O at P=3bar. For a stoichiometric mixture, the equilibrium conversion is 69.1% and the preferred conversion 86.4 %.

This is a highly nonideal system, which was analyzed by Bessling (1997) using the reactive distillation line concept. Figure 3 shows the high boiling azeotrope formic acid with water and the low boiling azeotrope methyl formate with methanol. Both azeotropes are nonreactive azeotropes. All reactive distillation lines start at that high boiling azeotrope and end at the low boiling azeotrope compare Figure 4.

In the preferred reactive distillation the conversion is higher compared to the equilibrium conversion. According the tangent criterion, reactive distillation is very suitable for the production of methyl formate. The production of formic acid from methyl formate and water is not so suitable for reactive distillation, because the preferred is in this case lower than the equilibrium conversion.

Conclusion

Reactive distillation is a very complex hybrid technology. Using the methods and ideas behind the preferred reactive distillation makes it possible to determine whether a reactive distillation is a desirable alternative. This can be accomplished either by determination of the preferred conversion by calculating a reactive flash or experimentally by the measurement of the simultaneous vapor liquid equilibrium.

Using this rule, it is possible for the first time, to determine the suitability of a system taking the chemical equilibrium and vapor liquid equilibrium simultaneously into account.

Nomenclature

K   Equilibrium constant
nc   number of components
P   pressure
v   reflux ratio
x_i liquid phase composition
X_i transformed liquid phase composition
y_i vapor phase composition
Y_i transformed vapor phase composition

Greek Symbols

v_i stoichiometric coefficient of component I

v_T sum of stoichiometric coefficient ($v_T = \sum_{i=1}^{nc} v_i$)

Ψ_i chemical compound, numbered from i=1 to nc

Subscript

B   Bottoms
D   Distillate
F   Feed
FA   Formic acid
I   component I
LB   Low boiler
MB   Medium boiler
MeOH Methanol
MF   Methyl formate
min   minimum
n   stage n of a distillation column
P   reference component for transformation
pref preferred

Superscripts

T   transformed

References

Barbosa, D. and M. F. .Doherty (1987). A new set of composition variables for the representation of reactive - phase diagrams. *Proc. R. Soc. Lond.*, **A 413**, 459–464.

Barbosa, D. and M. F. Doherty (1988a). The influence of equilibrium chemical reactions on vapor-liquid phase diagrams. *Chem. Eng. Sci.*, **43**, 3, 529-540.

Barbosa, D. and M. F. Doherty (1988b). The simple distillation of homogeneous reactive mixtures. *Chem. Eng. Sci.*, **43**, 3, 541-550.

Barbosa, D. and M. F. Doherty (1988c). Design of minimum reflux calculations for single-feed multicomponent reactive distillation columns. *Chem. Eng. Sci.*, **43**, 7, 1523-1537.

Bessling, B., G. Schembecker and K. H. Simmrock (1997). Design of processes with reactive distillation line diagrams. *Ind. Eng. Chem. Res.*, **36**, 8, 3032-3042.

Bessling, B., J. M. Loening, A. Ohligschaeger, G. Schembecker and K. Sundmacher (1998a). Investigations on the synthesis of methyl acetate in a heterogeneous reactive distillation process. *Chem. Eng. Tech.*, **21**, 5, 393-400.

Bessling, B. (1998b). *Zur Reaktivdestillation in der Prozeßsynthese*. PhD Thesis, University of Dortmund.

Stichlmair, J. (1988a). Distillation and rectification, *Ullmann´s Encyclopedia of Industrial Chemistry*, **B3**, VCH Verlagsgesellschaft mbH, D-6940 Weinheim.

Stichlmair, J. (1988b). Zerlegung von Dreistoffgemischen durch Rektifikation. *Chem.-Ing.-Tech.*, **60**, 10, 747-754.

Stichlmair, J., Offers, H., and R.W. Potthoff (1993). Minimum reflux and minimum reboil in ternary distillation, *Ind. Eng. Chem. Res.*, **32**, 2438-2445.

OPTIMUM DESIGN OF MASS EXCHANGE NETWORKS USING PINCH TECHNOLOGY

Duncan M Fraser
Department of Chemical Engineering
University of Cape Town
Rondebosch 7701, South Africa

Nick Hallale
Department of Process Integration
UMIST
Manchester, M60 1QD, UK

Abstract

This paper discusses a new method for targeting the minimum capital and total costs of mass exchange networks. This method makes it possible to optimise total cost with respect to minimum driving force ahead of design. The method is applied to four literature problems which had previously been optimised using a mathematical programming approach and shows that substantial improvements are possible. Networks can be designed to closely approach the optimised total annual cost target. Designs for the four problems using the new approach have total annual costs which achieve this and are consistently lower than those of previous optimum designs. The techniques developed have been extended to a wide range of mass exchange problems, including retrofit applications, in which an impact diagram is developed which may be applied to effluent reduction in existing processes.

Keywords

Mass exchange networks, Pinch technology, Optimisation, Capital cost targets, Impact diagram.

Introduction

El-Halwagi and Manousiouthakis (1989) first applied pinch technology to mass exchange network synthesis (MENS), where mass exchangers are used to transfer components from process streams to Mass Separating Agents (MSAs). These authors introduced the use of a minimum composition difference, ε, which is analogous to the minimum approach temperature in heat exchanger network synthesis (HENS). They showed how specifying the value of ε locates the mass transfer pinch, which is a thermodynamic bottleneck for mass transfer between streams. This allows a target for the minimum flow rate of external MSA required by a network to be determined. This target is analogous to the energy target in HENS. Avoiding the transfer of mass across the pinch ensures that the MSA target is met in design.

However, until recently, there have been no targets for capital costs. The use the minimum number of units was recommended in an attempt to minimise the capital cost. However, this does not necessarily give a minimum cost. The absence of capital cost targets also meant that optimisation of the total cost (operating plus capital) could not be achieved by targeting alone.

There have been two main approaches aimed at minimising the total cost. The first approach used MSA targeting combined with repeated network design and costing over a range of ε values in order to trade off capital and operating costs (El-Halwagi and Manousiouthakis, 1990). The second approach applied mixed-integer non-linear programming (MINLP) to optimise a hyperstructure of possible mass transfer options (Papalexandri *et al*, 1994). Both approaches have some drawbacks, with the most significant one being the fact that neither can guarantee optimality. This paper describes the results of a new, pinch-based approach that has been developed by the current authors. This approach not only determines the optimum but also generates designs that closely approach it, using an insight as a basis.

Capital Cost Targets

The new approach is based on the recent development of capital cost targets for mass exchange networks (Hallale and Fraser, 1998). This method is based on a new graphical representation termed the y-x composite curve plot (Figure 1). This is a plot of rich stream composition versus lean stream composition and consists of a composite operating line and the mass transfer equilibrium line. This new representation allows targets to be predicted

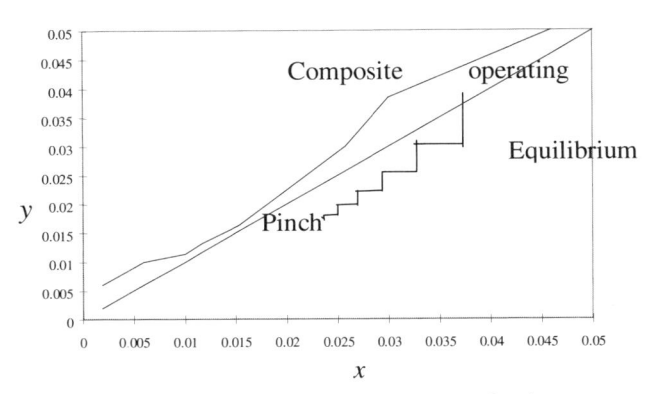

Figure 1. The y-x composite curve plot for a mass exchange network.

for the minimum number of stages for stagewise systems (as demonstrated in Figure 1), as well as for the minimum height in continuous-contact systems. Details are given elsewhere (Hallale, 1998; Hallale and Fraser, 1998). These targets can be used with cost correlations in order to predict a minimum capital cost target.

Total Cost Targeting

The real strength of the new targets is the ability to optimise the total network cost ahead of any design. This is achieved by targeting both MSA and capital costs and then adding them (on an annual basis) to yield a total annual cost (TAC) target. This is repeated over a range of range of ε values in order to locate the optimum as shown in Figure 2 (which is the optimisation for the coke-oven gas sweetening problem to be dealt with in the next section). Special design techniques have been developed (Hallale, 1998) which then allow these optimised targets to be closely approached in design. These techniques are largely insight-based and driven by the user.

Test Problems

This method has been applied to four literature problems which had previously been optimised using MINLP (Papalexandri et al, 1994). These problems cover a wide range of MENS problems: the copper recovery problem involves two MSAs, the coke-oven gas sweetening problem involves multiple components, the dephenolisation in coal conversion problem involves simultaneous mass exchange and regeneration network design, and the rayon waste desulphurisation problem involves reactive mass exchange and non-linear equilibria.

The results for the coke-oven gas sweetening problem will be shown here to illustrate how they differ from the MINLP results. The objective of this process is to remove two acid gases (H_2S and CO_2) from two process streams (coke oven gas and Claus tail gas) using two available MSAs (one a process MSA and the other an external MSA), and the design objective is to do this at a minimum cost. Full details of the solution of this problem may be found in Hallale and Fraser (1998).

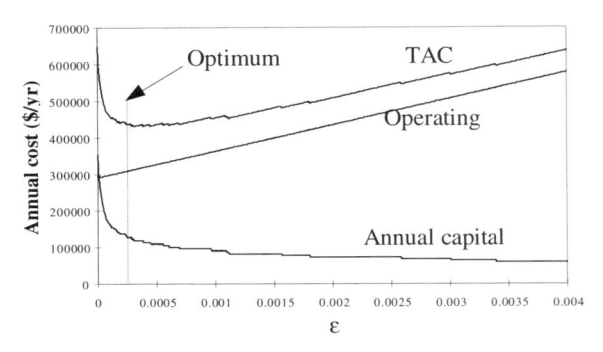

Figure 2. Optimisation of a mass exchange network before design (coke-oven gas sweetening problem).

Targets

Table 1 shows first the comparison between the MINLP results for these four problems and the optimum targets determined by our methods. It will be noted that the targets are significantly better than the MINLP results in three of the four cases. For the copper extraction problem the results are identical. In the rayon waste desulphurisation problem the target is only 0.25% of the MINLP result.

Designs

Table 1 also shows the results achieved with designs using our methods, which are seen to closely approach the targets, with the largest deviation being 3.2%.

For comparison, Figure 3 shows the network design for the coke-oven gas sweetening problem which was generated by Papalexandri et al (1994) using MINLP. This design has a TAC of $918 000/yr.

However, application of the new methods gives a target TAC of $413 600/yr and the design shown in Figure 4. This design has a TAC of $427 000/yr – less than half the cost of the previous optimum design.

As can be seen clearly, the two designs have very different structures and so it would not be possible to evolve the MINLP design to the network shown in Figure 4.

The new design also uses two more units and demonstrates that minimising the number of units does not necessarily give the best design.

Such improvements are certainly consistently achievable, as is summarised in Table 1. One of the reasons why the MINLP designs are not optimal appears to be limitations imposed on minimum composition differences in matches.

Extensions

The results presented in Table 1 indicate that the methods developed extend from systems with a single component being removed by a single MSA to systems with multiple components being removed by multiple MSAs with overlapping regions of operation (provided they are compatible and do not interfere with the equilibria of the other components), as well as to systems where regeneration of the MSA is included and systems with non-linear equilibria. These problems also include both stagewise and continuous contacters.

Another area of application for these techniques is to retrofit problems. This situation is more constrained in that there is already an existing network of mass exchangers in place. The approach here is to compare the existing process with the ideal target, on the basis of the size of mass exchange equipment used or required at the flowrate of MSA in the process. This gives the effieicncy of the existing process, α.

In determining what could be achieved, a target volume vs flowrate curve can be generated, as shown in Figure 5 for a gold extraction process. One would want to reduce the MSA requirement by adding extra mass exchange volume. A retrofit path is then chosen, either at the value of the current process efficiency, or one which approaches the target by using an incremental efficiency, Δα, of unity, as shown in Figure 5.

The retrofit path chosen can then be costed to give a diagram of operating savings vs investment. In the gold extraction example the savings also equated to a reduction in effluent, and this is plotted against investment in Figure 6. This diagram shows both a thermodynamic limit to the

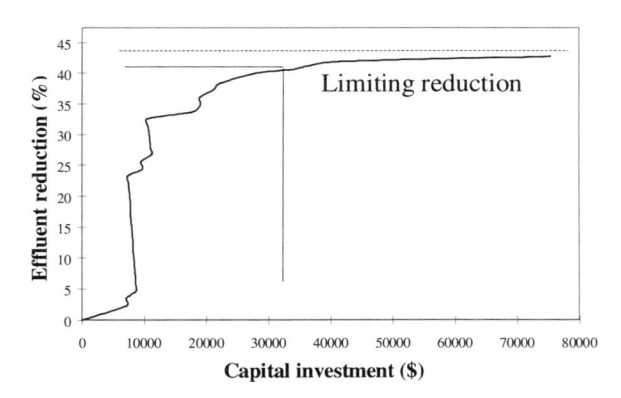

Figure 6. Impact diagram showing effluent reduction vs capital investment.

savings/reduction as well as a point of diminishing return on investment.

Conclusion

This paper has shown how capital cost targets for MENs can be combined with operating cost targets to produce total annual cost as a function of the minimum composition difference. The result is an optimum TAC at a particular minimum composition difference.

This method shown to produce targets sometimes considerably less than previously-reported optimum designs, as well as designs which closely matched these optimum targets. It was also noted using the minimum number of units does not guarantee optimality.

Extension of these techniques to retrofit problems leads to an impact diagram which is useful in setting effluent reduction targets.

Acknowledgement

The financial support of the SA Council for Mineral Technology (MINTEK), the SA Foundation for Research Development (FRD), and the University of Cape Town (UCT) is gratefully acknowledged.

References

El-Halwagi, M. M. and V. Manousiouthakis (1989). Synthesis of mass exchange networks. *AIChE J.*, **35**, 8, 1233-1244.

El-Halwagi, M. M. and V. Manousiouthakis (1990). Automatic synthesis of mass exchange networks with single-component targets. *Chem. Eng. Sci.*, **45**, 9, 2813-2831.

Hallale, N., (1998). *Capital Cost Targets for the Optimum Synthesis of Mass Exchange Networks*. PhD Thesis, University of Cape Town.

Hallale, N. and D. M. Fraser, (1998). Capital cost targets for mass exchange networks - a special case: water minimisation. *Chem. Eng. Sci.*, **53**, 2, 293-313.

Papalexandri, K. P., E. N. Pistikopoulos, and C. A. Floudas, (1994). Mass exchange networks for waste minimisation: a simultaneous approach. *Trans. I. ChemE.* (Pt A), **72**, 279-294.

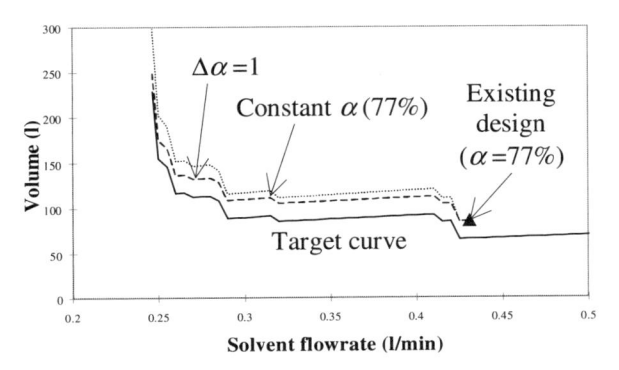

Figure 5. Retrofit path on volume-flowrate diagram.

Table 1. Comparison of Results from New Approach with those of MINLP.

Problem	TAC with MINLP ($/yr.)	Optimum TAC target ($/yr.)	TAC achieved ($/yr.)	Approach to target (%)	Reduction cf MINLP (%)
Copper extraction	49 000	49 000	49 000	0.0	0
Coke-oven gas sweetening	918 000	413 600	427 000	3.2	53.5
Dephenolisation in coal conversion	957 000	692 000	706 000	2.0	26.2
Rayon waste desulfurisation	11 273 500	28 000	28 000	0.0	99.75

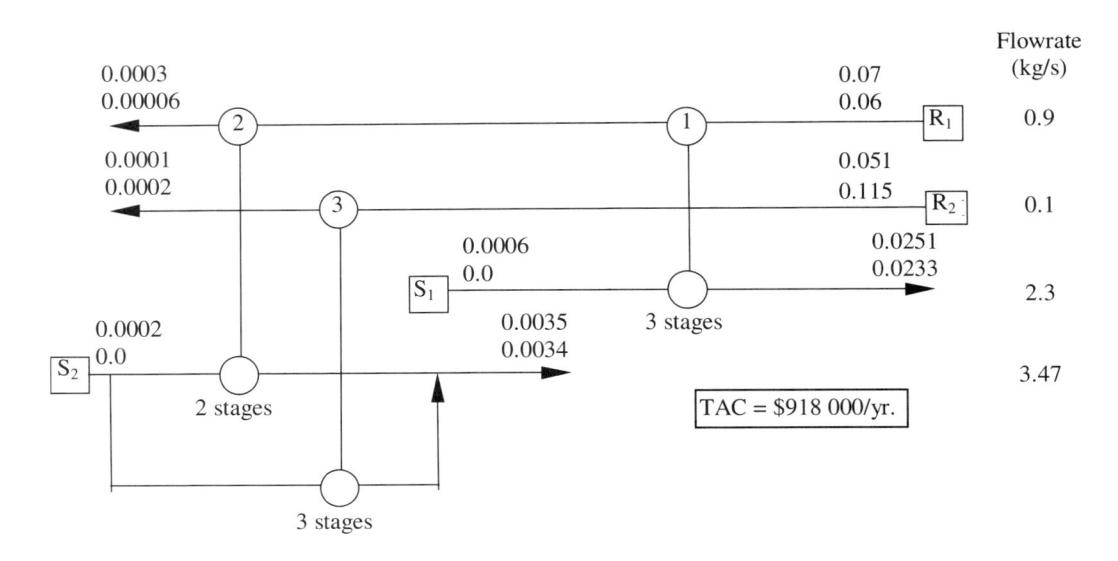

Figure 3. Optimum MINLP design for coke-oven gas sweetening problem (from Papalexandri et al, 1994).

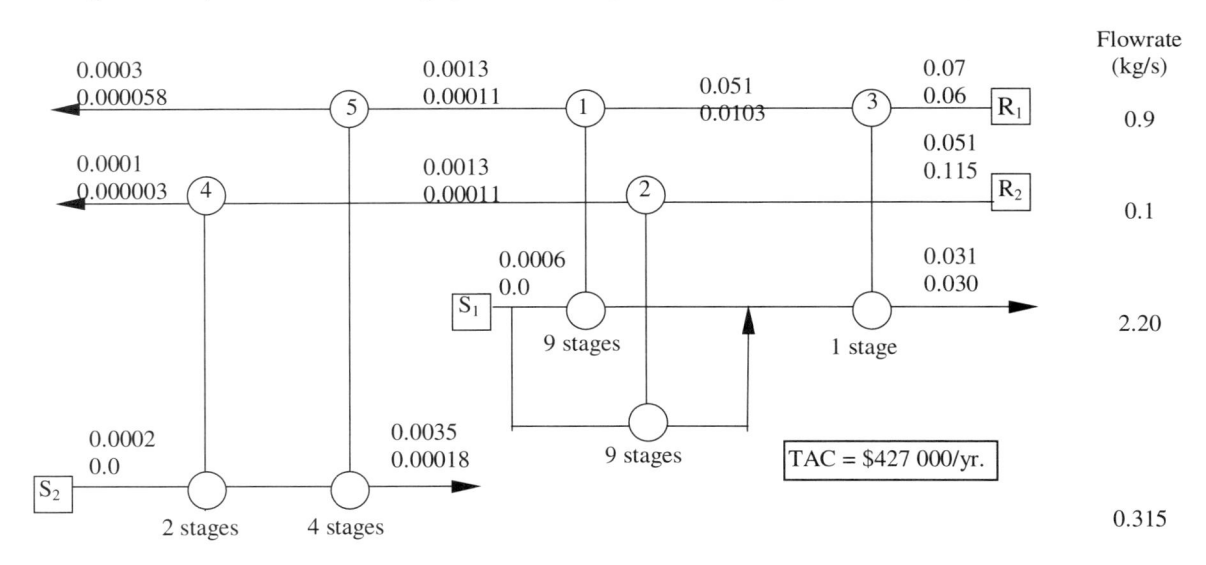

Figure 4. Design for coke-oven gas sweetening problem using new methods.

AUTOMATED GENERATION OF REACTION PRODUCTS

Haifa Wahyu, Ramachandran Lakshmanan, and Jack W. Ponton
Department of Chemical Engineering
University of Edinburgh
Edinburgh, Scotland, UK

Abstract

Automated generation of chemical species and chemical reactions has been developed using a simple combinatorial approach of free radicals and the building up of components from functional groups. This paper describes this approach and illustrates its application on side reaction generation of a vinyl chloride process plant. This paper also features the application of the program comparing the results obtained from experimental works and that of this work.

Keywords

Reaction path synthesis, Side reactions, Structural representation.

Introduction

For many processes of industrial interest there is often a large amount of information available on the main reaction or reactions. The number of possible main reactions is relatively small and the plausible alternatives quite obvious. This suggests that applying automatic synthesis methods to discover main reactions is unlikely to be profitable.

Possible side reactions and their products are much less widely explored. There are many more of these than the main reactions, and there is normally little intrinsic interest in them. However, when a process is transferred from a bench scale batch operation to full scale, substances which were present in undetectable quantities, or which were never identified because their presence was never suspected, may appear in significant amounts. Continuous operation, and in particular recycle, may cause build up of byproducts whose properties prevent their removal with identified byproducts or sidestreams.

This paper presents work on a systematic technique for process synthesis at the input-output block and recycle stages in the Douglas hierarchy (Douglas et al, 1995). The synthesis activity focuses on generating side reactions in a chemical process by creating plausible radicals and functional groups to build up a range of chemical compounds. Reactions are screened using stoichiometric and thermodynamic feasibility.

Background

Recently a number of workers have returned to the topic of synthesiing detailed chemical reaction networks. Nagel & Stephanopoulos (1995) and Prickett & Mavrovouniotis (1997a, 1997b) developed a computational language for reaction and molecular representation, covering both chemical representation and reaction networks in chemical reaction synthesis. The objective is to provide a generic computer language that allows flexibility in the generation of the types of molecules and reaction. By contrast, the present work does not attempt a rigorous representation of 'real' chemistry, but reverts to a heuristic approach as described below.

Automated generation using an empirical but systematic approach requires the following facilities which were adapted from Nagel's work (Nagel et. al, 1995) :

- A means of defining chemical structures. These are associated with structural formulae, functional groups, atoms and bonds.
- Definition of reactive behaviour of chemicals. The reactive behaviour of chemicals is

determined by functional groups which means the tendency of the atom bonds to cleave and form radicals.
- Definition of reactions and pathways. Reactions contain information about the species determining the reactants and the products. Pathways are the routes taken to generate certain chemicals.

Representation of Structure

Models for chemical structures require the capability to represent a compound uniquely. The representation must also be convenient to be manipulated. Molecules are represented in linear notation as standard characters, somewhat based on SMILES (Weininger, 1988). The compactness of the SMILES representation is convenient as a descriptive language, but requires further interpretation to extract precise molecular information. For this reason, we did not use SMILES notation.

As it is required to estimate chemical properties based on group contribution, molecules are built up directly on functional groups rather than atom to avoid additional recoding. For example

- ethane will be written by embedding two CH_3's, that is: CH3CH3
- ethylene is CH2:CH2:, double bond is denoted by colon possessed by both groups
- triethylamine equals CH3CH2[CH3CH2]NCH2CH3 with the ethyl branch enclosed in square brackets

See Wahyu (1996) for details. The group contribution method adopted is the Lydersen technique as improved by Joback (Reid et al 1987).

Method

The method involves three main stages summarized below.

Species Generation

Species generation uses the concept of free radicals, although we here use the term rather more widely than in the strictly correct chemical sense. To accommodate the different characteristics of each component class, sets of rules regarding radical formation and reactions have been developed (Isaac, 1975 and Finar, 1975).

Simple examples of radicals are methyl, phenyl and hydroxyl. Radicals can be produced by several ways, such as thermolysis, photolysis or radiolysis and redox reaction. Fragmentation also occurs in all possible ways predominated by those leading to the most stable free radicals. Stabilities of bonds affect the splitting patterns, the most stable being aromatics while the least being alcohols. Typical reactions of radicals are dimerization, disproportionation and exchange, hydrogen addition to an unsaturated system and oxidation and reduction. The mechanisms of radical formation of certain component classes are listed below.

- Alkanes undergo abstraction of hydrogen atom, fragmentation of methyl, ethyl, propyl .. etc
- Alkene undergo abstraction of hydrogen atom, formation of single bond and fragmentation of alpha and beta bond
- Arenes undergo the formation of C_6H_5, C_6H_4 and so on

Alpha and beta bonds are the position of carbon atom relative to the position of double bond or functional groups.

There are two possible approaches to combine radicals. Either rules may be used to allow any certain combination, or all combination which satisfy the elementary rules of valency may be allowed. In the latter case the feasibility screening procedures described below will eliminate impossible compounds. We prefer the second approach although it is computationally more expensive as it ensures that all possible species will be tested.

Stoichiometric Feasibility

The second step involves a combinatorial approach to determine reactions that might feasibly produce the species that are generated in the first stage. Each reaction consists of reactant and product species with stoichiometric coefficients. The atom balances for the reactions generate a set of linear equation in the stoichiometric coefficients. The solution is found by a linear equation solver. A reaction which is considered plausible if:

- a solution exists, and
- the coefficients are small whole numbers.

In fact the threshold free energy can be used as a 'tuning parameter' for the method. Positive free energy do not mean that the product will never be found, merely that the equilibrium constant will be small and so the amount of product will be small. A rule of thumb value of ± 40 kJ/mol has been suggested (May and Rudd, 1976). However it is precisely small amounts of byproducts which are at issue. Furthermore, the free energies of species are estimated using group contribution method, which can be subject to significant error.

One approach to this problem has been to start with a tight bound, e.g. strictly negative for free energy to limit the number of byproducts generated. The bound can then be relaxed and the program re-run, generating more species.

Thermodynamic Feasibility

Screening out those reactions which are thermodynamically unlikely under the plausible reaction conditions (Kyle, 1992 and Denbigh, 1997). Free energy (ΔG) of the reaction is determined from the free energy of reactants and products estimated using a group contribution method. It is calculated at temperatures in the range of 300 to 1500 K. If a negative free energy is identified then the reaction is deemed to be possible.

Implementation

The above stages form what we will refer to as a cycle. Series of cycles are created automatically using a build-tree program. The numerical procedures of the cycle are implemented in Fortran 90 while the build-tree program is written in LISP. The purpose of the build-tree program is mainly for user interaction and visualisation of species generation and the reactions involved. This program can be used to analyse the extent of the species generation and also to examine the reaction pathways.

Case Studies

Comparison with Experiments

The approach has been compared with experimental work on partial wet oxidation of p-coumaric acid (Mantzavinos et al, 1996 Herrera et al, 1998) and alkylation of toluene with ethanol (Walendziewski et al, 1996).

All components found in experimental work were generated by the program. Figure 1 shows some of the results of p-coumaric acid oxidation reactions. The first level generates 2 feasible reactions producing the following 4 surviving species, that $OHC_6H_4CH=CH_2$, OHC_6H_4CHO, $CHOCOOH$ and CO_2. Each species is reacted further to produce the next generation of components.

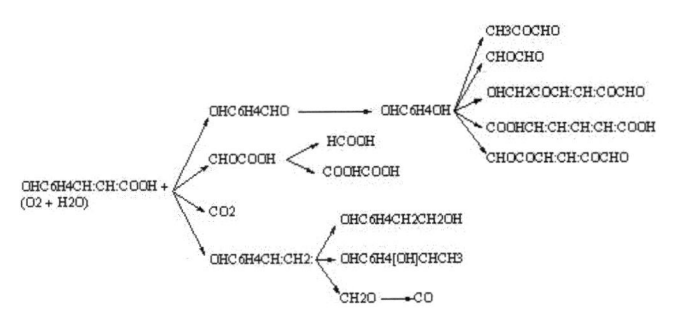

Figure 1. Species tree of the coumaric acid oxidation reaction.

However, the tree diagrams also contain some species which are not mentioned in the publications. These species could be intermediate species which are then transformed into other species or they occur in trace amount (minor species) so their occurences are not reported.

Vinyl Chloride Monomer

A further study on the application of the program is to investigate side reaction generations in vinyl chloride process. The layout of the plant and the main reactions are given below.

There are 3 main reactions in separate reactors:

- $CH_2=CH_2 + 2HCl + \frac{1}{2}O_2 \rightarrow ClCH_2CH_2Cl + H_2O$
- $CH_2=CH_2 + Cl_2 \rightarrow ClCH_2CH_2Cl$
- $ClCH_2CH_2Cl \rightarrow CH_2=CHCl + HCl$

The process contain two recycles. One of these spans two reactors and provides a possible place for light byproducts to accumulate. The other, around the pyrolysis reactor, is a likely sink for heavies.

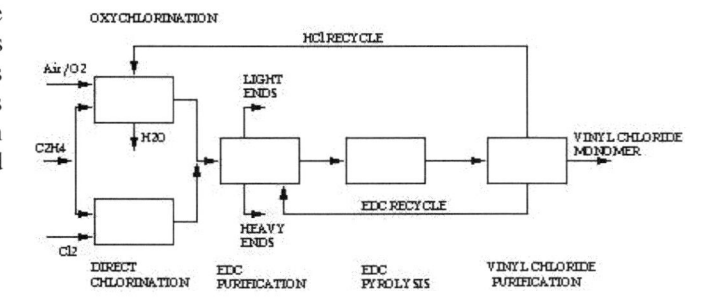

Figure 2. Vinyl chloride process block flowsheet.

The following list shows the ethylene dichloride (EDC), intermediates and some of byproducts having close boiling point and relative volatility generated by the programs.

Components	BP, °C	Rel. Vol.
HCl	-85	112.16
CH_3CH_2OH	78.5	5.96
H_2O	100	4.37
$CH_3[OH]CHCH_3$	82.4	2.43
CCl_4	76.5	1.93
$CH_2=CHCH=CHCH_2CH_3$	73.0	1.59
$ClCH_2CH_2CH_2CH_3$	78.4	1.50
$CH_3CH_2[Cl]CHCH_3$	68.0	1.49
$CH_3CH_2[OH]CHCH_3$	99.5	1.38
$ClCHCCl_2$	87	1.28
C_6H_6	80.1	1.19
$ClCH_2CH_2Cl$ (EDC)	**83.5**	**1.0**
$ClCH_2CHO$	85.0	0.92
Cl_3CCHO	97.9	0.823
Cl_2CHCHO	90.0	0.822
$CH_2=[Cl]CCH_2Cl$	94	0.71
$OHCH_2CH_2CH_3$	97.4	0.657
$CH_3[ClCH_2]CHCl$	96.4	0.656
$Cl[CH_3]CCl_2$	74.1	0.37

A RADFRAC simulation was used to investigate the separability of the byproducts from EDC into the 'lights' and 'heavies' streams. The system is highly nonideal and a number of azeotropes appear to be formed. Clean separation of the main intermediate seems unlikely to be possible.

Thus some oxidized and chlorinated compounds will be carried over to the pyrolysis section where further chlorination occurs. In order to initiate the chain reaction, CCl_4 is injected into the pyrolysis reactor. The components carried over from the previous reactors and the separation section will react to form new species. Some of the species have close boiling points relative to vinyl chloride monomer and so will increase the difficulty of the final high purity product separation.

However, the main difficulty is that precisely because there is a high purity final product, any heavy impurities, including maximum boiling azeotropes formed from new byproducts, will all enter the EDC recycle stream and ultimately accumulate in the plant.

Conclusions

The studies of oxidation and alkylation and have shown favourable results when compared with experiments. The side reaction study on vinyl chloride monomer shows that many components are generated beside the main product. Most of these components are heavily chlorinated and oxidized, and separation simulation confirms that they are likely to accumulate in the recycles, a situation confirmed by operators of such plants.

The work has demonstrated the ability of the programs to trace chemical compounds by employing the concept of free radicals and building up components from functional groups.

Acknowledgment

Haifa Wahyu would like to express her gratitude to the Islamic Development Bank, Jeddah, Saudi Arabia, which has provided funding and support so that the pursuit of a PhD degree is made possible.

References

Douglas, J.M. and G. Stephanopoulos (1995). Hierarchical approach in conceptual process design: framework and computer-aided implementations. *AIChE Symp. Ser.*, **91**, 183-197.

Finar, I.L. (1975). *Organic Chemistry Vol II : Stereochemistry of Natural Products*, 5th edition. Longman, 48-62.

Herrera, F., Pulgarin, C., Nadtochenko, V. and J. Kiwi (1998). Accelerated photo-oxidation of concentrated p-coumaric acid in homogeneous solution. Mechanistic studies, intermediates and precursors formed in the dark. *App. Cat. B: Environ.*, **17**, 141-156.

Isaac, N.S., (1975). *Reactive Intermediates in Organic Chemistry*. John Wiley Sons, London, 294-374.

Mantzavinos, D., Hellenbrand, R., Metcalfe, I.S. and A. G. Livingstone (1996). Partial wet oxidation of p-coumaric acid: oxidation intermediates, reaction pathways and implications for wastewater treatment. *Wat. Res.*, **30**, 12, 2969-2976.

May, D. and Rudd, D.F., (1976), Development of solvay clusters of chemical reactions. *Chem. Eng. Sci.,* **31**, 59-69.

Nagel, C. and G. Stephanopoulos (1995). Inductive and deductive reasoning: The case of identifying potential hazards in chemical processes. *Adv. Chem. Eng.*, **21**, 187-255.

Nagel, C.J., Han, C. and G. Stephanopoulos (1995). Modeling languages: Declarative and imperative descriptions of chemical reactions and processing systems. *Advances in Chemical Engineering – Intelligent Systems in Process Engineering. Part I : Paradigms from Product and Process Design, AICHE Symp. Ser.*, **21**, 1-91.

Prickett, S.E. and M.L. Mavrovouniotis (1997a). Construction of complex reaction systems. 2. Molecule manipulation and reaction application algorithms. To be published.

Prickett, S.E. and M.L. Mavrovouniotis (1997b). Construction of complex reaction systems. 3. An example: alkylation of olefins. *Comp. Chem. Eng.*, **21**, 12, 1325-1337.

Reid, R.C., Prausnitz, J.M. and B.E. Poling (1987*). The Properties of Gases and Liquids*, 4ed. McGraw Hill, New York.

Wahyu, H., Lakshmanan, R. and J.W. Ponton (1999). Automated generation of reaction products. To be submitted.

Walendziewski, J. and Trawczynski, J., (1996). Alkylation of toluene with ethanol. *Ind. Eng. Chem. Res.*, **35**, 3356-3361.

Weininger, D., (1988). SMILES, a chemical language and information system. 1. Introduction to methodology and encoding rules. *J. Chem. Inf. Comp. Sci.*, **28**, 31-36.

PROPERTIES OF SECTIONAL PROFILES IN REACTIVE SEPARATION CASCADES

Steinar Hauan and Kristian M. Lien
Norwegian University of Science and Technology
N-7034 Trondheim, Norway

Jae W. Lee and Arthur W. Westerberg
Carnegie Mellon University
Pittsburgh PA 15232, USA

Abstract

This paper addresses the fundamental differences between reactive and non-reactive composition profiles in VLE separation cascades. Specifically we demonstrate how the use of generalized difference points may be used to sketch the trajectories of seemingly complex profiles by hand. In systems where the local separation behaviour is C-shaped we show how the number of structurally different profile shapes is limited to just five basic profile types and how the actual type encountered in a specific cascade may be derived from very simple geometric arguments.

Keywords

Reactive VLE separation, Difference points, Composition profile properties.

Introduction

The sectional composition profiles of reactive separation cascades frequently exhibit complex behaviour. Examples include profiles intersecting themselves, new types of tangent pinch points and counter-intuitive "twists and turns". In this paper we aim to capture and understand these aspects from a mathematical and geometric viewpoint.

We base our approach on the Lewis-Matheson (Lewis, 1932) procedure previously extended to extractive and reactive cascades (Hauan et al, 1998) where the equations describing stage-to-stage compositions may be rewritten as a sequence of linear combinations of alternating thermodynamic steps and non-reactive material balances. Initially using a simplified set of equations to represent vapour-liquid equilibrium allows us to isolate and identify the contributions to cascade composition profiles from reaction and separation alone as well as their interactions. We then perform calculations in real VLE systems, both ideal and non-ideal for a selected set of chemical reactions. The analysis will be limited to systems where the local separation behaviour, or qualitative structure of residue curve maps, is C-shaped (Serafimov, 1971). This is a commonly occurring elementary topology found in all ideal systems and in selected simple distillation regions for many non-ideal systems. A more detailed description, discussion along with more results and references is found elsewhere (Hauan, 1998).

Theory

From material balances, the composition profiles in extractive and reactive separation cascades were derived by Hauan et al. (1998) and shown to be governed by normalized linear combinations of difference point arising from (side) feeds, products and chemical reactions. For a single chemical reaction, the composition of the overall cascade difference point in the rectifying section, δ^r_R, moves between the composition of the distillate product and the reaction difference point, $\delta_{(r)} = v/\Sigma v$.

$$\delta^r_R = a_r \cdot x_d + (1 - a_r) \cdot \delta_{(r)} \quad (1)$$

$$a_r = \frac{D}{D - \xi \Sigma v} \quad (2)$$

Here ξ is the cumulative molar turn-over by reaction from stage one throughout and including stage n. We will refer to the (extended) line between x_d and $\delta_{(r)}$ as the *locus of difference point compositions*. The composition of the vapour entering stage n from below, y_{n+1}, is found on a straight line between the liquid composition on stage n, x_n, and the cascade difference point.

$$y_{n+1} = \alpha \cdot x_n + (1-\alpha) \cdot \delta_R^r \qquad (3)$$

$$\alpha = \frac{L_n}{V_{n+1}} \qquad (4)$$

The straight balance lines allow us to apply the lever-arm rule independently to reaction and separation (Lee et al, 1998). Under the assumption of phase equilibrium, the composition profile of the rectifying section may be constructed through alternating steps thermodynamics with reaction and separation balance lines once the reaction distribution ($\xi(n)$) and internal reflux ratio (α) has been set.

The structural impact of simultaneous reaction and separation depends critically on the properties of the VLE separation vector field and its inverse. Figure 1 shows the composition geometry of a non-reactive rectifying section with a total condenser. The product composition (x_d) is identical to the composition of the vapor leaving the first stage (y_1), which again is in equilibrium with the liquid on this stage (x_1). By material balance, we know the composition of the vapor entering stage 1 from below, y_2, must be found on the straight line connecting y_1 and x_1. The vapor composition y_2 is again in equilibrium with liquid composition x_2. However, without knowledge of the specific VLE system in question, we cannot predict the direction in which x_2 will be found. More precisely, we may not in general make any assumptions about the relative orientation of consecutive equilibrium vectors, ($y_n - x_n$) versus ($y_{n+1} - x_{n+1}$).

In a reacting system, the picture becomes even more complex. Figure 2 shows the composition geometry for the first stage of a rectifying where chemical reaction takes place. The first VLE step is identical to Figure 1 with x_d equal to y_1 and in equilibrium with x_1. With chemical reaction on stage 1, construction of straight material balance lines changes compared to the non-reactive case: Instead of the composition of y_2 being located in between x_1 and x_d, it is now found on the line between x_1 and the moving cascade difference point composition as given by Equation 1. This direction not only depends on the set of chemical reactions taking place, but also on sectional amount of reaction turnover (Equation 2).

In order to study the full geometry of reaction and separation interactions, we need to evaluate all possible directional combinations of VLE behaviour and loci of cascade difference points. However, as only changes in relative orientation may give rise to new structural variants, it is sufficient to consider the effect of a continuum of reaction directions for each elementary VLE cell (Serafimov, 1971).

In ternary diagrams, locally C-shaped behaviour is by far the most commonly occurring case and covers all ideal

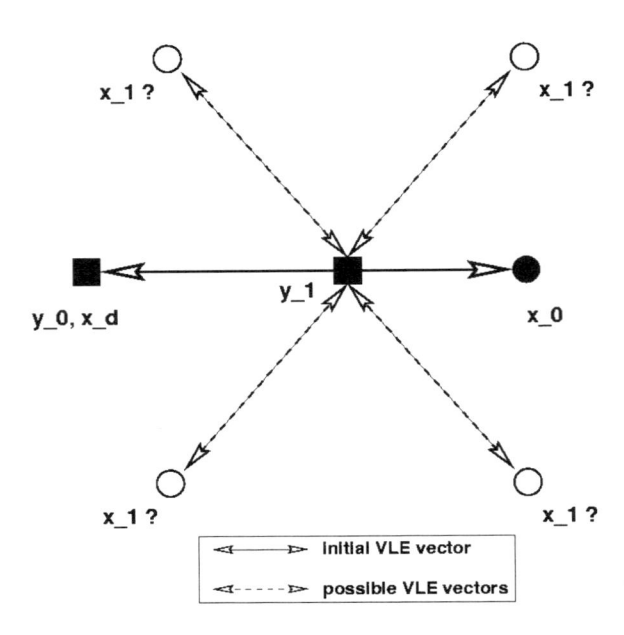

Figure 1. Composition geometry for a non-reactive rectifying section: Possible directions for the VLE separation vector on stage 2.

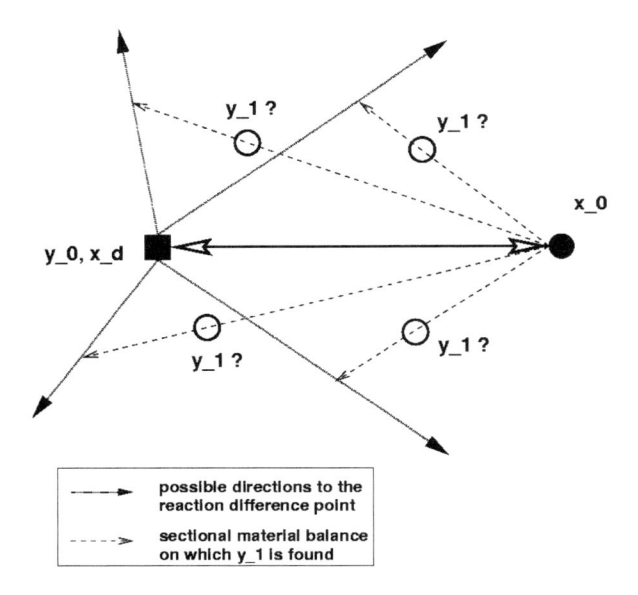

Figure 2. Composition geometry of a reactive rectifying section: Possible locations of the vapour composition from stage 1 depending on the locus of cascade difference points.

mixtures as well as many simple distillation regions in non-ideal systems. In order to completely isolate the effect of reaction-separation interactions from those arising from differences in local separation behaviour along non-reactive and reactive profiles, the initial analysis of composition profiles were performed in an artificially parametric vector field (Hauan, 1998) before being verified against real thermodynamics. Figure 3 shows how

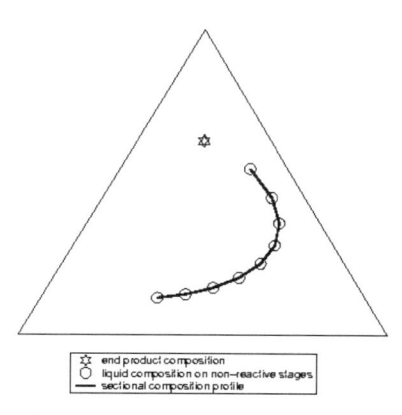

Figure 3. Schematic illustration of a C-shaped liquid composition profile in a non-reactive rectifying section.

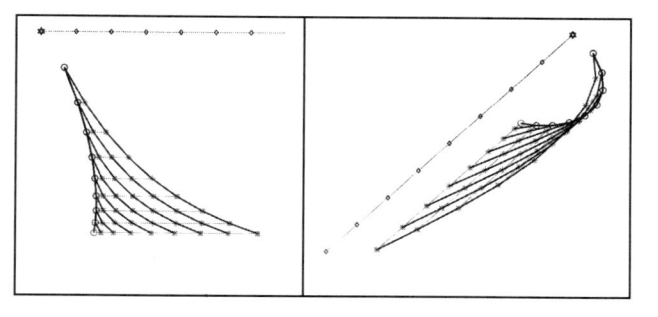

Figure 4: Sample liquid composition profiles in a parametric VLE vector field with a variable number of reactive stages.

a sample composition profile with artificial VLE closely resembles the profiles known from ideal mixtures.

With the non-reactive composition profile fixed with an arbitrary orientation compared to the figure axis, we computed reactive profiles for all possible loci of the cascade difference points. Figure 4 shows two example sets of liquid composition profiles when the reaction stoichiometry (and thus direction of the difference point locus) is fixed, but the internal reaction distribution varies. The hexagon represents the composition of the cascade end product while open circles correspond to stage-wise liquid compositions when no chemical reaction takes place. The fully drawn lines with asterisks correspond to composition profiles when the number of reactive stages is varied from one to (n-1) where n is the number of stages in the cascade section. The internal reflux ratio is the same for the whole family of profiles and the scaled reaction turnover per stage is constant as indicated by the overall cascade difference points (diamond) along the straight line toward the reaction difference point. Finally note that the dotted iso-stage lines drawn between compositions of identically sized cascade sections with a variable number of reactive stages. A very large set of such calculations with complete parametric variation was the initial basis for the geometric considerations in this paper.

The most obvious geometrical result follows trivially from Equations 1 and 2. As the number of subsequent stages with positive reaction turnover goes to infinity, the reactive profiles will always approach the locus of cascade difference points. However, this will usually happen as mathematical artefacts and take place outside the valid composition space. In the vicinity of attainable cascade products, the number of basic profile variants turns out to be only five. These profiles are illustrated in Fig. 5a-e and described below. Note that in the parametric vector field there are no axis or directional labels as our focus is on the structural differences in shape – or relative directions - between reactive and non-reactive profiles. The notion of generalized difference points allows us to extract the key geometric features:

Type 1: The reactive and non-reactive profiles do not cross each other, but have the same C-shape. The more chemical reaction occurring, the higher the curvature for the resulting profiles.

Type 2: At least one reactive profile crosses the non-reactive profile once from the concave side.

Type 3: The reactive and non-reactive profiles do not cross each other, but have different shapes. There are two different subtypes dependent on the relative direction of the reactive profile for a small and large number of stages:

(a) For a large number of stages, the reactive profile always has a vectorial component in its initial direction. However, for certain combinations of curvature introduced by reaction and separation, these profiles may also be S-shaped, i.e. have an opposite vectorial component for an intermediate number of stages.

(b) The reactive profile starts out with a vectorial component in the direction of the initial non-reactive profile. However, as the number of stages is increased, the sign of this vectorial component is reversed and the reactive profile makes a (partial) turn before bending away from the non-reactive profile.

The transitional state between Type 3a and 3b occurs when the locus of cascade difference point is normal to the initial direction of the non-reactive profile. This distinction determines how quickly a sectional profile will intersect the composition simplex, i.e. usually turn infeasible.

Type 4: As for Type 3b, the reactive profile starts out in the direction of the non-reactive profile, but quickly reverses the sign of this vectorial component. It then crosses the non-reactive profile once from the convex side and bends off in the same direction as the non-reactive

400   SEPARATION AND REACTION SYSTEMS

(a) Type

(b) Type

(c) Type

(d) Type

(e) Type

(f)

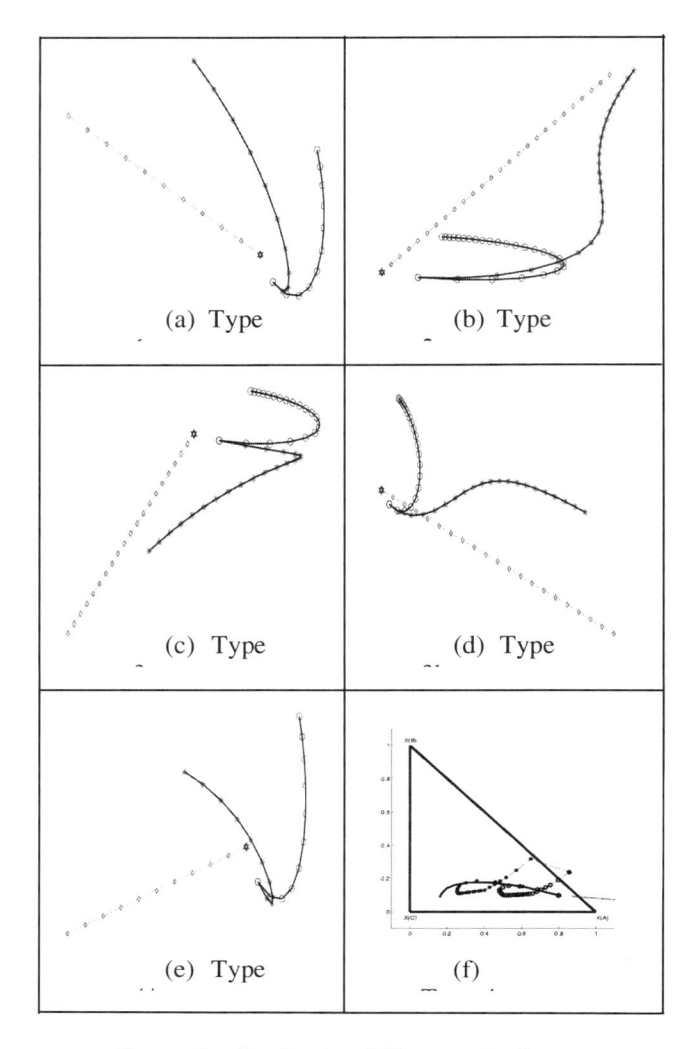

Figure 5. Qualitative differences in shape between non-reactive (open circle) and reactive (asterisk) composition profiles found by rotating the line of cascade difference points (diamond.)

profile. Calculations with real VLE mixtures reveal this profile type to be artificially created by our choice of as independent of composition (Hauan, 1998). In real systems, the transition between profiles of Type 3b and 1 occur via another structure not attainable with a parametric definition of F, but shown in Fig. 5f.

The different profile types are obtained by considering only the direction to all possible reaction difference points. This information is fully captured by an (NC-1) dimensional hyper sphere in the composition exterior. Based only upon the directions of the separation vector ($\mathbf{y} - \mathbf{x}$) for the first and last non-reactive cascade in a given cascade, Fig. 6 shows the region in which a locus of cascade difference point yields each profile type. Again, the figure orientation is chosen to illustrate the underlying geometric features and not to represent any particular VLE system. Finally, note that the composition profiles shown sometimes cross the locus of cascade difference points.

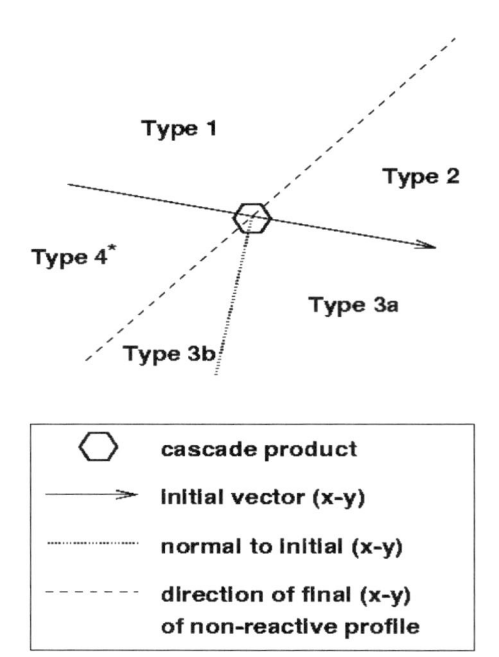

Figure 6. Regions of cascade difference point loci relative to tangents to x-y pairs along the non-reactive profile yields the structurally different types of reactive profiles in the parametric vector field.

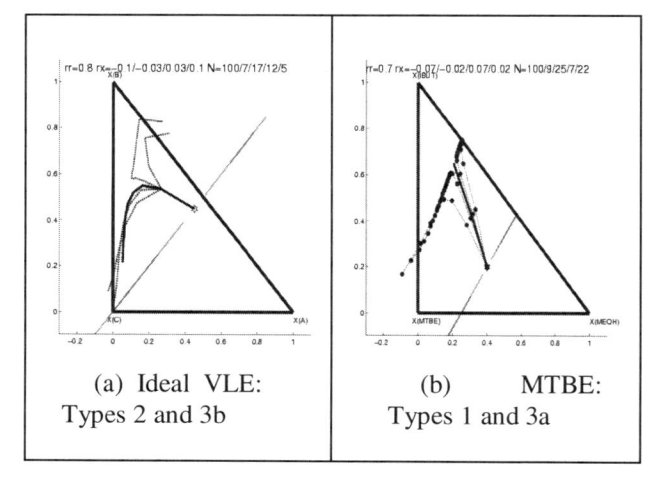

(a) Ideal VLE: Types 2 and 3b

(b) MTBE: Types 1 and 3a

Figure 7. Sectional non-reactive (fully drawn) and reactive (dashed) cascade profiles in real VLE systems for different reaction directions and distributions.

This is only possible in extractive and reactive cascades if such a locus traverses the composition interior in its path between two non-physical difference point compositions in the composition exterior

Two sample calculations in real VLE systems, one ideal and one highly non-ideal, are shown in Fig. 7. In both cases, the fully drawn line represents the non-reactive stage-by-stage composition profile for a suitably chosen reflux ratio starting from the composition marked with a hexagon. The four dashed lines, all starting at the same composition as the non-reactive profile, represent liquid stage-by-stage calculations with the same internal reflux ratio as the non-reactive case, but with different amounts and distribution of forward and backward chemical reaction. In Fig. 7b the actual compositions are marked with filled circles and we also note that the non-reactive profile encounters a pinch.

For both Fig. 7a and 7b, the straight (dotted) line through the hexagon represents the locus of actual cascade difference points in the system. Albeit less idealized and distinct, all eight reactive profiles fall into the theoretical classification types shown in Fig 5.

Conclusion and Significance

For a given case where the physical equipment involved has been decided upon, the main questions are to how to set up, initialise and solve the resulting models. Knowledge about what profiles to expect would most likely help us with initial composition estimates throughout a process unit. If a column simulation initiated from the chosen end products did not converge because both profiles exited the valid composition space without intersecting, their shapes may tell us if we need more or less initial reaction in each section. Knowing the basic profile types and the conditions under which they arise may indicate the need for structural variation as an alternative to changing the internal reaction distribution.

Using non-reactive stages at either end or in the middle of the column may identify "connecting paths" in composition space between products and (parts of) existing reactive profiles. Description and visualization of the difference between profiles resulting from reaction, separation or mixing alone - and combinations thereof through the use of generalized difference points - is believed to be important when it comes to assessment of process structure. Further, the visual results uncovered may enhance our understanding of reactive composition profiles and assist in sketching them by hand as well as prepare for future computer-aided design procedures.

References

Hauan, S. (1998). *On the Behaviour of Reactive Distillation Systems*. PhD thesis, Norwegian University of Science and Technology. Dr.ing Dissertation No. 1998:121.

Hauan, S., Ciric, A.R., Westerberg, A.W and K.M. Lien (1999). Difference points in extractive and reactive cascades. I Basic properties and analysis. In press, *Chem. Eng. Sci.*

Lee, J.W., Hauan, S., Lien, K.M. and A.W. Westerberg (1999). Difference points in extractive and reactive cascades. II - Generating design alternatives by the lever rule for reactive systems. In press, *Chem. Eng. Sci.*

Lewis, W. and G. Matheson (1932). Design of rectifying columns for natural and refinery gasoline. *Ind. Eng. Chem.*, **24**, 494.

Serafimov, L.A., Zharov, V.T. and V.S. Timofeyev (1971). Rectification of multicomponent mixtures I. Topological analysis of liquid-vapor phase equilibrium diagrams. *Acta Chemica Acad. Sci. Hung.*, **69**(4), 383-396.

FEED ADDITION POLICIES FOR TERNARY DISTILLATION COLUMNS

Brendon Hausberger, Diane Hildebrandt, David Glasser, and Craig McGregor
Centre for Optimisation, Modelling and Process Synthesis
University of the Witwatersrand
WITS 2050, South Africa

Abstract

Previous work using Attainable Region analysis has shown that for binary packed columns, there exists a single, optimal feed point. The feed location is known to be the point where the column composition is the same as that of the feed. In such systems it was shown that there is no advantage in the use of distributed feeds. This result is also known to apply to staged columns and is easily seen from a McCabe Thiele diagram.

However, for ternary and higher order systems it is unlikely that any composition in the column profile will match the feed composition. This implies that the advantages that were seen for a single feed in binary distillation in the form of minimising stages, and reducing utility loads, may or may not be of relevance in multi-component systems. Thus we must look further to determine the optimal feed policy. We will identify the optimal feed addition policy for a ternary system, using the fundamental processes of mass transfer and feed addition.

The system to be examined will consider a case of constant relative volatility, and the effects of changing feed addition policies. This allows us to determine the optimal feed addition policy for a ternary system. We will present the effects of distributed feeds, and examine the implications of the Attainable Region analysis on the design of packed and staged columns.

This behaviour is shown graphically in a triangular mole fraction space, with the combination effects of distributed feeds and separation processes being of particular interest

Keywords

Distillation, Feed addition policies, Column profiles.

Introduction

The number of trays in a distillation column can be reduced through the optimal placement of the feed tray. Most of the current design rules for the placement of the feed tray are based on heuristics developed by Fenske (1932), Ashakah (1979), Kirkbridge (1944) and Hengestebeck (1961).

It has only recently been proved that an optimal feed addition policy for binary systems exists. This result shows that the optimal feed tray for non-azeotropic packed binary distillation lies at the point where the composition of the liquid portion of the feed at the column pressure matches the liquid composition in the column at that stage. (McGregor et al 1997) This feed policy has certain clear logical advantages over other possible feed policies in that the energy and component back mixing effects are minimised at the feed location.

This result however relies on the feed composition existing in the column, and when we attempt to extend this result to ternary or higher component mixtures, it is evident that the result is no longer necessarily applicable. In multicomponent distillation, the column profile does not usually pass through the feed composition. An exception to this is the column profile that pinches from both the rectifying and stripping sections at the feed composition. This configuration, due to the infinite number of stages required, is an unattainable operating design.

None of the previous research in this field addresses the question of whether to make use of a single feed tray or distributing the feed stream over several trays, a configuration we have referred to as distributed feed. While there is some work in the field of reactive distillation where there are multiple feeds of different compositions, there has been little or no discussion on the topic of distributed feed configurations in distillation when there is only one feed composition. In this paper we will examine whether a distributed feed offers any advantages over a single feed configuration.

Effect of Distribution of Feed over Two Trays

We will consider a ternary ideal mixture with constant relative volatilities of 2 and 1.5 with respect to the third component respectively. For the sake of a fair comparison, we will fix the feed and product compositions and flowrates at the following values:

Table 1. Stream Flowrates and Compositions for Examples.

Stream	Mole Fractions			Flowrate
Component	1	2	3	mole
Feed	0.41	0.5	0.09	20
Distillate	0.81	0.11	0.08	10
Bottoms	0.01	0.89	0.10	10

Furthermore, we will consider only a liquid feed, and consider only feed addition in the rectifying section of a column.

When we examine a column with a reflux ratio approaching the minimum reflux ratio of 1.53 as calculated using the Underwood (1948) method, and a reboil ratio of 3.53, as derived from the feed stage mass balance. From rigorous calculations, it can be shown that the column will have 26 stages, 5 above the single feed point and 21 below. These profiles are superimposed on the residue curve map for the system in Figure 1. With the addition of a feed on the 2nd and 3rd stages, it is possible to reduce the number of stages required for the separation to 23. As we know that operations near minimum reflux are inefficient in their utilisation of stages, it may not be surprising that we can do better under these conditions.

However, if the feed flowrate is not carefully specified, this addition policy can lead to an increase in the number of required stages, up to 32 or even to a

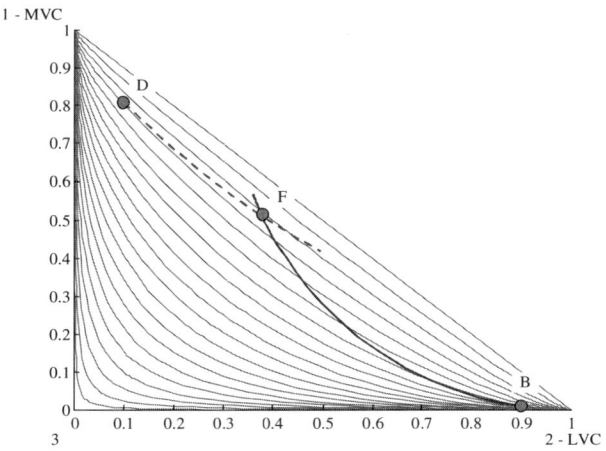

Figure 1. Column profiles for single feed minimum reflux column.

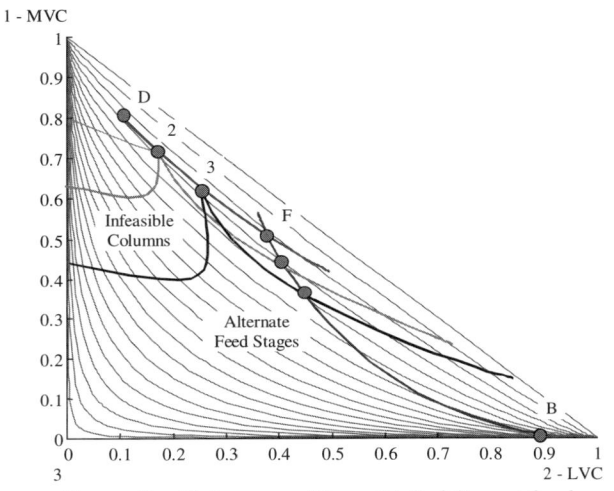

Figure 2. Column profiles with 2 different feed flowrates show off stage 2 of the column.

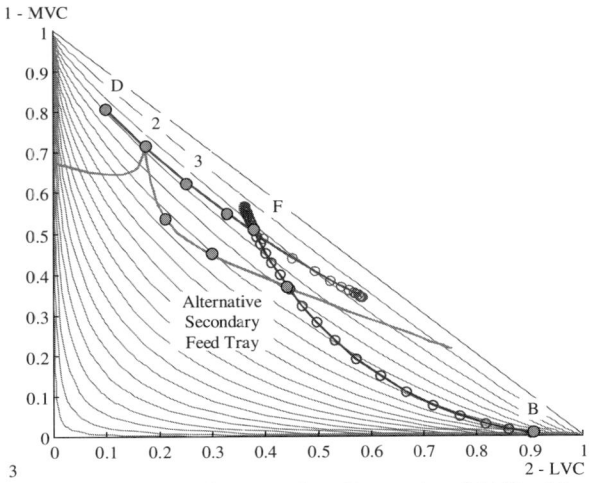

Figure 3. Column with reflux ratio of 1.99 with feed addition on stage 2 and 6.

column that does not produced the specified product, as shown in Figure 2. Later discussion will show that the increase of feed flowrate causes a rotation of the column profile, which may have several benefits, as well as some not so desirable effects.

However most columns, even for an easy separation are not designed to operate near minimum reflux, unless utility costs are of major concern, so for this result to be useful, the same benefits would need to be evident further from the minimum reflux operating condition. Making use of the design heuristic of an operating reflux of 1.3 times that of the minimum reflux, we arrive at a single feed column with a reflux ratio of 1.99, and 22 equilibrium stages of which 5 are in the rectifying section, and 17 stripping stages. This result is shown as the profiles in Figure 3.

With the addition of a second feed point on the 2nd, 3rd or 4th stages with a flow of as little as a tenth of the feed flowrate, we can reduce the number of stages by 3.

This shows that for just the addition of one extra feed tray, at a point that in no way has been optimised, we have been able to reduce the number of stages. Thus we have been able to demonstrate that this is a reproducible result that may lead to reduction in the number of stages in many systems, under various operating conditions.

Effect of Feed Distribution on Operating Regions.

Let us now consider the local effects of this feed addition on the column profile. For this work we have made use of the differential description of the distillation column, as defined by Doherty (1982), that is:

$$\frac{\partial \mathbf{X}}{\partial n} = \frac{V}{L}\mathbf{X} - \mathbf{Y}^* + \frac{D}{L}\mathbf{X}_D - \mathbf{X} \quad (1)$$

When we introduce a feed/s to the system, this equation includes a feed addition vector ($X-X_F$) to become:

$$\frac{\partial \mathbf{X}}{\partial n} = \frac{V}{L}\mathbf{X} - \mathbf{Y}^* + \frac{D}{L}\mathbf{X}_D - \mathbf{X} + \frac{F}{L}\mathbf{X} - \mathbf{X}_F \quad (2)$$

Or

$$\frac{\partial \mathbf{X}}{\partial n} = \frac{V}{L}\mathbf{X} - \mathbf{Y}^* + \frac{\Delta}{L}\mathbf{X}_\Delta - \mathbf{X} \quad (3)$$

Where

$$\mathbf{X}_\Delta = \frac{D\mathbf{X}_D - F\mathbf{X}_F}{D - F} \quad (4)$$

and

$$\Delta = D - F \quad (5)$$

It can be shown that when $F = D + B$ this equation describes the rate of change of the liquid composition with respect to increasing stage number in the stripping section.

This feed term has the effect of rotating the direction of the column profile vector between that of the rectifying section vector of a column when there is no feed addition, and the direction of the feed vector when we have an infinite amount of feed relative to the flows in the column.

So how does this affect the column composition profiles? If we examine Figure 4, we can see that we are no longer restrained to operate in the same leaf structure as described by Castillo and Towler (1997) for a single feed column. By moving outside of the envelope of the residue curve passing through the distillate, we have been able to extend the operating leaf. Also we have been able to adjust the pinch point by modifying the reflux ratio, thus enlarging the possible operating region further. This may have several ramifications so without performing rigorous calculations, how can we can identify the region of attainable column profiles.

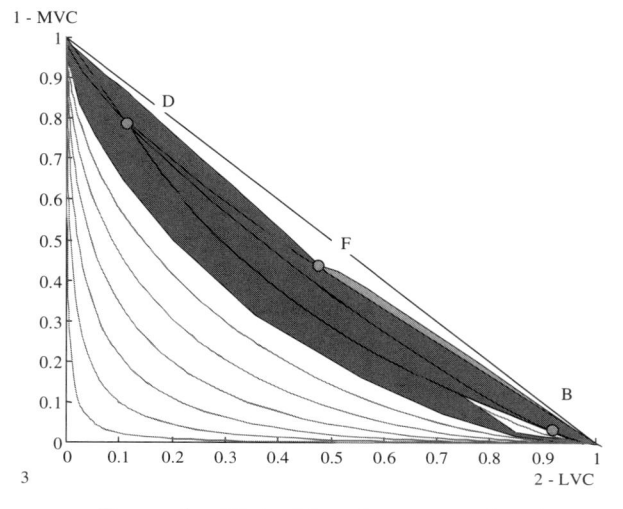

Figure 4. Attainable column operational compositions.

Figure 4 shows the effect of this change on the pinch curve. The vertical shaded region represents the rectifying leaf, the horizontally shaded region the stripping leaf, and the solid gray region the extension due to the change in the pinch curve. The dotted region indicates the extension due to the movement outside of the residue curve As can be seen this has extended the region enclosed by the pinch curve and the residue curve passing through the distillate point, and hence the attainable composition region.

If we extend this investigation, we can see that the leaf of attainable compositions with any columns, bounded by the residue curve passing through the product, and the pinch curve defined by that product point, has been extended not only by the amount of extension of the pinch curve, but additionally by the extension gained in the movement of the profiles below the residue curves in the direction opposite to the feed. To try and explain this extension let us examine the vector behaviour around the point. If we examine equation 2, we can see that we have two main vectors in effect at this point. The standard vector describing the rectifying section of a distillation column, and a another vector defining a separation of feed.

When we examine the directions of these vectors, it is easy to see the reason for the expansion of the composition region. This gives a possible route for the extension for side draw compositions particularly in Petlyuck column configurations.

Conclusions

Here we have shown the local effects on a column profile of a distributed feed, and used these to explain how the utilisation of a distributed feed can lead to an increased region of attainable compositions. This may well be of advantage when side draws are required, but the purity of the composition at the point of side draw may not be sufficiently high for product specifications for a particular component. We have also shown how this can adjust the number of stages required for a separation. Depending on the feed flowrate selected, and the feed point, this can lead to either an increase or in some cases a decrease in the number of stages.

On a fundamental level using only the number of stages as an indicator this would seem to have some advantage for a distillation column, however a full economic analysis would be necessary to establish the pricing effects of multiple feed points and control difficulties against the gains in reducing the number of stages.

Nomenclature

B: Bottom Product Molar Flowrate [mol/s]
D: Distillate Product Flowrate [mol/s]
F: Feed Molar Flowrate [mol/s]
L: Liquid Molar Flowrate [mol/s]
n: Number of stages
V: Vapour Molar Flowrate [mol/s]
X: Vector containing Liquid Mole Fractions
Y*: Vapour composition in equilibrium with X
Y: Vector containing Vapour Mole Fractions

Subscripts

B: Relating to Bottoms Product
D: Relating to Distillate Product
F: Relating to feed

References.

Akashah, S.A., Erbar, J.H., and R.N. Maddox (1979). *Chem. Eng. Comm.*, **3**, 461.

Fenske, M.R. (1932). Fractionation of straight run gasoline, *Ind. Eng. Chem.*, **24**, 482

Hengestebeck, R.J., *Distillation: Principles and Design Procedures*. Reinheld,1961.

King, C.J. (1980). *Separation Processes*, 2nd Edition. McGraw Hill, New York.

Kirkbridge, C.G. (1944). Process design procedure for multicomponent fractionators. *Pet. Ref.*, **23**, 87

Underwood, A.J.V. (1948). Fractional distillation of multicomponent mixtures. *Chem. Eng. Prog.*, **44**, 603.

Castillo, F.J.L. and G.P. Towler (1997). *Synthesis of Homogeneous Azeotropic Distillation Sequences*, A thesis submitted to University of Manchester, Institute of Science and Technology.

UNCERTAINTY CONSIDERATIONS IN THE REDUCTION OF CHEMICAL REACTION MECHANISMS

Marianthi G. Iepapetritou
Rutgers University
Piscataway, NJ 08854

Ioannis P. Androulakis
Exxon Research and Engineering Co.
Annandale, NJ 08801

Abstract

The objective of this work is to investigate the effects of uncertainty in the predictions of reduced kinetic mechanisms derived from complex networks such as the ones appearing in environmental and combustion models. Kinetic model reduction is achieved by employing the mathematical programming methodology proposed by Androulakis (1999). Flexibility analysis of the kinetic model results in quantification of the uncertainty space where the model is valid based on the simultaneous variation of different uncertain parameters that correspond to different initial conditions. The importance of the proposed approach lies in the fact that by quantifying the validity of a given reaction network under varying parameters, possible discrepancies between experimental observations and modeling predictions could be minimized. Examples are provided to illustrate the application of the proposed methodology.

Keywords

Kinetic modeling, Mechanism reduction, Flexibility analysis, Uncertainty.

Introduction

Detail modeling of chemically reacting systems is an important factor in the analysis of chemical reactors. By-product distribution, pollutant formation and process operation/optimization are greatly influenced by the details of the chemistry involved. Having accurate models that correctly predict the chemical pathways involved often result in complex models involving a large number of reacting species and reaction steps. Keeping track of species evolution in such large systems is a highly demanding computational tasks. It becomes therefore imperative to identify alternative representations of the detailed kinetic mechanisms that reduce the computational workload. One can possibly identify various types of mechanism reduction which are summarized in Tomlin *et al.* (1997). The approach taken by Androulakis (1999) reduces the number of reacting species and reactions while maintaining the structural integrity of the system. Let the detailed network be composed of N_R reactions and N_S species. Let us assume that detailed reaction network is defined as $S_{NR} = \{R_1, R_2, ..., R_{NR}\}$. A reduced network that maintains the structural integrity of the original one is composed of N'_{NR} reactions belonging to a sub-set of the original set: $S'_{NR} \subset S_{NR}$. The reaction pathways remain unaltered only some branches have been removed. The reduction is based on the minimization of user-defined objective that measures the accuracy with which the reduced network describes the time evolution of key observables. In the work of Androulakis (1999) this is defined as the integer optimization problem described through equations (1)-(7).

In this formulation "detailed" is the network for which all λ_r's are equal to 1. "K" is a user-defined subset of the set of all reacting species (molecular as well as free radicals). The error measure "L" defines the approximation error of the trajectories of the key observable quantities for the interval of interest. The temperature of the system, T, as well as the ignition time, τ, are also treated as target observables. The weight factors, ω, are appropriately selected so as to represent the biases of the decision maker in the way different observables are to be considered in the objective. The form of the objective can greatly affect the result of the reduction as discussed in Androulakis (1999). In terms of the actual simulation it is assumed that we study adiabatic plug-flow reactors of a given length Z.

$$\min_{\lambda \in I^{N_r}} \sum_{r=1}^{N_r} \lambda_r \quad (1)$$

subject to:

$$L = \sum_{k \in K} \omega_k \int_0^Z \left(\frac{y_k^r(z) - y_k^d(z)}{y_k^d(z)} \right)^2 dz$$

$$+ \omega_T \int_0^Z \left(\frac{T^r(z) - T^d(z)}{T^d(z)} \right)^2 dz \quad (2)$$

$$+ \omega_\tau \int_0^Z \left(\frac{\tau^r(z) - \tau^d(z)}{\tau^d(z)} \right)^2 dz \leq \varepsilon_R$$

$$\frac{dy_s^p(z)}{dz} = \sum_{r=1}^{N_R} \lambda_r \alpha_{rs} R_r \quad (3)$$

$$\frac{dT^p}{dz} = \sum_{s=1}^{N_S} H_s^g \frac{dy_s^p(z)}{dz} \quad (4)$$

$$\frac{d\tau^p}{dz} = \frac{\rho}{c_0} \quad (5)$$

$$R_r = K_F^r e^{-E/RT} \prod_{s=1}^{N_S} \left[X_s^p \right]^{\alpha'_{rs}}$$

$$- K_R^r e^{-E/RT} \prod_{s=1}^{N_S} \left[X_s^p \right]^{\alpha''_{rs}} \quad (6)$$

$$K_F^r = k_F^r e^{-E_F^r/RT}, \quad K_R^r = k_R^r e^{-E_R^r/RT} \quad (7)$$

$$\lambda \in I^{N_R} = \{0,1\}^{N_R}$$
$$s = 1, \ldots, N_s, r = 1, \ldots, N_R, p = \{detailed, reduced\}$$

Given the initial conditions, mol fractions and temperature, base-case profiles are obtained by simulating the detailed mechanism and recording the species and temperature profiles. The above-mentioned formulations aims at identifying the minimum number of reactions $\sum_{r=1}^{N_r} \lambda_r \leq N_R$ that would achieve an adequate representation of the original network. The mixed integer dynamic optimization problem is solved within a branch and bound framework as explained in Androulakis (1999). All thermophysical properties are estimated using **CHEMKIN II** (Lee et al., 1990), and the NLP relaxations are solved using an SQP implementation of the **NAG** Fortran Library (NAG, 1995).

Flexibility Analysis

Once a reduced mechanism has been identified, a very important question concerns the range of validity of it. In other words, if a given set of conditions has been used so as to perform the reduction, i.e., initial conditions, how will the reduced mechanism behave in conditions other than the nominal ones? We propose to address this problem by utilizing the concepts of flexibility analysis (Biegler .et all., 1997). It is assumed that the inlet conditions (compositions of key reactants, temperature of incoming stream) are uncertain parameters whose nominal values, θ^N, are known (values used for the reduction), and are subject to expected deviations $\Delta\theta^+, \Delta\theta^-$ in the positive and negative directions respectively. Evaluation of the flexibility index, F, corresponds to inscribing within the region of feasibility, R, a scaled hyper-rectangle T which may be expressed in terms of the non-negative scalar variable , δ, defined in Swaney and Grossmann (1985) as : $T(\delta) = \{\theta \mid \theta^N - \delta\Delta\theta^- \leq \theta \leq \theta^N + \delta\Delta\theta^+\}$. For a reduced mechanism the feasible region R is defined as the range of initial conditions (uncertain parameters) that maintain the accuracy with which a reduced mechanism described the time evolution of key observables, i.e., $L \leq \varepsilon$. The *flexibility index*, F, is then defined as follows:

$$F = \max \delta$$
subject to:
$$\forall \theta \in T(\delta)\{\exists x \mid f(d, x, \theta) \leq 0, h(d, x, \theta) = 0\}$$
$$T(\delta) = \{\theta \mid \theta^N - \delta\Delta\theta^- \leq \theta \leq \theta^N + \delta\Delta\theta^+\}$$

where

$$\theta = \{T(\delta), y_s(0)\}, x = \{R_r, y_s, T\}$$
$$f = \{(2)\}, h = \{(3), (4), (5)\}$$

The direct solution of the above problem is in general very difficult since it involves the max-min-max constraint as shown in Swaney and Grossmann (1995). In this paper we adopt the approximate solution approach based on the *vertex enumeration* procedure (Swaney and Grossmann, 1995). The vertex problems to de solved are defined as in (8)-(17). The coefficients λ_r^p are all equal to 1 when p = detailed, and are equal to the solution vector coming from the minimization of (1) when p = reduced. Constraint (16) enforces that the summation of the initial mol fractions should equal to 1, and (17) avoid trivial solutions in which initial conditions are selected that lead to non-reactive mixtures.

$$\max \delta_j \qquad (8)$$

subject to:

$$L = \sum_{k \in K} \omega_k \int_o^Z \left(\frac{y_k^r(z) - y_k^d(z)}{y_k^d(z)} \right)^2 dz$$
$$+ \omega_T \int_o^Z \left(\frac{T^r(z) - T^d(z)}{T^d(z)} \right)^2 dz \qquad (9)$$
$$+ \omega_\tau \int_o^Z \left(\frac{\tau^r(z) - \tau^d(z)}{\tau^d(z)} \right)^2 dz \leq \varepsilon$$

$$\frac{dy_s^p(z)}{dz} = \sum_{r=1}^{N_R} \lambda_r^p \alpha_{rs} R_r \qquad (10)$$

$$\frac{dT^p}{dz} = \sum_{s=1}^{N_S} H_s^g \frac{dy_s^p(z)}{dz} \qquad (11)$$

$$\frac{d\tau^p}{dz} = \frac{\rho}{c_0} \qquad (12)$$

$$R_r = K_F^r e^{-E/RT} \prod_{s=1}^{N_S} \left[X_s^p \right]^{\alpha'_{rs}}$$
$$- K_R^r e^{-E/RT} \prod_{s=1}^{N_S} \left[X_s^p \right]^{\alpha''_{rs}} \qquad (13)$$

$$K_F^r = k_F^r e^{-E_F^r/RT}, K_R^r = k_R^r e^{-E_R^r/RT} \qquad (14)$$

$$y_s^p(0) = y_s^N + \delta_j(y_s^{V_j} - y_s^N)$$
$$T^p(0) = T^N + \delta_j(T^{V_j} - T^N) \qquad (15)$$

$$\sum_{s=1}^{N_S} y_s^p = 1 \qquad (16)$$

$$T^p(Z) - T^p(0) \geq \Delta T_{min} \qquad (17)$$

The flexibility index is then defined as: $F = \min_{j \in V} \delta_j$.

Computational Results

Three chemical networks are analyzed. Detailed description of the reduction and it characteristics is given in Androulakis (1999). Here it is assumed that the reduced mechanism are given and we are concerned with analyzing the flexibility of them with respect to initial conditions.

H_2 Combustion – Model I (20 reversible reactions, 8 reacting species). Based on the results presented in Androulakis (1999) a reduced mechanism containing 9 reactions was determined. The nominal conditions used for reducing the mechanism were:

$$H_2 = 0.12, O_2 = 0.88, T_o = 1000K.$$

A total of 8 vertices are considered and the results are:

$$\delta_j = \{0.31, 0.45, 0.75, 0.85\}.$$

Therefore the flexibility index is F=0.31 which describes the feasible region:

$$H_2 = \{0.0827, 0.393\}, O_2 = \{0.6069, 0.9172\}, T = \{969, 1031\}$$

Fig. 1 depicts predicted profiles with the detailed and reduced mechanisms for the nominal conditions and for the extreme point of the feasible region. The agreement is excellent pointing to the fact that the reduced mechanism can predict the chemistry at conditions quite different than the nominal ones. Notice that the reduced mechanism not

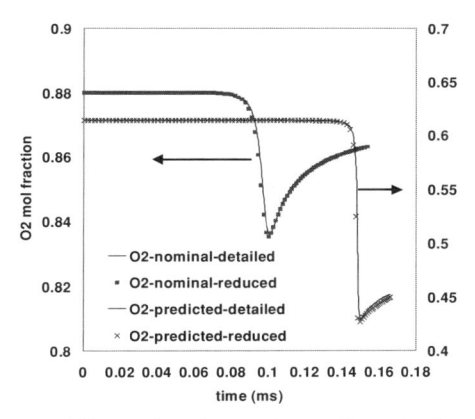

Figure 1. Comparison of nominal and predicted composition profiles.

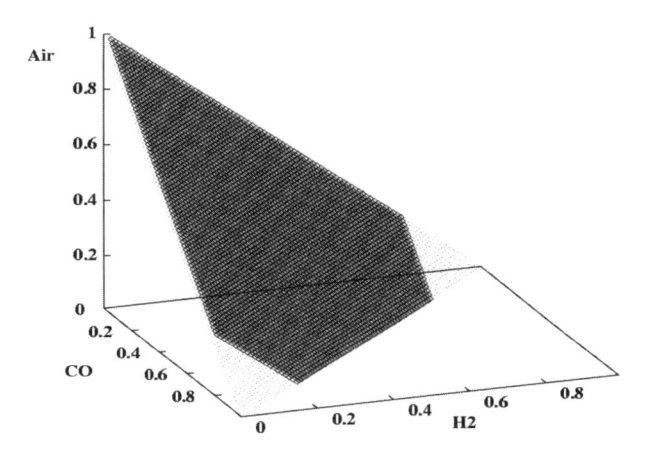

Figure 2. Feasible composition space for H_2/CO/Air combustion.

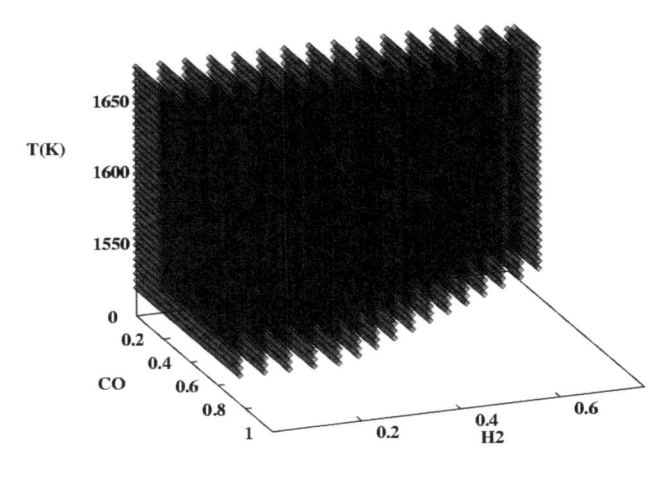

Figure 3. Temperature-Composition feasible space for H_2/CO/Air combustion.

only predicts the substantially different temperature rise, but also the radically different ignition times.

H_2 Combustion Model II (46 irreversible reactions, 9 reacting species). A reduced mechanism containing 18 reactions was identified in Androulakis (1999). It is an improved mechanism over Model I. the feasible domain, for the same nominal point, provided a feasible domain of flexibility index F=0.30.

CO/H_2/Air Combustion (47 reversible reactions, 12 reacting species). The nominal conditions used were:
$$H_2 = 0.05, O_2 = 0.189, CO = 0.095, N_2 = 0.711, T(K) = 1600$$
The flexibility formulation had to incorporate an additional constraint, namely: $\frac{y_{N_2}(0)}{y_{O_2}(0)} = 3.762$ to account for the fact that Air is used for the combustion and therefore we can not treat O_2 and N_2 independently. A total of 32 vertices are considered and the flexibility index was determined to be F=0.77. Fig. 2 and 3 depict the feasible region where the reduced mechanism predicts accurately the detailed one.

Conclusions

Flexibility analysis was used to determine the range of validity of reduced mechanisms obtained by a method proposed in Androulakis (1999). The vertex enumeration methodology was used to determine the flexibility of the reduced mechanisms. The results obtained illustrate not only the range of validity of the reduced mechanisms but also allow for a systematic procedure for characterizing reduced kinetic networks. Several questions remain in terms of addressing the flexibility of dynamic models: (a) Application of active set strategy methodology (Dimitriadis and Pistikopoulos, 1997). In this case model will involve non-convexities that will lead to sub-optimal solutions. (b) Consideration of uncertainty at the first mechanism reduction stage in order to increase the range of validity for the generated reduced mechanism and possibly to identify the different mechanisms that are valid in different initial conditions (Ierapetritou et al., 1996). (c) Consider uncertainty into kinetic constants and the effects into kinetic model reduction. Further work in this direction in currently under way.

Nomenclature

N_R = number of reactions in detailed mechanism
N_s = number of species in the detailed mechanism
λ_r^p = binary variable denoting the existence of rxn r
α_{rs} = stoic. coeff. of specie s in rxn r
y = mass fraction
X = molar concentration V_j = vertex point

References

Androulakis, I. P. (1999). Kinetic mechanism reduction based on an integer programming approach. *AIChE J.*, submitted for publication.

Biegler, L. T., I. E. Grossmann, and A. W. Westerberg (1997). *Systematic Methods of Chemical Process Design.* Prentice Hall.

Dimitriadis, V. D. and E. N. Pistikopoulos (1995). Flexibility analysis of dynamic systems. *Ind. Eng. Chem. Res.*, **34**, 4451.

Ierapetritou, M. G. and J. Acevedo and E. N. Pistikopoulos (1996). An optimization approach for process engineering problems under uncertainty. *Comp. Chem. Eng.*, **20**, 703.

Lee, R. J., F. M. Rupley and J. A. Miller (1990). *Chemkin II: A Fortran Chemical Kinetics Package for the Analysis of Gas-Phase Chemical Kinetic*. Sandia Report SAND90-8009.

NAG Fortran Library, Mark 17, 1st Edition. The Numerical Algorithms Group Limited.

Swaney, R. E., and I. E. Grossmann (1985). An index for operational flexibility in chemical process design. Part I – Formulations and theory. *AIChE J.*, **31**, 621.

Tomlin, A. S., T. Turanyi, and M. J. Pilling (1997). Mathematical methods for the construction, investigation and reduction of combustion mechanisms. In M. J. Pilling (Ed.), *Low-Temperature Combustion and Autoignition, Chemical Kinetics*, **35**.

COMPUTATIONALLY EFFICIENT DYNAMIC MODELS FOR THE OPTIMAL DESIGN AND OPERATION OF SIMULATED MOVING BED CHROMATOGRAPHIC SEPARATION PROCESSES

Karsten-U. Klatt and Guido Dünnebier
Department of Chemical Engineering,
University of Dortmund
D-44221 Dortmund, Germany

Abstract

Simulated moving bed chromatographic separation processes are an emerging technology, especially in the field of fine chemicals and pharmaceutical products. Because of the complex process dynamics, the choice of the optimal design and operating parameters is a challenging task the solution of which demands reliable and computationally effective simulation models. The SMB process models reported so far in the literature are either computationally very expensive, or very simplifying approaches. The purpose of this contribution is to present recent results for the generation of dynamic SMB process models. We here distinguish chromatographic processes by the type of adsorption isotherm and proceed from the ideal model for chromatographic columns while increasing the model complexity as far as necessary. In order to demonstrate the capabilities of the resulting simulation models, a model-based optimization for a SMB chromatographic separation is presented at the end of this paper.

Keywords

Chromatographic separation, Simulated moving bed, Dynamic simulation, Model-based optimization.

Introduction

Chromatographic processes provide a powerful tool for the separation of multi-component mixtures in which the components have different adsorption affinities. Typical applications are found in the pharmaceutical industry and in the production of the fine chemicals. In this area, the simulated moving bed (SMB) technology is becoming an important technique for large-scale continuous chromatographic separation processes since it provides the advantages of a continuous countercurrent unit operation while avoiding the technical problems of a true moving bed. The SMB process is realized by connecting several single chromatographic columns in series. The countercurrent movement is approximated by a cyclic switching of the inlet and outlet ports in the direction of the fluid stream. Thus, the process shows mixed continuous and discrete dynamics.

The successful design and operation of SMB separation units depend on the correct choice of the design parameters and the operating conditions, especially of the flow rates and the number and dimensions of the single columns in each section of the process. Because of the complex dynamics, the choice of the optimal parameters with regard to safe and economical operation and keeping of the product specifications at any time is not straightforward. For this challenging task, a detailed and reliable dynamic model of the process is necessary which includes the continuous dynamics of the single columns as well as the discrete events resulting from the cyclic operation.

The SMB models reported so far in the literature are either complex and computationally very expensive or very simplifying approaches which neglect the non-ideal behavior of the chromatographic columns and the complex dynamics of the process. However, due to the interdependence of the design and operating parameters the operability and controllability aspects and thus the complex dynamics of the process have to be taken thoroughly into account. Therefore, the purpose of this contribution is to present recent results for the design and implementation of a simulation model for SMB processes which on the one hand is computationally effective and on the other hand correctly describes the dynamics of the process which is relevant for process optimization and control.

The SMB Process

The simulated moving bed chromatographic process is the technical realization of a countercurrent adsorption process, approximating the countercurrent flow by a cyclic port switching. It consists of a certain number of chromatographic columns in series (see fig. 1) while the countercurrent movement is achieved by sequentially switching the inlet and outlet ports one column downwards in the direction of the liquid flow after a certain time period τ. In the limit of an infinite number of columns and infinitely short switching periods, the SMB operating mode comprises a real countercurrent process.

For economic reasons, a small number of columns is often desirable. The stationary regime of this process is a cyclic steady-state (CSS), in which in each section an identical transient during each period between two valve switches takes place. The CSS state is practically reached after a certain number of valve switches, but the system states are still varying over time because of the periodic movement of the inlet and outlet ports along the columns.

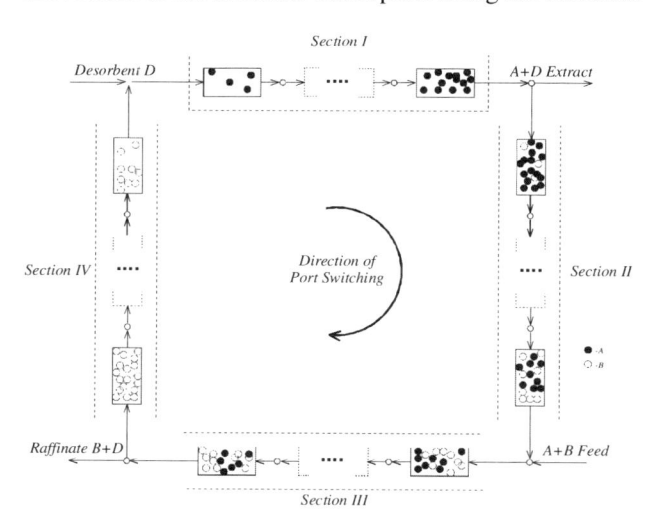

Figure 1. Scheme of a SMB process.

Modeling the SMB Process

There are two different approaches to built a mathematical model for the SMB process. The one tries to approximate the SMB process by the equivalent continuous true moving bed process (TMB model). The TMB model is fairly suitable for describing the steady-state operation of processes with three ore more columns in each section (Lu and Ching, 1997), but for processes with fewer columns (which are more and more utilized in industrial applications in order to reduce the investment costs) the conformity becomes poor. Furthermore, the TMB model does not describe the dynamics of the process which is mandatory for model-based control purposes.

Therefore, we here focus on the second approach which is building a realistic model of the SMB process by connecting dynamic models of the single chromatographic columns while considering the cyclic port switching. The fluid velocities and the inlet concentrations for each section can be calculated by mass balances around the inlet and outlet nodes (Node model, Ruthven and Ching, 1989). The mathematical modeling of single chromatographic columns has been extensively described in the literature by several authors. The different approaches can by classified by the phenomena they include and thus by their level of complexity. Fig. 2 shows this classification schematically.

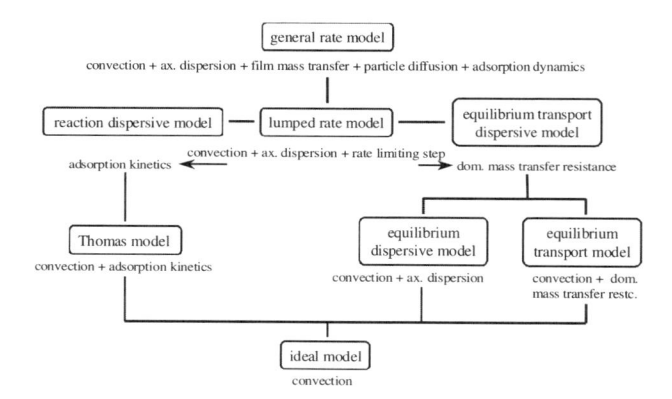

Figure 2. Classification of column models.

Most dynamic SMB models reported in the literature so far use an equilibrium transport dispersive column model. It is based on the adsorption equilibrium isotherm and a linear driving force approach for the mass transfer from bulk to solid phase. This results in a set of PDEs for the bulk phase and ODEs for the solid phase. They use finite difference, finite element or collocation methods to solve the system of model equations (see e.g. Lu and Ching, 1997; Strube and Schmidt-Traub, 1996; Wu *et al*, 1998). They computation times of these approaches are often within the range of the real process time, which is several minutes per switching period. Thus, they are not well suited for online optimization and control purposes. The purpose of the work presented here was to generate computationally more efficient simulation models by a bottom up strategy. We proceed from the ideal model which only includes convection and adsorption phenomena and increase the complexity as far as necessary to achieve sufficient accuracy. From the mathematical point of view, it is useful to distinguish chromatographic processes by the type of adsorption isotherm. Processes with linear isotherms result in a much simpler mathematical structure of the equations.

The Ideal Model for Linear Isotherms

Neglecting both mass transfer resistance and axial dispersion results in the following balance equation for the bulk phase concentration of each component

$$\frac{\partial c_i}{\partial t} + \frac{1-\varepsilon}{\varepsilon}\frac{\partial q_i}{\partial t} = -u\frac{\partial c_i}{\partial x}, \quad (1)$$

where u is the linear velocity of the fluid in each section of the process. The solution of this PDE yields in the case of a linear adsorption isotherm ($q = K \cdot c$) a constant speed of propagation for each single species

$$w_i = \frac{u}{1 + K_i(1-\varepsilon)/\varepsilon}. \quad (2)$$

Based on this expression (Zhong and Guiochon, 1996) proposed a dynamic SMB model which only contains algebraic equations to calculate the location of the adsorption and desorption fronts over time. The computational time is almost negligible, but from fig. 3 it is obvious that this model gives an insight into the qualitative behavior but shows unrealistic shock fronts and is not accurate enough for design and control purposes.

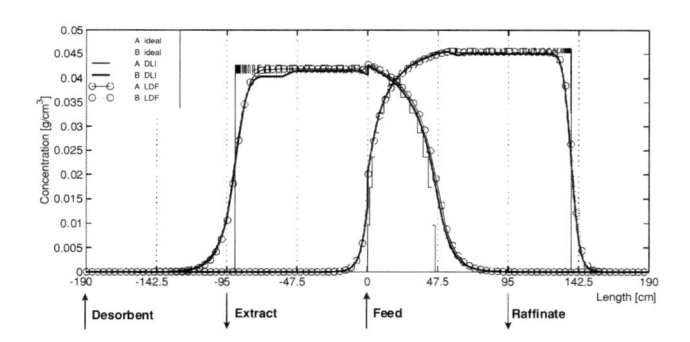

Figure 3. SMB Fructose/Glucose separation (axial concentration profiles at CSS).

The Dispersive Model for Linear Isotherms

In order to overcome the shortcomings of the ideal model we include mass transfer resistance and axial dispersion in the next step. (Van Deemter et al., 1956) have shown that in case of a linear isotherm these two effects are additive and can be incorporated into a single parameter, the apparent dispersion coefficient D_{ap}. This results in the following balance equation

$$(1 + k'_i)\frac{\partial c_i}{\partial t} + u\frac{\partial c_i}{\partial x} = D_{ap}\frac{\partial^2 c_i}{\partial x^2}. \quad (3)$$

(Lapidus and Amundsen, 1952) proposed a closed form solution of this type of equation for a set of general initial and boundary equations by double Laplace transform. From this, we generate the dynamic SMB model by connecting the solution for each single column by the respective node model (see Dünnebier et al., 1998, for details of the implementation). Fig. 3 shows the simulation results for a sugar separation in an 8 column SMB plant. The DLI model shows promising correspondence with the more complex LDF simulation model proposed by (Strube and Schmidt-Traub, 1996) which has been experimentally verified, while the computation times could be reduced by two orders of magnitude.

The Nonlinear Wave Model

Neglecting all non-ideal effects but assuming nonlinear adsorption equilibrium, the balance equations are similar to eqn. (1), but because of the competitive adsorption isotherms the hyperbolic PDEs are now strongly coupled. In the case of competitive Langmuir isotherms this system can be solved by using the theory of nonlinear wave propagation. The solution for standard Riemann problems as they occur in batch operation has been widely documented in the literature. Because of the more general initial and boundary conditions in the SMB operating mode, we extended this solution procedure by incorporating the front tracking approach (Wendroff, 1992), see (Dünnebier and Klatt, 1998) for details of the implementation. This results in a computationally very efficient simulation model which however shows the same lack of accuracy as the ideal model for the linear case (see fig. 4).

The General Rate Model

In order to generate both an accurate and computationally efficient dynamic model also in the case of general nonlinear adsorption isotherms we first followed the same approach as in the linear case by analyzing a model where the non-idealities are lumped into a single parameter. However, a closed-form solution is no longer possible in the nonlinear case and the numerical solution using standard techniques for the spatial discretization did not improve the computational efficiency substantially. Fortunately, there exists a very effective numerical solution for the complex general rate model

$$\frac{\partial c_i}{\partial t} = D_{ax}\cdot\frac{\partial^2 c_i}{\partial x^2} - u\frac{\partial c_i}{\partial x} - \frac{3(1-\varepsilon)k_{l,i}}{\varepsilon\, r_p}(c_i - c_{pi}(r_p))$$

$$(1-\varepsilon_p)\frac{\partial q_i}{\partial t} + \varepsilon_p\frac{\partial c_{pi}}{\partial t} = \varepsilon_p D_{p,i}\left[\frac{1}{r^2}\frac{\partial}{\partial r}\left(r^2\frac{\partial c_{pi}}{\partial r}\right)\right] \quad (4)$$

incorporating arbitrary nonlinear isotherms proposed by (Gu, 1995) for a single batch column. A finite element formulation is used to discretize the fluid phase and orthogonal collocation for the solid phase. We extended this approach to SMB processes and the simulation results are completely consistent with those of the complex dynamic SMB models reported in the literature so far while reducing the computational time by almost two orders of magnitude. Fig. 4 shows the simulation results for a SMB aromatics separation compared to the wave theory model.

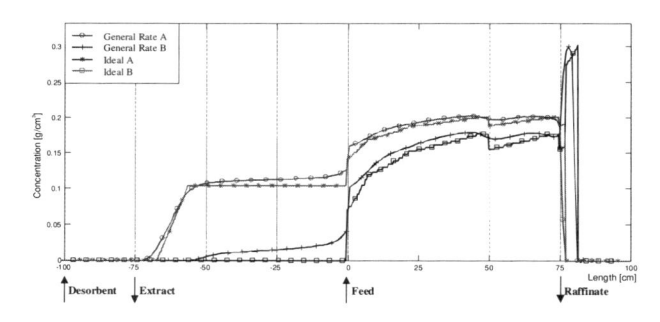

Figure 4. SMB separation, nonlinear isotherms (axial concentration profiles at CSS).

Process Optimization Example

The dynamic process model is used within the optimization algorithm to calculate the cyclic steady state of the process. The process can then be optimized with respect to the desorbent consumption as a measure for the operating costs or with respect to total separation costs if the necessary cost data is available. In the first case, the operating parameters (flow rates of desorbent Q_D, extract Q_E and recycle Q_{Rec}, and switching period τ) can be optimized for a given separation task on a given plant, in the second case, one can either determine the optimal plant design for a given separation or the optimal throughput for a given plant. In all cases, the product requirements are formulated as additional constraints and the resulting NLP can be solved with a standard SQP algorithm.

The example presented here is the separation of phenyalanine and tryphtophan. Estimated model parameters and the reference operating conditions are taken from (Wu et al., 1998). Because of the nonlinear adsorption equilibrium (Langmuir isotherms), the *general rate model* is used within the optimization framework. Due to the lack of cost data, the process is optimized with respect to the desorbent consumption for the plant for the feed flow rate given in the original work. The separation is optimized for three different product requirements. In the first run, the obtained purities for the original operating point are taken as constraints. The following runs are used to study the effect of higher or lower product requirements. The results are shown in tab. 1. A single optimization takes from 10 to 20 h cpu time on a PC PII 266 with around 150 function evaluations until convergence. The cyclic steady state for this system can be most efficiently determined by dynamic simulation within 75 switching periods in average with a mass balance error of less than 0.5%.

Table 1. Optimization Results.

	Original	Run 1	Run 2	Run 3
Pur_{Ex} [%]	99.5	99.5	99.0	98.0
Pur_{Raf} [%]	98.3	98.3	99.5	98.0
rel. Q_D [%]	100	**60.9**	**61.8**	**41.7**
Q_D [cm^3/s]	0.8275	0.5043	0.5110	0.3452
Q_F [cm^3/s]	0.2500	0.2500	0.2500	0.2500
Q_E [cm^3/s]	0.77989	0.4683	0.5081	0.3489
Q_{Rec} [cm^3/s]	0.2695	0.1489	0.1485	0.1423
τ [s]	1337.4	2096.7	2557.0	2675.2

Conclusions

The DLI model and the particular implementation of the general rate model represent very efficient and accurate dynamic simulation models for SMB processes which can be incorporated into an optimization and control framework. For the presented example, the model-based optimization predicts a potential saving of nearly 40% compared to the original operation for the same separation task, whereas even higher product qualities can be achieved with lower cost.

References

Dünnebier, G., I. Weirich and K.-U. Klatt (1998). Computationally efficient dynamic modeling and simulation of simulated moving bed chromatographic processes with linear isotherms. *Chem. Eng. Sci.*, **53**, 2537-2546.

Dünnebier, G. and K.-U. Klatt (1998). Modeling of chromatographic separation processes using nonlinear wave theory. *Proc. of IFAC DYCOPS-5*, Corfu, 521-526.

Gu, T. (1995). *Mathematical Modeling and Scale Up of Liquid Chromatography*. Springer, New York.

Lapidus, L. and N. Amundsen (1952). Mathematics of adsorption in beds IV. The effect of longitudinal diffusion in ion exchange and chromatographic columns. *J. Phys. Chem*, **56**, 984-988.

Lu, Z.P. and C.B. Ching (1997). Dynamics of simulated moving bed adsorptive separation processes. *Sep. Sci. Technol.*, **32**, 1993-2010.

Ruthven, D.M. and C.B. Ching (1989). Counter-current and simulated moving bed adsorption separation processes. *Chem. Eng. Sci.*, **44**, 1011-1038.

Strube, J. and H. Schmidt-Traub (1996). Dynamic simulation of simulated moving bed chromatographic processes. *Comp. Chem. Eng.*, **20**, S641-S646.

Van Deemter, J., F. Zuiderweg and A. Klinkenberg (1956). Longitudinal diffusion and resistance to mass transfer as causes of nonideality in chromatography. *Chem. Eng. Sci.*, **5**, 271-280.

Wendroff, B. (1993). An analysis of front tracking for chromatography. *Acta Appl. Mathematicae*, **30**, 265-285.

Wu, D.J., Y. Xie, Z. Ma and N.-H.L. Wang (1998). Design of simulated moving bed chromatography for amino acid separations. *Ind. Eng. Chem. Res.*, **37**, 4023-4035.

Zhong, G. and G. Guiochon (1996). Analytical solution for the linear ideal model of simulated moving bed chromatography. *Chem. Eng. Sci.*, **51**, 4307-4319.

RIGOROUS OPTIMAL DESIGN OF A PERVAPORATION PLANT

J. I. Marriott, E. Sørensen and I. D. L. Bogle
Department of Chemical Engineering
University College London
London WC1E 7JE, UK

Abstract

This paper introduces a new method for the rigorous optimal design of membrane separation systems which is illustrated with a pervaporation case study. A detailed model of a hollow-fiber membrane unit has been developed, in which the spatial variation of the variables is efficiently approximated using orthogonal collocation. The model is applied to a case study involving the dehydration of an ethanol/water mixture by pervaporation, showing excellent agreement with published experimental results. This study also considers the optimal design of a pervaporation pilot plant. An algorithmic approach to optimal design, based on the detailed hollow-fiber unit model, is proposed. A superstructure for a pervaporation plant is developed that enables all the degrees of freedom in the design to be considered. Using this method, a pilot plant case study is evaluated. The design study is formulated as an *MINLP* optimization problem and is solved using a branch and bound method. The new pilot plant design compares favorably with the actual plant design (Tsuyumoto *et al.*, 1997).

Keywords

Membrane separation, Pervaporation, Detailed dynamic model, Optimal design, Orthogonal collocation.

Introduction

During the last thirty years, the search for alternatives to traditional energy intensive separation methods such as distillation has led to the introduction of processes based on membranes. Membrane technology is used for a wide range of separations from particle-liquid separations, such as reverse osmosis, to gaseous and liquid-liquid separations. To achieve the desired separation, a large membrane area is often required. In an industrial plant, this area is supplied in a modular configuration for which there are a number of variations. Historically, plate and frame modules have been used, but more recently, spiral-wound and hollow-fiber modules have attracted a lot of interest as these offer much higher packing densities.

Pervaporation is an emerging membrane technology used for the separation of liquid mixtures. In this process the membrane forms a semi-permeable barrier between the liquid feed and a low pressure gaseous product. Consequently, the heat of evaporation must be supplied to the permeating material, typically resulting in a feed stream temperature drop.

In the past two decades, a large amount of work has been published describing models which characterize the separative properties of membranes (Ho and Sirkar, 1992). In comparison, there are far fewer models describing the whole membrane module, furthermore, these models make many simplifying assumptions. Typically, isothermal conditions, approximate membrane characterization, and steady-state conditions are assumed (e.g. Coker *et al.*, 1998; Krovvidi *et al.*, 1992). As a result of better solution algorithms and improved computer power such assumptions are no longer necessary. This paper describes a new model for a general hollow-fiber module which is more detailed than those presented so far.

At UCL, the use of detailed models in determining the optimal design of separation systems is being investigated. Though unit models have previously been used to optimize the operating conditions of a given membrane system (Ji *et al.*, 1994; Tessendorf *et al.*, 1999), full structural optimization has only been carried out using approximate models (El-Halwagi, 1992; Qi and Henson, 1998; Srinivas and El-Halwagi, 1993). Unfortunately, inaccuracies in modeling the membrane modules will lead to the development of sub-optimal plant designs with the possible over (or under) prediction of plant performance and a lack of generality due to implicit assumptions. The use of approximate models for pervaporation is particularly inappropriate due to the high dependence of permeability on temperature which, if ignored, will cause large inaccuracies. Therefore, in this paper, we consider the application of detailed models to the optimal design of membrane systems, illustrating our method with a pervaporation case study.

Mathematical Model

This section describes a detailed model developed to simulate hollow-fiber membrane modules. The principal model consists of a number of interacting but independent sub-models. A plug flow sub-model is used to describe conditions on both the feed and permeate sides of the membrane, whilst the flux of material through the membrane is calculated from a local transport sub-model. Fluid thermodynamic and physical properties are calculated assuming ideal mixture behavior. Pure component properties (including activity coefficients and enthalpies) are determined from empirical correlations. Figure 1 illustrates flow through a single fiber of a pervaporation module.

Plug Flow Sub-model

This model considers uni-directional liquid or gas flow, and is applicable to fiber-side and shell-side flow. The model is derived from dynamic mass, heat and momentum balances, for which the following assumptions are made:

- Dispersal effects are negligible.
- Radial concentration variations are negligible.
Frictional pressure losses are described by the Hagen-Poiseuille relationship.[1]

In this model, a number of common modeling assumptions have been disregarded. These include isothermal flow, assumed concentration gradients, and steady-state conditions.

Local Transport Sub-model (membrane characterization)

The flux of material through the membrane is calculated as a function of concentration, activity, and temperature. It is assumed that:

- The temperature of the membrane is equal to the liquid temperature.
- The resistances of the porous support and any concentration boundary layers are negligible.

---

[1] For shell-side flow an appropriate superficial cross-sectional area must be used.

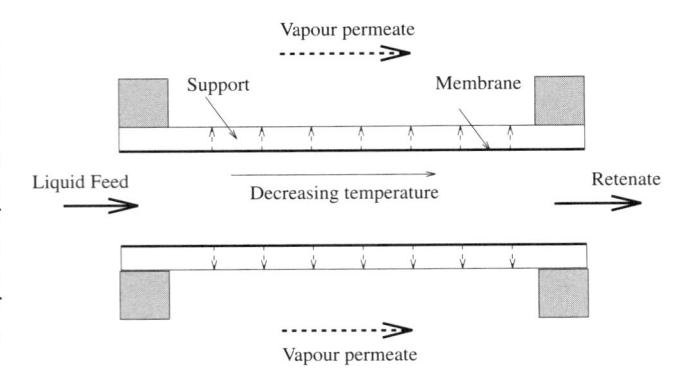

Figure 1. Flow through a single fiber.

Typical models used to characterize the mass flux through the membrane are given by Ho and Sirkar (1992).

Solution Method

The model is solved using the gPROMS simulation software (Process Systems Enterprise Ltd., 1998) on an IBM RS 6000. To approximate the spatial variation of the distributed variables both finite difference and orthogonal collocation methods have been investigated. Simulation studies, using this model, have shown that a finite difference approach requires 20 times more discretization points than orthogonal collocation to achieve the same accuracy. The model has been tested for a number of pervaporation and gas separation systems showing excellent results and efficient calculation. Typical CPU solution times are 2-3 seconds for steady state calculations using orthogonal collocation.

Optimal Design Methodology

The optimal design of any plant can be determined by the algorithmic solution of a superstructure optimization problem. The superstructure should enable all the degrees of freedom in the design to be investigated. In this section, we present a superstructure for a pervaporation process, and later we will illustrate its application with a case study. By making a few basic assumptions, it will be shown how the problem size can be greatly reduced.

Problem Size

A membrane separation system usually consists of a large number of membrane units as well as ancillary equipment such as heat exchangers, pumps, and compressors. A process superstructure can be rigorously generated by considering all possible units and allowing stream connections between all of these units. This can result in a very large number of optimization decision variables, making solution very difficult. However, as similar membranes operated in parallel will perform identically, it is possible to reduce the size of the model and thus the computation requirements. To this end, a stage is defined as a number of identical modules connected in parallel with an optional feed heat exchanger,

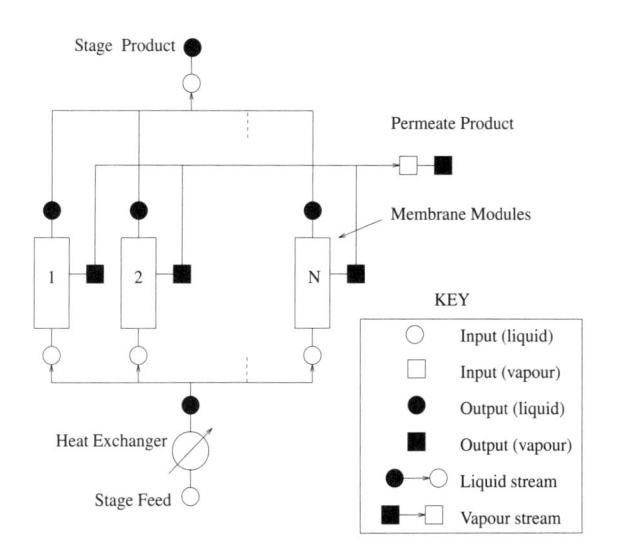

Figure 2. A single stage.

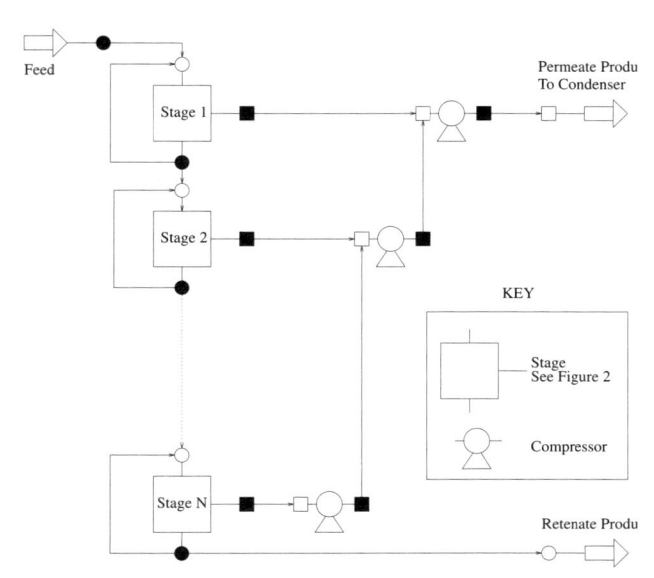

Figure 3. Pervaporation superstructure.

see Fig. 2. Consequently, only one unit model is required to describe any number of modules operated in parallel. This approach does not reduce the accuracy of the model but does rely on a few basic assumptions:

- All modules in parallel are identical.
- Feed temperatures and pressures are equal.
- Split of inlet flowrates is equal.

These assumptions are reasonable as it is uncommon to connect dissimilar modules in parallel. Furthermore, if the first two conditions are satisfied, then the third condition corresponds to optimality.

Pervaporation Process Superstructure

The process superstructure considered in this paper is illustrated in Figure 3. Due to the nature of pervaporation, for which vacuum conditions must be maintained on the permeate side, the following assumptions can be made:

- Vapor and liquid streams are not mixed.
- Compressors are connected in series.
- Compressor pressure at stage n ≥ compressor pressure at stage n+1.

For a system with N stages and thus N possible compressors, there are N^2 permeate stream alternatives. As these streams can not be considered simultaneously, a very large binary problem must be solved. Hence, the final two assumptions simplify the design of the compressor system, and greatly reduce the problem size.

Solution Method

The problem formulation contains linear equations in integer variables as well as non-linear equations in continuous variables. It is therefore a *MINLP* problem. Binary variables describe whether the heat exchanger and compressor associated with a given stage exist, and integer variables describe the number of modules within a stage. Several different algorithmic approaches to the solution of *MINLP* problems have been developed over recent years (see Grossmann and Kravanja, 1995). In this case, a manual branch and bound solution method has been selected due to its ease of application and high transparency. This requires that the integer and binary variables are relaxed to bounded continuous variables, and that a set of *NLP* sub-problems are solved. *NLP* problems are solved relatively easily within the gPROMS/gOPT (Process Systems Enterprise Ltd., 1998) simulation and optimization environment. However, a guarantee of global optimality is not possible.

To satisfy a defined objective, the optimal recycle flows, heat exchanger temperatures, and permeate pressures must be calculated. Additionally, the number of modules and, the existence of the compressor and heat exchanger must be established for each stage. This requires six decision variables per stage.

Application to a Pervaporation Case Study

To assess the validity of the optimal design procedure, a pervaporation case study has been selected. Tsuyumoto *et al.* (1997) recently presented the design of a pilot plant for the dehydration of an ethanol/water mixture. Their plant uses nine hollow-fiber modules based on a new type of polyion complex membrane to purify 100kg/hr of 94 wt% ethanol to a purity in excess of 99.5 wt% ethanol. The plant also contains four heat exchangers, two compressors and a condenser system.

Model Verification

The accuracy of the model presented in this paper is assessed for a single hollow-fiber module at two different flowrates. The results, summarized in Table 1, show

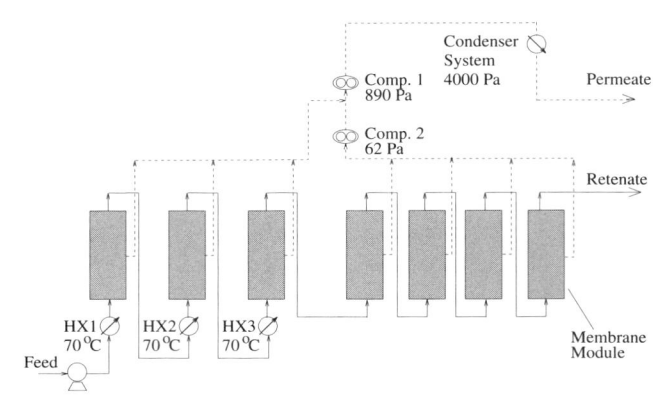

Figure 4. Optimal process flowsheet.

excellent agreement with the experimental data reported by Tsuyumoto *et al.* (1997).

Optimization

For this study, the objective is to find the plant design that minimizes production costs - subject to design constraints. The production costs are the sum of the operating costs and the depreciated capital costs. Note that, due to the small scale of the system, size independent capital costs have been used.

At this stage, we further simplify the optimization by assuming that the capital costs for piping and pumping are negligible in comparison to the cost of the major equipment items.

Using the procedure described in this paper, the optimal plant design (Fig. 4) has been found. The new design compares favorably with the actual plant design (Tsuyumoto *et al.*, 1997), with a total cost saving of 11%. It contains fewer modules (seven compared to nine) and less heat exchangers (two compared to four), leading to a reduction in capital costs. Operating costs are also lower as a result of a similar total heat input but lower compressor duties. The main emphasis of the new design is that the temperature is kept high throughout the process by adding all the heat in the initial stages, where most of it is otherwise lost, due to the higher rate of permeation. It is this constant high temperature that enables the number of modules and the compressor duties to be reduced.

Table 1. Simulation and Experimental Results.

Parameter	Case 1	Case 2
Feed Flowrate, kg/hr	44.8	248.5
Feed Concentration, wt% eth	94.03	96.81
Product Concentrations, wt% eth.		
This Study	97.29	97.44
Tsuyumoto *et al.* (1997) Simulation	97.29	-
Tsuyumoto *et al.*(1997) Experimental	97.17	97.43

Conclusions

Using the superstructure developed in this paper an improved design for a pervaporation pilot plant has been found. In contrast to previous membrane optimization studies, this study is based on an accurate membrane unit model that has been verified against published experimental data. The model enables efficient and rigorous modeling of a general hollow-fiber module.

If a similar number of stages are considered, then the application of the model to full-scale systems is straight-forward. However, when large membrane areas are required the method should be adapted to include stages of modules in series as well as in parallel.

Future work includes the removal of the assumptions used in the development of the hollow-fiber model, and the application of the optimal design procedure to larger and different membrane systems.

References

Coker, D.T., B.D. Freemann, and G.K. Fleming (1998). Modeling multicomponent gas separation using hollow-fiber membrane contactors. *AIChE J.*, **44**, 1289-1302.

El-Halwagi, M.M. (1992). Synthesis of reverse osmosis networks for waste reduction. *AIChE J.*, **38**, 1185-1198.

Grossmann I.E., and Z. Kravanja (1995). Mixed-integer nonlinear programming techniques for process systems engineering. *Comp. Chem. Eng.*, **19**, S189-S204.

Ho, W., and K. K. Sirkar (1992). *Membrane Handbook*. Van Nostrand Reinhold, 1st ed., New York.

Ji, W., A. Hilaly, S.K. Sikdar, and S. Hwang (1994). Optimization of multicomponent pervaporation for removal of volatile organic compounds from water. *J. Membr. Sci.*, **97**, 109-125.

Krovvidi, K.R., A.S. Krovvali, S. Vemury, and A.A. Khan (1992). Approximate solutions for gas permeators separating binary mixtures. *J. Membr. Sci.*, **66**, 103-118.

Process Systems Enterprise Ltd. (1998). *gPROMS Advanced User's Guide*, London.

Qi, R., and M.A. Henson (1998). Optimal design of spiral-wound membrane networks for gas separations. *J. Membr. Sci.*, **148**, 71-89.

Srinivas, B.K., and M. M. El-Halwagi (1993). Optimal design of pervaporation systems for waste reduction. *Comp. Chem. Eng.*, **17**, 957-970.

Tessendorf, S., R. Gani, and M.L. Michelsen (1999). Modelling simulation and optimization of membrane-based gas separation systems. *Chem. Eng. Sci.*, **54**, 943-955.

Tsuyumoto, M., A. Teramoto, and P. Meares (1997). Dehydration of ethanol on a pilot-plant scale using a new type of hollow-fiber membrane. *J. Membr. Sci.*, **133**, 83-94.

INTEGRATING CHEMISTRY PROCESS DEVELOPMENT WITH DOWNSTREAM CONSIDERATIONS

David C. Miller and James F. Davis
Ohio State University
Columbus, OH 43210

Abstract

We present a prototype Process Design Decision Support System (PDDSS) which applies concepts of concurrent engineering to the chemical process industry by integrating early experimental chemistry development and engineering analysis. We developed a formalism to represent the laboratory experimental data that is then entered into the PDDSS. The formalism closely follows the form and terminology used to report reactions in the chemistry literature. The PDDSS uses the chemistry description to develop a list of required functions which it uses to query an engineering knowledge database to determine devices capable of carrying out the required functions. The devices which are identified form the basis of a process topology which can then be used for various types of engineering analysis. The models and knowledge associated with each device in the database allow for an early analysis of cost, environmental, regulatory, and safety issues that may become important as the project moves from experimental process research to engineering design. By coupling experimental chemistry development with interactive, engineering-based evaluation, the PDDSS accelerates the overall development process and enables the underlying chemistry to better meet financial and processing goals. The PDDSS is unique in its approach by integrating downstream engineering considerations with laboratory-scale chemical synthesis.

Keywords

Preliminary process evaluation, Process chemistry critique, Process topology, Decision support, Engineering knowledge database.

Introduction

With recent advances in computer technology, more information can be processed and more calculations performed than ever before. As discussed at the recent FOCAPO '98 conference, many view this computer power as the key that will allow them to unlock the full potential of a given chemical process by using increasingly complex and detailed models of the process. Unfortunately, these complex and detailed models require a large amount of input information to be useful—information that is not always immediately available.

This is especially true in the early stages of the development of complex specialty chemicals and pharmaceuticals. Since many of these compounds never proceed past research and development, fully characterizing each compound would not be an appropriate use of resources. Instead, only when it becomes clear that the product is likely to be commercially viable should the additional information be obtained for more detailed analysis and simulation. At the same time, we recognize many benefits from beginning engineering analysis while a compound is undergoing laboratory development. In particular, engineering and manufacturing concerns can be used to focus process chemistry development on issues that may not otherwise be identified until much later in the development cycle, where the process changes are significantly more costly. In addition, development time can be shortened by identifying and correcting issues that would traditionally not arise until beginning engineering process design. Thus, in order to successfully and rapidly commercialize new, structurally complex, site selective molecules, the CPI must utilize engineering analysis to focus the on-going experimental development of the process chemistry.

This paper describes a prototype Process Design Decision Support System (PDDSS) which seeks to integrate process chemistry research with engineering analysis. The usefulness of the PDDSS has been evaluated using industrial case studies. We discuss the knowledge framework and operation of the PDDSS in the context of a simple case study drawn from literature.

Reaction Representation Formalism

Engineering analysis in the process domain almost always centers around a process flowsheet. This is a convenient method to identify major unit operations, process streams, and the interrelations between them. Typically, significant information is required to develop a process flowsheet. Indeed, most automated analysis programs require this information to be fully specified before they can provide useful results (e.g, Batch Design Kit, etc.). Such requirements inhibit the use of these tools for early process analysis since much of the required information has not yet been determined.

Instead of trying to force definition of the unknown data, we have developed techniques which make use of the information that is immediately available for process analysis. Thus, our analysis approach first develops a type of process flowsheet which we refer to as a "process topology" to recognize its lack of detail found in a typical flowsheet. It includes information on the material streams (i.e., approximate stream constituents and mass) and the equipment (i.e., what it is intended to do and what components are involved in the operation). The process topology forms the foundation for an engineering analysis of the process chemistry.

In order to develop the process topology, we begin with the chemical reaction description as recorded by the process chemists. We have formalized a syntax for the process recipes so they can be used as a reaction language to enter the information into the computer. Our formalism is strongly based on the general format used to report chemical reactions in standard chemistry journals such as *The Journal of Organic Chemistry*.

Unlike the formalisms of Nagel *et al.* (1995) and Prickett and Mavrovouniotis (1997) which attempt to capture detailed molecular and reactivity information, our formalism is concerned only with capturing the recipe. Thus, our representation does not require the identification of reactive sites on the molecule. Instead, it is sufficiently abstract that a reaction is merely described as a sequence of chemicals being added and another eventually being collected at the other end.

The simple, highly abstract representation is fully reflective of the information that is normally reported in the course of synthetic organic research. It is limited, but it is readily available. The more mechanistic representations, on the other hand, require significantly more information to define a chemical reaction. In some cases, additional experimentation may be required to obtain all the information required. From an analysis perspective, the more mechanistic representations allow for much more detailed reaction analysis since it has more information front loaded into the system. Our representation, however, is not intended for such a detailed reaction analysis. Instead, it provides only enough information to create an initial process topology which can then be used as the basis for engineering analysis.

Analysis Methodology

We describe our methodology using the synthesis of florfenficol described by Schumacher *et al.* (1990) as shown in Figure 1. In particular, our example focuses on the first reaction.

In the figure, the bold numbers are used to refer to the molecules in this document. Within the reaction descriptions (below) they are referred to as CaseStudy-X, where X is the number in bold. The numbers in bold italics refer to the numbers used in the original paper. The names of the reactions are used in this document to track the reactions when they are entered into the PDDSS prototype.

Developing a Process Topology

The first step is to enter the reaction description using the appropriate syntax and keywords. For the first reaction, this is shown in the first column of Table 1. Once the reaction information has been entered, the system begins reading through the reaction descriptions to identify the functions that are involved with each reaction step and to build an ordered list of functions. For those functions that involve adding or removing chemicals, the PDDSS queries

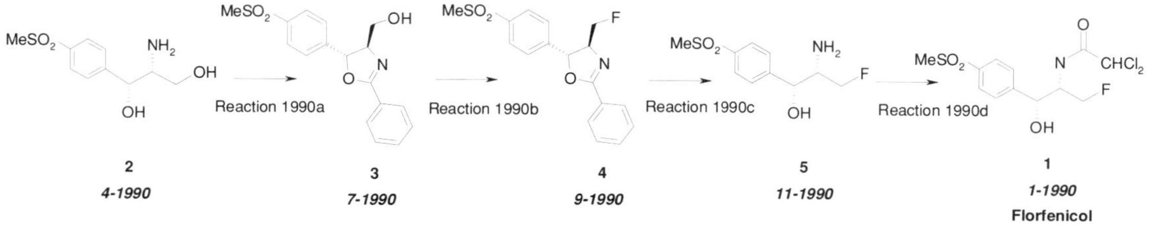

Figure 1. Reaction pathway developed by Schumacher et al. (1990).

a chemical information database to pull that data into the system.

When the system has finished parsing the reaction information, it has simultaneously developed an ordered list of the functions involved with each chemical transformation. Associated with mass transfer functions are the chemicals entering or leaving the system. This ordered list is shown in the second column of Table 1.

Using this ordered list of functions, the PDDSS begins to query the database to find devices that can be used to achieve the functions that were determined to be needed based on the chemistry. Our approach recognizes that often a single device will be used for multiple purposes. In our example, a reactor is initially selected for its ability to handle the Material Loading function derived from the *charge reactor with* keywords. The database is then queried to determine whether this device is capable of performing each subsequent function until a new device is required. In this case, the reactor can perform the additional Material Loading functions as well as the Agitation and Heat Transfer functions. The query for the Filter function, however, returns false indicating that a new device is required. Another query is then performed to find an appropriate device for this function.

Once an annual production basis is specified, the resulting mass requirements for each chemical are determined. Waste streams are also identified, and their masses calculated. This allows material and waste costs to be calculated. In addition, since we have determined the volume of material passing through each device, we can estimate the processing costs associated with the chemistry. This cost is currently estimated using a proprietary empirical correlation but any appropriate model can be used.

Employing Critics

In order to critique a process topology, we employ independent critics. Critics act much like an outside expert. They are external to the data structures developed for the process topology; however, they have complete access to the information contained therein. The application of one critic is typically not dependent on the application of another. In this sense, they are independent. However, the data structures permit a critic to make use of the results of a previous critic.

Each critic is designed to evaluate a particular aspect of the design. These can range from material issues (i.e., regulatory and hazard information associated with the chemicals in the process) to equipment issues (i.e., whether a particular piece of equipment is expected to operate properly or whether special construction materials must be employed) to cost issues (i.e., processing cost, waste cost, material cost and overall cost). The critics share a similar approach. They begin by navigating the links between data structures searching for those data types that indicate there may be something for it to examine.

The results of the critic analysis can be viewed at any level desired. Thus, analysis of the chemicals may indicate some regulatory issues that need to be resolved. This information is reflected not only directly in the chemical data structure, but also in the data structure of the equipment in which the chemical is being use, the data structure of the original chemical reaction description, and the data structure of the function serving as a bridge between the chemistry and engineering worlds. The results of equipment-based and function-based critics are similarly associated with the other data types so as to facilitate access to the information from multiple points of view.

While this is straightforward for critics which rely only on information contained directly within their own data type, such as a regulatory critic, other critics require information from the other data types. For example, an equipment-based critic, although it is concerned with the operation of a particular device, cannot effectively critique the device unless it has access to information about its functions and the chemicals involved. Thus, a distillation critic requires certain physical property information in order to determine the effectiveness of a particular distillation. Similarly, a waste critic requires information of all the functions of all the equipment in order to determine where waste is generated. It then requires chemical information to determine the amount and type of waste.

Table 1. Original Text from the Reaction is Description Shown with Resulting Function and Link to Chemical Information if Applicable.

Reaction Description	Ordered Function List	Device
Charge reactor with glycerol (10 g)	Material Loading	Reactor
add K2CO3 (0.43 g)	Material Loading	
add CaseStudy-2 (5.0 g)	Material Loading	
Heat to 115 C	Heat Transfer - heating	
add benzonitrile (3.5 g)	Material Loading	
Stir for 840 min	Agitation	
Cool	Heat Transfer - cooling	
add water (5 g)	Material Loading	
Filter	Filter	Filtration system
Wash with water (2 g)	Filtrate Washing	
Wash with MeCl2 (2 g)	Filtrate Washing	
Dry	Drying	Dryer
Collect CaseStudy-3 (6.4 g)	Material Collection	

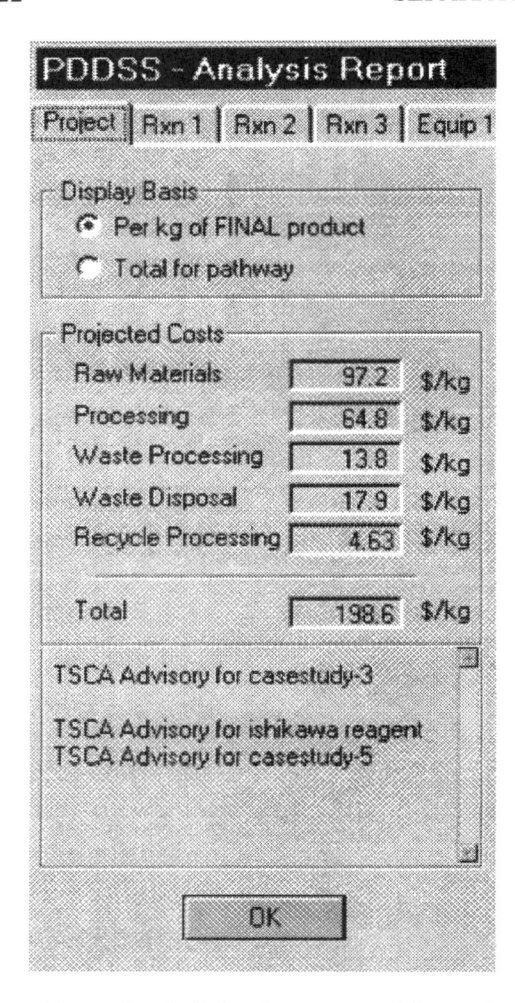

Figure 2. A dialog box report of the analysis of the synthesis of florfenicol.

This knowledge structure allows for critics to be independently developed based on the type of critique they perform. Since access is equally easy for any of the available data types, the same overall structure can be employed for all the critics. Thus, an equipment-based critic iterates through the equipment structures and uses the links to functional and chemical information to perform its critique. The results of the criticism can be stored in all the data structures associated with the equipment.

Analysis report

The system then reports the results of the analysis. Figure 2 shows one dialog box which reports the results of various mass balance and economic evaluations. The radio buttons allow selection of the display basis, either dollars per kilogram of final product or total dollars for the given production basis. The lower box shows results from the TSCA critic which determines the regulatory status of the chemicals involved in the process.

Other critics provide results in a similar format. For example, the cost critic provides a rough estimate of which areas of the chemical synthesis account for the largest portions of the cost (shown below).

reaction 1990a	Material Costs	% Total Cost: 27.34
reaction 1990bc	Material Costs	% Total Cost: 15.41
reaction 1990bc	Processing Costs	% Total Cost: 13.70
reaction 1990d	Processing Costs	% Total Cost: 11.41

The output from this critic indicates that material costs account for 27% of the total cost of production. Thus, the chemist knows that this is an area where significant overall improvement to the process could be achieved.

Additional features of the PDDSS include (1) exporting results to Excel for further analysis, formatting, printing, etc., (2) searching the Federal Register for additional regulatory information on the chemicals being used.

Summary

In summary, we have described a prototype PDDSS which provides a mechanism for integrating engineering analysis with process chemistry development. This analysis can accelerate the overall development cycle by helping to quickly screen synthetic process alternatives and identify areas which require more in-depth experimental analysis. In its prototype form, the PDDSS demonstrates how reaction information can be entered into the system and used to develop process topology for analyzing and critiquing the underlying chemistry.

The PDDSS is currently being evaluated by industrial chemists for application to their development process. They are providing helpful feedback regarding the presentation of analysis results and the user interface. We are also working with them to develop additional critics. As this interaction continues, we foresee the PDDSS becoming a practical tool used in industry.

Acknowledgements

We gratefully acknowledge support provided by Dow AgroSciences and the National Defense Science and Engineering Graduate Fellowship program.

References

Nagel, C., C. Han and G. Stephanopoulos (1995). Modeling languages: declarative and imperative descriptions of chemical reactions and processing systems. In G. Stephanopoulos and C. Han (Eds.), *Intelligent Systems in Process Engineering Part 1: Paradigms from Product and Process Design.* Vol. 21 of *Adv. in Chem. Eng.*, Academic Press, New York, 1-91.

Prickett, S. E. and M. L. Mavrovouniotis (1997). Construction of complex reaction systems – I. Reaction description language. *Comp. Chem. Eng.* **21**, 1219-1235.

Schumacher, D. P., J. E. Clark, B. L. Murphy and P. A. Fischer (1990). An efficient synthesis of florfenicol. *J. Org. Chem.* **55**, 5291-5294.

SYNTHESIS OF OPTIMAL DISTILLATION SEQUENCES FOR THE SEPARATION OF ZEOTROPIC MIXTURES USING TRAY-BY-TRAY MODELS

Hector Yeomans and Ignacio E. Grossmann
Carnegie Mellon University
Pittsburgh, PA 15213

Abstract

This paper describes a Generalized Disjunctive Programming (GDP) model for the synthesis of distillation sequences using rigorous design equations. The model is obtained systematically from the State Equipment Network (SEN) representation of superstructures, and results from the separation of three component mixtures illustrate its robustness and computational efficiency.

Keywords

Distillation sequences, Tray-by-tray models, Superstructure, MINLP, Disjunctive program.

Introduction

The optimal synthesis of distillation continues to be a central problem in the design of chemical processes, due to the high investment and operating costs involved in these systems. The recent trends in this area have been to address models of increasing complexity through the use of mathematical programming. Examples of these models include the short-cut models by Novak et al. (1996) and Yeomans and Grossmann (1998b), and the rigorous tray-by-tray models by Bauer and Stichlmair (1998) and Smith and Pantelides (1995). The high degree of nonlinearity and the difficulty of solving the corresponding MINLP optimization models, however, have prevented these methods from becoming tools that can be readily used by industry. For instance, a common problem that is experienced with rigorous models is when the columns are "deleted", as then the equations describing the MESH equations become singular.

This goal of this paper is to present a Generalized Disjunctive Programming (GDP) model for the synthesis of rigorous distillation sequences, that can avoid the existing pitfalls of MINLP optimization models. The case of separation of zeotropic mixtures is addressed, and its potential use for azeotropic distillation will be discussed.

Problem Definition

The objective of this paper is to generate an optimization model for the design of optimal distillation systems. Given is a feed stream with known composition required to be separated into essentially pure component product streams. The model has the following characteristics: (1) it is based on rigorous calculations, (2) ideal or non-ideal VLE equilibrium equations, (3) covers only simple column configurations.

Synthesis Framework

The tray-by-tray optimization model was systematically derived by the application of the synthesis framework proposed by Yeomans and Grossmann (1998a). The framework consists of three steps: (1) Generation of a superstructure of all possible flowsheet alternatives based on the State Task Network (STN) or State Equipment Network (SEN) representations. (2) Modeling of the superstructure using Generalized Disjunctive Programming (GDP; Raman and Grossmann, 1994; Turkay and Grossmann, 1996). (3) Solve the GDP model with a modification of the Logic-Based Outer Approximation Algorithm (Turkay and Grossmann, 1996).

The first step of the synthesis framework requires the identification of three key elements of any synthesis problem: states, tasks and equipment. These elements are assembled in a flowsheet, and linked to one another depending on the choice of representation (SEN or STN). Each of these representations can be translated into a unique mathematical programming model in GDP form, which is then solved with special purpose algorithms. To tackle the problem of interest, the SEN superstructure representation was used.

Superstructure Representation

Consider the separation of a three component mixture, where A,B and C represent the components ordered by decreasing relative volatility. There are four tasks that can be identified for this case: the separation of A from BC, of AB from C, of A from B and the separation of B from C. The minimum number of equipment units required to perform these separations is two, given that sharp splits are required and only one separation path is selected.

Considering a superstructure with the minimum number of equipment units, Figure 1 shows the SEN representation for the problem. Mixers and splitters are permanent equipment with permanent tasks, while the distillation columns represent permanent equipment with conditional tasks.

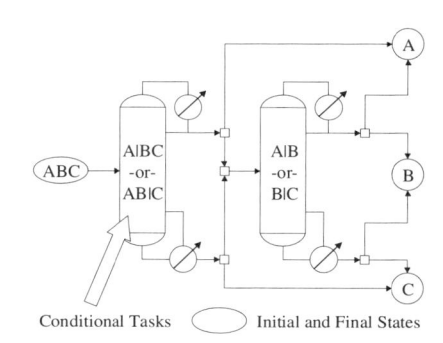

Figure 1. Sample SEN superstructure.

At this point the superstructure in Figure 1 is valid for aggregated, short-cut or rigorous models. If rigorous models are used, another discrete decision to make is the selection of the number trays in the column. This decision is not explicitly linked to the choice of a task, so it is possible to model a single column as a superstructure of smaller equipment units –the trays– that can be represented also as a SEN. For this case a tray can perform one of two tasks: VLE Mass exchange or no mass exchange as seen in Figure 2.

Because only simple column design is used, the feed tray, reboiler and condenser tray are considered equipment with one permanent task, but they can become conditional if complex columns configurations are included.

GDP Model for SEN Representation

The second stage of the synthesis framework requires the modeling of the superstructure representation as a Generalized Disjunctive Programming (GDP) problem. In this model, each discrete choice of a task or equipment is represented as a disjunction. The equations and constraints that apply whenever the equipment or task exists, are grouped with brackets in each disjunction. The equations and constraints that apply when a task or equipment does not take place are also grouped in brackets in the same disjunction. The *OR* logic operator ($\vee$) denotes the discrete choice between equations, and a set of boolean variables (Y) indicates a choice and propagate its effect to the rest of the problem by means of Logic Relationships ($\Omega(Y)=True$).

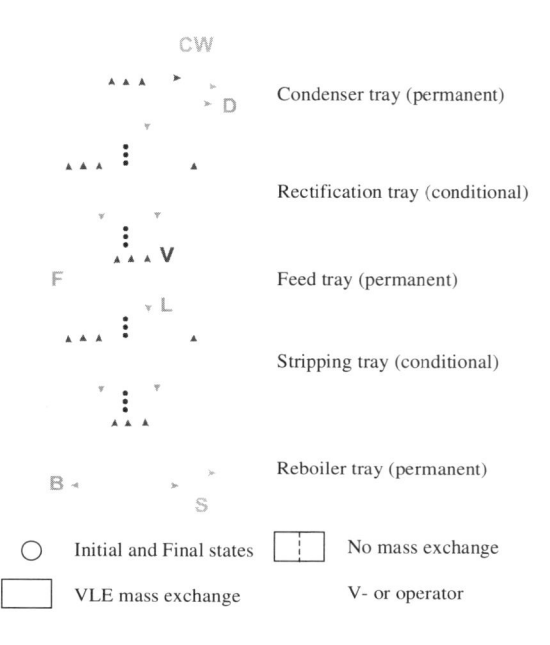

Figure 2. Superstructure for rigorous column.

The following model is based on the superstructure shown in Figure 1, but it can easily be extended to superstructures with more columns and tasks. The variable definition for the model can be found at the end of this paper, and the following set definitions were used: C is the set of i components to be separated; COL is the set of available columns j; TS is the set of available trays n; TM is the set of conditional trays (TM$\subseteq$TS); NRT, NCT and NFT are the reboiler, condenser and feed trays, respectively. These are the permanent trays.

$$\min \sum_{j \in COL} \{f(NT_j, Dc_j) + \alpha QR_j + \beta QC_j\} \quad (1)$$

$$\text{s.t.} \quad F_2^i = D_1^i + B_1^i$$
$$NT_j = \sum_{n \in TM} STG_n + 3 \quad (2)$$
$$DC_j = f(T_j^V, P_{n,j}, R_j, VAP_{n,j}) \quad n \in NCT$$

$$\left.\begin{array}{l}F_j^i + L_{n+1,j}^i + V_{n-1,j}^i - L_{n,j}^i - V_{n,j}^i = 0 \\ \sum_{i \in C}\begin{pmatrix} hF_j^i + hL_{n+1,j}^i + hV_{n-1,j}^i \\ - hL_{n,j}^i - hV_{n,j}^i \end{pmatrix} = 0 \\ F_j^i = FED_j x_{F,j}^i \\ hF_j^i = f(T_{n,j}^L) \end{array}\right\} n \in NFT \quad (3)$$

$$\left.\begin{array}{l} V_{n-1,j}^i - L_{n,j}^i - D_j^i = 0 \\ \sum_{i \in C}\left(hV_{n-1,j}^i - hL_{n,j}^i - hD_j^i\right) = QC_j \\ D_j^i = DIS_j x_{n,j}^i \\ D_j^i = R_j L_{n,j}^i \\ hD_j^i = f(T_{n,j}^L) \end{array}\right\} n \in NCT \quad (4)$$

$$\left.\begin{array}{l} L_{n+1,j}^i - B_j^i - V_{n,j}^i = 0 \\ \sum_{i \in C}\left(hL_{n+1,j}^i - hV_{n,j}^i - hB_j^i\right) = QR_j \\ B_j^i = BOT_j x_{n,j}^i \\ hB_j^i = f(T_{n,j}^L) \end{array}\right\} n \in NRT \quad (5)$$

$$\left.\begin{array}{l} L_{n,j}^i = LIQ_j x_{n,j}^i \\ V_{n,j}^i = VAP_j y_{n,j}^i \\ \sum_{i \in C} x_{n,j}^i = 1 \\ \sum_{i \in C} y_{n,j}^i = 1 \\ f_i^L = f_i^V \\ hL_{n,j}^i = f(T_{n,j}^L) \\ hV_{n,j}^i = f(T_{n,j}^V) \end{array}\right\} n \in TS \quad (6)$$

$$\left.\begin{array}{l} L_{n+1,j}^i + V_{n-1,j}^i - L_{n,j}^i - V_{n,j}^i = 0 \\ \sum_{i \in C}\left(hL_{n+1,j}^i + hV_{n-1,j}^i - hL_{n,j}^i - hV_{n,j}^i\right) = 0 \end{array}\right\} n \in TM \quad (7)$$

$$\begin{bmatrix} Y_{n,j} \\ f_i^L = f(T_{n,j}^L, P_{n,j}, x_{n,j}) \\ f_i^V = f(T_{n,j}^V, P_{n,j}, y_{n,j}) \\ T_{n,j}^L = T_{n,j}^V \\ STG_n = 1 \end{bmatrix} \vee \begin{bmatrix} \neg Y_{n,j} \\ x_{n,j}^i = x_{n+1,j}^i \\ y_{n,j}^i = y_{n+1,j}^i \\ T_{n,j}^L = T_{n+1,j}^L \\ T_{n,j}^V = T_{n-1,j}^V \\ f_i^L = 0 \\ f_i^V = 0 \end{bmatrix} \quad n \in TM \quad (8)$$

$$\begin{bmatrix} Y_{A|BC}^1 \\ x_{n,1}^A \geq \zeta \\ D_1^A \geq \mu F_1^A \end{bmatrix} \vee \begin{bmatrix} Y_{AB|C}^1 \\ x_{n,1}^B \geq \zeta \\ D_1^B \geq \mu F_1^B \end{bmatrix} \quad n \in NCT$$
$$\begin{bmatrix} Y_{A|B}^2 \\ x_{n,1}^A \geq \zeta \\ D_1^A \geq \mu F_1^A \end{bmatrix} \vee \begin{bmatrix} Y_{B|C}^2 \\ x_{n,1}^B \geq \zeta \\ D_1^B \geq \mu F_1^B \end{bmatrix} \quad n \in NCT \quad (9)$$

$$\begin{array}{l} Y_{A|BC}^1 \Leftrightarrow Y_{B|C}^2, \quad Y_{AB|C}^1 \Leftrightarrow Y_{A|B}^2 \\ Y_{n,j} \Rightarrow Y_{n-1,j}, \quad Y_{n-1,j} \Rightarrow Y_{n,j} \end{array} \quad (10)$$

$F, FED, D, DIS, B, BOT, L, LIQ, V, VAP \in R^+$
$R, T, P, NT, QR, QC, x, y, f, STG \in R^+$
$hD, hB, hF, hL, hV \in R$
$Y = \{True, False\}$

Equation (1) is the objective function, a nonlinear cost function in terms of the number of trays, column diameter and duties of reboiler and condenser. (2) defines the overall column interconnection, as well as the costing variables. Equations (3), (4) and (5) are the mass and energy balances for the permanent trays (feed, condenser and reboiler, respectively). The block in (6) represents all the equations that are valid for both permanent and conditional trays. (7) are the mass and energy balances for the conditional trays. The disjunction in (8) indicates the discrete choice for conditional trays: whenever VLE takes place, fugacities are defined and liquid and vapor temperatures are equal; if the choice is not VLE, then compositions and temperatures in the tray will depend on adjacent trays, while the fugacities of liquid and vapor are set to zero. The disjunctions in (9) enforce the discrete choice of task selection for each column, based on purity and recovery specifications for the key component recovered at the top of each column. Finally, (10) includes the logic relationships that hold in the superstructure.

Numerical Examples

The model described above was solved with a modified Logic-Based Outer Approximation algorithm (Yeomans and Grossmann, 1998b). The algorithm was implemented with the GAMS modeling environment (Brooke et al., 1992), on a Pentium II 300 PC.

Table 1. Results for the Separation of C3, C4, C5.

Parameter	Value
Continuous Variables	1952
Discrete Variables	56
Constraints (NLP)	2107
Max Trays per column	30
Objective Value	M$ 2.2573
CPU Time	43 min. 46 sec.
OA Iterations	6

Two numerical examples were used to test the model. The first one requires the separation of a mixture of butane (C4), pentane (C5) and hexane (C6) into pure components. The second example is for the separation of a mixture of benzene, toluene and o-xylene into pure components. Both systems were modeled with ideal equilibrium, and reasonable bounds on the number of trays required for the separation. The objective function is the present cost of the equipment and utility costs.

Figure 3 shows the optimal configuration obtained, and Table 1 shows relevant computational information. The results from the model were confirmed with the commercial simulator PROII, with very good agreement.

80% of the CPU time was for the master problem since it has more than 10,000 variables and more than 6,000 constraints. Because of the nature of disjunctive programs, the size of the NLP subproblems was considerably reduced, compared to MINLP models.

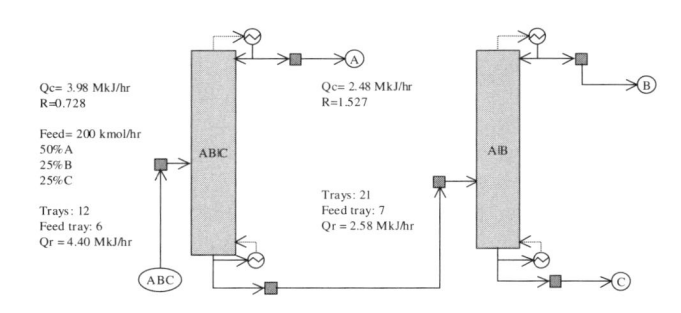

Figure 3. Optimal design for separation of ABC.

In the second example for the separation of benzene, toluene and o-xylene, the optimal solution separates the most abundant component first, the o-xylene, in a 36-tray column. The mixture of benzene and toluene is then separated in a 26-tray column. The optimal net present cost is M$1.30, and the solution was obtained in 5 OA iterations, with a total CPU time of 2.3 hrs, on an HP C-110 workstation. This high CPU time is due to the bound in the number of trays per column, which was set up to 50.

Conclusions

A mathematical programming model for the design of distillation sequences with tray-by-tray models was presented. The model was derived systematically, according to the synthesis framework proposed by Yeomans and Grossmann (1998). Two examples that have been tested suggest that the proposed method is robust and efficient for modeling the separation of ideal and zeotropic mixtures. It is important to remark that even though this model has not been tested for the separation of azeotropic mixtures, it can potentially solve these problems, provided an appropriate superstructure is developed. The main significance of this work is that the numerical difficulties of MINLP models produced by disappearing column sections and flows can be overcome.

Notation

B^i_j = Bottoms flow of species i in column j, kmol/hr
BOT_j = Total bottoms flow in column j
DC_j = Column diameter, ft.
D^i_j = Distillate flow of species i in column j
DIS_j = Total distillate flow in column j
f_i = Fugacity of liquid or vapor of species i
F^i_j = Feed flow of species i into column j
FED_j = Total feed flow into column j
hB^i_j = Liquid enthalpy of i in the bottoms, kJ/kmol
hD^i_j = Liquid enthalpy of species i in the distillate
hF^i_j = Liquid enthalpy of species i in feed of column j
$hL^i_{n,j}$ = Liquid enthalpy of species i in liquid stream
$hV^i_{n,j}$ = Vapor enthalpy of species i in vapor stream
$L^i_{n,j}$ = Liquid flow of species i out of tray n, in column j
$LIQ_{n,j}$ = Total liquid flow out of tray n in column j
NT_j = Number of trays in column j
QC_j = Condenser heat load of column j, kJ/hr
QR_j = Reboiler heat load of column j
R_j = Reflux ratio for column j
$STG_{n,j}$ = Binary variable indicating a stage uses VLE
$T_{n,j}$ = Temperature of liquid or vapor in tray n,K
P = Pressure (stage or column), bar
$x^i_{n,j}$ = Mole fraction of species i in the liquid phase
$y^i_{n,j}$ = Mole fraction of species i in the vapor phase
μ = Recovery fraction with respect to feed
ς = Purity specification (fraction)
α, β = Utility cost coefficients

References

Bauer, M.H. and J. Stichlmair (1998). Design and economic optimization of azeotropic distillation processes using mixed-integer nonlinear programming. *Comp. Chem. Eng.*, **22**, 541.

Brooke, A., D. Kendrick and A. Meeraus (1992). *GAMS – A User's Guide.* Scientific Press, Palo Alto.

Novak, Z., Z. Kravanja and I.E. Grossmann (1996). Simultaneous synthesis of distillation sequences in overall process schemes using an improved MINLP approach. *Comp. Chem. Eng.*, **20**, 1425.

Raman, R. and I.E. Grossmann (1994). Modeling and computational techniques for logic based integer programming. *Comp. Chem. Eng.*, **18**, 563.

Smith, E.M.B. and C.C. Pantelides (1995). Design of reactor/separation networks using detailed models. *Comp. Chem. Eng.*, **19**, S83.

Turkay, M. and I.E. Grossmann (1996). A logic based outer-approximation algorithm for MINLP optimization of process flowsheets. *Comp. Chem. Eng.*, **20**, 959-978.

Yeomans, H. and I.E. Grossmann (1998a). A systematic modeling framework of superstructure optimization in process synthesis. Accepted for publication, *Comp. Chem. Eng.*

Yeomans, H. and I.E. Grossmann (1998b). A disjunctive programming method for the synthesis of heat integrated distillation sequences. AIChE Annual Meeting, Miami.

ECONOMIC RISK MANAGEMENT FOR DESIGN AND PLANNING OF CHEMICAL MANUFACTURING SUPPLY CHAINS

G. E. Applequist, J. F. Pekny, and G. V. Reklaitis
School of Chemical Engineering
Purdue University
West Lafayette, IN 47907-1283

Abstract

A company faces economic risks when it decides among products to introduce and facilities in which to produce them. Uncertainties in the demands for products result in the risk that plant investments will not pay off. A model for the design and planning problem accounts for the economic loss of overproduction relative to uncertain market demands, with uncertainty in price as well. The expected investment value and its variance, as a measure of risk, are calculated exactly by polytope volume integration without the need for Monte Carlo simulation. This method of integrating over uncertain variables allows for a variety of possible demand and price distributions. A decision between design alternatives is based on the mean and variance of the investment returns. The company should expect a risk premium at least as good as the alternative investments in the financial market. The key decisions about production rate, capacity location, and customer service are made on this basis. The new ideas in this work are demonstrated on a design case study based on an agricultural chemical manufacturing example.

Keywords

Plant design, Planning, Risk management, Supply chain, Uncertainty.

Introduction

The problem of economic uncertainty is a rich area of research in methods for the design and operation of chemical facilities. Ongoing efforts to account for uncertain factors in design and planning are motivated by the risk of economic loss where large investments are to be made. A major challenge in formulating these problems is to decide on the kind of objective function and sets of constraints which appropriately capture the needs of the decision maker. The existing work contains recurring themes: expected values of objective functions, chance constraints, and scenarios. A beneficial viewpoint would be to treat engineering investments and financial investments as comparable alternatives.

A special feature of the risks in the supply chain is the interdependence of uncertainties: variation may be associated with the time of year, the type of product, the region of the customer, or any number of other effects. It would be beneficial to model uncertainties that fit this description, directly calculate their effects on return and risk, and use that information to guide the solution of design and operational problems.

Planning and Design with Uncertainty

Grossmann and Sargent (1978) treated demand parameters with uniform distributions, with the goals of minimizing expected costs while meeting demands and keeping the plant feasible over the range of uncertainties. The difficulty of continuous distributions was avoided with the introduction of discrete *scenarios*, or combinations of discrete samples of all uncertain parameters. Subrahmanyam (1996), using the discrete scenario approach and the expected NPV as the objective, proposed risk constraints with a probabilistic guarantee index, i.e. a bound on the probability that a demand goal is violated. Another approach takes uncertain demands as normal distributions and integrates an objective function to obtain the expected value while utilizing cumulative probability distributions to formulate chance constraints. Petkov (1997) presents a formulation with multivariate-normal demands and a deterministic-equivalent transformation, introducing chance constraints and penalties for underproduction in order to control shortfalls of production relative to the uncertain demand.

New Ideas and Methods for Risk Management

Previous work on production problems has utilized various measures of economic risks, but there seems to be a need for a real-world basis of comparison for choices among alternatives in design and operation under market uncertainties. A reasonable approach would embody the policy of *only making engineering investments that are better than alternative financial investments*.

Risk Premiums

It has long been observed that financial securities generally offer an increased rate of return in exchange for higher volatility. The capital-asset pricing model (CAPM), invented by Sharpe (1970), Lintner (1969), and others, partly states that risks, measured by the ratio of a security's volatility to overall market volatility, are rewarded by an increased rate of return. The *risk premium* is the increase in return in exchange for an increase in variance. Fig. 1 illustrates the historical evidence for the risk-premium idea. The various classes of investments define a line in the space of expected return and standard deviation. In the long term, investors can expect the returns and risks of diverse portfolios to average out along the line.

The annual return on an investment in period t is assumed to take the form of

$$r_t(\alpha,\sigma) = r_o + \alpha\sigma + \varepsilon_t\sigma$$

where the mean is r_o, the standard deviation is σ, ε_t is a standard normal random variable, and α is the risk premium. The ε_t are assumed independent for different periods (years). The parameters fit to the data in Fig. 1 by least squares are $r_o = -0.057$ and $\alpha = 2.34$. The market risk premium, 2.34, is a benchmark for evaluating other investments. A favorable design investment should pay a higher risk premium than the financial market.

Mean and Variance of Investment Returns

Having an expression for market returns and their uncertainty, one can find the variation in the value of a series of investments. A financial investment consists of outlays of money over a period of time, and the value of the returns on the investment in a period t depends on the

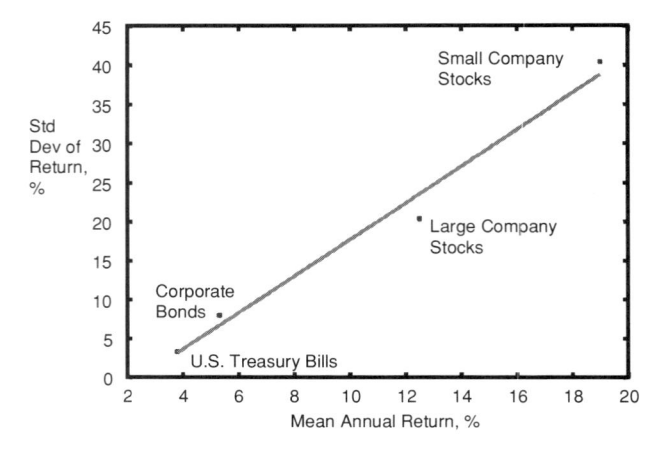

Figure 1. Historical risk-return relationship from data over 1926-1996 from Center for Research in Security Prices.

distribution of the rate r_t. Assume a series of annual investments $\underline{I}$ which earn dividends and a series of additional income $\underline{E}$ from the present to time n. The dividends and income have a total present value

$$R(\underline{I},\underline{E},\alpha,\sigma) = \sum_{t=0}^{n} I_t \exp(-\rho n)\left[\exp(\sum_{s=t}^{n} r_s(\alpha,\sigma))-1\right] + \sum_{t=0}^{n} E_t \exp(-\rho t)$$

where ρ is a discount factor. When the returns from year to year are assumed to be independent and normal, the expected value $E(R)$ and variance $V(R)$ are calculated analytically as functions of $\underline{I}$, $\underline{E}$, α, and σ.

The revenues R of a proposed investment in production facilities may be considered as the *dividends* of a capital investment which has an associated stream of production costs. The present value of these revenues may be compared directly to the dividends expected from market investments, as shown in the analogy in Table 1.

Table 1. Analogy of Manufacturing and Financial Investments.

	Manufacturing Investment	*Financial Market Investment*
I	Capital investments	Investments in securities
R	PV of sum of revenues	PV of sum of dividends
E	Production costs	Management expenses
Mean and variance calculations	Revenue model with polytope volume integration	$E(R(\underline{I},\underline{E},\alpha,\sigma))$ and $V(R(\underline{I},\underline{E},\alpha,\sigma))$
α	Risk premium for comparison with market	Market risk premium from historical data, $\alpha=2.34$
	comparison with market	premium from historical data, $\alpha=2.34$

The benefit of the analogy is to help decide whether a plan for design and production offers a good return relative to risk: The plan is acceptable when α is greater than the financial market premium. The additional ideas required are:

- A model of revenues under uncertainty.
- Methods to obtain the mean and variance of the revenues.
- The solution for a risk premium for comparison with the market.

Revenue model

Revenue from a product sale is generally defined as the price times the lesser of (i) the quantity of product ready to sell or (ii) the demand, the maximum quantity needed by customers. The revenue due to product j sold to customer k at time t is defined as

$$R_{jkt} = p_{jkt}\, min(q_{jkt}, D_{jkt})$$

The quantity of product available to sell is q_{jkt}, the uncertain price is p_{jt}, and the uncertain demand is D_{jkt}. This model assumes a type of product which cannot be held over to the next period if there is excess production. (The case of holding excess production as inventory is a model extension to be discussed in Future Work.)

Forecast Distributions

In practice, demand distributions are generally forecast from historical data for similar products or markets. A range of uncertainty, and also some measure of the relative sources of the uncertainty, is typically the most information available. A way to incorporate this information is to express uncertainties as weighted sums of independent distributions. Demands and prices are composed of uniform random variables u ranging from 0 to 1. A weight f is applied to the u associated with each product j, each customer k and each time interval t.

$$D_{jkt} = \overline{D}_{jkt}(1 + f_j \cdot (u_j - \overline{u}_j) + f_k \cdot (u_k - \overline{u}_k) + f_t \cdot (u_t - \overline{u}_t))$$

$$p_{jkt} = \overline{p}_{jkt}(1 + f_j \cdot (u_j - \overline{u}_j) + f_k \cdot (u_k - \overline{u}_k) + f_t \cdot (u_t - \overline{u}_t))$$

The price and demand are perfectly correlated in this case, meaning that higher demands are forecast to bring higher prices. Different prices and demands with a common index, such as a particular product k, are correlated. Hence for any forecast, all that is required is the expected value and ranges associated with various effects. This formulation of uncertainties facilitates the exact polytope volume integration to obtain moments of the revenue functions, as discussed below.

Polytope Integration

The expected value of a revenue, $E(R_i)$, a square $E(R_i^2)$, or of a cross product $E(R_i\, R_j)$ are all integrals over the independent uniform variables u_j, u_k, and u_t. The expected value integral over a vector of uniform variables $\underline{u}$ is

$$E(R_i) = \int_0^1 d\underline{u} \cdot p_i(\underline{u}) \min(q_i, D_i(\underline{u}))$$

Introducing two new dimensions of integration z_1 and z_2,

$$E(R_i) = \int_0^1 d\underline{u} \int_0^{p_i(\underline{u})} dz_1 \int_0^{\min(q_i, D_i(\underline{u}))} dz_2$$

This is the volume of the polytope defined by the constraints

$0 \le \underline{u} \le 1$
$0 \le z_1 \le p_i(\underline{u})$
$0 \le z_2 \le q_i$
$0 \le z_2 \le D_i(\underline{u})$

The calculation of expected values of squares and cross products is similar but extends to higher dimensions.

Lasserre (1983) provides a recursive algorithm for finding the polytope volume, given the description of the constraint matrix. A software implementation of Lasserre's method by Büeler et al. (1998) made this calculation possible. Matrix sizes formulated for this problem ranged from $n=5$ for $E(R_i(\underline{u}))$ to $n=10$ for cross terms $E(R_1(\underline{u}_1)\, R_2(\underline{u}_2))$.

Obtaining the Risk Premium

The integration methods described above contribute to the exact calculation of the total expected revenue $E(R_{tot})$ and variance $V(R_{tot})$. Given the series of capital costs $\underline{I}$ and expenses $\underline{E}$ required for production, one can subsequently calculate a corresponding risk premium α as the solution of

$$E(R_{tot}) = E(R(\underline{I}, \underline{E}, \alpha, \sigma))$$
$$V(R_{tot}) = V(R(\underline{I}, \underline{E}, \alpha, \sigma))$$

which is a nonlinear system of equations in the unknowns α and σ, to be solved within the domain of the historical σ data.

Consider a plant investment of $1000 required today in order to produce 100 units of a product in one year with an operating cost of $50. Assume a uniformly distributed demand of mean 100, price of mean 2.00, and price and demand ranging from -20% to 20% relative to the mean. By polytope volume integration, the revenue R has E(R) = 185.3, V(R) = 3940. By the solution of the equations for α and σ, the risk premium is α = 3.30. Fig. 2 shows the mean-variance space of the returns on this investment, with a curve for the returns of financial market investments, the point for the plant investment considered, and the range of effects of changing the magnitude of uncertainty between 10% and 40%. For the given forecast, the plant investment is favorable, as it lies to the lower right of the curve. However, for a high enough uncertainty parameter, the plant investment has a worse risk premium than the alternative market investment.

MILP for Design and Planning

To evaluate the capital costs and expenses for any production goals, it is necessary to have a basic model of the supply chain design. A mixed-integer linear programming formulation incorporating the general features of chemical plant design and planning is used to evaluate the capital costs and expenses for various production goals. The problem is to plan for the equipment purchases, materials, and operating costs required to produce several products in different plants and ship them via warehouses to customers in different locations. The production is aggregated at discrete time intervals over a number of years. The plant capacity is underestimated by dedicating equipment to each task.

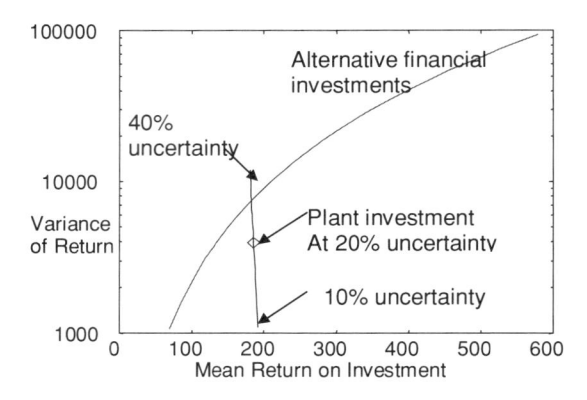

Figure 2. Return and risk of a simple plant design.

Decomposing the Decisions for Risk Measurement

Because of the strong dependence of risk upon many decision variables, there is a need for a decomposition approach which finds favorable solutions to subproblems, and considers combinations of solutions to obtain an overall plan. The following approach is used in the case study presented here.

Level 1: For each product and customer combination jk find a profile of production q_{jkt} over time which has the highest risk premium. Consider only profiles where q_{jkt} is 0 or the mean demand D_{jkt}. Obtain capital costs and expenses by solving the LP relaxation of the MILP, meaning fractional equipment is allowed.

Level 2: For each product j, consider the binary alternatives of choosing or rejecting the orders for each customer k. Investments and expenses are summed using the results the subproblems from Level 1, the mean and variance of revenues are calculated, and the risk premium is found. Among the binary combinations, choose the one with the best risk premium.

Level 3: Consider the binary alternatives of producing or rejecting each product. Find risk premiums as in Level 2. The combination having the highest risk premium is the recommended overall solution. Impose the selected production levels in the full MILP and solve, resulting in a design and planning solution.

When choices are made in binary fashion within the superproblems, the previously calculated mean and variance terms are reused, and the new variance terms involving nonzero production variables need to be calculated.

Case Study Results

The ideas for measuring and managing risks in the chemical manufacturing supply chain are applied here to a case study based on agricultural chemical industry data. It consists of planning for equipment needed to produce three products in batch processes in two plants and ship them through two warehouses to either of two customers. The MILP formulation contains 3785 variables (176 integer), 3549 equations, and 7801 nonzero coefficients. For this study, hypothetical uncertainties are introduced for prices and demands. Tables 2 and 3 show the forecast profiles of the mean demand and price. Table 4 shows the uncertainties as fractions of the mean, used as weights of the uniform random variables u.

Table 2. Mean Demand (kg/year).

Product	t = 1	t = 2	t = 3
P1	7,500,000	8,000,000	8,500,000
P2	5,000,000	5,500,000	6,000,000
P3	0	3,000,000	3,000,000

Table 3. Mean Price ($).

Product	t = 1	t = 2	t = 3
P1	4.50	5.00	5.20
P2	4.00	4.00	4.00
P3	0	8.00	9.00

Table 4. Demand and Price Uncertainties as Fractions of the Mean.

j	f_j	k	f_k	t	f_t
P1	0.05	C1	0.10	1	0.05
P2	0.05	C2	0.08	2	0.10
P3	0.10			3	0.20

The following tables show the results of the different levels of the decomposition. From Level 1, the production profiles, their associated capital costs and expenses, revenues, and their risk premiums are shown in Table 7. Tables 5 and 6 show the revenues and risk premiums resulting from combinations by customers (Level 2) and then by products (Level 3).

Overall the solution time for the current implementation required 6654 sec. on a HP 9000/J210 at 120MhZ, with 99.9% of the time required for polytope volume computations. The time requirements are mainly due to the higher dimension polytopes arising from the cross products of revenue terms with no random variables in common.

The recommended solution is to satisfy the demands of customers C1 and C2 for product P1, as its risk premium is the greatest at 7.9. The solution of the MILP for equipment selection and planning results in costs of $101 M, of which $15.2 M are capital. Hence it favorable to make this investment to capture an expected revenue of

Table 5. Customer Combination Solutions (Level 2).

Product	Customer Combination	Expected Revenue (M$)	Variance of Revenue (10^{12} ^{2})	Risk Premium
P1	C1	98.2	969	6.77
P1	C2	97.9	1039	6.52
P1	C1 C2	196.0	2742	7.91
P2	C1	54.8	325	2.73
P2	C2	54.8	328	2.73
P2	C1 C2	110.0	9418	3.20
P3	C1	42.0	210	3.14
P3	C2	42.1	191	3.31
P3	C1 C2	84.2	570	3.80

Table 6. Product Combination Solutions (Level 3).

Product Combination	Expected Revenue (M$)	Variance of Revenue(10^{12} ^{2})	Risk Premium
P1	196	2742	7.90
P2	110	941	3.20
P1 P2	306	5644	5.30
P3	84	569	3.80
P1 P3	280	4477	6.58
P2 P3	194	2170	3.96
P1 P2 P3	390	8038	5.40

$196 M, with the standard deviation of $52.3M, over three years.

The results show that every combination of solutions produces a risk premium above the market risk premium of 2.34. The results of the higher level problems, in Tables 6 and 7, are useful in gauging the investments for every combination of products with customers. Product P1 has the most favorable effect on a the risk premium of a portfolio of products, while product P3 is the second best, despite its product-specific uncertainty of 0.10 which is twice as much as the uncertainty specific to P1 or P2. Possible reasons why product P2 results in the lowest premiums may be its lower price forecast, higher raw material costs, and higher capital costs relative to the forecast revenue.

With variations in problem data and forecasts, the results for other industrial situations are likely to produce a broad range of risk premiums. With problem solutions potentially falling above and below the risk premium in the financial market, it will be possible to strongly discriminate the most favorable supply chain decisions.

Conclusion

Several ideas from different domains are combined in this work together with ideas specific to a chemical manufacturing supply chain situation. This is the only work as yet to formulate revenues functions which depend on several random variables and can be integrated exactly to obtain the mean and variances of revenues. It has been shown how these ideas are brought together with the notion of a risk premium which could be compared to financial markets as a rational basis for design decisions. A method is demonstrated for decomposing the overall problem into different levels where the risks premiums are evaluated and some elements are passed to the next level while others are rejected. For a simple supply chain with design elements based on actual tasks, recipes, and equipment, the results indicate a preference for investments in capacity for a particular product.

Future work

Opportunities for future work are numerous, and are now being pursued. An important feature to be developed is allow storage of product for sale at any time in the future, meaning the inventory is uncertain. The effect on the revenue function is that sales include current production plus available inventory, resulting in a coupling of revenue terms and all of their underlying uncertainties. Another important new idea is to evaluate future plant investments as options, analogously to stock options in the financial market. The future introduction of a product or expansion of capacity are decisions depending on current choices and are exercised only if a profit is expected. An option value should be used instead of considering committed production levels at future times.

Acknowledgments

This work greatly benefited from the public code for the Vinci software package for polytope volume computations, by B. Büeler, A. Enge, and K. Fukuda of ETH, Zurich, Switzerland.

References

Büeler, B., Enge, A. and K. Fukuda. (1998) *Exact Volume Computation for Polytopes: A Practical Study*. Technical Report, Institute for Operations Research, ETH Zürich.

Grossmann, I. E. and R. W. H. Sargent. (1978) Optimum design of chemical plants with uncertain parameters. *AIChE J.*, **24**, 1021-1028.

Lasserre, J. B. (1983) An analytical expression and an algorithm for the volume of a complex polyhedron in R^n. *Journal of Optimization Theory and Applications* **39**(3) 363-377.

Petkov, C. D. and S. B. Maranas (1997). Multiperiod planning and scheduling of multiproduct batch plants under demand uncertainty. *Ind. Eng. Chem. Res.*, **36**, 4864-4881.

Subrahmanyam, S., Pekny, J. F., and G. V. Reklaitis (1994). Design of batch chemical plants under market uncertainty. *Ind. Eng. Chem. Res.*, **33**, 2688-2701.

Subrahmanyam, S. (1996) *Issues in the Design and Planning of Batch Chemical Plants*. PhD thesis, Purdue University, West Lafayette, Indiana.

Table 7. Time Profile Solutions (Level 1).

Product	Customer	Production Profile (1E6 kg) t=1/2/3	Capital Investment (M$/yr) t=0/1/2/3	Expense Profile (M$/yr) t=0/1/2/3	Expected Revenue (PV M$)	Variance of Revenue (PV 10^{12} $\$^2$)	Risk Premium
P1	C1	7.5 / 8.0 / 8.5	5.76 / 0 / 0 / 0	0 / 15.5 / 15.5 / 15.5	98.2	969	6.77
P1	C2	7.5 / 8.0 / 8.5	5.75 / 0 / 0 / 0	0 / 15.5 / 15.5 / 15.5	97.8	1039	6.52
P2	C1	5.0 / 5.5 / 6.0	28.6 / 0 / 0 / 0	0 / 11.6 / 11.6 / 11.6	54.8	325	2.73
P2	C2	5.0 / 5.5 / 6.0	28.6 / 0 / 0 / 0	0 / 11.6 / 11.6 / 11.6	54.8	328	2.73
P3	C1	5.0 / 3.0 / 3.0	12.3 / 0 / 0 / 0	0 / 8.3 / 8.3 / 8.3	42.0	210	3.14
P3	C2	5.0 / 3.0 / 3.0	12.3 / 0 / 0 / 0	0 / 8.3 / 8.3 / 8.3	42.0	191	3.31

PROCESS PLANT ONTOLOGIES BASED ON A MULTI-DIMENSIONAL FRAMEWORK

Rafael Batres and Yuji Naka
Tokyo Institute of Technology
Yokohama 226-8503, Japan

Abstract

This work is motivated by the goal of improving integration of software components involved in a concurrent process engineering environment. We provide a description of ontologies that are based on a multi-dimensional formalism that represents in an explicit way physical, behavioral, and operational aspects of the plant. Ontologies are formal descriptions that define the vocabulary with which queries and assertions are exchanged among autonomous and collaborative software tools.

Keywords

Ontologies, Knowledge and data exchange, Simulation model interchange, STEP, Agents.

Introduction

To compete in the ever-increasing global markets, as well as to meet increasingly tighter safety and environmental constraints, process industries are being compelled to develop safer, and more operable and reliable plants and processes that result in high-quality products in shorter time and less cost. This has motivated the development of a conceptual framework for concurrent process engineering (McGreavy et al., 1995; Lu, et al., 1997) that integrates life-cycle activities, information and software tools from the very beginning of the design process. Integration of information requires knowledge and data to be shared and exchanged easily at all stages of the life-cycle, as well as between life-cycle stages, such as between design and operations. To support such integration, an unambiguous description of the worldview is needed that both computers and humans can understand. The specification of such a worldview is referred to as an ontology.

This paper proposes an approach in which process engineering knowledge and data are organized in ontologies based on a multi-dimensional framework that represents in an explicit way physical, behavioral, and operational aspects of the plant. We believe that ontologies based on this formalism (our worldview) will facilitate the sharing and exchange of data, knowledge, and models between tools such as CAD systems, simulation software tools, operating procedure synthesis programs, and control systems. The multi-view capabilities of ontologies are utilized to formally describe the objects of each dimension and their relationships.

Ontologies

In philosophy, Ontology is the study of questions about the existence of particular kinds of objects. In engineering, an *ontology* can be defined as a formal and explicit representation of unambiguous definitions of objects, concepts and the relationships that hold them.

We construct an ontology by defining terms such as classes of objects, their taxonomy, relations, and axioms. As in object-orientation, a class represents a category of similar terms that share a set of properties. A relation describes the association between two or more terms. Axioms are usually expressed as first order logic expressions that help to provide a more precise meaning of the terms and constrain their interpretation. In computer readable and computer processable ontologies, all the terms of the ontology ensure that applications pass legal data for semantic consistency with the definitions in ontologies. Without semantic checking and validation, misinterpretation of data could lead to catastrophic consequences. For example, without the adequate definition of "pipe", "line" and "segment" people or software components may not be able to exchange and share information in a consistent way.

In other words, an ontology provides the data structure and semantics that ensures the validation of data or knowledge that is to be exchanged. This requires two ontological commitments: the ontology should be consistent to the conceptual framework behind it and software components should be configured to be consistent with the ontology in order to share information and knowledge.

A Multi-dimensional Conceptual Framework

In order to support different aspects of the design, we have proposed a multi-dimensional formalism in which physical, behavioral, and operational aspects or dimensions of the plant, process and product are explicitly represented (Batres et al., 1999a). This formalism has been used as a conceptual framework for the development and implementation of a number of ontologies. Objects from each dimension can be classified into proposed, intended, assumed (what if?) or actual. The basic important concepts of these ontologies are described below.

Plant Structure Ontology

The plant structure ontology belongs to the physical dimension and refers to the description of the components of which the plant is built, their topology, their mereology and their spatial representation. The taxonomy and relations of this ontology are shown in Fig. 1. The most basic components are *elementary-structural-entities*. Assemblies of *elementary-structural-entities* are *composite-structural-entities*, which can be *functional* or *operational*. Instances of *composite-structural-entities* contain one or several other *composite-structural-entities*, which contain one or several *elementary-structural-entities*. Definitions of mereological and topological relations are complemented by axioms such as:

$\forall x,y \ \text{part\_of}(x,y) \Rightarrow \neg \text{part\_of}(y,x) \land \neg(\exists x \ \text{part\_of}(x,x))$

$\forall x,y \ \text{is\_connected}(x,y) \Rightarrow \text{component}(x) \land \text{component}(y) \land \neg\text{subpart\_of}(x,y) \land \neg\text{subpart\_of}(y,x)$

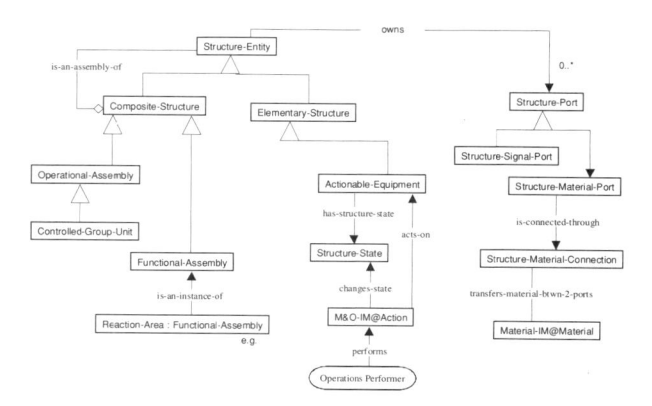

Figure 1. Structure taxonomy and relations.

Material Ontology

Construction and process materials belong also to the physical dimension. *Material* (Fig. 2) is defined as a sample of matter or substance. It has *material-properties* that include *specific-enthalpy*, *fugacity*, etc. A *material-phase* is a class of material and part of a certain amount of material that is in contact with one or more other phases all of which have uniform and homogeneous properties in each of them. Instances of *process-material* represent fluids and solid-materials that can be contained in or transported by instances of equipment. A material may have associated one or more instances of *component-in-mixture-properties* that contain information about components in a mixture. In this ontology pressure is the force per unit area that a material exerts on its surroundings while temperature is considered as a property that results from molecular interactions as in the kinetic molecular theory.

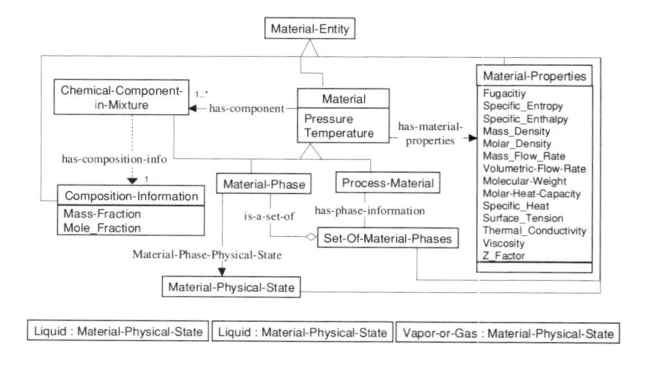

Figure 2. Material taxonomy and relations.

Behavior Ontology

The behavior ontology describes the conceptualizations about entities with properties that come out as the result of the interaction of instances of the plant structure ontology and instances of process material (Fig. 3). The state of the behavior that takes place in a physical component is described by *state-descriptors*. The state-descriptors contain a number of attributes that are updated by a simulation engine. The behavior that takes place in an instance of a structure is formulated through the use of another kind of behavior entity called *metamodel*. Metamodels can be interconnected via ports for transferring energy and mass flow. In this context, connecting ports is comparable to combining sets of equations that are assembled to mimic phenomena that occurs under the equipment boundaries. Metamodels that are not composed of other metamodels are called *protomodels*. Examples of protomodels are bond-graph elements (Karnopp and Rosenberg, 1975). For a given piece of equipment such as a distillation column, designers can combine multiple metamodels or assign an existing one to represent the proposed, assumed or actual behavior and not only functional information (intended behaviors).

Examples of such metamodels are descriptions of physicochemical phenomena such as flashing, mass flow, energy flow or chemical reactions. The main reason that motivates us to represent phenomena this way is because it allows us to create a library of reusable models that can be exchanged without modifying the structural or operational knowledge.

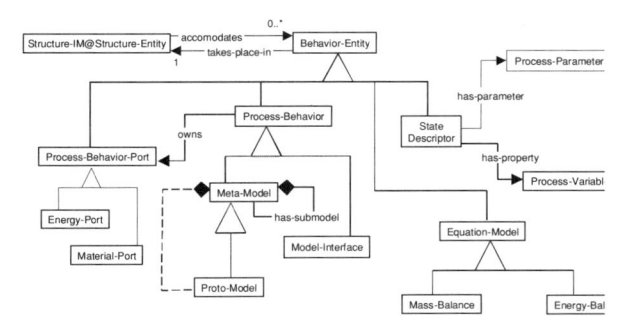

Figure 3. Behavior taxonomy and relations.

Management and Operation Ontology

The management and operation ontology specifies classes, axioms and relations to describe objects associated to a variety of control and operational tasks. These tasks run from management, planning and scheduling, and plant-wide control through to local advanced and regulatory control.

From business plans to plant operations, engineers and management people deal with *plans* to achieve economic objectives constrained to a number of factors. Plans are composed of *activities* that can be decomposed further into other plans. Activities are carried out over a time interval by a number of *activity-performers* that accomplish *actions*. Actions are deliberate and occur in a discrete fashion. An action can initiate another action or instruct *actuators* that directly operate over flows of materials, products, energy, information or money. At the operation and control level an action changes the state of the structure (such as opening a valve) and the behavior comes out from such changes.

Mappings between Dimensions

Mappings between dimensions are accomplished through relations such as *accommodates, transports, contains-process-material, conveys, etc*. From example, *Contains-process-material* is used to describe an instance of elementary-structural-entity that stores or contains process-material. Its inverse relation *is-contained-in* maps zero, one or more instances of process-material to an instance of elementary-structural-entity. For example, reference to the volume of the bulk of material in tank-1 is done with a sentence like *the volume of the process-material-inventory that is-contained-in tank-1*. Another example is the relation *Accommodates,* which maps an instance of elementary-structural-entity to zero, one or more instances of behavioral-entity (behavioral-entity is a parent class for metamodels and state-descriptors). Its inverse relation is *taking-place-in*. Similar definitions are given for other relations.

Validation of the Multi-dimensional Ontologies

A testbed of an engineering environment was developed in order to validate and refine the ontologies. The testbed is composed of a number of tools that are distributed in a networked environment. A schematic editor implemented in G2 (Gensym, 1997) is used for the development of PFD and P&ID diagrams. A semi-quantitative dynamic simulator and an operating procedure planning modules are responsible of startup and shutdown operations (Fig. 4). A number of other programs that implement topological and mereological algorithms are implemented

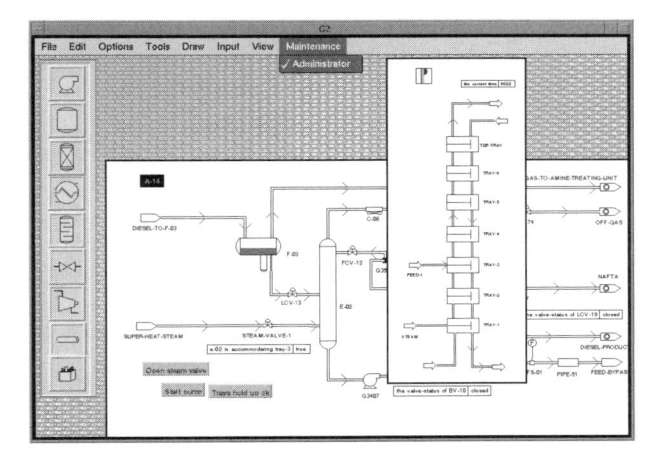

Figure 4. A stripper (instance of elementary-structural-entity) and its associated meta-models.

in CLIPS (CLIPS, 1997), and Java.

To illustrate the information flow between the tools Fig. 5 shows an example of a data transaction between a *simulator* and the operations planning module. The exchanged data are instances of the classes that are defined in their respective ontologies. In this example, the operations planner sends a message asking the simulator for the current state of the process material that is contained in tank-1. The simulator then sends a message back to the operations planner with the required data. Compiling information from other sources, such as the plant topology and mereology, the operations planner uses its planning algorithms and inference mechanisms to determine that valve-1 should be opened. Consequently, the operation planner sends a message to the simulator that instructs it to open valve-1.

A software component that exchanges messages using a predefined protocol over the network is called an agent (Batres *et al.*, 1999b). An agent communication protocol adds another semantic level for interoperability. With such a protocol it is possible to make a statement such as *(open valve-1)* a question (is valve-1 open?) or an assertion (valve-1 is open).

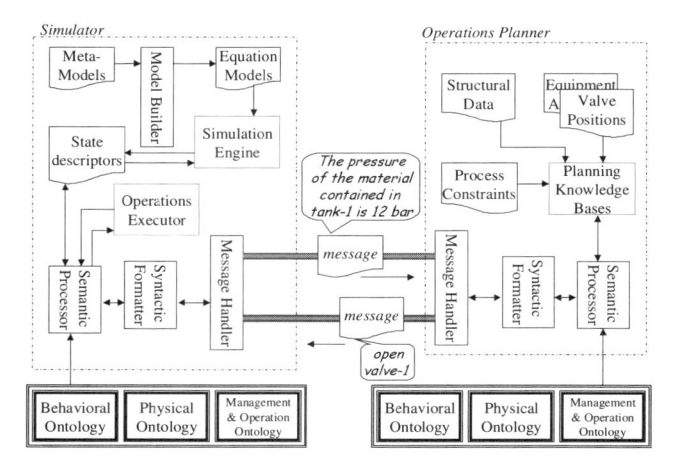

Figure 5. An example of a system architecture that illustrates the use of the multi-dimensional ontologies.

Constructing the Ontology

In order to construct the ontologies, we decided to use the Stanford ontology server (Farquhar, Fikes and Rice, 1997). The Stanford ontology server (Fig. 6) can be utilized remotely from a web browser and includes a number of tools for editing, browsing, and consistency checking as well as an HTML report generator. The editing tool allows developers to create and update classes,

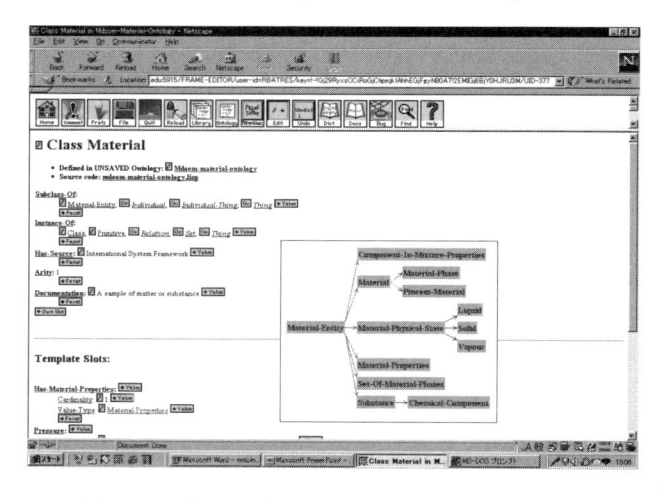

Figure 6. Screendump of the Stanford Ontology Editor showing the definition of material.

sub-classes, attributes, relations and axioms. The editor automatically converts the user's input to a formal representation called Ontolingua. Ontologies stored in this format can be translated to several different languages, including CLIPS, KIF (Genesereth and Fikes, 1992) and CORBA IDL (Zhonghua and Duddy, 1996). The modeling approach allows the extension and construction of ontologies from already existing ontology libraries, which results in less time spent in modeling other related domains. With this tool, it is possible to create an ontology that references the terms defined in another ontology and vice versa. This represents a natural approach for the modeling of the ontologies based on the multi-dimensional conceptual framework.

Axioms are defined in KIF (Knowledge Interchange Format) on which the ontology server is based. KIF provides a lisp-like syntax for representing first-order logic sentences.

Related Work

Functional Representation

Much research has concentrated on capturing functional knowledge in the design process (Elsass et al., 1998). Functional representation is a formalism that hierarchically represents the functions (what the device is intended to do), behavior and structure of a device following a top-down approach. In functional representation the behavior is how the device does what it does, while the behavior described in this paper is centered on the phenomena of the material, which is the basis for model interchange. The behavior of the equipment is described in terms of causal process descriptions (CPD) that represent state transitions of the device. In this approach it is not clear how to guide the simulation through the state transitions provided by the CPD, something that the multi-dimensional formalism complements with the management and operation dimension.

STEP

One of the uses of an ontology is as a standard where the terms related to design activities and information objects are defined. This results in something similar to STEP (Standard for the Exchange of Product Model Data) (NIST, 1999) which is an ISO initiative. ISO 10303 is the official designation for this standard. In ISO 10303 *Application Protocols* (AP) are parts of the standard that define data models for a certain application domain. As in the case of ontologies, each AP defines classes of objects, their taxonomy, and relations. Local and global rules are used to constrain the validity of the values for each object. From the point of view of information modeling, ontologies make a commitment to an unambiguous representation of the terms of a specific domain of discourse, while STEP makes a commitment to a common data format.

There are three APs for exchanging process engineering data: AP 231 (Process Engineering Data - process design and process specification of major equipment), AP 227 (Plant Spatial Configuration) and AP221 (Functional data and their schematic representation for process plants). Of these, AP231 and AP221 are still under development. One difficulty faced by current STEP developers in the process-engineering domain is the inadequate harmonization between APs. In our opinion, the problem lies in the use of the concept of unit operation, which is a functional representation with both structural and behavioral attributes. We believe that an explicit separation between structure, behavior and operation as presented in this paper might be an alternative for harmonizing the standards for both design and operations (an actual behavior can be interchanged for an expected behavior).

Conclusions

This paper presents a general description of ontologies in which process engineering knowledge and data is organized in such a way that facilitates the exchange of data/knowledge between tools such as CAD systems and simulation software tools. These ontologies are based on a multi-dimensional framework that represents in an explicit way physical, behavioral, and operational aspects of the plant. We believe that the separation of behavior and physical structure will make it easier to exchange simulation models, while keeping intact the plant structure (visualized on the simulator screen) for consistency with the PFD and P&ID diagrams developed using intelligent CAD tools. Furthermore, the proposed ontologies avoid unnecessary modifications to the simulation models by representing operations and management information in a separate ontology, which greatly facilitates the utilization of the same behavior models during the simulation of different operation and control modes for a certain design alternative.

Acknowledgements

This work was funded by Japan Society for the Promotion of Science, Japan.

References

Batres, R., M. L. Lu and Y. Naka (1999a). A multidimensional design framework and its implementation in an engineering design environment. *Concurrent Engineering: Research and Applications*, **7**, 1, 43-54.

Batres, R., S. Asprey, T. Fuchino, and Y. Naka (1999b). A KQML multi-agent environment for concurrent process engineering. *Comp. Chem. Eng.*, **23**, S653-S656.

CLIPS (1997). *CLIPS 6.05 Reference Manual*.

Elsass, M. J., D. C. Miller, J. R. Josephson, and J. F. Davies (1998). A process plant knowledge repository for multiple applications. In *Foundations of Computer-Aided Process Operations*, J. F. Pekny and G. E. Blau, editors. *AIChE Symposium Series*, **94**, 487-493.

Farquhar, A., R. Fikes, J. Rice (1997). *Tools for Assembling Modular Ontologies in Ontolingua. Technical Report KSL-97-03*, Stanford University, Knowledge Systems Laboratory. Available on the World Wide Web from: ftp://ftp-ksl.stanford.edu/pub/KSL_Reports/KSL-97-03.ps.gz

Genesereth and R. Fikes (1992). *Knowledge Interchange Format Version 3.0 Reference Manual*. Technical Report Logic-92-1. Stanford University Logic Group. Available on the World Wide Web from: http://logic.stanford.edu/papers/kif.ps

Gensym Corporation, Inc. (1997). *G2 Version 5.0 Reference Manual*.

Karnopp, D., R. Rosenberg (1975). *System Dynamics: A Unified Approach*. John Wiley, New York.

Lu, M. L., R. Batres, H. S. Li, and Y. Naka (1997). A G2 based MDOOM testbed for concurrent process engineering. *Comp. Chem. Eng.*, **21**, S11-S16.

McGreavy, C., X. Z. Wang, M. L. Lu, Y. Naka (1995). A concurrent engineering environment for chemical manufacturing. *Concurrent Engineering: Research and Applications*, **3**, 4, 281-293

NIST (1999). *The STEP Project*. Available on the World Wide Web from: http://www.nist.gov/sc4/www/stepdocs.htm

Zhonghua Y. and K. Duddy (1996). CORBA: A platform for distributed object computing. *ACM Operating Systems Review*, **30**, No. 2. Available on the World Wide Web from: http://www.infosys.tuwien.ac.at/Research/Corba/archive/intro/OSR.ps.gz

MODEL.LA: A PHENOMENA-BASED MODELING ENVIRONMENT FOR COMPUTER-AIDED PROCESS DESIGN

Jerry Bieszczad, Alexandros Koulouris, and George Stephanopoulos
Massachusetts Institute of Technology
Cambridge, MA 02139

Abstract

A computer-aided modeling environment, named MODEL.LA, has been developed which integrates a physico-chemical phenomena-based modeling language, for representing chemical process models, and modeling logic, for constructing the underlying models. The fully declarative modeling language provides means for model development at the high level of chemical engineering knowledge. Modeling logic is used to detect model inconsistencies and incompleteness, and to automatically derive the requisite mathematical model from the phenomena-based model description. The MODEL.LA modeling environment provides fast, reliable generation and solution of physico-chemical phenomena-based models of steady-state or dynamic systems of arbitrary structure and spatial distribution, hierarchical levels of detail, and coexisting multi-context depictions.

Keywords

Phenomena-based modeling, Modeling environments, Hierarchical design, Dynamic simulation.

Introduction

Effective chemical process design supplements human knowledge and expertise with computational power. Design engineers articulate ideas about process structure, behavior, and operation in terms of models. These models are solved by a computer, enabling engineers to evaluate process economics, screen alternatives, and optimize operating conditions. Thus, a computer-aided tool that facilitates fast, flexible, reliable, and high-level model development is critical to the design process.

Current computer-aided modeling tools cannot adequately address these needs. Modular flowsheet simulators are relatively easy to use, but their blackbox, *unit operation-based* models are inflexible and often require a level of detail that is more or less than what is suitable at a particular stage of design. Alternatively, process designers employ an *equation-based* approach, where mathematical equations for each alternative must be derived and solved using a spreadsheet, mathematical software, or some programming language. Unfortunately, this method is error-prone, difficult, and time consuming.

In order to address these issues, a new paradigm, *phenomena-based* modeling, has been proposed which combines the ease of use of flowsheeting tools with the power and flexibility of an equation-based approach.

Phenomena-Based Modeling

The MODEL.LA language (Stephanopoulos *et al*, 1990) introduced the concept of physico-chemical phenomena-based process modeling. It represents models in terms of knowledge about the physical and chemical structure and behavior of a process. From this description, model equations are automatically generated. Other researchers (Marquardt, 1996; Perkins *et al*, 1996; Preisig, 1995) have proposed similar ideas. However, engineers have been unable to embrace the phenomena-based modeling methodology due to two key reasons:

1. The impact of assumptions regarding a phenomena-based model on the resulting

mathematical model has not been formally explained. Rather, one must rely on intuition to interpret these models.
2. The integration of these ideas into computer-aided modeling tools has not passed the conceptual prototype stage. Pantelides and Britt (1995) stress that the implementation and practical application of these ideas is essential for assessing their value.

These two issues have been resolved by development of the MODEL.LA computer-aided modeling environment. This paper presents an overview of this environment. It discusses the elements of the phenomena-based modeling language, the logic used to automatically derive mathematical models, and the structure of the software and external components that it interfaces to.

Modeling Language Elements

MODEL.LA is a fully declarative, high-level modeling language rooted in the principles of chemical engineering science. The modeling elements of the language are based on a natural characterization of chemical processes. These elements provide a rich vocabulary for describing the topological and hierarchical structure of lumped or spatially distributed processes, thermodynamic and physical characterizations of materials, and mechanistic characterizations of physico-chemical phenomena assumed to occur in the process. The resulting models provide explicit documentation of the underlying assumptions and are much easier to develop, edit, analyze, and reuse.

Topological Structure

The topological structure of a model defines the systems of interest and how they interact. These are represented by the modeling elements *modeled-units* and *fluxes*. Modeled-units represent control volumes. Fluxes represent the transport of material, energy, or chemical species between two interacting modeled-units.

Hierarchical Structure

The hierarchical structure of a model defines how composite modeled-units are conceptually decomposed into sets of more refined modeled-units. The state of a composite modeled-unit may be viewed abstractly or as an aggregate of its subunits. Hierarchical modeling allows the modeler to control the complexity of a model by lumping related units into abstract units; to generate models at multiple resolutions, depending on the level of detail desired; and to incrementally develop a process model during process design where the behavior of a process model at a given level dictates the refinements at a subsequent more detailed level (Douglas, 1988).

Two special cases of the composite modeled-unit are the *staged* and the *distributed* modeled-units. In a staged modeled-unit, a unit is refined into a series of identical subunits. Assumptions made for the representative subunit are automatically propagated to the other subunits. A distributed modeled-unit represents a process unit that has internal spatial distribution. It is modeled using a differential element subunit, along with boundary subunits for each of the distributed dimensions. The balance equations for such a unit are partial differential equations (PDEs). Differential fluxes for each distributed dimension may be added to the differential element subunit. Fluxes to the boundary elements determine the boundary conditions of the PDE balance equations.

Material Characterization

An elementary modeled-unit may be characterized by a *material-content*. A material-content is characterized by the *phases*, *species*, and *reactions* assumed to occur within it. A vessel *geometry* may also be declared. A geometry introduces relationships between the volume and height of a material-content, and allows the port position of fluxes to be specified.

A material-content represents one or more thermodynamic phases at equilibrium. Phases are characterized by the species and reactions within them. Species represent the chemical compounds present within a process. Reactions represent the chemical reactions assumed to occur within a process.

Phenomena-Based Mechanistic Characterization

A phenomena-based mechanistic characterization of the topological, hierarchical, and material modeling elements introduces additional modeling assumptions. The associated modeling elements include *transport-mechanisms* (for fluxes), *rate-laws* (for reactions), and thermodynamic *property-models* (for phases).

Modeling Logic

The *syntax* of the MODEL.LA modeling language is defined formally using a context-free grammar. Instances of models are represented as semantic networks (i.e., labeled digraphs). However, syntax and common-sense explanations of the modeling elements cannot adequately characterize the exact meaning of a phenomena-based model. To complete the specification of MODEL.LA, the *semantics* of the language have been defined formally in (Bieszczad, 1999) using the notation of first order predicate logic. Each task in the modeling activity is described as an logical operator that acts on the model digraph. The modeling logic specifies operators used for model construction, detection of model inconsistencies and incompleteness, and automated generation and explanation of the resulting mathematical model.

Model Construction

To avoid the use of static textual input files, logical operators for construction of a phenomena-based model are defined. The modeling activity is characterized as a sequence of hierarchical tasks (e.g., *refine-modeled-unit*, *characterize-phase-behavior, define-material-geometry,* etc.). In response to the modeler's decisions, these tasks operate on the model digraph to automatically generate the underlying MODEL.LA language-based description of the model. These operators allow the modeler to define the model interactively and gradually, provide feedback on the validity of assumptions, capture the rationale of decisions made, and provide an explicit record of the modeling activity.

Detection of Incompleteness and Inconsistencies

Logical operators are also defined to detect missing assumptions, which indicate model incompleteness, and structural or behavioral discrepancies, which indicate model inconsistencies.

Mathematical Model Generation and Explanation

The translation of the phenomena-based model description into a equation-based representation is also defined by a set of logical operators based on the principles of chemical engineering science. Conservation of mass and energy is expressed by balance equations for each modeled-unit. Equilibrium relationships are generated for the phases of each material-content. Thermodynamic and physical property relationships are derived for each phase and convective flux. Transport-mechanisms, rate-laws, and property-models include additional mechanistic and empirical relationships in the model. Geometry assumptions also introduce conditional equations into the model, where the presence and state of a flux out of a system is dependant on the height of the material and flux outlet position.

By recording and linking the sequence of operations used during model derivation to the mathematical model, it is also possible to explain the resulting variables, terms, and equations in terms of specific modeling assumptions.

Modeling Environment

In order to provide an experimental framework to test the phenomena-based modeling concepts, a PC-based computer-aided modeling environment has been developed which integrates the modeling language and logic of MODEL.LA. The three key elements of this environment are the *Model Generator, Properties Manager,* and the *Numerical Engine.*

Model Generator

The Model Generator incorporates the modeling logic operators into a software environment. It provides an easy-to-use interface for the graphical and interactive declaration of phenomena-based models. Hierarchical and topological structure are declared intuitively using flowsheets as illustrated in Fig. 1. Assumptions regarding materials and phenomena-based mechanistic characterizations are declared using context-sensitive menus and dialogs.

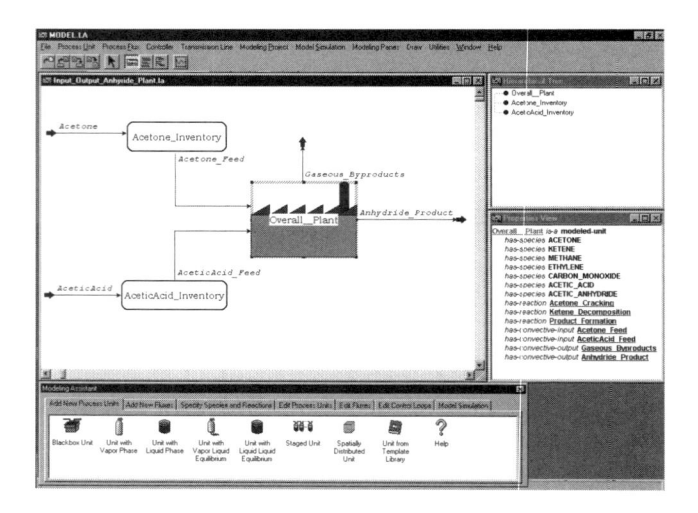

Figure 1. MODEL.LA user interface.

The model definition is checked for incompleteness and inconsistencies. Model equations are then generated based on the derivation context (e.g., steady-state or dynamic, level of resolution, intensive or extensive state characterizations, etc.) specified by the modeler. The modeler may also supplement the mathematical model by declaring *control structures* and *operating task schedules.*

Control Structures

Control structures may be added directly to a process flowsheet in order to introduce externally forced process dynamics. Control structures are constructed through declaration of *i.) process controllers,* which use *control laws* to determine the values of a set of manipulated variables from a set of measured variables, and *ii.) transmission lines,* which establish the linkage between variables measured or manipulated by a controller with process variables selected by the modeler.

Operating Task Schedules

The declaration of operating task schedules introduces a hybrid discrete/continuous behavior to a process model. Schedules are constructed graphically by specifying a hierarchical network of *events* and *actions.* When a state or time condition associated with an active event comes true, the subsequent action is triggered, resulting in the discrete manipulation of some process variable. The sequence of events in a schedule may include loops, conditional branches, and parallel paths.

Properties Manager

The Properties Manager adds additional equations describing the thermodynamic and physical properties of process materials. These equations are based on the equation-of-state models selected for each phase and pure species component data from a database.

Numerical Engine

In the Numerical Engine, dialogs help the modeler select structurally consistent design variables and specify initial guesses. For dynamic models, an index analysis is also performed and the modeler is assisted in specifying consistent initial conditions. The mathematical model is then formulated as a gPROMS (Barton, 1992) input file, and solved using gPROMS. Results are read back into MODEL.LA and displayed in graphical or tabular format.

Software Interaction

The MODEL.LA environment is an integrated tool and interfaces to several other software components:

1. The Properties Manager accesses the DIPPR database which contains data on 36 constant and temperature dependant properties of over 1400 chemical species.
2. The Numerical Engine accesses the gPROMS equation-based solver to solve the generated integral, partial differential, differential, and algebraic model equations.
3. OLE Automation is used to load and display graphical results in Microsoft Excel.
4. Direct linkage on the Model Generator flowsheet to other gPROMS model definition files provides access to the open architecture of the solver. This may be used to send, retrieve, and incorporate runtime data from external applications during simulation.

Case Studies

The MODEL.LA modeling environment has been applied to several case studies, including the hierarchical design of plants for the hydrodealkylation of toluene and the production of acetic anhydride (Fig. 2). Rigorous dynamic distillation process models have been developed, along with models of dynamic 1-D and 2-D spatially distributed reaction and separation processes. These examples have shown that MODEL.LA provides fast, reliable, generation of models of static or dynamic systems of arbitrary structure, hierarchical levels of detail, and coexisting multi-context depictions.

Conclusions

The high-level phenomena-based modeling approach of MODEL.LA can have a unique impact on process design by: *i.)* accelerating process model development and greatly increasing the number of alternatives considered, *ii.)* allowing experts in varying backgrounds to readily contribute to a design in parallel, *iii.)* providing complete flexibility in process specification, since models are not limited to an existing library, *iv.)* facilitating evolutionary process development by allowing addition of detail in a hierarchical manner, *v.)* enabling process models to be used and reused in different contexts (e.g., process design, steady state or dynamic optimization, training, controller design, and *vi.)* retaining the knowledge and assumptions behind model development.

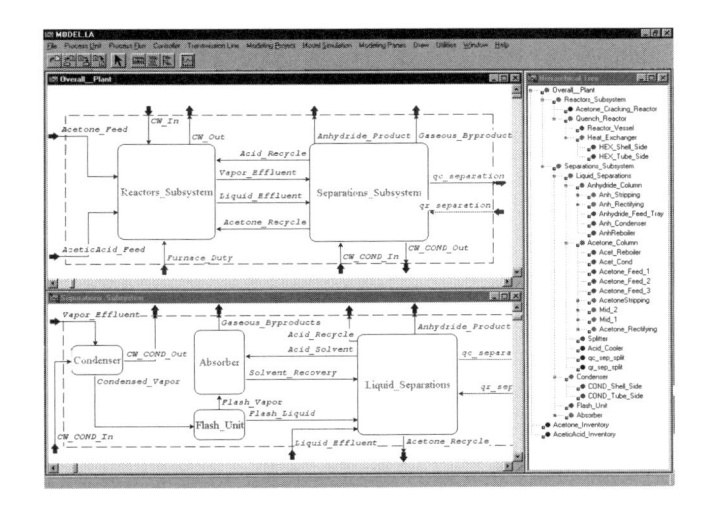

Figure 2. Acetic anhydride plant model.

References

Barton, P.I. (1992). *The Modelling and Simulation of Combined Discrete/Continuous Processes*. Ph.D. Thesis, Imperial College of Science, Technology and Medicine.

Bieszczad, J. (1999). *A Framework for the Language and Logic of Computer-Aided Phenomena-Based Process Modeling*. Ph.D. Thesis, Massachusetts Institute of Technology, in preparation.

Douglas, J. M. (1988). *Conceptual Design of Chemical Processes*. McGraw-Hill, NY.

Marquardt, W. (1996). Trends in computer-aided process modeling. *Comp. Chem. Eng.*, **20**(6/7) 591-609.

Pantelides, C. C. and H. I. Britt (1995). Multipurpose process modeling environments. In L. T. Biegler and M. F. Doherty (Eds.), *Foundations of Computer-Aided Process Design*, AIChE Symposium Series, **91**, 304, 128-141.

Perkins, J. D., R. W. H. Sargent, R. Vazquez-Roman and J. H. Cho (1996). Computer generation of process models. *Comp. Chem. Eng.*, **20**(6/7) 635-639.

Preisig, H. A. (1995). MODELLER—An object-oriented computer-aided modelling tool. In L. T. Biegler and M. F. Doherty (Eds.), *Foundations of Computer-Aided Process Design*, AIChE Symposium Series, **91**, 304, 328-331.

Stephanopoulos, G., G. Henning, and H. Leone (1990). MODEL.LA. A modeling language for process engineering—I: The Formal Framework. *Comp. Chem. Eng.*, **14**(8) 813-846.

INFORMATION MODELS FOR THE ELECTRONIC STORAGE AND EXCHANGE OF PROCESS ENGINEERING DATA

Neil L. Book
Department of Chemical Engineering
University of Missouri-Rolla
Rolla, MO 65409

Abstract

Information models are graphical and/or lexical constructs that define a body of data and its metadata (data about data). The metadata describe the relationships and dependencies amongst the elements in the body of data.

The Process Data eXchange Institute (pdXi), an Industry Technology Alliance of the American Institute of Chemical Engineers (AIChE), has sponsored the development of information models for process engineering data. The Standards for the Exchange of Product Model Data (STEP or ISO 10303) were adopted by pdXi as the methods and protocols for the storage and exchange of data captured by their models. The pdXi information models have been submitted to STEP and are under review to become a part of the international standard.

An application programming interface (API), software that executes the storage and exchange methods using the STEP protocols, has been developed for the pdXi information models. The API is designed to be used by individuals without extensive knowledge of the information models and/or the STEP methods and protocols. The API and the information models upon which the API is based enable the electronic storage and exchange of process engineering data.

Keywords

Information models, Process Data eXchange Institute, Standards for the Exchange of Product Model Data, Application programming interface.

Introduction

The Process Data eXchange Institute (pdXi) was formed following the FOCAPD '89 meeting as a Sponsored Reseach Program (now an Industry Technology Alliance) of the American Institute of Chemical Engineers (AIChE). It is a consortium of companies that provides funding and technical support to create an electronic storage and exchange system for process engineering data. The electronic data exchange system was developed using the methods and protocols of the Standards for the Exchange of Product Model Data (STEP or ISO 10303). The STEP standards (ISO, 1994a) require the development of information models for the data to be exchanged.

Information Models

Information models are graphical and/or lexical constructs that define a body of data and its metadata (data about data). The metadata describe the relationships and dependencies between the elements in the body of data. The Object Modeling Technique (OMT) of Rumbaugh et al. (1991) was used as the information modeling method. OMT is an object-oriented, graphical method for information modeling.

The pdXi Information Model

The OMT notation is easily learned so that individuals,

other than information scientists, can construct and/or evaluate information models that capture data in their domain of expertise. An OMT model of the process engineering domain was constructed and reviewed by representatives from the pdXi member organizations and by external domain experts. The model is large, containing over 500 classes, so the model was developed as 10 interlinked submodels: the Planning Level, Process Materials, Unit Operations, Physical Properties, Process Vessel, Material Transfer Equipment, Heat Transfer Equipment, Separation Tower Equipment, Simplified Geometry, and Metadata submodels. Revision 0 of the submodels was approved by pdXi in 1994 (Book, et al., 1994, AIChE, 1997a). The Planning Level submodel contains the classes at the highest level of abstraction, therefore, external references in the other submodels are ultimately resolved by classes in it. The Physical Properties submodel captures data describing the physical properties of substances. Substances can be pure substances, psuedochemical species, or mixtures. The Process Materials submodel captures data describing the quantities and flow rates of substances in a chemical process. Data associated with engineering models of process equipment, such as unit operations models in process simulators, are captured in the Unit Operations submodel. The Process Vessel, the Heat Exchange Equipment, the Material Transfer Equipment, and the Separation Tower submodels capture process engineering data describing these types of equipment. The Simplified Geometry and Metadata submodels are OMT models of data from other parts of STEP that are referenced in the pdXi model. The resolution of external references is not mandatory, so the submodels can be used individually to exchange data in their subdomain. There is a complete glossary that defines each class, each attribute, each role, and, for enumerated attributes, each enumeration in the model.

STEP

STEP is composed of a series of parts. There are parts that describe the methods and protocols that must be used, generic resources (data, such as units of measure and geometry, that are common to a number of domains), and application protocols (AP) where data from a domain, such as process engineering, are defined.

There are three candidate application protocols being developed for process data: AP 221, AP 227, and AP 231. The focus of AP 221 is the piping and instrumentation diagram. The three-dimensional representation of piping systems is the focus of AP 227. The pdXi models are the basis for AP 231.

An activity model (Sitton et al., 1993, AIChE, 1997b) for process engineering was developed using the IDEF0 notation (USAF,1981). The activity model describes the activities in process engineering and the flow of information between the activities. An activity model is a required component of an AP.

There are many stages that a candidate application protocol has to pass in order to become an official part of STEP. There are three stages that require international review and balloting: the committee draft (CD), the draft international standard (DIS), and international standard (IS). AP 231 has successfully completed the balloting as a committee draft (ISO,1998).

EXPRESS

EXPRESS (ISO, 1994b) is the information modeling language of STEP. EXPRESS is a lexical syntax (an EXPRESS model looks like a computer program) that has many, but not all, of the characteristics of object-oriented models. Graphical models are much easier to construct and review than lexical models so they are generally developed first and then translated to the lexical syntax. Translation of models from OMT to EXPRESS (or vice versa) is a relatively routine task.

Application Programming Interfaces

An application programming interface (API) is software that provides an interface between computer programs (applications) and data storage technologies. For example, structured query language (SQL) is an API for a database. With SQL, applications can create objects within the database, define attributes of the objects, and instantiate or query values of the attributes.

Three types of data storage technologies are used in STEP: 1) the STEP physical file; 2) the STEP working form; and 3) the PDES/STEP database (PSDB). The STEP physical file is an ASCII file for storing instances of data defined by an EXPRESS information model. Part 21 of STEP (ISO, 1994c) defines the mapping from an EXPRESS information model to records in the STEP physical file. The STEP working form is an in-memory representation of data defined by an EXPRESS model. The PSDB is a mapping of an EXPRESS model to objects and attributes in a database. Specific mappings from EXPRESS to STEP working form and to PSDB are not defined by STEP.

STEP contains a Standard Data Access Interface (SDAI) specification (ISO, 1994d) and bindings to programming languages that guide the development of an API for data in STEP data storage technologies.

The pdXi Prototype API

A prototype API was developed by pdXi and demonstrated at FOCAPD '94. The prototype API was created from public-domain software developed at the National Institute of Standards and Technology (NIST). The NIST software was developed for demonstration of the STEP methodology. The software included an EXPRESS compiler (Fedex+) and a toolkit that enabled the instantiation of STEP physical files through a graphical user interface (GUI). An EXPRESS compiler produces a working form from an EXPRESS information model. In the case of Fedex+, the working form (an EXPRESS working form) is a set of C++ classes and methods for storing instances of objects defined in the EXPRESS model. A STEP working form is created by populating the EXPRESS working form with data, either manually through the GUI or by reading a STEP physical file. A utility in the toolkit allowed the writing of a STEP working form to a STEP physical file. The pdXi prototype API was created by replacing the GUI with a C++ language binding. The prototype used a small portion of the pdXi information model (approximately 50 classes describing streams in process simulators) to demonstrate the exchange of process engineering data using STEP methods and protocols.

Creation of the C++ language binding involved writing code to provide functions to get and put an instance of each class in the information model. This task was performed manually for the small portion of the information model in the pdXi prototype API. However, an automation tool is required for information models that are large and/or frequently revised. An API code generator was developed and demonstrated (Khandekar and Book, 1998) that automated the creation of the C++ language bindings for any EXPRESS model. The code generator was used to recreate the pdXi prototype API (and located several bugs in the original) and to create an API for other portions of the information model. The existence of an API greatly simplifies the testing of an information model for effectiveness and completeness and, when generated automatically, allows for alternative models to be tested to determine the most efficient.

An interface to STEP databases (Gidh and Book, 1997) was also developed. Block Point Release (BPR) 3.2, a software toolkit developed at PDES, Inc. (PDES, 1992), was used as a pattern to develop the data definition language (DDL) to create objects and attributes (tables and fields in relational databases) in the database from an EXPRESS model. The BPR code is specific to the Oracle database. The BPR code was translated to Open Database Connectivity (ODBC) code so that any ODBC-compliant database can be used. Code that reads and writes data instances from the STEP working form to the database (also in ODBC) finalized the electronic data exchange system for EXPRESS information models shown in Fig. 1. Any application, database, physical file, or working form that has access to the API (with database and file utilities) for an EXPRESS model can exchange data captured by that model with any other application, database, physical file, or working form. If data captured by a model is exchanged from one computer system to another (physical file, database, or working form), both systems must have the API (with utilities) for the model.

The pdXi API

Based on the success of the prototype, pdXi awarded a contract to Simulation Sciences, Inc., to produce a commercial version of the API for the complete information model. The pdXi API produced by Simulation Sciences, Inc., is similar in structure to the pdXi prototype API, enabling data exchange between an application, a STEP working form, and a STEP physical file. (There is no interface to databases in the initial

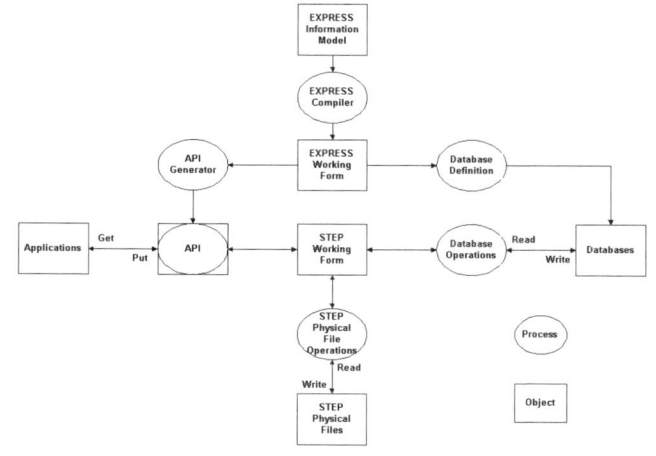

Figure 1. An electronic data storage and exchange system based on EXPRESS information models.

version of the API.) However, the pdXi API uses the C++ language bindings specified in Part 23 (ISO, 1995) of STEP (the bindings had not been specified when the pdXi prototype API was developed), has a STEP working form that is much more efficient, and has dynamic linked libraries (DLL) that enable the exchange of very large quantities of data from large EXPRESS models using computers with modest memories. Additionally, the STEP working form is in a platform-independent format that can be exchanged from one computer to another. The STEP working form is much smaller in size than the STEP physical file.

There are two levels in the pdXi API. The "low level API" has the C++ bindings specified by Part 23, thus, it is an SDAI. Use of the low level API requires extensive knowledge of the information model. The "high level API" has interfaces to C and FORTRAN and a set of convenience functions. The convenience functions provide the means for an application to exchange blocks of process engineering data with the STEP working form. There are convenience functions for material stream data, material flow data, component data, shell and tube heat exchanger data, separation tower data, stream lists, unit operation summaries, physical property tables, and unit operations data. The convenience functions were chosen to capture a large portion of the process engineering data that is frequently exchanged. In most cases, no knowledge of the information model is required to use the convenience functions. The User's Guide (Simulation Sciences, Inc., 1998a) and the Reference Manual (Simulation Sciences, Inc., 1998b) for the pdXi API describe the data structures that must be created within the application to store the data and the call to the function to put data from the data structure into the STEP working form or get data from the STEP working form into the data structure. The pdXi API completed beta testing and was released in March, 1998 (AIChE, 1998).

Extensions and Revisions to the pdXi Information Model

AP 231 and the initial release of the pdXi API are based largely on Revision 0 of the pdXi information models. The Metadata and Simplified Geometry submodels were replaced by the appropriate STEP generic resources and a revision of the Heat Exchanger Equipment submodel was included. However, the EXPRESS models in the Committee Draft of AP 231 have been extensively modified by the harmonization (integration with other application protocols and generic resources) and technical review processes. A second release of the API is planned for the approved CD of AP 231.

Submodels for Control Engineering Equipment (Smith and Book, 1999) and Piping System Equipment (Gidh, 1995) have been developed and a completely revised Planning Level submodel has been proposed to greatly enhance the efficiency for capturing batch and real-time operating data (Book and Sharma, 1998).

Conclusion

A revision of the pdXi information model has successfully completed balloting to become a committee draft application protocol (AP 231) of STEP. An application programming interface, based on Revision 0 of the pdXi information models, has been released. The API and its utilities use the methods and protocols of STEP to enable the exchange of process engineering data between computer programs, STEP working forms, and STEP physical files. An updated version of the API is planned.

References

AIChE, (1997a). *pdXi Release Volume I: Data Models*. AIChE.

AIChE, (1997b). *pdXi Release Volume II: Activity Model*. AIChE.

AIChE, (1998). *pdXi API Software*, AIChE.

Book, N. L., O. C. Sitton, M. R. Blaha, B. L. Maia Goldstein, J. L. Hedrick, R. L. Motard, and J. J. Fielding (1994). *The pdXi Data File Interchange Project Deliverables, Volumes I-XIV*. pdXi/AIChE.

Book, N. L., and A. Sharma, (1998). Information models for batch and real-time chemical process data, *Proceedings of FOCAPO 98, AIChE Symposium Series*, 320, 474.

Gidh, Y., (1995). *Development of an Application Programming Interface to STEP Databases for Chemical Process Software*. PhD Dissertation, University of Missouri-Rolla.

Gidh, Y., and N. L. Book (1997). Development of an API to STEP databases. *Proceedings of CHEMPUTERS* 5, McGraw-Hill, 157-180.

ISO, (1994a). *Process Data Exchange using STEP, Part 1: Overview and Fundamental Principles, International Standard*. U.S. Product Data Association, Fairfax, VA.

ISO, (1994b). *Process Data Exchange using STEP, Part 11: The EXPRESS Language Reference Manual*.

ISO, (1994c). *Process Data Exchange using STEP, Part 21: Clear Text Encoding of the Physical File Exchange Structure*.

ISO, (1994d). *Process Data Exchange using STEP, Part 22: Standard Data Access Interface*.

ISO, (1995). Part 23: *C++ Programming Language Binding to the Standard Data Access Interface Specification*. Working Draft, ISO TC 184/SC 4/WG 7 N 376.

ISO, (1998). Part 231:*Process Engineering Data: Process Design and Process Specifications of Major Process Equipment*. Committee Draft, ISO TC 184/SC 4/WG 3 N 745.

Khandekar, M. M., and N. L. Book, (1998). Application programming interfaces for EXPRESS informaiton models. *Proceedings of FOCAPO 98, AIChE Symposium Series*, 320, 480.

PDES, (1992). *Block Point Release 3.2 System Manual*. PDES, Inc., PTI018.04.00.

Rumbaugh, J., M. Blaha, W. Premerlani, F. Eddy, and W. Lorenson (1991). *Object-Oriented Modeling and Design*, Prentice-Hall, Englewood Cliffs, NJ.

Simulation Sciences, Inc., (1998a). *The pdXi API User's Guide*, Version 1.0-Beta. Simulation Sciences, Inc.

Simulation Sciences, Inc., (1998b). *The pdXi API Reference Manual*, Version 1.0-Beta. Simulation Sciences, Inc.

Sitton, O. C., N. L. Book, B. L. Maia Goldstein, J. L. Hedrick, R. L. Motard, and J. J. Fielding (1993). *pdXi Process Plant Activity Model*, Revision G, ISO Document TC 184/SC 4/WG 3 N 272.

Smith, V. A., and N. L. Book, (1999). Information models for control engineering data, This proceedings.

USAF, (1981). *ICAM Architecture Part II, Vol. IV—Function Modeling Manual (IDEF0)*, AFAWL-TR-81-4023. U.S. Air Force Wright Aeronautical Laboratories.

AN OBJECT-ORIENTED FRAMEWORK FOR PROCESS SYNTHESIS AND OPTIMIZATION

E. S. Fraga, M. A. Steffens, and I. D. L. Bogle
Department of Chemical Engineering, University College London
London WC1E 7JE, United Kingdom

A. K. Hind
Brewing Research International
Nutfield, Surrey RH1 4HY, United Kingdom

Abstract

The Jacaranda system is part of an object oriented framework written in the Java language. This framework has been designed to support a range of synthesis and optimization activities in process engineering. Through the use of object oriented modelling techniques and the dynamic extension capabilities of the Java run-time environment, the framework is easily adapted for solving problems in different areas. Java provides a portable extensible environment with extra capabilities for distributed computing, object persistence and access to native code when required. This paper describes some of the main classes in the Jacaranda system and presents case studies which demonstrate the generic nature of the framework for process synthesis.

Keywords

Process synthesis, Object oriented programming, Discrete programming, Optimisation, Simulation.

Introduction

Automated process synthesis methods are most often based on the definition of a superstructure, described as a mixed integer non-linear programme (MINLP). The MINLP is therefore usually defined explicitly and can be solved using a variety of methods. This general approach has been shown to be successful in a large range of problems but has some drawbacks:

- the difficulty in generating the superstructure manually,
- the size of the resulting MINLP, even for small problems, and
- the limitations due non-linear and non-convex models.

Alternative approaches which tackle one or more of these drawbacks have been proposed. For example, the P-graph method (Friedler et al., 1996) automates the generation of a superstructure. This procedure generates non-linear programmes so non-convexity is still an issue. An alternative is to simultaneously generate and search a superstructure. The CHiPS (Fraga & McKinnon, 1994) and Jacaranda (Fraga, 1998) systems combine graph search with discretization to address these problems:

- The superstructure is implicit in the search procedure and hence is neither explicitly required nor stored. No *a priori* decisions of the structures to include are required.
- Using dynamic programming with branch and bound techniques, the implementation is efficient, even on parallel computers (Fraga & McKinnon, 1995).
- Using discrete programming imposes no requirements on linearity or convexity. However, reasonable and appropriate discretization procedures are required.

Jacaranda is an object oriented implementation of the facilities provided by CHiPS. This paper describes this new OO architecture, illustrated with a set of case studies.

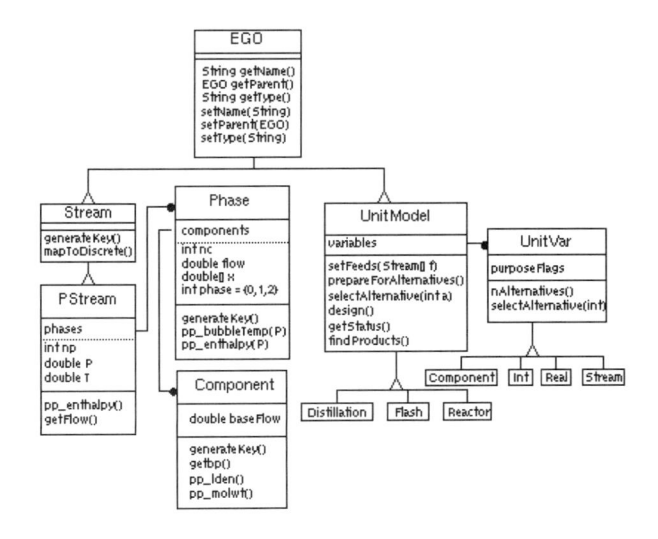

Figure 1. Stream and unit model class structure.

Jacaranda

Hendry & Hughes (1972) described a procedure for identifying the optimal separation sequence using a dynamic programming method. This procedure was automated by Johns & Romero (1979) by introducing discrete programming techniques. Fraga & McKinnon (1994) extended the underlying method to include the generation of ranked lists of solutions and designed parallel algorithms (Fraga & McKinnon, 1995). Recently, Fraga (1998) has extended the approach, with a new object oriented implementation, to cater for the uncertainties in early design. This implementation is based on the Java language (Sun Microsystems, 1999) which provides the basis for truly portable code with support for distributed and Internet based computing.

The search procedure in Jacaranda is based on implicit enumeration with depth first traversal to generate and search a graph which represents the superstructure. This superstructure is defined implicitly from the sets of raw materials, processing technologies and desired products. For a given set of streams, the procedure enumerates unit design instances which can process these streams. The outputs of a design are processed recursively. The recursion continues until streams which are either valid products or which cannot be processed any further are encountered. Discretization ensures that the search graph is finite, albeit possibly large.

The implicit enumeration procedure generates and traverses a state-task network where states are represented by streams and tasks by units. Therefore, there are two base classes in the object oriented implementation: Stream and UnitModel. These are abstract classes. Tackling new problems involves implementing new sub-classes. Figure 1 shows the basic class structure and subclasses for vapour-liquid systems. The Stream and UnitModel classes are themselves subclasses of the EGO (Eric's Generic Object) class. EGO objects provide the basic facilities for object persistence and problem definition. A detailed description of the object interfaces is given at http://www.ucl.ac.uk/~ucecesf/Jacaranda/.

The Stream Class

The search procedure relies on discretization both for ensuring a finite search space and for computational efficiency. The core of the discretization procedure is embodied in the Stream class and how it is sub-classed for specific types of problems. Two methods must be implemented: the mapToDiscrete method which takes a stream in continuous space and maps it to the nearest point in discrete space and the generateKey method which is used to generate a unique key string for each point in discrete space. The key string is used as a hash table index to identify previously solved sub-problems (Fraga, 1996). For the PStream class in Fig. 1, both methods are based ultimately on the discretization of component amounts. Each component has a base amount (baseFlow in the Fig. 1) which is the smallest amount of that component that will be represented exactly.

The UnitModel Class

Whereas the streams are used to label sub-problems in the search procedure, units are responsible for the size and shape of the search graph. For a given set of feed streams, units are used to generate new sub-problems based on the output streams of any feasible designs. The space of designs is defined by the values of unit design parameters which are represented by unit variable objects. These unit variables are manipulated by the search procedure which cycles through all combinations of allowed discrete values. The choice of discrete values is controlled by the user. The following algorithm illustrates how the enumeration procedure works.

```
for each unit u do
  if u.setFeeds(feedStreams) then
    u.prepareForAlternatives()
    n = u.nAlternatives()
    for a = 1 to n do
      u.selectAlternative(a)
      u.design()
      if u.getStatus() == OK then
        process the design...
      end if
    end for
  end if
end for
```

One advantage of the unit model class is that it can be used in simulation as well as synthesis. By not invoking the prepareForAlternatives method, specific alternatives can be simulated with differing feed streams or conditions.

In fact, given a particular design, the unit variables can be accessed directly and manipulated in continuous space. A discrete mapping is not necessary.

Problem Classes

The synthesis procedures are implemented in Problem classes. The pseudo-code shown below represents the base class which implements the basic features of enumeration, generation of ranked lists of solutions and multicriteria optimization:

```
class Problem
  init() { best = new List() }
  solve() {
    ie = new ImplicitEnumeration()
    while ie.hasMoreElements() do
      ienode = ie.nextElement()
      ienode.evaluate()
      best.insert(ienode)
    end while
  }
  inner class ImplicitEnumeration
    boolean hasMoreElements() { … }
    IENode nextElement() { … }
  end class
  inner class IENode
    evaluate() { … }
  end class
end class
```

The actual process synthesis class, PS_Problem, sub-classes the ImplicitEnumeration inner class to enumerate the allowable unit designs for a given set of feed streams. It also sub-classes the IENode class to evaluate each design and to recursively solve the sub-problems defined by the outputs of the unit. By sub-classing the enumeration and node inner classes, specific problem types can be tackled easily using the framework.

With discrete programming, it is possible to efficiently simultaneously generate, for each sub-problem, a set of lists, each list ranked according to a different criterion. Often, the unit design calculations are the most significant. Therefore, simultaneously collating results for multiple criteria can be efficient using this discrete implicit enumeration approach.

Case Studies

The description above indicates how the framework might be applied to the synthesis of separation sequences for vapour-liquid systems. The generic nature of the framework is now demonstrated by describing its application to a broad range of problems.

Bioprocess synthesis

In the biochemicals manufacturing industry, automated synthesis can be useful for achieving the responsiveness required in the global marketplace. Identifying the right process early on is critical due to the regulatory environment which requires the process designer to freeze the design once it has been approved. We have applied the synthesis procedure to several problems, including the production of Penicillin and Bovine Somatotropin (BST) (Steffens et al, 1999a, 1999b). In the latter case, the products form intracellularly as solid particles, or inclusion bodies. This problem is interesting because the modelling of the streams is complex and because the processing technologies include non-sharp separations (e.g. chromatography) and solid/liquid separation. Furthermore, the fermenter is modelled by a set of differential-algebraic equations. This type of problem uses the base synthesis problem class described above and requires the definition of consistent stream and unit models.

Heat Integration

The base synthesis procedure assumes that there is a decoupling between the solutions to sub-problems corresponding to the individual outputs of a given unit design. This assumption is violated in the case of heat integration. Therefore, the base procedure must be modified to handle this case: a problem sub-class is defined which introduces the concept of qualified problems. Qualifiers are used to specify that the solution to a particular sub-problem may, for instance, generate a given amount of excess heat at a specified temperature or that it may assume that the same amount of heat is available for free. The ImplicitEnumeration inner class is sub-classed to cycle through the combinations of qualifier assignment to different sub-problems. The rest of the class definition remains the same. Fraga & McKinnon (1999) present results based on a scalable implementation of this approach for large scale separation problems.

Recycle Structures

Identifying and generating recycle streams, especially for reaction/separation problems, also requires special handling. An iterative procedure has been designed which iterates on intermediate and partial results. The iterates include qualifiers which specify additional streams which are available for use anywhere in the process so long as the same process outputs equivalent streams. The approach is similar to converging on tear streams in sequential modular flowsheeting. This procedure is implemented as the IDP (Iterative Dynamic Programming) class which extends the QualifiedProblem class described above for heat integration. Qualifiers are used to indicate the availability of the recycle streams and the need to generate these same streams in the resulting process (Fraga, 1998).

Batch Processes

The framework can also support the use of dynamic modelling (e.g. the fermenter model in the bioprocess synthesis example above). Identifying the optimal malting schedule for barley is a typical example of a problem in the brewing industry (Holmberg et al., 1997), one that requires the use of dynamic models for all processing steps. The malting process consists of a series of steeping and drying stages, each modelled by systems of differential-algebraic equations. By defining a simple model for barley and the corresponding unit models, this

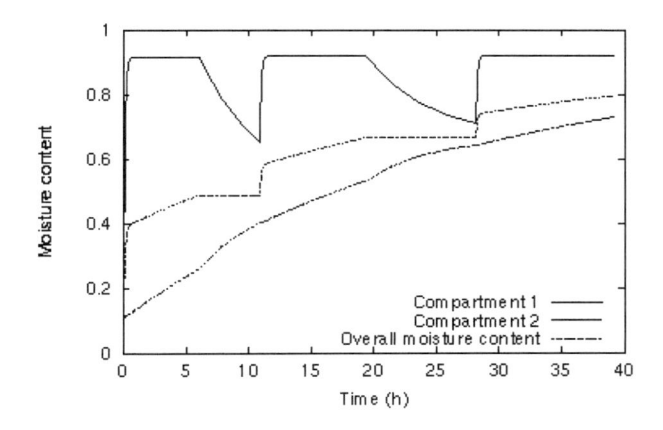

Figure 2. Profile of moisture uptake in malting.

problem can be solved in Jacaranda. The criterion for ranking solutions is the total time necessary to achieve desired product characteristics. The advantage of a discrete programming procedure is that there are no requirements on convexity, linearity or even continuity imposed on the models. This particular problem used native code (Fortran 90 & C++) for implementing and solving the dynamic models. The dynamic profile of moisture content for a solution is shown in Fig. 2. The saw-tooth behaviour of the content for compartment 1 corresponds to the steeping and drying stages.

Conclusions

The Jacaranda system provides an object oriented framework suitable for the automated synthesis of a wide range of processes. Specific types of processes can be designed by defining consistent stream and unit model definitions. A hierarchy of problem classes enables us to design processes with particular characteristics, such as heat integration and recycle structures. Furthermore, the underlying discrete programming nature of the synthesis procedure provides the ability to generate multiple ranked solutions and to tackle multiple criteria simultaneously and efficiently. The Jacaranda package is available for evaluation by contacting the first author.

Acknowledgements

The authors gratefully acknowledge the support provided by the EPSRC; part of this work was undertaken as part of the interdisciplinary research centre on Process Systems Engineering.

References

Fraga, E. S. (1996). Discrete optimization using string encodings for the synthesis of complete chemical processes. In C A Floudas and P M Pardalos (Ed.), *State of the Art in Global Optimization: Computational Methods & Applications.* Kluwer Academic Publishers, Dordrecht. pp. 627-651.

Fraga, E. S. (1998). The generation and use of partial solutions in process synthesis. *Chem. Eng. Res. Des.*, **76**, 45-54.

Fraga, E. S. and K. I. M. McKinnon (1994). CHiPS: A process synthesis package. *Chem. Eng. Res. Des.*, **72**, 389-394.

Fraga, E. S. and K. I. M. McKinnon (1995). A portable code for process synthesis using workstation clusters and distributed memory multicomputers. *Comp. Chem. Eng.*, **19**, 759-773.

Fraga, E. S. and K. I. M. McKinnon (1999). A scalable discrete optimization algorithm for heat integration in early design. In F. Keil, W. Mackens, H. Voß, and J. Werther (Eds.), *Scientific Computing in Chemical Engineering II – Simulation, Image Processing, Optimization, and Control.* Springer, Berlin. 306-313.

Friedler, F., J. B. Varga, E. Fehér and L. Fan (1996). Combinatorially accelerated branch-and-bound method for solving the MIP model of process network synthesis. In C. A. Floudas and P. M. Pardalos (Ed.), *State of the Art in Global Optimization: Computational Methods & Applications.* Kluwer Academic Publishers, Dordrecht. pp. 609-626.

Hendry, J. E. and R. R. Hughes (1972). Generating separation process flowsheets. *Chem. Eng. Prog.*, **68**, 71.

Holmberg, J., J.J. Hämäläinen, P. Reinikainen and J. Olkku (1997). A mathematical model for predicting the effects of the steeping programme on water uptake during malting. *J. Inst. Brew.*, **103**, 177-182.

Johns, W. R. and D. Romero (1979). The automated generation and evaluation of process flowsheets. *Comp. Chem. Eng.*, **3**, 251-260.

Steffens, M. A., E. S. Fraga and I. D. L. Bogle (1999a). Designing sustainable bioprocesses using multi-objective process synthesis. In F. Friedler and J. Klemeš (Eds.), *Proc. PRES'99.* Hungarian Chemical Society, Budapest. 613-618.

Steffens, M. A., E. S. Fraga and I. D. L. Bogle (1999b). Optimal system wide design for bioprocesses. *Comp. Chem. Eng.*, **23**, S51-S54.

Sun Microsystems Inc, 1999, *The Java Platform.* Available from: http://www.sun.com/java/.

A TASK AND VERSION-ORIENTED FRAMEWORK FOR MODELING AND MANAGING THE PROCESS DESIGN PROCESS

S. Gonnet, G. Mannarino, and H. Leone
INGAR (CONICET) - GIPSI (FRSF – Universidad Tecnológica Nacional)
3000 Santa Fe, Argentina
hleone@alpha.arcride.edu.ar

G. Henning
INTEC (Universidad Nacional del Litoral – CONICET)
3000 Santa Fe, Argentina
ghenning@intec.unl.edu.ar

Abstract

An object-oriented framework to support the modeling and management of the process design process is introduced. It naturally integrates the representation of both design tasks, and the outcomes that are achieved as the result of design activities, that are modeled as resources. Tasks are represented in terms of the goals they pursue and the different types of resources they are involved with. Mechanisms to explicitly represent task decomposition and temporal relationships among tasks are provided. Moreover, both the modeling of the alternative ways of performing a given design task and the representation of the different endings a design task may have are supported. The Version Administration System introduced in this paper provides an explicit mechanism to manage the different model versions being generated during the course of a design project as design tasks are executed.

Keywords

Management of the process design process, Task representation language, Versions' administration.

Introduction

In almost all areas of engineering design complexity prevents teams of designers from producing the desired artifact in just one step. Instead, the design process is separated into a number of phases, each corresponding to an increasing level of detail. Nevertheless, the design process is by no means linear, since phases are not separated but highly intertwined. The design process is a very complicated and creative human activity. Consequently, there is a need for tools able to capture and manage it in a useful form. Having such tools will allow the tracing of the design process and the analysis of its rationale, forming the basis for learning and future reuse. There is very little previous work in this area in chemical engineering. Westerberg *et al.* (1997) have developed design support tools based on the *n-dim* environment.

They have focused on supporting the management of information, a critical issue when addressing cooperative design. Bañares-Alcántara *et al.* (1995) have also made contributions to support the process design process. Their Epée environment makes possible the representation of the designer's intent and allows model traceability, reflecting to a certain extend the design history.

This contribution introduces a conceptual object-oriented framework that addresses the need outlined above for managing the design process. The framework proposes an explicit representation of each design task in terms of the goals it pursues and the different types of resources that participate in the task. Thus, it naturally integrates the representation of both design tasks, and the outcomes that are achieved as the result of design activities, that are represented as resources. Resources describe all the physical and conceptual entities that participate in a design project, such as the ones that model the evolving process design, the computer support tools being used, the adopted assumptions and simplifications, etc. Design tasks may operate on certain relevant perspectives of existing resources, by modifying, deleting and/or employing them. Design tasks may also create new resource entities. Mechanisms to explicitly represent task decomposition and temporal relationships among tasks are also provided. Moreover, both the modeling of the alternative ways of performing a given design task and the representation of the different endings a design task may have (i.e., successful, aborted, etc.) are supported. Details about the task-oriented modeling of the design process are given in the next section.

Having represented a design methodology by means of generic design tasks (i.e. task models), a particular design project is carried out by instantiating and executing these task models. As mentioned in the previous paragraphs, design tasks create, delete and/or modify resource perspectives that represent the structure, behavior and relevant characteristics of the chemical processing system being designed. At a given stage during the execution of the project, the states assumed by the set of relevant resources, from now on called model version, supply a snapshot description of the state of the artifact being designed. Consequently, it is necessary to have an explicit mechanism to administrate the different model versions that are being generated during the course of the design project. The mechanism proposed in this contribution, that is based on the situational calculus (Reiter, 1996), considers that a given model version represents a situation that can be reached from a previous one through the application of a certain design task. Details about this proposal are given in the section entitled Model Version Management. Due to lack of space small examples based on Prof. Douglas' hierarchical approach to the synthesis of chemical plants are presented (Douglas, 1985).

A Task-Oriented Representation of the Design Process

One of the main aims of the proposed framework is to support an explicit integration of the design process model and the representation of the evolving design artifact. This requirement establishes a difference with previous contributions (Han et al., 1995), that adopted an approach to the modeling of design tasks that has weak relationships with the modeling of the process being designed. According to the proposal discussed in this paper, the design process is envisioned as a series of design tasks that operate on resources by creating, deleting, modifying, and/or using them. Resources may be of different types: (i) people that participate in the design project (ii) information, such as assumptions, constraints, chemical reaction data, physical and chemical properties, etc., (iii) structural and behavioral models that describe the processing system being designed (Stephanopoulos et al., 1990), (iv) computer support tools such as equation-solving routines, optimization packages, etc. Resources are modeled according to a domain metamodel that is not described here due to a lack of space.

In a design project different kinds of design activities are executed: (i) Formulation or Specification tasks, that establish the scope of future design activities, (ii) Synthesis tasks, that generate structural components, (iii) Analysis or Evaluation tasks, (iv) Decisions' tasks, (v) Coordination tasks, etc. As it will be shown in the following paragraphs, all types of design tasks are modeled in the same fashion, according to a task metamodel proposed in our group (Mannarino et al., 1999). This feature establishes a distinction with other contributions (Bañares-Alcántara et al., 1995) that consider decision related tasks differently from other design tasks. Finally, it is important to stress that according to the characteristics of the proposed framework there is no need for a Planning Module to control the execution of tasks. A task will be able to be executed whenever both its preconditions are fulfilled and its temporal relationships with other tasks are satisfied.

The Task Metamodel

The task metamodel introduced by Mannarino et al., (1999) in the Coordinates' language prescribes the way design tasks can be modeled. According to this metamodel, a *Task Model* is used to represent a set of activities in terms of a set of resources that participate in different ways in order to achieve the tasks' goals. As only certain aspects or characteristics of a *Resource* may be of interest to a given task, a particular perspective of the *Resource* is actually viewed by the *Task*. For example, the task *SolveMaterialBalance* only views the *Material Perspective* of the involved *Stream* resources.

A *Task* is related to a *ResourcePerspective* by means of a *task-resource-link* that reflects the role that the *Task* plays in relation to the *ResourcePerspective*. The following roles have been considered in the proposed metamodel: *creates/eliminates* (non-renewable resources), *produces/consumes* (renewable resources), *modifies*, *uses*, *employs* (exclusive usage). The structure of the *task-resource-link* encapsulates the characteristics of the *Resource Perspective* that are relevant to the *Task*. These characteristics are expressed by resorting to the concept of *State*, which represents a "snapshot" of the *Resource Perspective* at a given time. A *ResourcePerspective* is linked to its possible states through the *resource-state* relationship. When a *ResourcePerspective* participates in a *Task*, three states need to be identified to express how the *Resource Perspective* behaves before, during and after executing the *Task*. Thus, (i) the *initial-state* relationship references the state the resource needs to assume in order to participate in the task, (ii) the *final-state* relationship references the state the resource assumes when the task has finished, and (iii) the *intermediate-state* encapsulates the evolution of the resource during the task execution. Based on this definition, the preconditions of a *Task* will be given by the initial states of a subset of its associated *ResourcePerspectives (PreStates)*. Similarly, the final states of the *ResourcePerspectives* (*PostStates*) provide the task's postconditions.

Having introduced the idea of *state* of a *Resource Perspective* it is now possible to describe its different roles in relation to a design task. It is said that a *ResourcePerspective* is *used* if the state it needs to be in order to participate in the *Task* is the same to the one it assumes when the *Task* has finished. The *creates* relationship represents the fact that a new *Resource Perspective* (and therefore, a new *Resource*) appears in the domain as a consequence of the *Task* execution. The link *eliminates* is the inverse of the *creates* one. Finally, the *modifies* link indicates that the *ResourcePerspective* suffers a change of state due to its participation in the *Task*.

As seen in the previous paragraphs tasks relate among themselves indirectly by means of the resources they operate on. However, two tasks can be directly linked through explicit temporal relationships (Allen, 1983). For example, when establishing the initial specifications of a design project the tasks *GetProductsSpecification* and *IdentifyPossibleRawMaterials* should be done *before* the task *IdentifyPossibleReactionSchemes*.

The fact that a *Task* may have different endings is explicitly represented by resorting to the *TaskMode* concept. Consider, for example, the task *Synthesize GeneralSeparationStructure* that is modeled as part of Prof. Douglas' hierarchical procedure. This task has three different modes since three possible general separation structures have to be considered provided the reactor exit may be (a) liquid, (b) vapor and (c) liquid and vapor. Each different *TaskMode* is associated to a distinctive set of final states (*PostStates*) of its related *Resource Perspectives* (Mannarino et al., 1999).

Tasks can be described at different abstraction levels, according to the complexity of the design activity that is being modeled. Hence, a task can be decomposed into subtasks. However, as there may exist alternative ways of disaggregating a given *Task*, the *TaskDecomp* concept is introduced. A *TaskDecomp* encapsulates a particular way of decomposing a Task, under a specific *TaskMode*.

Model Version Management

As it was introduced in the previous section, a *Task Model* can be used to represent the generic design tasks associated to a particular synthesis and design methodology. Thus, at a class level, generic design models are represented. However, when the design activities of a particular project are to be modeled, generic tasks have to be instantiated and executed. The instantiation process produces specific occurrences of the *Tasks* specified at the class level. On the other hand, the execution of tasks cause changes on the states of their associated *Resource Perspectives*.

When a design project is carried out, the states assumed by the set of relevant *Resource Perspectives* supply a snapshot description of the state of the processing system being designed. This description, called *ModelVersion*, will be modified each time a design task is executed. The model version management approach introduced in this paper provides an explicit mechanism to administrate the different model versions being generated during the course of a design project, as design tasks are executed.

In order to manage the evolution of the states of *ResourcePerspectives*, the Version Administration System (VAS) proposed by Gonnet et al. (1998) was specialized. The three components' architecture depicted in Figure 1 shows the constituents of the VAS.

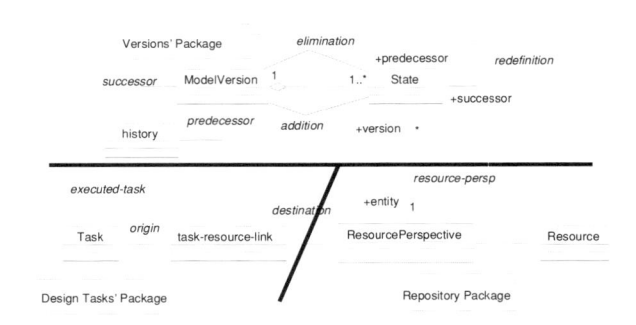

Figure 1. Model version administration architecture.

The *Repository* contains the resource entities that may evolve during a design project. So, resources that are only used or employed by design tasks (such as computer packages, human resources, etc.) are not part of the VAS system. On the other hand, entities representing processing systems, chemical reaction schemes, process streams, etc., that are modified, created or deleted by design tasks, will naturally be part of the *Repository Package*. In the *Versions' Package*, the evolution of the entities contained in the *Repository* is explicitly specified. The *ModelVersion* entity comprises the knowledge of the processing system being designed at a given time point. Therefore, while the different *ResourcePerspectives* compose the *Repository*, the states they can assume at different time points belong to the *Versions' Package*, and the tasks that actually transform the *ResourcePerspectives* are contained in the *Design Tasks' Package*.

According to the proposed framework, a new *ModelVersion* M_2 is generated as a consequence of the application of a particular *Task* T_1 to the components of a previous *ModelVersion* M_1. This is achieved by performing the following evaluation: $apply(T_1, M_1) = M_2$. Function *apply* is defined as follows

$$apply: \Phi \times M \to M \qquad (1)$$

where:

Φ: the set of all the design *tasks* instances ϕ.
M: the set of possible *ModelVersions*.

As mentioned in previous sections, *task-resource-links* define the effects of a *Task* over a set of *ResourcePerspectives*. Therefore, these links are the ones that express how a *ModelVersion* will evolve into another *ModelVersion*. In order to carry out this semantics the version administration primitive operations *add*, *delete* and *redefine* are introduced. The correspondence between the *task-resource-links* and these operations is expressed in the following table:

Table 1. Relationships between Task-Resource-Links and Operations.

Task-resource-link	Operation
creates	add
eliminates	delete
modifies	redefine

Thus, the procedural part of a design task ϕ is actually defined by the set of *add*, *delete* and *redefine* primitive operations that correspond to the *task-resource-links* associated with the task. So,

Then, the inductive definition of function *apply* is given by:

$$\text{Procedural part of } \phi = \begin{cases} \lambda \text{ empty sequence} \\ o \bullet \phi \text{ where } o \text{ is an operation} \end{cases} \quad (2)$$

$$\text{apply}(\lambda, m) = m$$
$$\text{apply}(o \bullet \lambda, m) = m', m \neq m' \quad (3)$$
$$\text{apply}(o \bullet \phi, m) = \text{apply}(\phi, \text{apply}(o \bullet \lambda, m))$$

The *add* operation generates a new *Resource Perspective* entity in the *Repository* and its corresponding *State* at the *ModelVersion*, as expressed by the *addition* relationship in Fig. 1. The *delete* operation eliminates a *ResourcePerspective State* from the *Versions' Package*. Finally, the *redefines* operation denotes the relationship between one *State s* that belongs to a *ModelVersion* m_i and *s'*, which represents the new state a particular *Resource Perspective* assumes in a model version that is a successor of m_i. The *successor state axiom* (Reiter, 1996) is used to determine the states of *ResourcePerspectives* that belong to a particular *ModelVersion m*. Provided the predicate *belong(s, m)* evaluates to **True** whenever the state *s* belongs to the *ModelVersion m*, eq. (4) holds:

$$\forall \phi, s, m \ Belong(s, apply(\phi, m)) \Leftrightarrow \\ ((Belong(s, m) \lor (add(s) \in \phi)) \quad (4) \\ \land \ delete(s) \notin \phi)$$

The following expression denotes the relationship among the *redefinition* link (Fig. 1), its associated *redefinition* predicate and the operation *redefine*:

$$\forall \phi, s, s', m_i, redefinition(s, s') \land belong(s, m_i) \land \\ \neg belong(s', m_i) \land belong(s', apply(\phi, m_i)) \Leftrightarrow \quad (5) \\ redefine(s, s') \in \phi \land \neg belong(s, apply(\phi, m_i))$$

Conclusions

A task and version-oriented framework to support the modeling and management of the process design process is presented. It truly integrates an explicit representation of design tasks with the model of the evolving design artifact. The proposed Version Administration System provides an explicit mechanism to properly manage the different model versions generated during a design project.

References

Allen J.F. (1983). Maintaining knowledge about temporal intervals. *Comm. ACM*, **26**, 832-843.
Bañares Alcántara, R. J. King, and G. H. Ballinger (1995). Extending a Process Design Support Systems to Record Design Rationale. *AIChE Symposium Series No. 304*, 332-335.
Douglas, J.M. (1988). *Conceptual Design of Chemical Processes*, McGraw-Hill, New York.
Gonnet S., R. Holzer, Melgratti H. and H. Leone (1998). Administración de versiones de modelos en una herramienta de soporte para el análisis y diseño orientados a objetos. *Proceedings of the IV CACIC*, Neuquén, Argentina.
Han, Ch., G. Stephanopoulos and J. Douglas (1996). Automation in Design: The conceptual synthesis of chemical processing schemes, *Intelligent Systems in Process Engineering*, G. Stephanopoulos, Ch. Han (Eds.). Academic Press.
Mannarino G., G. Henning and H. Leone (1999). Coordinates: a framework for enterprise modeling. *Information Infrastructure Systems for Manufacturing*, J.J. Mills and F. Kimura (Eds.). IFIP-Kluwer Academic Publishers, 379-390.
Reiter R (1996). *Knowledge in Action: Logical Foundations for Describing and Implementing Dynamical Systems*. available from: http://www.cs.toronto.edu/cogrobo/
Stephanopoulos G., G. Henning and H. Leone (1990). MODEL.LA.: A modeling language for process engineering. Part I: The formal framework. Part II: Multifaceted modeling of processing systems, *Comp. Chem. Eng.*, **14**, S813-S869.
Westerberg, A., E. Subrahmanian, Y. Reich, S. Konda and the n-dim group (1997). Designing the process design process. *Comp. Chem. Eng.*, **21**, S1-S9.

THE VALUE OF DESIGN RESEARCH: STOCHASTIC OPTIMIZATION AS A POLICY TOOL

Timothy Lawrence Johnson and Urmila M. Diwekar
Department of Engineering and Public Policy
Carnegie Mellon University
Pittsburgh, PA 15213

Abstract

The use of stochastic optimization in decision analysis and design research, while considerable in its potential, has not been fully realized in practice. The complexities of policy problems – especially the multiplicity of conflicting goals and the need to act with incomplete information – combine with analytical and computational demands to discourage its use. Yet related techniques can play a valuable role even in contexts where optimization itself may be of secondary importance. This paper describes a new and efficient stochastic annealing-nonlinear programming algorithm for combinatorial optimization through its application to a complex design problem: the vitrification of processing wastes from the production of nuclear fuels at the US government's Hanford site. Specifically, the optimization framework of previous analyses is extended to incorporate variance as an attribute in its objective function. The augmented algorithm produces results that are more robust than those of traditional techniques, and facilitates evaluation of important dimensions of the waste remediation effort – in particular, the value of devoting resources to research aimed at reducing uncertainty. The analysis presented here identified the predictive error of the glass property models as the most significant source of uncertainty. Not all sources of uncertainty are consequential; the ability to distinguish between those sources that are important and those that may be tolerated is therefore a valuable contribution.

Keywords

Stochastic optimization, Combinatorial optimization, Stochastic annealing, Vitrification, Nuclear waste.

Introduction

Despite their analytical complexity and computational demands, stochastic optimization methods have a valuable role to play in decision analysis and design work. Recently developed methods, for instance, facilitate the examination of uncertainty. Questions important to decision makers, such as how conservative one should be when information is limited or where resources should be allocated in order to reduce uncertainty, can be addressed in an optimization context using these techniques.

This paper illustrates the application of a new stochastic annealing-nonlinear programming (STA-NLP) framework in answering such questions. Specifically, the value of research in reducing select sources of uncertainty in the blending and vitrification of wasted generated in the production of nuclear fuels at the US government's Hanford, WA site is examined. The post-production wastes are currently stored underground in 177 tanks of varying sizes and ages, many of which are known to be leaking. Current plans call for the tank wastes to be removed and preprocessed, selectively combined, and mixed with glass formers – or "frit" – prior to vitrification and permanent disposal. Blending takes advantage of the fact that the tank contents differ; as the frit components are already present in the tank wastes, selective blending reduces the need for additional material, therefore decreasing the amount of glass produced. In addition, blending reduces the proportion of limiting components in a given waste stream, increasing the probability that the glass property constraints will be met and vitrification will succeed (see Johnson and Diwekar, in press, for an overview of blending and the Hanford problem).

Blending involves a two-stage decision process: assignment of tanks to a given blend, and assessment of the frit needed for each configuration. The dual nature of this process is typical of combinatorial optimization problems, where the levels of continuous variables must be determined in tandem with a set of discrete decisions. The dependency of these decisions prevents their decoupling. Complicating the Hanford problem, however, are imperfect knowledge of the tank contents and empirical uncertainties in the glass property models governing the frit requirements. When one also takes into account the combinatorial explosion that results when even a subset of tanks is to be divided into a limited number of discrete blends, as well as the nonlinearities and nonconvexities in both objective and constraint functions, the use of mathematical programming techniques becomes intractable as a means determining an optimal tank-blend assignment (see Birge, 1997).

The Hanford blending problem therefore poses a challenge to the use of stochastic programming. First, traditional methods have difficulty incorporating probabilistic events and uncertain parameters in a way that recognizes their full importance. Analytical methods, for instance, are limited to a few functional forms and relatively simple problem structures. Numerical methods, such as Monte Carlo or Hammersley Sequence Sampling, typically propagate expected values through an algorithm, with little attention paid to the variance of the sample values. Hence, while the *expected* values of decision variables will satisfy model constraints, there is no guarantee that the more realistic *sample* values will do so when used in place of their means. Robustness, however, is an important design consideration.

Second, the multiobjective nature of the blending problem demands the use of an optimization technique that is flexible enough to facilitate a full analysis of uncertainty. Uncertainties in both the tank waste composition and the empirical glass property models must be considered. As decision makers are typically interested in finding solutions that not only meet design criteria, but that are also robust to unknowns, the ability to examine the trade-offs that inevitably arise in reducing both cost and risk is essential. The STA-NLP algorithm introduced here permits such an analysis. In addition, the framework outlined below facilitates the inclusion of a third attribute: the value of time devoted to reducing uncertainty. Better characterization of the tank wastes and glass property models, for instance, would result in lower frit requirements and a greater probability that vitrification succeeds; research, however, carries opportunity costs that need to be minimized.

The following section describes a new STA-NLP framework designed to meet these challenges. The subsequent section applies the framework to an analysis of the conflicting objectives of the Hanford blending problem. Specifically, the trade-off between minimizing the costs of processing and disposal of the tank wastes and limiting the opportunity costs of research aimed at reducing uncertainty is examined. The objectives are described, and results are discussed. A conclusion summarizes the analysis.

The Augmented STA-NLP Framework

The numerical stochastic optimization technique described and employed here is derived from a more general stochastic annealing-NLP framework (Chaudhuri and Diwekar, 1996). Applied to the blending problem, the framework consists of three nested loops:

1. An inner sampling loop that calculates probabilistic values of the tank waste component mass fractions.
2. A continuous variable decision loop that determines frit requirements for a given tank-blend configuration, based on the sampled waste mass fractions and subject to a series of empirical glass property constraints.
3. A discrete variable decision loop that selects tank-blend configurations to minimize the expected frit mass and its associated sample variance.

This section provides a brief description of each level in the STA-NLP framework, starting from within.

The sampling loop utilizes an efficient sampling procedure to calculate probabilistic values of the waste component mass fractions for each tank. Sampling is performed over empirical distributions using data provided by Hanford scientists. The difficulty in acquiring samples from the non-homogeneous mixture of wastes that resides in the tanks, as well as inter-component reactions and the poorly documented history of the Hanford tank contents, result in a large estimated mass-fraction sample variance (var_{samp}). Normal distributions characterize this uncertainty; sample size increases as the discrete decision loop approaches an optimum, reducing the computational burden during the algorithm's initial exploration of the solution space (Chaudhuri and Diwekar, 1996).

The continuous variable decision loop takes the probabilistic values of the waste component mass fractions and determines the frit requirements for each blend in a given tank configuration. Empirical glass property models (Narayan, Diwekar and Hoza, 1996) constrain the proportion of each waste component in a blend, and therefore govern frit requirements. A nonlinear programming algorithm (NLP) determines the expected frit mass. As the frit requirements are mostly based on expected values of the waste component mass fractions, minimization of the variance in frit requirements (var_{frit}), calculated when sample values are used in place of their mean, becomes an important design objective.

The width of the NLP (inequality) constraint bounds in the continuous variable loop captures a third important source of uncertainty in the blending problem: the prediction error of the empirical glass property models. As this error increases, the bounds become progressively tighter; hence, the range across which the component mass fractions in a given tank-blend configuration can vary becomes more restrictive. To ensure that the constraints are met, frit must be added to "average out" limiting components in the waste feed. Less uncertainty in the empirical glass property models, and correspondingly liberal constraint widths, are therefore desired.

Finally, the discrete decision loop employs a stochastic annealing algorithm (Chaudhuri and Diwekar, 1996) to search for tank configurations that minimize the expected frit mass of each blend, its variance, and other variance-related terms as described in the following section. This loop governs the STA-NLP framework, which terminates when a stopping criterion – related to its rate of convergence – is met.

Variance as a Blending Objective

The three sources of uncertainty identified in the previous section – var_{frit}, var_{samp}, and the glass property model prediction error reflected in the constraint bound widths – have meaningful technical interpretations and capture important dimensions of the blending design problem. Previous analysis (e.g., Narayan, Diwekar and Hoza, 1996; see also Johnson and Diwekar, in press, and references therein) have been unable to treat the implications of these terms on more than an informal basis. The expansion of the blending problem objective function to capture multiple – and often conflicting – attributes is therefore a novel development.

As the costs of vitrification are directly proportional to frit requirements, the optimization techniques employed by earlier studies sought to minimize expected frit mass. The logical extension of this work is to consider robustness and incorporate the minimization of variance in expected frit requirements (var_{frit}). The first augmented objective function therefore becomes:

$$\text{minimize } E[frit] + w_1 * var_{frit} \quad (1)$$

where $E[frit]$ is the expected value of the frit mass, var_{frit} is scaled to the same order of magnitude, and w_1 is the weight placed on the frit mass variance (when frit requirements are calculated using individual sample waste mass fractions in place of their expected values). Table 1 presents optimization results for increasing values of w_1. Note that as the stress on minimization of var_{frit} increases, the frit requirements become greater, while the proportion of constraint violations declines. (Note that this analysis examines three blends of four tanks each. Initial remediation efforts at Hanford will focus on a small subset of tanks to prove the vitrification concept.)

Variation in expected frit mass could be reduced more directly through better characterization of the tank wastes, while constraint violations could be minimized by improving the prediction accuracy of the empirical glass property models, thereby relaxing the constraint widths. These research activities, however, carry opportunity costs. The STA-NLP framework facilitates examination

Table 1. Expected Value versus Variance Minimization.

w_1	$E[frit]$ (kg)	$\sqrt{Var_{frit}}$	% Violated Constraints
1.0	10075	190	6
5.0	10647	138	2
10.0	11558	118	1

of the trade-off between minimizing the costs of vitrification and those due to the reduction of uncertainty.

Assuming that uncertainty is inversely related to time spent on research, but that this reduction is characterized by diminishing marginal returns, an exponential function provides a good first-order approximation to the relationship between time spent on research and the variance of the waste mass fraction sample distributions:

$$uncertainty \Leftrightarrow var_{samp} \propto \exp(-time) \quad (2)$$

$$\text{or, } time \propto -\ln(var_{samp}). \quad (3)$$

Note that uncertainty in the glass property models (the prediction error) is inversely related to the constraint bound width, as described above. Hence,

$$time \propto -\ln(constr.\ width)^{-1} = \ln(constr.\ width) \quad (4)$$

Incorporating these terms into an augmented objective that seeks to reduce the cost of vitrification ($E[frit]$), the risk of vitrification failing (var_{frit}), and the time devoted to research ($-\ln[var_{samp}]$ and $\ln[constr.\ width]$) yields the following function:

$$\text{minimize } E[frit] + w_1 * var_{frit} - w_2 * \ln(\Sigma var_{samp})$$
$$+ w_3 * \ln(\Sigma constr.\ width) \quad (5)$$

where the w_i are weights as before, all terms are scaled to the same order of magnitude, and the summations are taken over the tank waste components and the constraints, respectively. A parametric analysis of the weights in Eqn. (5) yields the results summarized in Table 2.

Note that as less emphasis is placed on characterizing the tank wastes (i.e., as w_2 increases) variation in frit mass increases and constraint violations become more numerous. This effect is not surprising: a change in the variance of the waste component sampling distributions leads to a proportionate shift in the frit variance – and a

Table 2. The Trade-offs in Reducing Costs.

Focus	E[frit]	Var$_{frit}$	% Violated Constraints
Robustness (large w$_1$)	Increases	Decreases	Decreases
Min. tank characteriz-ation research (large w$_2$)	Increases	Increases	Increases
Min. time devoted to property models (large w$_3$)	No Change	Decreases	Increases

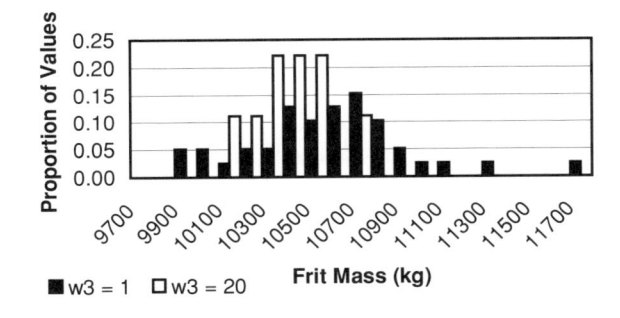

Figure 1. Parametric analysis of w$_3$.

similar impact on both the expected frit mass and the extent of the constraint violations (see Table 1).

Compared to changes in w_2, however, the variance in frit mass decreases while the percentage of constraint violations increases with larger weights on w_3, the constraint width coefficient; greater uncertainty in the glass property models translates into narrower constraint bounds, and a smaller range across which frit requirements may vary without consequence (see Fig. 1). This impact on process robustness leads to the conclusion that improvements in the glass property models should take priority over efforts to reduce uncertainty in the tank waste composition.

Conclusion

The STA-NLP algorithm described here facilitates a more realistic analysis of questions pertaining to uncertainty than competing optimization strategies, and recognizes the multiattribute nature of the trade-offs that arise when knowledge is incomplete. The preceding analysis identified the predictive error of the glass property models as the most significant source of uncertainty – an issue that previous approaches to the Hanford blending problem could not formally address.

The augmented STA-NLP framework is also more robust than equivalent mathematical programming techniques. The inclusion of variance in the objective function helps ensure that constraints are met by actual values of the decision variables – not just their expectations.

Development of the algorithm, however, is a work in progress. Future efforts will be aimed at enhancing efficiency through better sampling schemes, applying the model to larger-scale problems, incorporating a more realistic representation of uncertainty, and checking for nonconvexities in the multiattribute objective space. Such work will help make stochastic optimization a more useful technique in design and decision analysis applications.

Acknowledgements

The authors would like to thank Dr. Jared Cohon of Carnegie Mellon University for his perspective and advice. This work was sponsored in part by the National Science Foundation, under grant CTS-9613561.

References

Birge, J. R. (1997). Stochastic programming computation and applications. *INFORMS J. Comp.*, **9**, 111-132.

Johnson, T. L. and U. M. Diwekar (in press). Hanford waste blending and the value of research: Stochastic optimization as a policy tool. Submitted to *J. Multi-Criteria Decision Analysis*.

Narayan, V., U. M. Diwekar and M. Hoza (1996). Synthesizing optimal waste blends. *Ind. Eng. Chem. Res.*, **35**, 3519-3527.

Chauduri, P. and U. Diwekar (1996). Process synthesis under uncertainty: A penalty function approach. *AIChE J.*, **42**, 742-752.

DECOMPOSITION ALGORITHMS FOR NONCONVEX MIXED-INTEGER NONLINEAR PROGRAMS

Padmanaban Kesavan and Paul I. Barton
Department of Chemical Engineering
Massachusetts Institute of Technology
Cambridge, MA 02139

Abstract

A rigorous decomposition approach to solve general nonconvex MINLPs is presented. The proposed algorithms consist of solving an alternating sequence of the Relaxed Master Problem (a mixed integer linear program) and two nonlinear programming problems (NLPs; Primal and Primal Bounding problems respectively). A sequence of nondecreasing lower bounds and valid upper bounds are generated by the algorithms which converge in a finite number of iterations. A Primal Bounding Problem (convex NLP) is solved at each iteration to derive valid outer approximations of the nonconvex functions. Two decomposition algorithms are presented in this paper. On termination, the first algorithm yields the global solution, and the second finds a rigorous bound to the global solution of the original nonconvex MINLP. A distributed memory parallel algorithm (DMPA) variant which will drastically reduce the elapsed time for nonconvex MINLP is also presented.

Keywords

Nonconvex mixed-integer nonlinear programming, Decomposition algorithms, Global solution, Distributed memory parallel algorithm.

Introduction

A general class of optimization problems that involve integer and continuous variables can be defined as:

$$\min_{\mathbf{x},\mathbf{y}} f(\mathbf{x},\mathbf{y})$$
$$\text{s.t. } \mathbf{g}(\mathbf{x},\mathbf{y}) \leq \mathbf{0} \quad (P)$$
$$\mathbf{x} \in X \subset \mathbf{R}^n$$
$$\mathbf{y} \in U \subset Y \subset \mathbf{R}^q$$

where X is a nonempty, compact, convex set defined by set of linear inequalities. U is a finite discrete set defined by $U = \{y : y \in Y, \text{ integer}\}$ and Y is a convex polytope. $Y = [0,1]^q$ is assumed in the theoretical development of this paper. Finite sets of integer and discrete values may always be represented by sets of binary variables. Significant advances have been made in solving problem P by exploiting the special problem structure that results under the assumption that the functions $f : \mathbf{R}^n \times \mathbf{R}^q \to \mathbf{R}$ and $\mathbf{g} : \mathbf{R}^n \times \mathbf{R}^q \to \mathbf{R}^m$ are convex (e.g., Generalized Benders Decomposition (GBD) (Geoffrion, 1972) and Outer Approximation (OA) (Duran and Grossman, 1986; Fletcher and Leyffer, 1994). Equality constraints can be represented by a pair of inequalities which then conform to the form defined by P.

Decomposition algorithms to solve problem P when the functions f and $\mathbf{g}$ are nonconvex have been limited due to the difficulty in constructing valid support functions/linearizations. In particular, the support functions/linearizations derived by GBD or OA can cut off portions of the feasible space of problem P and convergence to a suboptimal solution is most likely. A rigorous decomposition strategy in which valid linearizations are derived at each iteration by solving a novel Primal Bounding Problem is presented here.

Construction of Subproblems

The decomposition strategy presented here is based on the construction of the following subproblems:

Lower Bounding Convex MINLP

The solution of the convex MINLP represents a valid lower bound to the global solution of P.

Construction of a valid lower bounding convex MINLP requires convexification and underestimation of nonconvex functions f and $\mathbf{g}$ in $X \times Y$. In particular, the convex envelopes of f and $\mathbf{g}$ (m constraints) are not necessary and any valid convex underestimator is sufficient. A detailed discussion on constructing convex underestimators when f and $\mathbf{g}$ are separable in the binary and continuous variables and linear in the binary variables is presented by Kesavan et al. (1999). These procedures can be directly extended to construct convex underestimators of $f:X \times Y \to \mathbf{R}$ and the vector valued function $\mathbf{g}:X \times Y \to \mathbf{R}^m$.

Let $L_1(\mathbf{x}, \mathbf{y})$ and $\mathbf{L}_2(\mathbf{x}, \mathbf{y})$ represent the convex underestimators of $f(\mathbf{x}, \mathbf{y})$ and $\mathbf{g}(\mathbf{x}, \mathbf{y})$ respectively. The lower bounding convex MINLP is:

$$\min_{\mathbf{x},\mathbf{y}} L_1(\mathbf{x}, \mathbf{y})$$
$$\text{s.t. } \mathbf{L}_2(\mathbf{x}, \mathbf{y}) \leq \mathbf{0} \quad \text{(P1)}$$
$$\mathbf{x} \in X, \mathbf{y} \in U$$

The following assumptions on P1 are necessary:

1. The functions L_1 and $\mathbf{L}_2$ are once continuously differentiable at $(\mathbf{x}^i, \mathbf{y}^i)$, where $\mathbf{y}^i \in U$ and $\mathbf{x}^i$ is the KKT point of the subproblem (convex NLP) obtained by fixing the binary variables in P1.
2. A constraint qualification holds at the solution of every NLP subproblem obtained by fixing the binary variables in P1.

Primal Problem

Nonconvex NLP obtained by fixing the binary variables (say to $\mathbf{y}^j$) in P (referred to as NLP($\mathbf{y}^j$)).

Primal Bounding Problem

Convex NLP obtained by fixing the binary variables (say to $\mathbf{y}^j$) in P1 and provides a valid and tighter lower bound to the Primal Problem for each binary realization than that provided by the current Relaxed Master Problem (referred to as NLPB($\mathbf{y}^j$)).

Master Problem

A Mixed-Integer Linear Program (MILP), the solution of which represents a valid lower bound to the global solution of P.

P1 is a convex MINLP and a Master Problem can be derived using outer approximation as presented by Fletcher and Leyffer (1994).

Relaxed Master Problem

A MILP, the solution of which represents a valid lower bound to that subset of U not yet explored by the algorithm.

An integer cut (Balas, 1972) that excludes the currently examined binary realization is added to the Relaxed Master Problem at each iteration where the corresponding Primal Bounding Problem is feasible. The integer cut is necessary since the constraint based on UBD as described by Fletcher and Leyffer (1994) for convex MINLPs will not exclude the currently examined integer realization. In the case of infeasible Primal Bounding Problem at any iteration, the constraints that exclude the binary realization ($\mathbf{y}^j$) is derived by solving the feasibility problem F($\mathbf{y}^j$) (Fletcher and Leyffer, 1994). Note that if the Primal Bounding Problem is infeasible, then the corresponding Primal Problem is also infeasible.

The Relaxed Master Problem solved at iteration k is:

$$\min_{\mathbf{x},\mathbf{y},\eta} \eta$$
$$\text{s.t. } \eta < \text{UBD}$$

$$\left. \begin{array}{l} \eta \geq L_1(\mathbf{x}^j, \mathbf{y}^j) + \nabla L_1(\mathbf{x}^j, \mathbf{y}^j)^T \begin{pmatrix} \mathbf{x} - \mathbf{x}^j \\ \mathbf{y} - \mathbf{y}^j \end{pmatrix} \\ \mathbf{L}_2(\mathbf{x}^j, \mathbf{y}^j) + \nabla \mathbf{L}_2(\mathbf{x}^j, \mathbf{y}^j)^T \begin{pmatrix} \mathbf{x} - \mathbf{x}^j \\ \mathbf{y} - \mathbf{y}^j \end{pmatrix} \leq \mathbf{0} \end{array} \right\} \forall j \in T^k$$

$$\mathbf{L}_2(\mathbf{x}^i, \mathbf{y}^i) + \nabla \mathbf{L}_2(\mathbf{x}^i, \mathbf{y}^i)^T \begin{pmatrix} \mathbf{x} - \mathbf{x}^i \\ \mathbf{y} - \mathbf{y}^i \end{pmatrix} \leq \mathbf{0}, \forall i \in S^k \quad (M^k)$$

$$\sum_{i \in B^j} y_i^j - \sum_{i \in NB^j} y_i^j \leq |B^j| - 1, \forall j \in T^k$$

$$B^j = \{i : y_i^j = 1\} \quad NB^j = \{i : y_i^j = 0\}$$

$$\mathbf{x} \in X, \mathbf{y} \in U$$

where,

$T^k = \{j \mid j \leq k : \text{NLPB}(\mathbf{y}^j) \text{ is feasible and } \mathbf{x}^j \text{ is the optimal solution to NLPB}(\mathbf{y}^j)\}$

$S^k = \{j \mid j \leq k : \text{NLPB}(\mathbf{y}^j) \text{ is infeasible and } \mathbf{x}^j \text{ solves F}(\mathbf{y}^j)\}$

Algorithms

Algorithm 1: Global solution of nonconvex MINLPs

Initialize:
1. Iteration counter $k = 0$, $l = 1$, $T^0 = \emptyset$, $S^0 = \emptyset$, $U^0 = \emptyset$.
2. UBD $= +\infty$, UBDPB $= +\infty$.
3. Integer combination $\mathbf{y}^1$ is given.

REPEAT

IF ($k = 0$ or (M^k is feasible and $M^k <$ UBDPB)) **THEN**

REPEAT
Set $k = k+1$

1. Solve the Primal Bounding Problem (NLPB($\mathbf{y}^k$)). If NLPB($\mathbf{y}^k$) is infeasible, solve the feasibility problem F($\mathbf{y}^k$). Let the solution be $\mathbf{x}^k$.

2. Linearize the objective and active constraint functions of P1 about $(\mathbf{x}^k, \mathbf{y}^k)$. Set ($S^k = S^{k-1}$ and $T^k = T^{k-1} \cup \{k\}$) or ($S^k = S^{k-1} \cup \{k\}$ and $T^k = T^{k-1}$) as the case may be.

3. If NLPB($\mathbf{y}^k$) is feasible and $L_l(\mathbf{x}^k, \mathbf{y}^k) <$ UBDPB, update $\mathbf{x}^* = \mathbf{x}^k$, $\mathbf{y}^* = \mathbf{y}^k$, $k^* = k$ and UBDPB $= L_l(\mathbf{x}^k, \mathbf{y}^k)$.

4. Solve the current relaxation M^k (solution η^k) of P yielding a new integer assignment $\mathbf{y}^{k+1}$.

UNTIL $\eta^k \geq$ UBDPB or M^k is infeasible.

ENDIF

IF (UBDPB $<$ UBD) **THEN**

1. Solve the Primal Problem NLP($\mathbf{y}^*$) and find the global minimum. Set $U^l = U^{l-1} \cup \{k^*\}$. If NLP($\mathbf{y}^*$) is feasible, let the solution be $\mathbf{x}_p^k$, and if $f(\mathbf{x}_p^k, \mathbf{y}^*) <$ UBD, update $\mathbf{x}_p^* = \mathbf{x}_p^k$, $\mathbf{y}_p^* = \mathbf{y}^*$ and UBD $= f(\mathbf{x}_p^k, \mathbf{y}^*)$.

2. If $T^k \setminus U^l \neq \emptyset$, update UBDPB $=$ min $(L_l(\mathbf{x}^m, \mathbf{y}^m))$, $m \in T^k \setminus U^l$ (UBDPB corresponds to the Primal Bounding solution (x^s, y^s)). Update $\mathbf{x}^* = \mathbf{x}^s$, $\mathbf{y}^* = \mathbf{y}^s$, $k^* = s$. Set $l = l+1$. Otherwise, set UBDPB $= +\infty$.

ENDIF

UNTIL UBDPB $\geq$ UBD and $\{M^k$ is infeasible or $\eta^k \geq$ UBD $\}$. The global solution of P is given by the current UBD, $\mathbf{x}_p^*, \mathbf{y}_p^*$.

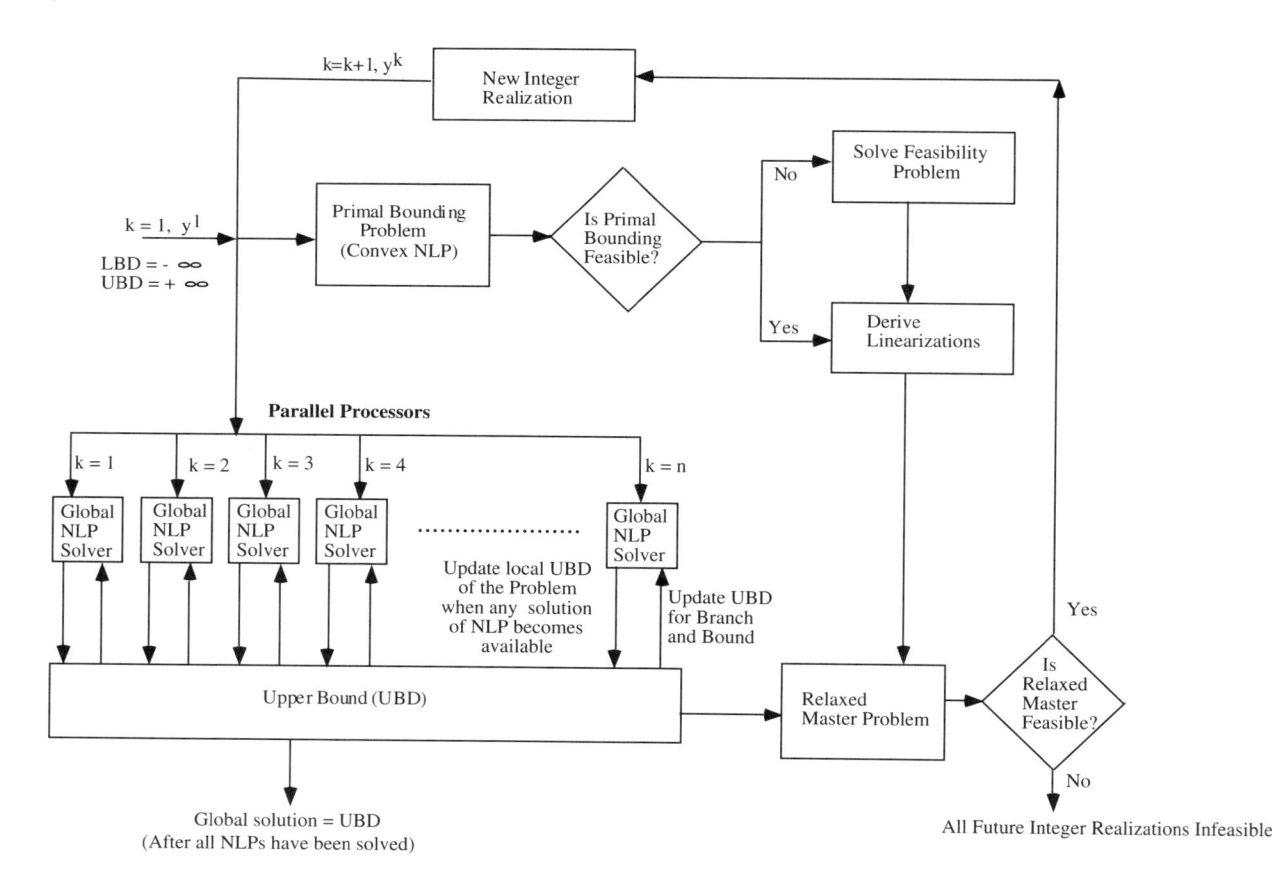

Figure 1. Distributed Memory Parallel Algorithm to determine the global solution of nonconvex MINLPs.

Algorithm 2: Rigorous Bound to the global solution of nonconvex MINLPs.

A rigorous bound to the global solution of P can be attained if the Primal Problem (nonconvex NLP) at each iteration is solved locally (global minimum not necessary) in Algorithm 1. The lower bound on termination is given by the minimum of the Primal Bounding solutions and the bound to the global solution is UBD – LBD. This algorithm is similar to that presented by Kesavan *et al.* (1999) for separable nonconvex MINLPs.

The convergence and optimality properties of Algorithms 1 and 2 directly follow from that presented by Kesavan *et al.* (1999) for separable nonconvex MINLPs.

Distributed Memory Parallel Algorithm (DMPA)

The linearizations for the Relaxed Master Problem in Algorithms 1 and 2 are derived from the solution of the Primal Bounding Problem. The solution of the Primal Problem is used only to update the UBD in both the algorithms. Therefore, the Primal Problem and the Relaxed Master together with the Primal Bounding Problem can be solved independently of one another. This leads to an efficient DMPA algorithm as shown in Fig. 1 which we anticipate will drastically reduce the elapsed time to solve P.

Preliminary Results

The heat exchanger network optimization problem as formulated by Yee and Grossman (1991) is considered as an example to analyze Algorithm 2. A detailed discussion is presented in Kesavan *et al.* (1999). A rudimentary implementation was done on a HP-C160 single processor machine with 128 MB RAM and running HP-UX B.10.20. The Relaxed Master Problems were solved using CPLEX via GAMS (Brooke *et al.* . 1988) and the Primal and Primal Bounding Problems were solved using Minos 5.4 (Murtagh and Saunders, 1987) via GAMS (Brooke *et al.* . 1988). The total CPU time required by the solvers alone was 85.2 seconds and the solution attained coincides with the global solution previously reported (Adjiman *et al.*, 1998).

Conclusions

A rigorous decomposition strategy and algorithms to solve general nonconvex Mixed-Integer Nonlinear Programs (MINLPs) has been presented in the current paper. A distributed memory parallel variant (DMPA) for efficient implementation of the proposed decomposition strategy has also been discussed. Preliminary results indicate that the proposed decomposition strategy is very efficient as regards to the computational time required to solve nonconvex MINLPs (relative to other rigorous algorithms reported in the literature).

Acknowledgements

This work was supported by the National Science Foundation under grant CTS-9703623.

References

Adjiman, C. S., C. A. Schweiger, and C. A. Floudas (1998). Mixed-Integer nonlinear optimization in process synthesis. In D. Z. Du and P. M. Pardalos (Eds.), *Handbook of Combinatorial Optimization*. Kluwer Academic Publishers Inc., Dordrecht. (in press).

Balas, E, and R. Jersoslow (1972). Canonical cuts on the unit hypercube. *SIAM J. Appl. Math.*, **23**, 61-79.

Brooke, A, D. Kendrick, and A. Meeraus (1988). *GAMS: A Users Guide*. Scientific Press, California.

Duran, M. A., and I. E. Grossman (1986). An outer-approximation algorithm for a class of mixed-integer nonlinear programs. *Math. Prog.*, **36**, 307-339.

Fletcher, R, and S. Leyffer (1994). Solving mixed-integer nonlinear programs by outer approximation. *Math. Prog.*, **66**, 327-349.

Geoffrion, A. M (1972). Generalized Benders decomposition. *J. Opt. Theor. Appl.*, **10**, 237-262.

Kesavan, P, R. J. Allgor, and P. I. Barton (1999). Outer approximation algorithms for separable nonconvex mixed-integer nonlinear programs. *Math. Prog.* (submitted).

Murtagh, B. A., and M. A. Saunders (1987). *Minos 5.1 Users Guide*. Technical report SOL 83-20R, Stanford University.

Yee, T. F., and I. E. Grossman (1991). Simultaneous optimization models for heat exchanger network synthesis. In I. E. Grossman (Ed.), *Chemical Engineering Optimzation Models with GAMS*, CACHE Design Case Studies, Series 6.

METAMODELING – A NOVEL APPROACH FOR PHENOMENA-ORIENTED MODEL GENERATION

Andreas A. Linninger
Department of Chemical Engineering,
University of Illinois at Chicago
e-mail: linninge@uic.edu

Abstract

This article discusses the phenomena-oriented modeling (POM) paradigm of TechTool, a computer-aided environment for large scale chemical process models. POM is a computer-based methodology for the automatic generation of mathematical representations from physical and chemical phenomena of relevance in chemical process engineering. Interpretation of the high-level design language leads to an equivalent set of system equations. Different modeling detail such as conceptual design, analysis of distributed systems and process dynamics require distinct modeling concepts and interpretation of the underlying phenomena. The *meta-modeling* feature allows engineers to define different instances of modeling languages in order to address specific modeling scope and objectives.

Keywords

Phenomena-based modeling, Model generation.

Introduction

Several distinct approaches to computer-aided modeling and design can be distinguished. The block-oriented flowsheet simulators constitute the state-of-the-art for many applications in the chemical industry, e.g. AspenPlus, Hysys, ProSim and others. For non-standard modeling tasks in teaching as well as academic and industrial research and development, equation-oriented modeling (EOM) languages enjoy increasing popularity, e.g. Piela, 1991; Barton and Pantelides, 1993; Abacuss, 1999. These latter tools offer ample opportunities for composing process models by means of mathematical expressions. In the industrial practice, several drawbacks are often noted: (i) long development times, (ii) lack of inner organization, (iii) requirement to employ experts rather than casual users. It seems that EOM tools provide satisfactory assistance for the mathematical solution of process models; support of the computer-aided model formulation itself is still missing.

Phenomena-oriented modeling (POM) is an initiative to address model generation from a system theoretical approach and aims at computer-aided support of the problem formulation. The focus lies on the specification of a phenomenological description without directly encoding detailed mathematical expressions. The main conjecture is that the concatenation of all relevant phenomena, such as reactions, material and heat transfer concepts, etc., leads to a complete model of the system. The associated equations can then be generated by an automated computer system through the appropriate interpretation of the phenomenological description. Model building by phenomena selection offers several advantages:

- The organization and composition of models is easily understood.
- Phenomena condense the information of the underlying system in highly compressed form.
- Adaptation, reuse and modifications of models can be accomplished through rearrangement and/or aggregation of phenomena rather than re-editing of the numerous underlying constitutive equations.

The scheme of Fig. 1. illustrates the mapping between the physical system, the phenomena-based modeling

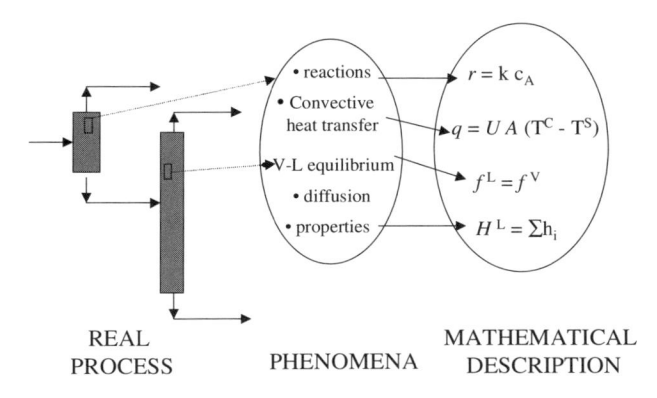

Figure 1. Schematic for phenomenological modeling.

elements and associated system equations. POM entails abstract descriptions of a system via high-level expressions for the physical and chemical phenomena of relevance. Models built from phenomena are associated with a set of formal mathematical expressions, i.e. system equations.

Section 2 gives an overview of prior work in model building and simulation. From this review, we will deduce a new proposition for the organization of a computer-aided modeling architecture entitled meta-modeling. Section 3 will discuss the main features of this new approach. Section 4 highlights shortcoming of the methodology. We will close with conclusions.

Earlier Approaches

The graphical interface of the flowsheet simulators for the continuous petrochemical systems can be interpreted as a pictorial unit-operations centered design language. The vocabulary matches the expectations of practitioners in the petrochemical industry ideally. While the work within the scope of the model library offers substantial convenience, e.g. distillation column, the graphical language does not support adaptation of unit models to new situations. This can only be done resorting to purely mathematical abstractions and programming. As an example take the gaseous stream containing a solid particulate phase. In traditional simulators, the reader will find concepts like the particle size distribution or pore structure missing and very hard to integrate.

BatchDesign-Kit (BDK) presented an operation-centered paradigm for the design of batch recipes for fine chemicals and pharmaceuticals, e.g. Linninger et al., (1998b) In BDK, batch recipes are constructed entirely through selection of operational tasks. The BDK operation centered language was designed after the natural language used by lab chemists and chemical engineers. The describing system equations are generated dynamically depending on the physical and chemical properties of the mixture. The model equations for each operation cannot be systematically modified at present.

Stephanopoulos presented the pioneering work in the area of systematic computer-aided modeling. The Model.la system (Stephanopoulos et al., 1990) gave the first account of a comprehensive modeling library of physico-chemical phenomena for the chemical engineering domain. The Model.la II design environment (Bieszczad et al., 1998) addresses the conceptual design of continuous process models in a phenomena-oriented spirit. Another advanced computer-aided modeling environment, MODKIT, was developed by Marquardt and his group (Lohmann and Marquardt, 1996; Lohmann, 1998). Both environments have graphical interfaces and can generate gPROMS code given a phenomenological description of a process model. MODKIT also offers computer-aided guidance for model evolution. It includes mechanisms for the automatic tracking of the model building phases, i.e. model history.

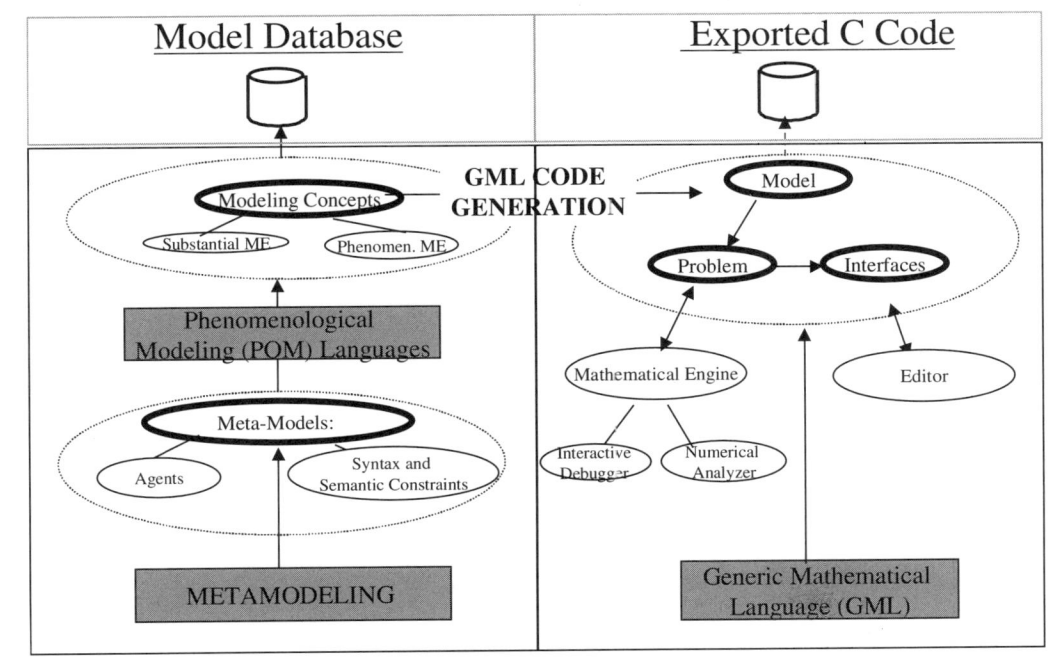

Figure 2. Overview of the system architecture.

In all these computer-aided efforts, a fixed set of modeling elements serve as the building blocks for model generation. It is worth mentioning that phenomena-based languages have a narrower range of applicability than EOM systems due their compressed information content. More informal POM languages can therefore be applied for specific domain with explicit user expectations. In general situations, a POM language may not offer adequate language elements with a suitable mapping onto the describing equations. In consequence, it cannot be expected that a unique modeling language will make phenomena-oriented modeling feasible. The multitude of modeling environments developed in different communities of science and engineering gives evidence of this hypothesis.

Meta-Modeling

The understanding of the evolutionary character of modeling paradigms leads us to proposition of the *meta-modeling* concept. Fig. 2. gives an overview of the TechTool architecture whose detailed description is well beyond the scope of this brief paper. It will allow for the gradual adaptation of the modeling paradigm as well as for the ad-hoc invention of entirely "new" concepts. Interpretation of the modeling elements leads to equivalent mathematical expressions at the highest level of abstraction. In this new modeling philosophy, process models will be represented by the following features: (i) a *vocabulary* of phenomena that represents the building blocks of the modeling language. (ii) An analogous *mathematical representation* that capture the model behavior in terms of differential and algebraic equations. (iii) A *model paradigm* that embeds the modeling concepts, the information for their consistent interpretation, as well as a set of specific rules that enforce syntactic and semantic constraints of the language.

Modeling Elements – The Vocabulary of the POM Language

Modeling elements constitute the words in the formal POM language and may be specialized by attributes or parameters. A conceptual language for chemical engineering composed of two classes of modeling elements was proposed in Linninger et al., 1998a: It follows Marquardt's definitions of substantial and phenomenological complexity (Lohmann and Marquardt, 1996). Substantial modeling concepts possess a specific physical or logical scope and usually describe separate physical entities. Most substantial modeling elements are characterized by extensive properties or delineated geometric dimensions. These concepts pertain to physical units and their connectivity, i.e. devices and links. Phenomenological complexity stems from the behavior of a substantial element. Phenomenological modeling elements refer to mathematical relations that determine the state or state transition of a substantial modeling element. Typically they involve generic mathematical expressions

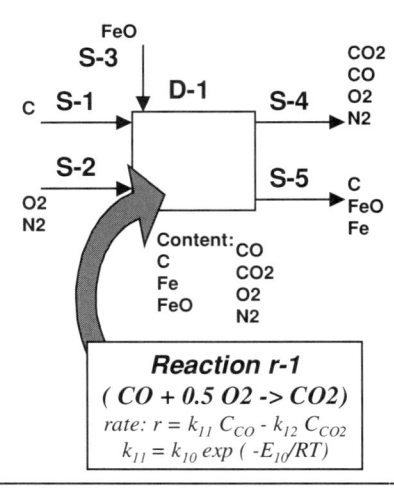

Figure 3. Addition of a reaction model.

such as reactions, transport phenomena, balance equations, and procedures for physical properties calculations.

Model Interpretation

The *model interpretation* step maps each of the modeling elements, into an equivalent mathematical expression. In the proposed environment, *procedural agents* accomplish this task. Each agent generates the system equations for an entire class of concepts giving rise to similar interpretation for members of same class. The objects themselves contribute the specific part of the information, while the agents supervise the generic code generation. A single phenomenon may give rise to different expressions depending on the presence of other concepts. Context-sensitive interactions are unavoidable, unless one wants to create a language that is as rigid as its mathematical abstractions.

This is best illustrated by a concrete example. Consider a modeling language that recognizes kinetic reaction models. Selection of a particular phenomenon, e.g. homogeneous first order *reaction* into the context of a *reactor vessel* triggers the execution of a *reaction agent*. This agent automatically updates the code, i.e. equations describing the reactor by building the specific reaction term into the vessel by using the pseudo code of Fig. 3. Similarly, the same agent produces the correct equations for another object in the reaction hierarchy, e.g. second-order reaction. In the latter case, the agent extracts the specific reaction term from the reaction template again.

Finally, model interpretation produces equation-based representations using the generic mathematical language (GML) described in Linninger, 1998. The mathematical aspects of a models can later be refined using a new object inheritance framework as described in Linninger et al.,

1998a. In effect, all models generated using any paradigm in the present methodology are compatible at the level of system equations.

Syntax and Semantics Consistency

The meta-level also provides means for the examination of model consistency. The syntax of the language can be enforced by specific rules that delineate the intention and appropriate utilization of the respective modeling element. The semantics of a model is more difficult to examine. It requires an assessment of the model's meaning. Fig. 4 shows that the semantics consistency checks of model are not trivial. Fig 4a shows a loop in a reaction network. Such a closed loop would make no sense in a practical reactor configuration and thus violate the semantic meaning. When adding a feed stream to first vessel and a purge stream, see Fig. 4b, the network may indeed represent a valid situation.

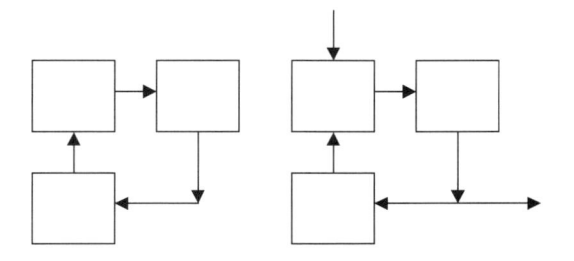

Figure 4 (a) and (b) - reactor network with loop.

We propose a *semantic rules concept* to address this problem. This feature is a rule-base which allows language designers to implement a set of first-order logical statements to prevent future misuse of this modeling concepts. These rules will be fired on any model instance and flag semantic rule violations of evolving applications.

Evolution of a POM language

Meta-modeling creates two levels of users. *Application designers* are modelers who implement process models using an instance of a POM language. *Paradigm developers* may augment or modify the paradigms of this language using the tools of the meta-level. Advanced languages add new features and are therefore backward compatible. For revisions of the same paradigm two outcomes are possible: Paradigms, which partially revise earlier model interpretations, can be incompatible with previously defined model applications. If the old interpretation was at fault, loading the model with the new paradigm fixes a modeling error. If the original behavior was desired, then the new and the old versions implement alternative interpretations of a physical phenomenon. Both versions may be needed in a multi-language modeling environment for the different levels of model detail and depend on the maturity of an emerging model, e.g. steady state versus dynamic view.

Conclusions

The necessity for meta-modeling arises from the different model scopes and the understanding of the transient nature of the real systems in practice. The approach presented in this article differs from the earlier work in that it makes the evolution of the adequate modeling information formalisms integral part of the modeling methodology. Instead of a careful design of the ultimate modeling language, we propose to design a system, which supports the ad-hoc evolution and generation of modeling paradigms. We acknowledge that these steps are implicit part of a human cognitive approach to modeling and in such are nothing new. The ability for designers to dynamically reconfigure the "rules" of their modeling environment and a framework to do this in a structured and formal way represents a new avenue in computer-aided model generation.

Acknowledgements

Financial support of this project by VAI is gratefully acknowledged.

References

ABACUSS; (1999). Available from: http://web.mit.edu/afs/athena/org/c/cheme/www/People/Faculty/Barton_Paul.html

Barton P.I. and Pantelides, C.C. (1993). gPROMS – A combined discrete/continuous modeling environment for chemical processing systems. *Sim. Ser.* **25**, 3, 25.

Bieszczad, J., Kolouris, A., Stephanopoulos, G., Foss, A., Geurts, K., and P. Goodeve (1998). Phenomena-based modeling using Model.la – Enabling students and computers to communicate in the language of chemical engineering. Paper, 172h, AIChE Meeting, Miami.

Linninger, A. A. (1998). Towards integrated Computer-Aided Process Engineering. Paper 240i, AIChE Meeting, Miami.

Linninger, A., A., Krendl, H., and H. Pinger (1998a). An initiative for integrated computer-aided process engineering. *Proc. FOCAPO '98*, Snowbird, Utah.

Linninger, A. A, Salomone, E., Ali, S.A., Stephanopoulos, E., and G. Stephanopoulos (1998b). Pollution prevention for production systems of energetic materials. *Waste Management*, **17**, 2/3, 165-173.

Lohmann B.; and W. Marquardt (1996). *Rechnerdemonstration European Symposium on Computer-Aided Process Engineering.* ESCAPE-6, Rhodes, Greece.

Lohmann, B. (1998). Ansaetze zur Unterstuetzung des Arbeitsablaufes bei der rechnerbasierten Modellierung verfahrenstechnischer Prozesse. PhD Thesis, D82, RWTH Aachen.

Piela, P., Epperly, T., Westerberg, K., and A. Westerberg (1991). ASCEND. An object-oriented computer environment for modeling and analysis: The modeling language. *Comp. Chem. Eng.*, **15**, 1, 53-72.

Stephanopoulos, G., Henning, G., and H. Leone (1990). MODEL.LA. A modeling language for process engineering I - The formal framework, *Comp. Chem. Eng.*, **14**, 8, 813 – 869.

INTEGRATED USE OF CAPE TOOLS - AN INDUSTRIAL EXAMPLE

H. H. Mayer and H. Schoenmakers
BASF AG
D-67056 Ludwigshafenm, Germany

Abstract

CAPE, Computer Aided Process Engineering, is currently applied in industry as a common tool in certain phases of the lifecycle of a process. The understanding of CAPE is that it may be applied in every phase of the lifecycle as an integrated tool. In the contribution the status of present applications is compared with the possible application fields under different aspects, especially the availability of suitable software for computing tools, for databanks and for interfaces on the one hand and of the economy of any application on the other hand. It turns out that presently the application is established in the phases of process development and project management and is coming up in the phase of plant operation. Needs for further development exist for tools, for example for the insertion of units containing solids in process simualators and for the interfaces between different tools and databanks.

Keywords

CAPE, Project lifecycle, Software application.

Introduction

One common definition of CAPE is the familiar ESCAPE-Definition by Lien (1996):

"Cape is the application of a systems modeling approach to the study of processes. as an integrated whole, from the viewpoint of development, design and operation."

This definition is a global one as it should be for a definition that tries to cover the whole field of possible applications.

In industrial reality the application of CAPE is determined by the developments of science and software tools under the aspect of practical benefits. Emphasis is centered there where real advantages are reached in time or quality or both. In addition the development trends and wishes may differ depending from the viewpoint of the three "players" that are on the market on the field of CAPE: The industry, the software companies and the universities.

The difference between the global approach of the definition and the industrial practice shall be described here.

Process Lifecycle and Application of CAPE

As stated in the definition, Cape tools should be applied from the very beginning of process concepts to the last shut down, that means from the cradle to the grave of a plant.

At the beginning of a process lifecycle CAPE application may contribute to product design, may continue with support for the chemical reaction path and the process synthesis, may fasten and improve the process design and enable the project team to make project management more effective. As well the running process may be enhanced by using CAPE for process control and even for handling commercial data of the plant

The main CAPE elements are:

- Databanks for basic data like physical properties, reaction data, cost data etc.,
- simulation or design tools for explicit use in different applications, from short cut to rigorous,

- data management tool(s) to combine and to connect the other elements including the management of the interfaces.

A broad offer of commercial tools is on the market and in addition in house tools exist in many chemical companies. Yet in the phases of a process lifecycle the availability and the quality for practical use of tools differ largely:

Product Properties

Product properties are the application field for molecular modeling tools. This work is done at some universities. It demands a tremendous amount of work to get results and the results till now describe only ideal systems. These tools are a typical "specialist software" for a few engineers or chemists with special knowledge.

So the application of CAPE tools in this phase is actually limited but it can be expected that the molecular modeling methods will improve strongly in the future.

Process Design and Development

Among the numerous tools for the *design step of a process* two packages exist that try to solve this problem in a totally different way:

The first one is the Mayflower package of Profs. Doherty and Malone from the University of Massachusetts, Amherst that is commercialized with Hyprotech as HYSIS.Concept. Its method are residue curves map calculations and the results may be used to design separation sequences for a fluid process.

The second one is the PROSYN-system of GHN in Dortmund/Germany, see Schembecker (1996), that uses the expert system approach. Here a process is designed on the basis of documented experience supported by numerical. Its results are suggestions for a process structure for a certain range of process options.

These approaches underline that there is no single tool for solving together the complete procedure, the exact solution and the complete range of possible processes. At least there is no tool suitable for the "normal" process engineer. Thus the design stage of a process can be supported by CAPE tools, but not as an integrated process starting with databanks containing the information needed and finishing with transferring the complete results through an interface to the next step in the work chain.

This next step is the process development. Here different tools have to be distinguished:

(a) Databanks for physical properties, costs and other basis data,
(b) Process simulation programs for mass and energy balances and operation data of the units,
(c) Design tools for equipment, specially heat exchangers, flares, fired, heaters etc. using explicit models or, just recently and, growing, CFD-tools,
(d) Data management tools for documentation containing the management of the interfaces.

For (a) - (c) a lot of commercial programs from software suppliers and in house programs from different chemical companies exist. The reason is that CAPE has its origins in process development.

It would need an excessive amount of space to comment the market for all the available software packages, and so only some remarks shall be made:

(a) The databanks are sufficient for the properties of fluids and gases, but users often have the difficulty not to have access to internal measurements of chemical companies. Thus estimation methods for physical properties like UNIFAC have to be applied.
(b) Many process simulators are available which in general have a good quality and user friendliness. They solve the normal balancing problems for fluid processes, but yet in a unsatisfactory way for processes containing solids. Some of them don't have optimal solving algorithms. The sequential modular approach for example seems to be more suitable for getting good starting values than for effective calculation of parameter variations. Here the equation based approach is much better.
(c) The design tools are complete for heat exchangers. For other devices exists at least one special supplier with one important exception: no software is on the market for the exact sizing of distillation columns. A growing application field is Computational Fluid Dynamics, CFD. Tools exist, but they again are "expert tools", not yet suitable for general use.
(d) Here the situation is completely different. There are some offers for data managament tools - ZYQAD from the UK that is now with ASPEN Tech and LOGOCAD from Germany (see LogoCAD) for example - but in practice they are rarely used because the implementation takes a lot of time and cost. Later in my presentation I will comment on this.

More as the phase of process design the phase of process development can be supported efficiently by CAPE-tools. Remaining limitations and lacks will be described later on.

Project Engineering and Management

In this phase of a process lifecycle an enormous amount of data has to be managed. These data come from process simulators and design software as an input from process development and - data specifically produced during the project engineering phase - from different CAD-tools, piping data and many listings of detail equipment. All these data have to be combined to get a plant documentation that is good enough for erecting the plant and for approval procedures of the authorities.

A large offer of software is available for all these separate tasks. The main program types are:

(a) databanks for the physical properties of the components, as described for the process development phase,
(b) CAD tools for flowsheets, for the electrical and control schemes, for the civil engineering documentation, for isometrics of piping, for the underground plans,
(c) design software for heat exchangers, for the piping data, for safety valves etc., etc.,
(d) databanks for any other information, often combined with CAD-software

These CAD-flowsheets with databank connections - "intelligent flowsheets"- are offered within the software packages of every deliverer. But very few independent software tools for intelligent flowsheets are available that - at least in principle - allow to combine the data from different sources.

Nevertheless an efficient support is actually possible in the phase of project engineering and management.

Production, Plant Operation

In the phase of plant operation CAPE-support may happen on the fields of trouble shooting and process control. In addition data for the economic evaluation of a plant may be gained. The main source of information are the data from the control systems. The extraction of these data needs special software like PI (Process Information System) or SIM21 from Aspen Tech. Than the data have to be validated. This can be done with data reconciliation software like VALI from Belsim im Belgium, RT-OPT from Aspen Tech or DATACON from SimSci. After the extraction and the validation of process data procedures like optimization can be startet. For the purpose of process control different software can be used like stationary or dynamic process simulators or special optimization software. Of course these tools have to be adapted to the special plant, so that a lot of work is connected with such applications.

It can easily be recognized from this enumeration that data management again is the main problem in this phase of a process' lifecycle.

Market

CAPE tools can support the product in the market in principle by transporting market data back to the process to influence capacities, quality data, product mix in order to optimize the process under the actual market situation. Few software is available. Some aspects are covered by batch process simulation tools like SIMPLE++.

It should be stated, that a lot of single tools are available but they cover not yet the whole range of possible support during process lifetime completely. The data management between these different elements in most cases is more complicated than the generation of data for a special phase of the lifecycle.

Practice in BASF as an Example

Till now the CAPE-potential was summarized under the view of estimated potential and support that is possible with available software. The next question is, how the industry uses CAPE in reality and how the experiences are.

As an example, how a big chemical company realizes CAPE-support, the practice of BASF shall be described:

BASF uses CAPE-tools consequently in the phases of process development and process engineering and management. Application during the phase of plant operation is developing. Other phases of process lifecycle are only partly concerned, the reasons are mainly the lack of software, that is the case with product development, or the bad relation between the estimated amount of work necessary for installation and the expected benefit. Another reason is that in different divisions different interests exist and CAPE is typically exceeding divisions limits. The coordination of these interests and the effort to bring them to an agreement may be difficult.

Process Development

Elements of the application of CAPE tools in the phase of process Development are:

- Process synthesis, inhouse programs
- Physical Properties Databank for pure components and mixtures including internal measurements, measurement data from the commercially available databanks,
- Physical Properties Parameter Databank with validated parameters for the main properties of pure components and mixtures for direct use in the simulation programs,
- Programs for thermodynamic calculations,
- Flowsheet simulator, equation based inhouse program CHEMASIM, see Hegner (1985),
- Equipment design for columns (inhouse programs), heat exchangers (inhouse programs), heat exchangers (commercial programs), other equipment (mainly inhouse programs), cost estimation (inhouse program), data reconciliation (commercial program) and CFD-tools.

All these programs are - more or less - included in a common software environment, called ProcessNet that is based on mainframes. This environment has common interfaces and thus it is possible to use the same set of physical properties to transfer input data and results from one program to another and to do the documentation of the results. Besides the ProcessNet the ASPEN PLUS program system is applied in BASF with some interfaces to the ProcessNet. It is used in parallel to the in house simulator CHEMASIM.

Project Engineering and Management

The main CAPE elements used in BASF in this phase are:

- Flowsheet simulation with ASPENPLUS,
- CAD programs, mainly INTERGRAPH-tools for the different applications, for example the P&ID's,
- Different programs for piping data, for equipment design, for the handling of data for machines and apparatuses, for the handling of electrical and control data, for civil engineering applications, for cost estimation etc..

Interfaces to transfer and to extract data from these programs for use in other programs or as technical information for the equipment lists exist partially. Normally the results are transferred in the form of flow sheet information from process simulators or they are extracted as technical data sheets. The interface between the Process Net and the programs used in the project engineering phase is twofold, either the ASPEN PLUS based interfaces or in house interfaces are applied. A concept for the integrated handling of date is under development.

CAPE in Other Phases of the Process Lifecycle

Some efforts are made to use CAPE tools in BASF for control purposes. Dynamic simulation is used to design difficult control schemes or to answer specific questions concerning process dynamics.

Some data reconciliation projects have been started with the aim to enable process control.

All these activities actually are isolated applications of CAPE tools, not integrated in the sense of the CAPE definition.

Summarizing the BASF position it has to be stated that in BASF CAPE activities focus on the phases of process development and of project engineering phase. During plant operation the application is growing. Within these phases these activities cover all the needs for databanks, computational power and internal data transfer With respect to data transfer between the two main process phases for CAPE applications they are more or less integrated and there are some interfaces for transfer and communication.

Even if a high standard is still reached there is a need for further development, described for example by Bessling (1997) and Mayer (1998). Of special importance is the inclusion of solids containing unit operations inprocess simulators and the offer of better data management tools. Some funded projects like CAPE OPEN are on the way (http//www.quantisci...).

Practice in other Chemical Companies, an Overview

The amount of CAPE tools used depends on the strategic orientation of the company. If engineering work is sourced out to a large degree, the contractors need CAPE tools, not the contracting company. Thus an integrated application of CAPE tools in the sense of the above definition can be expected only in companies carrying out most of their engineering work inside. This is true mainly for the big chemical companies, it doesn't concern the petrochemicals in the same way. So it can be expected that the main users of integrated CAPE tools will be chemical companies above a certain size that have service units for chemical engineering tasks and - still more favourable for CAPE application - do the project management mainly inside the company.

Conclusion

The CAPE-idea is to use CAPE tools in every phase of the lifecycle of a process and to transfer data from each phase to the following one for further use. It has been demonstrated that in every phase of the lifecycle possible applications exist and that a lot of tools are available. These tools do not cover the whole field, specially software for integrated tools and data management is not as much developed as computing software. But a lot of applications is possible. In industry only some of these applications are realized. The reasons are workflows, the division of engineering work and the economical justification. In industry CAPE tools are applied only, where benefits can be identified.

References

Bessling, B., B. Lohe, H. Schoenmakers, S. Scholl and H. Staatz (1997). CAPE in process design - Potential and limitations, *Comp. Chem. Eng.*, **21**, S17-S21.

Hegner, B. and H Schoenmakers (1985). Chemasim - experience with BASF's simultaneous process simluator. *Inst. Chem. Eng. Symp. Ser.* **92**, 365-375.

http//www.quantisci.co.uk/CAPE-OPEN

Lien, K.and T. Perris (1996). Future directions for CAPE research. *Comp. Chem. Eng.* **20**, 1551.

LogoCAD, product of INNOTEC Systemtechnik GmbH, Loher Straße 1, D-58332 Schwelm, Germany.

Mayer, H.H. and H. Schoenmakers (1998). Application of CAPE in industry – Status and outlook. *Comp. Chem. Eng.*, **22**, S1061-S1069.

Schembecker, G. and K.H Simmrock (1996). Heuristic numeric process synthesis with PROSYN. *AIChE Symp. Ser.*, **92**, No. 312, 275

A GENERAL FRAMEWORK FOR CONSIDERING DATA PRECISION IN OPTIMAL REGRESSION MODELS

Mordechai Shacham
Ben Gurion University of the Negev
Beer-Sheva 84105, Israel

Neima Brauner
Tel-Aviv University
Tel-Aviv 69978, Israel

Abstract

Construction of optimal (stable and of the highest possible accuracy) regression models comprising of linear combination of independent variable/s and their nonlinear functions is considered. It is shown that estimates of the experimental error which are most often available for engineers and experimental scientists, can be utilized for obtaining the optimal regression model and/or identifying the main causes preventing further improvement of a stable regression model. An orthogonalized variable based stepwise regression procedure (SROV), which is used for selection of the optimal model and for regression diagnostic, is presented. In this procedure, two indicators based on experimental error estimates are employed to analyze the stability of the regression model, while the error propagation in the various linear on nonlinear terms is estimated using perturbation techniques. The use of the proposed indicators and algorithm enables obtaining stable regression models that are much more accurate than the ones that have been published in the literature.

Keywords

Optimal regression model, Collinearity, Stepwise regression, Precision, Noise.

Introduction

Precise modeling and regression of experimental data becomes increasingly important as computer-based modeling, design, and control of chemical processes becomes more widespread. The accuracy of the models depends on the precision of the experimental data that is used for regression of the model's parameters. However, the precision also critically depends on the selection of appropriate explanatory variables (independent variables and their functions). If an insufficient number of explanatory variables are included in the equation, the regression model will not be able to reach the accuracy limit dictated by the precision of the experimental data. On the other hand, if redundant variables are included, the model becomes unstable (ill conditioned), whereby adding or removing experimental points from the data set may drastically change the parameter values. The derivatives of the dependent variable are not represented correctly and extrapolation outside the region, where the measurements were taken, yields absurd results even for a small range of extrapolation. Shacham and Brauner (1997) and Brauner and Shacham (1998a,b) provide several examples where regression models published in the chemical engineering literature are grossly inaccurate and/or unstable. The objective in regression of experimental data should be attaining an optimal regression model, where optimal model is defined as the most accurate (minimal variance) stable model. The most frequent causes of inaccuracy and/or ill-conditioning in regression are the following: non optimal or inadequate model (not all the influential explanatory variables are included in the model and/or non-influential variables are included); excessive errors in the data (as in the presence of the outlying measurements); presence of collinearity among the explanatory variables; and high sensitivity of the algorithm used for calculating the model parameters to numerical error propagation.

In view of the possible existence of several different causes for non-optimality, ill-conditioning or statistical invalidity of the regression model, the selection of the most adequate, optimal model should proceed in an iterative manner. An initial pool of explanatory variables, which can potentially be included in the regression model, is prepared. Then stepwise regression to identify the variables which should be included in an optimal, statistically valid model is carried out (an algorithm, which has low sensitivity to numerical error propagation must be used for regression). The cause/s that limit the accuracy and the number of explanatory variables that can be included in the model (like collinearity, outlying measurements and influential variables missing from the initial pool) are identified and remedial actions are taken (if possible). Then the process is reiterated to check whether further improvement of the model is possible.

There are several stepwise regression algorithms and programs available (for details see, for example, Neter *et al.*, 1990). These are however, rarely used by the engineering and experimental sciences communities. One reason is that existing programs are based on statistical indicators, where the validity of these indicators depends on certain conditions the data must fulfill. It has been long recognized that it is very difficult to find real data that meet all the necessary statistical requirements. Furthermore, existing stepwise regression programs overwhelm the user with statistical information, making him uncertain whether results obtained with data and model that do not meet all the statistical requirements can be considered statistically valid.

We have developed a stepwise regression procedure based on orthogonal variables (SROV). The two principal indicators used in this procedure are based on estimates of experimental errors and the use of the error propagation formula, rather than on statistical principles. These indicators are used to identify the optimal regression model and for diagnosing the main causes that limit the model accuracy.

Stepwise Regressing Using Orthogonalized Variables (SROV, Shacham and Brauner, 1999)

A standard linear regression model can be written:

$$\mathbf{y} = \beta_0 + \beta_1 \mathbf{x}_1 + \beta_2 \mathbf{x}_2 \cdots + \beta_n \mathbf{x}_n + \varepsilon \quad (1)$$

where $\mathbf{y}$ is an N-vector of the dependent variable, $\mathbf{x}_j$ (j = 1,2,$\cdots n$) are N vectors of explanatory variables, $\beta_0, \beta_1 \cdots \beta_n$, are the model parameters to be estimated and ε is an N-vector of stochastic terms (measurement errors). It should be noted that an explanatory variable could represent an independent variable or a function of one or more independent variables.

A certain error (disturbance, imprecision, noise) in the explanatory variables is also considered. Thus, a vector of an explanatory variable can be represented by $\mathbf{x}_j = \hat{\mathbf{x}}_j + \delta \mathbf{x}_j$, where $\hat{\mathbf{x}}_j$ is an N-vector of expected value of $\mathbf{x}_j$ and $\delta \mathbf{x}_j$ is an N-vector of stochastic terms due to noise. If estimates of the experimental errors are available, these can be used for $\delta \mathbf{x}_j$ and ε. Otherwise, it is usually assumed that the data is correct up to the last decimal digit reported. In such cases, the average rounding error can be used. If functions of the independent variables are used, or data transformation is carried out, the error propagation formula is used to calculate the resultant $\delta \mathbf{x}_j$. The vector of estimated parameters $\hat{\boldsymbol{\beta}}^T = (\hat{\beta}, \hat{\beta}_1, \cdots \hat{\beta}_n)$ is often calculated via the least square error approach by solving the normal equation: $\mathbf{X}^T \mathbf{X} \hat{\boldsymbol{\beta}} = \mathbf{X}^T \mathbf{y}$, where $\mathbf{X} = [\mathbf{1}, \mathbf{x}_1, \mathbf{x}_2, \cdots \mathbf{x}_n]$ is the data matrix and $\mathbf{X}^T \mathbf{X} = \mathbf{A}$ is the normal matrix. This method is subject to accelerated numerical error propagation due to collinearity (see for example, Brauner and Shacham, 1998a). The following SROV procedure is much less sensitive to numerical error propagation and as such, is more appropriate to be used in a general-purpose stepwise regression program.

The SROV procedure consists of successive stages, where at each stage one of the explanatory variables, say $\mathbf{x}_p$, is selected to enter the regression model. The explanatory variables, which have already been included in the regression model (at previous stages), are referred to as *basic variables*, and the remaining explanatory variables are the *non-basic variables*. At each stage, the non-basic variables and the dependent variable are first updated, by subtracting the information that is collinear with the basic variables. This updating generates a dependent variable and non-basic variables which are orthogonal to the basic variables set. When progressing from stage k to stage $k+1$ in the stepwise regression procedure, the parameter estimate corresponding to the explanatory variable $\mathbf{x}_p$ (selected as a basic variable at stage k) is obtained by:

$$\hat{\beta}^{k+1} \equiv \hat{\beta}_p = \frac{\mathbf{y}_k^T \mathbf{x}_p}{\mathbf{x}_p^T \mathbf{x}_p}; \quad \mathbf{x}_p \equiv \mathbf{x}^{k+1} \quad (2)$$

The updated (residual) values of the dependent variable $\mathbf{y}^{k+1}$, and the orthogonal components (residuals) of the non-basic variables are obtained by:

$$\mathbf{y}^{k+1} = \mathbf{y}^k - \beta_p \mathbf{x}_p \quad (3)$$

$$\mathbf{x}_j^{k+1} = \mathbf{x}_j^k - \mathbf{x}_p \left(\frac{(\mathbf{x}_j^k)^T \mathbf{x}_p}{\mathbf{x}_p^T \mathbf{x}_p} \right) \quad (4)$$

The model variance at this stage,

$$s^2 = \frac{(\mathbf{y}^{k+1})^T (\mathbf{y}^{k+1})}{\upsilon} \quad (5)$$

is the sum of squares of errors divided by the degrees of freedom: $\upsilon = N - n - 1$. Smaller variance indicates a better fit of the model to the data. The confidence interval, $\Delta\beta_p$ on a parameter estimate can be defined:

$$\Delta\beta_p = t(\upsilon, \alpha) \sqrt{s^2 (\mathbf{x}_p^T \mathbf{x}_p)} \quad (6)$$

where $t(\upsilon, \alpha)$ is the statistical t distribution corresponding to a desired confidence level, α and s is the standard error of the estimate. Clearly, if $|\hat{\beta}_p| < |\Delta\beta_p|$, then the zero value is included inside the confidence interval, Thus, there is no statistical justification to include the associated term in the regression model.

Criteria Used for Selection of a Basic Variable

The strength of the linear correlation between an explanatory variable $\mathbf{x}_j$, and a dependent variable $\mathbf{y}$ is measured by

$$YX_j = \mathbf{y}^T \mathbf{x}_j \quad (7)$$

where $\mathbf{y}$ and $\mathbf{x}_j$ are centered and normalized to a unit length. The value of $|YX_j|$ is in the range [0,1]. In case of a perfect correlation between $\mathbf{y}$ and $\mathbf{x}_j$ ($\mathbf{y}$ is aligned in the $\mathbf{x}_j$ direction), $|YX_j| = 1$. In case $\mathbf{y}$ is unaffected by $\mathbf{x}_j$ (the two vectors are orthogonal), $YX_j=0$. The inclusion of a variable $\mathbf{x}_p$, which has the highest level of correlation with $\mathbf{y}$ in the basic set ($|YX_p|$ is the closest to one) will affect the maximal reduction of the variance of the regression model. Therefore, the criterion $\mathbf{x}_p = \mathbf{x}_j \{\max | YX_j|\}$ is used to determine which of the non-basic variables should preferably be included in the regression model at the next stage, provided that the following *CNR* and *TNR* tests are both satisfied. The *CNRj* measures the signal-to-noise ratio of YX_j, and is defined by:

$$CNR_j = \left\{ \frac{|\mathbf{y}^T \mathbf{x}_j|}{\sum_{i=1}^{N} \left(|x_{ij}\varepsilon_i| + |y_i \delta x_{ij}| \right)} \right\} \quad (8)$$

Note that the denominator of eq. (8) represents the error in YX_j as estimated via the error propagation formula. A value of $CNR_j \gg 1$ signals that the correlation between x_j and $\mathbf{y}$ is significantly larger than the noise level. Thus, an accurate value of YX_j can be calculated. But when $CNR_j \leq 1$, the noise in YX_j, as affected by $\delta \mathbf{x}_j$ and ε is as large as, or even larger than $|YX_j|$. If this is the case, no reliable value for $|YX_j|$ can be obtained and the respective variable should not be included in the regression model.

The *TNR_j* measures the signal-to-noise ratio in an explanatory variable $\mathbf{x}_j$. It is defined in terms of the corresponding Euclidean norms (Brauner and Shacham, 1998a):

$$TNR_j = \frac{\|\mathbf{x}_j\|}{\|\delta \mathbf{x}_j\|} = \left\{ \frac{\mathbf{x}_j^T \mathbf{x}_j}{\delta \mathbf{x}_j^T \delta \mathbf{x}_j} \right\}^{1/2} \quad (9)$$

A value of $TNR_j \gg 1$ indicates that the (non-basic) explanatory variable $\mathbf{x}_j$, contains valuable information. On the other hand, a value of $TNR_j \leq 1$ implies that the information included in $\mathbf{x}_j$, is mostly noise. Thus it is practically collinear with the basic variables and should not be added to the model.

For calculation of CNR_j and TNR_j, ε, and $\delta \mathbf{x}_j$ values must be provided by the user. These values are based on the estimated experimental error or precision of the measurement and control devices. In the first stage of the regression, the values provided by the user applied. In subsequent stages, those estimates are updated using numerical perturbation. To this aim, the regression is carried out using two data sets in parallel, the original and perturbed data sets. The differences between the updated values of the explanatory variables (eq. 4) and the dependent variable (eq. 3), obtained with the original and perturbed data sets provide the estimates for the updated values of ε and δ xj at each new stage. (For more detailed explanation, see Shacham and Brauner, 1999). The selection of new variables (from among the non-basic variables) to be added to the basic variables in the SROV procedure stops when for all the non-basic variables either $CNRj \leq 1$ or $TNRj \leq 1$.

The SROV procedure consists of two phases. In the first phase, an initial (nearly optimal) solution is found. In the second phase, the variables are rotated in an attempt to improve the model. In the 0th stage of phase 1, the dependent and explanatory variables are centered (except for models which do not include a free parameter where centering is avoided). The first variable selected to enter the basis is determined using equations (7), (8) and (9) and the concepts that were explained in the previous paragraph. Then the calculation of the corresponding parameter value β^1 and the updating of the dependent and the remaining explanatory variables (and the corresponding errors) are carried out. In all subsequent stages of phase 1, the selection of additional variables to enter the basis, and the updating of the dependent and explanatory variables, is performed the same way as in the 0th stage. As noted above, the stopping criteria of phase 1 is $CNRj \leq 1$ or $TNRj \leq 1$ for all the remaining non-basic variables.

When the correlation between the original explanatory variables is weak (they are nearly orthogonal) the regression model found in phase 1 is the optimal with a minimal variance (and the sum of squares of errors) value. Howe- ver, if there is a considerable collinearity among the expla- natory variables, the order in which they enter the basis may change their effect on the reduction of the variance. In such cases, rotation of the variables can lead to a solu- tion with a smaller variance. Thus, in phase 2, the variables in the basis are rotated so that each of them is tested versus the non-basic variables and reselected as the last one to enter the basis. If a new variable enters the basis during rotation, a new rotation cycle (a new phase) is started. Before starting a new phase, all the variables are set back to their original values. Only the order at which they entered the basis in the previous phase is retained.

Regression Diagnostics and Model Improvement

After an optimal regression model has been found the indi- cators of the SROV procedure can be used for further diag- nostics. The analysis is based on comparison of the values of the statistical indicators (mainly the confidence interval/parameter value ratio) with the values of the data precision-based indicators (*CNRj and TNRj*). If both types of indicators rank the model obtained as the optimal one, further improvement of the model is possible only by providing more precise data of **X** and/or **y**. When precision based indicators show that more terms can be added to the model (improving further its accuracy) but when these terms are added the associated parameters values are not significantly different from zero, this usually indicates that the dominant limiting factor is not the precision of the data and other reasons must be sought. Several possible reasons were listed in the "Introduction" section. The real reason can be identified using the precision-based indicators and residual plots. Some examples for model improvements achieved using the proposed method are discussed in the next section.

Benefits of Data-precision Considerations in Optimal Regression Models

Using the SROV procedure, we have solved large-scale problems for linear, polynomial, quadratic and other regression models containing linear combination of various nonlinear terms. Due to space limitations, detailed examples that demonstrate the advantages of the proposed method could not be included in the paper. Only brief descriptions of some of the examples that have been solved are included (detailed descriptions will be presented in the conference).

One of the examples was presented by Wold *et al* (1984). They assumed that collinearity limits the number of independent variables that can be included to two out of eight independent variables, and attempted to stabilize the model using ridge regression, principal component regre- ssion and other techniques. These attempts yielded absurd results, where some of the parameters could be positive or negative, depending on the stabilization technique used. The techniques presented here eliminated collinearity as possible cause but identified an inappropriate model and existence of outlyers as the causes of ill conditioning. Identifying a more appropriate nonlinear model and removing the outlyers enabled reduction of the variance by more than an order of magnitude in a stable regression model.

Employing the SROV procedure on nitrogen's vapor pressure data from Wagner (1973) for a polynomial model has shown that the order of a stable polynomial model is not necessarily limited (to six, for example, as postulated by many). If very high precision data in a wide tempe- rature range are available, a high order, stable, polynomial model can be used to obtain the highest possible accuracy. For the nitrogen vapor pressure data, for example, a 14th order polynomial had to be used and still collinearity did not pose a limitation.

Using heat capacity data, it has been shown that in polynomial regression it can often be advantageous to include in the model non-consecutive powers of the inde- pendent variable. In regression of heat capacity data for solid 1-propanol, for example, it was found that the optimal model, which contains 5 non-consecutive powers of the independent variable, is a stable model which is an order of magnitude more accurate than the 3rd order poly- nomial that was recommended for this property in the literature.

Considering data precision in regression of expe- rimental data enables identifying the real reasons that limit the accuracy and stability of regression models. Using the methods described in this paper can help to obtain the most accurate and stable regression model for experimental data. This, in turn, can improve considerably the precision and reliability of process calculations and computations.

References

Brauner, N. and M. Shacham (1998a). Identifying and removing sources of imprecision in polynomial regression. *J. Math. Comp. Sim.*, **48**, 77-93.

Brauner, N. and M. Shacham (1998b). Role of range and precision of the independent variable in regression of data. *AIChE J.*, **44**, 603-611.

Neter, J., Wasserman, W. and M.H. Kutner (1990). *Applied Linear Statistical Models.* Irwin, Burr Ridge.

Shacham, M. and N. Brauner (1997). Minimizing the effects of collinearity in polynomial regression. *Ind. Eng. Chem. Res.*, **36** (10), 4405-4412.

Shacham, M. and N. Brauner (1999). A stepwise regression procedure based on data precision and collinearity considerations. Submitted for publication.

Wagner, W. (1973). New vapor pressure measurements for argon and nitrogen and a new method for establishing rational vapor pressure equations. *Cryogenics*, **13**, 470-482.

Wold, S., Ruhe, A., Wold, H., and W.J. Dunn III (1984). The collinearity problem in linear regression. The partial least squares (PLS) approach to generalized inverses. *Siam. J. Stat. Comp.*, **5**, 735-743.

REDEFINING THE PROCESS SIMULATION FLOWHSEET: A COMPONENT BASED APPROACH

Carl Spears and Vince Tassone
Hyprotech Ltd.
Calgary, Alberta T2E 2R2, Canada

Abstract

With the ever increasing demand placed on process simulators to deliver more sophisticated, more detailed functionality, so has the size and complexity of the programs. Although, software development approaches based on object oriented designed programs have gone a long way to help manage large software development projects, a need still exists to continually simplify the software development process. Recently, a move has been made towards the development of componentized/distributed simulation software systems allowing for easier development and integration of engineering technology into the process simulator. This has been accomplished through the adoption of current off the shelf PC based software development technology such as MFC/ATL COM and OLE/Active X technologies. The end product being a component based process simulator with open architecture allowing for customization of the process simulator. A unit operation palette now becomes a "suite" of components or an "ocx-palette," the engineering user builds a flowsheet composed of unit operation controls as one would build a VB application from its toolbox of controls. Flowsheet development or model building is then likened to program/software development and each block of code (the unit operation) is connected via stream blocks (decision blocks). The benefactors of this approach are all, that is, both developers and end user benefit. The user being allowed to customize the entire suite of process unit operations, plug in a customized valve operation for instance based on the original supplied by the software vendor or build an extension to augment the current engineering controls suite. In this work we present an Active X designed evaporative crystallizer (unitop) control found in Hysys Process Simulator. As well we present a cooling crystallizer built as a customized extension unit operation (OLE Automation) and linked to SPS (Separation Processes Services) solids-handling software design package.

Keywords

Active X, COM, Process simulator, Open architecture, Flowsheet, Controls.

Introduction

The development of today's commercial process simulators have become increasingly complex, if not through developing software technology then through continued added functionality. The result being "main" programs upwards of a million lines of code and beyond. In terms of programming the simulator, this has led to long periods of code compilation, difficult debugging and the inevitable slowing down of added new functionality to end user. Objected Oriented Design (OOD) programs have gone some way down the path of organizing program structure, however it hasn't solved the problem of programs becoming increasingly bigger, or decreased their size in any way when adding new functionality. Also, programming applications for the Windows operating environment with added user-written modules has become a challenge. Recently, developments in software programming technology has led to the development of COM technology in response to the need to componentize or simplify the programming structure to more manageable and efficient sizes. Applying this approach to the development of software programs, the commercial process simulator may be viewed as being composed of COM objects, for example Unit Operation COM objects, Thermodynamic COM objects, Reaction COM objects and in its purest sense, one may even choose to add a specific engineering method, say one's own viscosity equation without needing to know the EoS being implemented or the type of units (SI, Imperial etc.) having been chosen, or other pertinent physical property methods.

In the following paper we discuss two examples of COM object implementation as applied to HYSYS process simulator, the first a unit operation COM object with a Hysys GUI and the second, again a unit operation COM object using MFC GUI tools and then implemented within a generic PFD/Flowsheet test container.

In each of the examples, Hysys V. 2.1 Build 1350 Process

Examples

Example 1

SPS Cooling Crystallizer Extension, File types: *.dll, *.edf

In this approach, Hysys acts as the container exposing internal objects, the user or programmer utilizing these objects to build a unit operation COM module or "extension." This is illustrated below,

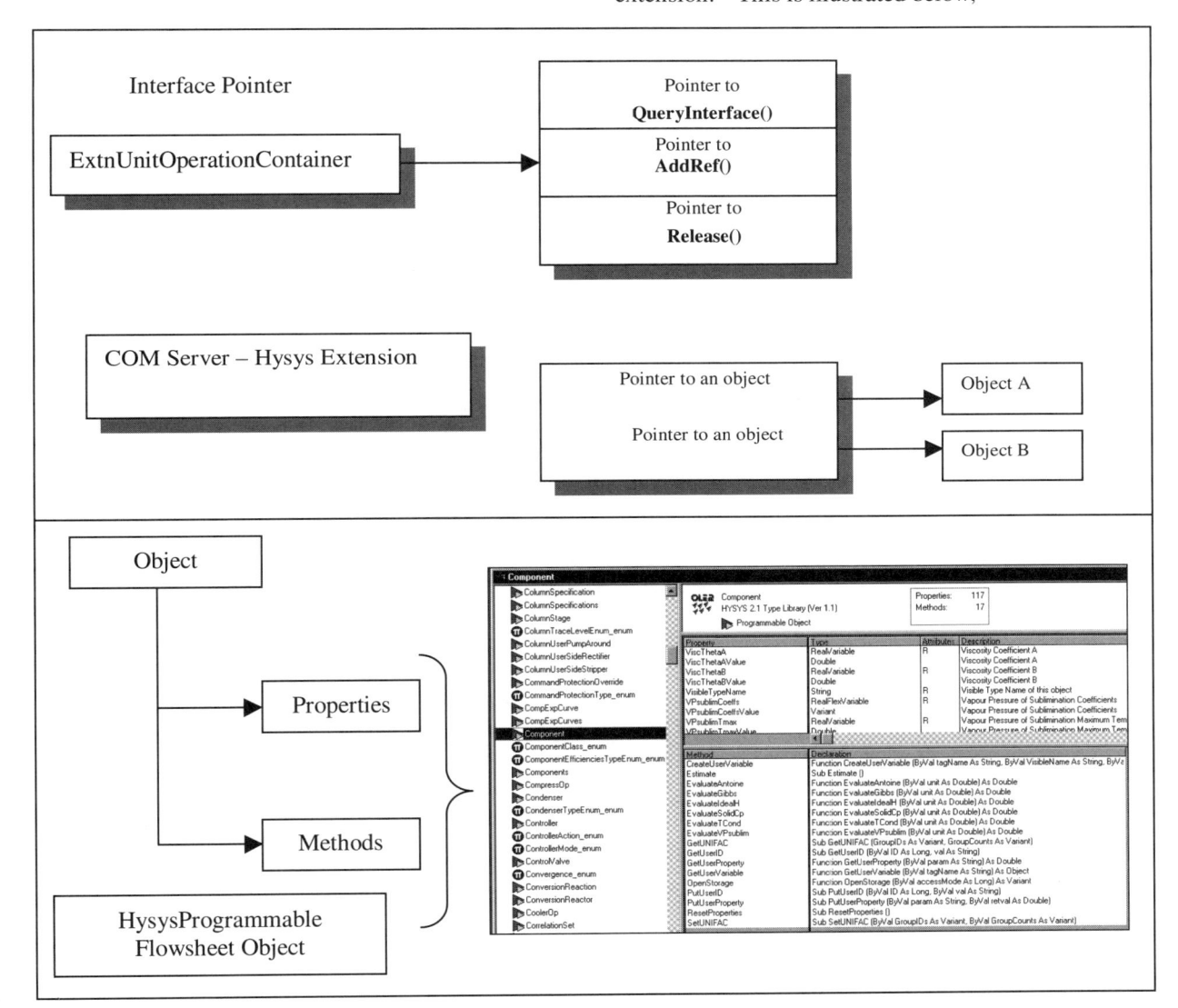

where each object obtained from Hysys container exposes methods and properties.

With this framework, a COM object or a Hysys extension can be readily set up. What remains is the graphical user interface GUI. In the SPS Cooling Crystallizer, a specific user interface is provided by the container, Hysys' container. The user is then given the tools necessary to implement the required functionality, such as units handling, serialization, subviews, worksheet matrices providing a look and feel of the simulator and allowing the programmer to concentrate on adding engineering methods as opposed to developing these tools themselves. In contrast to the Active X approach discussed in the next example, the user is required to develop the specific tools, but does have the option of utilizing GUI technology of their choosing, for example Microsoft COlePropertyPage, CPropertySheet or third party GUIs.

In the design of the SPS Cooling Crystallizer, the objective was to integrate differing technologies into a single process simulation environment, Hysys. Hysys acting as the flowsheet providing design data and flowsheet models with varying operating conditions. The COM object is written in VB (Visual Basic) the material and energy balances while a link to SPS Software representing different solids handling programs (*.exe) is written using an ATL COM object. Here the "best" tools for the job were chosen where simplicity and efficiency of implementation were the primary criteria for platform selection.

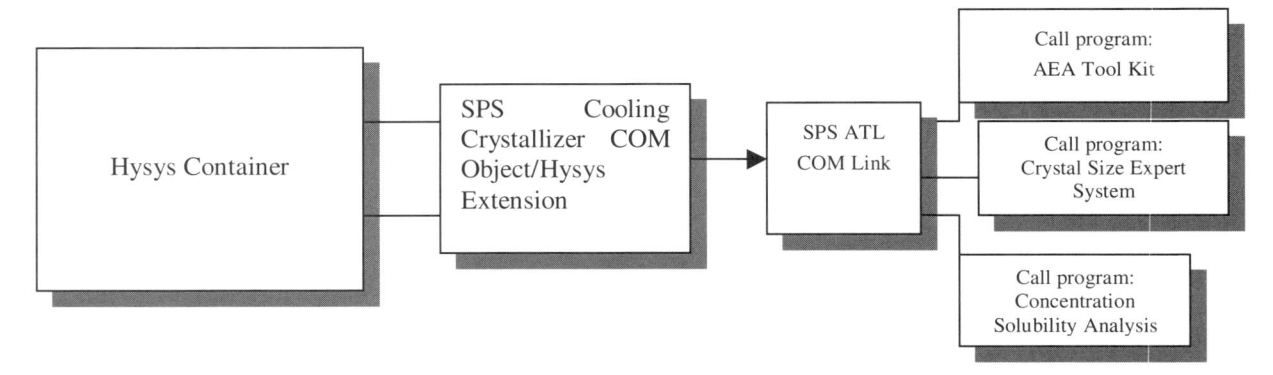

The many different software technologies integrated into a single simulation environment is shown in the screen capture below, the look and feel of Hysys is maintained through a standardized Hysys GUI interface, as well utilizing MFC VB treeview and dialog box tools. An ATL COM link calls SPS Crystal Size Expert where material and energy balance calculations are performed in the COM Server (Hysys Extension).

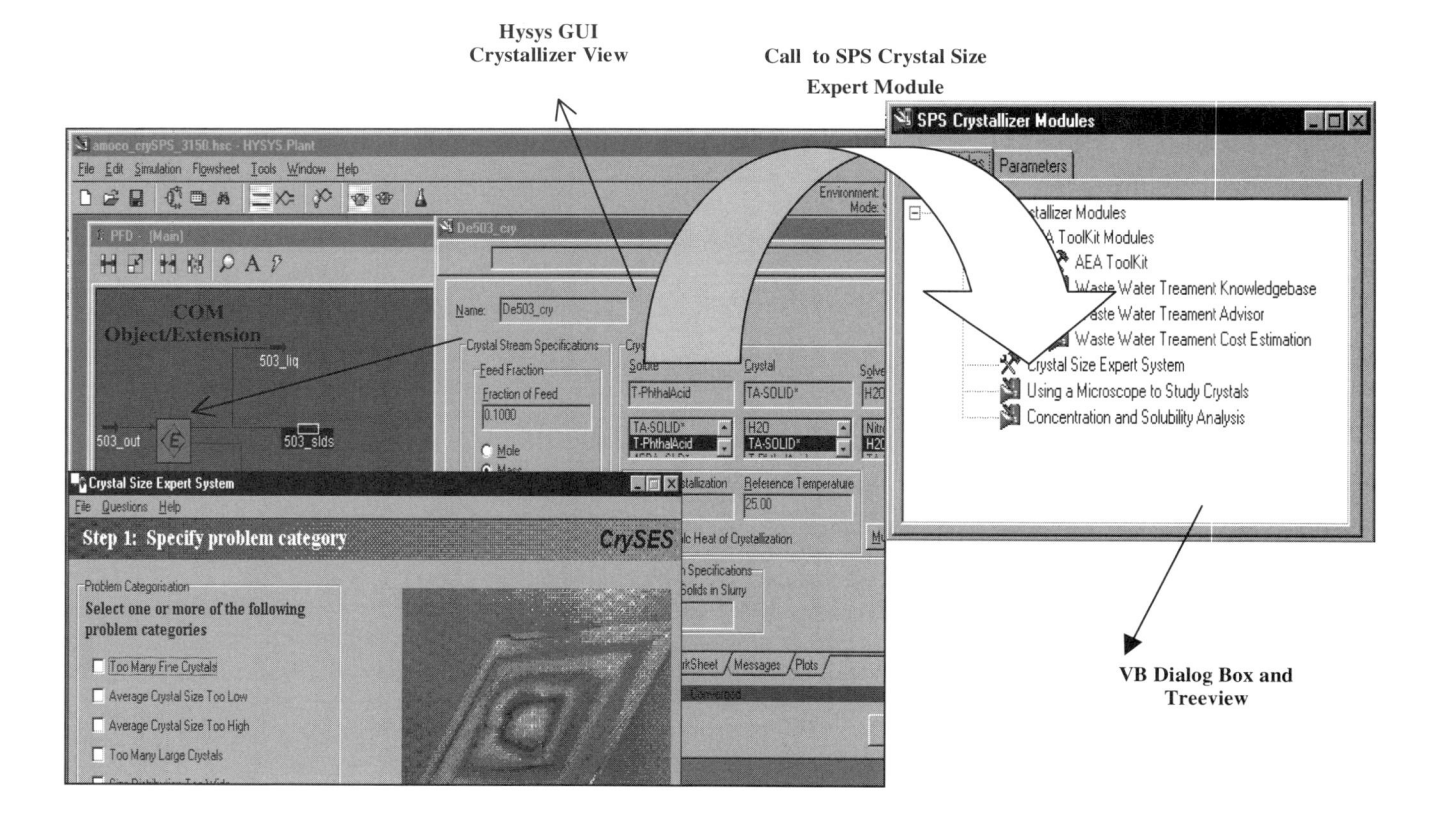

Example 2

An Active X Evaporative Crystallizer, File type: *.ocx

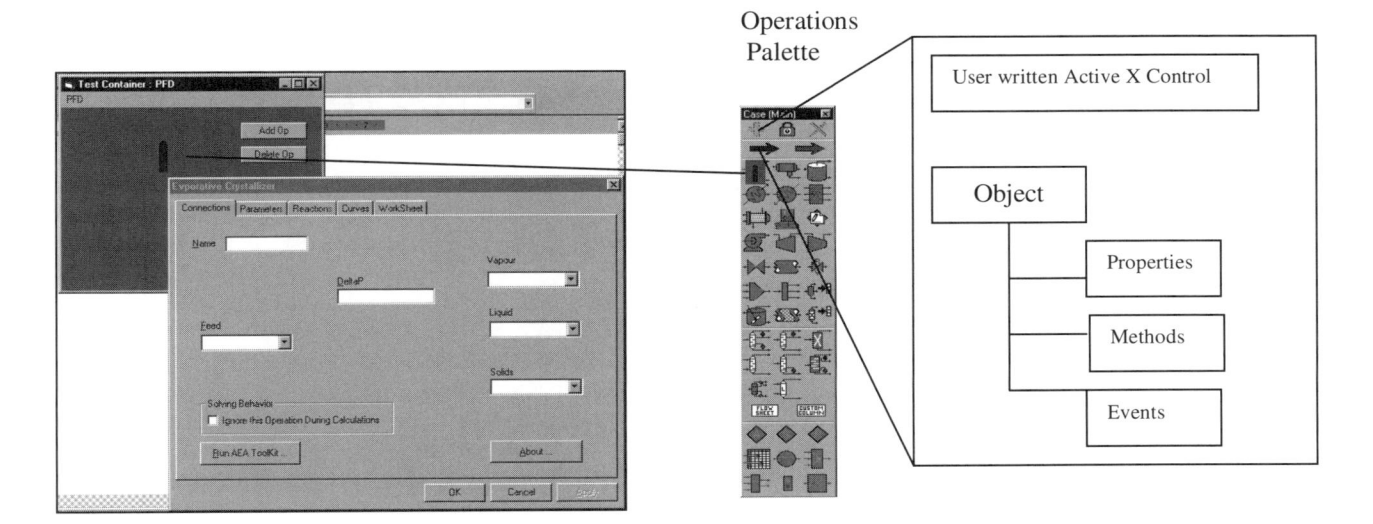

In an Active X approach, a unit operation becomes portable or autonomous, the programmer only needing to know what callbacks are necessary to import and export data to the simulation container. In this sense then, the Active X approach fulfills the "COM" philosophy. However, this approach is not without its problems. Here, data processing, units handling, serialization, views and subviews all need to be programmed. Specific controls developed for the simulator may not necessarily be available to the programmer. The disadvantage of course being time consuming interface development. On the positive side, the user being able to use generic off the shelf development tools. In this example an evaporative crystallizer has been setup and implemented in a generic VB PFD/Flowsheet container. The unit operation being able to draw itself in any container supporting its methods and properties. In an Active X control, events also need to be handled such as mouse clicks, updated displays when text or numeric data have changed. The end product being a flowsheet simulator of controls, Active X controls.

Conclusions

Two similar but different COM object approaches are presented in the context of Hysys process simulator. Although Active X controls present a different path to user customization, specifically designed user interfaces sometimes provide the user with "ready-made" tools to allow easier implementation of engineering technology within a simulation environment. Off-the shelf software development technology is certainly with its merits, however, where specific customized engineering technology has been developed, off-the-shelf development tools may sometimes require more work than having to use the tools designed specifically for the engineering simulation task at hand.

Nomenclature

ATL	Active Template Library
COM	Component Object Model
GUI	Graphical User Interface
MFC	Micrsoft Foucndation Class libraries
SPS	Separation Processes Services
Unitop	Unit Operation (e.g. Separator, Heat Exchanger, Crystallizer)

FUEL-ADDITIVES DESIGN USING HYBRID NEURAL NETWORKS AND EVOLUTIONARY ALGORITHMS

Anantha Sundaram, Prasenjeet Ghosh
Venkat Venkatasubramanian, and James M. Caruthers
School of Chemical Engineering
Purdue University
West Lafayette, IN 47907

Daniel T. Daly
The Lubrizol Corporation
Wickliffe, Ohio

Abstract

Fuel-additives play an important role in deposit reduction on the valves and combustion chamber of the automobile. They reduce cold-start problems, emissions and improve fuel-efficiency. The testing and design of fuel-additives is an expensive and lengthy procedure. A hybrid first-principles neural-network model for fuel-additive performance prediction was developed in this effort. This model products additive performance directly from the structure of the additive and outperformed existing models based on basic structural descriptors. The design of fuel-additive structures was accomplished using an evolutionary algorithm with problem specific representation and genetic operators. Some results from the model and the design algorithm are discussed in this paper.

Keywords

Computer-aided design of materials, Fuel-additives, Hybrid phenomenological neural-network models, Evolutionary algorithms.

Introduction

An additive is a substance added to another in small quantities to improve current performance or provide for a desired effect. Additives are added to gasoline to provide improved performance or to correct deficiencies. Gasoline additives have been historically used as combustion modifiers, anti-oxidants, corrosion inhibitors, anti-icing components as well as deposit control detergents(Gibbs, 1990). The primary focus of this paper is on the class of fuel-additives that act as deposit controllers in the gasoline. Specifically, this effort is on fuel-additives that control the deposit formation on the intake-valves of the automobile. Figure1 shows the schematic of the intake-valve and its surrounding components in an automobile. The intake-valve forms the opening into the combustion chamber. The fuel-injection nozzles spray gasoline directly on the intake-valve. When the valve opens it draws in a mixture of fuel and air into the combustion chamber where it is burned to supply power to the automobile. Over a period of time both during and shortly after engine operation, deposits tend to form on the surface of the intake-valve (Kalghatgi, 1990, Lacey et al., 1997). These deposits have been documented to affect driveability, cold-start efficiency, knock characteristics and emissions (Grant and Mason, 1992, Houser and Crosby, 1995).

The US Environmental Protection Agency (1996) has adopted a standard test to determine the deposit forming tendency of fuel package (gasoline + additives) before approving commercialization of the package. This is the ASTM Standard (1995) BMW-IVD test. The test uses a 4-cylinder 1985 BMW vehicle, operated over the road for a total of 16,093 km. The daily test cycle consists of 10\% city, 20\% suburban and 70\% highway mileage with an overall average speed of 45 mph. A fuel package must produce an average deposit of less than 100mg/valve for certification (Lacey et al., 1997). This is the performance measure of a fuel-additive.

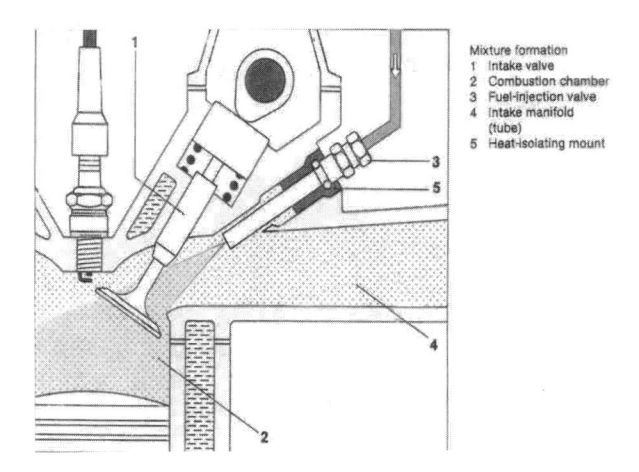

Figure 1. Intake-Valve and manifold.

The basis of any product design problem is the forward problem that involves the prediction of product performance from its components (Venkatasubramanian et al., 1995). The main function of the fuel-additive is the prevention of deposition on the valve by scavenging the deposit forming pre-cursors of the fuel. The mechanism of intake-valve deposit (IVD) formation is a complex process with a variety of influential factors such as additive and fuel chemistry, operating conditions and flow properties of the different components. Most of these parameters are not measured. This makes a purely first-principles model difficult if not impossible to determine. The high noise level in the BMW engine-test data (Arters et al., 1997) makes a purely statistical approach inaccurate and physically unreliable. A hybrid approach that combines the best possible phenomenological description of fuel-additive performance with the non-linear capabilities of neural-networks is proposed for additive performance prediction.

Hybrid Model for Fuel-Additive Performance Prediction

The hybrid model was developed along the following major steps

(i) Determination of a functional description of the fuel-additive in terms of key performance imparting components in the structure.
(ii) Development of a kinetic model that determines quantitatively the dynamic degradation of the functional components under given operating conditions.
(iii) Prediction of the time dependent "activity" of the additive in the fuel, in terms of its ability to sustain its deposit removal abilities.
(iv) Prediction of the IVD performance of the additive using a neural-network model to correlate the "amount of active additive" at different points in time to the experimental engine test data.

The main functional components of the fuel-additive structure were: (a) a core component that reactively bound to the deposit forming pre-cursors of the fuel (b) several branches of functional groups that prolonged the duration deposit removal activity of the additive (c) a transitional component that provided stability to the structure and held the core to the branches of the additive. A kinetic model for the degradation of the different components in the additive was developed and is given by Equation (1).

$$d\underline{X}/d\tau = \mathbf{K} \bullet \underline{X} \qquad (1)$$

Here, $\mathbf{K}$ is the matrix of rate constants describing the first-order irreversible reactions for additive degradation and $\underline{X}$ is the vector of concentrations of additives at different levels of degradation. Using this model, the distribution of the structure of the additive was determined as a function of time. A group contribution method (Barton, 1991) was employed to determine the cohesive energy density of the additive in the fuel from its structure and the given set of fuel characteristics. In combination with the degradative model, this gave a time-varying, quantitative description of the activity of the additive in the fuel. This information was then used as input to a neural-network model whose output was the IVD performance of the additive in the given fuel. This model is shown in Figure 2.

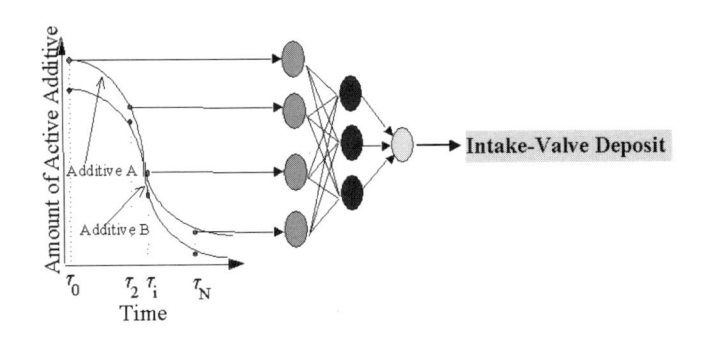

Figure 2. Hybrid model for IVD prediction.

This hybrid model was applied to the BMW engine test database provided by the Lubrizol Corporation. The data consisted of 92 engine-test results along with the structures of the additives and the fuel characteristics used in each. The results from different models are compared in Figure 3. The phenomenological model for IVD prediction used just a single descriptor (the activity of the additive at time = 0) and outperformed all other multivariate statistical models based on just the structural features of the additives such as the counts of different functional groups. This was clearly due to the mechanistic description of additive degradation and activity in the fuel included in the phenomenological model.

Design of High-Performance Fuel Additives

The hybrid approach gives a good model for the prediction of the IVD performance of a fuel-additive given its structure and operating conditions such as the fuel characteristics and temperature. The design problem involves the construction of optimal fuel-additive molecules given desired IVD requirements. This is a non-linear combinatorial optimization problem involving the use of a black-box objective function (due to the use of the NN). Widely used techniques such as knowledge-based enumeration (R.Gani and Fredenslund, 1993), mathematical programming (Maranas, 1996) and graph theoretic approaches (Skvortsova et al., 1993) will not work for this scenario, because of the above problem.

No. of Input Descriptors	Projections (PLS/PCA/None)	NN Architecture Transfer Function [No of Hidden Neurons]	RMSE (mg) in Testing (CV)
36	None	Tan-Sigmoid [7]	273*
4	PLS	Log-Sigmoid [3]	105*
3†	None	Tan-Sigmoid [3]	142
7	PCA	Tan-Sigmoid [3]	200
6††	None	Tan-Sigmoid [3]	172
1**	None	Radial Basis [4]	124

† Extracted from PLS factors, †† Extracted from PCA factors
* Large standard deviation across cross-validation data
** Phenomenological descriptor at time=0

Figure 3. Model Comparison for IVD prediction.

In this approach genetic algorithms (GAs) (Holland, 1975, Goldberg, 1989) which are search procedures based on Darwin's evolutionary model were adapted to the fuel-additive domain. Previous work in computer-aided molecular design has demonstrated GAs to be flexible in representing the rich underlying chemistry (Venkatasubramanian et al., 1996). Moreover, they are also robust to nonlinearities and are powerful algorithms for global search. The forward and inverse problem strategies for fuel-additives design are shown in Figure 4. The following were the main components of the evolutionary algorithm.

Representation: A fuel-additive was primarily composed of a group of functional components described previously. Differences between additive structures stemmed from differences in the chemical composition (in terms of the functional groups) and the topology of the functional components. Both of these affected the activity of the additive. Hence an additive was represented in an object-based manner, containing these functional components as the underlying objects. The representation also contained information on their connectivity.

Genetic Operators: There were rules concerning the connectivity and topology of the fuel-additive structure. These were chemical valence or connectivity constraints on specific components that avoid synthetic infeasibilities in the additive molecule. Moreover, every additive consisted of at least one of all three functional components in it. Genetic operations define the moves to navigate the search space of solutions that consisted of valid fuel-additive molecules.

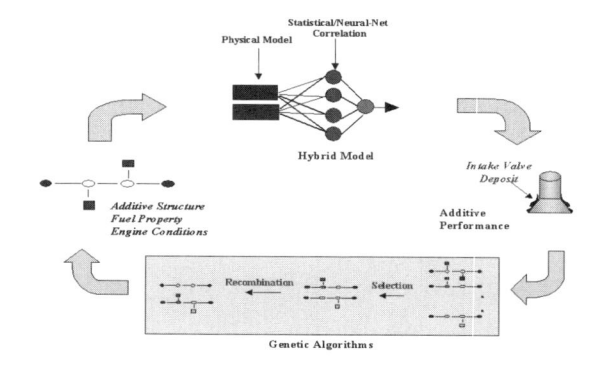

Figure 4. Designing for optimal fuel-additives.

Crossover and mutation are the usual genetic operators for evolutionary algorithms. Crossover involves large scale but random exchange of information across two members of the current set of solution candidates. Mutation involves the random but a small change to a solution candidate. To avoid the creation of infeasible or invalid fuel-additive structures from valid ones, these operators were customized to obey rules of construction. Four different constrained operators were employed in both crossover and mutation depending upon the component or region of application along the additive structure. A feasible mutation operation in the transitional component that connects the core of the additive to the branches is depicted in Figure 5. In this operation, the components connected to the mutated region were re-arranged into a feasible configuration such that none of the feasibility constraints were violated. Other operators were implemented along similar lines.

Initialization and evolution: The algorithm was initialized with a single structure that was randomly but feasibly constructed from a given set of functional components. A desired IVD objective was set as the goal. The initial population consisted of members that were copies of this structure. Each structure was evaluated using the hybrid model to determine its IVD performance and assigned a fitness value between zero and one quantifying its degree of desirability. Parents were chosen from the population based on their fitness and allowed to undergo crossover and mutation operations stochastically and according to a pre-determined frequency. The resulting offspring formed the next generation and the process of fitness evaluation, selection and recombination was continued. The evolutionary algorithm was employed to search a space of solutions that was roughly about a million in combinatorial size. This was used as a case study to examine the efficiency of the algorithm. With function evaluations at a premium this efficiency is crucial. The evolutionary algorithm stepped through 25 generations each containing a pool of 25 fuel-additive structures. A crossover frequency of 0.60 and a mutation frequency of 0.40 were used. The objective function was to design fuel-additive structures that, based on the hybrid forward model, were predicted to accumulate less than 10 mgs.

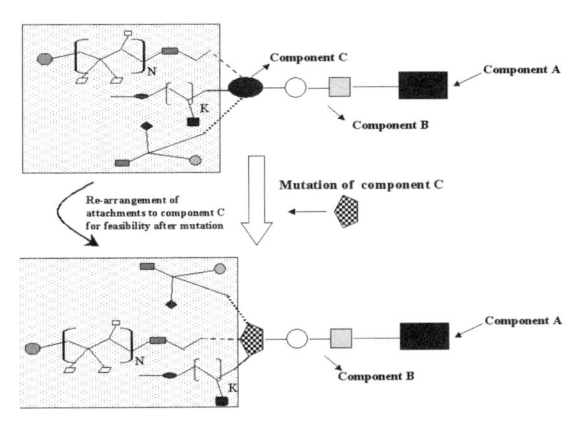

Figure 5. Mutation of a transitional component.

The neural-network model was trained from a set of HONDA generator engine tests with a predicted RMSE error on cross-validation of 30 mg. This was quite close to the quality of the engine tests whose repeats differed on an average of about 24mgs. The fitness function used is given by Equation (2)

$$F=1.0; IVD_{predicted} < IVD_{limit}$$
$$F=\exp(-\alpha\{(IVD_{predicted}-IVD_{limit})/IVD_{limit}\}^2); \text{otherwise} \quad (2)$$

All additive structures were assumed to be present at the same dosage initially. The results of three different runs of the evolutionary algorithm are summarized in Figure 6.

The evolutionary method was successful in constructing several known good performers as well as novel ones. The fitness and predicted IVD performance of the different additive structures are shown in the figure. In the figure, a structure is recognized as novel when all the functional components comprising it had never occurred in combination in any of the databases (BMW/HONDA). Structures, some of whose components had been examined in combination in the databases were characterized as variants of known structures. Even with a very small sampling of the search space (about 625 out of a possible one million), the evolutionary algorithm was successful in identifying diverse structures that met or were close in meeting the set objectives. Indeed, most of the best structures found in each of the runs were never encountered before. One fuel-additive molecule (III-1) in the Figure-1, comprised completely of frequently used components but never before tried in combination. This was not only a novel structure but was also seen to possess good synthesis potential.

Run	Rank/Identifier	Fitness	Predicted IVD (Hybrid-NN Model)	Structural Description
I	1, I-1	0.997	11.4 mg	Novel Structure. Rarely used components
I	2, I-2	0.996	11.5 mg	Novel Structure. Similar to best structure, different core.
I	6, I-6	0.993	12.0 mg	Variant of structure found in the BMW database.
II	1, II-1	0.999	10.1 mg	Novel Structure. Different from I-1. Infrequently used transitional component.
II	2, II-2	0.989	12.6 mg	Slight variant of additive structure found in BMW and HONDA databases.
II	4, II-4	0.983	13.2 mg	Minor variation of structure II-2 above. Slight modification of the core
III	1, III-1	1.00	8.9 mg	Novel Structure. Different from I-1 and II-1. Commonly used components.
III	2, III-2	0.994	11.9 mg	Variant of III-1. One of the transition and branch components different
III	3, III-3	0.993	12.1 mg	Variant of structure III-2 above. Slight modification of core. Contains an additional branch.

Figure 6. Evolutionary Search for Fuel-Additives.

Conclusions

A new hybrid first-principles neural network framework for fuel-additive performance prediction was developed and demonstrated to perform accurately even with limited and noisy data. Evolutionary algorithms with customized constrained operators were developed for the inverse design problem and proved to be successful in identifying novel, optimal candidates that also possess characteristics desirable in synthesis.

References

Arters, D. C., Schiferl, E. A. and D. T. Daly (1997). *SAE Technical Paper Series*, **SP-1277,** 67-80.
Barton, A. M. (1991).
Gibbs, L. M. (1990). *SAE Technical Paper Series: Lubricants and Fuels*, **99**(4), 618-638.
Goldberg, D. E. (1989). *Genetic Algorithms in Search, Optimization and Machine Learning*. Reading, Mass.
Grant, L. J. and R. L. Mason (1992). *SAE Technical Paper Series: Lubricants and Fuels*, **101**(4) 1221-1230.
Holland, J. H. (1975). *Adaptation in Natural and Artificial System*. Ann Arbor.
Houser, K. R. and T. A. Crosby (1995). *SAE Technical Paper Series: Lubricants and Fuels*. **103**(4), 1432-1451.
Kalghatgi, G. T. (1990). *SAE Technical Paper Series: Lubricants and Fuels*, **99**(4), 639-667.
Lacey, P. I., Kohl, K. B., Stavinoha, L. L. and R. M. Estefan (1997). *SAE Technical Paper Series: Lubricants and Fuels*, **106**(4), 880-891.
Maranas, C. D. (1996). *Ind. Eng. Chem. Res.*, **35,** 3403-3414.
R.Gani and Fredenslund, A. (1993). *Fluid Phase Equilibria*, **82,** 39-46.
Skvortsova, M. I., Baskin, I. I., Slovokhotova, O. L., Palulin, V. A. and N. S. Zefirov (1993). *J. Chem. Inf. Comp. Sci.*, **33,** 630-634.
Venkatasubramanian, V., Chan, K. and J.M. Caruthers (1995). *J. Chem. Inf. Comp. Sci.*, **35,** 188-195.
Venkatasubramanian, V., Sundaram, A., Chan, K. and J.M. Caruthers (1996). In *Genetic Algorithms in Molecular Modeling* (Devillers, J., Ed.). London, 271-302.

THE EMERGING DISCIPLINE OF CHEMICAL ENGINEERING INFO TRANSFER

Thomas L. Teague and John T. Baldwin
AIChE - Process Data eXchange Institute (pdXi)
New York, NY 10016

Abstract

Chemical Engineering Info Transfer is the exchange of process engineering technical information among the many people and organizations that conceive, design, construct, supply to and operate process plants. In current practice, Info Transfer is largely accomplished through the exchange of paper documents. Despite the large advances in process engineering software, in the course of daily work, engineers must still spend a significant portion of their time interpreting technical information developed in one software package and re-keying the information into other software packages to accomplish their work. Because current info transfer practice is so labor-intensive, automating the exchange of process technical data offers the promise of significant economic benefits to the process industry, estimated to be on the order of hundreds of millions, or even billions of dollars, annually. Electronic exchange of process technical information will also be essential for the successful introduction and use of e-commerce business practices by the process industry in the Internet age.

The idea for an industry-wide approach to automating process technical data exchange originated at the 1989 FOCAPD conference. As a result, the Process Data eXchange Institute (pdXi) was formed in 1991 as an AIChE Industry Technical Alliance. pdXi has discovered that achieving an implemented, internationally-recognized standard for process data exchange is a much more challenging technical and management problem than originally thought. Despite these challenges, pdXi is on the threshold of accomplishing its objectives in the next 1-2 years. This paper will outline the emerging discipline of Chemical Engineering Info Transfer, pdXi's accomplishments to date, and future outlook.

Keywords

Process engineering, Electronic data exchange, Process information, Information management, Information representation, Data models, STEP, ISO standards, Data sheet, PFD.

Introduction

It is generally acknowledged that 30-50% of an engineer's work time is spent finding and manipulating technical data. Often the data is found on paper such as computer program printouts, equipment lists, equipment data sheets or drawings (PFD's and P&ID's). Valuable engineering time is spent re-entering the data into one automated system or another. Considering the many times that the same process information is exchanged between people in owner, contractor, and supplier companies over the life cycle of a process facility, the current cost of info transfer is very high.

Chemical engineers have long studied the phenomena associated with transfer – heat transfer, mass transfer and momentum transfer. A new discipline is now emerging: Info Transfer. *Chemical Engineering Info Transfer is the exchange of process engineering technical information among the many people and organizations that conceive, design, construct, supply and operate process plants.*

Chemical Engineering Info Transfer is not new. Info Transfer has traditionally been carried out in the form of paper documents such as design basis memoranda, PFD's and P&ID's, equipment data sheets, major equipment lists, line summaries, etc. Process and equipment calculations were automated 20-30 years ago, but drawings and data sheets were still prepared manually. The production of engineering documents has migrated to electronic tools over the last 10-15 years, but the information content in these documents can still only be understood through human interpretation or by costly custom-built interfaces as illustrated in Figure 1.

designs. According to a DuPont study reported by Halford and Betteridge (1996) the ability to do electronic Info Transfer over a plant's life cycle could result in savings of several tens of millions of dollars. Broad, industry-wide application of electronic data exchange could result in annual savings of hundreds of millions, or perhaps billions of dollars, while significantly changing the work practices of process engineering. For a typical large company which does process engineering projects, hundreds, and probably thousands, of engineering work-hours will be saved annually by reducing, if not eliminating, the current practice of manual data entry and re-entry of data.

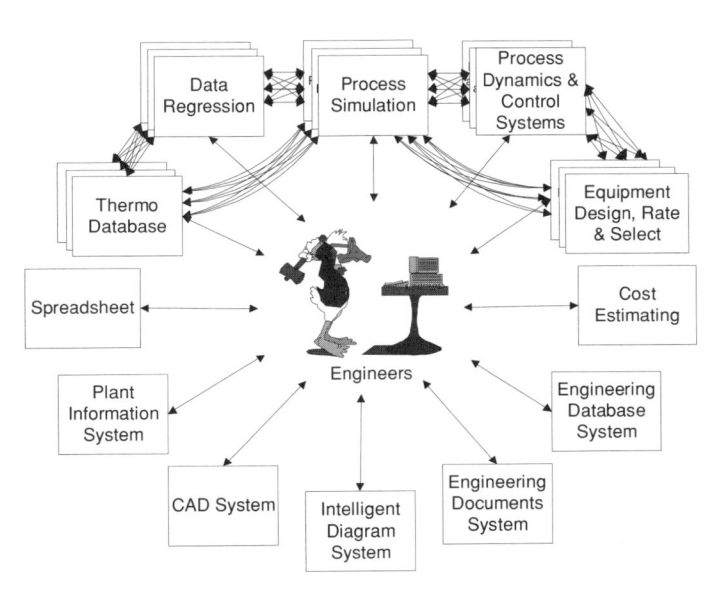

Figure 1. Current practices for chemical engineering info transfer.

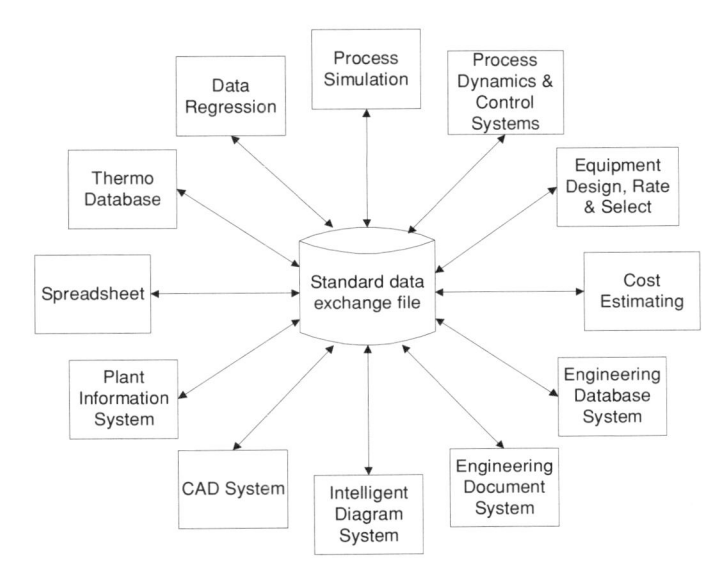

Figure 2. Standard approach to chemical engineering info transfer.

To get a more detailed description about the various paper-based process data exchanges associated with today's engineering practice, we refer you to our earlier paper (Baldwin, Teague and Witherell, 1998) and the description "Pat's Project." The schedule delays and labor costs associated with Info Transfer processes constrain the effectiveness of current engineering work process. If it were possible to automate the manual transcription of data from one software package to another, significant benefits would accrue to the process industry. In order to automate Info Transfer between software, a common descriptive language for Chemical Engineering terminology is needed together with a common electronic format to unambiguously transmit information from one software package to another. Such automation of the Info Transfer process is illustrated in Figure 2.

Benefits of Info Transfer

Automation of Info Transfer offers the opportunity to significantly reduce project schedules and labor costs, while reducing data transcription errors and allowing engineers additional time to create improved process

The Discipline of Info Transfer

Figure 2 illustrates a simple idea that has proven to be more difficult to achieve in practice than originally thought. Through its participation in the development of ISO STEP standards, pdXi has discovered that a complex, disciplined series of steps are needed to achieve implemented, standards-compliant data exchange for the process industry.

Developing an ISO STEP Standard

The international forum for technical data exchange standards is the International Standards Organization (ISO), Technical Committee 4, Standard for the Exchange of Product data (STEP). STEP has many parts to it, which are further described at U.S. Product Data Association web site (https://www.uspro.org). Other industries such as the automotive and aerospace industries have been long-time participants in STEP. The process industries have been active in STEP since about the 1993-1994 time frame. pdXi has been sponsoring the development of one of the parts of STEP known as an Application Protocol, or AP. pdXi is sponsoring AP231: Process Engineering Data:

Process Design and Process Specifications for Major Process Equipment. There are two other parts under development by the process industry, AP221: Functional Data and Schematic Representation for Process Plant (sponsored by USPI-NL, PISTEP and EPISTLE in Europe), and AP227: Plant Spatial Configuration (sponsored by the US-based PlantSTEP consortium.

To develop an effective InfoTransfer process using an ISO STEP standard, the following steps are required.

Step 1. *Define the business need* by developing a business activity model and a set of usage scenarios that involve economically important data exchanges. Review and agree upon the formal usage scenarios and business activity model. Draft conformance classes, based on the usage scenarios should be developed at this stage and refined in subsequent steps.

Step 2. *Define the industry vocabulary* by developing a formal object-oriented engineering data model of the data involved in the data exchange usage scenarios identified in Step 1. This data model includes:

a) Class definitions that name and describe engineering objects,
b) Class attributes that name and describe detailed data and data values associated with the engineering objects,
c) Relationships that name and describe the relationships between classes, e.g., one object is a kind of another object, one object is associated with many other objects, etc., and
d) Formal descriptions and detailed text definitions of all objects, attributes and relationships in the model.

Review and agree upon the formal data model that describes the data from the process engineering viewpoint. The "Working Draft" standards document is produced at this stage to facilitate international review and an international ballot is taken to progress to the next step. Formal comments are raised against the document as part of the international ballot process.

Step 3. *Validate the industry vocabulary* for completeness and to establish business value by developing working demonstration software implementations using a "working" file format. pdXi is currently considering the use of eXtensible Markup Language (XML), which is an emerging Internet standard (www.w3C.org/XML), for the Step 3 preliminary software validation implementations. From experience we know that developing working software implementations will improve the quality of the data model developed in step 2 as a part of developing the STEP standard in Step 4.

Step 4. *Standardize the industry vocabulary* by producing revisions to the "Working Draft" STEP documents produced in Step 2 to produce the "Committee Draft" document, which reflects the implementation experience learnings of Step 3. Additionally, we must develop a mapping between the engineering data model produced in Step 2 and the STEP Integrated Resources data model, which is a highly abstract data model used across many industries. This model uses very general terms such as product, property and representation. The STEP Integrated Resources data model must then be described with an object-oriented computer language (EXPRESS) that can be processed with STEP-compliant computer programs. This mapping from engineering domain data models to the STEP Integrated Resources model is illustrated in Figure 3.

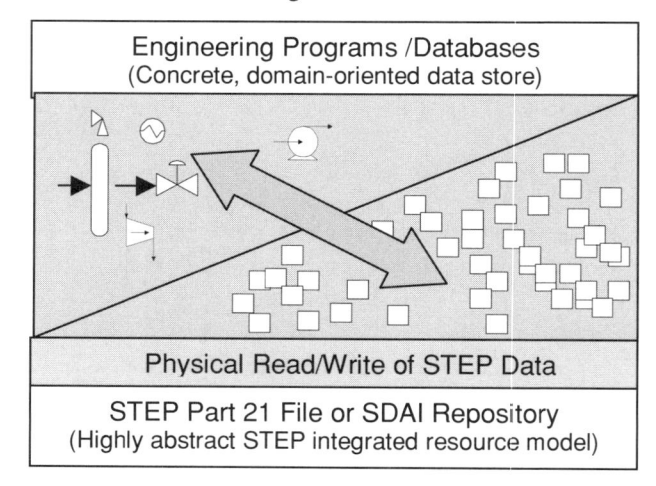

Figure 3. The data mapping challenge for STEP implementations.

The Committee Draft (CD) standard document is then issued for another round of international review and comment. This is the current stage of the pdXi-sponsored AP 231 standard.

Step 5. *Implement the standard* with a software demonstration project that illustrates how to use the STEP standard. In addition formal issues and comments raised by international reviewers against the Committee Draft need to be addressed. The Draft International Standard (DIS) document is then issued for international review, comment and ballot as before. Subsequently, address formal issues raised against the DIS document and issue the Final Draft International Standard (FDIS) document for international review, comment and ballot as before. Finally the International Standard (IS) is issued.

It should be readily apparent that the production of an international STEP standard is a lengthy, labor-intensive process. We estimate that the process industry, through pdXi, has invested well over $1.5 million to date in reaching the Committee Draft stage of the AP231 standard.

Software Implementation of the Standard

A standard is only useful and beneficial to the industry, however, when it is implemented through the software that engineers use on a daily basis. Throughout its existence, pdXi has tried to promote implementations so that actual data exchange benefits will be obtained from its data modeling results. pdXi showed, in two software implementation prototypes, that STEP Part 21 neutral file data exchanges can be built successfully. The first such prototype was demonstrated at the FOCAPD 94 conference. The second prototype was developed by Simulation Sciences and was completed in 1998. Both of these prototypes were based on the original pdXi data model that was completed in the 1991-1994 time frame.

As part of the AP 231 development since 1995, the pdXi data model has further evolved, undergoing significant revisions to address the comments from the international community.

At this point, pdXi is ready to start software implementations - both for Steps 3 and Step 5. pdXi believes it is critically important to implement the draft standards for two reasons. First, we believe we will learn things through practical implementations to improve the quality of the standards. Secondly, we believe that practical implementation will begin to deliver the business value of automated Info Transfer sooner than would be possible otherwise.

The XML implementations proposed for Step 3 of the overall process are expected to be relatively straightforward. First, pdXi will need to map the engineering level data model developed in Step 2 into the appropriate XML Schemas and Document Type Definitions (DTD's). Then, using publicly available XML parsing tools, it will be possible for any software owner (including spreadsheet users) to read and write XML files using the pdXi data model vocabulary. The Document Object Model (DOM) programming interface provides a standard programming interface for applications to read and write XML files.

STEP-based implementations are, unfortunately, more difficult to achieve. The primary difficulty is that, to build a software interface, the developer must hand-code the data mapping between the way the information is stored in the specific software program or database and the highly-abstract STEP integrated resources model illustrated in Figure 3.

This mapping complexity issue introduces a significant activation energy barrier of requiring the software developer to fully understand STEP methodology and the STEP integrated resources data model in order to implement standards-compliant interfaces. Commercial software utilities are available to help to physically read and write Part 21 files and to provide STEP Data Access Interfaces (SDAI) to commercial database products. However, there are no commercial tools to handle the data-mapping problem between the chemical engineering domain data model and the abstract STEP integrated resources model. How do we solve this problem?

One approach to the problem is to have each software owner become familiar with the STEP methodology and the detailed data mappings that are described in the AP 231 standards document. The difficulty with this approach is that the learning curve on STEP methodology is quite substantial, estimated to be on the order of a work-year. In chemical reaction terms, this "activation energy" is so large that the forward reaction (STEP implementations) cannot occur at reasonable rates at reasonable temperatures. In other words, it is economically infeasible for software owners to build standards-compliant interfaces without a catalyst of some sort to lower the activation energy. Therefore, one challenge remains for pdXi, which is to find a way to catalyze STEP-based software implementations. A possible solution to this problem may be for pdXi to develop a shared-cost software tool that maps the engineering view XML implementations from Step 3 to the STEP Integrated Resources model Part 21 file needed for Step 5.

Because the scope for AP 231 is very large, both the XML and the STEP software implementations need to be built incrementally, focusing first on the high value usage scenarios. In the remainder of 1999, pdXi is focusing on XML implementations of process stream data and heat exchanger data, which are conformance classes 1, 9 and 10 of the AP 231 draft standard.

Working software data exchange implementations are expected to provide immediate business value, while simultaneously improving the quality of the AP 231 standard. When completed, the final AP 231 standard will also have been demonstrated in practical applications.

Conclusions

A simple, but powerful idea of an industry-wide standard approach to automating process data exchange was born at FOCAPD 89. During the past 8 years, pdXi has discovered the emerging new discipline of Chemical Engineering Info Transfer. Developing an international, industry-wide data exchange protocol has turned out to be anything but simple. Rather it has consisted of a number of complex steps, requiring much longer and costing more that originally thought. The good news is that the process is now better understood and pdXi is on the threshold of completing practical implementations this year and next. Successful application of the Chemical Engineering Info Transfer discipline will enable software tools to exchange process engineering data automatically, rather than using today's labor-intensive paper-exchange process. The successful practice of automated Chemical Engineering Info Transfer will enable the process industry to save hundreds of millions of dollars annually, significantly improve the engineering work process and position the process industry to move fully into the era of electronic commerce in the new millennium.

References

Baldwin, John T., Teague, Thomas L., and W. David Witherell (1998). Info Transfer: An Emerging ChE Discipline. *AIChE CAST Newsletter*.

Halford, Joe, and L.B. Betteridge (1996). Economic impact of computing data standards. Paper 89a, Session 89, AIChE Spring Meeting.

A PRINCIPAL VARIABLE APPROACH FOR BATCH PROCESS DESIGN BASED ON A BLACK BOX MODEL

Po-Feng Tsai, Shi-Shang Jang, and M. Subramaniam
Department of Chemical Engineering
National Tsing-Hua University
Hsin Chu, Taiwan

Shyan-Shu Shieh
Chang Jung University
Tainan County, Taiwan

Abstract

Batch process is highly suitable for the manufacturing of small amount and high value added products such as specialty chemicals. However, for a batch process, a first principle model is still difficult to be derived because of complexities of chemical reaction and transport phenomena in the plant itself. It is hence desirable to implement an experiment oriented black box model, such as an artificial neural network (ANN). Naturally the initial conditions should be defined as the inputs of the ANN model. However, it is always possible to take transitional measurements during a batch cycle and these measurements may be useful to predict product quality. If all the parameters such as initial conditions and transitional measurements are included as variables for the black box model, then the dimensionality of the system increases. Also this over dimension may lead to over fitting the ANN model. In this work, we implement the principal curve transformation to reduce the input variables of the system. While carrying out principal curve transformation, the important information contained in the variables should not be missed. Very often, the information embedded in the transitional measurements is more useful in predicting product qualities. Hence, this information should be contained in principal variables. The principal variables obtained are in turn implemented as the inputs of the ANN model. Two examples are presented. One is a process whose input variables are non-linearly dependent among themselves where the 8 original input dimension is reduced to 2 new (principal variable) input dimension and the other is a realistic batch reactor with actual experimental data of a batch reactive distillation process with five original input dimension is reduced to 3 new input dimension to demonstrate the effectiveness of the principal variable approach.

Keywords

Non-linear principal curve analysis, Transitional measurements, Data compression, Non-linear dependent variables, Principal variable approach.

Introduction

Now a days the artificial neural network (ANN) is playing a very important role in dealing with systems, especially the highly non-linear systems. In the chemical engineering field, we apply this technique to construct a system model with inputs and outputs, called training data. With the increasing non-linearity nature of the system, we deal with more training data which are required to provide more information about the system. On the other hand, we do not want to increase our data points in data space unless we have to, because adding one more data point means adding one more experiment and increasing the cost. Unfortunately, in most chemical engineering processes, like batch reactor operations, the dimension of input variables is usually high that we have to reduce large data set for constructing a neural network model to prevent over fitting and make it work.

Usually, if we analyze the large number of input variables of non-linear complex processes, there may be some unknown relations among themselves. If we can process these variables by certain simple "compression" algorithm to reduce the dimension into a lower one, that means removing the dependence among themselves of these input variables, before training a neural model, it will make ANN to get trained easily with out over fitting with minimum number of data points.

In another case, if the information included in the input variables is not enough, the ANN model will not predict well, even after much training iteration because of the large error that cannot be successfully reduced. If we can add some extra information about the process, like dynamic measurements of variables during process proceeding, into input variables to increase the information content without increasing the dimension of input variables, we can also make the ANN model work with limited data points.

Hence to reduce the large number of input variables, we propose Non-Linear Principal Curve Algorithm with out loss of information content of the input variables that deal with complex, highly non-linear system.

Non-Linear Principal Curve and Reduction of Variable Dimensions

We first consider the relations of the inputs and outputs of a system and express them as the following equations:

$$\overline{\overline{Y}}_{ope*m} = F(\overline{\overline{X}}_{ope*n}) \qquad (1)$$

$$\overline{\overline{Y}} = [\overline{y}_1 \ \overline{y}_2 \ \overline{y}_3 \ ...... \ \overline{y}_m] \qquad (2)$$

$$\overline{\overline{X}} = [\overline{x}_1 \ \overline{x}_2 \ \overline{x}_3 \ ...... \ \overline{x}_n] \qquad (3)$$

where $\overline{\overline{X}}$: Input variable matrix
$\overline{\overline{Y}}$: Output variable matrix
ope : Operations
n : no. of input variables
m : no. of output variables

or

$$\overline{y}_1 = f_1(\overline{X}_1); \quad \overline{y}_2 = f_2(\overline{X}_2);$$

$$\overline{y}_3 = f_3(\overline{X}_3) \quad \text{etc.} \qquad (4)$$

where $\overline{X}_1$, $\overline{X}_2$, $\overline{X}_3$ ... are the rows of the input variable matrix of different batch operation.

If the dimension of the input variables, $\overline{\overline{X}}$, is high and also we have to add some more dynamic measurements to train a ANN model, then the dimension of the input data increases. So to reduce input dimension it is necessary to remove some dependent variables before training ANN.

The information loss should be prevented when we compress the input variables and even more careful during adding some extra information about the process or system.

First, we use the "dynamic terms" which means the measurements of variables during the process proceeding and these can be expressed as following equations:

$$\overline{\overline{D}}_{ope*num} = G(\overline{\overline{X}}_{ope*num}) \qquad (5)$$

where $\overline{\overline{D}}$: Dynamic term matrix
num : product of no. of samples and no. of dynamic variables.

Every row of dynamic term matrix, $\overline{\overline{D}}$, is a vector containing all samples of dynamic terms. We list all of them in a vector for every single operation, $\overline{D}_1$, $\overline{D}_2$, $\overline{D}_3$ ..., to form the dynamic term matrix, $\overline{\overline{D}}$.

$$\overline{\overline{D}} = [\overline{D}_1 \ \overline{D}_2 \ \overline{D}_3 \ ......]^T \qquad (6)$$

And then, we combine $\overline{\overline{X}}$ and $\overline{\overline{D}}$ to form our new input data matrix:

$$\overline{\overline{X}}_{new} = [\overline{\overline{X}} \ \overline{\overline{D}}] \qquad (7)$$

Now we apply Non-Linear Principal Curve Analysis (NLPCA) to compress the input data matrix as per the following steps:

Analysis (NLPCA) to compress the input data matrix as per the following steps:

(1) First, we define a principal curve to be a straight line formed by the first principal component as its direction vector and pass through the mean point of the data.

(2) Use the principal curve algorithm (Hastie and Stuetzle, 1989) and tune the factors of the principal curve for smooth factor to get an independent variable, λ and an estimated input data matrix, X_{est}.

(3) Define a residual matrix as per the following equation (8),

$$\overline{\overline{R}} = \overline{\overline{X}}_{new} - \overline{\overline{X}}_{est} \qquad (8)$$

and then set

$$\overline{\overline{X}}_{new} = \overline{\overline{R}}$$

(4) If the error becomes reasonable, we can stop the algorithm, otherwise go back to step 1 to get more independent variables.

We use these new independent variables, $\lambda_1, \lambda_2 \ldots \lambda_k$, to build a new data set instead of the original one formed by $\overline{\overline{X}}$. That means we map $\overline{\overline{X}}$ into a lower dimension space, $\overline{\overline{\lambda}}$, for $k \leq n$.

Illustrative Examples

Example 1: Input Variables are Non-linearly Dependent

In this example, we simulated a system whose input variables are non-linearly dependent with each other. We generated a data set (36 training data and 5 testing data) with 5 and even more (8) input variables, one output variable and 5 dynamic terms. The system is expressed as per equations (1) ~ (7), besides, there are also some non-linear relations between inputs. We considered only 2 degrees of freedom for the new input variables. The non-linear relation between original input and new input is as given below:

$$x_1 = f_1(x_a, x_b) \,;\, x_2 = f_2(x_a, x_b)$$

$$\ldots\ldots;\, x_n = f_n(x_a, x_b)$$

where $x_1, x_2 \ldots x_n$ are original input variables and they are functions of two independent unknown new variables, x_a, x_b.

Table 1. Simulated Output Variable Prediction for 5 Original Input Variables and also for 2 New Input Variables (5th test data is an extrapolated point and not listed).

Predicted output for $x_1\ldots x_5$ original inputs.	Predicted output for x_a & x_b new inputs.	Actual output

We obtained two λs for the new input data and the results are shown in Table 1.

-0.7828	0.1417	0.1814
0.8200	1.3080	1.2064
-1.8140	0.0865	0.0861
103.2000	99.4300	100.1651

Table 2. Simulated Output Variable Prediction for 8 Input Variables.

Predicted output for $x_1\ldots x_8$ original inputs.	Predicted output for x_a & x_b new inputs.	Actual output
47.15	31.52	29.63
44.58	37.77	31.82
149.90	188.7	180.05
241.60	241.5	266.58
146.10	154.4	158.09

Example 2: Reactive Distillation Model

In this reactive distillation experiment, the reaction is as shown below:

$$CH_3OH + CH_3COOH \xrightleftharpoons[]{H^+} CH_3COOCH_3 + H_2O$$

We have 5 input variables namely Methanol feed, Acetic acid feed, reflux ratio and temperatures at beginning and at 30 minutes operation and only one output variable which is product yield. There are totally 46 experimental data for our training and 6 extra experimental points to test our new model.

When we trained the ANN model using the 5 inputs, the training error could not be reduced to an acceptable range. If we use the 3 main operation input variables, 2 Feeds and reflux ratio, it is getting worse.

The training errors and results of the test set are given below:

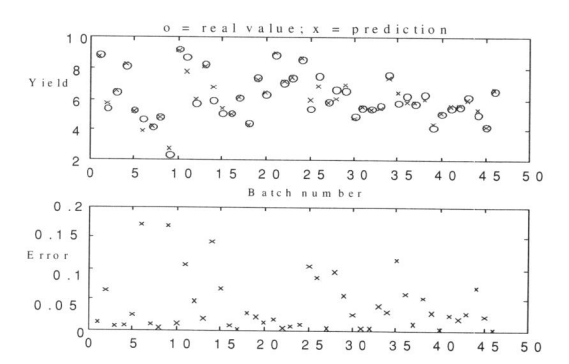

Figure 1. The predictions and errors for the original 5 inputs.

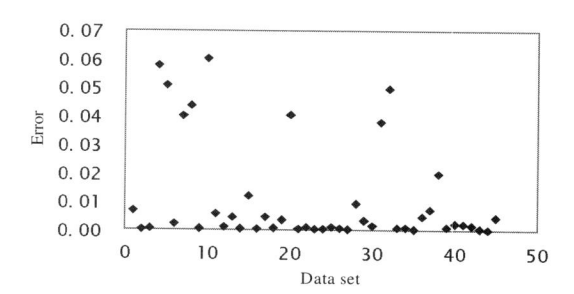

Figure 2. The training errors for 3 new independent variables (generated by NLPCA).

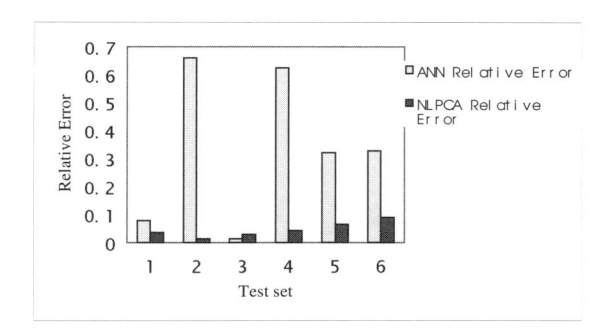

Figure 3. Comparison of relative error for the original inputs and new variables (generated by NLPCA).

Conclusions

Non-Linear Principal Variable Approach has been applied for the reduction of input variables with dynamic measurements for a system whose input variables are non-linearly dependent among themselves. On simulation it has been observed that the prediction of ANN output for the two new principal variable input is more accurate to actual output comparing to 5 and also 8 original input predictions.

Also the same algorithm is applied to actual experimental data of reactive distillation process. It has been observed that the prediction of ANN for reduced principal variable inputs is more accurate to actual output comparing to the original 5 or 3 inputs. Hence Non-Linear Principal Variable Approach is a promising technique to reduce the dimensions of input variables for ANN model development and also avoiding the over fitting of ANN model.

References

Hastie,T., and W. Stuetzle (1989). Principal curves. *J. Amer. Stat. Assoc.*, **84** (406), 502-516.

William S. Cleveland (1979). Robust locally weighed regression and smoothing scatter plots. *J. Amer. Stat. Assoc.*, **74**, 829-836.

Dong, D., and T. J. McAvoy (1996). Nonlinear principal component analysis – based on principal curves and neural networks. *Comp. Chem. Eng.*, **20**, 65 –78.

Dong, D., and T. J. McAvoy (1996). Batch tracking via nonlinear principal component analysis. *AIChE*, **42**, 2199-2208.

Mark A. Kramer (1991). Nonlinear principal component analysis using autoassociative neural networks. *AIChE*, **37**, 233-243.

Paul Nomikos and John F. MacGregor (1994). Monitoring batch processes using multiway principal component analysis. *AIChE J.*, **40**, 1361-1375.

Jolliffe (1986). *Principal Component Analysis.* Springer-Verlag, New York.

A SAMPLING TECHNIQUE FOR CORRELATED PARAMETERS IN NONLINEAR MODELS FOR UNCERTAINTY AND SENSITIVITY ANALYSIS

Victor R. Vasquez and Wallace B. Whiting
Chemical and Metallurgical Engineering Dept.
University of Nevada-Reno
Reno, NV 89557-0136

Abstract

A novel approach called Equal Probability Sampling (EPS) is presented for sampling the parameter space of nonlinear regression models with correlation among the parameters. The approach has been used for analyzing uncertainty and sensitivity in thermodynamic models. Uncertainty and sensitivity analysis for simulation and design of industrial processes are becoming increasingly important. The (EPS) method produces more realistic results in uncertainty analysis than methods based on other sampling techniques such as Latin Hypercube Sampling (LHS) or Shifted Hammersley Sampling (SHS). The EPS method is based on resampling to obtain uniform coverage over level sets of the objective function used to obtain the parameters of the model. The existence of unfeasible situations is substantially reduced with EPS. It can be extended to any regression model describing other kinds of physical applications and can be used as a better tool to estimate more reliable safety factors in the design and simulation of chemical industrial process.

Keywords

Sampling techniques, Monte Carlo simulation, Nonlinear models, Thermodynamics, Liquid-liquid equilibria, Statistical methods.

Introduction

Process design and simulation of chemical processes are strongly dependent on thermodynamic models. The uncertainty associated with these models is an important factor for risk analysis and performance studies, and this topic becomes very important for decision making in process safety and economic profitability analysis. Normally, in actual process design and simulation operations, the associated uncertainty is covered using safety factors, which can increase costs and investment without a quantitative measure of the avoided risk.

Uncertainty analysis can be used for studying the safety factors involved in the design. Decreasing their magnitude increases the efficiency and financial attractiveness of the global process. Alternately, the safety factor may be increased to attain a specific quantitative level of safety. Specifically, the uncertainty associated with thermodynamic models derives from their parameters and modeling errors (see Vasquez and Whiting, 1998a, 1998b, 1998c, 1999a).

The parameter-uncertainty depends on the parameterization chosen, and the random and systematic errors present in the experimental data. A special concern when performing the uncertainty analyses for regression models is the sampling approach used to obtain representative values from the parameter space.

Traditional methods are based on stratified sampling over individual parameter distributions, and their correlation structures are approximated through the use of pairing procedures. Examples of these methods are Latin Hypercube Sampling (LHS) (see Iman and Shortencarier, 1984) and Shifted Hammersley Sampling (Kalagnanam and Diwekar, 1997). In terms of the objective function used to regress the parameters, these techniques generate a first-order approximation in a Taylor's series expansion for the level sets. For nonlinear models (thermodynamic models belong to this category), the level sets are poorly estimated using first-order approximations when the parameter-effects and intrinsic nonlinearities are significant. If the parameter space is not sampled properly, the values for the stochastic variables generated may not represent correctly the physical problem, producing either unlikely results or unfeasible situations.

The main goal of this work is to introduce and apply the EPS sampling approach, which combines an improved level set estimation method with a new resampling scheme along the estimated level set, to obtain more realistic output distributions for performance assessments studies.

Vasquez and Whiting (1998b,1999b) have applied successfully EPS to study uncertainty and sensitivity of thermodynamic models in different process performance scenarios. The results show narrower cumulative frequency distributions than the ones obtained using traditional sampling methods, a consequence of sampling over the improved level sets obtained from the EPS technique.

Equal Probability Sampling and Uncertainty Analysis

The usual nonlinear regression model can be represented as:

$$y_t = f(x_t, \theta) + \varepsilon_t \qquad t = 1, \cdots, n \qquad (1)$$

where $y = (y_1, \cdots, y_n)^T$ is a vector of observations, $x_t = (x_{t1}, \cdots, x_{tk})$ are the control variables, $\theta = (\theta_1, \cdots, \theta_p)^T$ is a vector of unknown parameters, and ε_t are the errors, which are assumed to be independent normally distributed random variables, $\varepsilon_t \approx N(0, \sigma^2)$. The function f is supposed to be known and continuously differentiable in θ. The optimum value of θ, defined as θ^*, is obtained by minimizing the objective function $S(\theta) = \sum_{t=1}^{n}(y_t - f(x_t, \theta))^2$. The log-likelihood function for y is $L(\theta) = (2\pi\sigma)^{-n/2} \exp(-S(\theta)/2\sigma^2)$, and so the level sets of $S(\theta)$ are also level sets of the likelihood function. The minimum volume confidence region for θ at any given confidence level is clearly bounded by one of these level sets. Each level set is a closed hypersurface that divides the parameter space into areas of higher and lower likelihood, with each point on the level set being equally likely.

The probability distribution of $S(\theta)$ is stratified into **N** intervals of equal probability, and then the inverse image of these intervals form **N** shells of equal probability in the parameter space. From each of the **N** shells, a resampling scheme is used to obtain uniform coverage.

Fig. 1 shows hypothetical level sets of an arbitrary objective function $S(\alpha,\beta)$ defining ellipses of equal probability for the parameters α and β. In this Figure, the EPS sampling technique will take randomly one pair of values for α and β from each elliptical shell, covering in a stratified way the probability distribution for $S(\alpha,\beta)$. In this case, the use of pairing procedures to keep correlation structures coupled with sampling techniques such as LHS and SHS will produce similar results because an ellipsoidal region approximates the parameter space according to the following equation:

$$S(\theta) = S^* + \left(\theta - \theta^*\right)^T \frac{1}{2} H^* \left(\theta - \theta^*\right) \qquad (2)$$
$$\approx S^* + \left(\theta - \theta^*\right)^T \frac{1}{2} V_\theta^{-1} \left(\theta - \theta^*\right)$$

where $\mathbf{H}^*$ is the Hessian of $S(\theta)$ at $\theta = \theta^*$, and $\mathbf{V}_\theta$ is the covariance matrix of θ.

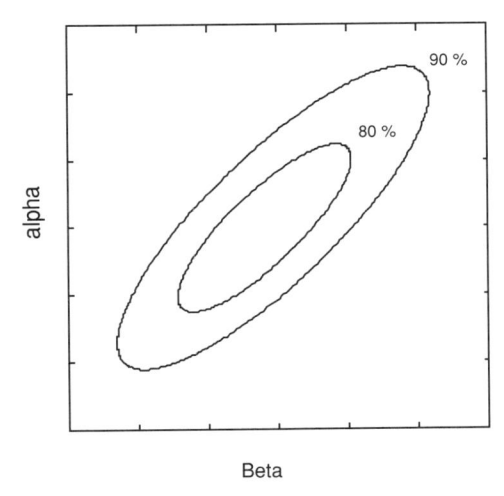

Figure 1. Hypothetical level sets of the objective function $S(\alpha,\beta)$ defining ellipses of equal probability for the parameters α and β.

A Basic Example: The Fieller-Creasy Problem

The nonlinear model for the two-sample case can be written as follows (Cook and Witmer, 1985; Potocký and Ban, 1992):

$$f(x_i, \theta) = \theta_1 x_i + \theta_1 \theta_2 (1 - x_i) \qquad (3)$$

where x_i is an indicator variable that takes values according to the population being studied. The level set estimation at 95% confidence level is presented in Fig. 2, expressed as a region centered at $\left(\theta_1^*, \theta_2^*\right)$ for the optimal values of $\theta_1^* = 3$ and $\theta_2^* = 0$. This transformation is represented in Fig. 2 as $k_1 = \theta_1 - \theta_1^*$, and $k_2 = \theta_2 - \theta_2^*$.

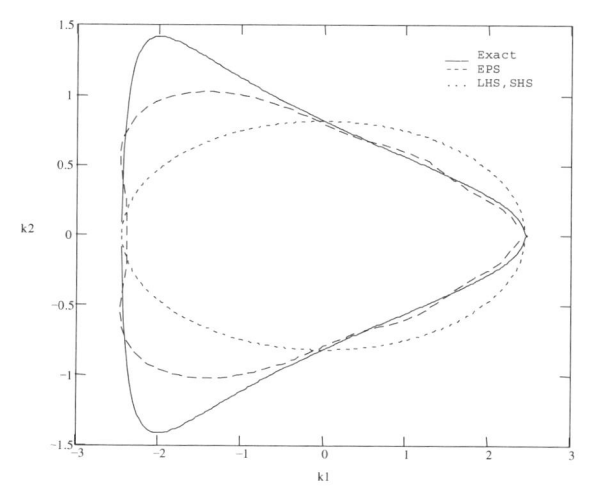

Figure 2. 95% Level sets for the Fieller-Creasy model for k_1 and k_2, where $k_1 = \theta_1 - \theta_1^*$ and $k_2 = \theta_2 - \theta_2^*$. The LHS and SHS results are using the pairing procedure of Iman and Conover (1982).

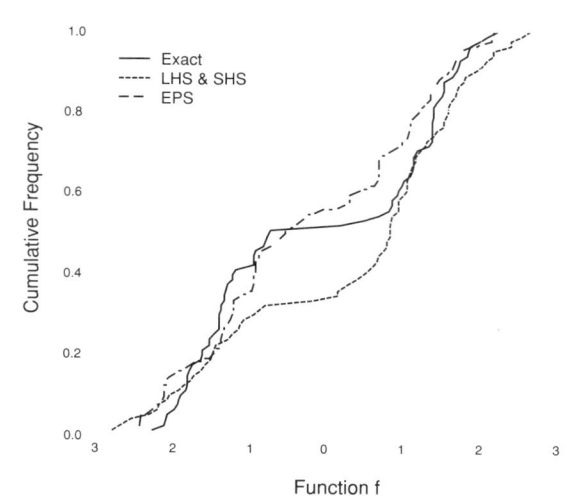

Figure 3. Comparison of EPS, LHS-SHS, and sampling over the exact confidence regions on the cumulative frequency curve estimation for the Fieller-Creasy model evaluated at $x = 0$.

We can see that about 32% of the confidence region defined by the exact level set is not explored by LHS and SHS, and about 5% is overestimated. With EPS about 6.5% of the confidence region is not sampled or explored and there is no significant overestimation. Additionally, Fig. 3 presents a comparison of EPS, Iman-Conover-SHS, Iman-Conover-LHS, and sampling over the exact confidence regions on the cumulative frequency curve estimation for the Fieller-Creasy model evaluated at $x = 0$ for the optimum parameters aforementioned. These results were obtained by first generating 100 samples of the parameters θ_1 and θ_2 using each of the approaches mentioned. Then each set of parameters θ_1 and θ_2 were evaluated in the Fieller-Creasy model at $x = 0$ and the results are presented in Fig. 3 as cumulative probability distributions. This methodology is commonly used when performing uncertainty analysis of stochastic models. When sampling over the exact and approximated level sets, the confidence regions have to be stratified uniformly before sampling because each random set of parameters over a given level set is equally likely. This is what we called the resampling scheme mentioned before.

Notice from the results generated using the EPS approach that they are closer to the ones obtained from sampling over the exact confidence regions than using LHS or SHS sampling techniques. This conclusion is the product of having a better representation of the parameter space before the sampling and the idea of using the level sets of the objective function for sampling purposes.

A Thermodynamics Example: The NRTL model

A case study based on the diisopropyl ether (1) + acetic-acid (2) + water (3) liquid-liquid system is presented. Experimental data from Treybal (1981) were used to regress the binary interaction parameters for the NRTL equation with $\alpha = 0.3$, using the ASPEN PLUS[1] process simulator. The regression results are presented in Table 1. One hundred samples were generated using both EPS and LHS sampling techniques. To illustrate and study the amplitude of the conditions chosen by the sampling methods, the equilibria based on the extreme values for the composition of component two were plotted for each technique. These results are presented in Fig. 4, which shows a high overestimation of the parameter space produced by the LHS technique, including an unfeasible extreme upper binodal curve. The EPS parameter space estimation is more reasonable from a practical standpoint for both binodal curves.

Vasquez and Whiting (1998b) performed an uncertainty analysis study using both LHS and EPS sampling approaches on this system for a liquid-liquid extraction operation.

As expected it was observed that the distribution from EPS is narrower in accordance with the results of Fig. 4. A more realistic or practical interpretation of the uncertainty analysis can be done using EPS than using the LHS approach. The existence of unfeasible simulations is substantially reduced through the use of the EPS technique.

Table 1. Binary Parameters b_{ij} and b_{ji} Regressed for the NRTL Model for the System Diisopropyl-Ether (1) + Acetic-Acid (2) + Water(3). α equal to 0.3

I	J	b_{ij} (K)	σ (K)	b_{ji} (K)	σ (K)
1	2	-400.72	289.84	635.08	240.82
1	3	786.73	34.749	1454.12	50.68
2	3	-343.87	73.68	430.72	381.48

[1] ASPEN PLUS is a trademark of Aspen Technology, Cambridge, MA, USA.

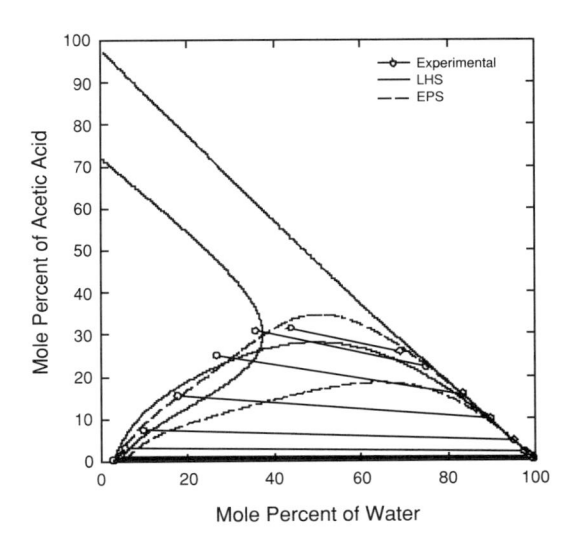

Figure 4. Extreme equilibrium binodal curves obtained for the sampling techniques LHS and EPS in the LLE prediction of the ternary system Diisopropyl Ether + Acetic Acid + Water at 25 °C. Predicted values by the NRTL.

Acknowledgments

This work was supported, in part, by U.S. National Science Foundation, grant CTS-96-96192.

Conclusions

The use of the EPS method generates a more practical interpretation of the uncertainty present in nonlinear regression models. Equal Probability Sampling produces uniform coverage by stratifying the model parameter space in regions with equal probability for the parameters. This provides sampling results and more importantly any correlation structure is automatically taken into account when the level sets of the objective function are constructed.

References

Cook, R. D., and J.A. Witmer (1985). A note on parameter-effects curvature. *J. Am. Stat. Assoc.*, **80**, 872-878.

Iman, R.L., and W.J. Conover (1982). A distribution-free approach to inducing rank correlation among input variables. *Comm. Stat.-Sim. Comp.*, **11**, 311-334.

Iman, R.L., and M.J. Shortencarier (1984). *A FORTRAN 77 Program and User's Guide for the Generation of Latin Hypercube and Random Samples for Use with Computer Models*. Report NUREG/CR-3624, National Technical Information Service, Springfield, VA.

Kalagnanam, J., and U.M. Diwekar (1997). Efficient sampling technique for optimization under uncertainty. *AIChE. J.*, **43**, 440-447.

Potocký, R., and T.V. Ban (1992). Confidence regions in nonlinear regression models. *Appl. Math.*, **37**, 29-39.

Treybal, R.E. (1980). *Mass Transfer Operations*. McGraw-Hill, New York.

Vasquez, V.R., and W.B. Whiting (1999). Evaluation of systematic and random error effects in thermodynamic models on chemical process design and simulation using Monte Carlo methods. *Ind. Eng. Chem., Res.*, Accepted for publication.

Vasquez, V.R., Whiting,W.B. (1998a). Effect of data type on thermodyamic model parameter estimation: A Monte Carlo approach. *Ind. Eng. Chem., Res.*, **37**, 1122-1129.

Vasquez, V.R., Whiting,W.B. (1998b). Uncertainty and sensitivity analysis of thermodynamic models using equal probability sampling. *Comp. Chem. Eng.*, accepted for Publication.

Vasquez, V.R., and W.B. Whiting (1998c). Uncertainty of predicted process performance due to variations in thermodynamics model parameter estimation from different experimental data sets. *Fluid Phase Equilibria*, **142**, 115-130.

Vasquez, V.R., Whiting,W.B., and M. Meerschaert (1999a). A sampling technique for correlated parameters in nonlinear regression models based on equal probability sampling (EPS). Submitted to *Risk Analysis*.

Vasquez, V.R., Whiting,W.B., and M. Meerschaert (1999b). Techniques for assessing the effects of uncertainties in thermodyamic models and data. *Fluid Phase Equilibria*, **158-160**, 627-641.

CHEOPS: A CASE STUDY IN COMPONENT-BASED PROCESS SIMULATION

Lars von Wedel and Wolfgang Marquardt
Lehrstuhl für Prozesstechnik
RWTH Aachen
52062 Aachen, Germany
{vonWedel, marquardt}@lfpt.rwth-aachen.de

Abstract

Process simulation has become a vital tool for assessing process design alternatives with respect to economic, environmental, and safety constraints. Commercial simulation packages are largely monolithic applications and rather inflexible. A new paradigm called component software promises to enable flexible environments which are extensible in a plug and play manner. This paper explores the use of components in process simulation and describes the architecture of the fully component-based, explorative simulation environment CHEOPS. It was developed as a part of the CAPE-OPEN project. Its open architecture and the interfaces developed permit interchangeable unit operation modules and thermodynamic property calculation methods. Further, different simulation strategies can be readily configured without recoding. A prototype has been built using CORBA as a middleware implementation.

Keywords

Process simulation, Open environment, Component software, CHEOPS, CORBA, CAPE-OPEN.

Introduction

The software engineering world is currently undergoing a major paradigm shift as it moves a step forward from object-orientation towards components (Orfali et al., 1996). Components are binary independent, standalone pieces that offer a public interface contract to their environment (Adler, 1995). Besides implementation independence, many component approaches offer also location and operating system transparency which makes components interoperable across computer system boundaries. An important advantage of components is the ability to be substituted dynamically.

This case study aims at exploiting the component approach and its plug and play nature for process simulation. A major issue is to explore to which degree of granularity components can contribute to the flexibility and extensibility of process simulation packages. To demonstrate the feasibility of these ideas, the simulator workbench CHEOPS (Component-based Hierarchical Open Process Simulator) has been developed.

The current version of CHEOPS is essentially a steady-state modular simulator. Hence, unit operation modules are the basic building blocks of the CHEOPS architecture. Following the idea of a component architecture, where the overall functionality is provided by interoperating pieces which can be exchanged (Mowbray, Malveau, 1997), we have experienced the possibility of a radical change of a process simulator architecture. Rather than setting up new models and simulation experiments within a static environment, one is now able to assemble a complete simulator from a set of interacting components. As an example, well-suited solution strategies such as sequential or simultaneous-modular as well as alternative convergence acceleration methods for torn streams can be selected from a dynamically extensible library.

This work is part of the EU-funded project CAPE-OPEN (CAPE-OPEN, 1998) which aims at defining standard interfaces for the major elements of a process simulator. As opposed to CAPE-OPEN where the integration of legacy systems was a requirement, this work intentionally started from scratch in order to overcome the limits of existing simulator architectures. Nevertheless, several components have been built reusing existing code.

The CHEOPS Architecture

CHEOPS has a multi-tiered component architecture (Mowbray, Malveau, 1997) which enables the substitution of components during compile time as well as during runtime. The interfaces are grouped into packages (Fig. 1). The core packages, *unit* and *thermo*, define interfaces for unit operations and thermodynamic calculation routines, respectively. The package *numerics* essentially defines only the variable interface which can be reused for several purposes throughout the whole architecture. Finally, the *executive* package defines the coupling to denote the connection between two coupled units as well as a solver interface which abstracts an arbitrary solution strategy for a flowsheet.

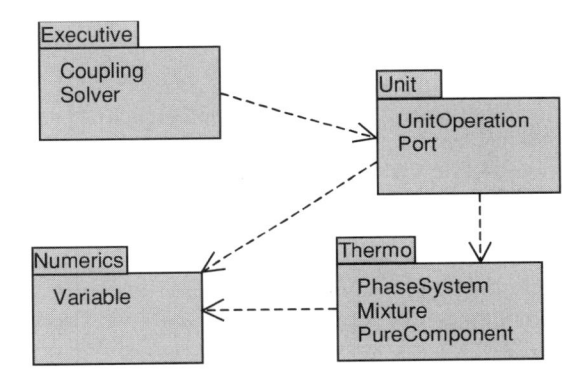

Figure 1. Organization of CHEOPS packages (UML package diagram).

Components as the architectural elements can be clustered into two parts: on the one hand there are components actually representing models such as a unit operation, a phase system, or a connection between two unit operation ports. These interfaces are derived from a general *cheops concept* interface. On the other hand there are components which represent services in the sense that they can be employed to execute a particular algorithm. A numerical solver is an example for such a service.

The component interfaces can be distinguished into horizontal and vertical interfaces. Horizontal interfaces are stable and reusable because they span several domains. An example is the variable interface which is reused in almost all other packages. Vertical interfaces, e.g. for a unit operation, are domain-specific and subject to changes due to extensions in scope and functionality.

The remainder of the paper explains the individual packages concerning numerics, thermodynamics, unit operations, and executive. Finally, the current implementation of CHEOPS is presented.

The Variable Interface

The *variable* is the most important concept defined in the numerics package. A variable represents a vector quantity of dimension 1.n. Beyond the actual value of a quantity, the variable interface also defines attributes for initial values, upper, and lower bounds. A status attribute indicates whether this variable is meant to be computed or given as an input by the user or specified by another component.

The variable interface is a core element in the overall architecture as it standardizes a unifying data exchange mechanism for process quantities among components. As an example, a solver can access a unit's stream and design variables through the same mechanism. The thermodynamics interfaces also rely on the variable concept for exchanging state information. Such horizontal and stable interfaces are a key issue for architectures not only in order to manage complexity but also to enable an evolution of the overall system (Brown et al., 1998). The variable concept has proven to be very valuable for this purpose.

The Design of Unit Operation Components

Conceptual Structure

As for any modular simulator, the unit operations are the core elements of CHEOPS. Roughly, the unit structure is described by a set of ports which represent the inlets and outlets of the unit (Fig. 2). A port is an abstraction of a unit operation's environment and thereby is an important means of encapsulation. As opposed to the classical stream view a port can rather be considered as one end of a generalized stream. Beyond the stream variables, which are associated to the ports and denote a flow of material, energy, or information across the unit operation boundary, a unit is directly linked to a set of design variables. During the calculation process of a unit operation, values are exchanged via the stream variables in the port and the design variables of the unit itself. Finally, a unit is associated with a phase system in order to calculate thermodynamic properties.

Simulation Contexts

In contrast to a strictly modular approach, a CHEOPS unit is not limited to an input/output formulation. A unit operation component supports several so-called *simulation contexts* which implement different formulations. Interfaces derived from a general simulation context interface abstract functionality for modular and equation-oriented formulations (Chen, Stadtherr, 1985). A modular simulation context provides only a computational method which triggers the classical input-output calculation whereas the equation-oriented simulation context can be queried for residuals and derivatives. The equation-oriented simulation context may provide additional, internal variables of a unit. The unit can optionally implement any of these interfaces. Simulation contexts must be considered as different views on the same unit operation component rather than independent objects. This simplifies changes of the simulation strategy on the fly.

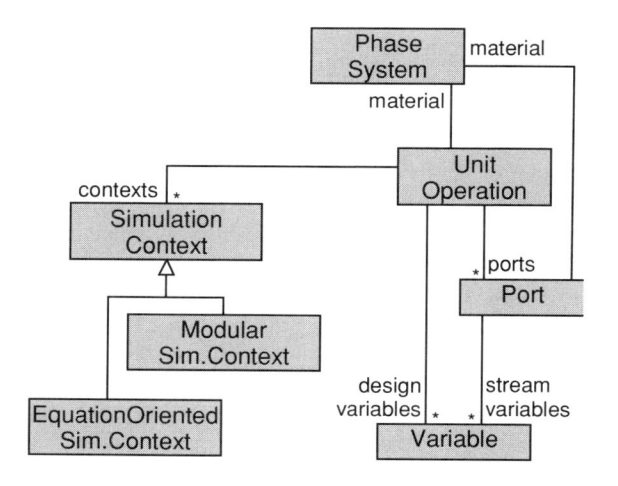

Figure 2. Conceptual object model of a CHEOPS unit operation (UML class diagram).

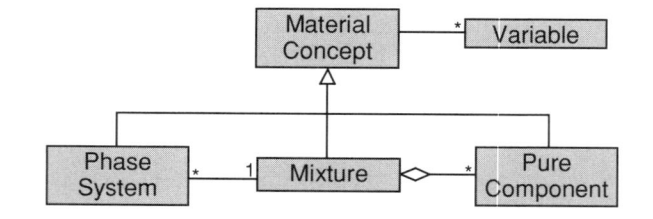

Figure 3. Conceptual object model of a thermodynamic system (UML class diagram).

Further Applications of Unit Operations

A unit operation component is a standalone object. Hence, heterogeneous tools using a totally different computational basis such as a CFD package or a neural net can be used to implement a unit. Within the CAPE-OPEN project, a wrapper around gPROMS has been built for demonstration purposes. It allows to plug-in a model written in the gPROMS language into CHEOPS. In this scenario, gPROMS (Barton, Pantelides, 1994) evaluates the model equations and provides residuals and derivative information through a CAPE-OPEN interface. An external solver must be used to compute the equation system in order to provide a modular simulation context.

Interfaces for a Thermodynamic System

The definitions of interfaces for a thermodynamics system can be split into two parts (Fig. 3). On the one hand there are (mixture and pure) components which represent abstract materials and their configuration. Phase systems, on the other hand, denote a concrete occurrence of a component in the flowsheet (and in the plant).

Phase systems are created from a mixture which acts as a factory (Gamma et al., 1995) and is subsequently associated with a unit or a port. The reason for also associating a phase system to a port becomes evident if e.g. an ideal flash unit operation is considered: though all ports may represent a single phase system, the material within the flash is (by definition) a multi phase system. Therefore, distinct instances have to be used.

Of course, all material components also exchange values of state quantities by the variable interface. All of these represent only intensive quantities. A port or a unit defines the context of a phase system by contributing the extensive quantities denoting a flux or a holdup, respectively. In combination, a thermodynamic system is fully determined.

Putting it All Together – The Executive

The executive package defines component interfaces which are required to perform a full simulation of a set of unit operations.

Couplings

First of all, a coupling is defined which denotes the connection between two ports. Each coupling denotes a directed edge in the flowsheet topology. Therefore, a set of unit operations and a set of couplings is sufficient to describe the full flowsheet topology. A coupling further offers two methods which are required for a sequential simulation strategy: a propagate method which is used if the coupling is not torn: it ‚copies' the values from one port to another, according to the direction of the coupling. If the coupling has determined to be torn, its converge method can be used in order to perform direct substitution or any accelerated method. The converge method returns an error which quantifies the quality of the guess. Couplings can be implemented as components on their own. This allows flexibility down to the level of convergence acceleration mechanism making it easy to compare and study their behavior.

Solvers

Furthermore, a solver interface is defined in the executive package. It can abstract different simulation strategies and expects as input a set of unit operations and couplings. The solver can be configured by a set of associated parameters. A sequential-modular and a simultaneous-modular solver have been implemented. The sequential solver employs a so-called graph service component in order to find a suitable partitioning and the torn couplings in the flowsheet. The simultaneous solver constructs a large equation system of all units and couplings that have to be taken into account. It has to interpret couplings as connection equations and reformelate the unit equations into a residual form based on the input/output formulation of a modular context (Chen,

Stadtherr, 1985). Further, perturbations of the unit module are required in order to compute linear relationships among inputs and outputs of the unit. An equation-oriented solver uses the equation-oriented context and can access residuals and derivatives directly. In both cases, the overall equation system is solved with a solver based e.g. on Newton's method.

Implementation of CHEOPS

The interfaces and components described so far have been implemented using CORBA as the underlying middleware layer. Packages and interfaces are mapped to CORBA IDL modules and interfaces, respectively. Components are implemented using C++ and Java and run on a SUN workstation.

A simple user interface on top of these components has been built in a Windows-based environment in order to set-up a flowsheet and to configure its individual parts (Fig. 4). On the left hand side the flowsheet topology can be graphically configured. On the right hand side registered components in various categories such as unit operations of thermodynamics are shown.

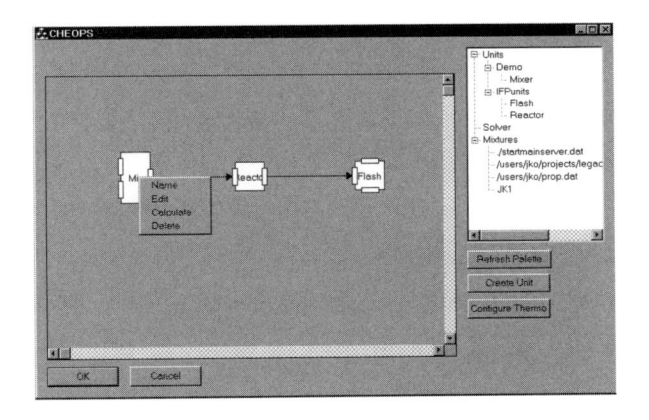

Figure 4. The CHEOPS user interface.

Currently, unit operations for a reactor, a flash and a mixer exist, where the former two have been implemented using existing FORTRAN code. A thermodynamic calculation package has been implemented as a wrapper around the IK-CAPE properties package (Fieg et al., 1995) which is written in FORTRAN. Due to the implementation independence components can be successfully employed to provide a modern interface for existing software.

Conclusions

A fully component-based, explorative simulator has been developed which permits flexible handling of different unit operation paradigms and solution strategies. In combination with architectural considerations, components have proven to be a very powerful technique and can considerably increase the flexibility of a simulation environment.

The flexibility of such an environment may, however, also lead to problems. The configuration of a set of components must be consistent in the sense that each can provide the functionality that is required in a certain context. As an example, a sequential-modular solver cannot be selected, if any of the units in the flowsheet does not support a modular simulation context. Further, consistency must be guaranteed e.g. among the phase systems employed by different units or for the stream variables in ports which are to be connected. This problem has not been solved so far. A possible solution might be to use a logical approach together with a detailed description of a components' facilities and services.

Future work will address dynamic simulation including consistent initialization. Both will fit seamlessly into the architecture by additional simulation contexts and solvers.

Acknowledgements

This work was partially funded by Brite/EuRam 3512. The authors gratefully acknowledge the fruitful discussions with all colleagues as well as CAPE-OPEN project members.

References

Adler, R. M. (1995). Emerging standards for component software. *IEEE Computer*, **28** (3), 68-77.

Barton, P., and C.C. Pantelides (1994). Modeling of combined discrete/continuous processes. *AIChE J.*, **40**(6), 966-979.

Brown, W.J., Malveau, R. C., McCormick, H. W. and T.J. Mowbray (1998). *Anti patterns – Refactoring software, architectures, and projects in crisis*. John Wiley and Sons, New York.

CAPE-OPEN (1996). *Conceptual Design Document*. Available on-line at http://bscw.quantisci.co.uk/pub/english.cgi/d145996/homepage.html.

Chen, H.S., and M. A. Stadtherr (1985). A simultaneous-modular approach to process flowsheeting and optimization. *AIChE J.*, **31** (11), 1843-1856.

Fieg, G., Gutermuth, W., Kothe, W., Mayer, H.H., Nagel, S., Wendeler, H., and G. Wozny (1995). A standard interface for use of thermodynamics in process simulation. *Comp. Chem. Eng.* **19**, S317-S320.

Gamma, E., Helm, R., Johnson, R., and J. Vlissides (1994). *Design Patterns: Elements of Reusable Software*. Addison-Wesley, Reading, Mass.

Mowbray, T. J., and R. C. Malveau (1997). *CORBA Design Patterns*. John Wiley & Sons, New York.

OMG (1999). *The Common Object Request Broker Architecture: Architecture and Specification*. The Object Management Group.

Orfali, R., Harkey, D., and J. Edwards (1996). *The Essential Distributed Objects Survival Guide*. John Wiley and Sons, New York.

A LAYERS ARCHITECTURE BASED PROCESS SIMULATOR

Naava Zaarur and Mordechai Shacham
Ben Gurion University of the Negev
Beer-Sheva 84105, Israel

Abstract

The implementation of layer-based, open-system architecture, in process simulators is studied. The simulators' various components: the user interface, the data base management system (DBMS) and the numerical solver are located on separate layers, where each layer communicates only with its adjacent neighbors and the operating system. The mathematical models of the process are constructed by the user aggregating individual equations stored in an object-oriented form in a database. Another database serves as physical properties library. The mathematical model of the process is translated to match the syntax requirements of the numerical solver and transferred to it for solution. The proposed simulator was implemented using Visual Basic for user interface, Access as DBMS and Matlab for numerical solution. Its operation is demonstrated for a reactive distillation process. It is concluded that layer-based architecture is preferable to the traditional architecture because of its flexibility in updating the model library and easy adaptability to frequent software and hardware changes. Expert and novice users can use the simulator effectively and beneficially.

Keywords

Process simulation, Layer-based architecture, Open-system architecture, Object oriented programming.

Introduction

The accelerating pace of technological development poses several difficult challenges to developers of process simulators. The growing competition in the chemicals market leads to shorter product life cycles, complex processes, and production of specialty materials and higher quality products. The meaning of those requirements is that new processes, which use lesser-known materials, have to be frequently developed. The process simulators that are currently available for the chemical industry are most appropriate to be used with well-known processes and materials. The models of the unit operations are invisible to the user; consequently he cannot verify that the model used is appropriate for his purposes. Adding new models to an existing simulator is a complex task that often requires programming. Adding thermophysical properties of new materials to the database is also a complicated, time consuming task.

The rapid change of computer hardware and software presents challenges of different nature. The user interface of the simulation program must be frequently modified to follow the changes in similar software packages and the changes dictated by a new operating system. Higher speed of the hardware allows incorporation of new, more rigorous solution techniques to the various programs. Because of the rigid architecture of the existing simulators it can be very expensive and time consuming to carry out the modifications needed in order to keep the program up to date.

An attractive approach to meet the challenges facing the process simulator developers is to adopt the "layered architecture". This type of architecture was developed during the early eighties to enable effective interconnection between various computers for distributed data processing (see, for example, Melendez and Petersen, 1999). The objective of using layered approach is to break down complex tasks into manageable subtasks. Different software and/or even hardware can be used to carry out the various subtasks. The great advantage of the layered architecture is that a layer can be modified, updated and even replaced without requiring any major changes in the other layers. Because of the independence of the various layers this architecture is often referred to as "open system" architecture.

Pantelides and Britt (1995) considered the possibility of using open architectures for interfacing between software originating from process modeling environment and software coming from other sources. They however saw a major obstacle for integrating various software products in the lack of standardization and nonexistence of appropriate communication protocols and suggested directing a major effort to standardization and establishment of communication protocols.

Assuming that modern operating systems provide the tools necessary to interconnect various software products we have started to develop an open architecture, layer based process simulator. In the rest of the paper the successful application of this new architecture in a prototype process simulator is described.

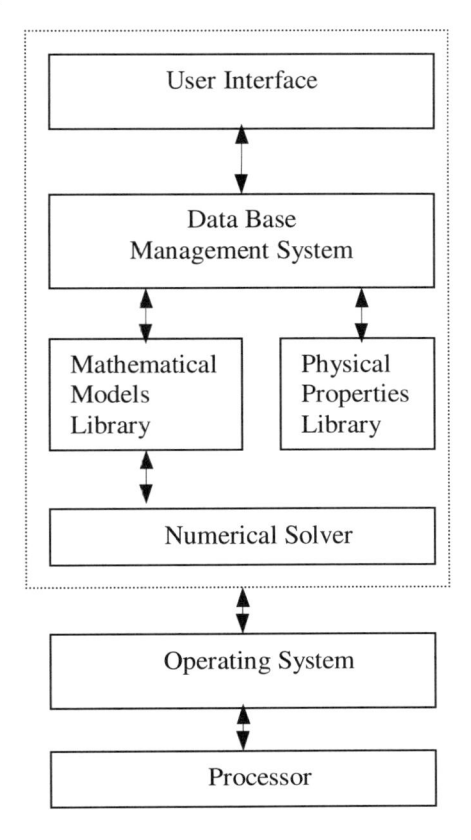

The Structure of a Layer Based Process Simulator

The structure of the proposed simulator is shown, schematically, in Figure 1. The main difference between this structure and the traditional process simulators' structure is in the restriction of the communication between the different parts of the program to well defined channels. In the proposed scheme each layer can communicate with the two adjacent layers and the operating system (OS). Data and commands can be transferred through the communication channels. While in principle many different OS can be used as computing environment for a layer based process simulator, for the prototype we have selected one OS that has enough means to support the required communication between the various layers. Brief description of the different layers follow.

The processor: The processor is involved in most of the stages of process simulation. It is envisioned that in the future various stages will be carried out by different processors (constructing the mathematical model of a distributed parameter system on a PC, for example, and solving the resultant set of differential equations on a super-computer) but for the prototype we have selected the PC as the common processor for the reasons that were described earlier.

OS: The OS must be compatible with the computer. For the prototype Windows 95[2] was selected as OS because of the many tools it provides for communication between different programs. It contains Application Programming Interface (API) functions intended for this purpose. Interfaces between various software products can be based on OLE (Object Linking and Embedding), DDE (Dynamic Data Exchange) or/and ActiveX.

User interface: This is the only layer of the program that the user communicates with, while the other layers are transparent to him. The user interface has a major role in determining the length of the learning curve for effective use of the process simulator. It can also help minimizing the time and effort spent on the technical details of the simulation of a particular process. For the prototype the user interface was built using Visual Basic[3]. The structure of the user interface will be discussed further in the next section.

Figure 1. Structure of a layer-based process simulator.

Data base management system (DBMS): This layer is responsible to aggregate the model equations, to retrieve physical and thermodynamic properties correlations and add user provided information in order to prepare the mathematical model for a particular simulation. It is also responsible for converting the data to the format required by the numerical solution package and storage and retrieval of the simulation results. Access[4] is used for data base management in the prototype simulator.

---

[2] Windows 95 is a trademark of Microsoft Corporation (http://www.microsoft.com)

[3] Visual Basic is a trademark of Microsoft Corporation (http://www.microsoft.com)

[4] Access is a trademark of Microsoft Corporation (http://www.microsoft.com)

Mathematical models' library: This is a database, which contains object-oriented representation of conservation principle based equations and constitutive equations for various unit operations. The equations are aggregated following the definition of the process by the user. Complete, previously defined models can be stored and retrieved and new equations can be added to the database by the user. Detailed description of the models' library will be provided in the following section.

Physical properties library: The library is a database that contains pure component properties and constants for correlation equations of various physical properties. Commercial databases such as DIPPR, DECHEMA etc. can be used in this layer as well as in-house databases. The user can add data that does not exist in the library or different from what is included in the library. For the prototype a database containing the data which is needed for the examples solved was prepared.

Numerical solver: Mathematical models of chemical processes can be categorized as systems of algebraic equations, systems of ordinary or partial differential equations and differential-algebraic systems. The selected numerical solver should contain programs for solving large systems of those types of equations. From among the programs that can be used to solve several types of equations Matlab[5] was selected as the numerical solver for the prototype simulator. The selection was based on the relative simplicity of interfacing Matlab with the data base layer of the simulator.

Assembling the Model Equations

The structure of the mathematical model's library is very similar to the three structure defined by Bogush and Marquardt (1995). The nodes are the equations, variables and parameters and the edges are "and" or "or" connections between the nodes. In the lowest layer there are the basic balance equations that are assigned according to the user's selection of a particular unit operation. The balance equations consist of functional terms, variables and constants. Some of the variables should be expressed as additional constitutive equations (heat and mass transfer rate equations, reaction rate etc.) Various constitutive equations are connected to the balance equations and the user can select which one of the equations is the most appropriate for his process. The constitutive equations may contain additional variables that can be either represented by further constitutive equations or can be treated as unknowns. The building of the "chain" of equations is continued until or the remaining variables are either unknowns or user-defined parameters.

Object-oriented techniques allow the representation of detailed information about the various equations. To each one of the equations in the database information strings are attached, which contain: 1. A unique index to identify the equation, 2. Equation type (differential or algebraic), 3. Definition of the variables and constants in the equation, 4. A brief explanation of what the equation represents and 5. Additional copy of the equation with Matlab syntax.

The generation of the model can be carried out completely automatically where the user's involvement is required only for selecting the appropriate constitutive equations from several options (when different options exist), and for defining parameter values. The second option is an automatic step-by-step model building where equations are added one by one to the model and the user has the option of inspecting the equation (and its explanation) before adding it to the model. This option allows the user to add his own equations to the model. The equations can be added using simple syntax similar to the mathematical definition of the equation.

The data base management system (DBMS) assembles the complete process model from the individual unit operations models. Only the pertinent model equations are passed to the DBMS. Two copies of the process model are assembled simultaneously. The first copy is the presentation copy, where the users can inspect/change the model equations. The second copy is the computational copy, which is built according to MATLAB's syntax rules. The difference between the two copies that in the first one the format of the equations is similar to the format they appear in textbooks for example, and in the second copy the syntax matches the requirements of the numerical solver. An indexed variable, for example, will appear with a subscript in the presentation copy while it will be included inside a do loop with the index inside parentheses in the computational copy. When the user introduces changes into one of the copies the other one is updated simultaneously. The complete process model can be saved for future modification and reuse.

Numerical Solution and Presentation of the Results

When the user selects the option to solve the model the complete process model is passed to the numerical solution layer in the form of three MATLAB m-files. The first file contains all the model equations. The physical and thermodynamic properties' correlation equations for the pertinent substances are included in the second file while the third file assigns numerical values to the user defined constants. There are pre-prepared m-files which serve as main programs and call the model equation files during the numerical solution process. The "main program" to be used depends on the type of the problem, whether it contains algebraic equations, ordinary differential equations or it represents a differential-algebraic system.

For steady state simulation, where all the equations are algebraic, for example, the main program assigns first initial estimates for all the unknowns. One of Matlab's built in functions or Newton-Raphson's method with numerical estimation of the derivatives can be used to solve the system of nonlinear algebraic equations.

After the solution has been found the results file is stored by the DBMS and the user can select through the

---

[5] Matlab is a trademark of The Math Works, Inc. (http://www.mathworks.com)

user interface the results that he wants to view or print. Results can be presented in tabular or graphical forms.

Esterification Reactor/Separator – An Example

Steady state simulation of a reactive distillation process is used, as an example, to demonstrate the feasibility and advantages of the proposed method. Reactive distillation was selected as example because it combines two processes of fairly high level of modeling complexity. This particular example is taken from Holland (1981). The column has 13 theoretical stages including a reboiler and a total condenser, and is operated at atmospheric pressure. The saturated liquid feed containing a mixture of acetic acid (A), ethanol (B) and water (C) enters the column on stage 6. The distillate rate and the reflux ratio are fixed. It is assumed that the reversible reaction: A+B↔C+D (where D is the product: ethyl acetate) occurs in the liquid phase in all stages.

In order to solve this example using the layer based process simulator, the user has to select first (assuming that the species present in the process have been defined previously) the option of "adding a new unit operation". From the unit operations menu "distillation" is selected, followed by the specification of the options: distillation column, including reboiler and full condenser, constant pressure, non-ideal thermodynamics and chemical reaction. At this point the pertinent material and energy balance equations are displayed (if the step-by-step model generation was selected). The user can inspect the equations and change them if necessary. By pressing "Next" the user requests the addition of the next equation to the model. In this case the stoichiometric equation is to be added next and the user is requested to specify the stoichiometric coefficients. This equation is followed by phase equilibrium, enthalpy mixing rule and mole fraction summation equations. Several types of expressions can be used to express the rate of the reaction. In this particular case and the rate of reaction is specified with respect to the ethanol where the rate constant changes with temperature according to the Arrhenius equation.

During construction of the model the DBMS has identified the parameters that must be defined by the user such as feed flow rate and composition, number of plates and feed plate, distillate rate, reflux ratio, Arrhenius equation constants etc. After finishing the construction of the model the user is prompted to define those parameters. When the model has been completed and all the parameters have been defined the option "Solve" is added to the available user options. Pointing on "Solve" initiates the writing of the m-files on the disk and solution of the model.

After Matlab has finished solving the problem, the option "viewing the results" becomes active. For the distillation column tabular or graphic display of tray temperatures, vapor and liquid compositions, vapor and liquid flow rates, stream properties as well as properties of the individual components can be selected for viewing or printing. The same type of data can be requested for the distillate and bottoms, and the duty of the reboiler and condenser can also be shown. The results obtained for this particular example are essentially the same as reported by Holland (1981) and others who have solved the same reactive distillation problem (see for example Venkataraman *et al*, 1990).

Discussion and Conclusions

It has been demonstrated that layer based, open architecture process simulators can be implemented with currently available operating systems and commercially available general purpose programs for data base management, numerical solution and user interface support. Using this new architecture for process simulators opens up a wide range of possibilities and it has many potential advantages.

The parts of the simulator, which must be the most flexible and the most easily updateable, namely the mathematical models' library and the physical properties' library, are built as databases. Therefore, it is very easy to add models of new unit operations or different versions of existing models and to add new physical properties. Adding and revising models and physical properties does not require any program changes only update of data in the databases and this can be carried out even by inexperienced user.

Adopting an open architecture process simulator to changes of hardware and software (changes that happen very frequently) should be much less expensive and time consuming than adopting simulators with the traditional architecture. The model and properties databases, which are the most expensive to develop, remain the same only the various general-purpose commercial programs should be replaced by their latest version. Some updating of the communication between the layers may also be required.

Users of various levels of expertise can make effective and beneficial use of the proposed simulator. Inexperienced users will probably rely on the automatic generation of the process model, using the equations stored in the database. The explanations attached to the model equations, that can be viewed when the model generation is carried out step-by-step, can be very beneficial to such user for understanding the principles on which a particular model is based. Experienced user can add new models or modify existing ones to represent his particular process more precisely.

In light of the many potential benefits of the layer-based open-architecture process simulator there is no doubt that it will gradually replace the simulators with the traditional architecture.

References

Bogush, R. and W. Marquardt (1995). A formal representation of process model equations. *Comp. Chem. Eng.*, **19**, S211-S216.

Holland, C. D. (1981). *Fundamentals of Multi-component Distillation*. McGraw-Hill, New York.

Melendez, W. A. and E. L. Petersen (1999). The upper layers of the ISO/OSI reference model (Part II). *Computer Standards and Interfaces*, **20**, 185-190.

Pantelides, C.C. and H.I. Britt. (1995). Multipurpose process modeling environments. *AIChE Symp. Ser.*, **304**, 128-140.

Venkataraman, S., Chan, W. K. and J. F. Boston (1990). Reactive distillation using ASPEN PLUS. *Chem. Eng. Prog.*, **86**(8), 45-54.

AUTHOR INDEX

A

Agrawal, Rakesh 381
Alger, Montgomery M. 163
Ali, Shahin 46
Alkaya, Dilek 125
Androulakis, Ioannis P. 406
Anselmo, Kenneth J. 125
Applequist, G. E. 427
Asenjo, Juan A. 306

B

Baldwin, John T. 482
Barbosa-Póvoa, Ana Paula 279
Barton, Paul I. 458
Basu, Prabir K. 284
Batres, Rafael 433
Bayer, Birgit 192
Bermingham, Sean K. 250
Bessling, Bernd 385
Biegler, L. T. 125
Bieszczad, Jerry 438
Bogle, I. D. L. 415, 446
Book, Neil L. 351, 442
Borland, John N. 31
Brauner, Neima 470
Braunschweig, Bertrand L. 220
Brennecke, Joan F. 371
Britt, Herbert I. 220

C

Carberry, John 26
Caruthers, James M. 478
Chakraborty, Aninda 355

D

Daly, Daniel T. 478
Dantus, Mauricio 360
Davis, James F. 419
Diwekar, Urmila M. 454
Doherty, Michael F. 163
Douglas, J. M. 265
Dünnebier, Guido 411

E

El-Halwagi, Mahmoud M. 367
Espinosa, José 342

F

Fidkowski, Zbigniew T. 381
Finlayson, Bruce A. 176
Floudas, Christodoulos A. 84
Fowler, Allan E. 1
Fraga, E. S. 446
Fraser, Duncan M. 389

G

Ganesan, K. D. 215
Ghosh, Prasenjeet 478
Glasser, David 311, 402
Gollapalli, Usha 360
Gonnet, S. 450
Gottschalk, Axel 364
Grievink, Johan 250, 324, 346
Grossmann, Ignacio E. 70, 423

H

Hallale, Nick 389
Harmsen, Jan 364
Hauan, Steinar 397
Hausberger, Brendon 402
Henning, G. 450
High, Karen 360
Hildebrandt, Diane 311, 320, 402
Hind, A. K. 446
Hinderink, Peter 364
Hooker, John 70
Huang, Y. L. 376
Huss, Robert S. 163

I

Iepapetritou, Marianthi G. 406
Iribarren, Oscar A. 306

J

Jang, Shi-Shang 338, 486
Jensen, Klavs F. 147
Johnson, Timothy Lawrence 454

K

Kesavan, Padmanaban 458
Kim, Sangtae 23
Klatt, Karsten-U. 411
Koulouris, Alexandros 438
Kramer, Herman J. M. 250
Kussi, J. S. 315

L

Lakshmanan, Ramachandran 393
Lee, Jae W. 397
Leimkühler, H.J. 315
Leone, H. 450
Lien, Kristian M. 397
Lin, Jun Hsien 338
Linninger, Andreas A. 46, 355, 462
Luyben, Michael L. 113

M

Malone, Michael F. 163
Mannarino, G. 450
Marquardt, Wolfgang 192, 494
Marriott, J. I. 415
Mateus, Ricardo 279
Mayer, H. H. 466
McGregor, Craig 311, 320, 402
Meeuse, F. Michiel 324
Miller, David C. 419
Mills, Patrick L. 147
Mockus, Linas 284
Montagna, Jorge M. 289, 294, 306

N

Nagy, Bert 215
Naka, Yuji 433
Neumann, Andreas M. 250
Neurock, Matthew 5
Nishitani, Hirokazu 298
Niwa, Tadao 298
Novais, Augusto Q. 279

O

O'Connell, John P. 5

P

Pantelides, Constantinos C. 220, 236
Papageorgiou, Lazaros G. 31

Pardalos, Panos M. 84
Pekny, J. F. 427
Perne, R. 315
Phimister, James R. 302
Pinto, José M. 306
Pistikopoulos, Efstratios N. 99
Ponton, Jack W. 393

Q

Quiram, David J. 147

R

Reklaitis, G. V. 427
Rosendall, Brigette M. 176
Ryley, James F. 147

S

Salomone, Enrique 46, 342
Sama, Sergei 220
Samsatli, Nouri J. 31
Schembecker, Gerhard 364
Schmidt, Martin A. 147
Schoenmakers, H. 466
Seferlis, Panagiotis 346
Seider, Warren D. 302
Shacham, Mordechai 470, 498
Shah, Ashish 215
Shah, Nilay 31
Sharif, Mona 31
Shieh, Shyan-Shu 338, 486
Sijben, Jo 364
Sinclair, Jennifer Lynn 138
Smith, David A. 26
Smith, Vernon A. 351
Sørensen, E. 415
Spears, Carl 474
Spriggs, H. Dennis 367
Stadtherr, Mark A. 371
Stefanović, Jelena 236
Steffens, M. A. 446
Stephanopoulos, George 46, 438
Stradi, Benito A. 371
Subramaniam, M. 338, 486
Sundaram, Anantha 478
Sylvester, Robert W. 26

T

Tassone, Vince 474
Teague, Thomas L. 482
Terrill, Daniel L. 329
Trainham, J. A. 265
Tsai, Po-Feng 486
Tyreus, Bjorn D. 113

V

van Rosmalen, Gerda M. 250
van Schijndel, Jan 99
vander Stappen, Michel L. M. 324
Vasquez, Victor R. 490
Vecchietti Aldo R. 289, 294, 306
Venkatasubramanian, Venkat 478
Verheijen, Peter J. T. 250, 324
Vinson, Jonathan M. 284
von Wedel, Lars 192, 494

W

Wahyu, Haifa 393
Watson, Brian A. 163
Westerberg, Arthur W. 265, 397
Wetzel, Mark D. 147
Whiting, Wallace B. 490
Wright, Stephen J. 58

X

Xu, Gang 371
Xu, Jianguo 333

Y

Yan, Q. Z. 376
Yang, Y. H. 376
Yeomans, Hector 423

Z

Zaarur, Naava 498

SUBJECT INDEX

A

Acquisitions, 1
Active X, 474
Adjoints, 236
Agents, 433
All solutions, 84
Application programming interface, 442
Attainable regions, 320
Azeotropes, 84
Azeotropic
   distillation, 302
   mixtures, 342

B

Batch
   distillation, 342
   facilities, 279
   plant design, 289
   process development, 31, 46
   processes, 46, 284
Bifurcations, 176

C

CAPE, 466
CAPE-OPEN, 220, 494
Capital cost targets, 389
CFD turbulence, 138
Chemical process
   design and development, 192
   modeling, 192
   simulation and optimization, 192
Chemical reactor, 176
CHEOPS, 494
Chromatographic separation, 411
Closed form modeling, 125
Collinearity, 470
Column profiles, 402
COM, 474
Combinatorial optimization, 454
Compartmental models, 250
Component software, 494
Composition profile properties, 397
Computational chemistry, 5
Computer supported collaborative work, 192

Computer-aided design, 46
   of materials, 478
Conceptual design, 113, 163, 324, 333, 364
Conceptual models, 342
Conference summary, 265
Continuation methods, 346
Control
   engineering, 351
   equipment, 351
Control
   structure selection, 346
Controllability, 360
Controls, 474
CORBA, 494
CRDT, 176

D

Data
   compression, 486
   models, 482
   reconciliation, 84
   sheet, 482
Decision
   support, 419
   making, 163
Decomposition algorithms, 458
Design, 279, 306, 311, 342
   and control, 113
   automation, 215
   by-analysis, 298
   engineering, 215
   methods, 364
   support system, 298
Detailed dynamic model, 415
Deterministic global optimization, 84
Development, 329
Difference points, 397
Discrete
   event system, 298
   programming, 446
Disjunctive program, 423
Distillation, 302, 402
   dynamics, 302
   sequences, 423
Distributed memory parallel algorithm, 458

Disturbance propagation models, 376
Dominant variables, 113
Draft tube baffle crystalliser, 250
Driving force cost, 333
Dynamic
    modeling, 338
    simulation, 411, 438

E

Education, 163
Electronic data exchange, 482
Emulsions, 324
Engineering knowledge database, 419
Environmental considerations, 46
Environmentally benign processing, 371
Evolutionary algorithms, 478
Experiment design, 338

F

Feed addition policies, 402
FIDAP, 176
Flexibility analysis, 406
Flowsheet, 474
    synthesis, 355
Fluidization, 138
Front-end integration, 215
Fuel-additives, 478

G

Gas-solid, 138
Generalized disjunctive programming, 294
Global solution, 458

H

Heat exchange, 333
    networks, 376
Hierarchical
    decomposition, 324
    design, 250, 438
High quality, 215
Hybrid phenomenological neural-network models, 478

I

Impact diagram, 389
Industrial radiant furnace, 176
Information
    management, 482
    modeling, 192
    models, 351, 442
    representation, 482
    silo, 23
    technology, 23
    theory, 338

Inherent safety, 26
Integer programming, 58, 84
Integrated design environments, 192
Integration, 215
Interface standards, 220
Intermediate storage tanks, 289
Internet, 23
Interoperability, 220
Interval analysis, 371
Intranet, 23
ISO standards, 482

K

Kinetic modeling, 406
Knowledge and data exchange, 433

L

Layer-based architecture, 498
Layout, 279
Life-cycle assessment, 26
Linear programming, 58
Liquid-liquid equilibria, 490
Lower cost, 215

M

Management, 311
    of the process design process, 450
Mass
    exchange networks, 376, 389
    integration, 367
Material resource planning, 315
MATLAB, 176
Mayonnaise and dressings, 324
Mechanism reduction, 406
Membrane separation, 415
MEMS, 147
Mergers, 1
Microchemical, 147
Microelectromechanical systems, 147
Microfabrication, 147
Microfluidic, 147
Microreactor, 147
Middle-vessel column, 302
Middleware, 220
Minichemical plant, 147
Minimum
    energy requirements, 381
    image convention, 236
Mixed integer
    linear programming (MINLP), 279, 294, 423
    programming, 70
Model generation, 462
Model-based optimization, 411
Modeling, 1, 31
    environments, 438

Molecular
  dynamics, 236
  simulation, 5
Monte Carlo simulation, 490
Multi-objective optimization, 355, 360
Multiphase, 138
Multiproduct, batch plants, 289, 294, 306

N

Next generation computer-aided process engineering, 220
Noise, 470
Nonconvex
  hull, 338
  mixed-integer nonlinear programming, 458
Noncovexities, 84
Nonlinear
  dependent variables, 486
  models, 490
  principal curve analysis, 486
  programming, 58, 70
Nuclear waste, 454

O

Object oriented programming, 446, 498
Ontologies, 433
Open
  architecture, 220, 474
  environment, 494
  form (equation oriented) modeling, 125
  system architecture, 498
Operability, 360, 381
Operating instruction synthesis, 284
Optimal
  design, 415
  regression model, 470
Optimization methods, 70
Optimization, 31, 58, 279, 306, 320, 329, 367, 389, 446
Orthogonal collocation, 415
Outsourcing, 1

P

Parameter estimation, 84
Parametric optimization, 346
Partial control, 113
Periodic boundary conditions, 236
Pervaporation, 415
PFD, 482
Phase equilibrium, 84, 371
Phenomena-based modeling, 438, 462
Physical properties, 5
Pilot plant, 329
Pinch technology, 389
Pipeless batch plant, 298

Planning, 427
Plant design, 31, 427
Plantwide control, 113
Pollution prevention, 26, 355, 367
Precision, 470
Preferred
  conversion, 385
  reactive distillation, 385
Preliminary process evaluation, 419
Principal variable approach, 486
Process
  analysis, 329
  chemistry critique, 419
  control, 99
  Data eXchange Institute, 442
  design, 26, 84, 99, 113, 333, 346
  development, 1, 311, 315
  engineering, 482
  information, 482
  integration, 125, 376
  operability, 99
  optimization, 99, 125, 333
  performance models, 306
  rate, 320
  simulation, 494, 498
  simulator, 474
  synthesis, 46, 70, 163, 311, 315, 320, 329, 364, 367, 446
  topology, 419
Production simulation, 298
Project lifecycle, 466
Property estimation methods, 5
Protein production, 306

R

Reaction
  engineering, 371
  path synthesis, 393
Reactive
  distillation, 385
  VLE separation, 397
Research, 329
Resource conservation, 367
Risk management, 427

S

Sampling techniques, 490
Scale-up, 250
Screening methods, 385
Sensitivity analysis, 346
Separation performance, 342
Shorter schedule, 215
Side reactions, 393
Simulated moving bed, 411

Simulation, 315, 446
    model interchange, 433
Software, 58
    application, 466
Solvent
    selection, 46
    substitution, 371
SQP algorithms, 125
Standards for the Exchange of Product Model Data, 442
Static controllability, 346
Statistical methods, 490
STEP standards, 351
STEP, 433, 482
Stepwise regression, 470
Stochastic
    annealing, 454
    optimization, 360, 454
Structural representation, 393
Structured products, 324
Supercritical fluids, 371
Superstructure, 423
Supply chain, 427
Sustainability, 26, 364

T

Tailored optimization, 125
Task representation language, 450
Ternary distillation, 381
Thermally coupled columns, 381
Thermodynamic efficiency, 381
Thermodynamics, 5, 490
Time-to-market, 31
Transitional measurements, 486
Transport, 176
Tray-by-tray models, 423

U

Uncertainty, 31, 355, 406, 427

V

Versions' administration, 450
Vitrification, 454

W

Waste, 26
    minimization, 360
    treatment, 46
Workflow management, 192